Trace Elements in Terrestrial Environments

Second Edition

Springer
New York
Berlin
Heidelberg
Barcelona
Hong Kong
London
Milan
Paris
Singapore
Tokyo

Domy C. Adriano

Trace Elements in Terrestrial Environments

Biogeochemistry, Bioavailability, and Risks of Metals

Second Edition

With 150 Illustrations

 Springer

QH 545 .T7 A37 2001

Domy C. Adriano
Savannah River Ecology Laboratory
University of Georgia
Aitken, SC 29802
USA

Adriano, D. C.

Trace elements in
terrestrial environments

Cover illustration: Courtesy of Dr. Laura Janecek and Mr. David Scott (SREL), an award-winning photographer.

Library of Congress Cataloging-in-Publication Data
Adriano, D.C.
    Trace elements in terrestrial environments: biogeochemistry, bioavailability, and risks
of metals/Domy C. Adriano.–2nd ed.
        p. cm.
    Rev. ed. of: Trace elements in the terrestrial environment. c1986.
    Includes bibliographical references.
    ISBN 0-387-98678-2 (hc: alk. paper)
        1. Trace elements—Environmental aspects.   2. Trace elements in nutrition.
3. Biogeochemical cycles.   I. Adriano, D.C. Trace elements in the terrestrial environment.
II. Title.
QH545.T7 A37 2001
577'.14–dc21   00-061263

Printed on acid-free paper.

Production coordinated by Chernow Editorial Services, Inc., and managed by MaryAnn
Brickner; manufacturing supervised by Erica Bresler.
Typeset by SPS Publishing Services (P) Ltd., Madras, India.
Printed and bound by Sheridan Books, Inc., Ann Arbor, MI.
Printed in the United States of America.

9 8 7 6 5 4 3 2 1

ISBN 0-387-98678-2   SPIN 10700898

Springer-Verlag   New York Berlin Heidelberg
A member of BertelsmannSpringer Science+Business Media GmbH

To my wife, Zena Gaviola Adriano, for her incredible understanding and patience throughout my professional career

and

To hundreds of loyal supporters and participants in the highly successful biennial series "International Conference on the Biogeochemistry of Trace Elements"

Preface

Based on citations in the literature, it is evident the first edition, entitled *Trace Elements in the Terrestrial Environment* (1986), met its primary objective, which was to provide students and professionals with a comprehensive book in many important aspects of trace elements in the environment. Indeed the extent of its use has exceeded my expectations. As a result of its usefulness and encouragement by colleagues in the field, I was compelled to write this edition following a similar format, but including new chapters on biogeochemistry, bioavailability, environmental pollution and regulation, ecological and human health effects, and risk and risk management and expanding the coverage to include freshwater systems and groundwater where appropriate. In addition to plants, which was the main biota of emphasis in the earlier edition, fish and wildlife and invertebrates (both terrestrial and aquatic) are discussed as necessary. The ecological and human health effects of major environmental contaminants, such as As, Cd, Cr, Pb, and Hg are also highlighted, along with relevant information on potential risks to the ecology and human health.

As in the first edition, the chapters are organized by element, which are grouped into "the big five" environmental metals, the essential elements (e.g. the traditional plant micronutrients, Zn, Cu, Mn, Mo, and B and now Ni, and Se, which is essential in animal and human nutrition), and other trace elements. It was challenging to find a solution to the daunting task of discussing emerging paradigms on bioavailability and how bioavailability can be influenced by the source term, chemical speciation, and the "driving" factors (i.e., pH, CEC, redox potential, and so on) that in turn can influence risk. Most challenging, however, were the areas of freshwater systems, with their attendant fishery and wildlife as well as human health effects of metals.

While the earlier edition was sparked by my mentors, especially Professor Al Page (University of California, Riverside, CA), but also Professor Parker Pratt (University of California, Riverside) and Dr. Larry Murphy (Kansas State University, Manhattan, KS), this edition is the result of encouragement by younger colleagues, foremost Professor Dr. Walter Wenzel (Universität für Bodenkultur, Vienna, Austria), Professor Dr. Jaco Vangronsveld (Limburgs Universitair Centrum, Diepenbeek, Belgium), and Dr. Ravi Naidu (CSIRO, Adelaide, Australia).

I trust that this book will enhance our understanding of the diverse roles, beneficial or detrimental, of trace metals in our society and environment.

Domy C. Adriano
Athens, Georgia
Aiken, South Carolina

Acknowledgments

An undertaking of this magnitude could have been painful without the support of the following people: my wife, Zena, for her continuous encouragement and sustainable patience; Dr. Mike Smith, our former director for allowing me to take my sabbatical leave in 1998–1999; Mr. Wouter Geebelen of Limburgs University (Belgium) for the artwork; Ms. Marie Heap and Ms. Pat Davis from SREL and Ms. Ilse Plesser of Limburgs University (Belgium) for typing parts of the manuscript; Mr. Brad Reinhart (SREL) for proofreading parts of the text; and Mrs. Jean Mobley, our librarian for ordering articles and books.

Although the majority of the chapters were not sent for external review before submission, the new chapter on bioabilability was reviewed by Drs. Tracy Punshon and Barbara Taylor (SREL) and the chapter on biogeochemistry was reviewed by Drs. Dan Kaplan (Savannah River Technology Center) and Vijay Vulava (SREL).

Hundreds of scientific articles were used to research the book, some of the most relevant were provided by the following colleagues in the United States, Europe, and Australia: Dr. Rufus Chaney (U.S. Department of Agriculture); Prof. Gary Pierzynski (Kansas State University); Dr. Bud Norvell (Cornell University); Dr. Bill Berti (DuPont Co); Prof. Bob Taylor (Alabama A&M University); Professor Andrew Chang (University of California, Riverside); Dr. Steve McGrath (Rothamsted Experiment Station, England); Professor Dr. Walter Wenzel (Universität für Bodenkultur, Vienna, Austria); Professor Dr. Jaco Vangronsveld (Limburgs Universitair Centrum, Belgium); Professor Bal Ram Singh (Agricultural University of Norway); Dr. Michel Mench (INRA, Bordeaux, France); Professor Dr. Alina Kabata-Pendias (IUNG, Pulawy, Poland); Professor Dr. Nicola Senesi (Universitá di Bari, Italy); and Drs. Ravi Naidu and Mike McLaughlin (CSIRO, Adelaide).

My sabbatical in 1998–1999 provided the venue conducive to this type of writing. Professors Wenzel and Vangronsveld not only provided me with the supplies and tools to be functional but they also served as a "sounding board" to discuss some of the emerging paradigms in trace metal research.

Often, I would walk with Dr. Wenzel at the Turkenshanz Park (my favorite park) by the University in Vienna to discuss "mind-boggling" things. To them, I am most grateful for such stimulating exchanges of ideas.

A special debt of gratitude goes to Dr. Laura Janecek and Mr. David Scott (SREL), an award-winning photographer, for their design and photography that is used on the cover of this book; also to Kathy Jackson and Barbara Chernow of Chernow Editorial Services in New York for their skillful editorial assistance throughout the production of the book.

Finally, my tenure at SREL, world renowned for its ecological research, enabled me to broaden and sharpen my vision on trace elements well into the twenty-first century. To my SREL colleagues, I am forever indebted for the inspiration to wander into the fields of ecology, environmental biology, risk assessment, and aquatic chemistry.

Domy C. Adriano
Athens, Georgia
Aiken, South Carolina

Contents

1
Introduction

All ethics so far evolved rest upon a single premise: that the individual is a member of a community of interdependent parts. His instincts prompt him to compete for his place in that community, but his ethics prompt him also to co-operate (perhaps in order that there may be a place to compete for). The land ethic simply enlarges the boundaries of the community to include soils, waters, plants, and animals, or collectively: the land.

Aldo Leopold (*A Sand County Almanac*)

Rapid population growth. Climate change. Global deforestation. Resource exhaustion. Biodiversity loss. They indicate the wide scope of serious global environmental problems that are becoming more complex and critical with each passing day. They will undoubtedly dominate the economic, social, and geopolitical agendas of developed as well as developing countries into the 21st century. Indeed, they became the centerpiece of the forum during the 1992 Rio Summit, i.e., the United Nations Conference on Environment and Development held in June 1992 in Rio de Janeiro, Brazil. From this conference—the largest meeting ever of heads of state—it became apparent that we lack knowledge on how human activities interact with the environment, and how to balance economic growth and environmental protection to ensure sustainable development.

As if heeding the Rio Summit promulgations, the National Research Council's Forum on Science and Technology identified the following six areas that merit increased attention and resources in environmental research and development (R&D) (National Research Council, 1997): economics and risk assessment, environmental monitoring and ecology, chemicals in the environment, the energy system, industrial ecology, and population. Although not singled out, metals are at the core of these issues. Inherent are the impact of rising population and growing affluence of society in the production and emissions of metals, cost/benefit ratio of decontaminating metal-contaminated areas, effect of fossil fuel combustion on the mobility of metals, bioindicators, recycling, and the like.

Major advances have been made in our understanding of the environmental biogeochemistry, biological effects, and risks associated with trace metals over the past two decades. The scientific community has learned a great deal about the sources, concentration levels in natural environments and biota, adsorption–desorption reactions, chemical speciation and bioavailability, bioaccumulation, exposure and risks, toxicology and ecotoxicology, and the transport and fate of trace metals. Additionally, major leaps have been achieved in analytical chemistry, environmental cleanup, and monitoring. But there is still a great deal to learn, particularly on metal behavior at the molecular level and at the interfaces between various environmental media. Also, particular attention should be directed toward understanding and predicting natural attenuation of trace metals in environments. Thus, a need is evident to address trace metal issues employing a multidisciplinary approach among physical, biological, and chemical sciences.

Although metals have become increasingly important as a pollutant group, only during the last three decades or so have they been widely acknowledged as potentially dangerous environmental toxins. While the commonly accepted micronutrient trace elements, i.e., B, Fe, Mn, Cu, Zn, Mo, and Cl, for higher plants were established mostly before 1940, those for animals were established over a longer period of time lasting until the 1980s (Adriano, 1986; Davies, 1992). In contrast to the much earlier findings of the essentiality of certain trace elements to plants and animals, environmental pollution by trace metals was highlighted much later—mostly in the 1950s and 1960s by the *Minamata* (methyl mercury poisoning) and *itai-itai* (cadmium poisoning) diseases in Japan (Hamada and Osame, 1996; Nogawa and Kido, 1996). In the 1980s and 1990s, lead poisoning in humans was acclaimed as a social disease due to its widespread occurrence. Today, more and more varied sources of metals are being discovered that can taint the quality of our food and drinking water resources. Very alarming is the widespread incidence of arsenic poisoning in the West Bengal region of India and in Bangladesh that has reportedly exposed millions of inhabitants (Chowdhury et al., 1997). The World Bank Group (1999) estimates that more than 20 million of Bangladesh's 120 million people are drinking contaminated water—perhaps the largest mass poisoning in history. High concentrations of naturally occurring As have already been found in water from thousands of tubewells, the main source of potable water, across more than half of Bangladesh's districts. The real extent of contamination is not exactly known as the majority of the country's wells still need to be tested.

Recently, a major mining incident of catastrophic proportion occurred in April 1998 in the Doñana, Spain (Meharg et al., 1999). Approximately 5 million cubic meters of acidic pyrite sludge from a collapsed tailing dam traveled 40 km down the Guadiamar tributary of the Guadalquivir River in southwestern Spain, affecting over 4400 ha. The sludge (pH 2) is enriched in Zn (0.8% DW), Pb (1.2% DW), Cd (0.1 g kg^{-1} DW), and As (0.6% DW). The Doñana (a Biosphere Reserve and a World Heritage Site) is perhaps the most important bird breeding and overwintering area in western Europe. It is home to many endangered and rare species, including the Spanish imperial eagle, marbled teal, and white-headed duck. This incident reminds us that accidents involving metals will generally be more serious than those involving organic chemicals because of the persistent, permanent nature of metals in the environment. Already, the heavy burden of the metals in the environment has alarmingly exposed birds and other wildlife populations, and the long-term impacts on wildlife diets will continue to be a serious concern (Meharg et al., 1999).

To safeguard the health of organisms and humans, society needs to know how trace metals behave in the environment, their pathways and fluxes, their health effects, and how to protect the general population from metal exposure. In several countries, this has resulted in the enactment of regulations and, in many cases, monetary expenditures to abate pollution.

1 Definitions and Functions of Trace Elements

Current universal interest in trace element research is being spurred by our needs to (1) increase food, fiber, and energy production; (2) determine trace metal requirements and tolerances by organisms, including relationships to animal and human health and diseases; (3) evaluate bioavailability of trace metals to organisms as influenced by biogeochemical processes and factors; (4) evaluate the potential bioaccumulation, biomagnification, and biotox-icity of trace metals; (5) understand trace metal cycling in nature, including their biogeochemistry; (6) elucidate the importance of trace metals in environmental health, including urban and indoor environments; (7) assess trace metal enrichment in the environment by recycling wastes; (8) discover additional ore deposits; (9) use certain trace metals in the high-tech industry; (10) comply with stringent state and federal regulations on releases of effluents (both aqueous and gaseous) to the environment; and (11) accomplish risk assessment and eventual risk reduction of metal-affected systems.

The trace elements that have been studied most extensively in soils are those that are essential for the nutrition of higher plants: B, Cu, Fe, Mn, Mo, and Zn. Similarly, those extensively studied in plants and feedstuff because of their essentiality for animal nutrition are As, Cu, Co, Fe, Mn, Mo, Zn, Cr, F, Ni, Se, Sn, and V. The term *trace element* is rather loosely used in the literature and has differing meanings in various scientific disciplines. Often it designates a group of elements that occur in natural systems in minute concentrations. Sometimes it is defined as those elements used by organisms in small quantities but believed essential to their nutrition. However, it broadly encompasses elements including those with no known physiological functions. Earth scientists generally view trace elements as those other than the eight abundant rock-forming elements found in the biosphere (i.e., O, Si, Al, Fe, Ca, Na, K, and Mg). It is a general consensus that an element is considered *trace* in natural media when present at levels of <0.1%. In biochemical and biomedical research, trace elements are considered to be those that are ordinarily present in plant or animal tissues in concentrations comprising <0.01% of the organism. In food nutrition, a trace element may be defined as an element that is of common occurrence but whose concentration rarely exceeds 20 parts per million (ppm) in the foodstuffs as consumed. It should be noted that some of the *nutritive* trace elements (e.g., Mn and Zn) may often exceed this concentration.

In this book, *trace elements* refer to elements that occur in natural and perturbed environments in small amounts and that, when present in sufficient bioavailable concentrations, are toxic to living organisms. Other terms that have been used and, for all practical purposes, considered synonymous for trace elements are: *trace metals, heavy metals, micronutrients, microelements,* and *minor elements.* The use of *micronutrients* usually is restricted to those elements required by higher plants (i.e., Zn, Mn, Cu, Fe, Mo, and B; Cl is also a micronutrient). The term *heavy metals* usually refers to elements having densities greater than 5.0 g cm^{-3} and denotes metals and metalloids that are associated with pollution and toxicity but also includes elements that are required by organisms at rather low concentrations. In this book, the *trace elements* to be considered are: arsenic (As), silver (Ag), boron (B), barium (Ba), beryllium (Be), cadmium (Cd), cobalt (Co), chromium (Cr), copper (Cu), fluorine (F), mercury (Hg), manganese (Mn), molybdenum (Mo), nickel (Ni), lead (Pb), antimony (Sb), selenium (Se), tin (Sn), thallium (Tl), vanadium (V), and zinc (Zn). The terms *trace elements* and *trace metals* are used interchangeably in this book.

In the plant kingdom, especially with higher plants, an element is considered essential if it meets the following criteria (Hewitt and Smith, 1974):

1. Omission of the element must directly cause abnormal growth, or failure to complete the life cycle, or premature senescence and death.
2. The effect must be specific, and no other element can be substituted in its place.
3. The effect must be direct on some aspect of growth or metabolism. Indirect or secondary beneficial effects of an element, such as reversal of the inhibitory effect on some other element, disqualify an element as essential.

In animal nutrition, the criteria for essentiality are somewhat similar (Underwood, 1977): (1) consistent significant growth response to dietary supplements of the element and this element alone, (2) development of the deficiency state on diets otherwise adequate and satisfactory, and (3) correlation of the deficiency state with the occurrence of subnormal levels of the element in the blood or tissues of animals exhibiting the response. In short, an element is essential if a reduction in its total daily intake below some minimal level consistently induces signs of deficiency, and its subsequent augmentation prevents and reverses the metabolic changes. The essentiality, benefits, and potential toxicity of trace elements to plants, animals, and humans are shown in Table 1.1. For higher plants, six of the trace elements displayed are essential; Co is required for N_2 fixation in bacteria and algae; Se and V are essential for *E. coli* and alga *Scenedesmus obliquus* (Hewitt and Smith, 1974). In animals, 13 of the elements are essential (Underwood, 1977). Except for As and V, elements required by animals are also essential for humans. [Note: Chlorine (Cl) is also a plant micronutrient; silicon (Si) and iodine (I) are also required in animal nutrition; and I is also essential in human nutrition.]

2 Production and Emission of Trace Metals

The ever-increasing production and demand for metals in developed and developing countries suggest the mounting probability of their dispersal and contact with the environment. A metal may be dispersed from the time its ore is being mined to the time it becomes usable as a finished product or ingredient of a product. In addition, increasing demands for fertilizers and soil amendments in high-production agriculture may enhance this probability. Land disposal techniques that seem promising for municipal wastes and other solid wastes may also increase the metal burden of the soil. Moreover, the enforcement of Clean Air and Clean Water Acts in developed countries, while ensuring cleaner air and water supplies, is likely to exacerbate the metal burden of the soil. Examples include the collection of

TABLE 1.1. Essentiality and potential effects of trace elements on plant, animal, and human nutrition.[a]

Element	Essential/beneficial to			Potential toxicity to			Comments
	Plants	Animals	Humans	Plants	Animals	Humans	
Ag	No	No	No	Yes	Yes	Yes	Phytotoxic
Al	No	No	No	Yes	Yes	Yes	Phytotoxic in low pH soils; toxic to fish in low pH lakes; relatively nontoxic to mammals
As	No	Yes	No	Yes	Yes	Yes	Phytotoxic before animal toxicity; similar geochemical behavior to P; carcinogenic; blackfoot disease (arsenicosis) in South Asia
B	Yes	No	No	Yes			Narrow margin, especially in plants; phytotoxicity is more prone in arid regions; relatively nontoxic to mammals
Ba	No	Possible	No				Insoluble; soluble forms—toxic
Be	No	No	No	Yes	Yes	Yes	Phytotoxic; carcinogenic
Bi	No	No	No	Yes	Yes	Yes	Relatively nontoxic
Cd	No	No	No	Yes	Yes	Yes	Narrow margin; bioaccumulative and phytotoxic; enriched in food chain; carcinogenic; itai-itai disease (Cd poisoning)
Co	Yes	Yes	Yes	Yes	Yes	Yes	Relatively phytotoxic; role in symbiotic N_2 fixation; carcinogenic
Cr	No	Yes	Yes	Yes		Yes	Cr^{6+} very toxic and mobile in soils; carcinogenic; Cr^{3+} relatively nontoxic to mammals
Cu	Yes	Yes	Yes	Yes			Easily complexed in soils; narrow margin for plants; immobile in soils; relatively nontoxic
F	No	Yes	Yes	Yes			Toxic to mammals in high doses; role in dental health; very mobile in soils
Fe	Yes	Yes	Yes			Yes	Fe deficiency in humans common worldwide; phytotoxic in low-pH soils; relatively nontoxic to mammals

Element						Remarks
Hg	No	No	No	Yes	Yes	Biomagnifies in aquatic food chain; a concern in newly established reservoirs; *Minamata* disease (Hg poisoning)
Mn	Yes	Yes	Yes			Wide margin; phytotoxic in low-pH soils; relatively nontoxic to mammals
Mo	Yes	Yes	Yes	5–20 ppm		High enrichment in plants; narrow margin for animals; molybdenosis in livestock
Ni	Yes	Yes	Yes	Yes	Yes	Very mobile in soils and plants; carcinogenic
Pb	No	No	No	Yes	Yes	Relatively nonphytotoxic; immobile in soils; human exposure to leaded gasoline, paint, and plumbing; young children most sensitive to Pb poisoning; a global social issue
Sb	No	No	No	Yes	Yes	Insoluble; relatively nonphytotoxic
Se	Yes	Yes	Yes	<4 ppm	Yes	Narrow margin for animals (selenocosis); interacts with other trace metals; similar geochemical behavior to S; Keshan and Kaschin–Beck diseases (Se deficiency)
Sn	No	Yes	No	Yes	Yes	Relatively nontoxic; very low uptake by plants
Ti	Possible	Possible	No			Insoluble; relatively nontoxic; possibly carcinogenic
Tl	No	No	No	Yes	Yes	Very mobile in plants; phytotoxic; highly toxic to mammals
V	Yes	Yes	No	Yes	Yes	Required by green algae; narrow margin and highly toxic in mammals; carcinogenic
W	No	No	No			Very mobile in plants; possibly phytotoxic
Zn	Yes	Yes	Yes			Wide margin; easily complexed in soils; similar geochemical behavior to Cd; may be lacking in some diets; relatively nontoxic to mammals

a Primarily referring to land-based plants and animals.
Sources: Extracted from Allaway (1968); Chang (1996); Hewitt and Smith (1974); Loehr et al. (1979); Luckey and Venugopal (1977); Miller and Neathery (1977); Underwood (1975, 1977); Van Hook and Shults (1976); Wood and Goldberg (1977); Zingaro (1979).

particulate matter in combusting coal (i.e., more fly ashes) and prohibition of ocean dumping of sewage effluent (i.e., more sewage sludge generated). Thus, it has become apparent that land will be the repository of choice for industrial by-products and hazardous wastes.

Increasing trends for the primary production of metals indicate society's need for technological advancement, to address the current emissions of huge quantities of some of the most environmentally important metals (e.g., Cd, Hg, Pb, Cr, Zn) into the biosphere (Table 1.2) (Appendix Tables A.1 to A.3). The major emitters appear to be smelting and refining for Zn, Pb, As, Cd, and Cu; energy production for Ni, Se, Hg, and V; and manufacturing for Cr and Tl.

Major contributors to the metal burden in soils include discarded manufactured products as in scrapheaps/landfills (As, Cr, Cu, Pb, Mn, and Zn), coal ashes (As, Cd, Pb, Mn, Hg, Mo, Ni, Se, V, and Zn), and agricultural and livestock wastes (As, Cu, and Zn). Although metals are generally enriched in coal combustion residues, only B, Mo, and Se, and possibly As, may be of some environmental concern (Adriano and Weber, 1998). Major contributors to the metal burden in aquatic systems include manufacturing (Sb, Cd, Cr, Cu, Mo, and Zn), electric power generation (As and Se), and domestic wastewater (As, Cd, Cr, Cu, Mn, Ni, and Zn). Indeed, the mobilization of metals from industrial activities is approximately 10 to 20 times greater for Cd, Cr, Cu, Pb, Hg, Mo, Se, and Zn than the releases from weathering of geologic materials (Appendix Table A.4).

Casting the information of Nriagu and Pacyna (1988) in a philosophical context, the toxicity of trace metals from emissions each year now exceeds that of all radioactive and organic pollutants, on the assumption that toxicity is measured as the amount of water that would be required to dilute

TABLE 1.2. Trends in the primary production of metals and the recent rate of global metal emissions reaching the soil (10^3 tonnes yr^{-1}).

Metal	Year			
	1930	1950	1980	1985
Al	120	1500	15396	13690
Cd	1.3	6	15	19
Cr	560	2270	11248	9940
Cu	1611	2650	7660	8114
Fe	80180	189000	714490	715440
Hg	3.8	4.9	7.1	6.8
Mn	3491	5800	26720	—
Ni	22	144	759	778
Pb	1696	1670	3096	3077
Sn	179	172	251	194
V	—	1.8	35	134
Zn	1394	1970	5229	6024

Sources: Extracted from Nriagu (1979); Nriagu and Pacyna (1988).

pollutant concentrations to safe drinking levels. Unlike organic pollutants, metals do not biodegrade, are generally not mobile, and therefore will reside in the environment for long periods of time.

Not all the news is bad, however. Atmospheric levels of Pb, Zn, Cu, and Cd have been dropping since the mid-1970s. Lead emissions will continue to fall worldwide because of regulations on leaded gasoline in the United States, Canada, Japan, and western Europe. However, emissions and environmental accumulation will continue in developing countries that have no controls on lead and other metals.

Nriagu (1984) indicated that in many instances the inputs from anthropogenic sources exceed the contributions from natural sources severalfold. Thus, it has become evident that human activities have altered the global biogeochemical cycles of trace elements.

3 Sources of Trace Metals

The United States Environmental Protection Agency (U.S. EPA) included 13 metals in their priority pollutants list: Ag, As, Be, Cd, Cr, Cu, Hg, Ni, Pb, Sb, Se, Tl, and Zn. Their natural origins are contrasted with their more diverse anthropogenic sources (Table 1.3). While parent rocks and metallic minerals dominate the natural sources, the anthropogenic sources range from agriculture (fertilizers, animal manures, pesticides, etc.), metallurgy (mining, smelting, metal finishing, etc.), energy production (leaded gasoline, battery manufacture, power plants, etc.), and microelectronics, to waste/scrap disposal. The pollutants can be released in gaseous (aerosol), particulate, aqueous, or solid form, depending on the industry. They can emanate from point or diffuse sources, as will be discussed in Chapter 4. Some of the more common sources of trace metals in the environment are discussed below.

3.1 In Agricultural Systems

Intensive land use in industrialized countries, especially in Europe and North America, has been stressing the soils with continuous and often heavy inputs of fertilizers, agrochemicals (pesticides), and soil amendments. For example, in 65% of areas surveyed in the European Union (EU), pesticide concentrations in groundwater exceed drinking water standards (Koshiek et al., 1994). Of the total landmass (~226 million ha) in the EU, about 57% is used for agriculture—of which 52% is arable and 37% is used for grass and green fodder production. The rest is used in fruit, wine grapes, and vegetable production.

The total metal load on agricultural soils is the sum of metal input from atmospheric deposition and the input from the addition of fertilizers, biosolids, metal-containing pesticides, and sometimes wastewater.

TABLE 1.3. Natural and anthropogenic sources and common forms in wastes of trace metals on the priority pollutant list.

Element	Natural source/or metallic minerals	Anthropogenic sources	Common forms in wastes
Ag	Free metal (Ag^0), silver chloride ($AgCl_2$), argentide (AgS_2), copper, lead, zinc ores	Mining, photographic industry	Ag metal, Ag–CN complexes, Ag halides, Ag thiosulfates
As	Metal arsenides and arsenates, sulfide ores (arsenopyrite), arsenite ($HAsO_2$), volcanic gases, geothermal springs	Pyrometallurgical industry, spoil heaps and tailings, smelting, wood preserving, fossil fuel combustion, poultry manure, pesticides, landfills	As oxides (oxyanions), organo-metallic forms, $H_2AsO_3CH_3$ (methylarsinic acid), $(CH_3)_2$-AsO_2H (dimethylarsinic acid)
Be	Beryl ($Be_3Al_2Si_6O_{16}$), phenacite (Be_2SiO_4)	Nuclear industry, electronic industry	Be alloys, Be metal, $Be(OH)_2$
Cd	Zinc carbonate and sulfide ores, copper carbonate and sulfide	Mining and smelting, metal finishing, plastic industry, microelectronics, battery manufacture, landfills/refuse disposal, phosphate fertilizer, sewage sludge, metal scrapheaps	Cd^{2+} ions, Cd halides and oxides, Cd–CN complexes, $Cd(OH)_2$ sludge
Cr	Chromite ($FeCr_2O$), chromic oxide (Cr_2O_3)	Metal finishing, plastic industry, wood treatment, refineries, pyrometallurgical industry, landfills, scrapheaps	Cr metal, Cr oxides (oxyanions), Cr^{3+} complexes with organic/inorganic ligands
Cu	Free metal (Cu^0), copper sulfide (CuS_2), chalcopyrite ($CuFeS_2$), mine drainage	Mining and smelting, metal finishing, microelectronics, wood treatment, refuse disposal/landfills, pyrometallurgical industry, swine manure, pesticides, scrapheaps	Cu metal, Cu oxides, Cu humic complexes, alloys, Cu^{2+} ions
Hg	Free metal (Hg^0), cinnabar (HgS), degassive from earth's crust and oceans	Mining and smelting, electrolysis industry, plastic industry, refuse disposal/landfills, paper/pulp industry, fungicides	Organo-Hg complexes, Hg halides and oxides, Hg^{2+}, Hg_2^{2+}, Hg^0

	Natural sources	Anthropogenic sources	Species
Ni	Ferromagnesian minerals, ferrous sulfide ores, pentladite	Iron and steel industry, mining and smelting, metal finishing, microelectronics, battery manufacture	Ni metal, Ni^{2+} ions, Ni amines, alloys
Pb	Galena (PbS)	Mining and smelting, iron and steel industry, refineries, paint industry, automobile exhaust, plumbing, battery manufacture, sewage sludge, refuse disposal/landfills, pesticides, scrapheaps	Pb metal, Pb oxides and carbonates, Pb-metal-oxyanion complexes
Sb	Stibnite (Sb_2S_3), geothermal springs, mine drainage	Microelectronics, pyrometallurgical industry, smelting	Sb^{3+} ions, Sb oxides and halides
Se	Free element (Se^0), ferroselite ($FeSe_2$), uranium deposits, black shales, chalcopyrite–pentladite–pyrrhotite desposits	Smelting, fossil fuel combustion, irrigation waters	Se oxides (oxyanions), Se–organic complexes
Tl	Copper, lead, silver residues	Pyrometallurgical industry, microelectronics, cement industry	Tl halides, Tl–CN complexes
Zn	Zinc blende (ZnS), willemite ($ZnSiO_4$), calamine ($ZnCO_3$), mine drainage	Mining and smelting, metal finishing, textile, microelectronics, refuse disposal/landfills, pyrometallurgical industry, sewage sludge, pesticides, scrapheaps	Zn metal, Zn^{2+} ions, Zn oxides and carbonates, alloys

Sources: Distilled from Adriano (1986); Ross (1994); Salomons and Forstner (1984); Smith et al. (1995); U.S. Bureau of Mines (1980).

3.1.1 Parent Rocks

This subject matter has been reviewed by Cannon (1978) and Mitchell (1964) and will not be discussed at length here. Free of human interference, the trace metal content of the soil is largely dependent on that of the rocks from which the soil parent material was derived and on the process of weathering to which the soil-forming materials have been subjected. The more aged and older the soil, the less may be the influence of parent rocks. The extremely variable nature of trace element concentrations in various soil-forming rocks is demonstrated in Appendix Table A.5.

3.1.2 Phosphate and Other Fertilizers

Phosphatic fertilizers and mineral sewage sludge are considered the most important sources of metal contamination in agricultural lands. This is particularly true for Cd. Fertilizers made from magmatic phosphates tend to have only negligible concentrations of Cd, whereas those from sedimentary phosphates tend to have high levels (Hansen and Tjell, 1983). Rock phosphate from Senegal contained unusually high Cd (≥ 7 mg kg^{-1}); those from Morocco and Tunisia were unusually high also (18 mg kg^{-1}), but that from the United States had only ≤ 6.5 mg kg^{-1} Cd (Hutton and Symon, 1986). Because of the potential risk in Cd enrichment of soils from long-term application of phosphate fertilizers, especially in sensitive soils, the levels of Cd in these fertilizers are now being constrained (Appendix Table A.6).

A micronutrient survey in the 1960s revealed that soils are deficient in plant-available forms of one or more of the micronutrients Zn, Mn, Fe, Cu, Mo, and B, and that applications are recommended in every state of the United States and every country of the world (Sparr, 1970). The approximate quantities of micronutrients sold for agricultural use in the United States in 1968, in thousands of tonnes, were Zn, 14.5; Mn, 10.6; Cu, 2.4; B, 2.5; and Mo, 0.8. Although micronutrient requirements by crops varied considerably from state to state, more vegetable crops needed these elements than did field, forage, or fruit and nut crops. The form, composition, and rates of these micronutrients commonly used in agriculture are available (Murphy and Walsh, 1972).

In addition to these micronutrient carriers, other commercial fertilizers contain small amounts of trace elements (Appendix Table A.7). Phosphatic fertilizers contain varying amounts of Zn, Cd, and other trace elements that originated from phosphate rock. Differences in trace element content of phosphate rocks mined in various areas are caused by impurities coprecipitated with the phosphates at the time of deposition. In general, phosphate rock from the western United States contains higher concentrations of most trace elements than phosphate rock from eastern phosphate deposits. Application of 500 kg ha^{-1} of diammonium phosphate (20-48-0) for 100 years using North Carolina phosphate rock contributed to the soil the following amounts, in kg ha^{-1}: Zn, 14.25; Cu, 0.05; Cd, 1.50; Cr, 9.75; Ni, 1.90; and Pb, 0.24 (Mortvedt and Giordano, 1977).

Because of the variation in Cd content in phosphatic rocks, annual Cd input to agricultural lands varies from country to country—3.5 g ha^{-1} yr^{-1} for Germany (Kloke et al., 1984) and 4.3 g ha^{-1} yr^{-1} in the United Kingdom (Hutton and Symon, 1986). An experimental plot at Rothamsted (United Kingdom) receiving superphosphate over a 96-yr period, received ~5 g ha^{-1} yr^{-1} (Rothbaum et al., 1986).

3.1.3 Pesticides

Pesticides (i.e., herbicides, insecticides, fungicides, rodenticides, etc.) are widely used in high-production agriculture for the control of insects and diseases in fruit, vegetables, and other crops. They can be applied by spraying, dusting, or soil application. Sometimes, seed treatment is used. Typical pesticides in use in Canada for several decades are shown in Appendix Table A.8. The amounts of elements applied vary according to the type of pesticide used, ranging from as low as 0.002 kg ha^{-1} of Hg from methyl mercurials to as high as 0.5 kg ha^{-1} of As and 2.3 kg ha^{-1} of Pb from lead arsenate per application (Frank et al., 1976). Such contributions can total several kilograms of certain elements in a given fruit orchard. It should be noted that metal-based pesticides are no longer in use.

3.1.4 Sewage Effluents and Wastewater

Municipality-owned wastewater treatment plants in the United States treat over 29 trillion liters of wastewater annually, generating approximately 8 million tonnes of biosolids (U.S. Federal Register, 1989). Under the right conditions, land application of wastewater can be a more effective and less energy-intensive treatment option than conventional systems, such as activated sludge process, trickling filter, or aerated lagoons. However, transportation costs, land use (agronomic, horticultural, silvicultural, etc.), soil type (i.e., infiltration rates), and public opinion may weigh heavily on its suitability.

Sewage effluents have been applied to land in Europe and Australia for nearly a hundred years, and at several sites in the United States, applications have been in progress for over 50 years (Carlson and Menzies, 1971). A good example of how treated sewage effluents can be used on the land is the system used at Pennsylvania State University (Kardos, 1970). The system involves using the water and nutrients in sewage effluents on forest and cultivated crops. Recycling nutrients to the land, restoring groundwater, preventing stream pollution, and eliminating the need to add commercial fertilizers to cropland are some of the benefits derived from such a system. This particular system holds promise for solving waste disposal problems in small villages and cities. The potential for recycling sewage effluents in cropland has been demonstrated also at a sewage farm in Braunschweig,

Germany, where, even after operation since 1895, the trace element burden of the soil is still within the tolerable limits with regard to their plant growth compatibility as proposed by the German Board of Environmental Protection (El-Bassam et al., 1979).

While N and other mineral nutrients derived from effluents are beneficial to plants, it is the water itself that may offer the most benefit, especially in regions where shortages occur for crop production.

A wastewater reclamation study (Sheikh et al., 1990) was conducted in Monterey, California to demonstrate the feasibility and safety of irrigating food crops (e.g., artichokes, celery, broccoli, lettuce, and cauliflower) with tertiary treated effluents. The treated water was applied over a 5-yr period in conjunction with standard agricultural practices (sprinkler and furrow irrigation, fertilization, and pesticide application). The results indicate excellent yields of high-quality produce and, more importantly, no bioaccumulation of heavy metals took place; in fact, far greater amounts of metals were added to the soil from the fertilization with conventional fertilizers.

Reviews on potential hazards from trace elements in wastewater applied to land have been conducted (Bouwer and Chaney, 1975; Knezek, 1972; Leeper, 1972). The same elements listed under sewage sludge (Cd, Zn, Cu, Ni, Pb, etc.) are important potential hazards if present in wastewater. In addition, B is considered potentially hazardous in any irrigation water if present at >0.75 ppm. At this concentration, phytotoxic symptoms have already been observed in some forest tree species (Neary et al., 1975).

3.1.5 Biosolids

Biosolids [sewage sludge, animal wastes, municipal solid waste (MSW), and some industrial wastes, e.g., paper pulp sludge] are an important group of soil amendments that are gaining more popularity in agricultural, forested, and reclaimed lands. In addition to supplying plant nutrients, the organic matter (OM) in biosolids enhances soil tilth, pore space, aeration, and water retention capacity. Biosolids have three major constituents that may affect the degree of safety in their use in agricultural lands: nutrients, potentially toxic metals, and pathogens. Only sewage sludge and animal wastes are discussed here.

As the residue of wastewater treatment, sewage sludge represents an agglomeration of contaminants originally present in the wastewater. Sewage sludge typically contains OM, trace elements, pathogens, organic chemicals, essential plant nutrients, and dissolved solids. At present, over 5.5 million dry tonnes of sewage sludge are produced annually in the United States (U.S. EPA, 1993) and comparable amounts are generated in western Europe (Davis, 1992). Of the total sludge amount generated in the United States, ~33% is destined for land application, ~33% for landfilling, ~16% for incineration, and the balance for surface and ocean disposal. Similar trends exist for large western European countries where, because of banning of ocean disposal, more and more sludge is being applied on land. It is very important to note that as wastewater treatment technologies improve, the quality of sewage sludge in terms of metal content should also improve. Also, aqueous discharges of metals from metal-dealing industries into the sewage system have been eliminated or severely limited in many countries. Thus, metal content data obtained more than two decades ago should be used with caution.

A number of factors must be considered when applying sewage sludge to agricultural lands: agronomic application rate, annual metal loading rate, cumulative metal load, type of crops, application mode, soil pH and cation exchange capacity (CEC), and site characteristics (see Chapter 4).

The elements of primary concern in sewage sludge include Cd, Zn, Cu, Pb, Se, Mo, Hg, Cr, As, and Ni, which may depress plant yield or degrade the quality of food or fiber produced when applied to soils in excessive amounts. The concentrations of trace elements in soils in areas treated with sewage sludge can be expected to increase with increasing amount of application or increasing concentrations of these elements in the sludge. The variable nature of trace element concentrations in sewage sludge has been reviewed (Page, 1974; Sommers, 1977). Regulatory guidelines for land application of sewage sludge have been formulated in the European Union (EU), Canada, Japan, Australia, and the United States. Comprehensive appraisals of the potential hazards of trace elements in sewage sludge to biota from land application of sewage sludge have been conducted in these countries (CAST, 1980).

The variable nature of trace element concentrations in sewage sludge is demonstrated in Appendix Table A.9. The concentrations are largely dependent on the type and amount of urban and industrial discharges into the sewage treatment system and on the amount added in the conveyance and treatment system. As urbanization and industrialization progress, the concentrations of most trace elements in sewage sludge can be expected to increase. As a general precaution, sludge should be analyzed for nutrients and trace elements before application on land to aid in calculating the fertilizer and metal loads for a particular situation.

A potentially major use of sewage sludge is in the reclamation and revegetation of drastically disturbed land (Sopper, 1993). When composted municipal sludge was applied properly according to present guidelines, no adverse effects on the soil, vegetation, or groundwater quality were observed. Application areas include strip-mine spoils, gravel spoils, coal refuse, clay–strip-mine spoils, iron ore tailings, abandoned pyrite mine spoils, and sites devastated by toxic fumes.

Synthesis of the literature on sludge application to croplands reveals the following results (Juste and Mench, 1992): sludge-borne metals remain in the zone of sludge application (0 to15 cm depth); phytotoxicity was rarely observed in grain crops; sludge application exhibited a positive effect on plant growth in 65% of the cases; Zn is the most bioavailable sludge-borne metal, followed by Cd and Ni, whereas Cr and Pb plant uptake was insignificant; harmful effects on certain legumes were attributed to the detrimental influence of metals on microbial activity in soils, as in N_2 fixation; and the concentrations of metals in plant tissue were influenced by the application rate (annual or cumulative), time following cessation of sludge application, and other factors.

Animal wastes have been of environmental concern because of their potential as a public nuisance (flies and malodor) and in contaminating the surface water and groundwater with nitrates, salts, and metals (Adriano, 1986). Recently, an outbreak of red tide disease in coastal waters off Maryland was being attributed to marine pollution by animal waste runoff from concentrated livestock operations inshore. This outbreak could threaten the existence of major seafood industries (fishery and shellfish) due to the high biochemical oxygen demand (BOD) and microbial (fecal coliform) contents of livestock wastes.

Concentrations of elements in animal wastes are variable, induced by the classes of livestock (cattle, swine, poultry, etc.). Within a class, additional variation is associated with age of the animal, type of ration, housing type, and even waste management practice. The trace elements in animal wastes originate from the rations and whatever dietary supplement is added. For example, in the production of milk and pork, Cu an Co may be added to the diet as 1% copper sulfate and as 1% cobalt iodized salts. Some swine and poultry are fed diets containing up to 250 mg kg^{-1} Cu and 100 to 200 mg kg^{-1} Zn in the absence of antibiotics. The manure from these animals is then 10 to 40 times higher in Cu and 4 to 10 times higher in Zn than normal. Application of these manures at N fertilization rate could add 3 to 6 kg Cu ha^{-1} yr^{-1} (Baker, 1974). Typical trace element composition of animal wastes is shown in Appendix Table A.10. Unlike sewage sludge, animal wastes are rarely high in potentially harmful trace elements. However, similar to sewage sludge, there are also land disposal problems associated with these abundant wastes, produced at a daily rate of about 3.4×10^6 tonnes in the United States. Foremost problems are likely to be excess N and salts (Adriano et al., 1971, 1973) and nutrient imbalance in plants (e.g., causing grass tetany in cattle with the use of poultry manure).

3.1.6 Coal Combustion Residues

Use of coal combustion to generate electric power has been increasing in developed countries to curtail dependence on foreign oil. The majority of the 556×10^6 tonnes of coal produced in the United States in the late 1970s was consumed by the electric utilities. As part of regulations to combat releases of particulate matter and greenhouse and acid rain gases, fly ash that otherwise could become fugitive dust and flue gas desulfurization wastes are captured each year by coal-burning power plants. Current total ash production by electric utilities is estimated at more than 105×10^6

tonnes yearly; 45% of this amount is fly ash (Adriano and Weber, 1998). Of the total ash produced in the United States, only about 30% is being used for cement making, concrete mixing, ceramics, and other products. Because there is no consistent trend of effective utilization, the accumulating fly ash becomes a continuing waste disposal problem. Excellent reviews on coal residues and their characterization, potential for utilization, and potential hazards to plants and/or animals have been conducted recently (Carlson and Adriano, 1993).

The concentration of trace elements in coal residues is extremely variable and depends on the composition of the parent coal, conditions during coal combustion, efficiency of emission control devices, storage and handling of the by-product, and climate. Typical concentrations of trace elements that can be expected from the combustion of fossil fuels are shown in Appendix Table A.11. When compared with elemental mobilization due to natural processes, such as weathering, some values are quite substantial. Particularly notable are the amounts of As, Cd, Mo, Se, and Zn.

Trace elements in coal residues that can bioaccumulate and, therefore, could be critical in the food chain are As, B, Mo, and Se (Adriano et al., 1980; Carlson and Adriano, 1993).

3.1.7 Atmospheric Deposition

In industrial countries, atmospheric deposition can be an important source of metals for soils and plants in agroecosystems. Indeed, nationally and regionally, this source may be the major source for metal input to agricultural plants and soils (Haygarth and Jones, 1992). This results from surficial deposition onto the foliage, or uptake and translocation of aerially borne metals entering the ecosystems in particulate form.

Volatile metalloids (e.g., Se, Hg, As, and Sb) can be transported in gaseous form or enriched in particles, whereas other metals (e.g., Cu, Pb, and Zn) are transported only in the particle phase. For example, Se input to agroecosystems in the United Kingdom from atmospheric deposition was estimated to be about two orders of magnitude higher than input from commercial fertilizers, i.e., 600 vs. 1.4 μg m^{-2} yr^{-1} (Haygarth and Jones, 1992). For Pb, about 90% of the total plant uptake can be attributed to atmospheric deposition rather than uptake from the soil. Likewise for Cd, atmospheric input can contribute as much as 50% or higher to the total plant concentrations of agronomic and horticultural crops (Harrison and Chirgawi, 1989a,b). The annual Cd deposition may vary from <0.1 g ha^{-1} in remote areas (northern Scandinavia and southern Spain, Italy, Greece) to over 2.4 g ha^{-1} in industrialized regions (e.g., central Poland) (Kohsiek et al., 1994). The impact of atmospheric deposition of industrial origin is apparent from the data for Norway where forest vegetation in the southern region showed higher enrichment ratios than for similar vegetation in the northern part, i.e., the former area being closer to the industrial hub of Europe (Berthelsen et al., 1995).

Smelting metal ores has been a major source of metal input in localized areas by the smelters. The emission impact can be noticed for several kilometers away from the stacks (Appendix Table A.12). In most cases, the ecological effects are caused not only by metal emissions but also by releases of acid-producing constituents (Dudka and Adriano, 1997).

3.2 Quantified Sources: A Case Study

An inventory of metal inputs to agricultural soils in England and Wales was constructed for several major metals from the following known major sources: atmospheric deposition, sewage sludge, livestock manure, fertilizers and lime, agrochemicals, irrigation water, and industrial by-products (Nicholson, 1998). The relative proportions of annual metal inputs to agricultural lands in England and Wales from all sources are indicated in Figure 1.1. For Zn, Cu, and Ni, 25 to 40% of total annual inputs (5247, 1821, and 225 tonnes for Zn, Cu, and Ni, respectively) to agricultural land were derived from animal manure, 33 to 45% from atmospheric deposition, and <15% from sewage sludge. By contrast, 71% of Pb (total 630 tonnes) was from atmospheric deposition and only 8% from animal manure. For Cd (39 tonnes), 50% of inputs were from atmospheric deposition and 34% from fertilizers (mainly phosphate fertilizers) and lime, with 11% from animal manure. The major source of Cr (574 tonnes total) to agricultural land was industrial by-product wastes (37%), with sewage sludge, fertilizers and lime, and atmospheric deposition all of equal importance ($\sim$20%). Over 50% of As (58 tonnes total) and over 90% of Hg (23 tonnes total) inputs were from atmospheric deposition.

It is important to note that although atmospheric deposition and animal manure are generally the major sources of metal inputs to agricultural land in terms of total quantities, loading rates to individual fields are relatively low. By contrast, metal loading rates in fields receiving sewage sludge or industrial wastes are generally high compared with livestock manure.

3.3 In Groundwater

Groundwater quality is very important to society because groundwater serves as a direct source of potable water from wells drilled into the aquifer, as a source of irrigation water, and as an outcropping to recharge surface reservoirs. Therefore, tainted groundwater can contaminate both terrestrial and aquatic food chains, as well as pose a direct threat to human health when used for drinking and cooking.

Since the 18th century, industrialization in Europe has proceeded without proper disposal of the generated wastes, until only recently (Kohsiek et al., 1994). Often, industrial fluids heavily contaminated with heavy metals and organics were injected via wells into the groundwater, leading to expensive

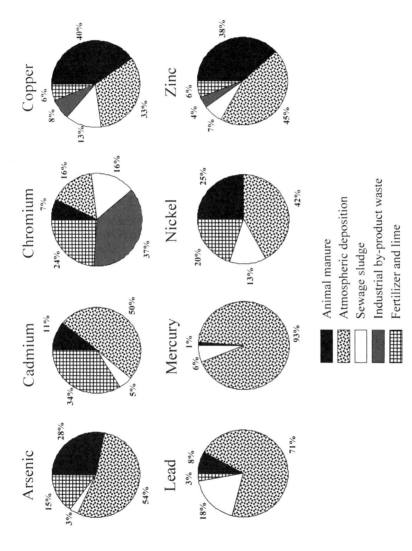

FIGURE 1.1. Relative proportions of the different sources of metal input on agricultural soils in Wales and England. (From Nicholson, 1998.)

TABLE 1.4. Typical sources of inorganic and organic substances for groundwater contamination.

Source	Inorganic contaminants	Organic contaminants
Agricultural areas	Heavy metals, salts (Cl^-, NO_3^-, SO_4^{2-})	Pesticides
Urban areas	Heavy metals (Pb, Cd, Zn), salts	Oil (petrol) products, biodegradable organics
Industrial sites	Heavy metals, metalloids, salts	Polycyclic aromatic hydrocarbons (PAHs), chlorinated hydrocarbons (trichloroethylene and tetrachloro-ethylene), hydrocarbons (benzene, toluene, xylene), oil (petrol) products
Landfills	Salts (Cl^-, NH_4^+), heavy metals	Biodegradable organics and xeno-biotics
Mining disposal sites	Heavy metals, metalloids, salts	Xenobiotics
Dredged sediments	Heavy metals, metalloids	Xenobiotics
Hazardous waste sites	Heavy metals, metalloids (more concentrated)	Concentrated xenobiotics
Leaking storage tanks	—	Oil (petrol) products
Line sources (motorway, railways, sewerage systems, etc.)	Heavy metals (Cd, V, Pb), salts	PAHs, oil products, pesticides

Sources: Extracted from Adriano (1986); Kohsiek et al. (1994).

contamination of soils and groundwater in and around industrialized towns of Europe. The sources of metal pollutants posing threats to groundwater quality can be categorized as shown in Table 1.4. The fate and pathways of metals to the groundwater are schematically depicted in Figure 1.2.

3.4 In Freshwater Systems

The major sources of surface water contamination include municipalities, agriculture, construction, and industry; more specific sources include wastewater, and runoff and erosion from upland areas (croplands, mining areas, etc.).

There are two types of wastewater: domestic and industrial. The first is a combination of human feces, urine, and graywater (from washing, bathing, and kitchen preparation), and the latter comes from industrial establishments. When wastewaters from some industries, schools, hospitals, airports, etc., are combined with domestic waste effluents for treatment, this becomes known as municipal wastewater. Most of the high-volume aqueous waste in the United States is treated in wastewater treatment plants prior to disposal. In the United States, >90% of all municipal wastewater and nonhazardous industrial wastewater as well as 10% of all hazardous wastewater is treated prior to disposal. Today, more than 15,000 wastewater treatment plants

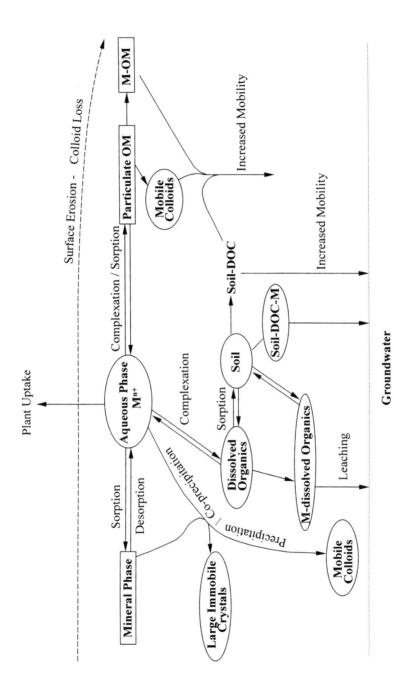

FIGURE 1.2. Fate and transport of trace elements in soil–groundwater systems. (Modified from Harter and Naidu, 1995.)

treat approximately 150 billion liters per day of wastewater in the United States (U.S. EPA, 1985). The chemical, paper, primary metal, and food processing industries discharge, in descending order, the most volumes of wastewater in the United States (Rhyner et al., 1995). The major pollutants in wastewater include toxic metals, biodegradable and recalcitrant organics, suspended matter, and pathogenic microorganisms. While wastewater treatment prior to disposal is the norm in developed countries, it is a rarity in developing countries.

Metals and other chemicals that are sorbed onto the soil solids can be transported by storm runoff (i.e., soil erosion) to riverine and lake ecosystems. Transported solids from croplands may contain metals, pesticides, phosphate, and other nutrients that are sorbed or complexed on clay minerals, humus, carbonates, etc. Likewise, transported solids from mining areas carry metals, salts, and other solutes that may enrich bodies of water not only with metals but also with acidifying constituents. The livestock industry is also a potentially major source of pollution, primarily in the form of runoff, especially during spring in temperate regions when snow deposited over the waste melts, soil infiltration is slow, and evaporation is slow.

Mining waste includes the materials that are moved to gain access to minerals (ores) and the tailings, slags, and residues that result from the processing of the materials. The low content of many minerals in ore deposits results in large quantities of earth disturbed in mining operations. Metals that often occur in the sulfide phases in the waste are exposed to the ground surface, subjecting them to oxidation and dissolution from the solid phases upon acidification. Tailings are the waste generated from the physical and chemical enrichment (beneficiation) process used to separate the valuable metals from the ore. Traditionally, mine waste and tailings have been disposed of as economically and expeditiously as possible. Ideally, the waste materials should be returned to the excavated area and the disturbed area restored or rehabilitated. But this is not always possible or practical, especially when a mine is still in operation (Rhyner et al., 1995). Estimates of mining waste generated in the United States range from 1 to 2 billion tonnes annually, with about one-half being mining waste and the rest tailings and dump/heap waste.

The metal industry can be an important source of trace elements in the environment from (1) the mining and milling operations with problems of grinding, concentrating, and transporting ores, and disposal of tailings along with mine and milling wastewater, and (2) the smelter–refinery process with problems of concentrate, haulage, storage, sintering, refining, atmospheric discharges, and fugitive dust. The proportion of trace elements released into the environment depends on the ores being processed. In the lead industry, Pb, Cu, Zn, and Cd can be released in substantial amounts (Dudka and Adriano, 1997; Wixson et al., 1973). In the case of Ni and Cu smelting, Co, Zn, Pb, and Mn, as well as Ni and Cu, can also enrich the surrounding environment significantly (Dudka and Adriano, 1997; Hutchinson and Whitby, 1974). Similarly, in smelting zinc ores, sizable releases of Zn, Cd, Cu, and Pb can occur (Dudka and Adriano, 1997; Jordan, 1975).

The impacts of atmospheric discharges from smelters can be detected within several kilometers from the point of release (Appendix Table A.12). In Great Britain, widespread contamination by Pb, Cd, and Zn in some mining areas has been found (Davies and Ginnever, 1979; Davies and Roberts, 1978).

Mining activities and spoils are concentrated in several large regions all over Europe, especially in England, France, Germany, former Czechoslovakia, Poland, Ukraine, and Russia (Kohsiek et al., 1994). The exploitation of these regions started over 100 years ago. Acid leachates, which are characteristically high in total dissolved solids and sometimes salinity, result from coal mining wastes that can contaminate surface water and groundwater.

3.5 In Forest Ecosystems

Atmospheric deposition is the main source of trace metals in forests. The initial source might be hundreds of kilometers away (diffuse source) from the impacted areas (Asami, 1984; Dudka and Adriano, 1997). For example, the documented contamination of ice in polar regions (e.g., Pb, Hg, Zn) illustrates the role of long-distance transport of contaminants virtually causing global contamination (Cheam et al., 1998; Murozumi et al., 1969; Rosman et al., 1993). In later chapters on specific elements in this book, ecological effects on forest ecosystems are discussed in sections on Nature for most of the elements.

3.6 In Indoor and Urban Environments

Worldwide, large cities and urban areas tend to have the worst environmental pollution problems due to high population density, energy usage, and industrial and transportation activities. The U.S. EPA has set National Air Quality Standards for carbon monoxide (CO), nitrous oxide (NO_2), ozone (O_3), particulate matter, sulfur dioxide (SO_2), and a metal, Pb. For Pb, the limit was set as a quarterly maximum of 1.5 $\mu g\ m^{-3}$ of air (U.S. EPA, 1994).

Indoor air contains a complex mix of volatile organic compounds (VOCs) and gases from fabrics, floor coverings, dyes, pesticides, refrigerants, and heating and cooking fuels. Lead can be added to this list of indoor pollutants; it has been shown to add to elevated blood Pb levels in humans in metropolitan areas, especially in old housing where leaded paints and lead plumbing were used (Thornton, 1991). Lead, a neurotoxin, comes primarily from the house dust and potable water. The dust originates from auto exhaust emissions (leaded gasoline) entering the windows and resuspension from floor coverings, such as carpet. Lead can be dissolved from lead pipes and solder, especially with hot soft water (low calcium content). For the toxicology of environmental Pb, see Chapter 10.

The greatest single source of air pollution in many countries is automobile exhaust. For example, nearly all the Pb in the air in the United Kingdom comes from the exhaust gases of petrol engines (Thornton, 1991). Roadside soils and vegetation have been shown to be contaminated with various trace elements primarily from auto emissions (Lagerwerff and Specht, 1970; Motto et al., 1970). These include Pb, Zn, Cd, Cu, and Ni, the more important being Pb as an additive from fuels and Zn from tires. Depending on the location and traffic intensity, contaminated zones can extend up to several hundred meters from the road.

In a survey of trace metal (As, Cu, Ni, Pb, Sn, and Zn) concentrations in sediments from the East and Gulf of Mexico coasts of the United States, it was concluded that highly elevated levels of metals, except As and Ni, are found at numerous sites located near large cities (Daskalakis and O'Conner, 1995). In addition, widespread use of solid waste incinerators in the United States and Europe over the last century suggests that solid waste incineration may have provided the dominant source of atmospheric Pb and other metals to many urban centers (Chillrud et al., 1999).

The contribution of the above sources of trace metals to contamination of the environment varies from country to country and from area to area within a country, depending on a myriad of factors, including climate, type of industrial activity, pollution abatement, and others.

References

Adriano, D.C. 1986. *Trace Elements in the Terrestrial Environment.* Springer-Verlag, New York.

Adriano, D.C., and J.T. Weber. 1998. *Coal Ash Utilization For Soil Amendment to Enhance Water Relations and Turf Growth.* EPRI Rep TR 111318, EPRI, Palo Alto, CA.

Adriano, D.C., P.F. Pratt, and S.E. Bishop. 1971. *Soil Sci Soc Am Proc* 35:759–762.

Adriano, D.C., A.C. Chang, P.F. Pratt, and R. Sharpless. 1973. *J Environ Qual* 2:396–399.

Adriano, D.C., A.L. Page, A.A. Elseewi, A.C. Chang, and I. Straughan. 1980. *J Environ Qual* 9:333–344.

Ainsworth, C., and D. Rai. 1987. *Chemical Characterization of Fossil Fuel Combustion Wastes.* EPRI Rep EA-5321, EPRI, Palo Alto, CA.

Allaway, W.H. 1968. *Adv Agron* 20:235–274.

Arora, C.L., V.K. Nayyar, and N.S. Randhawa. 1975. *Indian J Agric Sci* 45:80–85.

Assami, 1984. In J.O. Nriagu, ed. *Changing Metal Cycles and Human Health* (Dahlem Konferenzen). Springer-Verlag, Berlin.

Baker, D.E. 1974. *Proc Fed Am Soc Exp Biol* 33:1188–1193.

Berthelsen, B.O., E. Steinnes, W. Solberg, and L. Jingsen. 1995. *J Environ Qual* 24:1018–1026.

Berrow, M.L., and J. Webber. 1972. *J Sci Fed Agric* 23:93–100.

Bouwer, H., and R.L. Chaney. 1975. *Adv Agron* 26:133–176.

Cannon, H.L. 1978. *Geochem Environ* 3:17–31.

Capar, S.G., J.T. Tanner, M.H. Friedman, and K.W. Boyer. 1978. *Environ Sci Technol* 7:785–790.

Carlson, C., and D.C. Adriano. 1993. *J Environ Qual* 22:227–247.

Carlson, C.W., and J.D. Menzies. 1971. *BioScience* 21:561–564.

[CAST] Council for Agricultural Science and Technology. 1976. In EPA-43019-76-013, US EPA, Washington, DC.

Chang, L. 1996. *Toxicology of Metals*. CRC Pr, Boca Raton, FL.

Cheam, V., G. Lawson, J. Lechner, and R. Desrosiers. 1998. *Environ Sci Technol* 32:3974–3979.

Chillrud, S.N., R.F. Bopp, J.M. Ross, and A. Yarme. 1999. *Environ Toxicol Chem* 33:657–662.

Chowdhury, T.R. et al. 1997. In C.O. Abernathy et al., eds. *Arsenic: Exposure and Health Effects*. Chapman & Hall, London.

Daskalakis, K.O., and T.P. O'Conner. 1995. *Environ Sci Technol* 29:470–477.

Davies, B.E. 1992. In D.C. Adriano, ed. *Biogeochemistry of Trace Metals*. Lewis Publ, Boca Raton, FL.

Davies, B.E., and L.J. Roberts. 1978. *Water Air Soil Pollut* 9:507–518.

Davies, B.E., and R.C. Ginnever. 1979. *J Agric Sci Comb* 93:753–756.

Davis, R.D. 1992. *Water Qual Int* 3:22–32.

Domingo, L.E., and K. Kyuma. 1983. *Soil Sci Plant Nutr* 29:439–452.

Dudka, S., and D.C. Adriano. 1997. *J Environ Qual* 26:590–602.

El-Bassam, N., C. Tietjen, and J. Esser. 1979. In *Management and Control of Heavy Metals in the Environment*. CEP Consultants, Edinburgh, United Kingdom.

Frank, R., K. Ishida, and P. Suda. 1976. *Can J Soil Sci* 56:181–196.

Furr, K.A., A.W. Lawrence, S.S.C. Tong, M.G. Grandolfo, R.A. Hofstader, C.A. Bache, W.H. Gutenmann, and D.J. Lisk. 1976. *Environ Sci Technol* 7:683–687.

Hamada, R., and M. Osame. 1996. In L.W. Chang, ed. *Toxicology of Metals*. CRC Pr, Boca Raton, FL.

Hansen, J.A., and J.C. Tjell. 1983. In R.D. Davis et al., eds. *Environmental Efforts of Organic and Inorganic Contaminants in Sewage of Organic and Inorganic Contaminants in Sewage Sludge*. Reidel, Dordrecht, Netherlands.

Harrison, R.M., and M.B. Chirgawi. 1989a. *Sci Total Environ* 83:13–34.

Harrison, R.M., and M.B. Chirgawi 1989b. *Sci Total Environ* 83:35–45.

Hart, B.T. 1982. In P.G. Sly, ed. *Sediment–freshwater interaction, Developments in Hydrobiology*. Dr W Junk, The Hague, Netherlands.

Harter, R.D., and R. Naidu. 1995. *Adv Agron* 55:219–263.

Haygarth, and K.C. Jones. 1992. In D.C. Adriano, ed. *Biogeochemistry of Trace Metals*. Lewis Publ, Boca Raton, FL.

Hewitt, E.J., and T.A. Smith. 1974. *Plant Mineral Nutrition*. English Univ. Pr, London.

Hutchinson, T.C., and L.M. Whitby. 1974. *Environ Conserv* 1:123–132.

Hutton, M., and C. Symon. 1986. *Sci Total Environ* 57:129–150.

Jordan, M.J. 1975. *Ecology* 56:78–91.

Juste, C., and M. Mench. 1992. In D.C. Adriano, ed. *Biogeochemistry of Trace Metals*. Lewis Publ, Boca Raton, FL.

Kardos, L.T. 1970. *Environment* 10(2):10–27.

Kloke, A., D.R. Sauerbeck, and H. Vetter. 1984. In J.O. Nriagu, ed. *Changing Metal Cycles and Human Health* (Dahlem Konferenzen). Springer-Verlag, Berlin.

Knezek, B.D. 1972. *Heavy Metal Reactions in the Soil.* Mich State Univ Inst Water Res Tech Rep 30:2743, East Lansing, MI.

Kohsiek, L.H.M., D. Fraters, R. Franken, W.J. Willems et al. 1994. In M.H. Donker et al., eds. *Ecotoxicology of Soil Organisms.* Lewis Publ, Boca Raton, FL.

Lagerwerff, J.V., and A.W. Specht. 1970. *Environ Sci Technol* 4:583–588.

Leeper, G.W. 1972. *Reactions of Heavy Metals with Soil with Special Regard to Their Application of Sewage Wastes.* Contract DACW 73-73-C-0026. Dept of Army, Corps of Engrs, Washington, DC.

Loehr, R.C., W.J. Jewell, J.D. Novak, W.W. Clarkson, and G.S. Friedman. 1979. *Land Applications of Wastes,* vol. 2. Van Nostrand Reinhold, New York.

Luckey, T.D., and B. Venugopal. 1977. *Metal Toxicity in Mammals: Physiologic and Chemical Basis for Metal Toxicity.* Plenum Pr, New York.

McLaughlin, M.J., K.G. Tiller, R. Naidu, and D.P. Stevens. 1996. *Aust J Soil Res* 34:1–54.

Meharg, A.A., D. Osborne, D.J. Pain, and A. Sanchez. 1999. *Environ Toxicol Chem* 18:811–812.

Miller, W.J., and M.W. Neathery. 1977. *BioScience* 27:67–679.

Mitchell, R.L. 1964. In F.E. Bear, ed. *Chemistry of the Soil.* Reinhold, New York.

Mortvedt, J.J., and P.M. Giordano. 1977. In H. Drucker and R.E. Wildung, eds. *Biological Implications of Metal in the Environment.* CONF-750929. NTIS, Springfield. VA.

Motto, H.L., R.H. Dames, D.M. Chilko, and C.K. Motto. 1970. *Environ Sci Technol* 4:231–237.

Murozumi, M., T.J. Chow, and C.C. Patterson. 1969. *Geochim Cosmochim Acta* 33:1247–1294.

Murphy, L.S., and L.M. Walsh. 1972. In J.J. Mortvedt, P.M. Giordano, and W.L. Lindsay, eds. *Micronutrients in Agriculture.* Soil Sci Soc Am, Madison, WI.

National Research Council. 1997. *Environ Sci Technol* 31(1).

Neary, D.G., G. Schneider, and D.P. White. 1975. *Soil Sci Soc Am Proc* 39:981–982.

Nicholson, F. 1998. ADAS Gleadthorpe Res Center, Nottinghamshire, United Kingdom, personal communication.

Nogawa, K., and T. Kido. 1996. In L.W. Chang, ed. *Toxicology of Metals.* CRC Pr, Boca Raton, FL.

Nriagu, J.O. 1979. *Nature* 279:409–411.

Nriagu, J.O., ed. 1984. *Changing Metal Cycles and Human Health* (Dahlem Konferenzen). Springer-Verlag, Berlin.

Nriagu, J.O. 1990. *Environment* 32:7–33.

Nriagu, J.O., and J.M. Pacyna. 1988. *Nature* 333:134–139.

Oliver, B.G., and E.G. Cosgrove. 1975. *Environ Lett* 9:75–90.

Page, A.L. 1974. *Fate and Effect of Trace Elements in Sewage Sludge When Applied to Agricultural Lands.* EPA 670/2-74-005. U.S. EPA, Cincinnati, OH.

Ragaini, R.C., H.R. Ralston, and N. Roberts. 1977. *Environ Sci Technol* 8:773–781.

Rhyner, C.R., L.J. Schwartz, R.B. Wenger, and M.G. Kohrell. 1995. *Waste Management and Resource Recovery.* Lewis Publ, Boca Raton, FL.

Rosman, K.J.R., W. Chisholm, C.F. Boutron, J.P. Candelone, and U. Gorlach. 1993. *Nature* 362:333–335.

Ross, S.M. 1994. *Toxic Metals in Soil–Plant Systems*. Wiley, New York.

Rothbaum, H.P., R.L. Goquel, A.E. Johnston, and G.E.G. Mattingly. 1986. *J Soil Sci* 37:99–107.

Salomons, W., and U. Forstner. 1984. *Metals in the Hydrosphere*. Springer-Verlag, Berlin.

Sheikh, B., R.P. Cort, W.R. Kirkpatrick, R.S. Jaques, and T. Asano. 1990. *Res J Water Pollut Con Fed* 62:216–226.

Smith, L.A., J.L. Means, A. Chen, and M.D. Roper. 1995. *Remedial Options for Metal Contaminated Sites*. Lewis Publ, Boca Raton, FL.

Sommers, L.E. 1977. *J Environ Qual* 6:225–232.

Sopper, W.E. 1993. *Municipal Sludge Use in Land Reclamation*. Lewis Publ, Boca Raton, FL.

Sparr, M.C. 1970. *Commun Soil Sci Plant Anal* 1:241–262.

Thornton, I. 1991. In P. Bullock and P.J. Gregory, eds. *Soils in the Urban Environment*. Blackwell, London.

Underwood, E.J. 1975. In D.J.D. Nicholas and A.R. Egan, eds. *Trace Elements in Soil–Plant–Animal Systems*. Academic Pr, New York.

Underwood, E.J. 1977. *Trace Elements in Human and Animal Nutrition*. Academic Pr, New York.

U.S. Bureau of Mines. 1980. In *Minerals Yearbook, 1978–1979, vol 1: Metals and Minerals*. U.S. Dept of Interior, Washington, DC.

[U.S. EPA] United States Environmental Protection Agency. 1985. *Composting of Municipal Wastewater Sludges*. EPA 62514-85-014. U.S. EPA, Washington, DC. US-EPA, 1993. *Fed Register* 58:9248–9415.

US-EPA. 1994. *National Air Quality and Emissions Trends Report, 1993*. U.S. EPA, Research Triangle Park, NC. U.S. Federal Register. 1989. *Fed Register* 54(23):5745–5902.

Van Hook, R.I., and W.D. Shults. 1976. In *Effects of Trace Contaminants from Coal Combustion*. ERDA 77-64. NTIS, Springfield, VA.

Van Hook, R.I., D.W. Johnson, and B.P. Spalding. 1980. In J.O. Nriagu, ed. *Zinc in the Environment: Ecological Cycling*. Wiley, New York.

Wixson, B.G., E. Bolter, N.L. Gale, J.C. Jennett, and K. Purushothaman. 1973. In *Cycling and Control of Metals*. Natl Environ Res Center, Cincinnati, OH.

Wood, J.M., and E.D. Goldberg. 1977. In W. Stumm, ed. *Global Chemical Cycles and Their Alterations by Man* (Dahlem Konferenzen). Springer-Verlag, Berlin.

World Bank Group. 1999. *The Bangladesh Arsenic Mitigation Water Supply Project: Addressing a Massive Public Health Crisis*. World Bank Bangladesh Office.

Zingaro, R.A. 1979. *Environ Sci Technol* 13:282–287.

2
Biogeochemical Processes Regulating Metal Behavior

The soil is nature's purifying agent. The soil as a physical, biological, and chemical filter. The soil as a pollutant sink. These phrases signify the important role soils play in cleansing our environment of pollutants in terms of our food, surface water, and groundwater resources. The major physical, biological, and chemical processes that determine the fate and effects of environmental pollutants are discussed in this chapter.

There are several dynamic interactive processes, which in turn can be influenced by various biogeochemical factors, that govern metal behavior rendering the predictability of the fate and effects of trace metals in the environment rather cumbersome. These include abiotic and biotic processes and factors in heterogeneous environmental media. Therefore, a background understanding of the various biogeochemical processes, i.e., from the landscape to the molecular level, and relevant factors is in order.

1 Biogeochemical Cycles of Metals

The major accumulation pools and transfer pathways of metals within a terrestrial ecosystem are depicted in Figure 2.1. Meteorologic vectors (input and output) include airborne particulate matter, aerosols, dissolved substances in precipitation, and gases (e.g., gaseous forms of certain metals such as Hg, Se, Pb, etc.). Geologic vectors include dissolved and particulate matter moved by surface runoff and subsurface drainage. Biological flux results when substances gathered by animals in one ecosystem are deposited in another.

Within the ecosystem, the elements may be considered to be occurring in one of four basic compartments: (1) atmospheric, (2) organic, (3) bioavailable, and (4) mineral (soil and rocks). They can be classified as either mineral or organic (biomass and bioavailable compartments). The atmospheric compartment includes all elements, either in aerosol or gaseous form, both aboveground and belowground. The organic compartment includes all elements incorporated in living and dead biomass of plants and animals, including the roots and woody tissues of plants. The bioavailable compartment includes elements that are either in free ionic or complexed form, either dissolved in soil solution or sorbed on exchange sites of clays, humus, and oxides of Fe, Mn, and Al. The mineral compartment includes the primary (e.g., quartz and feldspars) and secondary (clays, allophane, sesquioxides) minerals.

The biogeochemical cycling of elements involves an exchange between the various compartments within the ecosystem. Fluxes across the ecosystems boundaries link individual ecosystems with the rest of the biosphere.

1.1 In Agroecosystems

In agroecosystems, trace metals cycle as visualized in Figure 2.2. In this conceptualization, the relative importance of the various transfer pathways varies considerably, depending on the element, plant species, soil type, site characteristics, management practices, and climate. There are basically two components: crops and soil. Practically speaking, there are two major routes for input of metals into agroecosystems: aerial (e.g., aerosol, particulate matter, resuspended and airborne dusts, etc.), and land (fertilizers, biosolids, other soil amendments, etc.). The output pathways can be represented primarily by losses through plant tissue removal for food, feedstuff, and fiber, and by leaching and erosion. The gaseous output pathway to the atmosphere for most trace metals is not important, except for those that undergo microbially mediated transformation, such as Hg and Se. Because of dynamic changes in outputs and inputs, agroecosystems are in a nonequilibrium state with respect to elemental cycling. Hence, an agroeco-

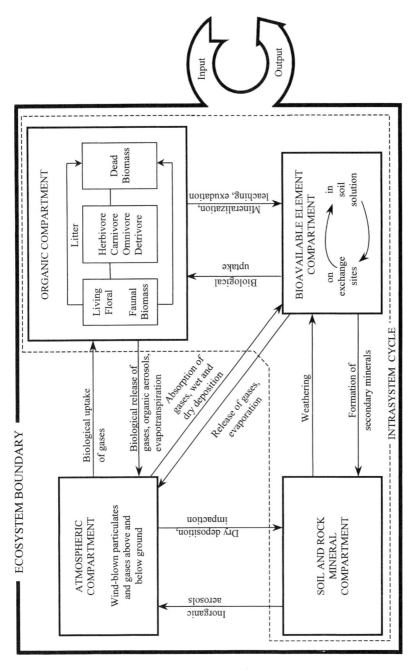

FIGURE 2.1. A generalized model depicting elemental relationships in terrestrial ecosystems. (Modified from Likens et al., 1977.)

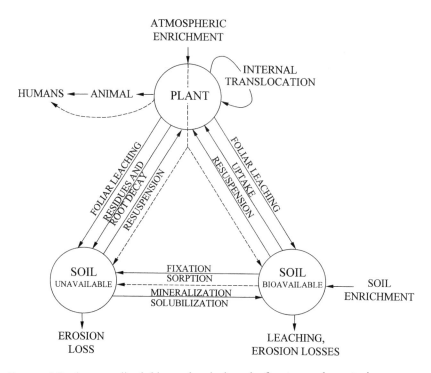

FIGURE 2.2. A generalized biogeochemical cycle for trace elements in agroeco-systems.

system can either be enriched, sufficient, or deficient in one or more elements with respect to agricultural productivity. Very seldom can they sustain optimum productivity for several years without supplemental addition of some micronutrients through fertilization or soil amendment.

The soil–plant system serves as an effective barrier against animal toxicity for certain trace metals (e.g., Ni, Cd) in that plant growth will cease or be greatly depressed before these elements could be taken up from the soil and bioaccumulated in concentrations that could be harmful to animals. The soil–plant system exerts an effective buffering action on the environmental cycling of trace elements. The amounts of trace elements listed previously in Table 1.1 (see Chapter 1) that are present in the root zone in the soil are several orders of magnitude higher than the amounts that can be removed by crop harvesting (Allaway, 1968), indicating the long residence time of these pollutants in the soil system.

1.2 In Forest Ecosystems

A conceptualized model for trace metal cycling in forest ecosystems (Fig. 2.3) is similar to that for agroecosystems in some respects. Again, the

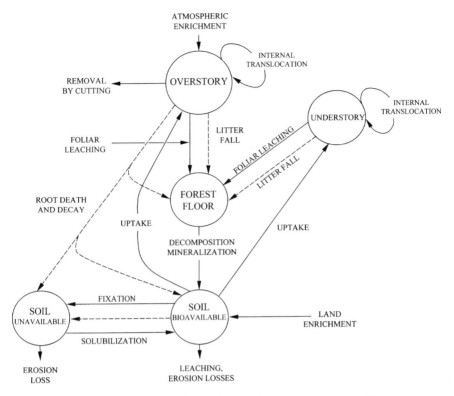

FIGURE 2.3. Generalized biogeochemical cycle for trace elements in forest ecosystems. (Modified from Van Hook et al., 1980.)

significance of various transfer pathways is dependent on factors similar to those that influence cycling in agroecosystems. But unlike agroecosystems, there is practically only one input pathway in established forest ecosystems, which is atmospheric deposition. Erosion seldom occurs in well-managed aged forests; however, leaching still occurs. Because of the presence of an understory and forest floor, there are more transfer pathways than in agroecosystems. In contrast to agroecosystems, aged and established forests are usually in a state of equilibrium with respect to elemental cycling.

Agricultural soils typically have higher buffering capacity against acidification (from acidic precipitation) and other airborne constituents than forest soils because of addition of acid-neutralizing amendments, such as lime and compost, to the former. Consequently, soil acidification has accelerated the transport (leaching) of essential nutrients (e.g., Ca and Mg) from the root zone of trees causing tree-dieback syndrome in regions of Europe and North America. But the declining atmospheric deposition of pollutants and the resiliency of the soil's buffering ability is good news for these impacted ecosystems.

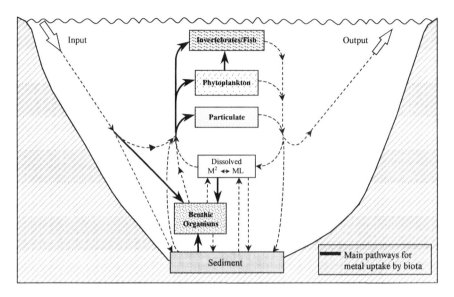

FIGURE 2.4. A simplified biogeochemical cycle for trace metals in freshwater systems. (From Hart, 1982.)

1.3 In Freshwater Systems

A simplified biogeochemical cycle for trace metals in freshwater systems is shown in Figure 2.4. The various transfer processes involved in the transformation and transport of metals from one compartment to another have been discussed in some detail by Hart (1982), and Salomons and Forstner (1984). The most important processes affecting the biogeochemistry of trace metals in freshwater systems include (1) wet and dry deposition of acidic and neutralizing compounds, (2) dilution and concentration during runoff and evaporation, (3) terrestrial neutralization or mobilization due to weathering, (4) anthropogenic influences, (5) recharge and detention, and (6) buffering and geochemical reactions occurring within the lake (Downs et al., 1998).

Trace metal inputs to a lake are eventually removed by deposition onto the sediments, where they are buried, or by transport out of the lake by outflow. The fraction of the metal inputs retained in sediments depends on the affinity of the metal ions for the settling particulates in the lake (Sigg, 1987). In freshwater lakes, the settling particulates consist mostly of materials of biological origin and inorganic constituents such as calcium carbonate ($CaCO_3$), Fe and Mn oxides, and silicate minerals. The role of these particulates in the transport and removal of metals from the water column depends on the extent of complexation of metal ions with the biological constituents, and Fe and Mn oxides; the latter are considered as effective conveyers of metals partly due to their large surface areas, with the carbonates playing an insignificant role as a carrier phase.

Humic substances are ubiquitous in the aquatic environment and constitute from 10 to 30% of dissolved organic carbon (DOC) in seawater to 70 to 90% in wetland waters (Thurman, 1985). The DOC originates from the decomposition and degradation of organic detritus. Humic substances typically range in concentration from 0.50 to 4.0 mg C L^{-1} in most lakes and rivers but can reach 10 to 50 mg C L^{-1} in wetlands and marshes (McCarthy, 1989). It has been well established that because of their polyelectrolytic nature, DOC complexes with or chelates trace metal ions in natural waters, thereby changing the physicochemical states of trace metals in solution (Florence and Batley, 1980; Mantoura et al., 1978). This complexation capacity of DOC therefore plays an important role in natural systems, influencing the mobility, bioavailability, and toxicity of trace metals by controlling their speciation.

2 Soil Constituents Relevant to Metal Reactions

As discussed in the previous section, there are two basic categories of compartments in a soil system: mineral (abiotic) and organic (biotic). Since the zone of most intense biological activity is in the surface soil, the emphasis is on this layer. Metal ions undergo a series of reactions involving both the aqueous and solid phases (Fig. 2.5), which vary in space and time. Thus, the chemical makeup of the soil solution is dynamic and is influenced by multiphase equilibria involving (1) the solid phase, i.e., phyllosilicates that include clay minerals such as kaolinite, illite, smectites, etc; hydrous oxides that include the hydrous Mn, Fe, and Al oxides, and the particulate OM; and (2) the aqueous phase consisting of water and dissolved

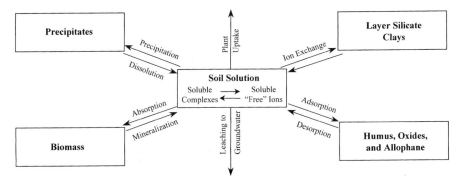

FIGURE 2.5. Schematic of key interactive processes in the soil system affecting the partitioning of trace metals between the aqueous and solid phases.

constituents (e.g., free metal ions, complexed metal ions, DOC, and other ligands). The major constituents of the soil are (see Fig. 2.5):

Primary minerals—are formed during the cooling of molten rock, are predominantly silicate minerals, feldspars, olivines, pyroxenes, amphiboles, and micas, and are considered as precursors of secondary minerals (Brady, 1990). They practically represent the sand fraction of soils. Soils of the temperate regions often have >50% primary minerals (Kabata-Pendias and Pendias, 1992). They are the least reactive soil constituent.

Secondary minerals—refer to advanced mineral products of weathering that are of clay size (i.e., <2 μm in diameter); also known as clay fraction or clay minerals; includes the layer silicates (e.g., 1:1 structure such as kaolinite, and 2:1 structure such as montmorillonite, illite, vermiculite, etc.), oxides of Fe, Al, and Mn, and the noncrystalline minerals (e.g., allophane). All have high surface area and are very reactive compared with primary minerals.

Humus—also known as soil organic matter (SOM); total of the organic compounds in soils exclusive of undecayed plant and animal tissues, their partial decomposition products, and the soil biomass (Stevenson, 1994). It is a heterogeneous mixture of products resulting from microbial and chemical transformations of organic residues. Its two components, humic acid and fulvic acid are very reactive. In most cases although humus is only a small fraction of the soil's total solid phase, it is very important in defining the surface properties of soils.

Soil biomass—organic matter present as live plant (roots), animal (e.g., invertebrates), and microbial tissues. One kilogram of surface soil may contain as many as 10 billion bacteria, 2 billion actinomycetes, 120 million fungi, 25 million algae, and a faunal population in the millions (Table 2.1). Added to this microbial biomass is the kilometer or so total length of roots from a single plant in the top meter of soils.

Precipitates—a generic term that includes the carbonates, phosphates, and sulfides. In agricultural and reclaimed soils, carbonates (e.g., calcite, dolomite), phosphates (e.g., hydroxyapatite or calcium phosphate), and

TABLE 2.1. Vertical distribution of different groups of microorganisms in soil ($\times 10^3$ g^{-1} soil).

Depth (cm)	Aerobic bacteria	Anaerobic bacteria	Actinomycetes	Fungi	Algae
3–8	7800	1950	2080	119	25
20–25	1800	379	245	50	5
35–40	472	98	49	14	0.5
65–75	10	1	5	6	0.1
135–145	1	0.4	—	3	—

Source: Eijsackers et al. (1994).

sulfides (as in rice paddy soils) are rather common, depending on the pH
and redox potential.

Colloids—a generic term for "mobile solid phases" that consist of very fine
particles that are between 0.001 and 0.1 μm in diameter. They can occur as
discrete layer silicate, oxide, or humus or as an association of them. Clay
surfaces can be coated with humic material and/or mineral oxides. Such
coating may block access of ions to reactive sites and, therefore, impede
cation exchange (Mingelgrin et al., 1977). Because of their small size,
colloids can be transported to groundwater with leaching water rather
easily. Soil clays and SOM are often referred to as soil colloids because their
sizes fall within, or close to, actual colloidal dimensions. Because of their
nature, colloids are often identified as effective "transporters" of contaminants in aquatic systems and in the vadose zone into the groundwater
(Kaplan et al., 1994, 1995; McCarthy and Zachara, 1989).

The above constituents are collectively known as the soil's solid phase.
Their composition and distribution vary from one soil type to another, even
within a soil profile, giving rise to spatial heterogeneity with respect to soil
properties.

Soil solution—also commonly known as the aqueous phase, and sometimes
referred to as soil water or soil moisture. Technically, it represents the field
moisture content at or below field capacity and is essentially an electrolytic
solution that contains dissolved substances, both organic and inorganic. It
also contains some gases from the atmosphere, and plant and microbial
activities (see Fig. 2.5). It is where biogeochemical speciation of metals
occurs. The interface between the solid and aqueous phases is the site of
most intense reactivity. In ordinary soils, the dominating cations in soil
solution include Ca^{2+}, Mg^{2+}, K^+, and Na^+, which are progressively
replaced by H^+ and Al^{3+} ions as soil acidity increases.

Speciation of metals in the solution phase can be facilitated by computer
programs such as SOILCHEM (Sposito and Coves, 1995), Geochem-PC
(Parker et al., 1995), and MinteqA2 (Allison and Brown, 1995), based on
the total concentrations of all elements and ligands in the soil solution.

Soils and sediments, and their clay fractions, may contain large amounts
of hydrous oxides of Fe, Al, or Mn that coat the clays and may form clay-
size particles themselves. Some of these oxides may exist as cutans,
concretions, or nodules, as well as colloids. These cementing compounds
play an important role when considering metal adsorption by soils. Jenne
(1968) proposed that some trace metal ions may be occluded and
coprecipitated with hydrous oxides of Fe and Mn. These amorphous and
partially crystalline hydrous oxides have an amphoteric cation exchange
capacity, especially Al, and may have a large capacity to adsorb cations.

It has been estimated that between 70 and 80% of the SOM in predominantly mineral soils consists of humic materials, namely humic acid, fulvic acid, and humin (Schnitzer, 1978). The humified material is the most active fraction of humus due to its high content of oxygen-containing functional groups including COOH, phenolic-OH, and C=O structures of various types. These functional groups are responsible for the complexing ability of humic and fulvic acids for trace metal cations like Cu^{2+}. Several studies indicate that humic and fulvic acids extracted from soils formed under differing geographic and pedologic environments (Argentina, Canada, Israel, Italy, Japan, and West Indies) have basically similar analytical characteristics and chemical structures (Chen et al., 1978; Schnitzer, 1977).

3 Biogeochemical Processes Regulating Metal Behavior

The various interactive biotic and abiotic processes that govern the behavior of metals in soils has been shown in Figure 2.5. The major phenomena include ion exchange (adsorption–desorption), solubilization (precipitation–dissolution), and absorption (assimilation or immobilization) by living biomass. Microorganisms and plant roots interact with the dissolved species, and microbial and root exudates can affect the solubility and eventual transport of these compounds. In essence, these processes largely determine the biogeochemical speciation of elements and influence their solubility, mobility, bioavailability, and toxicity. In turn, the predominance of any of these processes is influenced by the biogeochemical (biotic and abiotic) and environmental factors discussed below. Because of the dynamic and disproportional nature of the interchange of elements between the various phases, the system will remain in quasi-equilibrium under field conditions. There are other processes that can also influence the biogeo-chemical dynamics of metals—weathering, decomposition, acid–base reactions, and redox reactions—discussed in later sections.

3.1 Sorption

Considered as the most important process controlling the partitioning of metals between the aqueous and solid phases in soils, sorption may represent the combined effects of ion exchange, specific adsorption, (co)precipitation, and (surface)complexation on the transfer of ions from the aqueous phase to the solid phase. Thus, sorption can be broadly defined as the retention of chemical constituents through the transfer of ions from the aqueous phase to the solid phase. Because most metals occur in the cationic form in soils, cation exchange is emphasized here. There is also a corresponding anion exchange for metals that occur as oxyanions (e.g., Se, As, Cr, Sb, V, Mo).

When isomorphous substitution occurs in the clay and soil solution interface, cations of lower charge substitute for cations of higher charge (e.g., substitution of octahedral Al^{3+} by Mg^{2+} yields one negative charge unbalanced in the crystal lattice; similarly, tetrahedral Si^{4+} can be replaced by Al^{3+} leaving one unneutralized negative charge), leaving pH-independent permanent charge on the clay surface and creating exchange sites. Cations that are swarming in the soil solution are then attracted to the clay surface in order to maintain electroneutrality in the soil system. The cations are attracted to the negatively charged edges and surfaces of the particles by weak electrostatic bonds, the strength of attraction being proportional to the charge/radius ratio of the ions (i.e., ionic potential) (Appendix Table A.13). The attracted ions can be exchanged with the free cations in the solution. In acidic conditions, when the soil pH is below the PZC (point of zero charge) of the clay, positive charge develops and anions can be attracted to the surface. (Note: The PZC refers to the pH value at which there is no net charge on the exchange complex, i.e., equal amounts of + and – charges. For example, the PZCs are 4.0 to 5.0 and 7.0 to 8.0, respectively, for kaolinite and goethite.)

This phenomenon is known respectively as cation exchange (cation to cation) or anion exchange (anion to anion), which can be represented as

$$]-M^+ + M'^+ \rightleftharpoons]-M'^+ + M^+ \qquad \text{Cation exchange}$$

$$]+M^- + M'^- \rightleftharpoons]+M'^- + M^- \qquad \text{Anion exchange}$$

where] represents a fixed clay charge (+ or −), and M^+ and M^- representing cations and anions, respectively.

Exchange reactions are stoichiometric (e.g., 2 Na^+ ions exchanging with one attracted Ca^{2+} ion), rapid, and readily reversible. The order of exchange depends on the affinity of the cations on the charged site compared with their attraction for water molecules. The most abundant cations in soil solution (i.e., Ca^{2+}, Mg^{2+}, K^+, and Na^+) are only weakly retained by soils as exchangeable cations; in contrast, most trace cations, such as transition metal ions, are strongly retained by soils. Generally, Al^{3+} cations are the most difficult to exchange and Na^+ cations the least difficult, with other cations in between, in the order:

$$Al^{3+} > H^+ > Ca^{2+} > Mg^{2+} > K^+ > NH_4^+ > Na^+$$

Because the ions involved in the cation exchange process are reversibly adsorbed, they are referred to as exchangeable cations. The quantity of reversibly adsorbed cations per unit weight of adsorbent (e.g., clay) (e.g., cmol (+)/kg) is called the cation exchange capacity (CEC). In simple cation exchange, there is no change in surface charge or in the pH of the solution. One cation simply replaces an equivalent amount of another cation. Specific adsorption of cations differs in that some protons are displaced from the surface and therefore the pH is changed. In general, as the pH increases, the soil CEC generally increases due to an increase of pH-dependent surface charge.

Surface charge can also arise upon dissociation (deprotonation) of exposed hydroxyl (OH) groups present on the surface of Al octahedron and Si tetrahedron sheets in the clay lattice. In clay mineralogy jargon, the two types of edge surface hydroxyl groups are known as *aluminol* for the octahedral- and *silanol* for the tetrahedral-coordinated groups, respectively. The dissociation reaction is pH-dependent, increasing at high pH and decreasing at low pH, hence the term pH-dependent charge, or simply variable charge.

The protonation reaction at low pH is shown as

$$]-OH + H^+ \rightleftharpoons]-OH_2^+$$

The deprotonation at high pH is shown as

$$]-OH \rightleftharpoons]-O^- + H^+$$

The dissociated surface hydroxyl group can serve as a Lewis base toward metal cations (McBride, 1994). The metals may complex with one or possibly two of the deprotonated groups as follows:

$$]-OH + M^{n+} \rightleftharpoons]-OM^{(n-1)+} + H^+$$

$$2]-OH + M^{n+} \rightleftharpoons (]-O)_2M^{(n-2)+} + 2H^+$$

In acidic conditions, the exposed OH groups of the clay surface can be protonated by the addition of protons (H^+ ions) from the soil solution. Sorption of large concentrations of the H^+ ion creates positive charge. Thus, increased adsorption of metal ions with increased pH is due to both increased negative surface charge density and the increased concentration of the $M-OH^+$ species in soils. Adsorption of trace metal ions onto oxide minerals may be expected to increase from <20 to ~100% within a pH range of 3 to 7, except those that form oxyanions (Evans et al., 1995; Helmke and Naidu, 1996).

Essentially, soils consist of a mixed system with both permanent and variable charge surfaces. In such systems, the whole soil typically has a net negative charge. In variable charge soils (e.g., Oxisols, strongly weathered soils such as those in the tropics), increasing soil pH results in an increase in the net surface negative charge; decreasing the soil pH results in an increase in the net positive charge.

Organic C, particularly dissolved organic C (DOC), plays a major role in biogeochemical processes in a soil system. It interacts with mineral weathering, metal mobility and bioavailability, acid–base chemistry, solubility–dissolution of metal ions, and transport of metals. Dissolved organics that interact with soil constituents and trace metal ions are of two major types: (1) low-molecular-weight organic compounds (e.g., polyphenols, amino acids, sugar acids, etc.) and (2) water-soluble humic and fulvic acids. The humic and fulvic acids are more dominant, except in the rhizosphere where low-molecular-weight organics may predominate. The first type are continuously produced in soils through microbial activity. Root exudates consist of a variety of aliphatic acids, such as citric, oxalic, and tartaric acids, many of which complex with metal ions. The most abundant organic acids identified in tree root exudates include citric, fumaric, malic, malonic, and succinic acids (Smith, 1976).

Exudation of organic compounds by roots may influence metal ion solubility and uptake through their indirect effects on microbial activity, rhizosphere physical properties, and root growth dynamics, and directly through acidification, chelation, precipitation, and oxidation–reduction reactions in the rhizosphere (Marschner et al., 1989; Zang et al., 1991) of trace metals such as Cd (Mench et al., 1988; Mench and Martin, 1991). It has been further demonstrated that these low-molecular-weight compounds were able to influence the rate of Cd release from different soils and increase the solubility of Cd in bulk soils through the formation of soluble Cd–organic complexes (Cieslinski et al., 1998; Khrishnamurti et al., 1997). Subsequently, the complexed Cd could be more bioavailable for plant uptake.

The chemical reactivity of humus can be related to its total acidity value, which is the sum of titratable acidity from carboxylic (COOH) and phenolic hydroxyl (OH) functional groups. At about pH 3, the protons from the carboxylic groups start to dissociate, creating negative charge. The negative charge increases with increasing pH, and at about pH 9, the phenolic OH groups also dissociate. Because the intensity of the surface charge is sensitive to pH change, this is also known as pH-dependent or variable charge as described above, enabling humic molecules to perform several chemical reactions, such as adsorption of cations, complexation (or chelation) of metallic ions, and association with clay and oxide minerals. As with clay minerals, the negative surface charge must be balanced by cations present in the soil solution. The acidity from carboxylic and phenolic groups is mainly responsible for creating OM-associated CEC in soils.

Complexation reactions involving natural OM influence the behavior of metal ions, particularly when present in trace concentrations. Organic compounds in aquatic and terrestrial environments that may form complexes with metal ions may be grouped into three main classes (Senesi, 1992): (1) organic substances of known molecular structure and chemical properties, including biochemicals such as simple aliphatic acids, polysaccharides, amino acids, and polyphenols; (2) xenobiotic organics derived from agricultural, industrial, and urban activities; and (3) humic substances. Humic substances (HS) represent a significant proportion of total organic C in the global C cycle, constituting the major fraction in soils (70 to 80%) and the largest fraction of natural OM in streams, rivers, wetlands, lakes, sea, and groundwater (40 to 60% of DOC).

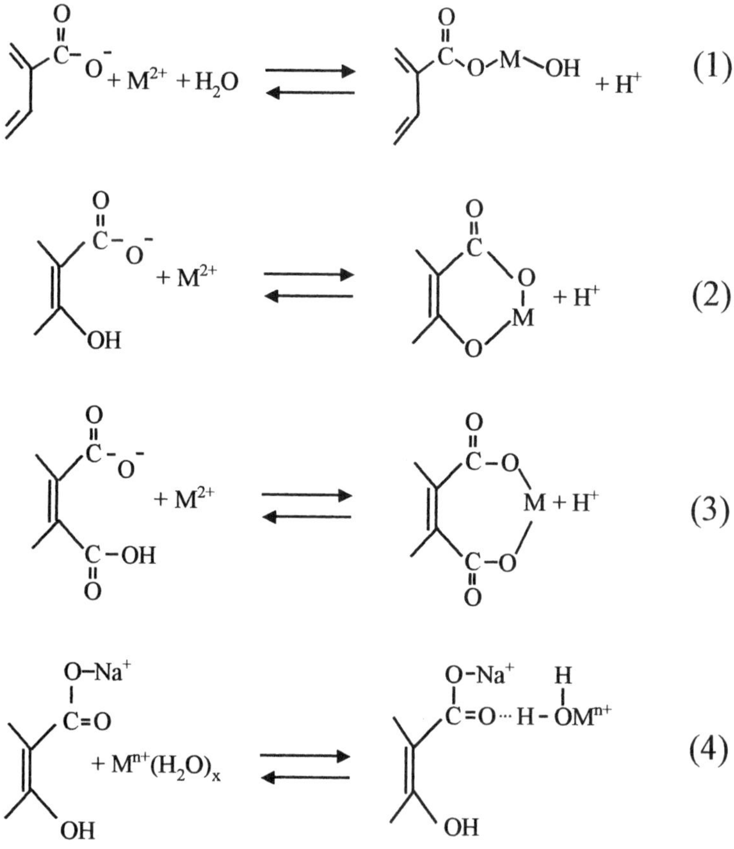

FIGURE 2.6. Mechanisms of complexation of trace metal ions with humic compounds in the environment. (Extracted from Senesi, 1992.)

Mechanisms of metal reaction with fulvic acid (FA) and humic acid (HA) are illustrated in Figure 2.6. The low-molecular-weight organic acids, and FA and HA, have basically the same functional groups. Using a divalent metal ion (M^{2+}) as an example, the complexation process is likely to occur with one or more, or simultaneously all four, reaction mechanisms (Senesi, 1992). In reaction 1, one COOH group reacts with one metal (M) ion to form a monodentate complex. In reaction 2, one COOH and one adjacent OH group react simultaneously with the M ion to form a bidentate complex or chelate (a complex is called a chelate when two or more functional groups of a single ligand are coordinated to a M cation). In reaction 3, two adjacent COOH groups react simultaneously with the M ion to form a bidentate chelate, and in reaction 4, a metal ion (M^{n+}) is dually bonded to the organics through electrostatic attraction and hydrogen bonding to a C=O group through a water molecule in its primary hydration shell. The last

TABLE 2.2. Common oxides of iron and aluminum found in soils.

Formula	Name	Oxyanion packing[a]	Comments
$\alpha\text{-FeOOH}$	Goethite	hcp	Common in temperate-region soils; gives soils characteristic yellow-brown color
$Fe_2O_3 \cdot nH_2O$	Ferrihydrite	Disordered	Common in temperate-region soils; reddish-brown to yellow-brown color
$\alpha\text{-Fe}_2O_3$	Hematite	hcp	Common in soils of hot climates, both humid and dry; reddish-brown color
$\gamma\text{-Al(OH)}_3$	Gibbsite	hcp (open packed)	Common in humid tropics; not common in cool temperate-region soils
$\gamma\text{-Fe}_2O_3$	Maghemite	ccp	Widespread but minor; may be product of oxidation–reduction cycles in soil; magnetic
Fe_3O_4	Magnetite	ccp	Widespread but minor; black, magnetic; may have origin similar to maghemite

[a]Abbreviations hcp and ccp refer to hexagonal and cubic closest packing, respectively.
Source: Extracted from McBride (1994).

reaction is of special importance if the cation has a high solvation energy that retains its primary hydration shell, thus forming an outer-sphere complex. Reactions 2 and 3 describe the formation of strong inner-sphere complexes. Inner-sphere complexes generally are more stable than outer-sphere complexes, because the latter cannot easily involve ionic or covalent bonding between the central group and ligand, whereas the former can (Sposito, 1989).

The trivalent metal cations Fe^{3+} and Al^{3+} form highly stable complexes with humus, each metal bonding to two or more functional groups. Mixed ligand complexes also can be formed within the soil system, in addition to simple metal complexes as discussed above. Among the oxides and hydroxides in the normal range of soil pH values, those for Al, Fe, and Mn are the most important (Table 2.2). Among the Fe compounds in soils, goethite ($\alpha\text{-FeOOH}$) is the one commonly found irrespective of climatic condition. Because it is the most thermodynamically stable of the iron oxides, it can be expected to be the dominant Fe oxide in clays. Aluminum in soil solution can isomorphically substitute for Fe, especially in highly weathered soils, such as Oxisols and Ultisols.

Among the aluminum minerals, gibbsite [$\gamma\text{-Al(OH)}_3$] is the most important in surface reactivity. Birnessite ($Na_{0.7}Ca_{0.3}Mn_7O_{14} \cdot 2H_2O$) is the most commonly found manganese mineral in soils. The metal oxides and hydroxides are considered as the "climax" mineralogy of soils that can form directly from the weathering of primary minerals or from the hydrolysis and desilication of clay minerals, such as kaolinite [$Al_4Si_4O_{10}(OH)_8$] and

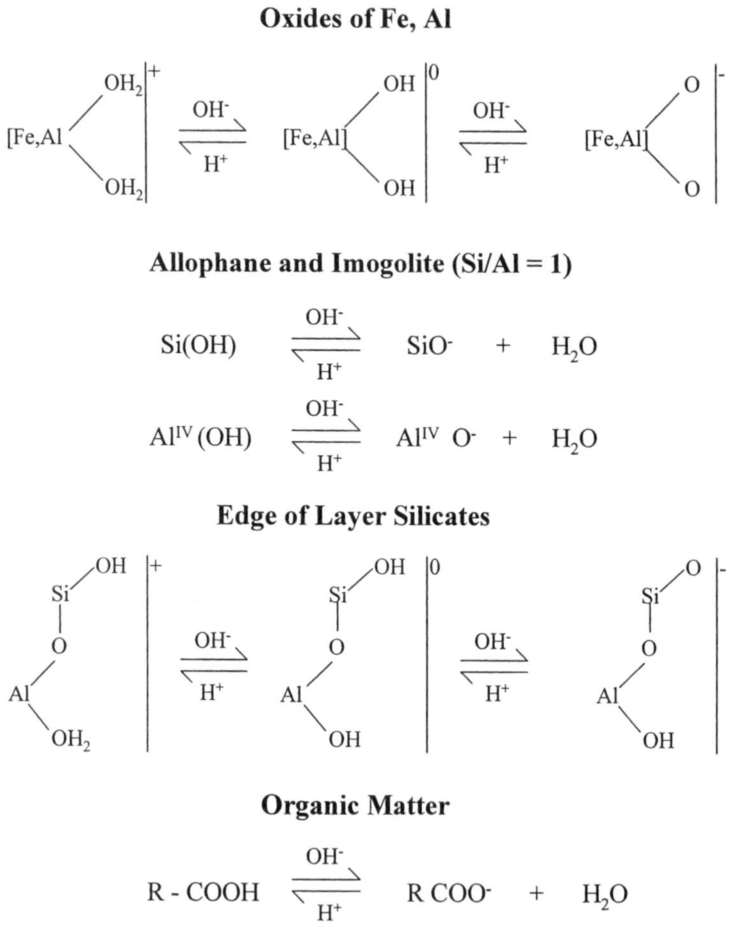

Oxides of Fe, Al

Allophane and Imogolite (Si/Al = 1)

$$Si(OH) \quad \underset{H^+}{\overset{OH^-}{\rightleftharpoons}} \quad SiO^- \quad + \quad H_2O$$

$$Al^{IV}(OH) \quad \underset{H^+}{\overset{OH^-}{\rightleftharpoons}} \quad Al^{IV} O^- \quad + \quad H_2O$$

Edge of Layer Silicates

Organic Matter

$$R - COOH \quad \underset{H^+}{\overset{OH^-}{\rightleftharpoons}} \quad R\,COO^- \quad + \quad H_2O$$

FIGURE 2.7. Mechanisms of charge generation on surfaces of soil constituents. Superscript IV on aluminum indicates fourfold coordination. (Modified from Wada, 1989.)

montmorillonite $[M_x Si_8 Al_{3.2} Fe_{0.2} Mg_{0.6} O_{20}(OH)_4]$, where **M** represents a monovalent interlayer cation (Sposito, 1989). Samples of surface adsorption sites for crystalline minerals (layer silicates as in 1:1 and 2:1 structures), noncrystalline minerals (oxides and allophane), and OM are indicated in Figure 2.7.

Two types of adsorption onto solid phases are known—the first is through the weak electrostatic, nonspecific attraction on pH-independent, permanent charges on the clay surface, where cations are reversibly sorbed through cation exchange, and the second is where cations are selectively and less reversibly sorbed as a result of bonding on pH-dependent, variable charge surfaces (e.g., oxides) and complexation with

OM functional groups. The first type of adsorption (sometimes known as physical adsorption) is facilitated by *cation exchange* to distinguish it from *specific adsorption* (sometimes referred to as chemisorption or surface complexation) for the latter. For example, Cu can be specifically adsorbed by layer silicate clays, oxides of Fe, Mn, and Al, and OM. In specific adsorption, ions are held much more strongly by the surface since these ions penetrate the coordination shell of the structural atom and are bonded by covalent bonds via O atoms (e.g., Cu–O–Al or Cu–O–Fe bond) or OH groups to the structural cations. As stated earlier, sorption can be broadly defined to include nonspecific and specific adsorption, and precipitation.

In ordinary soils, especially arable soils, the solubility of the more abundant elements (e.g., Al, Si, Fe, Ca, and Mg) can be limited by their precipitation, which occurs when the soil solution is supersaturated by these cations. The formation of the new solid phase (i.e., precipitates) (see Fig. 2.5) occurs when the solubility product for that phase has been exceeded. In normal soils, precipitation of trace metals is unlikely, but in highly metal-contaminated soils, this process can play a major role of immobilizing these contaminants. In fact, soil application of certain amendments (e.g., hydroxyapatite, a calcium phosphate compound) is now being done by the cleanup industry to immobilize Pb and other heavy metals (see Chapter 6).

Adsorption isotherms have been used to determine the nature and extent of affinity of metals for solid phases. The Langmuir adsorption isotherm was derived for the adsorption of gas molecules on solids (Langmuir, 1918), but has since been used to characterize the relationship between the adsorption of ions by solids and the concentration of ions in solutions (Boyd et al., 1947); the Freundlich isotherm has also been used to describe trace metal adsorption. The isotherm is simple to apply and gives parameters that can be interpreted in light of soil properties. Some adsorption isotherms by soils yield two or more linear portions of the Langmuir isotherm, indicating the presence of several types of adsorption sites.

3.2 Absorption

As indicated earlier, surface soils are teeming with dense and diverse populations of microorganisms and invertebrates (see Table 2.1). They play a myriad of important functions in soils—from the breakdown of plant litter to humification. Litter ingestion results in the intimate mixing of litter constituents in the gut of microfauna, further enhancing the mineralization process. Furthermore, soil fauna redistributes contaminants incorporated in the litter or deposited on the ground in the soil profile, a process known as bioturbation. In addition, microorganisms mediate the transformations of certain metals (e.g., Hg, Se, Sn, As, Cr) through oxidation–reduction and methylation–demethylation reactions.

Plant roots are also involved in soil processes. The rhizosphere (i.e., microzone around plant roots) is well known to attract dense microbial populations due to accessibility to energy sources (i.e., root exudates that include organic compounds such as sugars). The microbial population in the rhizosphere is about 10 to 100 times higher than in the surrounding bulk soil. Specific microbe–root associations, such as the vesicular arbuscular mycorrhiza (VAM) and ectomycorrhiza, play a dominant role in micronutrient uptake by plants.

Roots serve also as a cation exchanger, the degree of which depends largely on plant species. Organic acids in root exudates can complex metal ions into more mobile forms. Plant transpiration, a form of environmental pump, starts with the uptake of water by the roots that influence the soil solution dynamics. The combined living biomass immobilizes the metal ions once absorbed and assimilated, but upon death they are mineralized through decomposition.

3.3 Redox Reactions

Changes in the oxidation state of trace metals can occur depending on the redox condition. Mercury, As, Se, Cr, Mn, and Fe are examples of metals that are sensitive to redox changes (see individual chapters later in this book for redox transformations of these elements). The redox sequence in wetlands and other submerged soils (e.g., rice paddy) are indicated in Table 2.3. Redox potentials can vary according to the pH and temperature of the medium. Sediments in river systems may not be as anoxic (reduced) as those in lake sediments due to the "spiraling" effect of the moving water, which oxygenates the water column in the former.

A good example of the role of redox reactions on the behavior of metals is the transformation of Cd in rice paddies in the Jinzu Basin (Japan), site of *itai-itai* disease (Cd-poisoning) that occurred in the 1960s. The Cd came from mining effluents used to irrigate the paddies. The Cd remained largely unavailable as CdS for uptake by the rice plants during their early growth stage when the paddies were ponded to keep the weeds down. Cadmium, however, became bioavailable for uptake prior to the harvest when the paddies were drained to facilitate harvesting. During this period (a matter of few days) the paddy soil became oxic, oxidizing the precipitated CdS thereby dissolving the solid phase constituents into sulfate, Cd^{2+}, and other occluded metal ions, such as Zn^{2+} and Pb^{2+}. While the sulfide-bound metals were solubilized during this period, Fe and Mn became insoluble, further enhancing the uptake of Cd due to lack of competition with Fe and Mn ions for root absorption sites. Similar results can be expected when dredged polluted sediments from harbors and other waterways are spread on the ground. Metals that are sulfide-bound are similarly solubilized upon aeration of the sediments, rendering them more mobile and bioavailable to organisms.

TABLE 2.3. Oxidation–reduction reactions of primary importance in wetland soils, rice paddies, and sediments.

Element	Redox couple		Reaction	Redox potential for reactions (mV)[a]
	Oxidized species	Reduced species		
Oxygen	Oxygen	H_2O	$0.5O_2 + 2e^- + 2H^+ = H_2O$	700–400
Nitrogen	Nitrate (NO_3^-)	NH_4^+, N_2O, N_2	$NO_3^- + 2e^- + 2H^+ = NO_2^- + H_2O$	220
Manganese	Mn^{4+} (manganic: MnO_2)	Mn^{2+} (manganous: MnS)	$MnO_2 + 2e^- + 4H^+ = Mn^{2+} + 2H_2O$	200
Iron	Fe^{3+} (ferric: $Fe(OH)_3$)	Fe^{2+} (ferrous: FeS, $Fe(OH)_2$)	$FeOOH + e^- + 3H^+ = Fe^{2+} + 2H_2O$	120
Sulfur	SO_4^{2-} (sulfate)	S^{2-} (sulfide: H_2S, FeS)	$SO_4^{2-} + 8H^+ + 7e^- = 0.5S_2^{2-} + 4H_2O$	−75 to −150
Carbon	CO_2 (carbon dioxide)	CH_4 (methane)	$CO_2 + 8e^- + 8H^+ = CH_4 + 2H_2O$	−250 to −350

[a]Redox potentials are approximate values and will vary with soil pH and temperature.

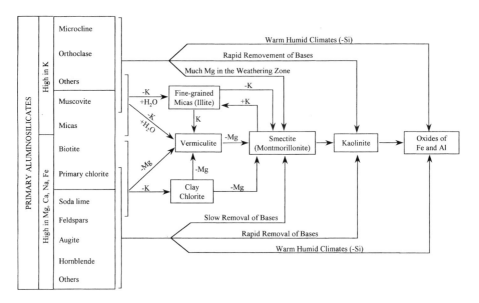

FIGURE 2.8. Sequence of natural weathering of primary minerals culminating in the release of trace and major elements including the micronutrients. (From Brady, 1990. Reprinted by permission of Pearson Education Inc., Upper Saddle River, NJ 07458.)

3.4 Weathering

Weathering refers to the disintegration and alteration of rocks and minerals by physical, biological, and chemical processes, or by *physico-biogeochemical* processes. Meteorological agents and living organisms assist in the weathering of primary minerals to form secondary minerals (Fig. 2.8), which results in the release of soluble constituents. Several of the solubilized materials include essential nutrients (e.g., P, K, Ca, Mg, Fe, Mn, Cu, Zn, etc.) that plants require for mineral nutrition. For example, feldspar and muscovite are sources for K, whereas biotite can enrich the soil with K and Fe.

Climate is perhaps the most important factor in the weathering of rocks and minerals. In the arid climate, *physical* weathering predominates and, due to the lack of water, chemical transformation is rather limited. In temperate and warm humid climates as in the tropics, both moisture content and temperature are more favorable for *biogeochemical* weathering, resulting in more clay formation. Biogeochemical factors of importance in the weathering process include biotic [e.g., decomposition of OM, respiration of living biomass (for the release of CO_2 and eventual formation of carbonic acid, H_2CO_3), and release of humic acids and other organic acids] and geochemical factors (e.g., oxidation–reduction, hydrolysis, carbonation, precipitation–dissolution, acid–base reactions, etc.).

3.5 Acidification and Buffering Capacity

Aside from natural weathering, there are two primary sources of acidification in soils: commercial fertilizers and acidic precipitation. Addition of biosolids (e.g., sewage sludge, especially those anaerobically digested) can also promote soil acidification. To promote optimal crop production, farmers regularly add N fertilizers in addition to the return of this nutrient to the soil from plant residues and livestock manures. The amount of N applied varies according to crop species and often is applied at least once yearly from different sources, such as anhydrous ammonia (NH_3), ammonium nitrate (NH_4NO_3), ammonium sulfate [$(NH_4)_2SO_4$], urea [$CO(NH_2)_2$], etc. It is not uncommon to fertilize crop fields with more than 100 kg ha^{-1} of elemental N per year in Europe and North America. Because plants can uptake and metabolize N in either the ammonium (NH_4^+) or nitrate (NO_3^-) form, the N fertilizers applied are usually in the NH_4^+ form. Once in the soil, this N form is transformed into the more mobile forms NO_2^- and NO_3^- by *Nitrosomonas* and *Nitrobacter*, respectively:

$$NH_4^+ + 2H_2O \rightleftharpoons NO_2^- + 8H^+ + 6e^-$$

$$NO_2^- + H_2O \rightleftharpoons NO_3^- + 2H^+ + 2e^-$$

Thus, nitrification can enhance the dissolution of minerals in soils by its generation of protons. Humic acids and other organic acids, formed during the decomposition of SOM, also contribute protons further enhancing the soil's dissolution potential.

Acidic precipitation (more commonly known as *acid rain*) originates from the anthropogenic releases of sulfur (SO_x) and nitrogen oxides (NO_x) when fossil fuels (e.g., coal, natural gas) are burned for energy production. Once these gases enter the earth's atmosphere, they rapidly react with moisture in the air to form sulfuric (H_2SO_4) and nitric (HNO_3) acids. This results in rainfall with pH <4.5 in many industrialized countries. When these acids are deposited, either by wet or dry deposition on land and freshwater systems, the protons from the dissociation of the acids can also promote acidification of the soils, sediments, and waters in these environments.

The degree of acidification of soils and sediments from N fertilizer addition, acidic precipitation, and OM decomposition depends on soil and sediment properties, especially their buffering capacity. Soils have the ability to resist changes in soil pH through their buffering capacity. The most resistant to acidification are soils in arid and semiarid regions due to their high alkalinity and the presence of free calcium carbonates ($CaCO_3$) (see Fig. 2.8). On the other hand, soils in the warm humid climate likely will have the least buffering capacity because soils are generally highly weathered and highly deficient of the base cations Ca^{2+}, Mg^{2+}, K^+, and Na^+. In between are soils in the temperate climate that are characteristically high in base cations.

In calcareous soils, free $CaCO_3$ can adsorb or precipitate metal ions depending on metal concentrations (McBride, 1994; Zachara et al., 1989). The affinity of Cd^{2+} for $CaCO_3$ is higher than that for Zn^{2+} or Cu^{2+} because of the similar ionic radii of Ca^{2+} and Cd^{2+} (0.10 nm and 0.095 nm for Ca^{2+} and Cd^{2+}, respectively), and the lower solubility of $CdCO_3$ ($K_{so} = 11.7$) than that for $CaCO_3$ ($K_{so} = 8.42$) can enhance the formation of a solid phase on the $CaCO_3$ surface. Thus, increasing the pH can enhance both adsorption and precipitation.

Arable soils typically are well buffered because of their high CEC arising from the clay minerals and SOM. Nevertheless, a test for lime requirement by soils is routinely conducted to advise farmers when to add lime to promote proper crop nutrition. In addition, an ample supply of base cations, primarily Ca^{2+} and Mg^{2+}, is ensured through the addition of lime. Potential environmental consequences can arise from acidification of soils when the predominant H^+ ions exchange with the base cations and acid cations (e.g., Al^{3+}) with respect to the nutrition of crops and trees. Consequences may include direct phytotoxicity from excess soluble Al or nutrient imbalances involving Ca and Mg.

4 The Driving Factors

A few major factors can be viewed as primary driving factors influencing the biogeochemical processes, and hence they are collectively termed master variables. Consider the biogeochemical process (*BP*) discussed in the previous section as a function of pH, CEC, redox potential, and other factors:

$$BP = f(\text{pH, CEC, redox, etc.})$$

where, *BP* is the dependent variable, f signifies function, and pH, CEC, redox, etc. are defining variables. These are not exclusive variables, as there are other biogeochemical and environmental factors that can influence the *BP*, mobility, and bioaccumulation of trace metals. The major factors are discussed in the sections below.

4.1 pH

In general, the retention capacity of soils for trace metals increases with increasing pH, with the expected maximum at around circumneutral. Exceptions are As, Mo, Se, V, and Cr, which are commonly more mobile under alkaline conditions. Accordingly, a decrease in plant uptake of B, Co, Cu, Mn, and Zn was observed when soil pH was increased from pH 5 to 8 (Hodgson, 1963). White et al. (1979) found that soil pH was an important factor in determining the relative Zn tolerance of several soybean cultivars. In a report of the Council for Agricultural Science and Technology (CAST, 1980), it was concluded that soil pH is the most critical factor in controlling

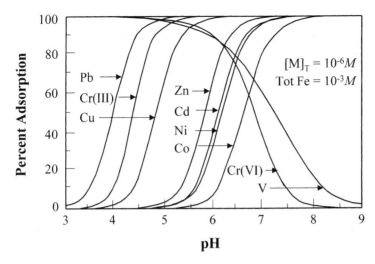

FIGURE 2.9. Modeled adsorption of certain trace metals onto hydrous ferric oxide. (From Evans et al., 1995, copyright © 1995 OPA (Overseas Publishing Association) N.V. Reprinted with permission of Taylor & Francis, Ltd.)

the plant uptake of Cd and Zn from sludge-treated soils. However, it was found that even at a soil pH of 6.5, the Cd added in many sludges is sufficient to increase Cd concentration in most crops.

The pH can be viewed as the master driver of all the driving factors because it can affect the surface charge of layer silicate clays, OM, and oxides of Fe and Al. In addition to the effect on the sorption of cations, which increases with increasing pH (Fig. 2.9), and complexation with OM, it also influences the precipitation–dissolution reactions, redox reactions, mobility and leaching, dispersion of colloids, and the eventual bioavailability of the metal ions.

The effect of soil pH on plant performance, both agronomic and horticultural, is illustrated in Figure 2.10. The optimal pH range for these crops is between 6 and 7. The number of plant species that may tolerate soil pH below 5.5 is limited to only a few agronomic (e.g., potatoes) and horticultural (e.g., azaleas, blueberries, etc.) species. Above pH 7, the risk of micronutrient deficiency, including Fe, Zn, Mn, and B, increases.

The effect of pH on chemical speciation of trace elements in soils and sediments is given in Table 2.4.

4.2 Cation Exchange Capacity

The CEC of soils is largely dependent on the amount and type of clay, OM, and the oxides of Fe, Al, and Mn. These soil components have different cation exchange properties. In general, the higher the CEC of a soil, the greater the amount of metals a soil can retain without potential hazards. For

THE pH SCALE

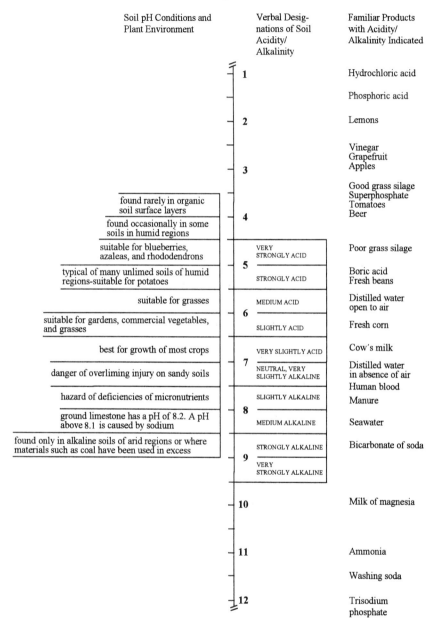

FIGURE 2.10. The pH scale indicating certain ranges of pH values that are optimal for growth of certain plants, including agronomic and horticultural crops.

TABLE 2.4. Expected dominant oxidation states and chemical species of trace elements in aqueous solution.[a]

Element	Acid soils/sediments	Alkaline soils/sediments
Ag(I)	Ag^+, $AgCl^0$	Ag^+
As(III)	$As(OH)_3$	AsO_3^{3-}
As(V)	$H_2As_4^-$	$HAsO_4^{2-}$
B(III)	$B(OH)_3$	$B(OH)_4^-$
Be(II)	Be^{2+}	$Be(OH)_3^-$, $Be(OH)_4^{2-}$
Cd(II)	Cd^{2+}, $CdSO_4^0$, $CdCl^-$	Cd^{2+}, $CdCl^-$, $CdSO_4^0$, $CdHCO_3^+$
Co(II)	Co^{2+}, $CoSO_4^0$	$Co(OH)_2^0$
Cr(III)	$Cr(OH)^{2+}$	$Cr(OH)_4^-$
Cr(VI)	CrO_4^{2-}	CrO_4^{2-}
Cu(II)	Cu^{2+}, $CuCl^-$	$CuCO_3^0$, $CuHCO_3^+$
Hg(II)	Hg^{2+}, $Hg(Cl)_2^0$, CH_3Hg^+	$Hg(OH)_2^0$
Mn(II)	Mn^{2+}, $MnSO_4^0$	Mn^{2+}, $MnSO_4^0$, $MnCO_3^0$, $MnHCO_3^+$
Mo(V)	$H_2MoO_4^0$, $HMoO_4$	$HMoO_4^-$, MoO_4^{2-},
Ni(II)	Ni^{2+}, $NiSO_4^0$, $NiHCO_3^+$	$NiCO_3^0$, $NiHCO_3^+$, Ni^{2+}
Pb(II)	Pb^{2+}, $PbSO_4^0$, $PbHCO_3^+$	$PbCO_3^0$, $PbHCO_3^+$, $Pb(CO_3)_2^{2-}$, $PbOH^+$
Sb(III)	$Sb(OH)_2^+$, $Sb(OH)_3$	$Sb(OH)_4^-$
Se(IV)	$HSeO_3^-$	SeO_3^{2-}
Se(VI)	SeO_4^{2-}	SeO_4^{2-}
Tl(I)	Tl^+	Tl^+
Tl(III)	$Tl(OH)_3^0$	$Tl(OH)_4^-$
V(IV)	VO^{2+}	oxidized to V(V) species
V(V)	VO_2^+, polyvanadates	$VO_2(OH)_2^-$, $VO_3(OH)^{2-}$
Zn(II)	Zn^{2+}, $ZnSO_4^0$,	$ZnHCO_3^+$, $ZnCO_3^0$, Zn^{2+}, $ZnSO_4^0$

[a]Because of greater solubility of humic compounds in alkaline media, metal–humic complexes (as with DOC) can be expected to also occur in this condition, in particular Cu, Hg, Pb, V, and Zn. Under certain conditions, some metals can also occur in methylated forms (e.g., Hg).

example, CEC along with soil pH, determine the loading rates for metals added with municipal sewage sludge (see Chapter 4). The CEC can be viewed as a general but imperfect indicator of soil components (i.e., clays, OM, and oxides) that may limit the solubility and mobility of metals instead of being a specific factor in the bioavailability of these metals.

Several trace metals (e.g., Co, Cu, Hg, Ni, Pb, Zn, etc.) exhibit rather high affinities for SOM. As discussed earlier, soluble and insoluble complexes between the metals and OM may form. Thus, indigenous SOM and the added animal manure, sewage sludge, compost, peat, and plant residues can bind trace metals in soils. Organic matter has both the cation exchange property and the chelating ability. It has been established that trace metals in sludge are taken up less readily by plants than are trace metals added to sludge as inorganic salts or trace metals added directly to soil in the same concentrations as present in sludge (Kirkham, 1977). The retaining power of OM for Cd was predominantly through its CEC property rather than its complexing ability (Haghiri, 1974).

The amount of clay in relation to the amounts of silt and sand determines the soil texture, which in turn influences the CEC of soils. In general, the higher the clay content, the higher the CEC. The CEC of soil is largely proportional to the surface area of individual components (Appendix Table A.14). For example, clay > silt > sand in CEC primarily because surface area decreases from clay to sand. Among the clays, the crystal structure determines the CEC; the 2:1 montmorillonite and vermiculite have much higher CEC values than the 1:1 kaolinite. Clay loam soils may have CECs of 4 to 58 meq/100 g, whereas sandy loams may only have 2.5 to 17 meq/100 g. As pointed out earlier, the oxides of Fe, Al, and Mn are important in sorbing and occluding various trace metals. These hydrous oxides can play a controlling role in the immobilization metals in soils and freshwater sediments (Gadde and Laitinen, 1974; Jenne, 1968).

4.3 Redox Potential

The moisture content of soils influences their retention for trace metals through oxidation–reduction reactions (see section 3.3). In oxidized soils, redox potential may range from about +400 to +700 mV. In sediments and flooded soils, redox potential may range from around −400 (strongly reduced) to +700 mV (well oxidized) (Gambrell and Patrick, 1978). Under reducing conditions, sulfides of metals can form. The metal-bearing sulfides are quite insoluble, so that metal mobility and bioavailability are considerably less than would be expected under oxidized soils. Elemental concentrations in solution extracted from sludge-treated soil indicate the decreased solubility of Cd, Cu, and Zn and increased solubility of Mn and Fe under reducing conditions (Bingham et al., 1976).

4.4 Type and Chemical Speciation of Elements

Certain physicochemical properties of elements (e.g., electronegativity and ionic potential) have some bearing on the biogeochemical behavior of metals (Appendix Table A.13). Electronegativity influences the order in which trace metals sorb on soil constituents. Stronger covalent bonds with oxygen atoms on the surface can be expected with the more electronegative metals. For some divalent metals, the bonding preference would be

$$Cu > Ni > Co > Pb > Cd > Zn > Mg > Sr$$

On the basis of electrostatic attraction, however, the strength of bonding is influenced by the ionic potential (charge/radius ratio), producing a different pattern:

$$Ni > Mg > Cu > Co > Zn > Cd > Sr > Pb$$

Likewise, the trivalent trace metals Fe^{3+} and Cr^{3+} would be preferentially sorbed in the presence of the divalent metals listed above.

Manganese oxides show particularly strong preference for the sorption of Cu^{2+}, Ni^{2+}, Co^{2+}, and Pb^{2+}; Fe and Al oxides preferentially adsorb Pb^{2+} and

Cu^{2+} most strongly among the divalent metals mentioned above (McBride, 1994). Adsorption of Pb^{2+} and Cu^{2+} in the hydrolyzed form is a plausible mechanism inducing the unusual selectivity of the Fe and Al oxides, since the two are most easily hydrolyzed among the metals above. The adsorption by the mineral phase proceeds as *chemisorption* in the acid range (at the low pH end of the isotherm), and as the pH is increased, metal ions cluster into metal oxide or hydroxide nuclei even before the surface is saturated. Ultimately, precipitation is formed as the pH is raised, removing all the hydrolyzed metals from the solution (at the high pH end of the isotherm).

The oxidation state and chemical species also influence the reactivity and mobility of metals in the environment. The following elements have multiple oxidation states of importance in soil and aquatic systems: As($-$III, 0, III, V), Cr(III, VI), Cu(I, II), Fe(II, III), Hg(0, II), Mn(II, III), and Se($-$II, II, IV, VI). For example, hexavalent Cr(VI) is much more mobile and toxic than the trivalent Cr(III).

Biogeochemical speciation of elements occurs in the soil solution (see Fig. 2.5). Here the ions exist as *free* (also termed *aquo* or aquated) or complexed either with inorganic (ion pairs) or organic ligands. Complexation occurs as an exchange reaction with the coordinated water molecules, which are exchanged with some preferred ligands. The principle of hard and soft Lewis acids and bases is useful to describe these reactions (Stumm and Morgan, 1981). A Lewis acid is any chemical species that employs an empty electronic orbital in initiating a reaction, while a Lewis base is any chemical species that employs a doubly occupied electronic orbital in initiating a reaction. Lewis acids and bases can be neutral molecules, simple or complex ions, or neutral or charged macromolecules. Accordingly, the proton and all

TABLE 2.5. Examples of hard and soft Lewis acids and bases found in soil solutions.

Lewis Acids
Hard acids:
H^+, Li^+, Na^+, K^+, Rb$^+$, Cs$^+$, Mg^{2+}, Ca^{2+}, Sr^{2+}, Ba^{2+}, Ti^{4+}, Zr^{4+}, Cr^{3+}, Cr^{6+}, MoO^{3+}, Mn^{2+}, Mn^{3+}, Fe^{3+}, Co^{3+}, Al^{3+}, Si^{4+}, CO_2
Borderline acids:
Fe^{2+}, Co^{2+}, Ni^{2+}, Cu^{2+}, Zn^{2+}, Pb^{2+}
Soft acids:
Cu^+, Ag^+, Au^+, Cd^{2+}, Hg^+, Hg^{2+}, CH_3Hg^+

Lewis Bases
Hard bases:
NH_3, RNH_2, H_2O, OH^-, O^{2-}, ROH, CH_3COO^-, CO_3^{2-}, NO_3^-, PO_4^{3-}, SO_4^{2-}, F^-
Borderline bases:
$C_6H_5NH_2$, C_2H_5N, N_2, NO_2^-, SO_3^{2-}, Br^-, Cl$^-$
Soft bases:
C_2H_4, C_6H_6, R_3As, R_2S, RSH, $S_2O_3^-$, S^{2-}, I^-

R represents an organic molecular unit; underline indicates a tendency to softness.
Source: Sposito (1984).

of the metal cations of interest in soil solutions are Lewis acids (Table 2.5). The Lewis bases include H_2O, oxyanions such as OH^-, SO_2^{2-}, PO_4^{3-}, COO^-, CO_3^{2-}, F^-, NO_3^-, and organic N, S, and P electron donors (see Table 2.5). Under the Lewis principle, hard bases prefer to complex hard acids, and soft bases prefer to complex soft acids, under comparable conditions of acid–base strength. Hard Lewis acid molecules are small, high oxidation state, high electronegativity, and low polarizability; the converse is true for soft Lewis acids. Similarly, a hard Lewis base has high electronegativity and low polarizability, and conversely for a soft Lewis base.

TABLE 2.6. Effects of soil/sediment factors on trace metal mobility/bioavailability.[a]

Soil factor	Causal process	Effect on mobility/ bioavailability
Low pH	Decreasing sorption of cations onto oxides of Fe and Mn	Increase
	Increasing sorption of anions onto oxides of Fe and Mn	Decrease
High pH	Increasing precipitation of cations as carbonates and hydroxides	Decrease
	Increasing sorption of cations onto oxides of Fe and Mn	Decrease
	Increasing complexation of certain cations by dissolved ligands	Increase
	Increasing sorption of cations onto (solid) humus material	Decrease
	Decreasing sorption of anions	Increase
High clay content	Increasing ion exchange for trace cations (at all pH)	Decrease
High OM (solid)	Increasing sorption of cations onto humus material	Decrease
High (soluble) humus content	Increasing complexation for most trace cations	Decrease/increase?
Competing ions	Increasing competition for sorption sites	Increase
Dissolved inorganic ligands	Increasing trace metal solubility	Increase
Dissolved organic ligands	Increasing trace metal solubility	Increase
Fe and Mn oxides	Increasing sorption of trace cations with increasing pH	Decrease
	Increasing sorption of trace anions with decreasing pH	Decrease
Low redox	Decreasing solubility at low redox potential as metal sulfides	Decrease
	Decreasing solution complexation with lower redox potential	Increase/decrease?

[a]Certain trace metals are more bioaccumulative in fish in aquo form (e.g., Cu^{2+}, Ag^+, etc.), whereas others in alkylated form (e.g., methyl Hg) are more toxic.

Chemical speciation is very important in evaluating the metal's mobility, bioavailability, and potential toxicity to organisms. For example, monomethyl Hg (CH_3Hg^+) is the most bioavailable and toxic to fish of all the Hg species in aquatic systems. This is also the Hg species that bioaccumulates in fish muscle, and when ingested by humans, results in virtually complete absorption by the gastrointestinal (GI) tract.

The effects of the different master variables on the mobility and bioavailability of trace metals are indicated in Table 2.6.

This chapter summarized the fundamental retention mechanisms and processes regulating the behavior of trace metals in aquatic and terrestrial systems. In turn, the major factors affecting these mechanisms and processes were discussed.

References

Allaway, W.H. 1968. *Adv Agron* 20:235–274.

Allison, J.D., and D.S. Brown. 1995. *MinteqA2/Prodefa2—A Geochemical Speciation Model and Interactive Processor*. Soil Sci Soc Am, Madison, WI.

Bingham, F.T., G.A. Mitchell, R.J. Mahler, and A.L. Page. 1976. In *Proceedings of the International Conference on Environmental Sensing and Assessment,* vol 2. Inst Elec & Electron Engrs, New York.

Bolan, N.S., R. Naidu, J.K. Syers, and R.W. Tillman. 1999. *Adv Agron* 67:87–140.

Boyd, G.E., J. Shubert, and A.W. Adamson. 1947. *J Am Chem Soc* 69:2818–2829.

Brady, N.C. 1990. *The Nature and Properties of Soils,* 10th ed. Macmillan, New York.

[CAST] Council for Agricultural Science and Technology. 1980. *Effects of Sewage Sludge on the Cadmium and Zinc Contents of Crops*. CAST Rep 83, CAST, Ames, IA.

Chen, Y., N. Senesi, and M. Schnitzer. 1978. *Geoderma* 20:87–104.

Cieslinski, G., K.C. Van Rees, G.S.R. Khrishnamurti, and P.M. Huang. 1998. *Plant Soil* 203:109–117.

Downs, S.G., C.L. McLeod, and J.N. Lester. 1998. *Water Air Soil Pollut* 108:149–187.

Eijsackers, H. 1994. Ecotoxicology of soil organisms: seeking the way in a pitch-dark labyrinth. In M.H. Donker et al., eds. *Ecotoxicology of Soil Organisms*. Lewis Publ, Boca Raton, FL.

Evans, L.J., G.A. Spiers, and G. Zhao. 1995. *Intern J Environ Anal Chem* 59:291–302.

Florence, T.M., and G.E. Batley. 1980. *CRC Crit Rev Anal Chem* 9:219–296.

Fuller, W.H. 1977. In EPA Rep. 600/2-77-020, Solid & Hazardous Waste Div., U.S. EPA, Cincinnati, OH.

Gadde, R.R., and H.A. Laitinen. 1974. *Anal Chem* 46:2022–2026.

Gambrell, R.P., and W.H. Patrick. 1978. In *Plant Life in Anaerobic Environments*. Ann Arbor Sci Publ, Ann Arbor, MI.

Haghiri, F. 1974. *J Environ Qual* 3:180–183.

Hart, B.T. 1982. In P.C. Sly, ed. *Sediment-Freshwater Interaction, Developments in Hydrobiology*. Dr W Junk, The Hague, Netherlands.

Harter, R.D., and R. Naidu. 1995. *Adv Agron* 55:219–263.

Helmke, P.A., and R. Naidu. 1996. In R. Naidu et al., eds. *Contaminants and the Soil Environment in the Australasia-Pacific Region.* Kluwer, Dordrecht, Netherlands.

Hodgson, J.F. 1963. *Adv Agron* 15:119–158.

Jenne, E.A. 1968. *Adv Chem* 73:337–387.

Kabata-Pendias, A., and H. Pendias. 1992. *Trace Elements in Soils and Plants.* CRC Pr, Boca Raton, FL.

Kaplan, D.I., D.B. Hunter, P.M. Bertsch, and D.C. Adriano. 1994. *Environ Sci Technol* 28:1168–1189.

Kaplan, D.I., P.M. Bertsch, and D.C. Adriano. 1995. *Groundwater* 33:708–717.

Khrishnamurti, G.S., G. Cieslinksi, P.M. Huang, and K.C. Van Rees. 1997. *J Environ Qual* 26:271–277.

Kirkham, M.B. 1977. In R.C. Loehr, ed. *Land as a Waste Management Alternative.* Ann Arbor Sci Publ, Ann Arbor, MI.

Kuo, S., and J.B. Harsh. 1997. In D.C. Adriano, ed. *Biogeochemistry of Trace Metals.* Applied Sci. Publ, Northwood, England.

Langmuir, I. 1918. *J Am Chem Soc* 40:1361–1403.

Logan, T.J., and S.J. Traina. 1993. In H.E. Allen et al., eds. *Metals in Groundwater.* Lewis Publ, Chelsea, MI.

McCarthy, J.F. 1989. In I.H. Suffet and P. MacCarthy, eds. *Aquatic Humic Substances.* American Chemical Society, Washington, DC.

McCarthy, J.F., and J.M. Zachara. 1989. *Environ Sci Technol* 23:496–504.

Mantoura, R.F.C., A. Dickson, and J.P. Riley. 1978. *Estuaries Coastal Mar Sci* 6:387–408.

Marschner, H. 1995. *Mineral Nutrition of Higher Plants.* Academic Pr, San Diego, CA.

McBride, M. 1994. *Environmental Chemistry of Soils.* Oxford Univ. Pr, New York.

Mench, M., and E. Martin. 1991. *Plant Soil* 132:187–196.

Mench, M., J.L. Morel, A. Guckert, and B. Gruillet. 1988. *J Soil Sci* 39:521–527.

Mingelgrin, U., S. Saltzman, and B. Yaron. 1977. *Soil Sci Soc Am J* 41:519–529.

Naidu, R., M.E. Sumner, and R.D. Harter. 1998. *Environ Geochem Health* 20:5–9.

Parker, D.R., W.A. Norvell, and R.L. Chaney. 1995. Geochem PC—A chemical speciation program for IBM and compatible personal computers. In *Chemical Equilibria and Reactive Models.* Soil Sci Soc Am, Madison, WI.

Salomons, W., and U. Forstner. 1984. *Metals in the Hydrosphere.* Springer-Verlag, Berlin.

Schnitzer, M. 1977. Recent findings on the characterization of humic substances extracted from soils from widely differing climatic zones. In *Soil Organic Matter Studies,* vol 2. Vienna (IAEA).

Senesi, N. 1992. In D.C. Adriano, ed. *Biogeochemistry of Trace Metals.* CRC Pr, Boca Raton, FL.

Sigg, L. 1987. In W. Stumm, ed. *Aquatic Surface Chemistry: Chemical Processes at the Particle–Water Interface.* Wiley, New York.

Smith, W.H. 1976. *Ecology* 57:324–331.

Sposito, G. 1984. *The Surface Chemistry of Soils.* Oxford Univ. Pr, New York.

Sposito, G. 1989. *The Chemistry of Soils.* Oxford Univ. Pr, New York.

Sposito, G., and J. Coves. 1995. Soil Chem on the MacIntosh. In *Chemical Equilibria and Reaction Models.* Soil Sci Soc Am, Madison, WI.

Stevenson, F.J. 1972. *BioScience* 22:643–650.

Stevenson, F.J. 1994. *Humus Chemistry,* 2nd ed. Wiley, New York.

Stumm, W., and J.J. Morgan. 1981. *Aquatic Chemistry*. Wiley, New York.

Thurman, E.M. 1985. *Organic Geochemistry of Natural Waters*. Kluwer, Boston.

Wada, K. 1989. In J.B. Dixon and S.B. Weed, eds. *Minerals in Soil Environments*. Soil Sci Soc Am, Madison, WI.

White, M.C., A.M. Decker, and R.L. Chaney. 1979. *Agron J* 71:121–126.

Zachara, J.M., J.A. Kittrick, L.S. Dake, and J.B. Harsh. 1989. *Geochim Cosmochim Acta* 53:9–19.

Zang, F., V. Romheld, and H. Marschner. 1991. *J Plant Nutr* 14:675–686.

Suggested References for Further Reading

Evans, L.J. 1989. *Environ Sci Technol* 23:1046–1056.

McBride, M.B. 1994. *Environmental Chemistry of Soils*. Oxford Univ. Pr, New York.

Stevenson, F.J. 1994. *Humus Chemistry,* 2nd ed. Wiley, New York.

3
Bioavailability of Trace Metals

Metals external to the organism are unlikely to cause any adverse effect in that organism but may do so once absorbed (or taken up) and assimilated. This implies that as a prelude step, metals have to come in contact with the organism to be of any benefit or consequence to that organism. In turn, metals have to be in a particular form to be able to enter an organism. In essence, for a contaminant to be assimilated, it will have to be *mobile* and *transported* and be *bioavailable* to the organism. Because bioavailability (also known as bioaccessibility) may have different meanings to different disciplines, it is prudent to define it according to the receptor organism.

1 Defining Bioavailability

A more generic definition of *bioavailability* is the potential of living organisms to take up chemicals from the food (i.e., oral) or from the abiotic environment (i.e., external) to the extent that the chemicals may become involved in the metabolism of the organism. More specifically, it refers to biologically available chemicals that can be taken up by an organism and can react with its metabolic machinery (Campbell, 1995), or it refers to the fraction of the total chemical that can interact with a biological target (Vangronsveld and Cunningham, 1998).

In ecotoxicology, *bioavailability* can be broadly defined as the portion of a chemical in the environment that is available for biological action, such as uptake by an organism (Rand, 1995). The conventional approach involves feeding groups of individuals known amounts of the chemical in question over a period of time. The portion of the total dose retained in the organism at the end of the feeding period is measured after withholding of food for sufficient time to allow complete clearance of the food from the gut (Newman, 1995). Bioavailability is then estimated by mass balance under the assumption that all chemicals measured in the organism are assimilated.

Bioavailability of sediment-associated contaminants can be defined as "the fraction of the total contaminant in the interstitial water and on the sediment particles that is available for bioaccumulation"; whereas *bioaccumulation* is the "accumulation of contaminant via all routes available to the organism" (Landrum and Robbins, 1990). As in other media, chemical measures of contaminant concentration in sediment do not always reflect the bioavailable fraction of sediment-bound contaminants (Landrum, 1989). Concentrations of metals on sediments can exceed those of the overlying water by 3 to 5 orders of magnitude (Bryan and Langston, 1992).

In pharmacology, *bioavailability* refers to the fraction of an orally administered dose (e.g., in diet) that reaches the systemic circulation (i.e., blood) of the animal (Gibaldi, 1991). This is basically the definition commonly used by mammalian toxicologists (Ruby et al., 1999). Applied to humans, oral bioavailability is defined as the fraction of an administered dose that reaches the blood from the GI tract. Oral (or dietary) bioavailability is an example of pharmacologically defined bioavailability (McCloskey et al., 1998). Measures of bioavailability are routinely determined for drugs in pharmacology and for chemicals pertinent to aquaculture with fish, mainly because of their fairly rapid metabolism or elimination. Bioavailability of CH_3Hg from oral solutions or commercial foods is high and has been reported to be between 70 and 90% in goldfish (*Carassius auratus*) (Sharpe et al., 1977) and between 50 and 70% in rainbow trout (*Oncorhynchus mykiss*) (Giblin and Massaro, 1973).

The above definitions center on a common condition that a contaminant or part of it should be free for uptake (i.e., movement into or onto) by the organism. In a risk-based context, the American Academy of Environmental Engineers (Linz and Nakles, 1996) defined *availability* as referring to the rate and extent that the chemical is released from the soil into the environment, i.e., air and water, or to bioavailability of the chemical to ecological and human receptors following dermal contact, ingestion, or inhalation (see Chapter 5). Thus, based on this context, the much broader environmental availability encompasses a physical component (i.e., mobility/transport by diffusion and mass flow) and a biological component (i.e., uptake and metabolism). Basically, this concept entails migration of chemicals into groundwater, surface water, or atmosphere (i.e., water and air quality), and bioaccumulation in organisms (i.e., toxicology).

2 Relevance of Bioavailability to Biological Response and Risk

It is essential to establish some important concepts pertinent to dose–response relationships in terrestrial and aquatic ecosystems. Bioavailability is the key driver defining the extent and magnitude of biological responses. A generalized dose–response curve for soils is displayed in Figure 3.1, whereby increasing concentration (dose) of a trace element exerts a biological effect. The effect can be in the form of yield of crop, or growth, or activity of an organism. The same concept can be applied to aquatic systems. Furthermore, this concept can be extrapolated to large mammals, including livestock and humans.

For trace elements that are essential for metabolism, there are essentially three parts of the curve: the *deficiency* range (biological activities increased in this range), the *buffering* (or normal) range (also known as sufficiency range where biological functions are somewhat optimal, have plateaued, and are sustainable over a wider range), and the *toxicity* range (where further increases in concentration inhibit metabolic functions and, depending on the severity, can be lethal to the organism). The shape of the curve and range of concentrations depends on a number of factors, including (1) type and nature (both physical and chemical) of metals, (2) sensitivity or tolerance of the receptor organism, and (3) nature and properties of soils (soil type, chemical properties, and buffering capacity). In aquatic ecosystems, especially freshwater systems, the buffering capacity is generally less limited than that of the soil system.

The risk associated with biological response occurs within the toxicity range. The threshold of toxicity is defined toxicologically in several ways, the first is the no observable adverse effect level (NOAEL) (or concentration, i.e., NOAEC) and the second is the lowest observable adverse effect level (LOAEL) (or concentration, i.e., LOAEC) (see also Chapter 6). To quantitate these thresholds, it is necessary to determine the highest concentration in a given medium that has no measurable effect (NOAEL),

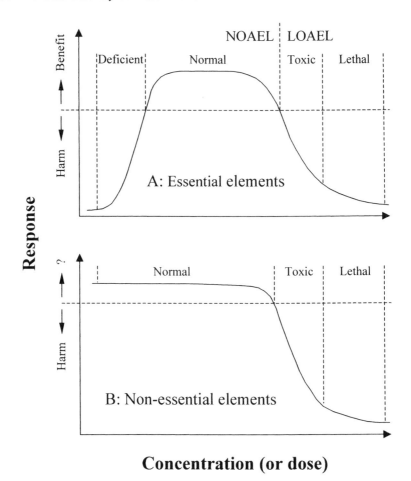

FIGURE 3.1. Dose–response relationships for (A) essential and (B) nonessential elements in higher plants and mammals including humans. Some nonessential elements may stimulate plant growth at low concentrations in the substrate.

or the lowest concentration that initiates the adverse biological effect (LOAEL). The LOAEL is a threshold level relevant to risk assessment. An example of an end point in polluted soil is the soil concentration of a particular metal that would induce, for example, a 10% reduction (or any given percentage in yield reduction deemed detrimental to crop quality or monetary return to the farmer) in crop yield.

3 Bioavailability in the Soil–Plant System

Because soils can be impacted by industrial emissions of metals, it is important to discuss some concepts pertinent to uptake and bioaccumulation. The sensitivity or tolerance of plants to excess metals is influenced by plant species and genotypes. Even among crops, sensitivity varies widely,

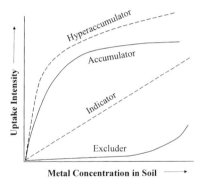

FIGURE 3.2. Relative uptake and bioaccumulation potential among plant species.

with members of the *Brassicaceae* family generally considered as the most tolerant in terms of accumulation. In general, plants can be grouped into three categories: *excluders, indicators,* and *accumulators* (Fig. 3.2). *Excluders* include members of the grass family (e.g., sudangrass, bromegrass, fescue, etc.) for their known insensitivity to metals over a wide range of soil concentrations; *indicators* include the grain and cereal crops (e.g., corn, soybean, wheat, oats, etc.), and *accumulators* include the mustard and *Compositae* families (e.g., lettuce, spinach, chard, etc.) and tobacco (Note: *Brassicaceae* members may respond better to elevated levels of metals in soil and in this regard they may be considered as a better indicator). There are extreme accumulators (known as *hyperaccumulators*) that seem to even thrive in heavily contaminated soils (or near ore deposits) and survive through a tolerance mechanism; in contrast, *excluders* survive through avoidance (or restriction) mechanism. *Indicators* are plant species that correspondingly respond to metal concentrations in soils, displaying linear curves, whereas *accumulators/excluders* display logarithmic curves (Baker, 1987).

3.1 Bioavailability in the Soil–Plant–Livestock Pathway

The extent of metal uptake by plants is important in the soil–plant–livestock pathway (Fig. 3.3) because of its potential implication to the consuming general population in terms of meat, dairy products, and organ tissues (e.g., kidney, liver, heart, etc.). For example, the quality of beef and milk from cattle depends on quality of the forage, which in turn depends on the quality of soil. In this case, the end point parameters would be the quality of the consumer products, i.e., beef and milk, if they are fit for human consumption. The main concerns in this pathway are elements that do not prove phytotoxic even if they bioaccumulate in plant tissue (e.g., Se, Mo, Pb, Cd, Hg) so that they are passed on to the animal. This is further confounded by the tendency of certain elements (e.g., Se and Mo) to have only narrow windows of safety in animal nutrition. On the other hand, elements that prove to be phytotoxic (e.g., Ni, Cu, Co, Zn, Tl) will not have the chance to present any consequential effect to the animal.

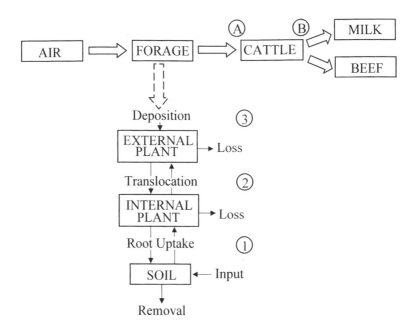

FIGURE 3.3. Schematic of metal transfer in the soil–plant–livestock pathway showing critical junctions for the bioavailability of metals. Soil–plant: (1) absorption by roots, (2) distribution within the plant parts, and (3) absorption by foliage. Plant–animal: (A) dietary intake (oral) and (B) absorption by intestinal tract.

As displayed in Figure 3.3, the forage is subject to contamination through foliar uptake from deposition and through root uptake. Bioavailability via foliar uptake can be important in localized instances such as in areas in the vicinity of smelters and emission stacks. Foliar spray of micronutrients is a common practice in agriculture to augment the available pool in the soil, especially if soil conditions are not conducive to solubilize these elements (e.g., calcareous, alkaline soils). Chelates are often added to foliar solutions to enhance bioavailablility of micronutrients (see individual chapters on Zn, Cu, Mn) to plants. By far, root uptake is the predominant route of entry of trace elements to plants. Thus, it is necessary to elucidate root uptake mechanisms.

3.2 Bioavailability in Plants Through Root Uptake

Metals need to come in contact with the roots as a prerequisite for their entry into the roots. This means localization of metals in the immediate proximity of the root, i.e., in the rhizosphere. The rhizosphere chemical environment is substantially modified over that of the bulk soil. This results from the metabolic activity of roots and their release of exudates, and the subsequent effects on type, number, and metabolic activity of associated microbial consortium. They undoubtedly influence the rates of solubilization and the equilibrium concentrations of individual ions associated with the soil solution, with subsequent effects on the bioavailability of both essential and contaminant ions for absorption by plants. The immediate *available* pool is the soil solution—the site of readily *bioavailable* forms of metals that are either in free ionic form or soluble complexes. This available pool is in quasi-equilibrium with other pools in solid phases, which is subject to change according to the influence of controlling master variables as discussed previously in Chapter 2.

It has been well accepted that metals generally accumulate in or on the surface of roots. Thus, this phenomenon gave rise to a popular connotation regarding the role of roots in plant nutrition—the *root barrier*—which simply refers to the restriction of plant uptake in the roots. Once trace element ions are absorbed into the root, mechanisms must be present to prevent their hydrolysis and sorption onto charge sites associated with structural surfaces and/or nonspecific chemical reactions with any of the myriad of trace and macroions being simultaneously stored, transported, or metabolized. For cations, these interactions are likely minimized by organic complexation reactions.

The ability of plants to control the behavior of nonnutrient cations, particularly polyvalent species, appears to involve two distinct processes after transport into the root. The first allows an element to remain chemically soluble after root absorption and during transport within the plant through interaction with low-molecular-weight ligands, which are normally employed to maintain the solubility of essential elements. The second process involves those metabolic and physiological phenomena that incorporate the element into either functional or sequestering metabolites. In both instances, the fate of a particular nonnutrient element depends on its chemical similarities (charge, hydrated radius, and redox requirements) to essential elements. These two processes are of key importance in altering the chemical form of elements within the plant (Cataldo et al., 1987). In order to shed some light on this concept, it is necessary to examine the root microscopically at the cellular level as displayed in Figure 3.4. In the root, the first and primary "barrier" to selective and regulated uptake of metal ions into the plant is the plasma membrane (or plasmalemma) of epidermal and cortical cells (Kochian, 1991; Marschner, 1986).

It is generally accepted that membrane transport of nutrient ions in higher plants exhibits multiphasic absorption isotherms at physiological concentrations and that membrane transport is an active process with some plant control over the rates of ion absorption in response to demand (Cataldo et al., 1987). The nonnutrient ions Ni (Cataldo et al., 1978), Cd (Cataldo et al., 1983), and Tl (Cataldo and Wildung, 1978) behave similarly, suggesting that mechanisms functioning for nutrient ion absorption also regulate absorption of these nonnutrients.

The plasma membrane and tonoplast are quite similar structurally to other biological membranes, consisting of a continuous lipid (e.g., phosp-

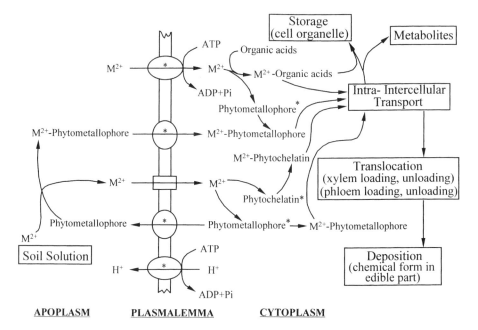

FIGURE 3.4. Plausible divalent metal ion transport systems in higher plants.
*Enhanced under certain conditions; phytometallophore or phytosiderophore.
(From Welch and Norvell, 1997.)

holipids, glycolipids, sterols, etc.) bilayer into which globular proteins are inserted. The lipid bilayer is relatively impermeable to ions; hence, the transport of ions into the cell is facilitated by specific transport proteins inserted into the lipid bilayer (Kochian, 1991). The different classes of transport proteins reside in the cellular membranes of all living organisms. These include ATPases that serve as primary active transport systems, tonoplast H^+-ATPases to facilitate active transport to another ion (secondary active transport), and ion channels that mediate passive ion transport. In short, the plasma membrane and tonoplast are the limiting membranes of the two major components of the cell—the cytoplasm and vacuole. They regulate the flow of metal ions into and within the cell through the action of specific ion transport systems in both membranes.

Because of the complexity of ion transport into, within, and from the root, the actual mechanisms of metallic ion transport across the various membranes are not exactly known. Welch and Norvell (1997, 2000) summarized plausible processes of ion (e.g., Cd^{2+}) transport into root cells as follows (see Fig. 3.4): (1) transport of divalent ions across the root-cell plasma membrane by a specific transporter protein; (2) entry as a phytometallophore (e.g., nicotianamine) or phytosiderophore complex via a phytometallophore–divalent cation transporter protein; (3) entry of divalent ions through a plasma membrane divalent cation channel (e.g., a Ca^{2+} channel) driven by the negative electrochemical membrane potential gradient resulting from operation of the plasma membrane H^+-translocating ATPase; and/or (4) transport by an ion carrier protein that normally transports Zn^{2+}, Cu^{2+}, or Fe^{2+}.

Once in the cytosol, certain divalent ions may bind to peptides and proteins that contain cysteine residues because these cations prefer sulfhydryl ligand groups to oxygen- or nitrogen-containing ligand groups (Welch and Norvell, 1997). The divalent ions may also be complexed by phytochelatins (a type of metallothionein) and organic acids, which facilitate the intracellular and intercellular movement of the ions.

Stated more simply, solute transport from outside the root to the central root xylem where the material is carried to the shoot takes place through two major pathways (Cataldo et al., 1987; Wu et al., 1999): the apoplastic (cell wall space between cell membranes) and the symplastic (crossing many cell membranes along the path). In the apoplastic pathway, the presence of the lipophilic Casparian strip at the root endodermis disrupts the apoplastic water flow and directs it to cross cell plasma membranes at least twice, where selective transport as well as passive permeation of the solutes occur (Tanton and Crowdy, 1972). Under normal plant physiological conditions, the Casparian strip–guarded pathway accounts for >99% of water flow through the roots (Hanson et al., 1985). Hydrophilic compounds tend to favor the apoplastic pathway, whereas lipophilic compounds favor the symplastic pathway. The root uptake rate and xylem steady-state concentration of a compound depend on the compound's lipophilicity.

Following the conventional definitions of bioavailability presented in section 1, trace element ions become officially bioavailable to the plant once they have crossed the "barrier" membranes in the roots, which then renders them available for transport and distribution to other plant tissues.

3.3 Indicators of Bioavailability in Plants

The readily soluble fraction of metals is generally considered to be bioavailable but there is a growing awareness that current methods of assessment of soluble and bioavailable fractions need reevaluation due to their variability both in space and time. These variations could arise from climate fluctuations or management practices. In agricultural soils, liming, irrigation, and fertilization could produce significant changes in the chemical composition of the soil solution.

Chemical extraction techniques are still the most commonly used method of estimating the fraction of a micronutrient or contaminant that is bioavailable in the short term. The content of a metal in soil solution (i.e., soluble) plus the "weakly adsorbed" content (i.e., exchangeable) provide a good measure of the plant-available amount. Several methods have been used to sample soil solution: suction cups, tensionless lysimeters, and centrifugation of soil samples. Single extractants that include $CaCl_2$, $Ca(NO_3)_2$, etc. have been used to extract mainly the exchangeable metals from soil (see next section). This extractable fraction generally correlates closely with plant uptake. Adsorption–desorption processes are important controls on the amount of metals in soil solution.

The use of organisms (e.g., worms) to assess bioavailability could provide a most reliable approach to assessment of bioavailability but there is still some concern about the most suitable organisms and about the extrapolation of results from the laboratory to the field.

Bioavailability of metals in soils can be manipulated by adjusting soil chemistry—such as additions of soil amendments (e.g., lime, biosolids, phosphate compounds, etc.)—inducing changes in the solubility and mobility of the metals.

3.3.1 Estimating Bioavailability Using Single Extractants

Much has been learned from soil chemists, horticulturists, and agronomists regarding the response of crops to micronutrient status in soils. They were among the pioneers to apply the "fractional" concept of estimating the bioavailable pool (in contrast to the total content) in soils based on fundamental principles of soil chemistry and verified by greenhouse and field research using crop indicators (Sims and Johnson, 1991). Thus, soil testing for plant micronutrients using single extraction techniques has become a standard practice to ascertain if a soil is deficient, borderline, or sufficient with regard to a given micronutrient for a particular crop (or group of crops). With agricultural lands now being allowed as receptors of certain type of biosolids (e.g., municipal sewage sludge and livestock manure), soil testing may also be employed to evaluate the potential of treated soils to accumulate excess metals, inevitably causing a breach of food chain quality, and phytotoxicity.

As discussed in later chapters on individual elements, in the section on "bioavailable" fraction, the major categories of micronutrient extractants currently in use are dilute acids (e.g., HCl, $H_2SO_4 + HCl$, NH_4OAc), and solutions containing chelating agents such as EDTA or DTPA. Based on extensive research and field trials, deficiency and sufficiency levels of micronutrients have already been established in the United States and abroad. Preferences of extractants, however, vary among regions and critical levels vary among crops. For example, in the United States, the DTPA solution is widely used in the Midwest and Great Plains, whereas the double-acid extractant (a Mehlich procedure variant) is widely used in the Southeast (Whitney, 2000).

There is no universal soil extractant that can be used to estimate the micronutrient and metal bioavailability to plants. This is because of the complexity of metal ion dynamics in the soil system and the interactive role of plant and environmental factors on the whole process.

3.3.2 Estimating Bioavailability Using Selective Extractants

Some concepts are needed to elucidate the dynamic relationship of metal ions in the aqueous and solid phases in soils. The sum of the two phases represents the total content; the sum of the soluble (in soil solution) form

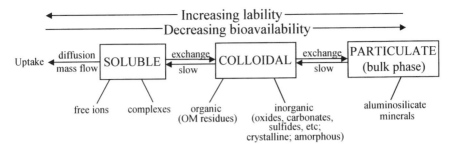

FIGURE 3.5. Trend of bioavailability (or lability) as influenced by the "operational" form of the metals in soils or sediments.

and the exchangeable form rarely exceeds 10% of the total content; in addition, the percentage in the soluble form is even much less (Shuman, 1991). For example, the percentages of total Zn and Cd in soils that received either inorganic fertilizer, farm yard manure, or sewage sludge compost that were in the exchangeable fraction (i.e., extracted with $CaCl_2$) ranged from 0.5 to 3.2% for Zn and from 4 to 18% for Cd (McGrath et al., 2000). As the most readily bioavailable amount in the solution phase and exchangeable phase is depleted by plant uptake, the equilibrium has to shift from the bulk solid phase to replenish the depleted pools, and if this replenishment is not enough as assayed by soil tests (as shown in Fig. 3.5, the rate of solubilization or desorption to the more bioavailable pool is usually very slow), then supplementary amounts need to be applied to the soil or directly to the plant by foliar spray. On the other hand, if the amount in the aqueous phase and exchangeable phase is excessive, as in heavily contaminated soils, the equilibrium would have to shift to the bulk solid phase. In heavily contaminated soils, the equilibrium can be induced to shift to the solid phase by adding certain soil additives as sorbent. For example, adding alkaline composted biosolids is an effective measure to mitigate excess metals in soil and defuse phytotoxicity. In this case, the metallic cations become bound mostly in the carbonate and OM fractions. Retention in this case is explained by the nature of the biosolid, i.e., alkaline pH due to addition of lime or lime-like material (e.g., alkaline kiln dust) and high OM content of the product. In Pb-contaminated soils, risks from Pb toxicity can be mitigated by addition of phosphate compounds (e.g., hydroxyapatite) which results in the formation of fairly stable complexes of lead–pyromorphite (see later chapter on lead).

The ultimate effect of the shift in equilibrium is reflected in bioavailability, and, therefore, potential risk. As featured in Figure 3.5, the lability (or mobility) of a metal follows a gradient of decreasing bioavailability from the soluble to the most nonlabile solid phase (i.e., the most tightly held, primarily within the silicate lattice—the so-called residual fraction). I propose the utilization of sequential fractionation technique to generate soil chemical indices as a way of evaluating the efficacy of remediation techniques to immobilize metals in soils. It has been our observation (Chlopecka and Adriano, 1997) as well as that of others (Geebelen, 2000; Mench et al., 1994) that the exchangeable fraction is usually higher in untreated contaminated soils than in remediated soils. These substantial reductions in the exchangeable fraction in treated soils are reflected as increases in amounts in other less labile fractions (i.e., carbonate, organic, residual, etc.). This led us to develop new chemical indices (or coefficients) that can be quantified to establish risk criteria using plant response data (Adriano et al., 2000). Due to the "operational" nature of selective extraction, such indices are still subject to a great deal of variation (Ahnstrom and Parker, 1999). Comprehensive testing, however, can overcome such shortcomings from this approach as was done with single extractants for micronutrient soil tests (Sims and Johnson, 1991). This would include more soil types, plant species, nature and extent of contamination, and possibly alternate methods to sequential extraction.

Despite some weaknesses (e.g., selectivity of extractants, readsorption, precipitation), inherent with the "operational" nature (i.e., not specific) of selective extraction (Chlopecka and Adriano, 1997; Shuman, 1991), it is still widely used as a means of characterizing the "forms" of metals (or their distribution among principal components) in a given medium, and delineating, although imprecisely, one pool from the other. More importantly, it can be helpful in understanding the dynamics of shift in form from one pool to the other, especially before and after remediation of contaminated soils. The method of Tessier et al. (1979) and its variants appear to be the most widely used.

In general, the method involves the selection of chemical reagents from the least to the most aggressive in a sequential fashion and from the least to the greatest extremes in temperature and stirring. There are prescribed extraction procedures for the following specific soil/sediment fractions (Shuman, 1991): (1) water soluble, (2) exchangeable, (3) sorbed, (4) organic, (5) oxide–crystalline or amorphous Fe and Mn oxides, (6) carbonate, (7) sulfide, and (8) residual. These various fractions (or phases) are briefly described: *soluble*—exists in soil solution either in free ionic or complexed form; *exchangeable*—sorption by electrostatic attraction (i.e., weakly adsorbed) to negatively charged exchange sites on colloidal particles; *sorbed*—adsorption on specific exchange sites on the surface of colloids (not exchangeable); *oxide*—specific adsorption on colloidal Fe/Mn oxides; *carbonate*—precipitation and/or occlusion in soils high in free $CaCO_3$, bicarbonate, and alkaline in reaction; *organic*—complexation with the organic fraction, chelated and/or organic bound; *sulfide*—very insoluble and stable compounds of metal-sulfides occurring in poorly aerated soils, such as rice paddy soils; and *residual*—fixation within the crystalline lattices of mineral (aluminosilicate) particles.

The number of phases determined varies typically from four to as many as six to seven per sample (Shuman, 1991). An example menu for a step-by-step sequential extraction of trace metals in soils has been applied by Keller and Vedy (1994).

TABLE 3.1. Concentration factors of trace elements in plant tissue.

Element	Soil (mg/kg)		Transfer factor			
	Range	Average	Grain	Vegetable	Root/tuber	Fruit
As	0.1–40	6	0.004	0.037	0.004	0.003
Cd	0.01–2	0.35	0.036	0.223	0.008	0.09
Pb	<1–300	19	0.002	0.0016	2E-5	14E-5
Hg	0.01–0.5	0.06	0.085	0.009	0.002	0.009
Se	0.01–1.2	0.4	0.002	0.015	0.042	0.021

Source: Data derived from Kabata-Pendias and Pendias (1994); Fergusson (1990).

3.3.3 Soil–Plant Transfer Coefficient as Indicator of Bioavailability

Data presented in Table 3.1 reflect soil values for certain trace elements from noncontaminated areas of the world. By dividing trace element concentrations in tissues for various plant types (organs) by soil concentrations from similar noncontaminated areas, the data for transfer coefficients (TC) were obtained. As expected, among the trace elements summarized, the bioavailability of Cd is highest to vegetables (TC = 0.223); Pb yielded the lowest bioavailability to vegetables (TC = 0.0016). However, in highly contaminated soils, bioavailability increases as indicated by much higher TC values that resulted from much higher percentages of bioavailable fractions (Kabata-Pendias and Pendias, 1994). For example, the TC values for Cd, Zn, and Tl are in the neighborhood of 1 to 10; for Cu and Ni, from 0.1 to 1; and for Pb and Cr, from 0.01 to 0.1. This reconfirms the relatively greater bioavailability of Cd, Tl, and Zn to plants.

4 The Plateau Concept in Bioavailability

It has been hypothesized that metal activity in soils and, therefore, plant uptake of metals from sludge-treated soils tend to approach a maximum as the metal loading rate increases (Chang et al., 1996, 1997; Corey et al., 1987). This phenomenon is known as the *plateau* concept for describing plant uptake and accumulation of potentially toxic metals from sludge-treated soils (the so-called "sludge protection" hypothesis).

The plateau concept is schematically presented in Figure 3.6. The plateau may be obtained by fitting the data to $Y = a + b(1 - e^{-cX})$, where Y and X denote the metal (e.g., Cd) content of plant tissue and soil, respectively, and a, b, and c are constants (see detailed discussion below). The existence of the plateau in plant uptake from sludged soils has been experimentally proven employing pot culture of several crops (broccoli, lettuce, Swiss chard, radish, etc.) using field stabilized sludge–treated soils with Cd content ranging from background to 12 ppm.

This is a very important concept because a main concern in long-term repeated applications of sludge is the accumulation of excess metals in soil and the eventual deterioration of plant quality and occurrence of phytotoxicity. If the land is used for food production, the accumulated metals in the soil may be transferred from the soil through the food plants to

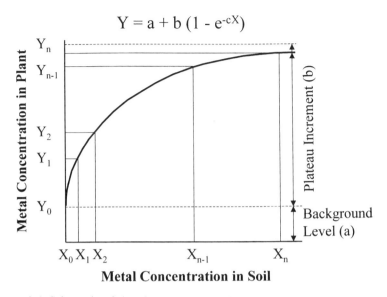

FIGURE 3.6. Schematic of the plateau concept of bioaccumulation of metals (e.g., Cd) by crops growing on sludged soils. (From Chang et al., 1997.)

consumers. The food chain transfer accounts for the most significant portion of a human's lifetime exposure to metals worldwide. Therefore, the metal loading rate for land application of sewage sludge must be controlled.

In sludge-treated soils, the characteristics of both sludge and soil affect the plant bioavailability of trace metals. More specifically, the phytoavailability of trace metals is determined by their concentrations (free ion activities) in the soil solution resulting from net reactions between the aqueous and solid phases of the sludged soil (Gerritse et al., 1983; Minnich et al., 1987; Sparks, 1984). Corey et al. (1987) postulated that "trace metal activity, and thus plant uptake of trace metals from sludged soils, tends to approach a maximum as the sludge rate increases." According to this postulation, the trace metal concentration in the plant tissue will gradually approach a plateau as the cumulative sludge input increases. They further assessed that "the estimates of plant bioavailability of the trace metals in sludge could be obtained from field or greenhouse studies with a specific variety of lettuce (or other crop that tends to accumulate the elements of interest) grown on sludge-amended soil" (Corey et al., 1987). Subsequently, this concept was adopted by the U.S. Environmental Protection Agency in justifying their decision not to set upper limits on pollutant loading in the *Standards for the Use and Disposal of Sewage Sludge* if pollutant concentrations in the sludge do not exceed the calculated limits (Chaney and Ryan, 1992; U.S. EPA, 1993). Under these auspices, sewage sludge may be applied on land without any restriction on cumulative pollutant loading if the concentration of each pollutant in the sewage sludge does not exceed the concentration for the pollutant as defined in the standards. Therefore, the plateau concept has far-reaching implications in land application of sewage sludge.

The various portions of the plateau curve can be mathematically explained as indicated in Figure 3.6. The Y and X axes of this graph represent the plant uptake of trace metal and the trace metal concentration of soil, respectively. The slope of the curve at any trace metal concentration of soil, n, may be calculated by the tangent to the curve at n by:

$$\Delta Y_n / \Delta X_n \tag{3.1}$$

where

$$\Delta Y_n = (Y_n - Y_{n-1}) \tag{3.2}$$

$$\Delta X_n = (X_n - X_{n-1}) \tag{3.3}$$

where, $n = 1, 2, 3, \ldots n$

The necessary and required condition for a plateau to occur would be that the slope of the curve $(\Delta Y_n / \Delta X_n)$ approaches 0 as the trace metal concentration of the soil increases. Therefore,

$$\Delta Y_1/\Delta X_1 > \Delta Y_2/\Delta X_2 > \Delta Y_3/\Delta X_3 \cdots \Delta Y_{n-1}/\Delta X_{n-1} > \Delta Y_n/\Delta X_n \to 0, \; n - \infty \tag{3.4}$$

Mathematically, the concept of a plateau may be depicted by the following equation (Chang et al., 1987):

$$Y = a + b(1 - e^{-cX}) \tag{3.5}$$

where Y = plant uptake of trace metal
  X = concentration of trace metal in sludge-treated soil
  a = background plant uptake of trace metal
  b = plant uptake plateau increment of trace metal
  c = rate constant for plant uptake of trace metal

If the plant growth is not affected by sludge application, the biomass harvested from soils containing different trace metal levels should not be significantly different from one another. Under this circumstance, the trace metal concentration of the plant tissue may be used as an indicator of the plant uptake. The dependant variable (Y) and parameters (a and b) in equation 5 may be expressed in terms of the concentration of trace metal in plant tissue.

The plateau concept is probably significant only for soils that are treated with sewage sludge or other sludge-like biosolids. The reason is the buffering (or sequestering) effect of the sludge constituents on the metal uptake by effectively complexing the metal ions by more or less specific adsorption that renders most of the metals nonbioavailable (i.e., bioavailability of metals from sludge is reduced due to the biosolid components, e.g., oxides of Fe and Mn, humic acids, phosphates, and sometimes carbonates from lime).

Is this plateauing effect real and infinite? The answer may depend, to a large extent, on the stability of the metal–OM complexation in sludged soils. Conceivably, the metals could be released from the OM bondage upon degradation (i.e., mineralization) of the OM itself—thus, the possibility of a *chemical time bomb* may exist. Indeed in contrast to the "sludge protection" hypothesis, the "sludge time bomb" hypothesis espouses that upon slow mineralization of sludge-associated OM, with subsequent acidification of the soil (if uncorrected), metals could transform into more soluble forms (McBride, 1995). Recently, however, McGrath et al. (2000) evaluated the two hypotheses using a long term field experiment (1942–61) at Woburn, England. Over a 23-yr period after 1961, when sludge was last applied, the extractability (using $CaCl_2$ solution) of Zn and Cd fluctuated, but neither decreased or increased consistently. The relationship between total soil and crop metal concentrations were linear, with no evidence of a plateau across the range of soil metal concentrations encountered. It was concluded that up to 23 yr after termination of sludge application, the extractability and bio-availability of Zn and Cd to crops (red beet, sugar beet, carrot, barley) have not decreased.

5 Bioavailability of Trace Metals in Aquatic Systems

As in soils, where plants vary widely in their uptake of metals (i.e., excluders, indicators, and accumulators), aquatic organisms also vary widely in their uptake of trace metals. They can be grouped into two major categories: *regulators* (or *excluders*) and *accumulators* (or *nonregulators*) (Mason and Jenkins, 1995). Simply, *regulators* are characterized by their low metal uptake; conversely, *accumulators* are characterized by their extreme metal uptake (Fig. 3.7). *Regulators* tend to control, hence the term regulator, metal accumulations and maintain their intracellular body burdens of metal within a narrow range over a broad concentration range of external metals. On the other hand, *accumulators* are those that are capable of adopting a system of detoxification based upon sequestration and tend to show elevated body burdens of metals even in noncontaminated environments (Mason and Jenkins, 1995).

As in the soil–plant system, a metal ion first has to approach the surface of a living aquatic organism before becoming bioavailable (Campbell, 1995). Also as in higher plants, the metal encounters a protective polysaccharide or glycoprotein layer (analogous to cell wall in higher plants). Moving inward, the metal will eventually meet the plasma membrane, the real barrier. The movement of metal across the lipid bilayer is relatively slow, and is generally considered to be the rate-limiting step during intake (Mason and Jenkins, 1995). The plasma membrane is generally characterized by its hydrophobic, phospholipidic nature and the presence of proteins, transport proteins, and/or ion channels that facilitate the transport of ions across the membrane (Campbell, 1995) (Fig. 3.8). Once across, the ions may interact with a wide array of binding sites, some of which are physiologically active sites where

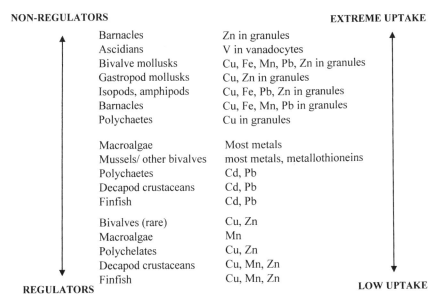

NON-REGULATORS   EXTREME UPTAKE

Barnacles	Zn in granules
Ascidians	V in vanadocytes
Bivalve mollusks	Cu, Fe, Mn, Pb, Zn in granules
Gastropod mollusks	Cu, Zn in granules
Isopods, amphipods	Cu, Fe, Pb, Zn in granules
Barnacles	Cu, Fe, Mn, Pb in granules
Polychaetes	Cu in granules
Macroalgae	Most metals
Mussels/ other bivalves	most metals, metallothioneins
Polychaetes	Cd, Pb
Decapod crustaceans	Cd, Pb
Finfish	Cd, Pb
Bivalves (rare)	Cu, Zn
Macroalgae	Mn
Polychelates	Cu, Zn
Decapod crustaceans	Cu, Mn, Zn
Finfish	Cu, Mn, Zn

REGULATORS   LOW UPTAKE

FIGURE 3.7. Aquatic organisms employing different strategies for the uptake, accumulation, and excretion of trace metals. (From Phillips et al., 1989.)

metals affect cell metabolism. Also, once within the cell, the metal may interact with a variety of intracellular sites, some resulting in metabolic consequences.

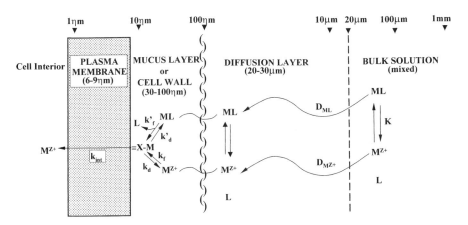

FIGURE 3.8. Schematic model of metal–organism interactions in aquatic systems. (M^{Z+} = free-metal ion; ML = metal complex in solution; M-X-membrane = surface metal complex; k_f, k_f' = rate constants for formation of the surface complex; k_d, k_d' = rate constants for dissociation of the surface complex; k_{int} = rate constant for "internalization" or transport of the metal across the biological membrane. Charges on ligand not shown for simplicity. (Modified by Campbell, 1995, from Tessier et al., 1994.)

The interactions of dissolved metals with biological surfaces, such as cell membranes (e.g., algal cellwalls and gill membranes of higher animals), detritus, and organically coated particles, in aquatic systems can affect the transport, bioaccumulation, and toxicity of metals. Biological surfaces are the most important substrate for metal binding in lakes and, in some cases, dissolved metal concentrations are controlled by adsorption to settling biological surfaces (Newman and McIntosh, 1991). For algae, metal ions must first adsorb onto the cell membrane before passing into the cell. The amount of metal bound to the surface thus directly affects the amount taken up. Higher organisms, such as zooplankton and fish, possess several mechanisms for metal uptake (e.g., ingestion of food and passage through membranes) and removal (e.g., defecation and removal of dissolved metals by the liver and kidney). The interactions that occur at biological surfaces in natural waters are very complex.

In aquatic systems, major factors affecting the dynamics and extent of contaminant bioaccumulation in organisms (e.g., fish) include (Allen et al., 1980; Campbell, 1995; Landrum and Robbins, 1990): (1) compound characteristics—solubility, chemical speciation, etc.; (2) sediment characteristics—organic C content, particle size distribution, clay type and content, CEC, pH, etc.; (3) water quality—pH, DOC, temperature, dissolved oxygen, alkalinity, etc.; (4) biological characteristics—organism behavior, modes and rates of feeding, source of water (i.e., interstitial vs. overlying water); and, (5) age and size (length) of organism. A large body of literature indicates that bioavailability or toxicity of trace metals is directly correlated to concentrations of free metal ions rather than to total or complexed metal concentrations.

The bioavailability of sediment-borne metals is related to the magnitude of metal sorption sites on the sediment. For example, the higher the organic C and oxides of Fe and Mn contents of the sediment, the lower the bioavailability. Under anoxic conditions, metals in sediments occur primarily as highly insoluble sulfides. Consequently, toxicities of metals in sediments under these conditions are strongly influenced by the sulfides. For this reason, normalization of metal concentrations to sulfide concentration has been used for estimating toxic impact in anoxic sediments (Ankley et al., 1991; Carlson et al., 1991; DiToro et al., 1990). Solid-phase sulfides in sediments are estimated with acid-volatile sulfides (AVSs). Procedurally, AVSs are the sulfides extractable with cold HCl. They are assumed to be primarily Fe and Mn sulfides. Contaminants in sediments can be transferred to the water column via a variety of processes, including diffusion and advection from sediments (often biotically mediated), sediment resuspension and release, and biotransfer through organisms that feed at the sediment–water interface (Mason and Lawrence, 1999).

Although the routes of metal absorption have been studied intensively, apparently no consensus exists on the relative importance of metal uptake from dietary or aqueous sources (Spry et al., 1988; Spry and Wiener, 1991). Some suggest direct uptake via the gills to be the primary route of metal accumulation in freshwater fish (Hodson et al., 1978; Pärt et al., 1985), whereas others indicate dietary uptake to be an important pathway of metal accumulation (Spry et al., 1988; Spry and Wiener, 1991). The absorbed metals are redistributed from the uptake sites throughout the body and accumulate in target organs (Handy, 1992). The pattern of metal distribution in the tissues may reflect the route of metal uptake in fish, i.e., metal concentration in the gills correlating with waterborne exposure, and metal content of the intestine reflecting dietary exposure (Harrison and Klaverkamp, 1989; Spry and Wiener, 1991).

Studies of Arctic char (*Salvelinus alpinus*) in two oligotrophic high-mountain lakes (the two lakes have pH 5.4 and 7.1, and alkalinity 1.3 and 100 μeq L^{-1}, respectively) in northern Austria indicate that in both sampling periods (winter, summer) the gills appear to represent a major route of Pb uptake in fish from low-alkalinity lakes (Koch et al. 1998). However, during the ice-covered period, the diet (oral) becomes an increasingly important additional source of Pb contamination.

Fish can metabolize trace metals both directly by absorption through the gills and skin of burrowing fish species (water pathway), or indirectly via the food chain where detritus feeders recycle sediment (sediment pathway), thus accumulating metals before being consumed by fish (Downs et al., 1998). The extent to which sediment is a source of the metals in fish is largely dependent on the particle size and OM content of the sediment. For example, Hg becomes strongly adsorbed onto small-size particles, such as silt and OM, resulting in lower fish bioconcentration factors for Hg (Watras et al., 1995). In southern Swedish lakes, the concentration of Hg in the sediment best reflect measured levels in pike (*Esox luscious*), which is a top predator (Hakanson et al., 1988; Lindqvist et al., 1991). The combination of acidification and increased loading of atmospheric Hg to the sediment had caused concentrations in pike to rise from background concentrations of 0.05 to 0.30 mg kg^{-1} to a new level of 0.50 to 1.0 mg kg^{-1}. Lindqvist et al. (1991) indicated that bioaccumulation of Hg in fish from sediment via the benthic food chain is probably of greater importance than the water pathway, implying that fluxes of Hg through the food chain at lower trophic levels are significant in determining Hg concentrations in fish. Methyl Hg bioaccumulation by benthic organisms is of concern because these organisms serve as the food for small fish and large invertebrates.

Organisms appear to have evolved a number of mechanisms to limit uptake of metals at the binding or transport stage (Mason and Jenkins, 1995). The most common are: (1) altering the chemical speciation of metal in the surrounding environment to reduce its bioavailability, (2) complexing the metal at the surface of the animal, (3) decreasing the permeability of the epithelial surfaces by introducing extracellular barriers, (4) reducing transport into the cell across the lipid bilayer, and (5) undertaking behavioral avoidance activity. Simply, the first strategy for an organism to avoid metal toxicity is to prevent the initial entry of the metal into the cell.

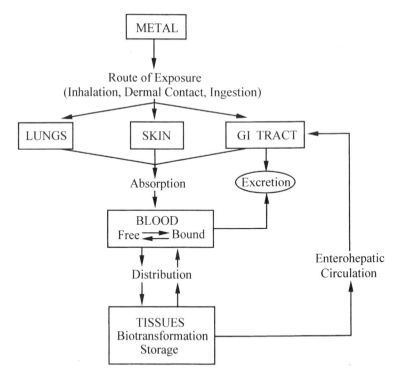

FIGURE 3.9. Schematic representation of the possible pharmacokinetic fate of metals in humans after exposure by inhalation, dermal contact, and ingestion. (Reprinted from Ballantyne, B., and J.B. Sullivan, Basic Principles in Toxicology. In J.B. Sullivan and B.R. Krieger, eds. *Hazardous Materials Toxicology: Clinical Principles of Environmental Health.* Copyright © 1992, with permission from Williams & Wilkins.)

6 Bioavailability of Trace Metals in Humans

As discussed later in Chapter 5, bioavailability of metals in humans can occur via three exposure pathways: inhalation, dermal contact, and ingestion (either direct or from the food chain).

The inhalation and dermal contact routes are more significant in an occupational setting, whereas ingestion (oral uptake of food, drinking water, and beverages) is the most relevant pathway for the general population. As previously defined, metal contaminants have to be absorbed through the skin, lungs, or gastrointestinal (GI) tract and become part of the systemic and lymphatic circulation systems (distribution through the blood and lymph systems) for them to be rendered bioavailable for metabolism (Fig. 3.9). Upon circulation, part is biotransformed and retained in certain storage tissues (e.g., bone, fat, soft tissue) and the rest is excreted in the form of feces, urine, expired air, and secretions.

In humans, bioavailability of metals and their compounds is very complex and is influenced by a magnitude of factors including physical and chemical forms (speciation), route of exposure (inhalation vs. ingestion vs. dermal contact), the individual, absorption and retention in the body, diet, and other interactions, including those with other elements, proteins, and other macromolecules (Chang, 1996).

Transfer of metals to the body following inhalation depends on deposition in the lungs of respirable particles of generally <10 μm in diameter (Bennett, 1986). Metals absorbed into the bloodstream are distributed to organs and tissues, the extent of which varies from one metal to another and largely depends on many factors including similarity to essential elements and filtration or storage locations (see Fig. 3.9). In this case, bioavailability can be expressed as the percentage of ingested materials absorbed through the intestinal wall into the systemic bloodstream.

Absorption (also known as fractional absorption) into the GI tract is believed to occur primarily through diffusion processes following the concentration gradient between the intestinal lumen, mucosa cells, and blood (Ewers and Schlipkoter, 1991). Absorption is induced by relatively rapid binding of metal ions to plasma proteins and blood cells as well as rapid elimination, and distribution and storage into various organs and tissues, creating a diffusion gradient towards the blood. This passive transport appears to be a main absorption mechanism for metals, in contrast to active transport (i.e., ions move against an increasing concentration gradient by means of metabolic energy) which dominates K^+ and Na^+ ion transport.

The rate of absorption varies among the metals, chemical species within a metal, and individuals (e.g., children vs. adults), and may be largely influenced by dietary factors as well as physiological condition of the individual (Table 3.2). In general, metals with low solubility in water are less readily absorbed than soluble compounds. Lipophilic organo-metallic compounds are usually absorbed to a greater extent than ionic species (Ewers and Schlipkoter, 1991). For example, methyl Hg and phenyl Hg compounds are almost completely absorbed, whereas only <10% of inorganic Hg is absorbed.

Dietary deficiencies of Ca, Cu, Fe, Zn, protein, and vitamin D have been shown to promote GI tract absorption of Pb and Cd (Rosen and Sorrell, 1978). The gastrointestinal absorption of Pb, Mn, and Hg(II) is greater in infants and young children than in adults. Also, fasting can promote the absorption efficiency of metals.

One of the most important factors determining the extent of bioavailability to any organism is the chemical speciation of the element. For example, in freshwater systems the free ionic forms of Ag and Cu are more toxic to fish in terms of metal–gill interaction than any other chemical species of these metals (Bergman and Dorward-King, 1997). In soils, the complexed-metal forms (e.g., $CdCl_2$, Zn–EDTA) generally proved to be more bioavailable to plants as a result of their greater solubility and mobility into the rhizosphere. The methylated forms of certain metals (e.g., methyl Hg) have the greatest bioavailability and therefore are the most bioaccumulative in fish and other aquatic vertebrates, and in turn are the most bioavailable when ingested by humans.

Table 3.3 indicates the general relative importance of chemical species of metals to organisms. Much qualitative evidence exists to indicate that

TABLE 3.2. Gastrointestinal absorption of metals in humans and animals.

Metal	Physicochemical form	GI tract absorption (%)	Factors affecting GI tract absorption
Aluminum		>5	Increased absorption in acidic diet
Antimony		~15	
Arsenic	Inorganic As(III) and As(V) compounds	<90	
	Organic As compounds in seafood	>80	
Barium	Insoluble Ba salts	>1	
	Soluble Ba salts	10–30	
Beryllium	Soluble Be salts	~20	
Cadmium	Soluble Cd salts	4–7	Increased absorption in case of Ca, Fe, and protein deficiency
Chromium	Cr(III) compounds	>1	
	Cr(VI) compounds	2–6	
Cobalt	Co oxide	>0.5	
	$CoCl_2$	~30	
Copper	Soluble Cu salts	50–70	Absorption regulated by Cu status in the body; increased absorption in case of Cu deficiency
Iron	Fe salts	2–20	Absorption regulated by Fe status in the body; increased absorption in case of Fe deficiency
Lead	Water-soluble Pb salts	~10–30	Absorption in infants up to 50%; increased absorption in case of Ca, vitamin D, and Fe deficiency
Manganese	Mn(II) salts	3–5	Increased absorption in conditions of Fe deficiency and in young infants/animals
Mercury	Elemental Hg	>0.5	
	$HgCl_2$	>10	
	Organic Hg compounds	<80	
Molybdenum	Mo(VI) compounds	40–80	
Nickel	Ni salts	~10	
Selenium	Water-soluble Se compounds	<80	
Silver	Water-soluble Ag compounds	10–20	
Thallium	Water-soluble Tl salts	<90	
Tin	Inorganic Sn(II) and Sn(IV) compounds	>5	
	Monoethyltin, dibutyltin, tributyltin, triphenyltin	~10	
	Triethyltin	<90	
Vanadium	Water-soluble V compounds	~2	
Zinc	Water-soluble Zn salts	50–80	

Source: Extracted and modified from Ewers and Schlipkoter (1991).

TABLE 3.3. Important environmental chemical species of trace metals with regard to their bioavailability and potential toxicity to organisms.

Metal	Dominant chemical species[a]		Most toxic species[b]
	Soil	Water	
Ag	Ag^+	Ag^+	Ag^+
As	AsO_4^{3-}	AsO_4^{3-}; AsO_3^{3-}	AsO_4^{3-}
B	$B(OH)_3$	$B(OH)_3$	$B(OH)_3$
Ba	Ba^{2+}	Ba^{2+}	Ba^{2+}
Be	Be^{2+}; $Be_xO_y^{2x-2y}$	Be^{2+}	Be^{2+}
Bi	Bi^{3+}?	Bi^{3+}?	?
Cd	Cd^{2+}	Cd^{2+}	Cd^{2+}
Co	Co^{2+}	Co^{2+}	Co^{2+}
Cr	Cr^{3+}	Cr^{3+}; Cr^{6+}	Cr^{6+}
Cu	Cu^{2+}	Cu^{2+}-fulvate	Cu^{2+}
Hg	Hg^{2+}; Hg^{2+}-fulvate	$Hg(OH)_2^0$; $HgCl_2^0$; CH_3Hg	CH_3Hg
Mn	Mn^{4+}; Mn^{2+}	Mn^{2+}	Mn^{2+}
Mo	MoO_4^{2-}	MoO_4^{2-}	MoO_4^{2-}
Ni	Ni^{2+}	Ni^{2+}	Ni^{2+}
Pb	Pb^{2+}	$Pb(OH)^+$	Pb^{2+}
Sb	$Sb^{III}O_x$?	$Sb(OH)_6^-$?	?
Se	$HSeO_3^-$; SeO_4^{2-}	SeO_4^{2-}	SeO_4^{2-}
Sn	$Sn(OH)_6^{2-}$?	$Sn(OH)_6^{2-}$?	?
V	$V^{IV}O_x$?	?	?
W	WO_4^{2-}?	WO_4^{2-}?	?
Zn	Zn^{2+}	Zn^{2+}	Zn^{2+}

[a]Does not account for ion-pairs or complex-ion species.
[b]Considers degree of bioavailability.
? = Most likely species.
Source: Modified from Logan and Traina (1993).

the total aqueous concentration of a metal is not a good predictor of its bioavailability, i.e., the metal speciation greatly affects its availability to aquatic organisms (Campbell, 1995). Another classic example of species importance is the case of Cr: Cr(VI) is nonessential and highly toxic, whereas Cr(III) is essential and nontoxic (see also later chapter on chromium). Thus, the challenge for environment scientists is to reduce the Cr(VI) to Cr(III), a process that may be mediated naturally in the rhizosphere (by the so-called phytoremediation) or by the introduction of reduction-promoting compounds such as Fe(II).

7 Factors Affecting Bioavailability

Because bioavailability in a generic sense is a function of the solubility and mobility of trace metals, factors regulating these processes also greatly influence the bioavailability of metals in organisms. These factors have been elucidated in Chapter 2. In soils, factors that are generally considered master variables include pH, OM, redox potential, oxides of Fe and Mn, etc; in aquatic systems (seawater, freshwater, and rivers), the master variables include pH, DOC, suspended particulate matter, ionic strength, alkalinity, and salinity (Turner, 1995). However, for clarity, they should be delineated into *external* and *internal* (in the organism) factors. For example, the *external* category may include pH, OM, redox potential, salinity, and oxides of Fe and Mn, whereas the *internal* category may include chemical speciation, pH, etc.

Geochemical approaches have been used to assess bioavailability of sediment-associated metals (Chen and Mayer, 1999). They were based on the principle that OM, oxides of Fe and Al, and acid-volatile sulfides (AVSs) act as solid metal-binding phases to prevent availability and toxicity of metals to organisms (DiToro et al., 1992; Van Gestel et al., 1995).

In view of the improper nature of total metal-based expressions of exposure, even in homogeneously prepared artificial soils, the need for surrogate measures of metal bioavailability in soils has never been greater. Weak electrolyte extractions seem to be very promising as an attendant method to assess ecotoxicological aspects of metal contamination in soils a la soil testing methods to ascertain micronutrient deficiency in crops. Indeed, Conder and Lanno (2000) observed that $Ca(NO_3)_2$-extractable Cd, Pb, and Zn in artificial soil tests relate well to toxic responses in earthworms (*Eisenia fetida*).

7.1 Importance of the Source Term

Most risk in humans from metals is associated with the forms of metals that are biologically available for absorption through the GI tract into the systemic (i.e., blood) circulation—in short, bioavailable to humans. In turn, bioavailability is heavily influenced by the source term of the metals, which entails the physical form as well as the geochemical form of the metals. For example, Pb and As exist in several physical forms (e.g., slag, tailings, calcine, waste ore, flue dust) and geochemical forms (e.g., oxides, sulfides, phosphates, etc.) at contaminated areas impacted by mining and smelting emissions (Ruby et al., 1999). These waste forms vary in their solubilities and geochemical stabilities in such a manner that they are unlikely to be bioavailable and, therefore, may pose only minimal risks to humans (Davis et al., 1993, 1994; Oliver et al., 1999; Rodriguez et al., 1999).

Most evidence indicates that ≥90% of ingested As and ~10 to 30% of ingested Pb is bioavailable, based on toxicological studies using pure As and Pb salts (U.S. EPA, 1990). Recent data have indicated that blood Pb levels in children living in urban areas (e.g., Cincinnati, Ohio) (Clark et al., 1991) and smelter communities (Butte, Montana) (BSBDOH and UCDEH, 1991) vary considerably, even when the Pb concentrations in the soil are considerable (Fig. 3.10). The differences are due to a number of variables—primarily the source of Pb in the soil, forms of available Pb, housing, and concentrations of Pb in house paint and house dust (Davis et al., 1994).

The geochemical factors that control Pb availability from mining waste are largely responsible for the lower potential bioavailability of Pb in mine waste soils compared with the urban or smelter-associated soils (Davis

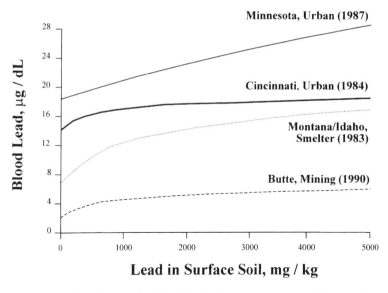

Lead in Surface Soil, mg / kg

FIGURE 3.10. Comparison of the blood lead dose response at mining, smelter, and urban sites in the United States. (From Davis et al., 1992.)

et al., 1994, 1993; Ruby et al., 1999). The differences in the solubilities of the Pb as effected by the differences on Pb geochemistry of different sites appear to have induced the variations in blood Pb responses (see Fig. 3.10). A number of geochemical factors decreased the solubility of Pb from mine waste, including the mineralogical composition, the degree of encapsulation in pyrite or silicate matrices, the nature of alteration rinds, and particle size. In the Butte, Montana site, the samples contain predominantly sulfide/sulfate and oxide/phosphate of Pb-bearing phases associated with mine waste. The sulfide/sulfate assemblage consists primarily of insoluble galena (PbS), altering to anglesite and plumbojarosite. For example, soils collected from the gardens of houses where children reside consist of 2% galena and 98% anglesite in this particular assemblage. Thus, the bioavailability of soil Pb is assumed to be associated with the solubility and dissolution kinetics of the soil/waste Pb in the GI tract.

The Pb dissolved is also controlled by physiological factors, including gastric pH, the rate of Pb dissolution relative to the residence time of Pb-bearing mine waste in the GI tract, and by *in vivo* sorption and precipitation reactions that may limit dissolved Pb concentrations (Ruby et al., 1993). Gastric pH ranges from a basal level of 1 to 2, to 4 to 6 after ingestion of food, which, with its attendant buffering capacity, appears to be one of the most important physiological factors in determining the amount of Pb solubilized from a mine waste. Lead that enters the fluid phase in a fasting stomach (assumed pH ~ 1.3 and high chloride activity) would be present at approximately equimolar concentrations of Pb^{2+} and $PbCl^{+}$. Another critical factor is the residence time in the stomach. The gastric contents are emptied completely into the small intestine within approximately 2 hr in humans when fed various test meals. On entering the duodenum, $NaHCO_3$ excreted with the pancreatic juice is mixed with the intestinal chyme, resulting in an increase in pH to approximately 7 in humans. Consequently, dissolved Pb concentrations are likely to decrease in the small intestine due to the adsorption to mineral and food particle surfaces with increasing pH and as a result of precipitation reactions.

To estimate the bioavailability of Pb, As, and other waste-associated metals, Ruby et al. (1993,1996) developed a physiologically based extraction test (PBET) for site-specific estimation of soil Pb bioavailability, based on the assumption that the form and solubility of Pb in a soil or mine waste will control its bioavailability in a test animal or in humans. Subsequently, this technique has been adopted for bioavailability evaluations of other metals relative to human health risk assessment (Oliver et al., 1999; Rodriguez et al., 1999; Ruby et al., 1999).

References

Adriano, D.C., A. Knox, W. Geebelen, J. Vangronsveld, and M. Mench. 2000. Univ of Georgia. Unpublished data.

Ahnstrom, Z.S., and D.R. Parker. 1999. *Soil Sci Soc Am J* 63:1650–1658.

Allen, H.E., R.H. Hall, and T.P. Brisbin. 1980. *Environ Sci Technol* 14:441–443.

Ankley, G.T., G.L. Phipps, E.N. Leonard, and J.D. Mahony. 1991. *Environ Toxicol Chem* 10:1299–1307.

Baker, A.J.M. 1987. *New Phytol* 106:93–111.

Bennett, B.G. 1986. In *Exposure Assessment for Metals in Carcinogens*. Intl Agency for Res on Cancer, Lyon, France.

Bergman, H.L., and E.J. Dorward-King. 1997. In *Reassessment of Metal Criteria for Aquatic Life Protection*. SETAC, Pensacola, FL.

Bhattacharya, B., and S.K. Sarkar. 1996. *Chemosphere* 33:147–158.

Bjornberg, A., L. Hakanson, and K. Landbergh. 1988. *Environ Pollut* 49:53–65.

Bryan, G.W., and W.J. Langston. 1992. *Environ Pollut* 76:89–99.

[BSBDOH and UCDEH] Butte, Montana, Department of Health and University of Cincinnati, Department of Environmental Health. 1991. *Butte-Silver Bow Environmental Health Lead Study*. Univ of Cincinnati, Cincinnati, OH.

Campbell, P.G.C. 1995. In A. Tessier and D.R. Turner, eds. *Metal Speciation and Bioavailability in Aquatic Systems*. Wiley, New York.

Carlson, A.R., G.L. Phipps, V.R. Mattson, P.A. Kosian, and A.M. Cotter. 1991. *Environ Toxicol Chem* 10:1309–1319.

Cataldo, D.A., and R.E. Wildung. 1978. *Environ Health Perspect* 27:149–159.

Cataldo, D.A., T.R. Garland, R.E. Wildung, and H. Drucker. 1978. *Plant Physiol* 62:563–565.

Cataldo, D.A., T.R. Garland, and R.E. Wildung. 1983. *Plant Physiol* 73:844–848.

Cataldo, D.A., R.E. Wildung, and T.R. Garland. 1987. *J Environ Qual* 16:289–295.

Chaney, R.L., and J.A. Ryan. 1992. *Water Environ Technol* 4:36–41.

Chang, A.C., A.L. Page, and J.E. Warneke. 1987. *J Environ Qual* 16:217–221.

Chang, A.C., A.L. Page, and J.E. Warneke. 1996. In D.C. Adriano, ed. *Biogeochemistry of Trace Elements*. Science Technol Letters, Northwood, England.

Chang, A.C., H.N. Hyun, and A.L. Page. 1997. *J Environ Qual* 26:11–19.

Chang, A.C., A.L. Page, and J.E. Warneke. 1997. In D.C. Adriano, ed. *Biogeochemistry of Trace Metals*. Lewis Publ, Boca Raton, FL.

Chang, L.W. 1996. *Toxicology of Metals*. Lewis Publ, Boca Raton, FL.

Chen, Z., and L.M. Mayer. 1999. *Environ Sci Technol* 33:650–652.

Chlopecka, A., and D.C. Adriano. 1997. *Sci Total Environ* 207:195–206.

Clarks, S., et al. 1991. *Chem Spec Bioavail* 3:163–171.

Conder, J.M., and R.P. Lanno. 2000. *Chemosphere* (in press).

Corey, R.B., L.O. King, C. Lue-Hing, and J.M. Walker. 1987. In A.L. Page et al., eds. *Land Application of Sludge: Food Chain Implications*. Lewis Publ, Chelsea, MI.

Davis, A., M.V. Ruby, and P.D. Bergstrom. 1992. *Environ Sci Technol* 26:461–468.

Davis, A., J.W. Drexler, M.V. Ruby, and A. Nicholson. 1993. *Environ Sci Technol* 27:1415–1425.

Davis, A., M.V. Ruby, and P.D. Bergstrom. 1994. *Environ Geochem Health* 16:147–157.

Derelanko, M.J., and M.A. Hollinger, eds. 1995. *CRC Handbook of Toxicology*. CRC Pr, Boca Raton, FL.

DiToro, D.M., J.D. Mahony, D.J. Hansen, and M.S. Redmond. 1990. *Environ Toxicol Chem* 9:1487–1502.

DiToro, D.M., J.D. Mahony, D.J. Hansen, and G.T. Ankley. 1992. *Environ Sci Technol* 26:96–101.

Downs, S.G., C.L. McLeod, and J.N. Lester. 1998. *Water Air Soil Pollut* 108:149–187.

Ewers, U., and H.W. Schlipkoter. 1991. In E. Merian, ed. *Metals and their Compounds in the Environment*. VCH, Weinham, Germany.

Fergusson, J.E. 1990. *The Heavy Elements: Chemistry, Environmental Impacts, and Health Effects*. Pergamon, Oxford.

Geebelen, W. 2000. Univ of Georgia, Savannah River Ecology Lab (unpublished data).

Gerritse, G.E., W. van Driel, K.W. Smilde, and B. van Luit. 1983. *Plant Soil* 75:393–405.

Gibaldi, M. 1991. *Biopharmaceutics and Clinical Pharmacokinetics*. Lea & Febiger, Philadelphia, PA.

Giblin, F.J., and E.J. Masarro. 1973. *Toxicol Appl Pharmacol* 24:81–89.

Hakanson, L., A. Nilsson, and T. Andersson. 1988. *Environ Pollut* 49:145–162.

Handy, R.D. 1992. *Arch Environ Contam Toxicol* 22:74–85.

Hanson, P.J., E.I. Sucoff, and A.H. Markhart. 1985. *Plant Physiol* 77:21–24.

Harrison, S.E., and J.F. Klaverkamp. 1989. *Environ Toxicol Chem* 8:87–97.

Hintelmann, H., R. Ebinghaus, and R.D. Wilkin. 1993. *Water Res* 27:237–246.

Hodson, P.V., B.R. Blunt, and D.J. Spry. 1978. *Water Res* 12:869–878.

Kabata-Pendias, A., and H. Pendias. 1994. *Trace Elements in Soils and Plants*. CRC Pr, Boca Raton, FL.

Keller, C., and J.C. Vedy. 1994. *J Environ Qual* 23:987–999.

Koch, G., M. Trend, and R. Hofer. 1998. *Water Air Soil Pollut* 102:303–312.

Kochian, L.V. 1991. In J.J. Mortvedt et al., eds. *Micronutrients in Agriculture*. SSSA 4. *Soil Sci Soc Am*, Madison, WI.

Landrum, P.F. 1989. *Environ Sci Technol* 23:588–595.

Landrum, P.F., and J.A. Robbins. 1990. In R. Bando et al., eds. *Sediments: Chemistry and Toxicology of In-Place Pollutants*. Lewis Publ, Chelsea, MI.

Lindquist, O., K. Johansson, M. Aastrup, and B. Tim. 1991. *Water Air Soil Pollut* 55:193–199.

Linz, D.G., and D.V. Nakles. 1996. In *Environmentally Acceptable Endpoints in Soil*. Am Acad Environ Engrs, Annapolis, MD.

Logan, T.J., and S.J. Traina. 1993. In H.E. Allen et al., eds. *Metals in Groundwater*. Lewis Publ, Chelsea, MI.

Marschner, H. 1986. *Mineral Nutrition in Higher Plants*. Academic Pr, London.

Mason, A.Z., and K.D. Jenkins. 1995. In A. Tessier and D.R. Turner, eds. *Metal Speciation and Bioavailability in Aquatic Systems*. Wiley, New York.

Mason, R.P., and A.L. Lawrence. 1999. *Environ Toxicol Chem* 18:2438–2447.

Mason, R.P., J.R. Reinfelder, and F.M.M. Morel. 1995. *Water Air Soil Pollut* 80:915–927.

McBride, M.B. 1995. *J Environ Qual* 24:5–18.

McCloskey, J.T., I.R. Schultz, and M.C. Newman. 1998. *Environ Toxicol Chem* 17:1525–1529.

McGrath, S.P., F.J. Zhao, S.J. Dunham, A.R. Crosland, and K. Coleman. 2000. *J Environ Qual* 29:875–883.

Meili, M. 1991. *Water Air Soil Pollut* 56:719–728.

Mench, M., J. Vangronsveld, V. Didier, and H. Clijsters. 1994. *Environ Pollut* 86:279–286.

Minnich, M., M.B. McBride, and R.L. Chaney. 1987. *Soil Sci Soc Am J* 51:573–578.

Newman, M.C. 1995. *Quantitative Methods in Aquatic Ecotoxicology*. CRC Pr, Boca Raton, FL.

Newman, M.C., and A.W. McIntosh, eds. 1991. *Metal Ecotoxicology: Concepts and Applications*. Lewis Publ, Chelsea, MI.

Oliver, D.P., M.J. McLaughlin, R. Naidu, L.H. Smith, E.J. Maynard, and I.C. Calder. 1999. *Environ Sci Technol* 33:4434–4439.

Pärt, P., O. Svanberg, and A. Kiessling. 1985. *Water Res* 19:427–434.

Phillips, D.J.H., and P.S. Rainbow. 1989. *Mar Environ Res* 28:207–210.

Rand, G.M. 1995. *Fundamentals of Aquatic Toxicology*. Taylor & Francis, Washington, DC.

Rodriguez, R.R., N.T. Basta, S.W. Casteel, and L.W. Pace. 1999. *Environ Sci Technol* 33:642–649.

Rosen, J.F., and M. Sorell. 1978. In J.O. Nriagu, ed. *Biogeochemistry of Lead in the Environment*. Elsevier, Amsterdam.

Ruby, M.V., A. Davis, T.E. Link, R. Schoof, R.L. Chaney, G.B. Freeman, and P. Bergstrom. 1993. *Environ Sci Technol* 27:2870–2877.

Ruby, M.V., A. Davis, R. Schoof, S. Eberle, and C.M. Sellstone. 1996. *Environ Sci Technol* 30:422–430.

Ruby, M.V., R. Schoof, et al. 1999. *Environ Sci Technol* 33:3697–3705.

Sharpe, M.S., A.S.W. DeFreitas, and A.E. McKinnon. 1997. *Environ Biol Fishes* 2:177–183.

Shuman, L.M. 1991. In J.J. Mortvedt et al., eds. *Micronutrients in Agriculture*. SSSA 4. *Soil Sci Soc Am*, Madison, WI.

Sims, J.T., and G.V. Johnson. 1991. In J.J. Mortvedt et al., eds. *Micronutrients in Agriculture*. SSSA 4. *Soil Sci Soc Am*, Madison, WI.

Sparks, D.L. 1984. *Soil Sci Soc Am J* 48:415–418.

Spry, D.J., and J.G. Wiener. 1991. *Environ Pollut* 71:243–304.

Spry, D.J., P.V. Hodson, and C.M. Wood. 1988. *Can J Fish Aquat Sci* 45:32–41.

Tanton, T.W., and S.H. Crowdy. 1972. *J Exp Bot* 23:600–618.

Tessier, A. 1994. *Chemical and Biological Regulations of Aquatic Systems*. Lewis Publ, Boca Raton, FL.

Tessier, A., P.G.C. Campbell, and M. Bisson. 1979. *Anal Chem.* 51:844–851.

Turner, D.R. 1995. In A. Tessier and D.R. Turner, eds. *Metal Speciation and Bioavailability in Aquatic Systems*. Wiley, New York.

[U.S. EPA] United States Environmental Protection Agency. 1990. *User's Guide for Pb: A PC Software Application of the Uptake/Biokinetic Model*. Dept Research and Development, Washington, DC.

U.S. EPA. 1993. *Fed Register.* 58:9248–9415.

Van Gestel, C.A.M., M.C.J. Rademaker, and N.M. Van Straalen. 1995. In W. Salomons and W.M. Stigliani, eds. *Biogeodynamics of Pollutants in Soils and Sediments*. Springer-Verlag, Berlin.

Vangronsveld, J., and S.D. Cunningham, eds. 1998. *Metal-Contaminated Soils: In-Situ Inactivation and Phytoremediation*. Springer-Verlag, Berlin.

Watras, C.J., K.A. Morrison, and N.S. Bloom. 1995. *Water Air Soil Pollut* 84:253–267.

Welch, R.M., and W.A. Norvell. 1997. In I.K. Iskandar et al., eds. *Proc 4th Intl Conf Biogeochemistry of Trace Elements, June, 1997, Berkeley, CA*. Univ California, Berkeley, CA.

Welch, R.M., and W.A. Norvell. 2000. Cornell Univ, personal communication.

Whitney, D. 2000. Kansas State Univ, personal communication.

Wolfe, M.F., S. Schwarzbach, and R.A. Sulaiman. 1998. *Environ Toxicol Chem* 17:146–160.

Wu, J., F.C. Hsu, and S.D. Cunningham. 1999. *Environ Sci Technol* 33:1898–1904.

4
Environmental Contamination and Regulation

Analyses show that certain soils in Switzerland are contaminated, in some cases to an appreciable extent. These are not cases of abandoned contaminated sites from old landfills but of land affected for many years by diffuse or isolated instances of contamination with pollutants not covered by law. Examples include land alongside roads carrying heavy traffic and vineyard soils contaminated with copper. In the case of soils used for agricultural and horticultural purposes, the high levels of contamination sometimes observed are attributable to the use of pollutant-containing waste-based fertilizers and plant treatment products.
USGrev, Bern, Switzerland, 1993

How can society minimize environmental contamination? Or maybe a better question is—what degree of contamination is acceptable?

Environmental contamination as a result of human activities is not a recent phenomenon. Cicero first related structural damage of buildings and statues in Rome to smoky rains of wood and charcoal burning about 2100 years BP (Eney and Petzold, 1987). Some of the detrimental effects by mining activities on human health had been recognized a long time ago. The Romans used slaves to extract cinnabar (a Hg-containing ore) at the Almaden mine in Spain. Due to acute Hg exposure, the miner's life expectancy was only about three years (Wren et al., 1995). Peat cores from a Swiss bog indicate that As, Sb, and Pb fluxes due to anthropogenic activities have been exceeding natural fluxes for more than 2000 years (Shotyk et al., 1998). The present enrichment factors in this bog are in the order of 20 times for As, 70 for Sb, and 130 for Pb. Modifications of the natural cycles of metals have led to a situation in which the inputs of metals in soils generally exceed the removal due to harvest of agricultural crops and the losses by leaching, volatilization, etc. (Jones, 1991; Van Driel and Smilde, 1990). The Industrial Revolution started in the mid-1800s and the large use of coal to produce energy caused the release of considerable amounts of gas, e.g., CO_2, SO_x, NO_x, and fly ash into the atmosphere. Since then, the biogeochemical cycles of inorganic contaminants (e.g., metals) naturally present in the environment have been largely accelerated by human activities. The conversion of the world's economy from coal to oil, initiated between the two world wars, enlarged the range of contaminants released in the environment to organic compounds, e.g., PAHs.

The development of the organic compound industry in this century led to the commercialization of enormous amounts of new substances, some of which were found to be toxic to the environment and animal life, e.g., PCBs entered the market in the late 1920s and DDT in the 1940s. In the last two decades or so, tremendous amounts of xenobiotics have been produced and released into the biosphere. It has been estimated that about 1000 new compounds are being synthesized annually and many times more chemicals are currently available in the market (Tolba and El-Kholy, 1992). The utilization of nuclear energy has introduced a large number of radionuclides of potential concern into the environment. After the atmospheric nuclear detonations that took place from the mid-1940s to 1960s, three major reactor accidents (Chernobyl, Ukraine, 1986; Three Mile Island, United States, 1979; Windscale, United Kingdom, 1957) have been the main sources of radionuclide emission. Urbanization and increasing affluence in western countries have resulted in increased generation of wastes. The dramatic accumulation of waste from chemical manufacturing, wastewater treatment, agriculture, mining and smelting, food processing, energy production, pulp and paper production, and other industries will induce waste–environmental interaction with potential consequences on soil and water quality. Implicit is the potential adverse effect on the food chain and drinking water. With the ever-expanding scientific database on the ecological and human health aspects of these contaminants, including metals, guidelines or standards on soil, drinking water, irrigation water, crop tissues, foodstuff, biosolids, and fertilizers are now becoming available.

1 Defining Contamination vs. Pollution

Although the terms *pollution* and *contamination* are considered synonymous and therefore interchangeably used in the literature, they do not necessarily connote the same meaning. Both terms should be viewed relative to background or baseline concentrations (also termed *normal* or *natural* concentrations). *Normal* may not designate the same meaning as *background* concentration, as the former term usually refers to the soil concentration that is not expected to induce phytotoxic or ecotoxicological effects; it is inclusive of contamination values. *Contamination* is used for scenarios where a metal is present in the environment, its concentration deviating from the background level, and is not causing any harm to any organism. *Pollution,* however, is more appropriate for situations where harmful effects can be detected. Therefore, pollution is a more intensive form of contamination. A widely used definition of pollution is "the introduction by man into the environment of substances or energy liable to cause hazards to human health, harm to living resources and ecological systems, damage to structure or amenity, or interference with legitimate uses of the environment" (Holdgate, 1979). Davies (1992) pointed out the necessity of distinction between contamination and pollution, and suggested the word *pollution* be reserved for levels of contamination that can cause some adverse effect on highly exposed, sensitive organisms, by the most sensitive pathway for contaminant transfer. Basically the entire global environment is contaminated due to long distance, i.e., thousands of kilometers away, transport of metals from their sources.

Environmental contamination by metals require the following (Fig. 4.1): source, metals themselves, transport of the metals, sink, and receptor targets. In pollution jargon, there are two types of pollutant sources: *point* sources and *diffuse* (or nonpoint) sources. Point sources refer to discrete and localized contamination processes. Natural point sources are represented, for example, by particular pedogenic substrates rich in specific trace elements (i.e., metalliferous soils developed on serpentinic rocks). Local accumulation of pollutants may originate also from human activities, such as landfilling, and mining and smelter processes. Nonpoint sources are related to diffuse processes or human activities that cover large areas. Atmospheric transport of volcanic emission and fossil fuel combustion particulates are one of the most common processes that disseminate pollutants in the environment. Agricultural practices represent another example of diffuse sources of contamination that involve large areas.

Point sources are generally responsible for high contaminant concentrations in small areas, whereas diffuse sources are influenced by dilution

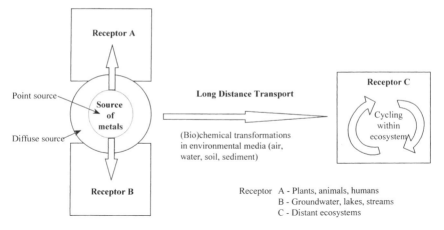

FIGURE 4.1. A simplified model of environmental metal pollution showing source–receptor relationships.

effects. Effects of emissions from diffuse sources in Europe and the United States have been detected even in remote areas, such as Antarctica. In the European area, the atmospheric transport of metals is a significant process: 30 to 90% of the metals emitted from each European country are deposited in other countries (Bartnicki, 1996).

Depending on the source, metals may not need any distant transporting to affect the environment. In Figure 4.1, this is represented by Receptors A and B that are in contact or in the vicinity of the source. Examples of Receptor A include individuals, populations, or communities of plants, animals, or people, whereas individuals or populations of invertebrate animals, fungi, and microbes in soils and groundwater exemplify Receptor B. In essence, the transport is vertical in nature as contrasted to a more or less horizontal transport to reach Receptor C, distant ecosystems.

There are scenarios where the source may also serve as the sink. Examples may include sludged soils, where research has shown that metals stay mostly in the application zone, i.e., 15 to 20 cm of the top soil (Juste and Mench, 1992). Although landfills may also serve as a sink, certain metals can be more mobile, eventually leaching to the groundwater because of oftentimes complex composition of waste and variable environmental conditions (pH, redox, etc.).

The cycling of metals varies among the types of receptors because of the extent of affected areas. Receptors A and B may represent small (<0.5 ha) to medium (>1 ha) size areas, whereas Receptor C may represent an ecosystem or the entire landscape (hundreds to thousands of hectares). Consequently, they may vary in the size and composition of target populations.

2 Processes and Factors Contributing to Soil Contamination

Chemical degradation of soils is a consequence of the post–Industrial Revolution. It is an environmental cost that society has to pay for increased energy use, modern agriculture, rapid population growth, and urbanization. Land contamination has been associated with the accumulation of unwanted waste by-products and major environmental perturbations resulting from accelerated human activities. Soil degradation through physical, biological, and chemical processes can reduce soil quality with potential consequences of soil erosion, low productivity, and soil and groundwater contamination (Fig. 4.2).

Although less area is involved compared with physically degraded (i.e., eroded) soils, chemically degraded soils are becoming important in terms of food production and quality (Adriano et al., 1997). Crops cultivated on soils of marginal quality may not only suffer in yield but also in quality, making them less competitive in the world market. For example, certain cereal and vegetable crops from eastern Europe may not be permitted in the European Union (EU) market because of questionable quality. Therefore, it has become apparent that polluted soils are not only a social and health issue but an economic issue as well (World Com. Environ. Develop., 1987).

The following are major biological and chemical processes in soil that may promote its chemical degradation: weathering, acidification, pollution,

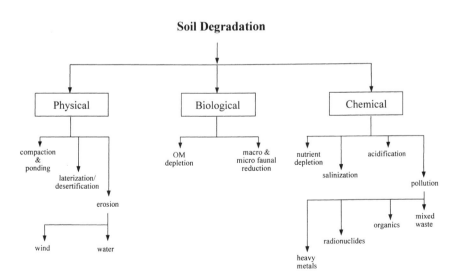

FIGURE 4.2. Processes that influence the physical, biological, and chemical degradation of soils. (Modified from Lal, 1998.)

organic matter buildup or depletion, nutrient buildup or depletion, and salinization.

Weathering in the pedosphere is a natural process that alters the chemical composition and structure of soil minerals, as well as the soil parent rocks (Dixon and Weed, 1989; Mortvedt et al., 1991). This results in the gradual loss of the alkaline earth cations (e.g., Ca^{2+} and Mg^{2+}) and the accumulation of insoluble compounds of Si, Al, and Fe. Over the long term, this process usually results in the shifting of soil reaction to more acidic conditions. Young, unweathered soils (i.e., those formed from basic parent rocks) have a greater capacity to buffer against acid constituents than older, more weathered soils, such as those in the tropics and in the southeastern part of the United States. Weathering also may result in the mineralization of metal-bearing rocks, which at times causes anomalous levels of certain elements, such as Zn, Cu, Ni, Co, Se, and others (see also Chapter 2).

In soil, acidification can be caused by weathering of certain soil constituents and by certain cultural practices. Soil acidification is a natural weathering process in humid environments, but is greatly accelerated by certain factors. When soils are perturbed to the extent that pyrite-containing strata are exposed to more aerobic conditions, acid-forming constituents such as H_2SO_4, are generated. This is the case in mine spoils generated by mineral ore exploration and coal mining. Soils also can become acidic from the applications of nitrogenous fertilizers that undergo nitrification and when organic matter is decomposed by soil microorganisms that produce organic acid by-products (Stevenson, 1982; Summer, 1991). Due to industrialization, acid rain can significantly contribute to soil acidification because of the direct input of acidic constituents, such as NO_x and SO_x compounds (Adriano and Johnson, 1989; Reuss and Johnson, 1986) (see also section 5.2.2 later in this chapter).

Due to urbanization and industrialization, many types of pollution are recognized: air, water, soil, and groundwater pollution from gaseous, liquid, and solid forms of chemicals. Pollution varies from country to country, but is generally more serious in developed countries with high population densities (Barth and L'Hermite, 1987; Hansen and Jorgensen, 1991; Harrison, 1990; Tolba and El-Kholy, 1992). Air pollution is a transboundary issue in North America, Europe, and other regions.

Nutrients and organic matter may build up or deplete in soil depending on farming practices (Stevenson, 1982). In intensive agriculture, such as in the midwestern U.S. Corn Belt, heavy fertilizer applications are required, especially in irrigated areas. This usually results in the leaching of excess nitrates into the groundwater (Powe, 1992; Williams, 1992). Organic matter from plant residues normally can build up in prairies and forest ecosystems, whereas loss of organic matter can occur when grasslands are burned or when forests are slashed and burned for shifting agriculture.

Under semiarid and arid conditions, salinization often occurs. It is frequently the result of poor irrigation management practices (Tolba and

El-Kholy, 1992; U.S. Salinity Lab Staff, 1954). Major factors inducing salinization include excessively high salt contents of irrigation water and improper leaching and drainage conditions of irrigated fields. This problem is the net result of salt accumulation on the soil surface due to an imbalance in the evaporation and leaching of salts in the soil profile.

Other chemical processes in soil, such as ion exchange, complexation, oxidation–reduction, sorption–desorption, precipitation–dissolution, and others, may also influence the severity of its degradation (Bolt and Bruggenwert, 1976; McBride, 1994; Sposito, 1989). In the context of soil protection, it is important to bear in mind that soil possesses some capacity to resist (buffer) drastic changes in soil reaction and pollutant load. Soils that are healthy, such as typical prairie soil and agricultural soil, have high buffering capacities.

The following are some of the more common practices that are conducive to chemical degradation of soil: fertilizer–pesticide application, land disposal of waste, landfilling–storage of waste, combustion of fossil fuels, mining–smelting of metal ores, disposal–releases of radioactive waste, storage of petroleum products, deforestation, and dredging.

High-production agriculture requires repeated applications of pesticides and fertilizers, increasing the potential to contaminate groundwater with pesticides and nitrates (Bar-Yosef et al., 1991; Greenwood et al., 1990; Powe, 1992; Racke and Coats, 1990; Williams, 1992). Once ingested, nitrates can be transformed to nitrites that become a health risk when converted to nitrosamine. Pesticide residues can taint the quality of crops and the quality of drinking water from the aquifer (Ekstrom and Akerblom, 1990). Lately, land users are resorting to recycling sewage sludge, livestock manure, and other resource materials, such as coal combustion residues (i.e., fly ash and flue gas desulfurization sludge). Potential problems with sewage sludge and livestock manure arise not only from their pathogen and excessive nutrient contents, especially N and P, but also from heavy metals (Adriano, 1986; Page et al., 1983). Cadmium is the metal of most concern from sewage sludge application on land; concerns with excessive Cu, Zn, and As may arise from swine and poultry manure application on land.

Modern landfills are engineered to contain chemical constituents by installing clay and plastic liners; however, old landfills often leak due to the absence of these barriers. Thus, old landfills are more prone to leak potentially hazardous constituents to underlying soil strata and ground-water (Andelman and Underhill, 1990; Fuller and Warrick, 1985; Suter et al., 1993).

Acid leachates and runoff water from the oxidation of pyrite-type material in coal arise from stockpiling of fossil fuel, especially coal high in S (Carlson and Adriano, 1993). Leachate and runoff water may have pH <3. In addition, the runoff water is usually high in soluble constituents, such as Fe, Mn, Ni, Zn, sulfate, and others (Anderson et al., 1993). These contaminants may adversely affect the quality of surface water and groundwater when they leach from the runoff storage basin. Leaking of petrol products, such as refined oil, from large storage tanks frequently occurs and can potentially contaminate the underlying groundwater (Cairney, 1993).

Drastically perturbed landscapes almost invariably result from mining of metal ores and coal (Davies, 1991). Deep geologic materials as well as subsoils are brought up to the surface along with pyrite-containing materials, resulting in an acidic soil environment. During the smelting and refining of metal ores, acidic substances and metal-bearing particulates escape into the atmosphere. Combustion of fossil fuels, such as coal and oil, for generating electric power results in the release of SO_x and NO_x compounds (Adriano and Johnson, 1989; Longhurst, 1990; Moldan and Schnoor, 1992). Most of these acid rain constituents, along with heavy metals, are deposited in nearby areas, but some are transported over long distances.

In 1986, the Chernobyl accident in the Ukraine demonstrated that severe ecological consequences can arise from failure of nuclear reactors and releases of their waste (Eisenbud, 1987; Kryshev, 1992). Today, massive programs are being planned to restore some health to the ecosystems of the Chernobyl area and the neighboring areas in the Ukraine and Belarus.

Sediments have to be dredged to render waterways commercially navigable. Dredged materials are often disposed of in adjoining areas that may overlie already contaminated land. Once exposed, certain chemical constituents in the sediment may undergo transformation, causing certain pollutants to become more toxic (Salomons and Forstner, 1988). For example, Cd in its nontoxic sulfide form in the sediment is transformed to the more toxic form once aerated upon spreading on land, resulting in the formation of H_2SO_4.

Depletion of soil OM may be caused by commercial logging and clear-cutting of forests (Allen, 1985). The tree biomass is burned for rapid cleanup and for rapid mineralization of plant nutrients. Under this practice, the soil may only be productive for a few years, after which it becomes unproductive due to nutrient depletion and soil erosion (Sumner and Miller, 1992). This practice is considered a major reason for the controversial global change issue.

The ever-intensifying exploitation of the land to meet the demands of the world's burgeoning population will confound the chemical stress on soil by the increasing trend to dispose of chemical wastes on land, as a result of banning other disposal options. Concomitant with society's demand for more and improved products is the ever-increasing waste, including hazardous waste, to be disposed of. Waste recycling technology may never catch up with production technology. Thus, more and more waste will accumulate on land, thereby endangering the quality of the soil, surface water, groundwater, and the atmosphere.

The following factors should ensure greater chemical stress on the soil: burgeoning population, more chemicals being produced, implementation of more stringent air and water pollution control measures, more incentives to recycle renewable materials onto land, difficulty in incinerating and ocean dumping of hazardous waste, land as a cheap disposal medium, expensive waste minimization technology, expensive environmental restoration, and poor public awareness.

3 Soil Contamination from Industrialization and Urbanization

3.1 Magnitude and Extent of Contamination

Environmental neglect by society since the dawn of the Industrial Revolution has resulted in severe contamination of the soil, groundwater, and surface water. For only about three decades or so has greater attention been given to high-stake chemical-related environmental issues relevant to the food chain and drinking water in North America and western Europe. Because of the potential risk to humans and other organisms from these contaminated sites, the EU, United States, Canada, Japan, Australia, and New Zealand have made some accounting of these affected areas. In these countries, surveys of the magnitude and extent of contamination have been underway for years. Indicated in Table 4.1 are the estimated number and areal extent of contaminated industrial and urban sites in the EU. In Table 4.2, the likely number of contaminated as well as severely (critically) contaminated sites in the EU and the United States are displayed. There are now over 1300 sites on the U.S. National Priority List (NPL) otherwise

TABLE 4.1. Estimated number of contaminated urban and industrialized areas in the European Union.

Source	Number	Amount of waste (tonnes)	Area (km^2)	Potentially contaminated area (km^2)
Industrial estates	12×10^6		10,000	16–40 thousand
Landfills/impound-ments	60 – 120 thousand	3 – 6 thousand	600 – 1200	900–7200
Fuel storage tanks	$3 – 6 \times 10^6$			250–4000
Mining waste dump	A few thousand	$17,000 \times 10^6$	250 – 500	350–5000
Line sources			10 – 25 thousand	1500–7500
Dredged sediment dumpsites	Hundreds			Hundreds
Hazardous waste sites	Hundreds			Hundreds
Estimated total contaminated area				20–60 thousand

Source: Kohsiek et al. (1994).

TABLE 4.2. Number of contaminated sites in selected countries in Europe and North America.

Country	Contaminated sites (total)	Contaminated sites (in critical condition)
Austria	3300	—
Germany	up to 100,000 +	10,000
Belgium	8300	2000
Italy	5600	2600
Netherlands	~110,000	4000
Denmark	20,000	3600
United States	37,600[a]	>1300 (NPL sites)
United Kingdom	up to 100,000	—
Norway	25,000	—

[a]As of 1993 reported to U.S. EPA, pursuant to CERCLA Section 103; NPL = National Priority List.
Source: Eijsackers and Hamers (1993); Overcash (1996).

known as Superfund sites. The projected costs to clean up these hazardous sites are staggering. The number of these hazardous sites in individual countries is expected to increase; for example, six more were added in the early half of 1998 to the NPL, as more surveys and risk assessments are conducted.

The toxic constituents in severely contaminated sites vary from just inorganics, including metals, to organics, and to mixed (both organics and inorganics) waste. The dominant metals at NPL sites are Pb (47% of the time), As (41%), Cr (37%), Cd (32%), Ni (29%), and Zn (29%) (Fig. 4.3).

Gasworks sites can also have a suite of inorganics and organics (Appendix Table A.15). This kind of setting may pose more challenges to the cleanup industry because of the varied nature of the waste. Where the food chain may not be affected, groundwater quality is usually compromised by old-fashioned disposal practices that enrich the inorganic contents of the water.

3.2 Generation and Disposal (or Uses) of Waste

3.2.1 Municipal Solid Waste

There are four basic categories of waste generated by industrial activities: combustible waste, solid waste, sludge and slurry waste, and wastewater (Fig. 4.4). The form of the waste depends on the type of the industry and can affect air, land, and water resources. For example, mining generates the most and primarily solid waste and because of the traditional disposal practice of this industry, i.e., piling (or waste heaps), the waste can affect air, soil, and groundwater quality.

The U.S. EPA has estimated that in 1990, $\sim 177.6 \times 10^6$ tonnes of municipal solid waste (MSW) were generated in the United States, of which 147.3×10^6 tonnes (83% of total) were disposed in landfills or incinerators,

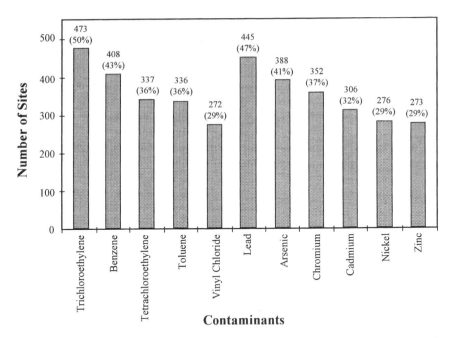

FIGURE 4.3. Frequency of the most common contaminants in all environmental matrices at U.S. NPL sites with Record of Decisions (RODs). (From U.S. EPA, Office of Solid Waste and Emergency Responses, ROD Information Directory, 1995.)

and the remainder composted or recovered (U.S. EPA, 1990). Municipal solid waste includes wastes, such as durable and nondurable goods, containers and packaging, food scraps, yard trimmings, and miscellaneous organic and inorganic wastes from residential, commercial, institutional, and industrial sources (Rhyner et al., 1995). The description of the usual components of MSW is given on Appendix Table A.16. About 2.0 kg per day per person of MSW is generated in the United States (U.S. EPA, 1992). Residential waste accounts for 55 to 65% of the total MSW generated, while 35 to 45% comes from commercial waste. Industrial waste, wastewater sludge, construction and demolition waste, agricultural waste, junked automobiles, and mining waste are not considered part of MSW.

The generation of MSW is apparently higher in the United States than in other developed countries as shown in Table 4.3. The MSW composition varies widely in different countries under differing climatic conditions (Rhyner et al., 1995).

3.2.2 Hazardous Waste

Hazardous waste refers to any solid, semisolid, liquid, or gaseous waste that cannot be handled by routine waste management methods because of

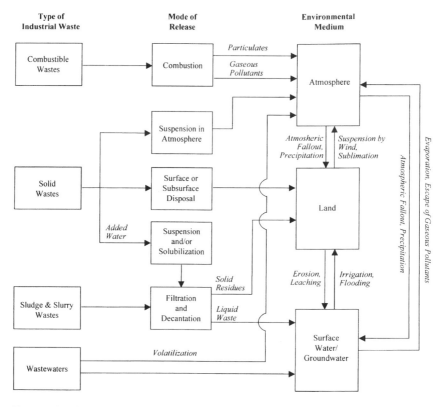

FIGURE 4.4. Sources and modes of release of metals into the environment. (Modified from Artiola, 1996.)

chemical, biological, or physical properties that present a significant threat to the health of humans and other organisms, or the environment (Rhyner et al., 1995). In some cases, hazardous waste may be a mixture of solid, liquid, or gaseous components. Substances that qualify as hazardous include explosives, flammables, oxidizers, irritants, corrosives, pesticides, acids, caustics, poisons, pathological waste, and radioactive waste.

In the United States, the RCRA regulations define hazardous waste as being either a listed hazardous waste, or a characteristic hazardous waste. A listed waste is one that appears on the U.S. EPA list of substances known to have hazardous and/or toxic properties. If a waste does not appear on the list, it may still be considered hazardous based on its characteristics and a laboratory test using the toxicity characteristics leaching procedure (TCLP). Furthermore, the regulations specify that a solid waste is a hazardous waste if it is a mixture containing a listed hazardous component along with other nonhazardous components or is a waste derived from the treatment, storage, or disposal of a listed hazardous waste (Rhyner et al., 1995).

TABLE 4.3. Generation of municipal solid waste (MSW) per capita in selected countries, including a comparsion with per capita economic output (GNP).

Country	Year of waste estimate	Per Capita Values		
		Annual MSW (kg)	Annual GNP value (U.S.$)	MSW/GNP (kg/$1000 U.S.)
Austria	1988	355	17,360	20.5
Belgium	1989	349	16,360	21.3
Canada	1989	625	19,020	32.9
Denmark	1985	469	20,510	22.9
England	1989	357	14,570	24.5
Finland	1989	504	22,060	22.8
France	1989	303	17,830	17.0
Germany (FRG)	1987	318	20,750	15.3
Greece	1989	259	5340	48.5
Italy	1989	301	15,150	19.9
Japan	1988	394	23,730	16.6
Luxembourg	1990	466	24,860	18.7
Netherlands	1988	465	16,010	29.0
New Zealand	1982	670	11,800	56.8
Norway	1989	473	21,850	21.6
Spain	1988	322	9150	35.2
Sweden	1985	317	21,710	14.6
Switzerland	1989	424	30,270	14.0
United States	1986	864	21,100	40.9

Source: World Resources Institute, *World Resources 1992–93,* Oxford University Pr, New York, 1992.

Over 100 substances listed as RCRA hazardous wastes are present in household products, such as cleaners, solvents, pesticides, paints, and household and automotive maintenance products. Examples of hazardous metallic materials and the products in which they may be found are shown in Table 4.4. The disposal of these products is exempt from hazardous waste regulations because of the small quantities generated by each household.

Hazardous waste and toxic waste are sometimes used interchangeably, except that RCRA guidelines make a clear distinction between them. According to the guidelines, certain waste streams are designated as hazardous if they satisfy any one of several characteristics, such as toxicity. (Note: Toxicity refers to the capacity of a substance to produce personal injury or illness to humans through ingestion, inhalation, or dermal contact pathway.) Therefore, toxic waste is essentially a subset of hazardous waste that is regulated on the basis of human toxicity. Examples of toxic wastes include heavy metals and many synthetic organic compounds, such as pesticides, PCBs, and solvents. Exposure to such a waste can result in acute or chronic health damage.

A waste may be defined as toxic because of its characteristics, or on the basis of a TCLP. In this test, a waste sample is treated with an acidic

TABLE 4.4. Hazardous metallic ingredients found in common household products.

Ingredient	Type of products found in
Arsenic(III) oxide	Paint (nonlatex antialgae)
Cadmium	Ni–Cd batteries, paints, photographic chemicals
Lead	Stain/varnish, auto batteries, paint
Mercury	Batteries, paint (nonlatex antialgae), fluorescent lamps
Silver	Batteries, photographic chemicals

Source: Modified from Rhyner et al. (1995).

solution to simulate leaching activity in a landfill. After 24 hr, the extract is analyzed to determine if the concentrations of toxic constituents exceed the levels in Table 4.5. The waste is considered hazardous if the concentrations exceed the limits.

3.2.3 Disposal of Solid Waste

Landfilling is the most commonly used waste disposal method in the United States and accounts for about 80% of the discarded MSW. Landfilling is also the predominant method of disposal in other developed countries, except Japan and some European countries (Appendix Table A.17). Landfills are a necessary component of any MSW management system. Waste reduction efforts that include recycling, incineration, and composting can reduce the volume of materials sent to a landfill, but residual materials from these activities will most likely require landfilling.

A modern sanitary landfill is an engineered site—designed and operated in such a manner as to minimize environmental impacts. The MSW is deposited in a confined area, spread in thin layers, compacted to the smallest practical volume and covered at the end of each working day. Certain types of industrial wastes may not require daily cover. The design, construction, and operation of the site are subject to state and federal regulations and performance standards.

TABLE 4.5. Regulatory levels for selected metals in wastes according to a leaching test.[a]

Contaminant	Maximum concentration (mg L^{-1})
Arsenic	5.0
Barium	100.0
Cadmium	1.0
Chromium	5.0
Lead	5.0
Mercury	0.2
Selenium	1.0
Silver	5.0

[a]TCLP test = Toxicity characteristic leaching procedure; procedure and levels promulgated March, 1990. (U.S. EPA, 1990.)

Landfilling in the United States has been decreasing because (1) old landfills that do not meet current design and management standards are being closed, (2) siting new landfills has become increasingly more difficult, and (3) the costs of site design, construction, operation, leachate and gas monitoring and collection, leachate treatment, management, and engineering favor the construction of large facilities (Rhyner et al., 1995). The difficulty encountered by municipalities in identifying and developing new landfills is not caused by a lack of land at a suitable location with suitable soil and hydrogeological conditions, but rather by public objection. People object to landfills because of the nuisance (e.g., dust, noise, traffic, odor), aesthetics, and potential environmental consequences (e.g., groundwater pollution from landfill leachate, migration of landfill gases to adjacent properties, and use of agricultural land). Subsurface conditions, such as soil type, underlying rock strata, and groundwater conditions are important factors for designing economically and environmentally safe landfills. The primary concern is the protection of groundwater quality from contamination. Accordingly, a database must be obtained regarding the distance from the bottom of the proposed fill to the groundwater, the soil type and other unconsolidated materials as well as bedrock beneath the site, the volume and direction of flow of the groundwater, and the existence of any impervious bedrock or clay layers between the fill and the groundwater.

Regarding hazardous waste, there are essentially only two methods for their disposal: land disposal and incineration. Landfills designed for MSW are not suitable for the disposal of hazardous waste, but some types of hazardous waste can be landfilled in specially designed sites known as *secure* landfills. These landfills are equipped with double liners, leakage detection equipment, leachate monitoring and collection, and groundwater monitoring equipment. Synthetic liners with a minimum thickness of 0.76 mm (30 mil) are mandated. Improvements in technology now allow liners to be put in place in very large sections, thus minimizing the number of joints. Some states allow the use of natural clays for liners, provided the hydraulic conductivity is less than 10^{-6} cm sec^{-1} and a leachate collection system is in place.

Incineration with appropriate emission control devices is the preferred method for destroying organics, particularly halogenated organic compounds. The predominant combustion products are CO_2 and H_2O, with small amounts of HCl or other halogen compounds. The last must be captured from the stack gases through reactions with lime or other alkaline compounds to produce nonhazardous salts.

The ever-increasing generation of municipal sewage sludge in developed countries in tandem with more stringent water quality regulations has switched emphasis on landfilling and agricultural utilization as the most preferred means of disposing of them. Data presented on Appendix Table A.18 indicate a zero option for ocean disposal due to water pollution issues as well as a zero option for incineration due to energy cost and air quality considerations. With methods to stabilize metals in these biosolids, such as composting under alkaline conditions, uses in agriculture and soil reclamation are very likely to become more attractive.

4 Contamination of Agroecosystems from Soil Amendments

The two major sources of metals in agricultural soils are biosolids, more specifically municipal sewage sludge, and phosphatic fertilizers. The main metal of concern in phosphatic fertilizer is Cd, whereas several metals from sludge are regulated with respect to contaminating the food chain and eventual exposure of humans. To minimize the risk to the public, concentrations of metals in phosphatic fertilizers and sludges as well as metal concentrations in the receiving soils are regulated. In addition, annual loading rates and total accumulative loads in soils are also regulated.

Interest in using municipal sewage sludge on cropland has increased considerably in the past two decades or so. The sludge is a valuable source of N, P, micronutrients (e.g., Cu, Fe, Mn, Zn), and OM that should benefit growing crops. In addition, sewage sludge disposal on agricultural land is an attractive option to other disposal options discussed above. However, concern over the potential hazards associated with the metal content of sludges (e.g., Cd, Cr, Ni, Pb, Zn) has sparked closer scrutiny of the material. Indeed, soil contamination by sludge-borne metals can result in decline of crop yields due to metal phytotoxicity or in enhancement of metal transfer into the food chain.

Two kinds of metal responses should be considered when applying sludges on land: (1) short-term effects due to the readily bioavailable forms of metals from the sludge, and (2) long-term effects resulting from the dynamics in metal speciation or other biogeochemical processes in the soil–plant system following sludge application or after application ceases. While short-term effects of metals are now quite understood, we are still in the dark with long-term field effects due to our lack of understanding of the long-term behavior of metals in the soil-plant system. Moreover, long-term field experiments represent the only comprehensive approach to precisely assess the cumulative and residual effects of metals as sludge progressively decays in soils.

A decade ago, Juste and Mench (1992) reported gleaning over 200 references on sludge field experiments alone. However, there are only a limited number of long-term field studies, i.e., those exceeding 10 yr in duration. Most of the well-known experiments are in the EU (France, Germany, Denmark, United Kingdom) and the United States. Some of these experiments are described in Table 4.6. The biosolid used generally consisted of sewage sludge of various types, such as liquid sludge,

TABLE 4.6. Maximum metal inputs (kg ha^{-1}) in selected long-term sewage sludge experiments.

Site	Experiment	Cd	Cu	Cr	Mn	Ni	Pb	Zn
				Metal (kg ha^{-1})				
Bordeaux, France	Bx1	26	158	65	5679		659	4882
	Bx2	641	170	64	234	1337	231	976
Nancy, France		9	132		122	37	197	746
Woburn, UK	Market garden	70	864	704		135	694	2158
St. Paul, MN USA		25						348
Riverside, CA USA		44	730	1180		301	936	2888
Joliet, IL USA	J1	58						1290
	J2	101						2358

Source: Modified from Juste and Mench (1992).

dehydrated sludge, and anaerobically digested sludge. In some cases, composted sludge was used. The metal contents of the sludge were always reported, but other parameters (pH, OM content, type of sludge process, etc.) were sometimes not mentioned. Among the metals, concentrations of Cd and Zn were always measured due to their large contents in sludges, their mobility, and because they represent the greatest risk of bioaccumulation in the food chain. Copper, Cr, Ni, Pb were less frequently included for different reasons: their bioavailability is generally lower and therefore deemed to be of lesser risk. Iron and Mn were often present in large amounts in sludges but were also less frequently measured because they are considered to be nontoxic. Other elements that are important to plant or animal nutrition, such as Co, Mo, Hg, Tl, V, As, B, Be, and Se, were at times not accounted for.

The maximum input of metals in the long-term field experiments from sludge application have been quantified in Table 4.6. The summary may provide some information for better understanding and interpreting the experimental data. These data indicate that metal inputs in the soil were greatly variable from one site to another. For example, the maximum input of Cd and Zn ranged from 9 to 641 kg ha^{-1} and from 348 to 4882 kg ha^{-1}, respectively. Such differences in metal inputs obviously induce marked differences in the metal behavior in soil and plants.

The generally high rates of Zn addition through sludge application underlines the potentially important role of Zn on the behavior of other metals that were incorporated into soils, especially that of Cd. Because of similarity in the biogeochemical properties of Zn and Cd, an excess of one might be for the good of the other in terms of bioaccumulation and food chain transfer.

The researchers from Table 4.6 reported that incorporation of the sludges into the soil generally resulted in marked increases of the soil metal contents. One major question arises with respect to the movement of metals in sludged soils. Do they remain near the soil surface in the zone of incorporation, or are they leached from this zone? Field studies have generally shown no difference in metal concentrations between control and sludged soils below 30 cm (Berti and Jacobs, 1998; Chang et al., 1982; Dowdy and Volk, 1983; McGrath and Lane, 1989; Sloan et al., 1998; Williams et al., 1985). However, few studies have also indicated the potential for metals to move from the application zone as indicated by low recovery (e.g., 47 to 83%) of sludge-applied Cd and Zn after application (Bell et al., 1991; Dowdy et al., 1991), or evidence of metal movement (i.e., as much as 13%) past the treated layer (Yingming and Corey, 1993).

The loss of metals from fields that received sludge may potentially represent an environmental risk based on the recent findings of Steenhuis et al. (1999). Using soil–liquid partition coefficient (K_d) values generated from an orchard site treated with sludge 15 years earlier, preferential flow model predicts that given sufficient time and rainfall, a substantial portion of less strongly adsorbed metals can leach out of the zone of incorporation. The rather rapid transport under preferential flow can be attributed to the inhibiting effect of metal–organo complexes (i.e., DOC promotes metal dissolution by forming soluble metal–organo complexes) on adsorption rendering metals more prone to leaching loss than previously thought.

Metal contents of crops grown on sludge-amended soils are generally a function of the annual sludge loading rate. This trend was especially observed in the Bordeaux (France) and United States experiments for Zn, Cd, and Ni (Hinesly et al., 1977; Juste and Mench, 1992). However, long-term changes in other soil parameters could have also occurred to affect the biogeochemical behavior of the metals. For example, an elevation of soil pH can reduce metal bioavailability, thereby reducing the metal concentrations in plant tissue in spite of a progressive rise in the total soil metal contents. The cumulative metal input to the soil is also a major factor determining the metal concentrations in plant tissues (Chang et al., 1987; Juste and Mench, 1992; Soon et al., 1980).

Perhaps unique with sludge-borne metals, metal concentrations in plant tissue may approach an ultimate level (i.e., a plateau) with the progressive metal input, indicating a threshold of the metal uptake by plant. In this respect, Chang et al. (1987) established a nonlinear regression model for Swiss chard, radish (leaves), and tubers grown in a long-term experiment:

$$Y = a + b(1 - e^{-cX})$$

where Y is the concentration in plant tissue, X is the total cumulative sludge loading, and $a, b,$ and c are parameters whose values are determined by an adjustment (see also Chapter 3).

Phytotoxicity is rarely observed even though large amounts of sludge-borne metals are added into the soil. Several hypotheses have been suggested to account for the relatively low phytotoxic effect of sludge-borne metals.

These include increased soil pH, formation of insoluble salts (e.g., phosphate, sulfate, and silicate salts), and metal sorption by iron and manganese oxides or OM. The protective nature of sludge against phytotoxicity was demonstrated when soybean plants grown on $CdCl_2$-amended soils exhibited Cd phytotoxicity symptoms, whereas plants grown on sludge-treated plots given equivalent Cd rates in the form of sludge were not affected (Heckman et al., 1987). Some antagonistic interaction among the sludge-metals may be another reason for their low phytotoxic effect.

Because of the threat to the quality of the food chain, sludged soils as well as the sludge itself are now regulated in the United States (Table 4.7) and EU countries (Appendix Table A.19). The primary parameters covered by the so-called 40 CFR Part 503 Standard include the annual loading rate, cumulative loading rate, and metal content of the sludge (Table 4.7).

Improvements in wastewater treatment technologies have already been achieved in North America and western Europe, which should subsequently allow higher application rates of sewage sludge on agricultural land without jeopardizing the quality of the food chain with respect to certain metals. For example, in the United Kingdom, substantial improvements in biosolid quality with respect to Cd, Pb, and Zn total concentrations in sludge between 1980 and 1990 have been observed (Matthews, 1999). However, no significant improvement in Ni and Cu were observed during the same period (Table 4.8). The same trend in sludge quality was observed for Germany (Bode, 1998) where federal requirements for effluent concentrations for the metal plating industry (e.g., Ruhr catchment area) have resulted in enhancement of biosolid quality.

TABLE 4.7. Metal limits relative to soil application of sewage sludge in the United States (U.S. EPA 40 CFR Part 503 Standard).

Metal	Ceiling concentration in sludge (mg kg^{-1})	Ceiling concentration in clean sludge[a] (mg kg^{-1})	Maximum cumulative loading (kg ha^{-1})	Maximum annual loading (kg ha yr^{-1})
Arsenic	75	41	41	2
Cadmium	85	39	39	1.9
Chromium	3000	1200	3000	150
Copper	4300	1500	1500	75
Lead	840	300	300	15
Mercury	57	17	17	0.85
Molybdenum[b]	75	18	18	0.90
Nickel	420	420	420	21
Selenium	100	36	100	5
Zinc	7500	2800	2800	140

[a]Metal concentration limits for biosolids designated as a Class A sludge.
[b]Limits for molybdenum stayed by court action in 1994.
Source: U.S. EPA (1993).

TABLE 4.8. Improvement in sewage sludge quality as measured by total concentrations for some metals in 1980 and 10 yr later in the United Kingdom. The sewage sludge was used in agriculture.

	mg kg^{-1} in dry solid				
	Zn	Cu	Ni	Pb	Cd
1980					
Min	143	20	5	25	0.1
Med	1002	440	45	260	7.0
Mean	1123	519	101	329	16.3
Max	4920	2900	615	3106	158
1990					
10%	454	215	15	70	1.5
50%	889	473	317	217	3.2
90%	1473	974	225	585	12

Source: Courtesy of P. Matthews (1999).

Phosphatic fertilizers applied to land may contain high levels of Cd. Fertilizers originating from magmatic phosphates tend to contain only negligible amounts of Cd, whereas those made from sedimentary phosphates tend to contain a range of concentrations, some of which may be very high (McLaughlin et al., 1995). In Australia and several countries, Cd contents in phosphate fertilizers are now being regulated (see also Chapters 1 and 8).

Annual inputs of Cd to agricultural soils from phosphatic fertilizers are as variable as the concentrations in the fertilizers themselves. A mean annual input of 3.5 g ha^{-1} yr^{-1} is reported for soils in Germany (Kloke et al., 1984). Inputs to soils in the United Kingdom are estimated to be 22 tonnes of Cd per year, a mean of 4.3 g ha^{-1} yr^{-1}. Cadmium inputs to an experimental plot at Rothamsted Experiment Station (United Kingdom) receiving applications of superphosphate over a 96-yr period were estimated to be 5 g ha^{-1} yr^{-1} (Rothbaum et al., 1986). However, in a study in California in which triple superphosphate was applied over a period of 36 yr, the concentration of Cd in the top 15 cm of the experimental plot was 1.2 µg g^{-1}, as compared with 0.07 µg g^{-1} in the control plot (Mulla et al., 1980). An approximately fivefold increase in the Cd concentration of the upper 2.5 cm of the soil profile was observed in a similar experiment (Williams and David, 1976).

The initial concern about Cd concentration in fertilizers in relation to the accumulation of Cd in food crops was raised (Schroeder and Balassa, 1963). The concentration of metals in the fertilizers is dependent upon its concentration in the parent rock and the processing technologies used to manufacture the product.

Several countries have expressed concern about Cd in phosphatic fertilizers, and Sweden has established a limit of 150 mg Cd kg^{-1} P, which was later lowered to 100 mg Cd kg^{-1} P. Since virtually all Cd in phosphatic fertilizers was derived from the phosphate component, the expression of the standard on the basis of P content allows normalization across different fertilizer products. Because different rock sources differ so widely in Cd concentrations (e.g., Florida, Kola, and South Africa are very low in Cd, while Idaho, Morocco, and Senegal are quite high in Cd) and because some countries have begun imposing strong limitations on Cd in fertilizer products, commerce has been affected. Currently, little agreement among countries regarding regulations of Cd in fertilizer products exists. For example, the Netherlands has imposed a substantially lower fertilizer Cd limit (25 mg Cd kg^{-1} P$_2$O$_5$) than Sweden (Landner et al., 1995). Technologies to remove Cd during phosphate fertilizer manufacture have been evaluated and different technologies may be required depending upon the Cd limitation imposed on the product.

4.1 The Rhine Basin Soil Contamination: A Case Study

The Rhine Basin in Europe provides an excellent case study of the complex relationship between economic activities and environmental quality (Stigliani et al., 1993). The region (Fig. 4.5) covers ~220,000 km^2 and comprises most of Switzerland, parts of Germany and France, most of the Netherlands, and all of Luxembourg. It is almost the size of Georgia, (United States), but has more population, ~50 million. The density of industrial activity has been one of the highest in the world, accounting for 10 to 20% of the total chemical industry in OECD (Organization of Economic and Cooperative Development) countries. Crop yields in the basin as well as agrochemical inputs have been among the highest in the world. A comprehensive historical analysis of heavy metal contaminants has been conducted with emphasis on Cd, Pb, and Zn, and accounting for inputs and outputs, and the balance for the above elements for the entire basin (Stigliani et al., 1993; Stigliani and Anderberg, 1994). However, only Cd will be highlighted here.

The three major sources of Cd to agricultural lands in the basin are atmospheric deposition, application of phosphatic fertilizer, and the spreading of sewage sludge. Once the Cd is in the soil, it can be transported out of the soil by plant offtake, surface runoff, and erosion. Since leaching of Cd was deemed minimal, it was not included in the calculation. The net accumulation of Cd in agricultural soils, determined by accounting for inputs and outputs, is shown on Table 4.9.

The net Cd inputs to the basin soil have been reduced by more than 50% between the early 1970s and the late 1980s. The most significant reduction has been in atmospheric deposition, which decreased by nearly 50 tonnes per year (80%) over this time period. In addition, there were more moderate reductions in inputs from phosphatic fertilizer (25% reduction) and sewage sludge (about 50% reduction). Whereas atmospheric deposition was the largest source of Cd to agricultural lands in 1970, by the mid-1970s phosphatic

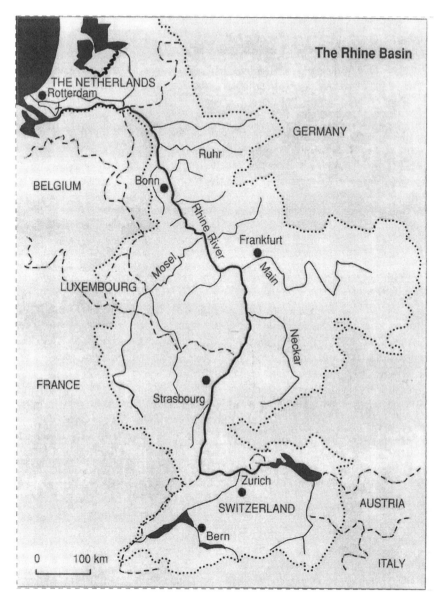

FIGURE 4.5. The Rhine Basin: an excellent model for a long-term assessment of accumulation in soil of environmentally important metals, including Cd, Pb, and Zn. (Reprinted with permission from Stigliani et al., 1993. Copyright © 1993 American Chemical Society.)

fertilizer became the dominant source. By the late 1980s it accounted for more than 70% of the total inputs (Stigliani and Anderberg, 1994).

Plant offtake of Cd increased mainly because of increases in crop yields over the 18-yr period. The yield of cereal crops in the basin is estimated to

TABLE 4.9. Inputs and outputs of cadmium (in tonnes yr^{-1}) to agricultural soils in the Rhine Basin, Europe.

Source	1970	1975	1980	1985	1988
Atmospheric deposition	+61.6	+38.0	+24.1	+15.4	+13.2
Phosphatic fertilizer	+47.7	+43.0	+43.7	+38.0	+35.9
Sewage sludge	+2.2	+2.3	+1.8	+1.1	+1.0
Plant offtake	-2.2	-2.7	-3.3	-4.0	-4.1
Runoff	-6.8	-4.2	-2.6	-1.7	-1.4
Erosion	-2.0	-2.1	-2.3	-2.3	-2.4
Net input (tonnes yr^{-1})	+100.5	+74.3	+61.4	+46.5	+42.2

Source: Modified from Stigliani and Anderberg (1994).

have increased from about 2.7 tonnes ha^{-1} yr^{-1} in 1970 to about 5.6 tonnes ha^{-1} yr^{-1} in the late 1980s. Surface runoff of Cd decreased because of decreases in its concentration in wet deposition. Erosion of Cd increased slightly, reflecting the slow increase in the total soil Cd content over time.

An important question is whether the cumulative Cd inputs to agricultural soils were causing significant increases in the soil concentrations of Cd above background levels and, if so, whether these increases could lead to unacceptably high levels of Cd in the crops grown in the basin. With this question in mind, the net inputs of Cd to agricultural soils since 1950 were estimated, using available information on historical phosphate use and calculating atmospheric deposition. Stigliani et al. (1993) had estimated that ~3000 tonnes of Cd have been added to the agricultural soils in the basin over the 40-yr period from 1950 to 1990. The net inputs were converted into soil concentration units, and the calculated increase in concentration was plotted over time as shown in Figure 4.6. Under the assumptions of the model, the average Cd concentration in agricultural soils has risen from approximately 360 g ha^{-1} in 1950 to about 700 g ha^{-1} in 1988, corresponding to an increase of 94% in the 38-yr period. The trends for the year 2000 and beyond have two scenarios: scenario A refers to a policy in which all Cd is removed from the phosphatic fertilizers; scenario B refers to a policy in which Cd is left in the fertilizer. The plateauing of scenario A indicates that the remaining input from atmospheric deposition is nearly balanced by the output of Cd from the system.

5 Unusual Alterations of Biogeochemical Processes

5.1 From Impoundment of Freshwater

The risk of Hg to humans from environmental rather than from occupational exposure has been recognized since the late 1950s and includes a number of scenarios relating to Hg from specific sources (D'Itri and D'Itri,

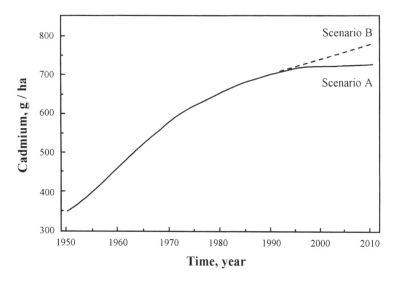

FIGURE 4.6. Projected cadmium accumulation in agricultural soils in the Rhine Basin. (Reprinted with permission from Stigliani et al., 1993. Copyright © 1993 American Chemical Society.)

1977). Mercury readily accumulates in tissues and is biomagnified through the aquatic food chain. Methylation of Hg has been accepted to be a key process controlling its bioavailability in aquatic ecosystems (Stokes and Wren, 1987).

A significant portion of the Hg deposited in the terrestrial system is immobilized in the organic fraction of upper soil horizons (Nater and Grigal, 1992). Concentrations of Hg in soil have been shown to be related to the OM in surface horizon and to some extent silt + clay content in the horizon. A small percentage, 0.03 to 0.04% annually, of Hg in the soil pool is transported via runoff from forest soil to surface waters (Aastrup et al., 1991). Others have estimated that ~4% of annual atmospheric input is transported to surface water and groundwater (Krabbenhoft et al., 1995). The terrestrial transport of Hg via runoff has been recognized as an important source of Hg to remote surface waters (Lee and Iverfeldt, 1991; Swain et al., 1992).

During the last 15 years, occurrences of elevated Hg levels in tissues of fish from regions considered to be remote from point or local sources of Hg have been documented. These appear to be related to (1) acidification of surface water (Lindqvist et al., 1984) and, (2) recent impoundments, usually in connection with hydroelectric dam construction (Bodaly et al., 1984).

In natural ecosystems, fish accumulate Hg concurrently via food ingestion and directly from the water (Stokes and Wren, 1987). The efficiency of Hg assimilation from food appears to vary among species. Controlled studies have shown that 68 to 80% of Hg ingested is assimilated by rainbow trout

(*Oncorhynchus mykiss*) (Phillips and Buhler, 1978; Rodgers and Beamish, 1982), about 20% by northern pike (*Esox lucius*) (Phillips and Gregory, 1979), and an efficiency of 80% has been used in uptake models for yellow perch (*Perca flavescens*) (Norstrom et al., 1976). Mercury assimilation across the gut can decrease above certain critical concentrations (Rodgers and Beamish, 1982).

Mercury uptake from the water is determined by the water concentration, fish metabolic rate, and bioavailability as determined by the ambient water characteristics. The rate of accumulation and toxicity of Hg and certain metals to fish are known to be reduced by increased water hardness or calcium levels (Carroll et al., 1979; Davies et al., 1976; Howarth and Sprague, 1978). The mechanism to account for this effect is thought to be increased gill permeability at low Ca levels (Spry et al., 1981) or competition between metals and Ca for cellular binding sites (Zitko and Carson, 1976). Observations from field studies also show correlations between fish Hg level and water Ca concentration (McFarlane and Franzin, 1980; Wren and MacCrimmon, 1983).

Other physiochemical conditions within lakes have been empirically related to Hg levels in fish with the inverse relationship with pH being the most prominent (Brouzes et al., 1977; Hakansson, 1980; Jernelov et al., 1975; Wren and MacCrimmon, 1983); Hg levels were higher in brook trout (*Salvelinus fontinalis*) from drainage lakes (pH <5.0) than in brook trout from seepage and bog-type lakes (pH >5.0). A thorough evaluation of data from over 200 lakes in Sweden suggests that Hg levels are elevated in fish within specific regions due to increased atmospheric deposition of Hg and lake acidification (Bjorklund et al., 1984).

The first documentation of Hg bioaccumulation in fish from impoundment was on the high Hg levels in fish from Willard Bay Reservoir, Utah in 1974 (Smith et al., 1974). Predatory fish in this shallow reservoir (pH ~8.1 to 8.3) contained total Hg in the range 0.27 to 7.3 mg kg^{-1}. This was attributed to geogenically (or pedogenically)-generated Hg. Another evidence of biomagnification of Hg was obtained in 1975 from another man-made reservoir (Lake Powell, New Mexico), where predatory fish commonly had Hg levels greater than 0.5 mg kg^{-1}, and some exceeding 1.0 mg kg^{-1} (Potter et al., 1975). This phenomenon was attributed to natural weathering of the basin. Soil that had deposited into another reservoir (Cedar Lake Reservoir, Illinois) was believed to be the original source of Hg in crappie (*Pomoxis spp.*) and largemouth bass (*Micropterus salmoides*) where Hg levels were >0.5 mg kg^{-1} (Meister et al., 1979).

Similar early observations on Hg bioaccumulation in fish from impoundments have been reported for several reservoirs in Canada. This includes Smallwood Reservoir, Labrador and La Grande Reservoir, Quebec (Boucher and Schetagne, 1983). Most fish from these reservoirs exceeded the federal limit of 0.5 mg kg^{-1} WW, which limited fish consumption by the public. The roles of sediments resulting from erosion and of OM as factors mediating the influx and methylation of Hg were offered as possible mechanisms (Boucher and Schetagne, 1983).

A major controlling factor for methylation of Hg in these reservoirs was a decrease in pH of surface water from about 6.5 to 6.0, with slightly lower (5.4 to 5.9) pH in subsurface water. Inundation of vegetation was also considered an important factor, since terrestrial plants, especially mosses, have a high metal-binding capacity, and may be loaded with Hg from atmospheric deposition. Another factor could have been the increased concentrations of nutrients after flooding, which could have enhanced methylation of Hg.

Subsequent, more comprehensive studies in other Canadian reservoirs (e.g., Cookson Reservoir, Saskatchewan; Churchill and Nelson River Basins, Manitoba) lend support to the reservoir paradigm (Bodaly et al., 1984; Waite et al., 1980). Preimpoundment and postimpoundment Hg levels for selected fish from Canadian reservoirs are indicated on Table 4.10. These data as well as data from nearby reservoirs indicate a 20 to 40% increase in Hg level after impoundment (Bodaly et al., 1984).

There is general agreement that methyl Hg results in higher levels of uptake into fish than does inorganic Hg. There is also general agreement that most (70 to 85%) of the Hg in fish muscle is methyl Hg. From the evidence presented to date it is plausible to assume that the source of Hg is the material already in the lake or river prior to flooding, or the flooded soils and vegetation, and that this Hg is either natural or at least background, even if not preindustrial levels (Lindqvist et al., 1984) since there is no immediate atmospheric source in these situations. The question of mechanism then becomes one of mobilization and methylation of existing Hg

TABLE 4.10. Preimpoundment and postimpoundment mean Hg levels (in mg kg^{-1} WW) in fish in lakes flooded by the Churchill River, Northern Manitoba, Canada.

Fish species	Lake	Preimpoundment	Postimpoundment
Lake whitefish	SIL, area 4[a]	0.05 (25)* – 1975	0.22 (16) – 1978
(Coregonus clupeaformis)	SIL, areas 2 and 6[a]	0.06 (50) – 1975	0.30 (17) – 1978
	IS[a]	0.15 (24) – 1975	0.32 (5) – 1978
	WK[b]	0.08 (1) – 1970	0.33 (28) – 1970
Northern pike	SIL[a]	0.29 (12) – 1971–73	0.47 (33) – 1976–79
(Esox lucius)	IS[b]	0.61 (5) – 1978	0.90 (26) – 1982
	WK[b]	0.91 (75) – 1979	1.13 (1) – 1982
Walleye	SIL[a]	0.25 (12) – 1971–73	0.40 (21) – 1975–79
(Stizostedion	IS[b]	1.52 (5) – 1978	0.79 (19) – 1982
vitreum)	WK[b]	0.34 (1) – 1970	0.89 (34) – 1981

*(N) = number of samples.
SIL = South Indian Lake, Churchill River, at point of diversion.
IS = Issett Lake, upper end of Notigi Reservoir.
WK = Waskwatin Lake, on Burntwood River Basin, Notigi Reservoir.
Sources: Extracted from (a) Bodaly and Heck (1979); (b) Bodaly et al. (1984).

resulting from the factors that change between preimpoundment and postimpoundment (Stokes and Wren, 1987). Also, evidence indicates that methylation is promoted by vegetation and/or already decaying vegetation or other OM in the flooded area. Methylation may occur in the previously exposed soil and vegetation, or in the material that is washed into the reservoir and becomes bed sediment or suspended particulate matter. In short, bacterial methylation of Hg can be enhanced under acidic or low-alkalinity conditions, or by increased availability of organic materials in newly flooded reservoirs (Bodaly et al., 1984). Under these conditions, even background levels of Hg in water, sediments, and soils can lead to elevated Hg concentrations in fish tissue.

5.2 From Acidic Precipitation

5.2.1 Effects on Freshwater Systems

Surface waters and soils have considerable capacity to neutralize acidic precipitation (i.e., buffering capacity). The primary buffers of soils are the basic cations: Ca^{2+}, Mg^{2+}, K^+, and Na^+, with Ca^{2+} being the most important. The decrease in a watershed's buffering capacity over time depends on several factors, including the rate of deposition, the nature of the soil, the size of the watershed, and the flow characteristics of the lake or groundwater. These variables often make it difficult to detect acidification while in progress. Surface water usually reflects the buffering capacity of the watershed soil because of outcropping and runoff of water through the soil profile. However, soils overlying granitic parent rock are commonly deficient in basic cations and therefore, have poor buffering capacity (Sparling, 1995). The Big Moose Lake in the Adirondack Mountains in the State of New York is a case in point where the watershed soil originated from granitic rock and where the historical record of acidification has been well documented (Fig. 4.7). This area has received some of the highest inputs of acid precipitation in North America because it is downwind from western Pennsylvania and the Ohio Valley, historically the industrial heartland of the United States. Contaminants carried by the westerly winds are trapped in the mountains and deposited via wet and dry deposition. The ecological effects of acidification on surface waters might take a long time even in sensitive areas, as illustrated in Figure 4.7, which indicates the historical trends in lake water pH (dashed line), SO_2 emissions upwind from the lake (solid line), and the extinction of different fish species. The lake pH remained nearly constant at around 5.6 over the entire period from 1760 to 1950. Then, within a span of 30 years, from 1950 to 1980, the pH declined more than one whole unit to about pH 4.5. The decline in pH lagged behind the rise in SO_2 emissions by some 70 years, and the peak years of S emission preceded the decline in pH by 30 years. The deposition rate is estimated to be ~2.5 g S m^{-2} yr^{-1} during the peak period. Thus, beginning around 1950,

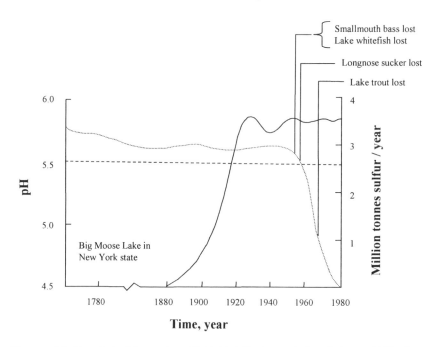

FIGURE 4.7. Trends in lake water pH (dotted line), upwind emissions of SO_2 from the U.S. industrial Midwest (solid line), and fish extinction for the period 1760 to 1980. (Modified from Stigliani, 1988.)

acid-sensitive fish species such as smallmouth bass (*Micropterus dolomieu*), whitefish (*Coregonus spp.*), and longnose sucker (*Catostomus catostomus*) began to disappear, followed in the late 1960s by the more acid-resistant lake trout (*Salvelinus namaycush*) (Stigliani, 1988). Because of the difficulty to precisely determine the mass balance of H^+ in natural waters due to complex interaction of strong mineral acids and weaker organic acids, the use of alkalinity or acid-neutralizing capacity (ANC) is more appropriate (Munson and Gherini, 1991):

$$ANC = (HCO_3^-) + (CO_3^{2-}) + (OH^-) + (\text{other } H^+ \text{acceptors}) - (H^+ \text{donors})$$

where reactants are in µeq L^{-1}.

For field applications, freshwater systems with ANC ≤ 200 are considered acid sensitive, ANC ≤ 100 are considered very sensitive, and ANC ≤ 0 are already acidic. The increases in the emissions of SO_x and NO_x since the onset of the Industrial Revolution have been a major causal factor in the acidification of surface waters and soils in parts of the United States, Canada, central Europe, and Scandinavia (Adriano and Havas, 1989; Husar et al., 1991). Accumulated sample data indicate that acidic precipitation has caused serious environmental problems in certain regions of North America and Europe. In the United States, surface water with low alkalinity (<200

μeq L^{-1}) extends along the eastern seaboard from Maine to Florida and then extends westward into Louisiana (Ormernik, 1985). A narrower belt also occurs in the Cascades of the Pacific Northwest, and patches of low alkalinity also occur in the Sierras of California, northern Midwest, and in the Rocky Mountains. The U.S. EPA coordinated a national survey of 2332 lakes and 558 streams during the mid-1980s, with part of the summary presented in Appendix Table A.20. The survey (Charles, 1991; Landers et al., 1988) indicated the following: (1) The lakes in the Adirondacks were the most affected by acid precipitation; soils primarily come from granitic gneiss that are more prone to acidify; 18% of the 1290 lakes larger than 4 ha in size were acidic (ANC $\leq$0) (Driscoll et al., 1991); acidity was strongly associated with deposited SO_4^{2-}; 24% of the lakes were fishless; lake size, depth, elevation, and pH determined the existence of fish; (2) Of all the states, Florida had the highest proportion of acidic lakes (i.e., of the 453 lakes surveyed, 22% had ANC $\leq$0, and 84% were sensitive (ANC $\leq$200); the rate of acidic deposition (H^+ + SO_4^{2-}) in the state had increased by 27% over the last 15 years; and (3) Eastern Canada received much of its air pollution from industrialized regions of the United States (Jeffries, 1991); about 1.7% of the 700,00 lakes in the region were acidic (ANC $\leq$0), and another 1.4 to 5.7% could be classified as sensitive.

Acidification of lakes and streams in North America and Europe has altered their trace metal chemistry (Newman and McIntosh, 1991) by: (1) increasing total metal concentrations (Al, Mn, Fe, Cd, Pb, Zn, etc.); (2) shifting the speciation of dissolved metals toward free aquo ions, typically the species most toxic to aquatic biota; and (3) reducing particulate metal concentrations in favor of higher dissolved levels.

Severe problems of surface water acidification were discovered in Scandinavia in the late 1960s. The acidification was largely caused by sulfur emissions from continental Europe and the United Kingdom, and drew attention to the existence of large south to north transport of air pollution (Brannvall et al., 1999). Today, there is great concern for contamination of the Arctic environment by atmospheric emissions derived from lower latitudes (Pacyna, 1995). Studies of ice cores from Greenland and lake sediments and peat in Sweden (Renberg et al., 1994) indicate that large-scale pollution of high latitudes by emissions from cultural centers in Europe has occurred for several thousand years, and there is also convincing evidence of early pollution from the British Isles and continental Europe (Shotyk et al., 1998). Despite this evidence, the common opinion is still that large-scale atmospheric pollution is a problem that started with the Industrial Revolution in the late 18th century and that concentrations observed in remote regions today represent the natural background values. This is a false picture, at least for Pb and Cu, but most likely for several other atmospheric pollutants, such as Hg.

The following are general observations on the effect of acidic deposition on surface water ecosystems (Adriano and Johnson, 1989; Sparling, 1995):

(1) Aluminum becomes mobilized and changes its species with reduced pH. This metal is readily leached from acidified watersheds and can accumulate to potentially toxic levels in water. Its toxicology in water is complex and depends on concentration, organism, pH, concentration of ligands that alter its bioavailability, and Ca levels. At pH greater than 6.0, Al solubility is low and most precipitates onto sediments or substrates. Below pH 5.5, Al solubility increases and more of it is in inorganic monomeric (Al^{3+}), hydroxide [$Al(OH)^{2+}$ or $Al(OH)_2^+$], or fluoride (AlF_3) forms. These inorganic forms are usually more toxic than organically bound Al, especially to fish; (2) Other metals may also be mobilized and become more toxic with reduced pH. Specifically, aqueous levels of Pb may increase because of concomitant deposition of Pb with acidifying constituents, whereas Cd may increase in water because of leaching from watershed soils. Evidence indicates that Hg levels in fish tissues are inversely related to the pH of water, as discussed above; (3) High H^+ concentrations can have direct toxic effects on aquatic organisms. The most common problem associated with low pH is impaired ability to balance Na^+, Cl^-, and K^+. Perturbed ion regulation has been demonstrated for benthic invertebrates, fish, and amphibians; (4) Base cations ameliorate the effects of H^+ toxicity. Calcium is typically low in acid-sensitive waters because the watersheds are mostly granitic with little limestone. Lethal pH's for benthic invertebrates and fish are often higher in very soft water (<100 to 150 μeq L^{-1} Ca) than in waters with greater Ca. Low levels of Ca increase plasma membrane permeability to ion and water exchange and enhance the toxic effects of H^+. Mollusks and crustaceans that incorporate Ca into their shells or exoskeletons are particularly sensitive to low levels of Ca and are often the first species to disappear with acidification; (5) Humic substances have complex relations with pH. At pH 4 to 5, these compounds are weak buffers that can help maintain pH. Surface waters, such as seepage lakes that are high in DOC are typically less sensitive to acidic deposition than are clearwater lakes. Humic substances may bind with nutrients, Al, and other metals and alter their bioavailability to organisms. Aluminum toxicity is often lower in water with moderate levels of DOC (5 to 10 mg L^{-1}) than with low DOC; (6) Lake clarity often increases with reduced pH. Increased transparency may be due to changes in the chemical composition of dissolved organics, or to flocculation of organoaluminum complexes rather than to decreased phytoplankton abundance. Greater clarity may allow photosynthesis at greater depths; and (7) Decreased species richness with acidity has been observed in phytoplankton, zooplankton, periphyton, macroinvertebrates, fish, and some amphibians.

5.2.2 Effects on Forest Ecosystems

During the past 30 years, declines in forest vigor and diebacks have been observed in Germany, central Europe, Scandinavia, the United States, and

Canada (Adriano and Havas, 1989; Adriano and Johnson, 1989; Irwin, 1989). Over 50% of the forests in Germany show some evidence of decline, with the greatest frequency occurring among fir forests in the southern part of the country. Beech, oak, pine, and spruce forests have also been affected. In North America, diebacks or reduced growth are obvious at high elevation for red spruce (*Picea rubens*) of the Appalachians, balsam (*Abies balsamea*) and Fraser firs (*Abies fraseri*) in New England and Canada, loblolly (*Pinus taeda*) and slash pines (*Pinus elliottii*) in commercial forests of the Southeast, and sugar maples (*Acer saccharum*) in the eastern United States and south-eastern Canada (NAPAP, 1991). Red spruce has been mostly affected by acidic deposition in the United States with declines observed in New England and upstate New York since 1960 (Hornbeck et al., 1986). Signs of unhealthy trees, which may dominate entire stands, include extensive loss of older needles, thinner crowns, yellowing of needles, attenuation of fine-root biomass, and reduced growth rates.

The causes of forest declines have not been clearly identified. However, several causes have been offered: climate, disease, insects, natural senescence, community dynamics, and other forms of atmospheric pollution, such as ozone. Johnson et al. (1986) found that more than 40 cases of red spruce die-offs have been reported since the early 1800s. Some of these cases were attributed to drought, outbreaks of spruce budworms and European sawflies, changes in winter temperatures, and unusually hot summers. However, airborne acidic compounds and heavy metals in soils were 3 to 8 times greater above 1000 m elevation, where most of the problems are now occurring, than below 800 m. Red spruce seedlings are more sensitive to low pH and elevated Al than are several other species of spruces and pines and are also affected by ozone at ambient levels.

Sparling (1995) summarized the several hypotheses that had been advanced to explain diebacks of conifers in North America and central Europe: (1) Aluminum antagonizes Ca uptake by roots and Ca transport in stems of seedlings; (2) complex high-elevation disease characterized by a combined effect of ozone, acidic deposition, and nutrient deficiencies; (3) red-needle disease of spruce in Germany caused by a parasitic foliant fungus that may be facilitated by ozone, NO_x, or sulfur dioxide; (4) general stress syndrome explained by reduced photosynthetic activity and growth by a combination of atmospheric pollutants; (5) increases in organic air pollutants such as aniline, ethylene, or dinitrophenol, which can reduce the growth of trees; and (6) excess N deposition, primarily in the form of ammonium or nitrates. This last process can exert multiple effects including (a) greater-than-normal growth of biomass may result in Mg, P, or K deficiencies; (b) direct inhibition on mycorrhizae or cuticle formation; (c) increased vulnerability of trees to pathogens or pests; and (d) impaired normal growth configuration of the trees, particularly stem/root ratios.

6 Regulatory Aspects of Contamination

6.1 Defining Background vs. Baseline Limits

The formulation of legislative guidelines pertinent to contamination of environmental media requires clear understanding and determination of certain parameters, such as background and baseline values. While established regulatory limits for safe levels of elements in drinking water and foodstuffs are available (in the United States, this is provided by the U.S. EPA and the U.S. Food and Drug Administration), those for soils, plants, rocks, and sediments are lacking. Thus, to determine whether a plant, soil, rock, or sediment contains a *high* or *unusual* quantity of a specific contaminant, it is necessary to determine what quantity is *normal* or *usual*. These levels are referred to as *background* or *baseline* measurements, but they are essentially different.

A *background* measurement represents natural levels of an element in natural media without human influence, representing an idealized situation, and is usually more difficult to measure than a baseline. True backgrounds may be obtained by sampling and analyzing items practically free of anthropogenic influence, such as deep glacial ice cores, tree rings, or layers of sediments from lake or ocean floors. An exotic background sample would be soil or geological material underneath the Great Wall of China.

A *baseline* measurement represents levels measured at some point in time and is generally not a true background. Baseline concentrations are usually expressed as a range, not a single value. A baseline is computed as two deviations about the mean. Two deviations include about 95% of the area of the distribution under a normal frequency curve. Thus, a baseline value for a contaminant may represent the upper limit of the normal range of concentration. In formulating regulatory or guideline levels, it is necessary to accurately determine the background levels, considering spatial heterogeneity so that unrealistically low values can be avoided.

Technically speaking, background levels of metals in soils are related to the parent (rock) material in the lower portion of the soil profile. Anomalous levels of metals in soil due to unusual influence by the parent material may interject some confusion to these definitions. A good example is displayed on Figure 4.8 for As depicting three scenarios: noncontaminated, anthropogenically contaminated, and geogenically-enriched soil profile. The soil is located in lower Austria, near a mining area. Because the geogenically enriched values far exceed the established baseline values, one may conclude a severe contamination. But the fact that these values also far exceed the anthropogenic values indicates that the unusually high values were induced naturally by geologic phenomena. Similar conditions may exist in areas where serpentine rocks predominate.

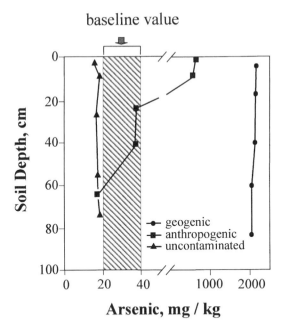

baseline value

FIGURE 4.8. Example of background, baseline, and geogenically enriched soil profile for arsenic. (Courtesy of Professor W. Wenzel, Universität für Bodenkultur, Vienna.)

6.2 Guideline Values for Metals

Since water quality issues predate the ones for soil quality, the development of guidelines for drinking water quality will be used here as a model. Approaches used to set the maximum permissible contaminant levels (MCLs) for drinking water are similar worldwide and the numerical limits for many pollutants are identical in many drinking water quality regulations due to the following: the mode of exposure via drinking water is unambiguous (direct ingestion); the exposed population is well defined; the dose–response relationships for the chemicals are universal, and the performance of water treatment processes is relatively consistent throughout the world.

The primary aim of the *Guidelines for Drinking Water Quality* (WHO, 1984) (Appendix Table A.21) is the protection of public health and thus the elimination, or reduction to a minimum, of constituents of water that are known to be hazardous to the health and well-being of the community. The following are the relevance of guideline values pertinent to drinking water quality:

1. A guideline value represents the level (a concentration) of a constituent that ensures an aesthetically pleasing water and does not result in any significant risk to the health of the consumer.

2. The quality of water defined by the *Guidelines for Drinking Water Quality* is such that it is suitable for human consumption and for all usual domestic purposes, including personal hygiene. However, water of a higher quality may be required for some special purposes, such as renal dialysis.

3. When a guideline value is exceeded this should be a signal: (i) to investigate the cause, with a view to taking remedial action; and (ii) to consult with public health authorities for advice.

4. Although the guideline values describe a quality of water that is acceptable for lifelong consumption, the establishment of these guidelines should not be regarded as implying that the quality of drinking water may be degraded to the recommended level. Indeed, a continuous effort should be made to maintain drinking water quality at the highest possible level.

5. The guideline values specified have been derived to safeguard health on the basis of lifelong consumption. Short-term exposures to higher levels of chemical constituents, such as might occur following accidental contamination, may be tolerated but need to be assessed case by case, taking into account, for example, the acute toxicity of the substance involved.

6. Short-term deviations above the guideline values do not necessarily mean that the water is unsuitable for consumption. The amount by which, and the period for which, any guideline value can be exceeded without affecting public health depends on the specific constituent involved.

It is recommended that, when a guideline value is exceeded, the surveillance agency (usually the authority responsible for public health) should be consulted for advice on suitable action, taking into account the intake of the substance from sources other than drinking water (for chemical constituents), the likelihood of adverse effects, the practicability of remedial measures, and similar factors.

7. In developing national drinking water standards based on these guidelines, it will be necessary to take account of a variety of local geographical, socioeconomic, dietary, and industrial conditions. This may lead to national standards that differ appreciably from the guideline values.

8. In the case of radioactive constituents, the term guideline value is used in the sense of *reference level* as defined by the International Commission on Radiological Protection (ICRP). Reference levels may be established for any of the quantities determined in the course of radiation protection programs, whether or not there are limits for these quantities. A reference level is not a limit and is used to determine a course of action when the value of a quantity exceeds or is predicted to exceed the reference level (ICRP, 1977).

In arriving at the guideline values for various substances in water, the total intake from air, food, and water for each constituent is taken into consideration from the best information available.

For the majority of the substances for which guideline values are proposed, the toxic effect in man is predicted from studies with laboratory animals (WHO, 1984). The accuracy and reliability of a quantitative prediction of toxicity in man from animal experimentation depend upon a number of factors, e.g., choice of animal species, design of the experiment, and extrapolation methods.

Information on the toxicity of chemicals is obtained from research in which the adverse effect occurs at considerably higher dosages than would be experienced in humans. When extrapolating from such animal data to humans, therefore, a safety factor must be introduced to provide for the unknown factors involved. The current doubts concerning both the biological and the mathematical reliability of methods of extrapolating from high doses to low doses necessitate the use of somewhat arbitrary safety factors, such as reduction by a factor of 100 or 1000.

These uncertainties arise from the nature of the toxic effects and the quality of the toxicological information. Other considerations are the size and type of the population to be protected and thus, under certain conditions, safety factors (or uncertainty factors) as high as 1000 may be necessary. However, assessment of the health risk to the population involves more than routine application of safety factors, and it must be emphasized that, strictly speaking, the extrapolation from animal experimentation applies only to the conditions of the particular experiment.

The existing methods of extrapolation from animal data to humans deal with exposures to single substances, whereas, in the human environment, a large number of hazardous chemicals and other factors may interact. In the special case of constituents possessing carcinogenic properties, a risk factor in arriving at the proposed guideline value is used. Owing to the considerable uncertainties in the available evidence, the proposed guideline values are, in many cases, deliberately cautious in character and, therefore, must not be interpreted as standards.

It should be reemphasized that guideline values are not strict standards that must be adhered to, but are subject to a wide range of flexibility and are provided essentially in an endeavor to protect public health and enable a judgment to be made regarding the provision of drinking water of acceptable quality. Thus, according to WHO's definition, the main purpose of developing water quality guidelines is to provide a basis for the development of standards based on the context of prevailing national or regional environmental, social, economic, and cultural conditions.

For example, the guideline (i.e., MCL) value of 0.05 mg L^{-1} for As set by the U.S. EPA for drinking water (i.e., 50 ppb As) has to be changed in view of widespread As poisoning in West Bengal, India, and Bangladesh from drinking water (see later chapter on arsenic). A good example of established regulatory standards to protect soil quality is the one for sludged soils, where, because of differences in scientific and sociopolitical considerations, disparity on contaminant levels in soils as well as loading rates exists between the Americans and Europeans. More interestingly, variations even among the EU members exist (Appendix Table A.19).

For soil protection purposes, environmental legislation relevant to concentrations of metals and organic compounds in soil has been formulated in several countries. Because public perception and political consideration, in addition to the scientific database, also play a role in such legislation, guidelines or standards that are fragmented and inconsistent are usually the norm. Such is the case within the EU and even among the states in the United States. Among developed countries, U.S. standards for acceptable levels of metals and organic contaminants are by far the most lax according to a 1999 report by Cornell University's Waste Management Institute (ES&T News, 2000).

Because of severity of contamination and potential risk to the public and the environment, several countries in North America and Europe, and Australia have adopted legislation for soil protection and remediation purposes. As will be discussed in Chapter 6, usually two to three tiers of values are established: soil protection (or reference) values, soil cleanup (or action, or intervention) values, and in between are the soil investigation (or threshold, or tolerable) values. The reference values sometimes adopt background values such as those displayed in Appendix Table A.22, but policymakers are more commonly adopting the baseline values. The reference value is defined as the maximum soil concentration of a contaminant that allows unlimited multifunctional use of the land. For example, the reference values for selected metals in soils in the United States are based on the 95th percentile from the frequency distribution (Holmgren et al., 1993; Ryan and Chaney, 1997). Similarly, in Germany, the 90th percentile values are used for mobile (NH_4NO_3- extractable) trace elements in soil as shown in Appendix Table A.23 (Pruess, 1995). Because of the importance of certain soil factors, values are computed or adjusted according to pH, OM content, and clay content of the soil (Adriano et al., 1997).

References

Aastrup, M.J., et al. 1991. *Water Air Soil Pollut* 56:155–167.

Adriano, D.C. 1986. *Trace Elements in the Terrestrial Environment.* Springer-Verlag. New York.

Adriano, D.C., and M. Havas, eds. 1989. *Acidic Precipitation, vol 1: Case Studies.* Springer-Verlag, New York.

Adriano, D.C., and A.H. Johnson, eds. 1989. *Acidic Precipitation, vol 2: Biological and Ecological Effects.* Springer-Verlag, New York.

Adriano, D.C., A. Chlopecka, D.I. Kaplan, H. Clijsters, and J. Vangronsveld. 1997. In R. Prost, ed. *Soil Contamination.* INRA 85. INRA, Paris.

Adriano, D.C., A. Chlopecka, and D.I. Kaplan. 1998. In P.M. Huang et al., eds. *Soil Chemistry and Ecosystem Health.* SSSA Spec Publ 52. Soil Sci Soc Am, Madison, WI.

Allen, J.C. 1985. *Biotropica* 17:15–27

Andelman, J.B., and D.W. Underhill. 1990. *Health Effects from Hazardous Waste Sites.* Lewis Publ, Chelsea, MI.

Anderson, M.A., P.M. Bertsch, and L.W. Zelazny. 1993. In R.F. Keefer and K.S. Sajwan, eds. *Trace Elements in Coal and Coal Combustion Residues.* Lewis Publ, Boca Raton, FL.

Angelone, M., and C. Bini. 1992. In D.C. Adriano, ed. *Biogeochemistry of Trace Metals.* Lewis Publ, Chelsea, MI.

Artiola, J.F. 1996. In I.L. Pepper, C.P. Gerba, and M.L. Brusseau, eds. *Pollution Science.* Academic Pr, New York.

Bar-Yosef, B., N.J. Barrow, and J. Goldshmid, eds. 1991. *Inorganic Contaminants in the Vadose Zone.* Springer-Verlag, Berlin.

Barth, H., and P. L'Hermite. 1987. *Scientific Basis for Soil Protection in the European Community.* Elsevier, Essex, United Kingdom.

Bartnicki, J. 1996. *Water Air Soil Pollut* 92:343–374.

Bell, P.F., B.R. James, and R.L. Chaney. 1991. *J Environ Qual* 20:481–486.

Berti, W.R., and L.W. Jacobs. 1998. *J Environ Qual.* 27:1280–1286.

Bjorklund, I., H. Borg, and K. Johansson. 1984. *Ambio* 13:118–121.

Bodaly, R.A., and R.E. Heck. 1979. In *Post Impoundment Increases in Fish Mercury Levels in the Southern Indian Lake Reservoir Manitoba.* Fisheries Marine Serv Manus Rep 1531. Dept. Fisheries Environment, Canada.

Bodaly, R.A., R.E. Heck, and R.J.P. Fudge. 1984. *Can J Fish Aquat Sci* 41(4):682–691.

Bode, H. 1998. Ruhrverband, Essen, Germany. Communication via. P. Matthews, Anglian Water Co, United Kingdom.

Bolt, G.H., and M.G.M. Bruggenwert, eds. 1976. *Soil Chemistry: A Basic Element.* Elsevier, Amsterdam.

Boucher, R., and R. Schetagne. 1983. *Repercussions de la Mise en eau des Reservoirs de la Grande 2 et Opinaca sur la Concentration du Mercure dans les Poissons.* Direction de L'Environment, Quebec, Canada.

Brannvall, M.J., R. Bindler, I. Renberg, O. Emteryd, J. Bartnicki, and K. Billstrom. 1999. *Environ Sci Technol* 33:4391–4395.

Brouzes, R.J., R.A. McLean, and G. Tomlinson. 1977. *The Link Between the pH of Natural Waters and the Mercury Content of Fish.* Domtar Res Centre Report, May 3, 1977.

Cairney, T., ed. 1993. *Contaminated Land: Problems and Solutions.* Lewis Publ, Boca Raton, FL.

Carlson, C.L., and D.C. Adriano. 1993. *J Environ Qual* 22:227–247.

Carroll, J.J., S.J. Ellis, and W.S. Oliver. 1979. *Bull Environ Contam Toxicol* 22: 575–581.

Chang, A.C., A.L. Page, and F.T. Bingham. 1982. *J Environ Qual* 11:705–708.

Chang, A.C., A.L. Page, and J. E. Warneke. 1987. *J Environ Qual* 16:217–221.

Charles, D.J., ed. 1991. *Acidic Deposition and Aquatic Ecosystems.* Springer-Verlag. New York.

Chen, J., F. Wei, Y. Wu, and D.C. Adriano. 1991. *Water Air Soil Pollut* 57–58:699–712.

Davies, B.E. 1992. In D.C. Adriano, ed. *Biogeochemistry of Trace Metals*. Lewis Publ, Boca Raton, FL.

Davies, M.C.R., ed. 1991. *Land Reclamation*. Elsevier, Essex, United Kingdom.

Davies, P.H., J.P. Goettl, J.R. Sinley, and N. Smith. 1976. *Water Res* 10:199–206.

D'Itri, P.A., and F.M. D'Itri. 1977. *Mercury Contamination: A Human Tragedy*. Wiley, New York.

Dixon, J.B., and S.B. Weed, eds. 1989. *Minerals in the Environment*, 2nd ed. Soil Sci Soc Am, Madison, WI.

Driscoll, C.T., et al. 1991. In D.F. Charles, ed. *Acidic Deposition and Aquatic Ecosystems*. Springer-Verlag, New York.

Dowdy, R.H., and V.V. Volk. 1983. In *Chemical Mobility and Reactivity in Soil Systems*, SSSA Spec Publ 11, Soil Sci Soc Am, Madison WI.

Dowdy, R.H., J.J. Latterrell, T.D. Hinelsy, and D.L. Sullivan. 1991. *J Environ Qual* 20:119–123.

Eijackers, H.J.P., and T. Hamers, eds. 1993. *Integrated Soil and Sediment Research: A Basis for Protection*. Kluwer, Dordrecht, Netherlands.

Eisenbud, M. 1987. *Environmental Radioactivity*. Academic Pr, Orlando, Fl.

Ekstrom, G., and M. Akerblom. 1990. *Rev Environ Contam Toxicol* 114:23–55.

Eney, A.B., and D.E. Pedzold. 1987. *Environmentalist* 7:95.

[ES&T] Environmental Science and Technology News. 2000. June issue.

Fuller, W.H., and A.W. Warrick. 1985. *Soils in Waste Treatment and Utilization*, vol. 1 and 2. CRC Pr, Boca Raton, FL.

Greenwood, D.J., P.H. Nye, and A. Walker. 1990. *Quantitative Theory in Soil Productivity and Environmental Pollution*. Royal Society, London.

Hakansson, L. 1980. *Environ Pollut* (B) 1:285–304.

Harrison, R.M. 1990. *Pollution: Causes, Effects, and Controls*. Royal Soc Chem, Cambridge.

Hansen, P.E., and S.E. Jorgensen, eds. 1991. *Introduction to Environmental Management*. Dev in Environ Modeling 18. Elsevier, Amsterdam.

Heckman, J.R., J.S. Angle, and R.L. Chaney. 1987. *J Environ Qual* 16:113–117.

Hinesly, T.O., R.L. Jones, E.L. Ziegler, and J.J. Tyler. 1977. *Environ Sci Technol* 11:182–188.

Holdgate, M.W. 1979. *A Perspective of Environmental Pollution*. Cambridge Univ Pr, Cambridge.

Holmgren, G.S.S., M.W. Meyer, R.L. Chaney, and R.B. Daniels. 1993. *J Environ Qual* 22:335–348.

Hornbeck, J.W., R.B. Smith, and C.A. Federer. 1986. *Water Air Soil Pollut* 31:425.

Howarth, R.S., and J.P. Sprague. 1978. *Water Res* 12:455–462.

Husar, R.B., T.J. Sullivan, and D.F. Charles. 1991. In D.F. Charles, ed. *Acidic Deposition and Aquatic Ecosystems*. Springer-Verlag, New York.

[ICRP] International Commission on Radiological Protection. 1977. ICRP Pub 26. Amm ICRP 1(3):1–53.

Irwin, J.G. 1989. *Arch Environ Contam Toxicol* 18:95.

Jeffries, D.S. 1991. In D.F. Charles, ed. *Acidic Deposition and Aquatic Ecosystems*. Springer-Verlag, New York.

Jernelov, A., L. Landner, and T. Larsson. 1975. *J Water Pollut Control Fed* 47:810–822.

Johnson, A.H., A.J. Friedland, and J.G. Dushoff. 1986. *Water Air Soil Pollut* 30:319.

Jones, K.C. 1991. *Environ Pollut* 69:311–325.

Juste, C., and M. Mench. 1992. In D.C. Adriano, ed. *Biogeochemistry of Trace Metals.* Lewis Publ, Boca Raton, FL.

Kloke, A., D.R. Sauerbeck, and H. Vetter. 1984. In J.O. Nriagu, ed. *Changing Metal Cycles and Human Health* (Dahlem Komferenzer). Springer-Verlag, Berlin.

Kohsiek, L.H.M., R. Franken, et al. 1994. In M. Donker et al., eds. *Ecotoxicology of Soil Organisms,* CRC Pr, Boca Raton, FL.

Krabbenhoft, D.P., et al. 1995. *Water Air Soil Pollut* 80:425–433.

Kryshev, I.I., ed. 1992. *Radioecological Consequences of the Chernobyl Accident.* Nuclear Soc Intl, Moscow.

Lal, R. 1998. In P.M. Huang, D.C. Adriano, T.J. Logan, and R. Checkai, eds. *Soil Chemistry and Ecosystem Health.* SSSA 52. Soil Sci Soc Am, Madison, WI.

Landers, D.H., W.S. Overton, R.A. Linthurst, and D.F. Brakke. 1988. *Environ Sci Technol* 22:128.

Landner, L., M.O. Oberg, and M. Aringberg-Laavatza. 1995. In *Cadmium in Fertilizers* (OECD workshop). Stockholm, Sweden.

Lee, Y.H., and A. Iverfeldt. 1991. *Water Air Soil Pollut* 56:309–321.

Lindqvist, O., A. Jernelov, K. Johansson, and H. Rodhe. 1984. *Mercury in the Swedish Environment.* Nat Swedish Env Protect Bd Rep SNV PM 1846.

Longhurst, J.W.S., ed. 1990. *Acid Deposition: Origins, Impacts and Abatement Strategies.* Springer-Verlag, New York.

Matthews, P. 1999. Anglian Water Intl Ltd, Cambridgeshire, United Kingdom, personal communication.

McBride, M.B. 1994. *Environmental Chemistry of Soils.* Oxford Univ. Pr, New York.

McFarlane, G.A., and W.G. Franzin. 1980. *Can J Fish Aquatic Sci* 37:1573–1578.

McGrath, S.P., and P.W. Lane. 1989. *Environ Pollut* 60:235–256.

McLaughlin, M.J., K.G. Tiller, R. Naidu, and D.P. Stevens. 1995. *Aust J Soil Res* 34:1–54.

Meister, J.F., J. Di'Nunzio, and J.A. Cox. 1979. *J Am Water Works Assoc* 71:574–576.

Moldan, B., and J.L. Schnoor. 1992. *Environ Sci Technol* 26:14–21.

Mortvedt, J.J., et al., eds. 1991. *Micronutrients in Agriculture.* SSSA 4. Soil Sci Soc Am, Madison, WI.

Mulla, D.J., A.L. Page, and T.J. Ganje. 1980. *J Environ Qual* 9:408–412.

Munson, R.K., and S.A. Gherini. 1991. In D.F. Charles, ed. *Acidic Deposition and Aquatic Ecosystems.* Springer-Verlag. New York.

Nakles, D.V., D.G. Linz, and T.D. Hayes. 1993. In *Management of Manufactured Gas Plant Sites Technology Seminar.* Remed Technol. Inc., Arlington, VA.

[NAPAP] National Acidic Precipitation Assessment Program. 1991. *1990 Integrated Assessment Report.* Natl Acid Precip Assess Prog, Washington, DC.

Nater, E.A., and D.F. Grigal. 1992. *Nature* 358:139–141.

Newman, M.C., and A.W. McIntosh, eds. 1991. *Metal Ecotoxicology: Concepts and Applications.* Lewis Publ, Chelsea, MI.

Norstrom, R.J., A.F. McKinnon, and A.S.W. Defreitas. 1976. *J Fish Res Bd Can* 33:248–267.

Ormernik, J.M., G.E. Griffith, and A.J. Kinney. 1985. In *Total Alkalinity of Surface Waters.* U.S. Environmental Protection Agency (EPA) Corvallis Environ Res Lab, Corvallis, OR.

Overcash, M. 1996. *Crit Rev Environ Sci Technol* 26:337–368.

Pacyna, J.M. 1995. *Sci Total Environ* 160–161:39–53.

Page, A.L., T.L. Gleason, J.E. Smith, I.K. Iskandar, and L.E. Sommers, eds. 1983. *Utilization of Municipal Wastewater and Sludges on Land*. Univ California, Riverside, CA.

Phillips, G.R., and D.R. Buhler. 1978. *Trans Am Fish Soc* 107:853–861.

Phillips, G.R., and R.W. Gregory. 1979. *J Fish Res Bd Can* 36:1516–1519.

Potter, L., D. Kidd, and D. Standiford. 1975. *Environ Sci Technol* 9:41–46.

Powe, J.F. 1992. In *International Symposium in Nutrient Management for Sustainable Production*. Punjab Agric Univ Pr, Ludiahana, India.

Pruess, A. 1995. In R. Prost, ed. *Contaminated Soils*. INRA 85. INRA, Paris.

Racke, K.D., and J.R. Coats, eds. 1990. *Enhanced Biodegradation of Pesticides in the Environment*. ACS Symp. Ser. 426, American Chemical Society, Washington, DC.

Renberg, I., M.W. Persson, and O. Emteryd. 1994. *Nature* 368:323–326.

Reuss, J.O., and D.W. Johnson. 1986. *Acid Deposition and Acidifications of Soils and Waters*. Springer-Verlag, New York.

Rhyner, C.R., L.J. Schwartz, R.B. Wenger, and M.G. Kohrell. 1995. *Waste Management and Resource Recovery*. Lewis Publ, Boca Raton, FL.

Rodgers, D.W., and F.W. Beamish. 1982. *Aquat Toxicol* 2:271–290.

Rothbaum, H.P., R.L. Gognel, A.E. Johnston, and G.E.G. Mattingly. 1986. *J Soil Sci* 37:99–107.

Ryan, R., and R.L. Chaney. 1997. In R. Prost, ed. *Contaminated Soils*. INRA 85. INRA, Paris.

Salomons, W., and U. Forstner. 1988. *Chemistry and Biology of Solid Waste-Dredged Material and Mine Tailings*. Springer-Verlag, Berlin.

Schroeder, H.A., and J.J. Balassa. 1963. *Science* 140:819–820.

Shacklette, H.T., and J.G. Boerngen. 1984. In U.S. Geological Survey (USGS) Prof Paper 1270. USGS, Washington, DC.

Shotyk, W., D. Weiss, P.G. Appleby, A.K. Cheburkin, and W.O. Vander Knaap. 1998. *Science* 281:1635–1640.

Sloan, J.J, R.H. Dowdy, and M.S. Dolan. 1998. *J Environ Qual* 27:1312–1317.

Smith, F.A., R.P. Sharma, R.I. Lynn, and J.B. Low. 1974. *Bull Env Toxicol* 12:153–157.

Soon, Y.K., T.E. Bates, and J.R. Moyer. 1980. *J Environ Qual* 9:497–504.

Sparling, D.W. 1995. In D.J. Hoffman, B.A. Rattner, G.A. Burton, Jr., and J. Cairns, Jr., eds. *Handbook of Toxicology*. Lewis Publ, Boca Raton, FL.

Sposito, G. 1989. *The Chemistry of Soils*. Oxford Univ Pr, New York.

Spry, D.J., C.M. Wood, and P.V. Hodson. 1981. In *Effect of Environmental Acid on Freshwater Fish with Particular Reference to the Softwater Lakes in Ontario, and the Modifying Effects of Heavy Metals*. Tech Rep Can Fish Aquat Sci 999.

Steenhuis, T.S., M.B. McBride, B.K. Richards, and E. Harrison. 1999. *Environ Sci Technol* 33:1171–1174.

Stevenson, F.J. 1982. *Humus Chemistry: Genesis, Composition, Reactions*. Wiley, New York.

Stigliani, W.M. 1988. *Environ Monitor Assess* 10:245–307.

Stigliani, W.M., P.R. Jaffe, and S. Anderberg. 1993. *Environ Sci Technol* 27:786–793.

Stigliani, W.M, and S. Anderberg. 1994. *Industrial Metabolism-Restructuring for Sustainable Development*. UN Univ Pr, Tokyo.

Stokes, P.M., and C.D. Wren. 1987. In T.C. Hutchinson and K.M. Meema, eds. *Lead, Mercury, Cadmium and Arsenic in the Environment.* SCOPE 31. Wiley, New York.

Summer, M. 1991. In G.H. Bolt et al., eds. *Interactions at the Soil Colloid-Soil Solution Interface,* Kluwer, Dordrecht, Netherlands.

Sumner, M.E., and W.P. Miller. 1992. *Am J Altern Agric* 7:56–62.

Suter, G.W., R.J. Luxmoore, and E. D. Smith. 1993. *J Environ Qual* 22:217–226.

Swain, E.B., et al. 1992. *Science* 257:784–787.

Tolba, M.K., and O.A. El-Kholy. 1992. *The World Environment, 1972–1992.* Chapman & Hall, London.

U.S. Salinity Laboratory Staff. 1954. *Diagnosis and Improvement of Saline and Alkali Soils.* USDA Handb. 60. U.S. Dept Agriculture, Washington, DC.

U.S. EPA. United States Environmental Protection Agency. 1990. *Fed. Register* 55(126):40 CFR, parts 261–302.

U.S. EPA. 1992. *Characterization of Municipal Waste in the US: 1992 Update.* EPA 530-R-92/019. EPA, Washington, DC.

U.S. EPA. 1993. *Fed. Register* 58:9284–9415.

U.S. EPA. 1995. In *Contaminants and Remedial Options at Selected Metal-contaminated Sites.* EPA 540. R-95/512. Washington, DC.

USGrev. 1993. *Botschaft zu einer Änderung des Bundesgesetzes über den Umwelts-chutz.* Eidg Drucksachen und Materialzentrale, Bern.

Van Driel, W., and K.W. Smilde. 1990. *Fert Res* 25:115–126.

Waite, D.T., G.W. Dunn, and R. J. Stedwill. 1980. In *Mercury in Cookson Reservoir (East Poplar River).* Sask Env Report, Saskatoon.

[WHO] World Health Organization. 1984. *Guidelines for Drinking-Water Quality, vol. 1: Recommendations.* WHO, Geneva.

Williams, C.H., and D.J. David. 1976. *Soil Sci* 121:86–93.

Williams, D.E., J. Vlamis, A.H. Pukite, and J.E. Corey. 1985. *Soil Sci* 140:120–125.

Williams, P.H. 1992. In *Proc Int Symp Nutrient Management for Sustainable Production.* Punjab Agric Univ Pr, Ludiahana, India.

[World Com. Environ. Develop.] World Commission on Environment and Development. 1987. *Our Common Future.* Oxford Univ Pr, Oxford.

[World Res. Instit.] World Research Institute 1992. *World Resources 1992–93.* Oxford Univ Pr, New York.

Wren, C.D., and H.R. MacCrimmon. 1983. *Can J Fish Aquat Sci* 10:1737–1744.

Wren, C.D., S. Harris, and N. Harttrup. 1995. In D.J. Hoffman et al., eds. *Handbook of Ecotoxicology.* Lewis Publ, Boca Raton, FL.

Yingming, L., and R.R. Corey. 1993. *J Environ Qual* 22:1–8.

Zitko, V., and W.G. Carson. 1976. *Chemosphere* 5:299–303.

5
Ecological and Health Risks of Metals

Some elements are paradoxically intriguing! Too little of them may not be enough for proper nutrition but a little bit more might be too much! This indicates that some elements may serve as a nutrient or a toxin.

The dissemination of metals in the environment from industrial activities ensures that a significant portion of the population is exposed to these contaminants. Depending on the degree of exposure and environmental setting, a fraction of the population may be at risk. Another fraction of the population may be at risk due to occupational exposure. Toxicological effects of metals to humans, particularly those of Cd, As, Hg, and Pb, have been well documented (see also individual chapters on these elements later in this book). Their ecotoxicological effects on agricultural, forest, and aquatic ecosystems are also well documented.

TABLE 5.1. Category of elements according to their environmental and toxicological relevance.

Category	Ion
Class A	Li^+, Na^+, K^+, Cs^+, Ca^{2+}, Ba^{2+}, Sr^{2+}, Mg^{2+}, Be^{2+}, and Al^{3+}
Class B	Ag^+, Au^+, Tl^+, Hg^{2+}, Bi^{2+}, Tl^{2+}, and Pb^{2+}
Borderline	Mn^{2+}, V^{2+}, Zn^{2+}, Ni^{2+}, Fe^{2+}, Co^{2+}, Cd^{2+}, Cu^{2+}, Sn^{2+}, Cr^{3+}, Fe^{3+}, As^{3+}, Sn^{3+}, and Sb^{4+}

Source: Extracted from Nieboer and Richardson (1980).

Trace elements may be grouped into three categories (Table 5.1) according to their biological and chemical reactivities, which in turn are based on their physicochemical properties. Elements belonging to class B (nitrogen/sulfur-seeking) may be expected to be more toxic than those in the borderline category which, in turn, are more toxic than the ones in class A (oxygen-seeking) (Nieboer and Richardson, 1980). The elements in class B and, to a lesser extent, the elements in the borderline (intermediate) category, are likely to be more bioreactive, resulting in biotoxicities, because of their greater tendencies to form more mobile (cation/ligand) complexes. Biotoxicity induced through metal exposure is the consequence of (bio)chemical reactions that disrupt biological processes by blocking the functional groups of biomolecules, displacing essential trace element ions in biomolecules, or modifying the active configuration of such molecules.

Biotoxicity can be induced directly on organisms through direct exposure to the toxic chemicals (e.g., in occupational settings via inhalation or dermal contact) or somewhat indirectly through indirect exposure to the chemicals (e.g., through food chain transfer and eventual ingestion). The main sources of exposure for humans are listed in Table 5.2, which indicates that exposure from occupational setting and food intake are the main routes for

TABLE 5.2. Main intake routes of trace elements to humans.

Element	Source					
	Occupational	Food	Water	Air	Dust	Smoke
Antimony	x	x				
Arsenic	x	x	xx^a		✓	✓
Cadmium	x	xx			✓	✓[b]
Chromium	x	x				
Lead	x	x	✓	✓	xx^c	✓[b]
Mercury	x	xx		✓		
Selenium	x	x	✓			✓
Thallium	x	x				
Zinc	x	x				

x = Major source, ✓ = Minor source.
[a]From drinking water.
[b]From cigarette smoking.
[c]For children.

most metals. There are no fewer than 45 different metals used in industrial processes, which may lead to their exposure to humans (Elinder, 1984). Food intake can potentially expose a much larger segment of the population, depending on the locality and food distribution pattern. For certain elements, the major source may vary according to circumstances. For example, drinking well water is the major route for As for millions of inhabitants in West Bengal (India), Bangladesh, and other As hot spots in the world.

Ingestion of Pb-laden dust or pica in inner cities by young children is reportedly the most important route for Pb (see later chapter on lead).

1 Metabolism and Toxicity of Essential vs. Nonessential Metals

Iron, Zn, and Cu are of considerable clinical interest (Elinder, 1984). Iron and Zn deficiencies in humans are rather common, especially in developing countries. Other essential elements (Mn, Se, Cr, Co, and Mo) receive less attention than the three above, with the exception of Se via Se-enriched drainage effluents (e.g., in the Central Valley, California) and groundwater. The body burdens and concentrations of essential metals in biological fluids are maintained at optimal levels through homeostasis, rendering humans less susceptible to long-term toxicity from ingestion of these metals.

The nonessential metals that have attracted the most attention worldwide are Cd, Pb, Hg, and As. Cadmium enters the body via ingestion and inhalation (see Table 5.2). Exposure via inhalation occurs primarily under occupational setting. Smokers are liable to additional exposure, since cigarettes and cigars contain Cd, which is inhaled while smoking. For the rest of the population, food ingestion is the major source of exposure. Fortunately however, the gastrointestinal (GI) tract absorption for Cd is limited to only ~6% (Table 5.3). Metallothionein, a low-molecular-weight and S-rich protein, plays a key role in the metabolism of Cd and other metals (Foulkes, 1982; Hutton, 1987). Cadmium has a very strong affinity for this protein and it is thought to play a key role in the bioaccumulation of Cd, especially in kidney and liver (Hutton, 1987). Once in the body, the elimination of Cd is very slow, with a biological half-life of >10 yr. The kidney is considered to be the critical organ for long-term Cd exposure and is the organ that first reveals signs of toxicity. One of the early signs of renal dysfunction is an increased urinary excretion of protein. Cd-induced proteinuria is generally considered to be characterized by the excretion of low-molecular-weight proteins, in particular α_2-, β_2-, and γ-globulins.

Most of the people presently at risk from environmental Cd poisoning are believed to be in Japan and central Europe, where the soils have become contaminated with Cd from various sources (Asami, 1997; Kloke et al., 1984). Emerging sites that can potentially offer some risk due to Cd input

TABLE 5.3. Major metabolic factors associated with environmental exposure to lead, mercury, cadmium, and arsenic (see also Chapter 3 on bioavailability).

	Arsenic	Cadmium	Lead[a]	Mercury[b] (organic)
Major routes of entry	Ingestion	Ingestion and inhalation (tobacco)	Ingestion and inhalation	Ingestion
Gastrointestinal absorption (%)	>80	~6	~10	~95
Organs of accumulation	Keratinous tissue	Kidney and liver	Bone, kidney, and liver	Brain, liver, and kidney
Major routes of excretion	Urine	Urine	Urine	Feces
Biological half-life	10–30 hr	>10 yr	~20 yr	~70 days

[a]For infants, GI absorption values of 25–50% are commonly reported.
[b]For inorganic Hg, GI absorption values of <20% are commonly reported.
Sources: Ewers and Schlipkoter (1991); Ferguson (1990); Hutton (1987).

from phosphatic fertilizers and sewage sludge are scattered throughout the world, including the Rhine Basin in Europe (Stigliani et al., 1993).

Like Cd, Pb enters the body mainly via inhalation and ingestion. Exposure to Pb via inhalation is caused primarily by occupational sources. The most important routes of environmental exposure include ingestion of Pb-contaminated food, wine, dust, and paint chips. The last two are particularly important for young children. The pronounced toxic effects of Pb are manifested as dysfunction in the production of heme, in the nervous system, and the renal system. Lead also inhibits several enzymes, especially ALA-dehydratase (ALA = aminolevulinic acid), which is crucial in the normal production of heme. Sites of bioaccumulation in the human body are bone, kidney, and liver (Fig. 5.1). Its GI tract absorption efficiency is similar to Cd and it also has a long biological half-life of ~20 yr (see Table 5.3). In the United States, the EPA in the 1980s had concluded that "there is a national health problem associated with exposure to environmental Pb for the general population, and, in particular, pre-school children" (U.S. EPA, 1986). Thus, the Centers for Disease Control in Atlanta, Georgia now recommends that all children 9 mo to 6 yr of age be screened for Pb at least once yearly. On a global scale, the actual number of people exposed to elevated levels of environmental Pb, which can result in Pb concentrations of 20 µg dL^{-1} (200 µg L^{-1}) or more, is well over one billion (Nriagu, 1988). It shows significant transplacental transfer (Hutton, 1987). In short, Pb is the greatest cause of global public health concern.

Inorganic Hg vapor is chiefly an industrial concern. It is effectively absorbed via the lungs after inhalation, passing through the blood–brain barrier, and subsequently bioaccumulating in the brain (see Fig. 5.1). Toxic effects of inorganic Hg (also known as *mercurism*) include tremor and a

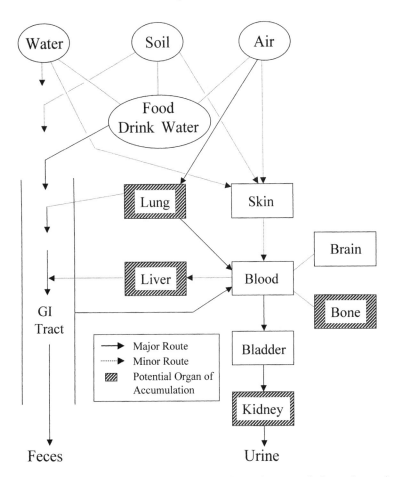

FIGURE 5.1. Exposure routes and potential sites of accumulation of metals in humans. Most of the body burden for lead is in bones. The brain has a high affinity for mercury.

mental disorder called erethism. The absorption of metallic Hg by the GI tract is virtually nil.

As already discussed in Chapter 2, inorganic Hg in the environment can be methylated, the form of Hg that bioaccumulates and biomagnifies in the aquatic food chain. Fish and seafood consumption constitutes the predominant route of human exposure to organic Hg. The GI absorption of methyl Hg is very efficient ($\sim$95%) and the residence time in the body is long ($\sim$70 days) (see Table 5.3). Its principal excretion route is via biliary excretion into the GI tract (Hutton, 1987). The brain is also the critical organ for exposure to organomercury compounds, but toxicity symptoms are different. Methylmercury intoxication is characterized by effects on the central nervous system, and the areas mainly affected are those associated with the sensory, visual, and auditory functions, and those involved in coordination (Hutton, 1987). The first sign of chronic organomercury poisoning is parasthesia. Later ataxia and deafness may develop. Organomercury can be transferred effectively from the placenta to the fetus (Hutton, 1987).

Since fish and seafood represent the primary exposure route for Hg, more specifically methyl Hg, most of the people likely at risk are those inhabiting small fishing villages and townships around the world.

The metabolism of As is rather complex. Several different species of inorganic and organic As exist, all of which appear to be metabolized differently (see later chapter on arsenic). Most soluble As compounds are effectively absorbed ($\geq 80\%$) by the GI tract and most of this As is rapidly eliminated via the urine. Chronic exposure to As may give rise to hyperkeratosis of the palms and the soles of the feet, together with hyperpigmentation, particularly in areas not exposed to the sun (Hutton, 1987). Skin tumors have also been commonly reported, often located on the hands and feet. The drinking of As-contaminated well water is accepted now as the most critical exposure of humans to As (see later chapter on arsenic). The recent outbreak of large-scale As poisoning in West Bengal (India) and Bangladesh from ingestion of contaminated groundwater implies that people residing in areas with As-contaminated groundwater are likely to be at risk (e.g., China, Chile, Argentina, Taiwan, India, Bangladesh, etc.).

A method of assessing the risk to the general population from exposure to a particular chemical is by making a comparison between the actual exposure and the exposure threshold for initial effects. This has been done for Cd, Pb, Hg, and As using data derived from dose–response analysis (see Table 5.4). Potential toxicity in humans from chronic exposure to Tl, Sn, V, and Sb has also been reported (Elinder, 1984).

2 Human Exposure to Metals

Metal contamination of the environment may lead to contamination of other media, potentially resulting in humans receiving a total dose via several exposure scenarios and through a number of exposure pathways (Chang et al., 1993; Ryan and Chaney, 1997).

The total dose represents the sum of doses resulting from each exposure pathway and may be calculated as:

$$Dose_{total} = Dose_{soil} + Dose_{food} + Dose_{water} + Dose_{air} + Dose_{dermal}$$

with units in $mg_{of\ metal} \cdot kg_{body\ wt}^{-1} \cdot day^{-1}$

where $Dose_{soil}$ = dose resulting from ingestion of contaminated soil and/or dust

$Dose_{food}$ = dose resulting from ingestion of contaminated food

$Dose_{water}$ = dose resulting from drinking contaminated water

$Dose_{air}$ = dose resulting from inhalation of contaminated air

TABLE 5.4. Estimated threshold values for chronic exposure to lead, methylmercury, cadmium, and inorganic arsenic.

Element	Effect	Estimated threshold	Reported values in the general population or background	Notes and authorities
Lead	ALAD[a] FEP	11 $\mu g\ dL^{-1}$, 15–20 $\mu g\ dL^{-1}$ (blood)	10–20 $\mu g\ dL^{-1}$ (blood)	Data refer to children. Value for ALAD is for a 10% prevalence rate
Methylmercury	Earliest signs and symptoms	20 $\mu g\ g^{-1}$ (hair), 80 $ng\ ml^{-1}$ (blood)	2 $\mu g\ g^{-1}$ (hair), 10 $ng\ ml^{-1}$ (blood)	Thresholds refer to nonpregnant adults. The threshold for the fetus is considered to be lower. Background values strongly influenced by fish consumption
Cadmium	β_2-m[b] cancer	150 $\mu g\ day^{-1}$ dietary intake	20–70 $\mu g\ day^{-1}$ dietary intake	Threshold refers to 50-yr daily intake needed to cause elevated urinary β_2-m in 10% of the population with average body weight of 70 kg
Arsenic (inorganic compounds)	Skin cancer	200 $\mu g\ L^{-1}$ (drinking water)?	2 $\mu g\ L^{-1}$ (drinking water)	Threshold refers to 5% prevalence after lifetime (70 yr) exposure

[a]ALAD = δ-aminolevulinic acid dehydratase activity.
[b]β_2-m = β_2-microglobulin.
Source: Data summarized by Hutton (1987).

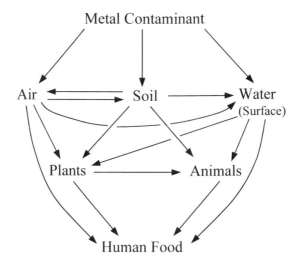

FIGURE 5.2. Routes of metal contaminants into human foods. (Modified from Niagu, 1990.)

Dose$_{dermal}$ = dose resulting from dermal absorption of chemicals from contaminated soil or water.

As already discussed above, the exposure from food (of both terrestrial and aquatic origin) consumption represents the most critical pathway for the general population (Fig. 5.2). Under certain situations, ingestion of metal-contaminated drinking water (as in India and Bangladesh), as in As, represents the most critical exposure pathway.

2.1 Exposure to Terrestrial Food Chain

The terrestrial human food chain, in general, is the most important exposure route of metals to the general population. The most common pathways for the terrestrial human food chain are (see also Chapter 6):

1. Contaminated soil → plant → human (more general)
2. Contaminated soil → plant → home gardener (relatively more isolated)
3. Contaminated soil → plant → animal → human (more general)
4. Contaminated soil → animal → human (more general)

The more common environmental settings for the above include sludged soils as well as home gardens in metropolitan areas. Sludged soils are given top priority in environmental regulation because of the ever-increasing popularity of sewage sludge as a soil amendment in North America and Europe. In addition, it is not uncommon to have high concentrations of potentially toxic metals in sewage sludge that can be transferred and bioaccumulated in food plants. Thus, a genuine threat to the health of humans exists.

The threat is relatively higher in areas where distribution of contaminated foodstuff is confined within the affected area, the extreme case being that for a home gardener. The threat can be minimized by following wise cultural/management practices that soil chemists have a menu for, such as the use of plant excluders of metals and regulating the pH and CEC of the soil to minimize plant uptake of metals (see Chapter 2). The threat from consumption of animal products mainly comes from ingestion of animal tissue, primarily kidney, liver, heart, and brain, that bioaccumulate metals. Milk, other dairy products, and carcass are seldom high in metal contents (except for radiocesium in dairy products).

The soil–animal (or sediment–animal) link can be a critical pathway for increasing the body burden of metals in animals that heavily rely on grazing. For example, ruminant animals grazing on grasses can have as much as 30% soil material in their diet (Table 5.5). This can lead to contamination in animal tissue in pastoral areas affected by sludge treatment or impacted by metal input from atmospheric deposition. Some wildlife species ingest substantial amounts of soil or sediment incidentally as they feed, exposing them to high concentrations of environmental contaminants (Beyer et al., 1994, 1998). The importance of this route of exposure has been demonstrated for mule deer (*Odocoileus hemionus*) ingesting metals and radioactive elements (Arthur and Alldredge, 1979), domestic animals feeding on sludge-treated fields contaminated with polychlorinated biphenyls (Fries, 1982), and wood ducks (*Aix sponsa*) ingesting Pb-contaminated sediment (Beyer et al., 1997). Ingestion of soil

TABLE 5.5. Estimates of percentage of soil-solids in the diet of various animal groups.

Animal type	Food type	% Soil in diet (DW)
Terrestrial feeding		
Carnivorous		
Mammals, birds	Mammals, birds, reptiles	4
Insectivorous		
Mammals, birds, reptiles	Insects, spiders	17
Herbivorous		
Mammals, birds, reptiles	Leaves, fruits, seeds	2
Rodents	Plant material	4
Mammals	Roots	10
Ruminants, macropods, birds	Grasses	30
Arboreal feeding		
Carnivorous		
Mammals, birds	Mammals, birds, reptiles	0.1
Invertebrates		
Mammals, birds, reptiles	Insects, spiders	0.1
Herbivorous		
Mammals, birds	Leaves, fruits, seeds, pollens, nectars	0.1

Source: Langley et al. (1998).

or dust is also the principal source of Pb to children in urban areas (Chaney and Ryan, 1994).

2.1.1 The *Itai-Itai* Disease: A Case Study

Contamination of soils with Cd worldwide from anthropogenic sources is well known. Perhaps no other metal poisoning incident has sensitized the scientific and clinical communities more than the itai-itai disease (the Japanese word *itai* means ouch or painful). This disease has been endemic among elderly women in the Jinzu River basin in Toyama Prefecture in Japan since World War II (Asami, 1997; Kobayashi, 1978; Nogawa and Kido, 1996; Tsuchiya, 1978). The pain results from unusual changes in bone, and the administration of large doses of vitamin D to affected individuals has been demonstrated to be effective in alleviating it.

The disease occurred mostly in multiparous farm women, middle-aged or older, who had lived for >30 years in this area. Starting in 1962, systematic surveys and research have been conducted by the Department of Health of the Toyama Prefecture Government. Subsequently, the Japanese Ministry of Health and Welfare and Japan Public Health Association conducted environmental research and epidemiological studies on the relationship between Cd pollution and *itai-itai* disease in the region. The results of these extensive studies were released in 1968 and indicated the following: renal tubular dysfunction was first caused by Cd poisoning, followed by the development of osteomalacia, resulting in *itai-itai* disease (Asami, 1997; Nogawa and Kido, 1996). The incidence and severity of the disease increased during periods of pregnancy, lactation, aging, Ca deficiency, and other mineral imbalances (Chaney, 1988; Flanagan et al., 1978).

Approximately 9.5% of rice paddy soils ($\sim 3 \times 10^6$ hectares) in Japan are contaminated with Cd, coming primarily as effluents from nonferrous metal mines. About 6000 ha of paddy fields in 54 areas in Japan have been designated as "agricultural land soil pollution policy" areas under the Japanese Land Soil Pollution Prevention Law because they are highly contaminated with Cd (Asami, 1997). The largest affected areas are in Toyama and Akita Prefectures, composing $\sim 53\%$ of the most severely affected areas. In addition to paddy fields, about 2.5×10^6 ha of upland fields, including orchards, were contaminated with Cd. Copper and As were also of concern in affected paddy areas. Consequently, maximum allowable limits of metals in soils have been set for Cd, Cu, and As. The requirements for the designation of "agricultural land soil pollution policy" area are based on the maximum allowable limit of Cd at 1.0 mg kg^{-1} DW in unpolished rice grown on the soil, on soil concentration of Cu at 125 mg kg^{-1} DW by 0.1 M HCl extraction and soil concentration of As at 15 mg kg^{-1} DW by 1 M HCl extraction. Because the relationship between the Cd concentration in rice and that in soil is affected by many factors, the maximum allowable limit of Cd in soil is based on the concentration of Cd in rice grown on the paddy soil.

However, the maximum allowable limit of Cd in unpolished rice (1.0 mg kg^{-1} DW), set by the Ministry of Health and Welfare in 1970, will still prove to be a hazardous level for human consumption.

Health effects from Cd poisoning were observed primarily in five areas (Table 5.6). Cadmium information from these areas indicates the mean Cd concentrations of the paddy soils ranged from 1.12 to 6.38 mg kg^{-1} DW. The maximum Cd concentration ever found in Japanese paddy soils was 930 mg kg^{-1}, with a range of 71 to 3304 mg kg^{-1} (Asami, 1997). The Cd in this case originated from wastewater from a factory dismantling Ni–Cd batteries.

The pollution source of paddy soils in the Jinzu River basin is the Kamioka Mine located about 40 km upstream of the Jinzu River from the polluted area. A survey of the concentrations of Cd, Zn, and Cu in the paddy soils of the basin and the unpolished rice from 3128 ha of this area from 1971 to 1976 was conducted. Heavy metal concentrations [mean (range)] in the paddy surface soils ($n = 1667$) extracted with 0.1 M HCl were 1.35 (0.18 to 6.88) mg Cd kg^{-1}, 59 (8.1 to 865) mg Zn kg^{-1}, and 6.3 (0.1 to 44.1) mg Cu kg^{-1}. Heavy metal concentrations in the unpolished rice ($n = 2570$) were 0.37 (0.00 to 5.20) mg Cd kg^{-1}, 25.2 (9.8 to 78.3) mg Zn kg^{-1}, and 4.5 (0.9 to 19.5) mg Cu kg^{-1}. When the amounts of the metals extracted by 0.1 M HCl were compared with total amounts, 67.0% of Cd, 17.0% of Zn, and 29.3% of Cu were extractable by 0.1 M HCl. Vertical distribution in the paddy soils of this area shows that the metals have moved to the lower layers, as the contamination history is very long.

Remediation of the paddy fields in the Jinzu River basin was accomplished by covering the surface soil with a 25-cm deep layer of noncontaminated soil and applying several soil amendments. After remediation, the Cd concentrations of the unpolished rice became lower than 0.1 mg kg^{-1}, the Cd concentration produced in unpolished paddy fields under the usual

TABLE 5.6. Cadmium concentrations of surface paddy soils and unpolished rice in the most severely affected areas in Japan.

Location	Soils (mg kg^{-1} DW)[a]			Unpolished rice (mg kg^{-1} DW)		
	n	Mean	Range	n	Mean	Range
The Jinzu River basin	544	1.12	0.46– 4.85	544	0.99	0.25–4.23
The vicinity of Ikuno Mine	19	6.38	3.90–12.16	19	1.06	0.79–1.44
The Sasu and Shiine River basins	69	5.6	1.0 –12.0	400	0.90	0.20–3.64
The Kakehashi River basin	122	3.11	1.01–17.70	122	0.81	0.41–2.84
Kosaka area	—	4.01	1.43–11.33	—	0.78	0.16–4.81

[a]0.1 M HCl extraction method was used.
Source: Asami (1997).

agricultural practice. Measures to decrease the emission of Cd and related metals have also been taken at the pollution source.

The prevalence of *itai-itai* disease and renal dysfunction in the Jinzu River basin, and its relationship to the mean Cd concentration of rice produced in each village has been reported (Nogawa and Ishizaki, 1979; Nogawa et al., 1983). Household unpolished rice was collected from 32 villages in the polluted area. As shown in Figure 5.3, there were no patients with *itai-itai* disease in the control area. The prevalences of the disease and renal dysfunction (proteinuria with glucosuria was used as an index of renal dysfunction) increased with the increase of Cd concentrations in rice. By 1993, 164 patients were certified by the pollution-related Health Damage Compensation Law as still suffering from the disease in the Jinzu River basin. In addition, the Toyama Prefecture Government has placed 388 suspected patients living in the designated area under medical care and observation.

The kidney is considered as the critical target organ for Cd toxicity for the general population. In order to define a maximum allowable intake of Cd, Nogawa et al. (1989) made an epidemiological study on the dose–response relationship of 1850 Cd-exposed and 294 nonexposed inhabitants in the Kakehashi River basin. β_2-microglobulin in urine was used as an index of the effect of Cd on health (renal dysfunction) and the average Cd concentration in locally produced rice was used as an indicator of Cd exposure. From the linear regression lines between the prevalence of β_2-microglobulin and total Cd intake, an average daily dose of about 440 µg day^{-1} in men and 350 µg day^{-1} in women is expected to cause a 50% response rate when 1000 µg β_2-microglobulin per liter of urine is used as the

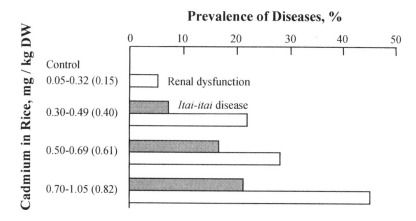

FIGURE 5.3. Prevalence of *itai-itai* (Cd poisoning) disease and renal dysfunction among females over age 50 in relation to average Cd concentration (mg kg^{-1} DW) in rice in affected villages in the Jinzu River basin. (Extracted from Nogawa et al., 1983.)

cut-off value (above which adverse effects on health may occur). The Cd intakes leading to 20, 10, and 5% response rates were 220, 150, and 110 µg day^{-1}, respectively, for men, and 200, 150, and 120 µg day^{-1} for women (Nogawa and Kido, 1996). The total lifetime Cd intake that can produce an adverse effect on health was calculated to be about 2000 mg for both men and women. If a total of 2000 mg Cd is consumed over 50 years, it corresponds to an intake of ~110 µg day^{-1}.

Recently, the amount of rice consumption by Japanese adults is decreasing. It was estimated to be 201 g day^{-1} per person in 1988 (Asami, 1997). It has been estimated that 10% of the population consumes twice the average quantity of a particular food class, and 2.5% consumes three times the average. Asami (1997) summarizes the Cd poisoning issue as follows: "Food-borne Cd is the major source of exposure for most people. Average daily intakes from food in most areas not polluted with Cd are 10 to 40 µg. In polluted areas, the value has been found to be several hundred µg per day. In nonpolluted areas, uptake from heavy smoking may equal Cd intake from food. An association between Cd exposure and increased urinary excretion of low-molecular-weight proteins has been noted in humans with a life-long daily intake of about 140 to 260 µg Cd, or a cumulative intake of about 2000 mg Cd or more."

In summary, the most important exposure route for Cd is ingestion of terrestrial foodstuff. A large segment of the general population could be potentially exposed to Cd by consuming foodstuff produced from treatment of arable soils with high Cd-sludge. Home gardeners consuming their own produce might also be a sensitive segment of the population.

2.2 Exposure to Aquatic Food Chain

In some areas, exposure to metals via the aquatic food chain may represent an important route to humans. There are certain cultures and economic circumstances that favor consumption of fish and other seafoods. Countries in Asia, Africa, and South America heavily depend on seafood as part of their staples. For example, the elevated concentration of Hg in the blood of Yanomamo Indians from South America compared with the rest of the population indicates either a local or an imported food effect (Ferguson, 1990). Similarly, Japanese living in South America exhibit higher Hg in the hair than the rest of the population, reflecting their heavy preference for fish consumption (Tusgane and Kondo, 1987). People inhabiting coastal areas are expected to have higher dietary intake of aquatic foodstuff. For cultural and economic reasons, certain aborigines in North America largely depend on these food items for their livelihood, e.g., the levels of Hg in the hair of Eskimos were higher compared with the rest of U.S. males (Ferguson, 1990).

The most common pathways for the aquatic human food chain are:

1. Contaminated water/sediment → plant → human (e.g., floating plants)
2. Contaminated water/sediment → plant → animal (aquatic) → human

3. Contaminated water/sediment → plant/animal → animal (waterfowl)
   → human
4. Contaminated water/sediment → animal → human

Undoubtedly, the aquatic food chain scenario is more complicated than its terrestrial food chain counterpart due to the role of the trophic level relative to the bioaccumulation of these contaminants. Both floating and emergent plant species can be important transfer vectors directly to humans (e.g., watercress), or indirectly through their ingestion by fish and waterfowl. Emergent plants, such as the water lotus (popular in Chinese cuisine), derive their elemental uptake mostly from the sediments and some from the water.

In several Asian countries (Philippines, India, China, Indonesia, Thailand, Vietnam, etc.) inhabitants who are literally subsistence farmers/fishermen rely heavily on aquatic foodstuff for their livelihood. For example, inhabitants along coastal areas hunt for fish and waterfowl not only for family consumption but also for commerce. Paddy rice farmers also gather fish and other aquatic organisms (e.g., crustaceans and snails) from paddies irrigated with canal water. These activities may also be a contributing factor to the severity of As poisoning in Bangladesh, where inhabitants gather fish and other paddy organisms for family consumption (Ullah, 2000). This segment of the population may represent the most potentially exposed individuals—similar to the home gardener scenario discussed in the previous section.

2.2.1 The *Minamata* Disease: A Case Study

The so-called *Minamata* disease was the first incident of epidemic intoxication occurring in a large area and affecting a large number of patients. It is known as *Minamata* disease because it occurred in Minamata, Japan. Minamata is situated at the southern end of Kumamoto Prefecture that borders on Kagoshima Prefecture. The disease was caused by methyl Hg borne from wastewater discharged from a chemical plant (acetaldehyde plant) into the sea facing Minamata (Hamada and Osame, 1996). The chemical was discharged into Minamata Bay at first, and the pollution was primarily restricted initially to areas along the bay. Later the methyl Hg was discharged outside the bay, thus spreading the pollution to wider areas. Methyl Hg was bioconcentrated in the food chain, including the fish and shellfish. In the era of the 1950s, fish was the main source of protein for the inhabitants in the bay area. Thus, virtually all inhabitants were exposed to Hg via the consumption of methyl Hg–contaminated fish and other seafood. The population of the bay was exposed to methyl Hg for a very long period of time since the 1950s so that by November 1993, 2256 patients had been officially identified in the Minamata area (Hamada and Osame, 1996). Although almost all the inhabitants were exposed below or above the threshold for Hg poisoning, fishermen and their families were more greatly affected since their consumption of fish is normally greater than the other segments of the population. For example, of the 487 patients in Kagoshima, 316 were fishermen vs. only 40 farmers. The others were from other occupations.

As mentioned earlier, ingested methyl Hg is almost completely absorbed in the GI tract and is degraded into inorganic Hg once in the body. For example, after high oral intakes of methyl Hg for 2 mo, the percentages of inorganic Hg per total Hg burden in the tissues were 7% in whole blood, 22% in plasma, 39% in breast milk, 73% in urine, and 16 to 40% in liver (WHO, 1990). A considerable amount (>80%) of Hg in the brain is reported to be in inorganic Hg form in patients with chronic *Minamata* disease (WHO, 1991). Although methyl Hg is distributed to all tissues within a short time (~4 days), its toxic effects are selective to the nervous system; the central nervous system and peripheral nerves are both affected. The most characteristic feature is the manifestation of lesions.

Ten years after the Minamata episode, a similar phenomenon occurred in Niigata, Japan. The wastewater was discharged into a river in Niigata instead of the sea in Minamata. Since river fish are consumed by fewer people compared with sea fish, methyl Hg poisoning occurred at a much lower scale in Niigata (690 total identified patients vs. 2256 in Minamata Bay residents).

Sensory impairment is the most basic symptom of Hg poisoning, which may still occur even in mildly affected patients (Bakir et al., 1973). The first symptoms include paresthesia or dysesthesia, generally initiated from the peripheral regions, i.e., fingers or toes. In severe cases, anesthesia may be manifested. Other signs and symptoms of Hg poisoning are given in the literature (Hamada and Osame, 1996) (see also the later chapter on mercury).

In summary, human exposure to Hg is virtually limited to consumption of seafood. Subsistence fishermen and persons heavily relying on seafood would represent the most sensitive segment of the population.

2.3 Exposure to Drinking Water

2.3.1 *Arsenicosis* in Bangladesh: A Catastrophic Case Study

In the early 1970s, most of Bangladesh's rural population took its drinking water from the surface ponds and nearly a quarter of a million children died each year from waterborne diseases (World Bank Group, 1999). The provision of tubewell water for 97% of the rural population has been credited with bringing down the high incidence of diarrheal diseases and contributing to a halving of the infant mortality rate. Paradoxically, the same wells that saved so many lives now pose a threat due to the unforeseen hazard of As. Until wells are tested to determine which are As-affected and which are not, millions of rural poor in Bangladesh face a dilemma: either continue the risk that they may be drinking As-contaminated water or return to surface water and risk diarrhea and other deadly waterborne diseases.

The problem of high levels of As in numerous shallow and deep wells was first detected in Bangladesh in 1993, and was subsequently confirmed after 1995 (World Bank Group, 1999). As more than 20 million of Bangladesh's 120 million people are exposed to As through drinking water, this has to be the largest mass poisoning in history. The As contamination in Bangladesh is unprecedented. There are examples of geologic As contamination in other parts of the world, including neighboring West Bengal in India, but they are relatively site specific and affect a limited number of people (also see later chapter on arsenic).

Arsenic is a silent killer. Undetectable in its early stages, As poisoning takes between 8 and 14 years to impact on health, depending on the amount of the metal ingested, nutritional status, and the immune response of the individual. Until a certain point, contamination may be treated and reversed by drinking As-free water. The effects of As poisoning can vary from skin pigmentation, development of warts, diarrhea, and ulcers during the initial stages. In the most severe cases, As poisoning causes liver and renal deficiencies or cancer that can lead to death.

Arsenic poisoning is difficult to detect because there is a lack of capacity and tools to diagnose it. Furthermore, only a minority of those suffering from *arsenicosis* can be easily identified from their skin condition. Thus, the majority of *arsenicosis* cases have gone undetected. To date, several thousands of patients in Bangladesh with As-related skin disease have been found in the first limited surveys. Available data on As-related mortality are scarce, but dozens of deaths due to As-induced skin cancer have been reported in the last few years. Because of the majority of the country's tubewells were installed in the past two decades, it is likely that many more people will start developing symptoms in the next few years.

The social consequences of the As crisis are far reaching and tragic. Because of the illiteracy and lack of information, many confuse the skin lesions caused by *arsenicosis* with leprosy. The most hard-hit villages where health problems have gripped a large population are treated much like isolated leper colonies. Within the community, As-affected people are barred from social activities and often face rejection, even by immediate family members. Women are unable to get married and wives have been abandoned by their husbands. Children with symptoms are not sent to school in an effort to hide the problem.

Although scientists in the region and around the world are investigating the problem, the exact cause of the contamination is still unclear. In some countries with As contamination, it was found that oxygen introduced into groundwater by the regular lowering of the groundwater table (e.g., through intensive agricultural irrigation with groundwater), triggered an oxidative process that dissolved As from geologic materials. This does not appear to be the case in Bangladesh. According to a preliminary hydrogeological study commissioned by the World Bank and funded by the British Department for International Development, the As in Bangladesh's groundwater is already dissolved and present under natural circumstances, thus making human-related causes unlikely (World Bank Group, 1999).

In summary, the most important exposure route for As is via ingestion of drinking water. The most sensitive segment of the population may be represented by people relying heavily on groundwater as a source of potable water in As hot spots around the globe.

2.4 Exposure via Dietary Intake

Collection of reliable data on the food consumption habits of a population is the most difficult problem facing any assessment of the dietary intake of natural and synthetic chemicals. This indicates the fundamental difficulty in the exposure assessment relative to dietary intake of contaminants, including those from homegrown produce. Several factors are necessary in estimating exposure to contaminants in food: (1) the concentration of contaminant in each dietary item, (2) the quantity of each dietary item consumed, and (3) the frequency and duration of consumption of each item.

In general, information on food consumption habits of a population can be obtained either by collecting information directly from individuals or households (i.e., surveys) or indirectly by gathering data about the movement and disappearance of foods in a nation or region (food balance approach) (Langley, 1996). As can be deduced from Table 5.7, food

TABLE 5.7. Regional food consumption patterns and global diet (g/person/day).

Food group	Middle Eastern	Far Eastern	African	Latin American	European	Global diet
Cereals	432	452	320	254	226	405
Roots/tubers	62	108	321	159	242	288
Pulses	21	27	18	21	9.3	23
Sugar/honey	95	50	43	104	105	58
Nuts/oil seeds	4.3	18	15	19	12	18
Vegetable oil/fat	38	15	24	26	48	38
Stimulants	8.0	1.5	0.5	5.3	14	8.1
Spices	2.3	2.0	1.6	0.3	0.3	1.4
Vegetables	193	168	74	124	298	194
Fish/seafood	13	32	32	40	46	39
Eggs	14	13	3.6	12	38	38
Fruit	220	94	85	288	213	235
Milk and products	132	33	42	168	338	55
Meat/offals	72	47	32	78	221	111
Animal oil/fat	0.5	1.5	0.3	5.0	10	5.3
Other	4.3	—	0.5	7.0	2.0	3.2
Total	1311	1062	1012	1311	1822	1520

Source: Chang and Page (1996).

TABLE 5.8. Estimated human exposure to selected trace elements via food intake.

	Exposure (µg day^{-1})				
Diet	As	Cd	Pb	Hg	Se
Middle Eastern	12	6	17	2	0.7
Far Eastern	12	6	17	2	0.8
African	9	4	12	1	1
Latin American	7	4	11	1	0.8
European	9	4	10	1	1
Global	12	6	16	2	1

Source: Chang and Page (1996).

consumption patterns vary considerably from region to region (Galal-Gorchev, 1991; Ryan et al., 1982). There are differences in terms of the total amounts consumed as well as type of food consumed. The Food and Agriculture Organization (FAO) recommends that the average food consumption data in FAO Food Balance Sheets (global diet) be used to estimate human exposure to contaminants via food intake (WHO, 1989). The food consumption described in the global diet does not resemble any specific diet (see Table 5.7) but rather a generic aggregation of diets from various regions. Grain/cereal, vegetables, root/tuber, and fruits account for ~76% of the total daily food intake according to the global diet. Thus, the potential for contaminants to be transferred to humans through other food groups, i.e., dairy/animal products, oil/fat, shortening, sugar/honey, fish/seafood, eggs, etc. is small (McKone and Ryan, 1989).

Based on literature data and regional food consumption patterns (see Table 5.7), Chang and Page (1996) estimated human exposure to certain trace elements via dietary intake (Table 5.8). Daily dietary intakes of As, Cd, Pb, Hg, and Se by adults have been reported to be in the range of 10 to 130, 10 to 120, 5 to 1700, ~30, and <10 to 5000 µg day^{-1} (Anke, 1986; Fergusson, 1990). Thus, it appears from the exposure estimation presented in Table 5.8 that if the soil (or food crops) is not contaminated, human exposure through food consumption is substantially below the provisional tolerable intake guideline of the Joint FAO/WHO Expert Committee on Food Additives (WHO, 1989). For example, the calculated values for Se appear to be insufficient to meet human nutritional requirement from the selected food groups, since the reported value is at 25 µg day^{-1}.

3 Ecotoxicology of Metals

A fairly new field of science, *ecotoxicology* developed from the traditional fields of toxicology and environmental chemistry that gave rise to environmental toxicology. When applied to the ecosystem setting, environmental toxicology virtually becomes ecotoxicology. Cairns and Mount (1990) defined *ecotoxicology* as "the study of fate and effect of toxic agents in ecosystems."

The traditional environmental toxicology refers to the study of effects, direct or indirect, of toxic agents on organisms. Thus, aquatic ecotoxicology refers to the study of such toxic effects on aquatic organisms (e.g., plants, animals, microorganisms).

Nowadays, ecotoxicology commonly includes the study of toxic effects at the cellular, individual, population, and community levels. However, the fast-progressing field of gene research has spurred the use of molecular techniques (e.g., in genotoxicity). In general, however, the vast majority of studies performed to date have been at the individual level (Adams, 1995). Furthermore, ecotoxicology integrates several disciplines, including environmental biogeochemistry, toxicology, and ecology, dealing with mechanisms, processes, and responses. In short, ecotoxicological research deals with the interactions among the organisms, the toxic agents (in this case metals), and the environment.

3.1 Basic Requirements in Ecotoxicology

There are several levels of biological organization in ecotoxicology research (Eijsackers, 1994; Newman, 1995): molecular level (e.g., RNA/DNA); cellular level (e.g., enzymes, hormones, ATP content, phytochelatins, metallothioneins); individual level (e.g., growth rates, organ functioning, behavioral responses); population level (e.g., lethality, reproduction, genetic drift); community level (e.g., species diversity, community structure); and ecosystem level (e.g., nutrient cycling, energy fluxes).

Figure 5.4 summarizes the present understanding of effects along the ecological spectrum of organization (i.e., molecular), which becomes poorer as the level of organization increases. The lower level effects are generally more sensitive (i.e., manifested at lower toxicant concentrations) than effects at higher levels of organization. They often respond more rapidly to a toxicant.

In the Soil Ecotoxicology Risk Assessment System (SERAS), tests are categorized into (Eijsackers and Lokke, 1992): (1) physical features, such as particle size, mineral composition, organic matter composition, etc.; (2) soil processes, such as carbon and nutrient cycling, nutrient transformation, mobility, etc.; (3) community structure, with features such as species abundance, species diversity, trophic structure, etc.; and (4) community functions, such as predation, mutualism, etc.

There are also basic criteria for ecotoxicological tests (Doelman and Vonk, 1994): relatively quick and cost effective, standardizable (reliable), practically applicable, relevant to environmental regulations, ecologically relevant to many ecosystems, discriminating from natural fluctuations, and relevant cause–effect relationship. In soil ecotoxicity research, the following indicators are generally used (Domsch, 1991): microbial biomass; population of bacteria, actinomycetes, and fungi; soil respiration; nitrogen transformation (e.g., ammonification, nitrification); and enzyme activities. McGrath (1994) pointed out that symbiotic N_2 fixation appears to be a reliable index of metal

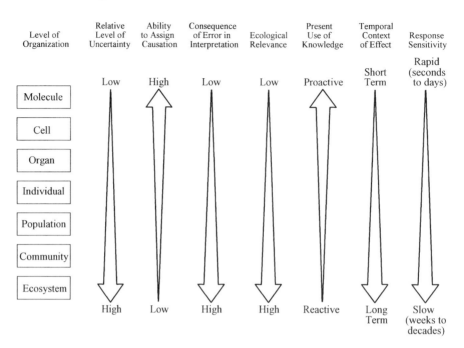

FIGURE 5.4. Features of ecotoxicological effects based on level of ecological organization. (Reprinted from Newman, *Quantitative Methods in Aquatic Ecology*. Copyright © 1995 Lewis Publishers, an imprint of CRC Press, Boca Raton, FL.)

stress in sludged soils. Doelman and Vonk (1994) as well as others (Breckle, 1997) strongly believe that microorganisms are especially suitable to serve as indicators of environmental pollution, including their possible role as an ecological warning system due to their ubiquity, size, and versatility, and their important role in the foodwebs and cycling of elements.

Soil fauna, such as earthworms, springtails, and beetles are also widely used in soil ecotoxicology research (Doelman and Vonk, 1994; Roth, 1992; Van Gestel and Van Straalen, 1994). Because earthworms are generally agreed to be beneficial to soil, their toxicity should be taken into account when assessing soil ecotoxicology (Beyer and Linder, 1995). In addition, species richness should be considered in soil assessment on the rationale that a decrease in species diversity should be construed as detrimental. The common parameters used include lethality, behavior, food choice, survival, hatching, growth, reproduction, or development of larval stages.

3.2 Case Studies in Terrestrial Ecotoxicology

Two types of terrestrial ecosystem studies dealing with soil ecotoxicity will be highlighted here: sludged-agricultural ecosystems and deciduous woodland ecosystems, the latter being impacted by industrial emission.

3.2.1 Arable Land Treated with Sewage Sludge

The long-term ecological effects of metals in sewage sludge added to soils are very difficult to assess as there are only a few such experiments that exist; thus, a lack of good long-term data exists. Less is known about the adverse long-term effects of metals on soil microorganisms than on crop growth and metal uptake. Recently, McGrath et al. (1995) summarized the evidence for impacts of metals on the growth of selected plants and on the effects of metals on soil microbial activity and soil fertility in the long term. They summarized the information from long-term controlled field experiments in the United States, United Kingdom, Sweden, and Germany. The following field experiments were included:

1. Woburn, England—began in 1942, where the plots received anaerobically digested sewage sludge from 1942 to 1961. The soil is a sandy loam with about 10% clay and 2% organic C. Soil pH was maintained at ~6.5, with regular dressing of lime.
2. Lee Valley and Luddington, England—began in 1968, where sewage sludge from various sewage works, contaminated mainly with Zn, Cu, Ni, or Cr was used in both locations for 4 yr. The Lee Valley soil is a heavy silt loam with 21% clay, 4% organic C, and pH of 5.6 to 5.9. That at Luddington is a sandy loam with 15% clay, 3% organic C, and pH of 6.5.
3. Gleadthorpe, England—started in 1982, used sewage sludge spiked with metals by adding metal salts to raw sewage and dewatered. The sludge was spiked with Zn, Cu, or Ni, with some plots spiked again in 1987. The soil is a sandy loam with 9% clay, and 1 to 2% organic C.
4. Ultuna, Sweden—began in 1956, used sewage sludge contaminated by sewage works in Uppsala. The sludge was added every 2 yr from 1956 to 1988. The site has postglacial soil with 35% clay, 1.2% organic C, and pH of 6.2.
5. Braunschweig, Germany—two field experiments began in 1980, where moderately contaminated or metal-spiked liquid sludge was added to the plots for 10 yr (1980 to 1990). Braunschweig I site was an old arable soil with pH of 6.0 to 7.0, and organic C of 0.8 to 1.5%; Braunschweig II was an ex-woodland soil with soil pH of 5.3 to 5.7, and 1.6 to 2.6% organic C. Both soils were silty loam with 5% clay.
6. Maryland, United States—two experimental sites at Fairland and Beltsville, Maryland, began in 1975 and 1976, respectively. Anaerobically digested sludge was applied to sandy loam (Fairland) with 6.4 to 6.9 pH, and to a fine sandy loam (Beltsville) with pH of 5.1 to 6.6.

On these field trials, either treatment of just inorganic fertilizers or application of farmyard manure (FYM) was used as a control.

The yields of white clover in a mixed grass/clover sward significantly decreased at the Lee Valley experimental plots predominantly contaminated by Zn 7 yr earlier (Vaidyanathan, 1975). Copper also decreased the clover yield, but to a lesser extent, whereas Ni and Cr had no effect. At Woburn, the yields of monoculture white clover decreased by 60% on sludged plots compared with FYM plots more than 20 yr after the sludge application (McGrath et al., 1987). Similarly, yields of white clover at Gleadthorpe decreased in Zn- and Cu-contaminated soils (Royle et al., 1989). At Luddington, red clover failed almost completely in 1985 from some of the most metal-contaminated plots (Jackson, 1985). From the above studies, it is impossible to determine whether the adverse effects on clover yield were due to direct phytotoxicity, or due to some effect on N_2 fixation, or on the *Rhizobia* in these sites. McGrath et al. (1987) however, showed that the yield decreases at Woburn contaminated plots were not due to phytotoxicity, as the clover yields could be boosted to the yields from the control (FYM plots) by adding N fertilizer. Thus, it became obvious that the yield reductions were due to the adverse effect on N_2 fixation. Since the metal levels in contaminated plots at Lee Valley, Luddington, and Gleadthorpe were similar to metal-contaminated plots in Woburn, it can be deduced that the yield decrements in the other sites were due to detrimental effect on N_2 fixation. Indeed, the *Rhizobium leguminosarum* biovar *trifolii* isolated from nodules on clover plants grown in metal-contaminated soils were found to be ineffective in N_2 fixation (McGrath et al., 1987).

In addition to adverse effects on *Rhizobia,* soil microbial biomass was only about half in the sludge-treated soils compared with the FYM (farmyard manure)-control plots (Brookes and McGrath, 1984). Metal concentrations at which microbial biomass was affected are shown on Table 5.9. Soil microbial biomass C and N were found to be reduced by about 60% in metal contaminated plots in Ultuna compared with the control FYM plots (Witter et al., 1993).

At the Woburn site, significant decreases in acetylene reduction activity (ARA, a test for nitrogenase activity) by N_2 fixers were reported in metal-contaminated soils. The reductions occurred at metal concentrations close to the EU upper limits for Zn and Cu, and 3 to 4 times the limit for Cd (see Table 5.9). Similar observations were obtained in Ultuna, Sweden where heterotrophic N_2 fixation was severely reduced (Martensson and Witter, 1990). Furthermore, reductions in N_2 fixation by heterotrophic bacteria were also observed in the old arable soil plots in Braunschweig, Germany. The N_2 fixation by phototrophic cyanobacteria (i.e., autotrophs that grow on the soil surface and utilize light to fix CO_2 in photosynthesis) was also adversely affected in Woburn, Ultuna, and Braunschweig (see Table 5.9) (McGrath et al., 1995).

The levels of metals in soil at which adverse effects on soil microbial parameters were observed at the above sites are summarized in Figure 5.5. The lowest observed adverse effect concentrations (LOAECs) for Zn and Cu that adversely affected microbial parameters were greater than the largest background (BG) concentrations for these metals in soil, except for

TABLE 5.9. Minimum concentrations of metals in soils at which negative effects on (A) yields of clover or the population of *Rhizobium leguminosarum* biovar *trifolii,* (B) N_2 fixation in cyanobacteria (blue-green algae), and (C) soil microbial biomass were observed.

Experimental site		mg kg^{-1} soil					
		Zn	Cd	Cu	Ni	Pb	Cr
				– A –			
Woburn, England[a]		180	6.0	70	22	100	105
Gleadthorpe, England[a]		281	—	150	—	—	—
Braunschweig I, Germany[b]		200	1.0	48	15	—	—
Braunschweig II, Germany[b]		130	0.8	27	11	—	—
				– B –			
Woburn, England[c]		114	2.9	33	17	40	80
Ultuna, Sweden[d]		230	0.7	125	35	40	85
Braunschweig I, Germany[e]	L sludge	42– 93	0.36–0.81	—	1.75– 4.5	—	—
	H sludge	132–305	0.58–2.38	—	5.4 –17	—	—
Braunschweig II, Germany[f]	L sludge	26– 81	0.4 –0.88	—	1.5 – 4.6	—	—
	H sludge	86–240	0.7 –2.15	—	3.7 –16	—	—
				– C –			
Woburn, England		180	6.0	70	22	100	105
Luddington, England		281	—	150	—	—	—
Lee Valley, England		857	—	384	—	—	—
Ultuna, Sweden		230	0.7	125	35	40	85
Braunschweig I, Germany		360	2.8	102	23	101	95
Braunschweig II, Germany		386	2.9	111	24	114	105

[a]Clover yields.
[b]Rhizobial population.
[c]50% Reduction in N_2 fixation.
[d]>50% Reduction in N_2 fixation.
[e]Braunschweig I: 30% and 70% reduction in N_2 fixation in the Low (L) and High (H) rate of sludge application, respectively.
[f]Braunschweig II: 25% and 100% reduction in N_2 fixation in the L and H, respectively.
Source: Extracted from McGrath et al. (1995).

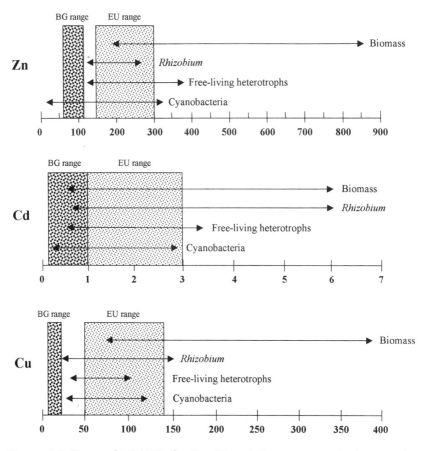

FIGURE 5.5. Range of LOAECs for Zn, Cd, and Cu on groups of microorganisms found in soils of six long-term field experiments, compared with soil background concentrations found in U.K. soils. BG range = background concentration range; EU range = range of limits for sludge-treated soils in the European Union. (Adapted from McGrath et al., 1995.)

cyanobacteria for which there were LOAEC values below the lowest BG level for Zn. For Cd, the LOAECs overlapped the highest BG level found in soils. McGrath et al. (1995) caution that the metal effects may indeed be additive since the metals occurred in the same matrix, or the effects may have been due to another metal rather than the specific ones indicated on the graph. They also pointed out the interactive influence of soil type, OM content, and pH on the LOAECs for each parameter. For example, the LOAECs for *Rhizobia* from contaminated soils in Woburn and Braunschweig were similar since both sites are loam soils, with low OM contents. However, Braunschweig II resulted in smaller LOAECs for *Rhizobia* than for Braunschweig I because the former soil had a lower pH (5.3 to 5.7) than the latter (pH 6 to 7). Heavier soils (i.e., more clayey), near-neutral soils (i.e., higher pH), or soils having high OM content may be expected to reduce the detrimental effects of metals to soil microorganisms by binding the contaminants and rendering them less bioavailable, thus giving high LOAECs.

3.2.2 Woodland Ecosystems Impacted by Industrial Emissions

Smelters were commissioned in the Avonmouth area, near Bristol, England starting in 1929 (Martin and Bullock, 1994). The site smelted Zn and Pb, and in 1969 a much larger smelter began operation where Pb and Zn ores were simultaneously smelted. The study site has two woodlands—Hallen Wood and Haw Wood are located at about 3 km from the smelting complex. The sites are primarily oak–hazel (*Quercus robur, Corylus avellana*) woodlands, located on clayey soils.

The woodland studies have utilized various biological monitoring techniques, including tree leaves (Little and Martin, 1972), grass (Burkitt et al., 1972), *Sphagnum* moss (Gill et al., 1975; Little and Martin, 1974), and woodlice (Hopkin et al., 1986). Results have shown that the highest deposition of heavy metals was close to the smelting facility and an exponential decline in deposition occurred with increasing distance from the source. The largest burden of metals was in the litter and the mineral soil, i.e., 98.7% Cd, 96.1% Pb, 98.9% Zn, and 98.1% Cu (Table 5.10). The amount of litter at Hallen Wood was 9 times greater than at a similar type of woodland 23 km away from the smelter (Coughtrey et al., 1979). This greater accumulation of litter on the impacted site is not due to greater annual inputs of OM on the forest floor but to interference in the decomposition process. At this site, the litter accumulation was closely related to both Cd and Zn concentrations in the litter. Indeed, the greater litter accumulation in heavily metal-impacted forest ecosystems has been a well known phenomenon (Freedman and Hutchinson, 1980; Strojan, 1978; Tyler, 1975; Wiener and Grodzinski, 1984). In one instance, litter far from the smelter in Pennsylvania respired at a rate 3.6 times that of the same mass

TABLE 5.10. Distribution of metals (%) in the Hallen Wood, Avonmouth, England forest ecosystem.

	Cd	Pb	Zn	Cu
Trees	1.25	2.94	1.03	1.82
Shrubs	0.04	0.98	0.06	0.12
Ground flora	0.03	0.02	0.03	0.02
Litter	10.75	37.24	9.49	14.37
Soil (0–31 cm)	87.93	58.82	89.40	83.76
Total, mg (metal) m^{-2}	4945	82162	285786	11062

Source: Extracted from Martin and Bullock (1994).

of litter near the smelter (Strojan, 1978b). In addition, weight loss (via litter bag studies) of litter near the smelter were only about half the expected rates (Strojan, 1978a).

Studies of litter/soil fauna (Table 5.11) indicate some marked differences in the taxa of invertebrates present in the contaminated site (Hopkin et al. 1985). In particular, millipedes (Diplopoda), earthworms (Annelida), and woodlice (Isopoda) were greatly reduced in population in the contaminated site.

Enzymatic activity in soil has been regarded as an indication of soil microbial activity. Thus, certain enzymatic activity has decreased in several occasions in direct relationship with soil metal concentrations (e.g., dehydrogenase enzyme) (Reddy et al., 1987; Tyler, 1975). Indeed, reduced populations of actinomycetes, bacteria, and fungi in ridgetop soils contaminated with Zn, Cd, and Pb from zinc smelters in Pennsylvania were found (Jordan and Lechevalier, 1975).

TABLE 5.11. The number of invertebrates in litter collected from Hallen Wood, Avonmouth, England compared with a relatively uncontaminated woodland (Wetmoor Wood). Data are presented as number per kg of litter.

	Hallen Wood (contaminated), no. kg^{-1}	Wetmoor Wood (control), no. kg^{-1}
Isopoda		
Oniscus asellus	3.9	14.8
Trichoniscus pusillus	0	112
Diplopoda		
Polydesmidae	0.6	58.5
Julidae	0	8
Glomeridae	0	15.6
Chilopoda		
Lithobiidae	7.8	86
Geophilomorpha	23	195
Arachnida		
Acari	9034	14,370
Aranae	17.4	60
Pseudoscorpionidae	14	50
Insecta		
Collembola	1457	6436
Coleoptera	63	36
Coleoptera (larvae)	8.4	3
Diptera (larvae)	321.4	215.6
Annelida		
Lumbricus rubellus	2	2.2
Lumbricus terrestris	0	21.5
Aporrectodea longa	0	3
Aporrectodea caliginosa	0	22
Octoclasium cyaneum	0	6.7

Source: Extracted from Martin and Bullock (1994).

3.3 Ecotoxicity in Aquatic Systems

The basic principle upon which all toxicity tests are based is the recognition that the response of living organisms to the presence (exposure) of toxic agents is dependent upon the dose (exposure level) of the toxic agent (Adams, 1995). Based on this dogma, aquatic toxicity tests are designed to describe a concentration–response curve when the measured effect is graphically plotted with the concentration. Acute toxicity tests are short term and are usually designed to evaluate the concentration–response relationship for survival; whereas, chronic tests are conducted over a significant portion of the organism's life cycle to evaluate sublethal effects, such as growth, behavior, reproduction, or biochemical effects, and are usually designed to provide an estimate of the concentration that produces no adverse effect. The American Society for Testing and Materials (ASTM) has published standard guides on how to perform acute toxicity tests for both freshwater and marine invertebrates and fish (Adams, 1995). The most commonly used test species in the United States include fathead minnows (*Pimephales promelas*), rainbow trout (*Oncorhynchus mykiss*), bluegill (*Lepomis machrochirus*), channel catfish (*Ictulurus puntatus*), waterflea (*Daphnia magna*), amphipod (*Hyalella azeteca*), midge (*Chironomus* spp.), duckweed (*Lemna* spp.), green alga (*Selenastrum capricornutum*), and shrimp (*Mysidopsis bahia, Penaeus* spp., *Palaemonetes pugio*). For chronic toxicity tests, the following organisms are often used: brook trout (*Salvelinus fontinalis*), fathead and sheepshead minnow, waterflea (*Daphnia magna*), zebrafish (*Brachydanio verio*), and mysid shrimp.

3.3.1 Ecotoxicity of Mercury and Silver

Water-breathing animals are normally more vulnerable to chemicals than animals that breathe air. The simple reason is toxic chemicals come into direct contact with very sensitive gills (Hogstrand and Wood, 1998). The fish gill is a complex, delicate organ that has multiple physiological functions. In addition to gas exchange, the gill is a central ionic regulatory organ, the major site of acid–base regulation, and the principal route for excretion of nitrogenous waste products. Thus, disruption of any of the gill functions could produce adverse effects, including death.

Mercury is highlighted here because of its uniqueness among trace metals to biomagnify in the aquatic food chain. In addition, reservoirs not known to have directly received Hg input are also known to have resident organisms that have bioaccumulated the metal, especially in newly established reservoirs. Spry and Wiener (1991) reported that when the whole-body concentrations of Hg in fish are in the range of 10 to 30 $\mu g\ g^{-1}$ (WW), toxicity can be expected. Freshwater invertebrates have a wide range of toxicity to Hg (Table 5.12). Concentrations of Hg acutely toxic (LC_{50}) to aquatic invertebrates range from 0.2 $\mu g\ L^{-1}$ for crayfish (*Procambus clarki*),

TABLE 5.12. Toxicity of inorganic mercury to freshwater invertebrates.

Organism	End point[a]	Concentration (µg L^{-1})	Test conditions
Crayfish	LC_{50} (96 hr)	7	flow
Waterflea (*Daphnia magna*)	LC_{50}	2.2	nr[a]
Waterflea (*Daphnia magna*)	LC_{50}	5	nr[a]
Snail (*Amnicola* spp.)			
Embryo	LC_{50} (96 hr)	2100	static
Adult	LC_{50} (96 hr)	80	static
Mussel (*Lamellidans maginalis*)	LC_{50} (72 hr)	3690	static
Stonefly	LC_{50}	2000	static
Caddisfly	LC_{50} (96 hr)	1200	static
Mayfly	LC_{50} (96 hr)	2000	static

[a]nr = Not reported.
Source: Adapted from Wren et al. (1995).

2.2 µg L^{-1} for waterflea, 2.0 mg L^{-1} for mayfly (*Ephemerella subvaria*), to up to 2100 µg L^{-1} for the snail (*Amnicola* spp.) (Wren et al., 1995). Furthermore, Hg concentrations as low as 0.04 µg L^{-1} inhibited reproduction in waterflea.

In freshwater fish, inorganic Hg is not as toxic as some other metals (i.e., Cu, Pb, Zn, or Cd). Standard 4-day LC_{50} for inorganic Hg range from 33 µg L^{-1} in 2 mo-old rainbow trout to 687 µg L^{-1} in adult white sucker (*Catostomus commersoni*) (Table 5.13). As can be deduced from Table 5.13, organic Hg is more toxic than inorganic Hg under short-term

TABLE 5.13. Acute toxicity of mercury to freshwater fish.

Organism	End point[a]	Concentration (µg L^{-1})	Test conditions
Inorganic mercury[b]			
Rainbow trout (51–76 mm)	LC_{50} (96 hr)[c]	33	flow
Brook trout	LC_{50} (96 hr)	0.3–0.9	nr[a]
Largemouth bass (embryo-larva)	LC_{50} (8 days)	140	static
Banded killifish	LC_{50} (96 hr)	110	static
Pumpkinseed	LC_{50} (96 hr)	300	static
Striped bass	LC_{50} (96 hr)	90	static
White perch	LC_{50} (96 hr)	220	static
Common white sucker	LC_{50}	687	nr
Carp	LC_{50} (96 hr)	180	static
Organic mercury			
Rainbow trout (fry)	LC_{50} (96 hr)	24	static
Rainbow trout (fingerling)	LC_{50} (96 hr)	42	static
Brook trout	LC_{50} (96 hr)	75	flow
Brook trout (yearling)	LC_{50} (96 hr)	65	nr
Lamprey	LC_{50} (96 hr)	48	flow

[a]nr = Not reported.
[b]Mercuric chloride.
[c]Mercurous nitrate.
Source: Adapted from Wren et al. (1995).

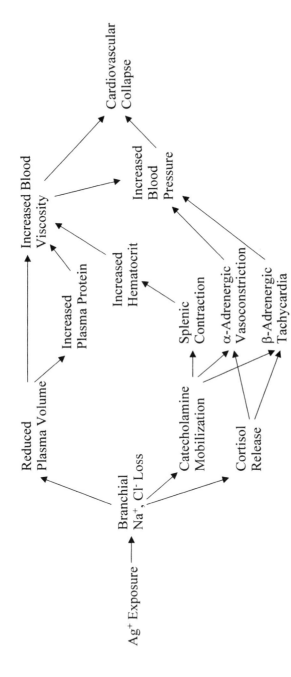

FIGURE 5.6. Likely etiology of acute silver toxicity in freshwater fish. (Reprinted with permission from Hogstrand and Wood, *Environmental Toxicology and Chemistry* 17:547–561. Copyright © 1998 SETAC, Pensacola, Florida, USA.)

exposures. For example, methyl Hg chloride was 7 times more toxic than mercuric chloride to 8-g rainbow trout; methyl Hg chloride was 20 to 40 times more effective than inorganic Hg in reducing plasma Na^+ and Cl^- concentrations in adult rainbow trout. In general, the potential for toxicity to aquatic fish and invertebrates from direct exposure to aqueous Hg is low (Wren et al., 1995). For example, the reported aqueous concentrations of total Hg and methyl Hg in low-alkalinity waters in the north-central United States, Sweden, and Ontario, Canada are typically 2 to 3 orders of magnitude less than aqueous concentrations proven to cause acute or chronic toxicity.

While the GI tract in most cases may be the dominant pathway for metal metabolism, it is usually branchial uptake of dissolved metals that is significant for acute toxicity in freshwater fish (Hogstrand and Wood, 1998). A sample of toxicological effects of metals, as exemplified by free Ag^+ ions, to freshwater fish is illustrated in Figure 5.6. Exposure to Ag^+ ions results in a net loss of Na^+ and Cl^- ions from the blood plasma. This osmolyte loss triggers a series of events that can eventually lead to fatally increased blood viscosity and arterial blood pressure. The final cause of death is likely cardiovascular collapse.

References

Adams, W.J. 1995. In D.J. Hoffman et al., eds. *Handbook of Ecotoxicology*. Lewis Publ, Boca Raton, FL.

Anke, M. 1986. In W. Mertz, ed. *Trace Elements in Human and Animal Nutrition*. Academic Pr, New York.

Arthur, W.J., and A.W. Alldredge. 1979. *J Range Manage* 32:67–71.

Asami, T. 1997. In D.C. Adriano, Z.S. Chen, S.S. Yang, and I.K. Iskandar, eds. *Biogeochemistry of Trace Metals*. Science Reviews, Northwood, United Kingdom.

Bakir, F., et al. 1973. *Science* 181:230–241.

Beyer, W.N., and G. Linder. 1995. In D.J. Hoffman et al., eds. *Handbook of Ecotoxicology*. Lewis Publ, Boca Raton, FL.

Beyer, W.N., E.E. Connor, and S. Gerould. 1994. *J Wild Manage* 58:375–382.

Beyer, W.N., L.J. Blus, C.J. Henry, and D. Audet. 1997. *Ecotoxicology* 6:181–186.

Beyer, W.N., D. Day, A. Morton, and Y. Pachepsky. 1998. *Environ Toxicol Chem* 17:2298–2301.

Breckle, S. 1997. In D.C. Adriano, Z.S. Chen, S.S. Yang, and I.K. Iskandar, eds. *Biogeochemistry of Trace Metals*. Science Reviews, Northwood, United Kingdom.

Brookes, P.C., and S.P. McGrath. 1984. *J. Soil Sci* 35:341–346.

Brookes, P.C., S.P. McGrath, D.A. Klein, and E.T. Elliott. 1984. In *Environmental Contamination*. CEP, Edinburgh.

Burkitt, A., P. Lester, and G. Nickless. 1972. *Nature* 238:327–328.

Cairns, J., and D.I. Mount. 1990. *Environ Sci Technol* 24:154–161.

Chaney, R.L. 1988. In J.R. Kramer and H.E. Allen, eds. *Metal Speciation: Theory, Analysis, and Application*. Lewis Publ, Chelsea, MI.

Chaney, R.L., and J.A. Ryan. 1994. In *Proceedings: Criteria for Decision Finding in Soil Protections: Evaluation of Arsenic, Lead, and Cadmium in Contaminated Urban Soils. Frankfurt, Germany, Oct. 9–11, 1991*.

Chang, A.C., and A.L. Page. 1996. In L. Chang, ed. *Toxicology of Metals*. CRC Pr, Boca Raton, FL.

Chang, A.C., A.L. Page, and T. Asano. 1993. In *Developing Human Health-Related Chemical Guidelines for Reclaimed Wastewater and Sewage Sludge Application in Agriculture*. World Health Organization (WHO), Geneva.

Coughtrey, P.J. 1978. *Cadmium in Terrestrial Ecosystems: A Study at Avonmouth*. PhD thesis, Univ Bristol, Bristol, United Kingdom.

Coughtrey, P.J., C.H. Jones, M.H. Martin, and S.W. Shades. 1979. *Oecologia* 39:57–60.

Doelman, P., and J.W. Vonk. 1994. In M.H. Donker, H. Eijsackers, and F. Heimbach, eds. *Ecotoxicology of Soil Microorganisms*. A SETAC Spec. Publ. Lewis Publ, Boca Raton, FL.

Domsch, K.H. 1991. *Toxicol Environ Chem* 30:147–152.

Eijsackers, H. 1994. In M.H. Donker, H. Eijsackers, and F. Heimbach, eds. *Ecotoxicology of Soil Microorganisms*. A SETAC Spec. Publ. Lewis Publ, Boca Raton, FL.

Eijsackers, H., and H. Lokke. 1992. *SERAS Soil Ecotoxicology Risk Assessment System: A Report from a European Workshop*. NERI, Silkeborg, Denmark.

Elinder, C.G. 1984. In J.O. Nriagu, ed. *Changing Metal Cycles and Human Health*, (Dahlem Konferenzen). Springer-Verlag, Berlin.

Ewers, U., and H.W. Schlipkoter. 1991. In E. Merian, ed. In *Metals and their Compounds in the Environment: Occurrence, Analysis, and Biological Significance*. VCH Publ, Weinheim, Germany.

Ferguson, J.E. 1990. *The Heavy Elements: Chemistry, Environmental Impact and Health Effect*. Pergamon Pr, New York.

Flanagan, P.R., J.S. McLellan, J. Haist, G. Cherian, H.J. Chamberlain, and L.S. Valberg. 1978. *Gastroenterology* 74:841–846.

Foulkes, E.C., ed. 1982. *Biological Roles of Metallothionein. Proc. Joint U.S.-Japan workshop (Cincinatti, Ohio)*. Elsevier, Amsterdam.

Freedman, B., and T.C. Hutchinson. 1980. In T. C. Hutchinson and M. Havas, eds. *Effects of Acid Precipitation on Terrestrial Ecosystems*. Plenum Pr, NewYork.

Fries, G.F. 1982. *J Environ Qual* 11:14–20.

Galal-Gorchev, H. 1991. *Food Additives Contam*. 8:793–806.

Gill R., M.H. Martin, G. Nickless, and T.L. Shaw. 1975. *Chemosphere* 4:113–118.

Hamada, R., and M. Osame. 1996. In L. Chang, ed. *Toxicology of Metals*. CRC Pr, Boca Raton, FL.

Hogstrand, C., and C.M. Wood. 1998. *Environ Toxical Chem* 17:547–561.

Hopkin, S.P., and M.H. Martin. 1985. *Bull Environ Contam Toxicol* 34:183–187.

Hopkin, S.P., G.N. Hardisty, and M.H. Martin. 1986. *Environ Pollut* 11:271–290.

Hutton, M. 1987. In T.C. Hutchinson and K.M. Meema, eds. *Lead, Mercury, Cadmium, and Arsenic*. SCOPE 31. Wiley, New York.

Jackson, N.E. 1985. *Research and Development in the Midlands and Western Regions*, 1985. Wolverhampton.

Jordan, M.J., and M.P. Lechevalier. 1975. *Can J Microbiol* 21:1855–1865.

Kloke, A., D.R. Sauerbeck, and H. Vetter. 1984. In J.O. Nriagu, ed. *Changing Metal Cycles and Human Health* (Dahlem konferenzem). Springer-Verlag, Berlin.

Kobayashi, J. 1978. In F.W. Oehme, ed. *Toxicity of Metals in the Environment.* Marcel Dekker, New York.

Langley, A. 1998. In R. Naidu et al., eds. *Contaminants and the Soil Environment in the Australasia-Pacific Region.* Kluwer, Dordrecht, Netherlands.

Little, P., and M.H. Martin. 1972. *Environ Pollut* 3:241–254.

Little, P., and M.H. Martin. 1974. *Environ Pollut* 6:1–19.

Martensson, A.M., and E. Witter. 1990. *Soil Biol Biochem* 22:977–982.

Martin, M. H., and R. J. Bullock. 1994. In S. M. Ross, ed. *Toxic Metals in Soil-Plant Systems.* Wiley, London.

McGrath, S.P. 1994. In S.M. Ross, ed. *Toxic Metals in Soil-Plant Systems.* Wiley, London.

McGrath, S.P., K.E. Giller, and P.C. Brookes. 1987. In *AFRC Report for 1986.* Harpenden, Herts, United Kingdom.

McGrath, S.P., P.C. Brookes, and K.E. Giller. 1988. *Soil Biol Biochem* 20:415–425.

McGrath, S.P., A.M. Chaudri, and K.E. Giller. 1995. *J Ind Microbiol* 14:94–104.

McKone, T.E., and P.R. Ryan. 1989. *Environ Sci Technol* 23:1154–1163.

Newman, M.C. 1995. *Quantitative Methods in Aquatic Ecotoxicology.* CRC Pr, Boca Raton, FL.

Nieboer, E., and D.H. Richardson. 1980. *Environ Pollut Ser B* 1:3–26.

Nogawa, K., and A. Ishizaki. 1979. *Environ Res* 18:410–420.

Nogawa, K., and T. Kido. 1996. In L. Chang, ed. *Toxicology of Metals.* CRC Pr, Boca Raton, FL.

Nogawa, K., Y. Yamaha, R. Honda, and T. Katoh. 1983. *Toxicol Lett* 17:263–266.

Nogawa, K., R. Honda, T. Kido, and H. Yamaha. 1989. *Environ Res* 48:7–16.

Nriagu, J.O. 1988. *Environ Pollut* 50:139–161.

Reddy, G.B., A. Faza, and R. Bennett. 1987. *Soil Biol Biochem* 19:203–205.

Roth, M. 1992. Metals in invertebrate animals of a forest ecosystem. In D.C. Adriano, ed. *Biogeochemistry of Trace Metals.* Lewis Publ, Boca Raton, FL.

Royle, S.M., N.C. Chandrasekhar, and R.J. Unwin. 1989. In *WRC York Symposium.* York, United Kingdom.

Ryan, J.A., and R.L. Chaney. 1997. In R. Prost, ed. *Contaminated Soils. INRA 85.* INRA, Paris.

Ryan, J.A., H.R. Pahren, and J.B. Lucas. 1982. *Environ Res* 28:251–302.

Spry, D.J., and J.G. Wiener. 1991. *Environ Pollut* 71:243–304.

Stigliani, W.M., P.R. Jaffe, and S. Anderberg. 1993. *Environ Sci Technol* 27:786–793.

Strojan, C.L. 1978a. *Oikos* 31:41–51.

Strojan, C.L. 1978b. *Oecologia* 32:203–212.

Tsuchiya, K., ed. 1978. In *Cadmium in Japan: A Review.* Elsevier, New York.

Tusgane, S., and H. Kondo. 1987. *Sci Tot Environ* 63:69–76.

Tyler, G. 1975. In T.C. Hutchinson, ed. *Proc Intl Conf Heavy Metals in the Environment,* Univ of Toronto, Canada.

Ullah, A. 2000. University of Dhaka, Bangladesh, personal communication.

[U.S. EPA] United States Environmental Protection Agency. 1986. *Air Quality Criteria Document for Lead.* U.S. EPA, Research Triangle Park, NC.

Vaidyanathan, L.V. 1975. In *Experiments and Development in the Eastern Region 1975.* Great Britain Agric Develop Advis Serv, Cambridge.

Van Gestel, C.A.M., and N.M. Van Straalen. 1994. In M.H. Donker, H. Eijsackers, and F. Heimbach, eds. *Ecotoxicology of Soil Microorganisms*. SETAC Spec. Public. Lewis Publ, Boca Raton, FL.

[WHO] World Health Organization. 1989. Offset Pub 87. WHO, Geneva.

WHO. 1990. *Environmental Health Criteria 101: Methylmercury*. WHO, Geneva.

WHO. 1991. *Environmental Health Criteria 118: Inorganic Mercury*. WHO, Geneva.

Wiener, J., and W. Grodzinski. 1984. In W. Grodzinski, J. Wiener, and P.F. Maycock, eds. *Forest Ecosystems in Industrial Regions: Studies on the Cycling of Energy, Nutrients, and Pollutants in the Niepolomice Forest, Southern Poland*. Springer-Verlag, Berlin.

Witter, E., A.M. Martensson, and F.V. Garcia. 1993. *Soil Biol Biochem* 25:659–669.

World Bank Group. 1999. *The Bangladesh Arsenic Mitigation Water Support Project: Addressing a Massive Public Health Crisis*. World Bank Bangladesh office.

Wren, C.D., S. Harris, and N. Harttrup. 1995. In D.J. Hoffman et al., eds. *Handbook of Ecotoxicology*. Lewis Publ, Boca Raton, FL.

Xian, X. 1989. *Plant Soil* 113:257–265.

6
Risk Assessment and Management in Metal-Contaminated Sites

How clean is clean? How clean is clean enough? These are indications of the changing cleanup paradigm being faced by the scientific community, policymakers, and industry. A decade or so ago policymakers believed that cleanup activities should use background (or baseline) values of contaminants in soils as an index of a successful cleanup (e.g., in the case of soil washing/flushing or excavation). Today, the cost and even the best available remedial technologies may not be able to achieve such rather wishful levels without destroying the integrity of the contaminated site. This has led to the current cleanup questions: *How clean is clean enough? In other words, how much risk reduction is acceptable without reaching the background level? Or can standards be tailored to specific land use?*

Environmental policies regarding water and air quality in Europe and North America have already been well developed even before addressing soil quality issues because they have been perceived as important sociopolitical issues. For example, national policies embodied in the Clean Air Act and Clean Water Act in the United States have already been in place for decades while a specific Clean Soil Act still needs to be legislated. A reason for this may be the clear visibility of polluted air (e.g., smog) and water (e.g., floating dead fish), whereas the consequences of polluted soils may not become apparent until, for example, plants die, which may take long periods to manifest. Another plausible reason is, perhaps the lack of meaningful dialogue among the scientific community, the public, and policymakers (including politicians) with respect to the soil's limited capacity for accepting pollutants.

1 Defining Soil Quality and Multifunctionality

Because of society's increasing awareness of the important role of soil on the health of plants, animals, and humans, *soil quality* has been defined in a number of ways. The following are the more commonly used definitions for soil quality, which are interchangeably used for soil health (soil health is preferred by the farming community, whereas soil quality is preferred by the science community).

Larson and Pierce (1991) define "*soil quality* as the capacity of a soil to function both within its ecosystem boundaries (e.g., soil map unit boundaries) and within the environment external to that ecosystem." They proposed *fitness for use* as a simple operational definition for soil quality and stressed the major functions of soil as a medium for plant growth, in partitioning and regulating the flow of water in the environment, and as an environmental buffer (see Chapter 2). Another definition for *soil quality* is the capability of a soil to produce safe and nutritious crops in a sustained manner over the long term, and to enhance human and animal health without impairing the natural resource base or harming the environment (Parr et al., 1992). More recently, *soil quality* is defined as the fitness of a specific kind of soil to function within its capacity and within natural or managed ecosystem boundaries, to sustain plant and animal productivity, maintain or enhance water and air quality, and support human health and habitation (Karlen

TABLE 6.1. Indices for soil quality and their relationship to soil functions (NRC, 1993).

	Soil functions			
Indicator	Promote plant growth	Regulate water flow	Buffer environmental changes	Promote biodiversity
Nutrient availability	Direct	Indirect	Direct	Direct
Organic matter	Indirect	Direct	Direct	Direct
Infiltration	Direct	Direct	Indirect	Direct
Aggregation	Direct	Direct	Indirect	Direct
pH	Direct	Direct	Indirect	Direct
Soil fauna	Indirect	Indirect	Indirect	Direct
Bulk density	Direct	Direct	Indirect	Indirect
Topsoil depth	Direct	Indirect	Indirect	Direct
Salinity	Direct	Direct	Indirect	Direct
CEC	Indirect	Indirect	Indirect	Indirect
Water holding capacity	Direct	Direct	Indirect	Direct
Soil enzymes	Indirect	Indirect	Indirect	Indirect
Soil flora	Indirect	Indirect	Indirect	Direct
Heavy metal bioavailability	Direct	Indirect	Direct	Direct

et al., 1997). Thus, it appears that soil quality can best be defined in relation to the functions that soils perform in ecosystems (Table 6.1).

The quality of soil resources historically has been closely related to soil productivity. Indeed, in many cases, soil quality and soil productivity have been used interchangeably (NRC, 1993). Other definitions relevant to *soil quality* include "healthy soil, an essential component of a healthy ecosystem, is the foundation upon which sustainable agriculture is built" and "soil health is the soil fitness to support crop growth without becoming degraded or otherwise harming the environment" (Acton and Gregorich, 1995).

Soil health degrades mainly by wind and runoff erosion, loss of OM, salinization, breakdown of soil structure, and chemical contamination. As such, soil quality cannot be measured directly much less quantitatively, but it can be deduced or estimated by measuring key indicators including physical, biological, and chemical properties such as pH, electrical conductivity, OM and nutrient content, and other indirect indicators, such as crop performance and surface water and groundwater quality. Thus, soil quality reflects the composite (holistic) picture of the soil physical, chemical, and biological properties and the processes that interact to determine its condition (see Table 6.1). In practice, soil quality is largely dependent on human activities. The ways man uses soil affects soil quality. Soil erosion strips away fertile topsoil, leaving the soil less hospitable to plants. Heavy farm equipment can compact the soil and impede its capacity to retain water. Loss of OM due to erosion or poor farming practices can seriously hamper the soil's ability to filter out pollutants (see Table 6.1).

For the purpose of soil protection, due regard must be directed to the topsoil, subsoil, and deeper layers and to their association with the groundwater. Because soil pollution decreases the availability of resources, it is imperative that soils be treated carefully in order to ensure sustainable ecosystems for future generations. To appreciate the role of soil in ecosystems it is necessary to enumerate their various functions (i.e., multifunctionality) for society and the environment; the most important include (Blum and Santelises, 1994): as a supporter of plants, animals, and humans; as a base for the production of food, fiber, and other biomass; as a controller of elemental and energy cycles in ecosystems; as a bearer of groundwater aquifers and mineral deposits; as a genetic reservoir (i.e., biodiversity); as a base for construction of buildings and various structures; and, as an archive of archaeology and natural history.

The word *function* refers to a set of soil properties or attributes that are essential for a certain type of land use. Implicitly included in the multifunctionality of soils is their ability to buffer inputs of environmental contaminants, such as acid precipitation and hazardous materials, including metals. Also implied is the importance of land use—the way soils should be utilized according to, in a broad sense, their *fitness* and *quality,* or more specifically, according to their level of contamination (e.g., the use of heavily contaminated soils should be restricted according to regulations). Thus, maintaining the multifunctionality of a soil indicates keeping all options for future land use open. A good multifunctional soil (1) poses no harm to any use by humans, plants, or animals; (2) can function without restricting the natural cycles; and (3) does not contaminate other parts of the environment (Vegter, 1995). Figure 6.1 illustrates the effect of soil contamination on the soil's

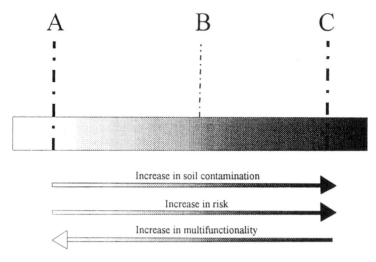

FIGURE 6.1. A, B, and C values (see text) as related to soil concentration of contaminants and soil multifunctionality.

multifunctionality. (Note: For a discussion of the A, B, and C values, see Section 3.1 Guidelines for Metal Levels in Soils.) In addition, society uses soils for the disposal of materials of anthropogenic origin, including hazardous waste.

Soils make it possible for plants to grow by mediating the biological, physical, and chemical processes that supply plants with nutrients and water. These processes buffer environmental changes in air quality, water quality, and global climate (Lal and Pierce, 1991). The soil serves as an incubation chamber for the decomposition of organic wastes, such as pesticides, sewage sludge, solid waste, and other organic materials. Depending on how they are managed, soils can be important sinks or sources of CO_2 and other gases that contribute to the greenhouse effect. Soils store, degrade, or immobilize nitrates, phosphates, metals, pesticides, radionuclides, and other substances that eventually can become air or water pollutants.

2 Potential Consequences of Polluted Soils

Physical, biological, and chemical processes may contribute to soil degradation. While much is known about physical processes of soil degradation, much less is known about the biological and chemical processes, especially pollution with heavy metals, radionuclides, organic chemicals, and mixed waste (i.e., inorganic and organic). Chemically degraded soils may result in an imbalance in the physical, biological, and chemical processes upon which soil productivity depends.

Although less acreage is involved compared with physically degraded soils, chemically degraded soils are becoming important in terms of food production and quality (Adriano et al., 1997a). Crops cultivated on soils of marginal quality may not only suffer in yield but also in quality, making them less competitive in the world market. For example, certain cereal and vegetable crops from eastern Europe may not be permitted in the European Union (EU) market because of questionable quality. Therefore, it has become apparent that polluted soils are not only a social and health issue but an economic issue as well (World Commission on Environment and Development, 1987).

The following are some potential major consequences of chemically degraded soils: poor physical–biochemical properties of soil, soil infertility, increased susceptibility of soil to erosion, diminished sustainability of the soil, contamination of the food chain, diminished quality of surface water and groundwater, lower biodiversity, human illness, and economic loss. Severely chemically degraded soils could manifest one or more of these consequences. The lesson from the collapse of the former Soviet Bloc countries where environmental quality was compromised for industrialization and militarization should remind us that sustainable development depends also on a sustainable environment.

3 Scientific and Legislative Basis for Soil and Water Cleanup

The life cycle of environmental contamination and cleanup is diagrammatically represented in Figure 6.2. It has basically four stages: first is the scientific perception stage; followed by a political perception stage; then by the legislative and policy action stage, and finally followed by the quality attainment stage. Databases obtained by scientists precipitate the concern over the issue. Then scientists can bring these concerns to the attention of politicians and the public, primarily through the media. Legislation then can be developed, enacted, and implemented with public and industry participation. It is apparent from Figure 6.2 that from the time scientific data are gathered to the time successful environmental laws are legislated, considerable time could elapse. Then more time would be required for the completion of remediation; the time varies according to the nature and extent of contamination, public demand, adopted technology, and cost. A suite of interim regulatory guidelines have emerged in North America and within the EU that are rather fragmented and inconsistent. Although most depend on professional judgments, some were likely formulated in the

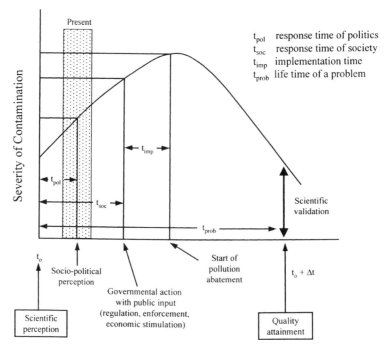

FIGURE 6.2. Diagram of the life cycle for environmental contamination and remediation.

absence of reliable technical data. Environmental research has been a priority for many countries and much information has and will continue to be obtained. Unfortunately, a considerable portion of this information is generated from an academic point of view without full recognition of the needs of policymakers and regulators. Conversely, the regulatory focus has often been inconsistent and lacking scientific credibility, which has impaired a coordinated approach. In many cases, cleanup goals do not appear achievable at all, regardless of the time allotted. This dilemma is exacerbated by the public's outcry that the environmental abuse resulting from past negligence be cleaned up rapidly and permanently, and a frequently voiced desire for zero contamination and zero risk, which is totally unrealistic even in most developed countries.

3.1 Guidelines for Metal Levels in Soils

In this book, a soil is considered contaminated according to the definition offered in Chapter 4. Background level corresponds to the content of metals in soils not effected by human activities. These values are available in a number of publications (Adriano, 1986; Alloway, 1995; Kabata-Pendias and Pendias, 1992; Merian, 1991) (Appendix Table A.24).

In reality, human influence on soil composition cannot completely be excluded due to the diffuse sources of anthropogenic contamination, e.g., air pollution, application of fertilizers, pesticides, etc. Their presence even in pristine areas and their ecological effects are well documented. Deposition of metals are evident in snow packs in the Arctic region and in humus layers in remote forest ecosystems (Lindberg et al., 1990). Background values may vary as a function of the locality from which a given soil is sampled. For example, metal concentrations in serpentine-derived soils can be highly toxic to animals and plants as a result of the naturally elevated contents of the parent rock from which the soil is derived. Metal concentrations in soils are also known to be affected by the clay and OM contents of soils and increase almost linearly as a function of them. Because of the spatial variability of metals in soils, background values may not serve as good reference values for legislative purposes. Therefore, it is not possible to arrive at a single background value for any of the metals. Perhaps the most extensive set of quantitative environmental standards is for water, like the ones developed by the U.S. EPA. Soil standards are fewer since implementing them is technically difficult.

In an effort to expedite remediation of hazardous waste sites across the United States in the absence of a national soil cleanup standard, many state agencies have developed their own cleanup standards that vary from those of other states (Bryda and Sellman, 1994). These variations are due to different assumptions, algorithms, and exposure scenarios. Despite this lack of consistency among the states' cleanup levels, some trends are apparent. In general, cleanup levels promulgated for industrial sites tend to be less stringent than those for residential sites. On the other hand, soil cleanup levels established to protect groundwater quality tend to be more stringent than those established, based on direct human exposure, for contaminated soils. In addition, carcinogens tend to be assigned more stringent levels than noncarcinogens.

At present, there is only one set of critical levels that apply to all the countries of the EU, those defined in Annex 1A of Council Directive 86/278/ EEC, which establishes limit values for concentrations of metals in soils that should not be exceeded when sewage sludge is applied in agriculture (Marmo, 1999; Reiniger, 1997). The directive has been implemented in the form of national laws, with some much lower nationally defined critical levels (e.g. Netherlands, Denmark, Sweden). In many cases, these critical values have been extended to soils in general and not only limited to the application of sewage sludge.

Several European countries have adopted soil protective policies having two- to three-tier (stage) systems. This "precautionary" principle is a common tenet in European environmental policy that aims to safeguard soil quality for a wide range of functions (Reiniger, 1997; Vegter, 1997). Alongside the precautionary (or protective) levels, separate critical values are established for the cleanup of contaminated sites, based on functional criteria and toxicological aspects. The first tier is generally known as the *reference* level, designated also as *guide, target, baseline,* or *preventive* values. Exceeding reference values indicates that the soil's multifunctionality might be imperiled, but no hazards may be expected. The next tier is the *investigation* level, designated also as *trigger, signal* or *threshold* values where potentially adverse effects on organisms are likely to occur (i.e., hazards possible). In essence, these values are still *tolerable*. When these values are exceeded, further site-specific investigations that include the relevant exposure pathways have to be conducted (equivalent to risk assessment), followed by consideration of remedial options. The *remediation* level, also referred to as *cleanup, action,* or *intervention* values are for severe contamination that warrants a definite need for remedial action, as dictated by risk assessment that confirms certain hazards exist. This tier system is also known in Europe as the A-B-C value system (see also Fig. 6.1). An example for Cd is displayed for Switzerland, Germany, and the Netherlands (Fig. 6.3).

After 10 years of cleanup experience, the A-B-C system for the Netherlands has recently been revised to include only A-C values (Vegter, 1995). The new C values in the Dutch guidelines were based on exposure routes and toxicological criteria. As more reliable technical data become available, the values or the tier system will be modified. Other countries, notably Canada, the United Kingdom, Germany, and the Netherlands have progressed further than the United States in setting soil standards for soil protection. The relative dearth of soil standards, let alone their inconsistency, tends to render the decision-making process more contentious.

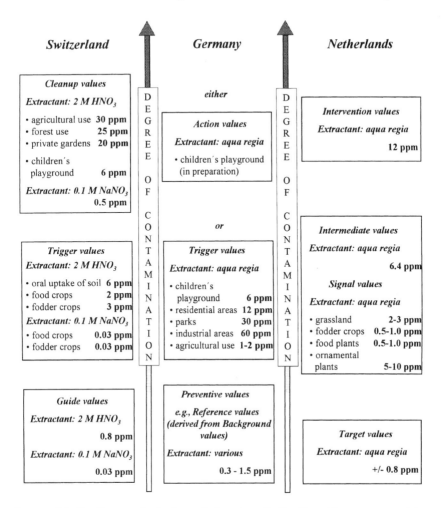

FIGURE 6.3. Quality standards for cadmium in soils for three European countries. All German values and Swiss trigger and cleanup values are provisional only. (Modified from Vollmer et al., 1997.)

Several attempts were made to adapt background values to certain local situations in order to obtain more viable reference values for regulatory decision. Data in Table 6.2 present an example of reference values for eight metals legally adopted for the region of Flanders, Belgium. The information on this table was based on several surveys of soils, collected throughout Flanders from sites remote from sources of metal pollution (Tack et al., 1995). The mean clay content was 9% and OM was 2%. The definition for the standard soil value, as presented in Table 6.2, was then deduced from these data.

TABLE 6.2. Reference (or target) values (mg kg^{-1} soil DW) of standard soil samples and correction factors (A, B, C) for the equation, legally adopted in Flanders, Belgium.[a]

Element	Reference value	A	B	C
As	19	14	0.5	0
Cd	0.8	0.4	0.03	0.05
Cr^{3+}	37	31	0.6	0
Cu	17	14	0.3	0
Hg	0.55	0.5	0.0046	0
Ni	9	6.5	0.2	0.3
Pb	40	33	0.3	2.3
Zn	62	46	1.1	2.3

[a]Besluit van de Vlaamse Regering houdende achtergrondwaarden, bodemsaneringsnormen en toepassingen van gereinigde grond. Ministry of Environment and Employment, Brussels, April 22, 1995. The standard soil sample contains 10% clay and 2% OM.

The reference (or target) values adopted correspond to the 90th percentile upper limit of background values for the standard soil. The reference values for soil samples not corresponding to the composition of the standard soil can be corrected as follows:

$$N(x, y) = N(10, 2) \times \frac{A + xB + yC}{A + 10B + 2C} \tag{6.1}$$

where N = reference value, when the clay content is x or 10%, and the OM is y or 2%, x and y are the respective clay and OM contents of the soil sample, A, B, and C are factors depending on the element in question presented in Table 6.2.

Similarly, the role of clay fraction and soil OM was taken into consideration by the Dutch in developing the reference values for metals (Vegter, 1995). In this case, a standard soil contains 25% clay and 10% OM.

Following the isolated case of metal pollution from irrigation water in the Jinzu River basin in Japan in the late 1950s, maximum allowable limits of heavy metals in agricultural soils were set by the Japanese Agricultural Land Soil Pollution Prevention Law for Cd, Cu, and As (Asami, 1997). These were based on the maximum allowable limits (MALs) of Cd in rice (i.e., 1 mg kg^{-1} in unpolished rice) and for Cu and As in paddy soil (125 mg Cu kg^{-1} DW by 0.1 M HCl extraction, and 15 mg As kg^{-1} by 1 M HCl extraction, respectively). Restoration of polluted paddy soils in Japan was accomplished by covering the contaminated surface soil with a 25- to 30-cm deep layer of noncontaminated soil that enabled the Cd content in the rice to fall below the desired 0.1 mg kg^{-1} level. Today, there are 54 sites in Japan having as much as 6000 ha of paddy fields identified as highly polluted by Cd according to the agricultural land soil pollution policy.

In Lekkerkerk, the Netherlands, a former waste dump was discovered in 1980 in a residential area (Theelen et al., 1995). This created a major public outcry that led to a legislative mandate for soil restoration in this country. Metals, inorganics, and a wide range of organic compounds were found. For each contaminant, three different values were initially adopted:

A—*Reference* (or target) value—soil is considered of good quality.
B—*Threshold* value (for pollution)—above which no biological or ecological damage is yet observed; soils with this level of pollution however, should be further monitored.
C—*Intervention* value—above which cleanup is mandated.

These A, B, and C criteria were recently revised (Vegter, 1997). Table 6.3 presents the values for metals adopted by the Dutch legislature in May 1994. The intervention values for soil remediation will be used to assess whether contaminated land poses serious threat to public health. These values indicate the concentration levels of the metals in soil above which the (multi)functionality of the soil for human, plant, and/or animal life is seriously compromised or impaired. Concentrations in excess of the intervention values correspond to serious contamination. The intervention values replace the old C values in the soil protection guidelines.

The present values (1) take into account both human toxicological and ecotoxicological considerations; (2) are based on considerations not only of the natural concentrations of the contaminants, which indicate the degree of contamination and its possible effects, but also of the local circumstances, which are important with regard to the extent and scope for spreading or contact; (3) are related to spatial parameters; they are regarded as having been exceeded and the soil is therefore being seriously contaminated if the soil mean concentration in at least 25 cubic meters of soil volume exceeds the intervention values; (4) are dependent on soil type, since they are related to the content of OM and clay in the soil; this is achieved by means of the

TABLE 6.3. Dutch target (also referred to as A value or reference value) and intervention (also referred to as C value) values for selected metals in soil.

Element	Soil (mg kg^{-1} DW)	
	Target value	Intervention value
Arsenic	29	55
Barium	200	625
Cadmium	0.8	12
Chromium	100	380
Cobalt	20	240
Copper	36	190
Lead	85	530
Mercury	0.3	10
Molybdenum	10	200
Nickel	35	210
Zinc	140	720

Source: Dutch Ministry of Housing, Spatial Planning, and Environment, The Hague, Netherlands (Vegter, 1997).

so-called soil type correction formula; and (5) are important for remedial and preventive policies. The target values (see Table 6.3) indicate the soil quality levels ultimately aimed for. These values are derived from the analysis of field data from relatively pollution-free rural areas, regarded as noncontaminated.

In practice, the strict application of the Dutch target values is difficult to achieve for several reasons, among which the most obvious are (1) essential parameters influencing the bioavailability of metals (e.g., metal speciation, soil properties such as pH, CEC, etc.) were not taken into account, and (2) the land use after restoration, based on the context of soil multifunctionality, was also not taken into account (Theelen et al., 1995).

In other European countries, the public has asked for similar legislation for soil restoration. Tolerable metal concentrations for agriculture and horticulture were published in Germany and in Switzerland (Vollmer et al., 1995), where they were included in Sections 33 and 39 of the Environmental Protection Law of 1983. In the United Kingdom, soil use after restoration was proposed as a criterion for determining the threshold value for redevelopment of contaminated sites (Dept. Environ., 1987). In Germany, threshold values were introduced for playgrounds, parks, parking areas, and industrial sites (Eikmann and Kloke, 1993). In agricultural and horticultural soils, lower threshold values were proposed when growing leafy vegetables than for fruit production, or for the cultivation of grain or ornamental plants. A similar approach is being proposed in Poland where agricultural and horticultural uses vary according to the severity of soil contamination (Kabata-Pendias and Adriano, 1995).

In recent environmental legislation of the Flemish region of Belgium, cleanup values for restoration that somewhat correspond to the intervention values vary with the intended land use for the remediated site (Table 6.4). These threshold values were defined using the Human Exposure to Soil

TABLE 6.4. Cleanup values (in mg kg^{-1} standard soil, DW) adopted in Flanders, Belgium, as a function of the intended use of remediated site.[a]

Element	Natural park	Agriculture	Residential area	Recreational area	Industrial area
As	45	45	110	200	300
Cd	2	2	6	15	30
Cr^{3+}	130	130	300	500	800
Cu	200	200	400	500	800
Hg	10	10	15	20	30
Ni	100	100	470	550	700
Pb	200	200	400	1500	2500
Zn	600	600	1000	1000	3000

[a]These five classes of land use are extensively described in the legislation "Besluit van de Vlaamse Regering houdende achtergrondwaarden, bodemsaneringsnormen en toepassingen van gereinigde grond." Ministry of Environment and Employment, Brussels, April 22, 1995. The standard soil sample contains 10% clay and 2% OM.

Pollution Model, which estimates the transfer of contaminants from soil to man by different pathways (i.e., by inhalation, ingestion, drinking water, animal or plant food, etc.) (Stringer, 1990). It was recently improved and several other models (Theelen et al., 1995) are proposed to assess the human risk of soil pollution.

An overview of cleanup goals (actual and potential) for both total and leachable metals in the United States are displayed on Appendix Table A.25. Based on inspection of the total metal cleanup goals, one can see that they vary considerably both within the same metal and among metals. Similar variation is observed in the actual or potential leachate goals. The observed variation in cleanup goals has at least two implications with regard to technology alternative evaluation and selection. First, the importance of identifying the target metal(s), contaminant state (leachable vs. total metal), the specific type of test and conditions, and the numerical cleanup goals early in the remedy evaluation process is made apparent. Depending on which cleanup goal is selected, the required removal or leachate reduction efficiency of the overall remediation can vary by several orders of magnitude. Second, the degree of variation in goals both within and among the metals, plus the many factors that affect mobility of the metals suggest that generalizations about the effectiveness of a technology for meeting total or leachable treatment goals should be viewed with some caution.

4 Risk Assessment

Contamination of soils from anthropogenic chemicals and their subsequent degradation has become a major concern because of the critical role of soil resources in promoting sustainable environment and economic development. Both anthropogenic inorganic and organic compounds in soils may not only adversely affect their production potential but also may compromise the quality of the food chain and the underlying groundwater. This scenario may require regulatory-driven risk assessment and evaluation of remedial technology to a specific site condition.

Risk assessment is a complex, multifaceted process that identifies the sources and pathways of chemical exposure and quantifies the risks resulting from such events. Suter (1993) defines *risk assessment* as an array of methodologies that has grown out of the actuarial techniques of the insurance industry and is concerned with the estimation of probabilities and magnitudes of undesired events such as human morbidity, mortality, or property loss. Environmental risk assessment deals with the risks that arise in or are transmitted through the air, water, or biological food chains to humans. Regulatory agencies and international bodies concerned about public health, food safety, and environmental protection are interested in assessing the exposure of the general population or subgroups to hazardous chemicals present in the environment. Results of risk assessments are often used as the basis for taking regulatory action for issuing advisory guidelines.

Risk assessment refers to the characterization of the potential adverse health effects of human exposures to environmental hazards. *Risk* is conceptually a random variable related to chance or peril. In this case, it denotes a predictable probability that an adverse effect would result from a given level of exposure to potentially toxic chemical substances. Risk assessment includes several aspects (Suter et al., 2000; U.S. EPA, 1998): description of the potential adverse health effects based on an evaluation of results of epidemiological, clinical, toxicological, and environmental research (*hazard identification*); extrapolation from those results to predict the type and estimate the extent of health effects in humans under given conditions of exposure (*dose–response assessment*); judgments on the number and characteristics of persons exposed at various intensities and durations (*exposure assessment*); and summary judgments on the existence and overall magnitude of the public health problem (*risk characterization*). Risk assessment also includes characterization of the uncertainties inherent in the process of inferring risk.

Hazard identification is a procedure by which potentially toxic chemicals in an affected site are identified, and the need for undertaking a risk assessment is evaluated. End points to be used to assess effect and data required to value the assessment need to be clearly identified. Data from epidemiological and environmental toxicological investigations usually are the starting point of the selection process. If a chemical is potentially toxic, as indicated by epidemiological or toxicological data, its occurrence in soil and fate during treatment should be examined to ensure it will not be a health hazard when the site is chosen for a certain land use.

While the hazard identification portion of the risk assessment can establish causality, it does not provide quantitative information on exposure. Exposure assessment defines the amount and availability of pertinent substances and should include the magnitude, frequency, and duration of exposure. The dose–response relationship quantitatively defines the effect or lack of effect with a clear statement of confidence of statements, i.e., statement of statistical confidence and power. This relationship is specific for a chemical and may be derived from data obtained in human epidemiological investigations, lifetime feeding studies of experimental animals, and toxicity assays of mammalian or bacterial cells. Depending on the nature of the data, the resulting dose–response relationships may vary considerably.

In a contaminated site, a toxic chemical may be transported by several media at the same time and is susceptible to various degradation mechanisms. Based on these environmental transport pathways, plausible exposure scenarios must be developed that describe the circumstances surrounding an exposure event. Exposure scenarios are more likely to be hypothetical events. It is essential that the exposure scenarios one develops reflect realistic situations. Each exposure scenario is associated with an identifiable risk group (exposed population). All individuals of the exposed population receive exposure at levels above the background and they must be a clearly identifiable segment of society. In its most complete form, an exposure analysis should determine the magnitude and duration of exposure for each scenario; the size, nature, and classes of the exposed population; and uncertainties of all estimates.

Because it is difficult to obtain environmental measurement data for the exposure analysis, model simulations are an accepted and normal part of the exposure assessment process. Two approaches have been used to describe the environmental transport of pollutants mathematically (Jones et al., 1991). Some models assume steady state equilibrium between concentrations of the pollutant in environmental compartments along an exposure route (*quasi-equilibrium models*). In this type of model, the partition of a chemical between two adjoining environmental compartments is described by linear transfer coefficient. The other type of model is designed to describe the spatial- and temporal-dependent behavior of a pollutant in the environment (dynamic models). Consequently, *dynamic models* are structured to approximate the kinetics in a real system and are applicable to all situations (e.g., acute, intermittent, and continuous exposures). They are however, impractical to use because spatial-, temporal-, and concentration-dependent model parameters are difficult if not impossible to define.

Prior to undertaking any form of risk assessment, there are several key issues to be considered (Langley, 1996): (1) What is the purpose of the risk assessment? (2) Who (or what) are to be protected? (3) What is the expected end point of the risk assessment? Is there any concern for cancer and/or noncancer outcome? (4) Is it to be a quantitative or qualitative assessment of risk? Is it the actual, calculated, or perceived risk? There are likely to be considerable differences between the types and an appreciation of each is necessary. If it is for quantitative risk assessment, there is a need to determine what quantitative level of risk is acceptable or tolerable (unless several risks are being compared in order to rank them rather than to consider the absolute level of risk). If a qualitative risk assessment process is used, what criteria will be used to categorize or interpret the results? (5) How is the risk assessment information going to be used? for regulatory purposes? for standard setting? for education? to provide advice? (6) Is there concern for current exposures, or past and future exposures? (7) Is there concern for the risks to individuals, groups of people, or for the population as a whole? If for individuals, are those typical individuals or those with worst case exposures? (8) Are there mixtures with the possible interactions between substances? and (9) What information is needed for sensible decision making?

Risk assessment has an increasingly important role in the remediation of hazardous waste sites. Both federal and state (or provincial) regulations call for the use of risk assessment techniques to help in identifying and prioritizing sites requiring remediation, developing remedial objectives and cleanup standards, and selecting the most appropriate remedy for a particular location. Protection of human health may not ensure adequate environmental protection thus, there has been an emphasis on the development of ecological risk assessment methods. These methods must consider a variety of receptors and end points at the population, community, and ecosystem levels as well as a suite of exposure pathways (Newman, 1995). In the assessment phase of remediation, the *diagnosis* and *prognosis* approach should be observed (Prost et al., 1997), i.e., the current degree of risk as well as the long-term status of sites with respect to the contaminants in question. Thus, the final choice of remedial technology largely depends on the nature and degree of contamination, the intended function or usage of the remediated site, and the availability of innovative and cost-effective techniques. The choice is further complicated by environmental, legal, geographical, and social factors. More often, the choice is site specific. For example, home gardens and agricultural fields in large rural areas that are contaminated may require a remedial approach different from that for smaller but more heavily contaminated areas. Similarly, large areas around old mining and smelter sites need an approach that differs from that of a heavily polluted area.

Risk assessment frameworks have been established in the United States by the Environmental Protection Agency (EPA). Regulatory agencies in Canada and the United States are developing suites of tests for use at contaminated sites (Linz and Nakles, 1997). Many of the agencies rely on a similar set of terrestrial and aquatic species in their ecotoxicity testing protocol. However, these frameworks do not specifically address human health concerns, although some genotoxicity tests are included.

Risk assessment is used as a basis for determining acceptable levels of exposure for toxic chemicals in order to prevent unnecessary excessive exposure (Suter et al., 2000; U.S. EPA, 1998). Risk assessment of metals generally follows the same methodology used by toxicologists, which is to identify the toxic properties and the dose–response relationship, followed by assessment of exposure and risk characterization (Fan, 1996). The actual exposures may be reflected in biological media, such as blood and urine, and in some cases, biomarkers in the biological system.

The Remedial Investigation/Feasibility Study (RI/FS) process for contaminated sites includes human health and ecological risk assessment (Fig. 6.4). Information gathered is used for the baseline risk assessment, development of preliminary remediation goals, and risk evaluation of remedial options. There are two risk assessment frameworks: an ecological assessment and a human health assessment.

There are many similarities and differences between the two frameworks (Linz and Nakles, 1997). A major difference is that the human health risk assessment framework addresses risks to people at the individual level, while the ecological risk assessment framework (Fig. 6.5) addresses key

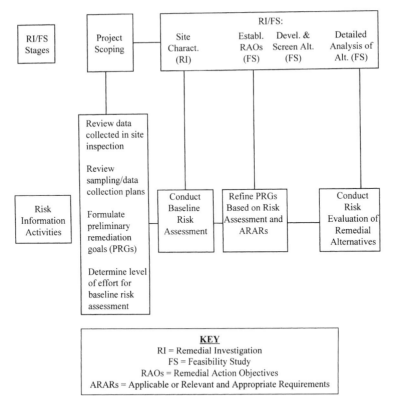

FIGURE 6.4. Risk assessment activities in the remedial investigation/feasibility study (RI/FS) process. (Reprinted from *Environmentally Acceptable Endpoints in Soil,* Linz and Nakles. Copyright © 1997, with permission from Elsevier Science.)

species typically at the population level. In addition, some of the end points used in ecological risk assessments may not be as readily understood to risk managers as the end points (e.g., cancer) used in human health risk assessments. Another major difference lies in how regulatory criteria are established. Under the ecological framework, a weight of evidence approach, along with an estimate of the potential for risk are used to develop regulatory decisions, whereas numerical estimates of acceptable cancer risk and noncarcinogenic hazards are used in the human health framework. Ecological risk assessments evaluate the likelihood that adverse ecological effects will occur or are occurring as a result of exposure to stressors related to human activities, such as dredging or filling of wetlands, or release of chemicals (Norton et al., 1995; U.S. EPA, 1992). The term *stressor* describes any chemical, physical, or biological entity that can induce adverse effects on *ecological components,* i.e., individuals, populations, communities, or ecosystems. Adverse ecological effects encompass a wide range of disturbances, from mortality in an individual organism to a loss in ecosystem function.

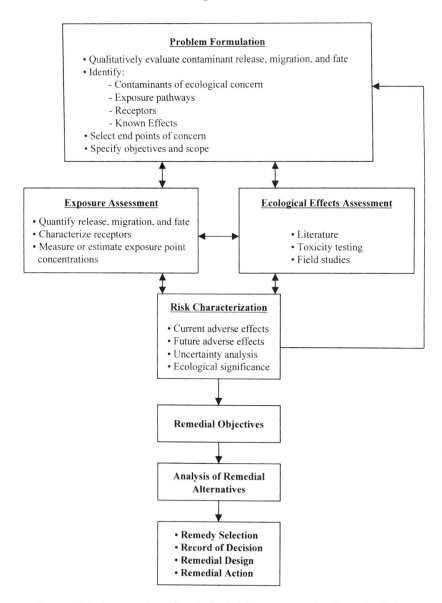

FIGURE 6.5. An overview of ecological risk assessment for Superfund sites.

In the Netherlands, a new methodology for assessing risk of contaminated soil has been developed (Theelen et al., 1995). The method is based on a scientific approach using risk analysis that includes exposure assessment as well as human and ecotoxicological data. This approach is based on the premise that soil contamination cannot be tolerated if risks to humans or ecosystems that exceed the *maximum tolerable risk* (MTR) levels are likely to occur.

The MTR level for humans in the Netherlands is determined by the *tolerable daily intake* (TDI), which can be estimated from epidemiological studies in human toxicology or derived from toxicological studies on experimental animals (WHO, 1987). Essentially, it describes the daily maximum tolerable exposure. Then, the exposure of humans from a contaminated site must be calculated. Such exposure calculation can be carried out using an exposure model, the C-SOIL model, developed by the Dutch National Institute of Public Health and Environmental Protection (Van den Berg et al., 1993). The overall Dutch intervention value is set by combining the human and ecotoxicological values (Theelen et al., 1995). By taking into account the contribution of the various pathways, one can calculate the TDI of a chemical. Then, the human toxicological intervention value is taken as equal to the concentration of the chemical in the soil or groundwater, resulting in an intake equal to the TDI.

For ecological risk assessment, a statistical approach is used in which a percentage of ecosystem species at risk is related to the chemical concentration in the soil (Denneman and Robberse, 1990; Van Straalen and Denneman, 1989). The median value, i.e., roughly 50% of the species at risk, is assigned to the ecological intervention value. This value is designated as HC_{50}, or the *hazardous concentration* for ∼50% of the species.

Risk assessment frameworks are also available for Germany, the United Kingdom, Switzerland, New Zealand, and Australia (Langley, 1996; Theelen et al., 1995; Vollmer et al., 1997). The German model (UMS) is getting ready for regulatory deployment. A risk-based approach is also being followed in the United Kingdom, where trigger values are calculated by the CLEA (Contaminated Land Exposure Assessment) model. The C-SOIL, UMS, and CLEA models are similar in that they calculate the acceptable daily exposure. The Australia–New Zealand framework has been heavily influenced by the U.S. and Dutch methodologies (Langley, 1996). Thus, with risk-based approaches being adopted in cleanup policies in the United States, Canada, the United Kingdom, the Netherlands, and Germany, risk assessment that integrates human and ecological end points has become the norm.

4.1 Utilization of Dose–Response Relationship

An assessment of dose–response relationships (Fig. 6.6) can reveal the derivation and importance of several parameters used in risk assessment. The *reference dose* (RfD) or *reference concentration* (RfC) is typically determined from data generated from a subchronic or chronic animal study and represents an estimate of the level of exposure to which the human population, including sensitive segments of the population, can be exposed

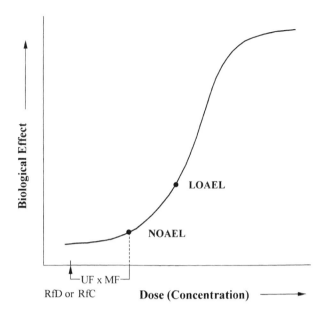

FIGURE 6.6. A dose–response curve from a typical toxicology study showing the relationship between reference dose (RfD) or reference concentration (RfC) with the no observable adverse effect level (NOAEL). UF, uncertainty factors; MF, modifying factor, LOAEL, lowest observed adverse effect level. (Reprinted from *Food and Chemical Toxicology*, Munro. copyright © 1981, with permission from Elsevier Science.)

during a lifetime without detrimental effects. The RfD and RfC are used by the U.S. EPA for noncarcinogenic health effects having a threshold level associated with chronic exposure to chemicals (Rubenstein and Segal, 1993). The RfD assesses only systemic health effects (i.e., health effects other than cancer and gene mutations). These effects are considered to occur only after a threshold exposure has been exceeded. The RfD serves as a benchmark for deriving a regulatory threshold level to protect exposed populations from adverse health effects from noncarcinogens due to ingestion of chemicals. The RfD is defined by the U.S. EPA (1988) as an estimate, with uncertainty spanning perhaps one order of magnitude, of a daily exposure to the human population (including sensitive subgroups) that is likely to be without an appreciable risk of detrimental effects over a lifetime of exposure. The RfD is a daily intake level and is a risk assessment tool.

The RfD is derived from the highest experimentally determined dose in humans or animals without adverse biological effect (NOAEL, no observed adverse effect level) divided by the product of several uncertainty factors (UF) and a modifying factor (MF) (Fig. 6.6). RfDs should not be used as a sharp demarcation between safe and unsafe exposures because of the uncertainties inherent in RfD derivation. The LOAEL (lowest observed adverse effect level, the lowest dose or exposure level that produces an adverse health effect) is the dose or concentration higher than the NOAEL on the curve (see Fig. 6.6).

The NOAEL is defined as the highest experimental dose of a chemical at which there is no statistically or biologically significant increase in frequency or severity of a toxicological effect between an exposed group and its valid control. Adverse effects are defined as any effects that result in functional impairment and/or pathological lesions that may affect the performance of an organism or ability of an organism to respond to new challenges. The NOAELs for several toxicological end points differ. The major end points in chronic as well as subchronic exposures are not mortality but nonlethal adverse effects, which can be defined by biochemical, hematological, or clinical measurements. In instances in which a NOAEL cannot be demonstrated, a LOAEL may be used to evaluate critical toxic end points.

In principle, the RfD (or RfC) is similar to the concept of the Food and Drug Administration's (FDA) *acceptable daily intake* (ADI) and the Occupational Safety and Health Administration's (OSHA) *permissible exposure limit* (PEL). For groundwater intended for drinking, *maximum concentration limits* (MCLs) are set of regulatory limits for each chemical of concern.

The sigmoidal shape of the dose–response curve in Figure 6.6 can be mathematically defined by the flexibly shaped Richards sigmoid model which has been applied to describe uptake–effect (growth) relationships in consumer organisms (e.g., waterfowl) (Brisbin and Newman, 1991).

4.2 Defining Environmentally Acceptable End Points

As part of an ongoing debate on "what concentration of contaminant is environmentally acceptable" relative to site closure, an *environmentally acceptable end point* (EAE) needs to be defined. The EAEs for soils are commonly defined as concentrations of chemicals or other indices of contamination (e.g., leachability, or biological response) that are judged acceptable by a regulatory agency or an appropriate entity based on established standards or guidelines, or following an analysis of site- or chemical-specific information and/or testing (Linz and Nakles, 1997). In either case, risk management decisions are made that account for various factors related to the site and the chemicals present at the site. The following factors need to be considered in arriving at EAEs for a given site:

1. Extent and nature of contamination
2. Current and future potential land use
3. Availability of chemical- and site-specific data
4. Risk to ecological or human receptors
5. Federal and/or state standards or guidelines

6. Feasibility
7. Cost
8. Public acceptance

Subsequently, the most commonly used approaches for establishing EAEs in soils are (Linz and Nakles, 1997) (1) site-specific target concentrations derived from site- and chemical-specific data, and risk-based analysis; (2) chemical- or medium-specific criteria or guidelines; (3) adaption of local, state, or regional background or baseline concentrations; and (4) state or federal limits developed using risk-based analysis.

4.3 Role of (Bio)availability in Risk Assessment

In a more generic sense, the term *availability* refers to (1) the rate and extent at which the chemical is released from the medium of concern into the environment (e.g., groundwater) or (2) the bioavailability of the chemical to ecological and human receptors through direct contact, inhalation, ingestion, or absorption (or uptake). In other words, the chemical's mobility and/or bioavailability can be used to express the chemical's availability.

Availability of chemicals may mean different things to different disciplines, as exemplified by (1) leaching to groundwater, (2) volatilization, (3) capillary rise to topsoil, (4) outcropping or lateral movement, (5) bioavailability to plants, (6) bioavailability to microorganisms, (7) bioavailability to invertebrates, (8) bioavailability to insects, (9) bioavailability to animals (fish, wildlife, etc.), and (10) bioavailability to humans (by dermal contact, inhalation, or ingestion).

The different bioavailability pathways have been discussed in Chapter 3 and range from absorption (or uptake) by roots and foliage for plants to the three main exposure pathways, i.e., dermal contact, inhalation, and ingestion, for humans.

The risk arising from metals largely depends on their bioavailability, which in turn depends on the form (i.e., more specifically, chemical speciation) in which they occur. This is because the risk to human health by metals cannot be solely assessed on the basis of the total concentration of the toxic metal in question, but rather on their physicochemical state (Kelly, 1988; Prost et al., 1997). The speciation of metals depends on a number of factors, including their behavior on the soil's solid phase, pH, redox potential, complexation with soil components, and others (Adriano, 1986; Alloway, 1995; McBride, 1994; Sposito, 1989). For example, the bioavailability and toxicity of Cr depends on its oxidation state: Cr^{6+} is highly toxic to organisms, whereas Cr^{3+} is much less toxic. Metals that form organometallic complexes have increased mobility and therefore, may become more bioavailable to organisms. However, metals in this form are not necessarily more toxic. For example, free Cu^{2+} ions are more toxic to aquatic organisms (e.g., algae) than Cu complexed by humic substances (Florence et al., 1984;

Hirose and Sugimura, 1985). In the case of Cd, $CdCl_2$ is more bioavailable to potato (*Solanum tuberosum*) than free Cd^{2+} ions (McLaughlin et al., 1994; Tiller et al., 1997).

Bioavailability pertains to the availability of a metal with respect to a specific organism and environmental conditions, which can be measured in terms of its effect on the organism. In the soil–plant system, it is manifested in the plant itself in the form of growth or physiological functions, such as shoot biomass, root development, respiration, etc. In acidic soils, most metals are known to have increased bioavailable species with the exception of a few elements, such as Mo, Se, and possibly As. The primary soil factor contributing to metal mobility and bioavailability to plants is pH, followed by OM (Adriano, 1986; Kabata-Pendias and Adriano, 1995; Page et al., 1983). These factors have not been consistently incorporated into soil criteria for soil remediation.

One of the most important site- and chemical-specific factors is the availability of the chemicals in the medium. The importance of chemical availability on risk management of contaminated sites is depicted in Figure 6.7, which shows the various exposure routes from contaminated soils. Chemical availability will determine the extent of exposure from all of the routes. This simply implies that the chemical would be inconsequential if only a small fraction of the chemical was available (i.e., mobile or bioavailable).

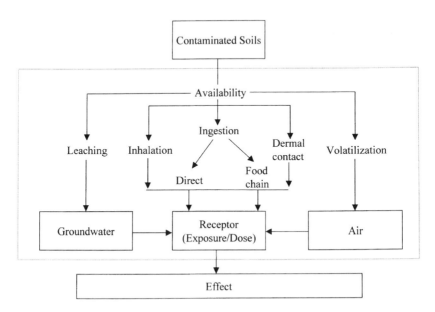

FIGURE 6.7. Role of (bio)availability in physical and biological effects, and risk assessment. (Reprinted from *Environmentally Acceptable Endpoints in Soil,* Linz and Nakles. Copyright © 1997, with permission from Elsevier Science.)

A more quantitative illustration of the importance of chemical (bio)avail-ability can be provided by examining a standard risk calculation (Equation 6.2) adapted for hydrocarbons and other organic chemicals in soils (Linz and Nakles, 1997).

$$Risk = CS \times \frac{EF \times ED \times IR \times ABS}{BW \times AT} \times CPF \qquad (6.2)$$

where CS = chemical concentration in the soil (mg kg^{-1})
 EF = exposure frequency (occurrences per day)
 ED = exposure duration (days)
 IR = soil ingestion rate from the site (kg of soil per day)
 ABS = ingestion absorption factor or bioavailability of chemical to human receptor following ingestion
 BW = body weight (kg)
 AT = averaging time (typical life span in days)
 CPF = cancer potency factor (mg kg^{-1} per day)

Typically, the absorption factor of an ingested chemical is assumed to be 100%, i.e., all of the chemical present in the soil enters the tissue and/or fluids of the body, which of course is unrealistic. If the above equation is rearranged (Equation 6.3), it is evident that there is an inverse relationship between an acceptable chemical concentration in the soil (CS) and the chemical bioavailability to the receptor (ABS) for any given risk ($Risk$):

$$CS = Risk \left[\frac{1}{ABS} \right] \left[\frac{BW \times AT}{EF \times ED \times IR \times CPF} \right] \qquad (6.3)$$

Thus, when chemical bioavailability is reduced 100-fold (i.e., from 100 to 1%) the chemical concentration that yields a specific risk will increase by an equivalent factor (e.g., 10 to 1000 mg kg^{-1}). This is illustrated in Figure 6.8. The computation could be performed to establish action goals for soil remediation, based upon the risk of cancer to humans resulting from soil ingestion.

4.4 Risk Assessment: A Case Study

Risk assessment is employed customarily to evaluate hypothetical events that have not happened. Uncertainties and errors of the assessment may be introduced at every step because of data deficiency, incorrect assumptions, and inappropriate computational models. In addition, results of a risk assessment are problematic; the reality of the events oftentimes cannot be verified experimentally. Uncertainties and errors in risk assessment, unfor-tunately, could lead to erroneous conclusions, unwarranted advisories, and unjustified restrictions for land use. To ensure that risk assessors are aware of the uncertainties and to minimize errors in selecting data, guidelines that

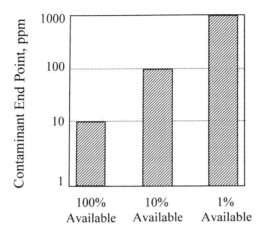

FIGURE 6.8. Importance of (bio)availability on the calculation of environmentally acceptable end points for contaminants in soils. (Reprinted from *Environmentally Acceptable Endpoints in Soil,* Linz and Nakles. Copyright © 1997, with permission from Elsevier Science.)

formalize protocols of risk assessment are necessary to promote consistency among various risk and exposure assessment activities (U.S. EPA, 1986a,b).

Basically, the same information protocols are needed to employ risk assessment in developing numerical limits for land application of biosolids (e.g., sewage sludge). In this case, the maximum permissible exposure of the receptor must be determined first, based on an acceptable level of "risk"; the exposure scenario for each pollutant transport route must be constructed; and the permissible pollutant loading (or input) through waste application must be calculated according to the mathematical relationships that define the transfer of the toxicant through the exposure pathways. The numerical limits of toxic chemicals in the drinking water guidelines are developed in this manner (NAS, 1977; WHO, 1984). In the case of land application of biosolids, assessment is far more complicated because risk of exposure to pollutants deposited in soil involves a multitude of exposure routes.

The U.S. EPA promulgated a regulation, 40 CFR Part 503, known as *Standards for the Use or Disposal of Sewage Sludge* to protect human health and the environment from any potential adverse effects of certain contaminants that may be present in sewage sludge (Ryan and Chaney, 1997; U.S. EPA, 1993). This is an example of a result of risk assessment conducted in its most comprehensive manner in which 14 pollutant transport pathways were involved. Its technical support document that formed the basis for deriving the standards is by far the most complete risk-based analysis on land application of sewage sludge ever undertaken. The regulation establishes requirements for the final use or disposal of sewage sludge in three circumstances: (1) sewage sludge applied on the land for beneficial purposes, (2) sewage sludge disposed on land by placing it in surface disposal sites, and

(3) sewage sludge to be incinerated. The standards for each end use or disposal practice consist of numerical limits on the pollutant concentrations in sewage sludge, management practices, operational standards, frequency of monitoring requirements, record-keeping requirements, and reporting.

The development of sludge contaminant loading limits consisted of the following general components:

Delineation of contaminants of concern in sewage sludge;

Identification of potential pathways for exposure and receptors (humans, soil biota, plants, and animals) to these contaminants through land application of sewage sludge;

Identification of effect end points (e.g., dose–response relationships) for the receptors and contaminants of concern;

Determination of the level of pollutant in the soil that would protect the highly exposed individual/receptor (HEI) from adverse effects for each exposure pathway and contaminant; and

Determination of maximum acceptable loading rate of each contaminant based on the most limiting value for all evaluated pathways.

From the maximum acceptable loading rates and the sludge concentration distribution of the National Sewage Sludge Survey (NSSS), the contaminant limits (cumulative soil pollutant application limit and maximum allowed sludge contaminant concentration) for the Part 503 Rule were determined.

4.4.1 Critical Contaminant and Exposure Pathways

In the final regulatory report, 14 organic contaminants and 10 inorganic contaminants were evaluated in the land application Part 503 risk assessment (U.S. EPA, 1985). The inorganic contaminants were As, Cd, Cr, Cu, Pb, Hg, Mo, Ni, Se, and Zn, and the organic contaminants were aldrin/dieldrin (total), benzene, benzo[a]pyrene, bis(2-ethylhexyl)phthalate, chlordane, DDT/DDE/DDD (total), heptachlor, hexachlorobenzene, hexachlorobutadiene, lindane, N-nitrosodimethylamine, polychlorinated biphenyls, toxaphene, and trichlorethylene. In the final development, 14 pathways of exposure were included for the Part 503 risk assessment (U.S. EPA, 1993). They are presented in Appendix Table A.26.

The pathways identified in this document described almost all the possible environmental routes of contaminants released through land application of wastes, with receptors of contaminants ranging from child and adult humans, livestock, plants, soil biota, to predators of soil biota. Mathematical models based on linear transfer between components were developed to describe the transfer of pollutants through each pathway.

Not all of the 14 pathways were evaluated for each contaminant because preliminary analysis (hazard identification) showed certain pathways were not critical for the compound, and/or data needed to make the calculation were not available. For example, only the organic contaminants were evaluated by the vapor pathway (i.e., pathway 13).

A separate risk assessment was conducted for certain pathways for managed lands (e.g., agricultural land that includes pasture and range land, and lawns and home gardens) and for unmanaged lands (e.g., forests and reclamation sites). Not every pathway was evaluated for each type of land. For example, the soil ingestion pathway (pathway 3) was not evaluated for unmanaged land because children are not expected to be exposed for a long period to sewage sludge applied to that type of land. For details of all pathways and data used in the risk assessment see the technical support document for sewage sludge land application, vol. I, PB93-11057 and vol. II, PB93-110583 (U.S. EPA, 1993).

4.4.2 Specific Exposure Pathway

Pathway 2 (sludge–soil–plant–home gardener) for Cd will be used as an example (Ryan and Chaney, 1997). Cadmium is selected because (1) it is one of the most mobile and bioavailable metals in the soil–plant system, (2) it is the metal of most concern in agroecosystems, and (3) its epidemiological role in *itai-itai* disease is well documented. This pathway depicts that sludge contaminants are taken up from the soil through plant roots, with direct adherence of sludge or soil to crop foliage being minimal as the small amounts of contaminants on the foliage are presumably washed off before consumption. The HEI for this pathway is the home gardener who produces a large portion of their dietary consumption of the plant food (potatoes, leafy vegetables, root crops, garden fruits, nondried legumes, and sweet corn) on sludged soils. It is also assumed that the soil is at the maximum cumulative application and that this practice is carried out for a lifetime.

For inorganics, a cumulative reference application rate (RP) is calculated according to the following equation (U.S. EPA, 1989b):

$$RP = \frac{\left[\frac{RfD \times BW}{RE} - TBI\right] \times 10^3}{\sum UC_i \times DC_i \times FC_i} \tag{6.4}$$

where RP = reference (allowed cumulative) application rate of contaminant (kg ha^{-1})

RfD = reference dose (mg/kg day^{-1})

BW = human body weight (kg)

TBI = total background intake rate of contaminant (mg day^{-1})

RE = relative effectiveness of ingestion exposure (unitless)

10^3 = conversion factor (µg mg^{-1})

UC_i = uptake response slope for the food group i (µg/g DW) (kg/ha)$^{-1}$

DC_i = daily dietary consumption of the food group i (g DW day^{-1})

FC_i = fraction of food group i assumed to originate from sludge-amended soil (unitless)

By assuming that the sludge is mixed into the plow layer (15 to 20 cm) of the soil, RP can be converted to a soil concentration (RCL) by the following equation:

$$RCL = \frac{RP}{MS \times 10^{-9}} \qquad (6.5)$$

where RCL = reference (allowed cumulative) soil concentration of contaminant ($\mu g \; g^{-1}$ DW)

RP = reference application rate of contaminant (kg ha^{-1})
MS = 2×10^9 g ha^{-1} = assumed mass of soil in upper 15 cm
10^{-9} = conversion factor (kg μg^{-1})

The variables in the equation utilized to calculate RP can be classified as dose (RfD) or exposure variables (all others) and are discussed below.

Risk Reference Dose (RfD)

The risk reference dose used in pathway 2 requires a maximum allowable dietary intake of a contaminant as a measure of the potential for adverse effect. The ADI (acceptable daily intake) is "the daily intake of a chemical which, during an entire lifetime, appears to be without appreciable risk on the basis of all the known facts at the time. It is expressed in milligrams of the chemical per kilogram of body weight (mg/kg)" (Lu, 1983). This value is developed to protect the more susceptible members of the population and thus allows greater protection for the majority of the population (Barnes and Dourson, 1988; Ryan et al., 1982; U.S. EPA, 1988). The U.S. EPA prefers the term reference dose (RfD) to avoid the connotation of acceptability. Doses less than the RfD are not likely to be associated with adverse health risks, and therefore, are less likely to be a regulatory concern. As the amount of the exposures exceeding the RfD increases, the probability of adverse effects in the exposed population increases, becoming a regulatory concern (Barnes and Dourson, 1988; Hallenbeck and Cunningham, 1986; U.S. EPA, 1988).

The RfD for Cd (0.001 mg Cd/kg body weight day^{-1} or 0.070 mg Cd/kg body weight day^{-1}) used in this assessment has some conservative assumptions associated with the calculation (Chaney and Ryan, 1986; Chaney and Ryan, 1994; Ryan et al., 1982). The errors identified suggest that daily intake could exceed 1000 μg Cd day^{-1} for a lifetime without injury. The best model for allowable Cd intake (Kjellstrom and Nordberg, 1978) estimated over 400 μg Cd could be ingested daily for a lifetime without causing tubular proteinuria. Even the malnourished peasant rice farmers could ingest 200 μg Cd day^{-1} without Cd health effects (Nogawa et al., 1978). Thus, the use of the current accepted RfD value in Cd is conservative and could be increased by a factor of 10 without concern. However, for this evaluation, the conservative 0.001 mg Cd/kg body weight day^{-1} (70 μg/day) was used.

As the human health end point (RfD) is for chronic lifetime exposure, the duration of exposure (DA) and exposure averaging time (AT) must also be considered as lifetime values. Thus, it has to be assumed that this level of exposure occurs for a lifetime. If this level of exposure is to occur for the 50- to 70-yr exposure period, the soil is presumed to be at the maximum cumulative application rate (RP) for the entire exposure period.

Body Weight (BW)

A body weight of 70 kg was used in the calculations and represents the average adult male (U.S. EPA, 1990b).

Total Background Intake (TBI)

Humans are exposed to Cd, even if no sewage sludge is applied to agricultural land. A value of 0.0161 mg Cd/day was used (U.S. EPA 1993). The sources included baseline levels (natural and anthropogenic) in drinking water, food, and air. When TBI is subtracted from the weight-adjusted RfD, the remainder defines the increment that may result from sewage sludge use or disposal without exceeding the threshold. Since conservative assumptions are used in determining the values for most of the variables in this risk assessment, more realistic values for this parameter were used.

Relative Effectiveness (RE)

The relative effectiveness of dose is assumed to be 1.0 for this calculation. As the exposed population (i.e., home gardener) has a diet rich in Ca, Fe, and Zn that would reduce absorption of Cd, RE should be less than one [see Chaney and Ryan (1994) for more detail].

Plant Response (UC)

Relationships between the concentrations of metals in growth medium and plants have been established from field and greenhouse studies, with different forms and carriers of metals, spiked and nonspiked sludges, having divergent results (Logan and Chaney, 1983; Page et al., 1987). These findings relate to the differences between salts and sludge and between growing plants in greenhouse pots vs. in the field (Chaney and Ryan, 1993). Because of the difficulty in extrapolating greenhouse pot results to field conditions, all available field data were utilized in the revised risk assessment (U.S. EPA, 1993). For Cd, the plant uptake data in the final Part 503 exposure assessment were based entirely on sludge field studies. Only pH has been shown to have a consistent significant effect of all the soil variables (OM content, CEC, soil texture, pH, etc.) on plant uptake of sludge-applied metals (Page et al., 1987). Thus, acid soil system is also well represented in the Part 503 data set.

As the UC data set for any food group (potatoes, leafy vegetables, nondried legumes, root vegetables, garden fruits, corn, grain, and cereal) appears to represent a log normal distribution, the geometric mean was used to best represent the distribution.

The choice of the values for FC and UC have significant effects on the allowable RP or RCL. The allowable value for RCL ranges from 2 mg Cd kg^{-1} (FC = 100% and UC = highest observed for each food group) to more than 100,000 mg Cd kg^{-1} (FC = 1% and UC = lowest observed for each food group) (Ryan and Chaney, 1993, 1994).

Dietary Consumption (DC)

The 503 Rule (U.S. EPA, 1989b) used the highest food consumption groups to represent the diet of individuals from birth to age 70 (the so-called *mega-eater*). This diet assumed that an individual had for their entire life the following consumption rates:

The teenage male (14 to 16 yr) for the food groups grains, potatoes, root vegetables, dairy, and dairy fat;

The adult male (25 to 30 yr) for the food groups legume vegetables, garden fruit, beef, pork, poultry, beef fat, poultry fat, and pork fat;

The adult female (25 to 30 yr) for the food groups lamb and lamb fat;

The adult female (60 to 65 yr) for the food group leafy vegetables; and

The adult male (60 to 65 yr) for the food groups beef liver, eggs, and beef liver fat.

The use of this *mega-eater* diet resulted in an overestimate of dietary consumption (DC) that makes it impossible to define the exposed population, leading to the conclusion that the exposed population did not exist and therefore, the *mega-eater* diet was a bounding estimate. A more reasonable exposure model averaged the dietary consumption rates across sex and age, calculating the Estimated Lifetime Average Daily Food Intake, used in the revised risk assessment (Ryan and Chaney, 1993, 1994; U.S. EPA, 1993).

Fraction of Food Produced (FC)

The subsistence gardener is assumed to produce 37% of their lifetime consumption of potatoes, 43% of their lifetime consumption of flour and cereal, and 59% of their lifetime consumption of all other garden food groups (leafy vegetables, roots crops, garden fruits, nondried legumes, and sweet corn). As discussed below in the HEI section, changes in the fraction of food originating from sludge-amended soil (FC) alter the expected size of the exposed population (HEI) and these conservative assumptions restrict the defined HEI population to only a few people.

Population at Risk (HEI)

The highly exposed individual (HEI) is of critical importance in exposure assessment. The HEI can be a human, plant, or animal that represents a living organism that has the maximum exposure to a given contaminant for a particular disposal scenario. The HEI must correspond to a very small, but statistically meaningful, percentage of the population before it is appropriate to create algorithms to attempt to quantify its exposure. Thus, information on the exposed population becomes critical as well as information on chemical concentration and time of contact data (duration of exposure) in completing the objective of exposure assessment (U.S. EPA, 1991).

Ryan and Chaney (1993) estimated that <0.005% of the population were lifetime subsistence gardeners who were utilizing soil that may have received sludge. This does not consider how many are at the RCL, thus possibly receiving the calculated exposure, or the number of gardeners who use good agronomic management practices and maintain near-neutral soil pH, thus receiving less than the calculated exposure. It was also calculated that <3% of current sludges could ever produce the required RCL, and that it would take 300 to 600 yr of continuous application at agronomic rates before the soil could become equal to sludge residual and thus reach the RCL.

The implication is that even if there exists a population of subsistence lifetime gardeners who use sludge, only a few could use sludges with sufficient Cd concentrations to allow their soil to reach the RCL, and it would take several lifetimes of continuous application at agronomic rates before the soil reached a concentration equal to the RCL. Since it is unlikely that continuous yearly applications would occur for these time frames and soil concentrations are not likely to reach the RCL, the HEI population that receives the projected exposure most likely does not exist in the United States. Thus, this is another conservative scenario.

Based on the algorithm used to calculate the allowable RCL, it is possible to use information on current sludge composition, assumed rate of application of sludge, and number of applications to calculate increases in dietary Cd ingestion, using the application rate of 10 tonnes ha^{-1} (U.S. EPA, 1993) and sludge composition data from the NSSS (U.S. EPA, 1990a), and if sludge OM had disappeared from the soil-sludge matrix (Ryan and Chaney, 1993). These calculations indicate that 80% of present-day municipal sludge could not cause a 10 µg day^{-1} increase in dietary Cd ingestion by the subsistence gardener and it would take 25+ yr of application at 10 tonnes ha^{-1} yr^{-1} for even the upper 1% of the sludges to cause that increase of dietary Cd. Furthermore, it would take 300 yr of application at 10 tonnes ha^{-1} yr^{-1} for the upper 1% of sludges to produce dietary increases in excess of the RfD. Thus, even if the lifetime subsistence gardeners use sewage sludge and their soil has reached equilibrium with the sewage sludge, their exposure is unlikely to reach the RfD in any year, and even less likely for 70 years.

In short, the model estimated that garden soils could reach 60 mg Cd kg^{-1} before HEIs (i.e., those who grow 59% of vegetables and fruits and

37% of potatoes they consume for 50 yr in a garden containing the maximum allowable cumulative load) would consume a lifetime average of 70 μg Cd day^{-1}, which is the WHO and U.S. EPA recommended maximum daily intake for Cd. Simply stated, allowing soil Cd to reach the RCL does not represent a risk, even in extreme local contamination.

5 Remedial Option and Technology

Should we clean up polluted soils? If yes, to what extent? There is no universal answer to this question. The risks associated with polluted soils vary from site to site and country to country according to the scientific database, public perception, political perception, national priority, and other factors. While severely contaminated soils may require some form of remediation, there may be instances in which remediation is not desirable. These include (1) the cost to society of cleanup far exceeds the expected benefits of cleanup in terms of human health and ecological sustainability, (2) the contaminated soil is not being used and has a low potential to be used in the future, (3) there are inexpensive substitutes for the contaminated soil in question, (4) the site will not be used after remediation because users will take some averting action, and (5) the contamination does not degrade soil and/or water quality to an unsafe or unhealthy level (NRC, 1993).

Remediation of soils contaminated with metals is a relatively new field of technology. It is advancing rather slowly due to regulatory pressure, public concern, and high cost. In addition, it has not been addressed aggressively because of (1) lack of guidance on the cleanup levels to be achieved; (2) more emphasis on toxic organics, explosive residues, and hydrocarbons than on metals; (3) assumption that metals applied to soils remain immobile; (4) lack of data on long-term ecotoxicological and health effects; and (5) insufficient knowledge on environmental fate and effects of metal species in the soil–plant–water system. The U.S. EPA is mandated by law, through section 121(b) of the Comprehensive Environmental Response, Compensation, and Liability Act (CERCLA) to select remedies that "utilize permanent solutions and alternative treatment technologies or resource recovery technologies to the maximum extent practical" and to prefer remedial actions in which treatment "permanently and significantly reduces the volume, toxicity, or mobility of hazardous substances, pollutants, and contaminants as a principal element." The underlying requirement is "permanent," which indicates that technologies be proven scientifically effective on a long-term basis.

Selecting an appropriate technology for soil remediation depends on the appropriate level of cleanup desired, the length of time allowed, forms and amounts of metals, site characteristics, and cost. In the United States, the initial guidance for cleanup level was based on the so-called EP (extraction procedure) toxicity test. This test involves leaching the soil or waste with weak acids, representing the worst-case scenario. In the absence of EP toxicity data, total metal concentration was used. Currently, the TCLP (toxicity characteristic leaching procedure) described in the 40 CFR 261.24 (1992) is being used (Smith et al., 1995). Needless to say, leaching the soil with acid solution does not represent the natural processes of metal release in soils.

Metals are prevalent at most Superfund sites. At sites with signed RODs (record of decisions), metals are the sole contaminants ($\sim$16%) or are found in combination with other contaminants, such as volatile or semivolatile organics ($\sim$49%) (U.S. EPA, 1997). Soil remediation techniques have been progressing since 1986 when the Superfund Amendments and Reauthorization Act (SARA) established coordinated research and development (R&D) programs and promoted technology demonstrations. In 1987, the U.S. Department of Defense (DOD) reviewed current engineering practices on treatment of soils contaminated with heavy metals (Kesari et al., 1987). Of the 21 technologies reviewed in the report, three technologies were selected for further R&D: microencapsulation, low-temperature thermal (roasting), and extraction. This report has recently been updated and a major R&D program for remediation of soil contaminated with metals was proposed (Bricka et al., 1994). The emphasis of the program is on (1) determination of metal partitioning, speciation, mobility, and mass transport relationships and mechanisms; (2) development and enhancement of metal recovery methods; and (3) development and enhancement of metal immobilization techniques. Currently, however, the remediation of soil contaminated with metals using *in situ* techniques, natural attenuation, extraction, and plant extraction (hyperaccumulators) is being advanced. In general, *in situ* remediation is more cost efficient when compared with traditional treatment methods, but relatively few alternatives exist for the *in situ* treatment of metals.

When the Superfund program was initiated, acceptable technologies then for hazardous waste and soil remediation were excavation, incineration, and land disposal. Specially constructed landfills with double liners were employed. In 1984, the U.S. EPA emphasized the in-place treatment of soil rather than removal and disposal or incineration, based on the premise that containment was only a temporary solution, that the magnitude of the problem was much larger than initially anticipated, and that incineration that leaves residue to be treated was not always necessary (U.S. EPA, 1987). Two years after the passage of SARA, there was a dramatic increase of new treatment technologies; 50% of RODs in 1987 selected some treatment for source control, while 70% of RODs did so from 1988 to 1990.

The following criteria for selecting a remedial option have been established by the National Contingency Plan (Lafornara, 1991): (1) the overall protection of human health and the environment; (2) compliance with applicable or relevant and appropriate requirements; (3) long-term effectiveness and permanence; (4) reduction of toxicity, mobility, and volume through treatment; (5) short-term effectiveness; (6) implementability; (7) state acceptance; (8) community acceptance; and (9) cost.

In general, there are three major categories of treatment of metal-contaminated sites: (1) *physical methods,* which simply restrict access to the contamination through containment or removal; (2) *chemical methods,* which attempt to alter contaminant speciation to either enhance mobility under various extraction scenarios or decrease mobility to reduce potential exposure hazards; and (3) *biological methods,* which attempt to use natural or enhanced biochemical processes to either increase contaminant mobility for extraction (e.g., phytoaccumulation) or reduce mobility by altering metal speciation. For example, natural geochemical processes such as precipitation, sorption, ion exchange, and even redox manipulation have been employed as remediation methods. In some cases, a combination of different types of treatment is employed. A few, selected, more commonly used techniques are listed, with brief descriptions in Table 6.5. The following is a general and brief description of selected techniques and their status, advantages, and limitations, when applicable.

5.1 Physical Treatment

Physical treatment of metal-contaminated soils includes the use of physical separation and washing, frozen-ground barriers, and thermal processes such as *in situ* or *ex situ* vitrification, incineration, high-temperature fluid-wall reactor, plasma-arc vitrification, cyclone furnace, and roasting (Iskandar and Adriano, 1997).

5.1.1 Vitrification

Vitrification is a solidification technique that involves heating of contaminated soils or mine spoil material by various means to the point of producing a melt that hardens into a glasslike material. Similar to the final product from a solidification process, the vitrified material has very low permeability, is resistant to weathering, and meets TCLP criteria. Heating methods include natural gas-fired burners, electric current, and plasma torches where temperatures can rise to as much as 2000 °C. The vitrification processes reduce the volume of the treated material by 25 to 35%. Vitrification is best suited for wastes that are difficult to treat, such as mixed wastes. A significant advantage to vitrification is that organic contaminants are oxidized by the high temperatures employed. A disadvantage is the possible need for emission control for waste gases.

5.1.2 Encapsulation

Encapsulation is a process that is employed for relatively small areas (e.g., smelter sites) and involves covering the site with an impermeable or very

TABLE 6.5. Selected remedial treatments for metal-contaminated sites being adopted by the cleanup industry.

Method	Brief description
Physical	
Vitrification (*ex situ*, or *in situ*)	Heating the contaminated medium to produce a glasslike, non-porous, nonleachable material
Encapsulation	Covering polluted site with impermeable layer
Washing (*ex situ*)	Extraction of contaminants with acid or chelate solution after particle separation
Flushing (*in situ*)	Leaching of contaminants with acid or chelate solution
Electrokinetics[a]	Migration of contaminant ions to electrodes induced by electric current
Groundfreezing	Formation of artificial, temporary barrier induced by cryogenic application
Chemical	
Neutralization	Neutralization of soil-waste acidity or alkalinity
Solidification	Addition of cementing agent to contaminated medium to produce hardened nonporous, nonleachable material
Stabilization[a]	Immobilization of constituents by adding materials to induce sorption/precipitation
Biological	
Phytoremediation[a]	Use of plants to bioaccumulate (phytoextraction) and volatilize (phytovolatiliation) contaminants
Revegetation (rehabilitation)	Site stabilization with vegetative groundcover against wind and water erosion

[a]These techniques are fairly recent and some, especially phytoremediation and certain stabilization techniques, are largely still experimental in nature.

slowly permeable layer, such as polyethylene or compacted clay, to minimize percolation of water through the site. This strategy is similar to that used for closing landfills. Wastes or affected soils are covered with a polymeric substance, such as polyethylene or asphaltic bitumen. Chappell and Willitts (1980) reported that the process effectively isolated metals, that leaching with acid produced a leachate with less than 1 ppm total metal, and that the permeability of the product was less than that for concrete.

The use of asphaltic product to form a base or a subbase has also been demonstrated (Brenner and Rugg, 1982). The best result was obtained when 60% waste was mixed with 40% asphalt/sulfur blend. The process is site specific and must be developed for each waste.

5.1.3 Soil Washing (Soil Extraction)

Soil washing can be effective in removing organics, metals, and low-level radionuclides. The process is designed to segregate and reduce the volume of contaminated soils/sediments. It involves high-energy contacting and mixing of excavated contaminated media with an aqueous-based washing solution in a series of washing units. Dilute acids or bases (Siller, 1993), chelating or complexing agents (Iskandar and Mobley, 1988; Kesari et al., 1987), and proprietary agents have been used. In some cases, the slurry is heated to about 40 to 50 °C. The concentration of the reacting agents and the number of sequential rinses vary depending on the metal concentration in the soil, cleanup levels desired, and physical and chemical properties of the medium. Sedimentation, centrifugation, or filtration may be necessary after each extraction process. Flocculating agents, such as aluminum sulfate (alum), may be used to assist in separation of fine particles, centrifugation, or filtration. Examples of removal efficiency by soil washing are given in Appendix Table A.27.

A variant of soil washing is its *in situ* version called *soil flushing,* which involves extraction of contaminants from soil using water or other suitable solutions. This technique could be effective in removing water-soluble species, such as Cr^{6+} (U.S. EPA, 1997).

5.1.4 Artificial Ground Freezing (Cryogenic Barriers)

Artificial ground freezing has been in use for at least 100 yr for groundwater control, tunneling, mine shafts, and other civil engineering construction (Iskandar and Adriano, 1997). During ground freezing, ice forms between soil particles, decreasing the soil permeability. Frozen soils become a barrier for water and chemical transport. Walls of frozen soil may serve to concentrate the contaminant in the unfrozen portion of the soil. Frozen barriers may also be used temporarily to isolate contaminated soils until other treatment methods can be employed.

5.1.5 Electrokinetics

Electrokinetics, also known as electrokinetic fencing, relies on the use of an anode and conditioning pore fluid placed in the soil at some distance from a cathode (U.S. EPA, 1997). A low-intensity electrical current induces movement of ions to their respective electrodes. Contaminants are mobilized in the form of charged species, particles, or ions. The removal efficiencies for Pb, Cr, Cd, and U ranged from 75 to 95% when initial soil concentrations were as high as 200 mg kg^{-1} (U.S. EPA, 1989a).

5.1.6 Considerations for Use

In summary, with the exception of physical separation, soil washing, soil flushing, filtration, and ultrafiltration, most of the physical technologies are based on thermal treatment. Although thermal treatment is effective in immobilizing metals and results in very low TCLP test, these processes are limited by their destructive nature and the high cost associated with energy, excavation, transportation, and backfilling. They may be attractive for treatment of special wastes, such as medical waste, organics, ash from incineration of industrial sources, nuclear waste, and soils contaminated

with very high levels of metals. For the remediation of soil contaminated
with low and medium concentrations of metals, chemical and biochemical treatment may be more cost effective and environmentally acceptable.
This is particularly true if the health risk and metal mobility in the soils are
not high.

5.2 Chemical Processes

Several methods for chemical treatment of metal-contaminated soils have
been developed. The main objective of these treatment processes is to
remove the metals or to decrease their availability to the food chain
and groundwater. The chemical treatment processes can be grouped into
neutralization, precipitation, solidification, stabilization, ion exchange, and
mobilization.

5.2.1 Neutralization

This process simply neutralizes the soil acidity or alkalinity of affected
media. It has been used in waste sites where batteries had leaked or had been
disposed of. The process is based on the general premise that metals, such as
Pb, are highly mobile at low pH. Lime and sludge were used to remediate a
dredge material disposal site in Delaware (Palazzo and Reynolds, 1991).
Prior to treatment, the pH was about 2.4. Extractable Cu, Zn, Cr(VI), Ni,
Pb, and Cd in the soil decreased significantly over the 16-yr study. The plant
tissue metal concentrations also decreased with time.

5.2.2 Solidification

This is probably the most widely used process for treatment of metal-
contaminated soils, sediments, and sludges. It has been used for treatment of
dredged material and sediments (Cullinane et al., 1986). The process
involves *ex situ* mixing the soil, sediment, or sludge with binder material
to change its physical and chemical properties. Then the metals and/or the
free water are immobilized by chemical interaction. Typical binders include
Portland cements, pozzolans, or thermoplastics (Pojasek, 1979).

In the cement-based techniques, the Portland cement and other additives,
such as fly ash, react with the soil to form a concrete-type material. In the
lime-based treatment, the addition of lime and silica to the soil form a hard
concrete-like material. Fly ash, cement kiln dust, and other materials can
also be used. The hardened material typically has a permeability on the
order of 10^{-9} to 10^{-4} cm sec^{-1} and is durable under freezing and
thawing cycles. The solidification process can increase volume by 10 to
30% and increase bulk density by up to 10%. Successful solidification
processes can reduce TCLP concentrations to acceptable levels. Overall,

the process prevents or substantially retards the interaction between the contaminants and the environment. Several commercial applications have already been demonstrated. The materials produced met the TCLP toxicity limit.

5.2.3 *In Situ* Stabilization

This process heavily relies on the formation of insoluble or sparingly soluble materials, by adding to contaminated soils rather inexpensive materials, such as lime and hydroxyapatite, and immobilization of metal ions by such agents as zeolites and clay-type minerals. Their use physically limits these techniques to areas of shallow contamination where such materials can be effectively applied and mixed with the affected soils. The ability of lime, the most widely needed amendment material in agriculture, to immobilize metals in soils is well known (see Chapter 2).

Zeolites are framework aluminosilicates consisting of extended three-dimensional networks of linked SiO_4 and AlO_4 tetrahedra. They possess interconnected channels or voids that form ideal sorption sites for alkaline-earth and transition metals. Zeolites derive their CEC from Al^{3+} substitution for Si^{4+}, with the size of the channel determining the type of exchangeable cation that is preferred (Breck, 1974).

Both greenhouse and field studies have demonstrated the ability of zeolites to reduce the uptake of Cs, Sr, Cu, Cd, Pb, and Zn in plants (Leppert, 1990; Mineyev et al., 1990; Mumpton, 1984; Rebedea and Lepp, 1994). However, zeolites are not effective in immobilizing transuranic species, such as uranyl (UO_2^{2+}), that are commonly found at sites with elevated Cs and Sr levels (Vaniman and Bish, 1995). To date, results from the use of zeolites as soil amendments to reduce plant uptake of radionuclides and metals have been inconsistent (Adriano et al., 1997b; Leppert, 1990; Mineyev et al., 1990). For example, clinoptilolite application to an acidic podzolic soil spiked with Zn, Pb, and Cd reduced acid-extractable Zn but not the concentrations of acid-soluble Pb and Cd (Mineyev et al., 1990). Clinoptilolite application also reduced Zn and Pb accumulation in barley grain, but had no effect on the yield. On the other hand, the application clinoptilolite (15 g kg^{-1} soil) significantly decreased exchangeable Zn, Cd, and Pb in contaminated soils (Chlopecka and Adriano, 1996, 1997a). For example, the exchangeable Zn concentration decreased from 237 to 189 mg kg^{-1} with this zeolite application.

In other studies, application of phillipsite, a natural zeolite, to contaminated soils from Canada, Poland, Taiwan, and the Czech Republic (with As, Cd, Pb, and Zn from mining, smelting, and other industrial activities) significantly enhanced the yield of maize and oats while reducing the plant uptake of Cd, Pb, and Zn (Chlopecka and Adriano, 1997b). Zeolite application also influenced the plant uptake of both macronutrients and micronutrients for the Czech soil, with an increase in plant tissue concentrations for Ca and Mg and a decrease of Mn from 933 to 256 mg kg^{-1}. In the Czech soil, zeolite addition reduced exchangeable Cd, Pb, and Zn by 43, 46, and 29% respectively, but increased the concentrations of those metals in the residual fraction. Greenhouse pot experiments using contaminated soils have also demonstrated the ability of synthetic zeolites to reduce metal phytotoxicity to maize (Rebedea et al., 1994).

Stabilization studies on metal-contaminated wastes and soils using apatite have focused mainly on Pb (Berti and Cunningham, 1997; Chen et al., 1997; Laperche et al., 1997; Ruby et al., 1994; Zhang et al., 1997). Ma et al. (1993, 1995) reported that before hydroxyapatite can be successfully used as a Pb-immobilizing material, three factors need to be considered: hydroxyapatite must immobilize Pb^{2+} in the presence of interfering cations, anions, and dissolved OM; the resulting products must be stable in the contaminated medium; and the reactions should be rapid.

Precipitation of not only Pb but also other metals can occur when phosphate compounds are added to a contaminated medium (Ma et al., 1993; Misra and Bowen, 1981). Ma et al. (1994) showed that hydroxypyromorphite [$Pb_5(PO_4)_3OH$] precipitated after the reaction of hydroxyapatite with Pb^{2+} in the presence of NO_3^-, SO_4^{2-}, and CO_3^{2-}, while chloropyromorphite [$Pb_5(PO_4)_3Cl$] and fluoropyromorphite [$Pb_5(PO_4)_3F$] formed in the presence of Cl^- and F^-, respectively. Lead removal results primarily from the dissolution of apatite followed by the precipitation of hydroxyl fluoropyromorphite. Minor otavite ($CdCO_3$) precipitation was observed in the interaction of apatite with aqueous Cd, but other sorption mechanisms, such as surface complexation, ion exchange, and the formation of amorphous solids are primarily responsible for the removal of aqueous Zn and Cd (Wright et al., 1995).

Apparently the pH under which a reaction between metals and apatite occurs plays an important role. Wright et al. (1995) reported that the immobilization of Pb was primarily through a process of apatite dissolution, followed by precipitation of various pyromorphite-type minerals under acidic condition, or the precipitation of hydrocerussite [$Pb_3(CO_3)_2(OH)_2$ or $Pb(OH)_2 \cdot 2PbCO_3$] and lead oxide fluoride (Pb_2OF_2) under alkaline conditions. Otavite, cadmium hydroxide [$Cd(OH)_2$, and zincite (ZnO)] were formed in the Cd or Zn system, respectively, especially under alkaline conditions; while hopeite [$Zn_3(PO_4)_2 \cdot 4H_2O$] might only precipitate under alkaline conditions. The selectivity order of heavy metal sorption by apatite depends on pH; at pH < 7, Pb $>$ Cd $>$ Zn, and at higher pH, Pb $>$ Zn $>$ Cd (Wright et al., 1995).

Recent studies on the stabilization of metals by phosphate minerals have focused on the reduction of plant uptake of metals, as well as their reduction in mobility. Apatite significantly improved plant growth and yield on highly contaminated soils (Chlopecka and Adriano, 1997b). Also, metal concentrations in plant tissues, such as leaves and roots, significantly decreased.

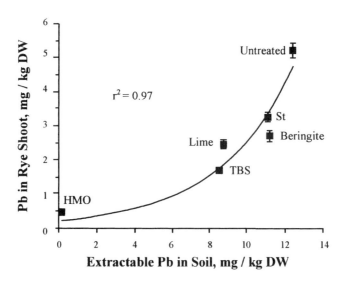

FIGURE 6.9. Relationships between soil Pb extracted by acetic acid and Pb concentrations in the shoots of ryegrass in contaminated soil amended with lime, beringite, phosphate (TBS), hydrous manganese oxide (HMO), and steel shots (St). (Reprinted from *Environmental Pollution,* Mench et al., copyright © 1997, with permission from Elsevier Science.)

For example, the Cd concentration in maize leaves decreased from 11.8 mg kg^{-1} in the control (Czech contaminated soil) to 5.3 mg kg^{-1} in apatite treatments. Similarly, Pb content in shoot tissue decreased with increasing apatite addition (Laperche et al., 1997). However, Pb and P contents in the plant roots increased as the quantity of added apatite increased. It was hypothesized that Pb accumulated in/on the roots as lead phosphate, since in the absence of phosphate, Pb was readily translocated from the roots to the shoots.

The effects of various materials added to contaminated soils on the bioavailability of Zn and Pb are illustrated in Figures 6.9 and 6.10. The soybean plants were grown on soil amended with lime, phosphate, biosolids, and various combinations of lime and biosolids (Pierzynski and Schwab, 1993). Similarly, ryegrass plants were grown on contaminated soil amended with lime, beringite, phosphate, hydrous manganese oxide, and steel shot (Mench et al., 1997). Both soils have the bioavailable metal concentrations altered by the stabilizing amendments.

5.3 Biological Processes

Potential use of plants and microorganisms for remediation of soils contaminated with metals has been gaining public and regulatory support. The concept of using vegetation involves growing specific plant species

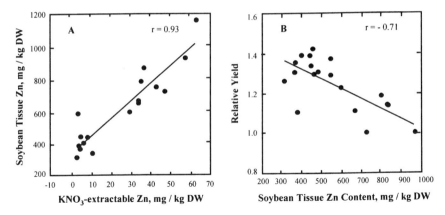

FIGURE 6.10. Relationships between KNO_3-extractable soil Zn and soybean tissue Zn concentrations (A), and soybean tissue Zn concentrations and relative yield (B) in a metal-contaminated alluvial soil. The variation in tissue Zn concentrations and extractable Zn were a result of changes in soil bioavailable Zn levels induced by various soil amendments (lime, phosphate, biosolids, and various combinations of lime and biosolids) without changing the total soil Zn concentration. (From Pierzynski and Schwab, 1993.)

known as *hyperaccumulators* on metal-contaminated soils and harvesting the plants at the end of the growing season or at a certain growth stage at which time the rate of metal uptake has become minimal. The harvested plants are then dried, incinerated with ashes disposed in a secured landfill, or extracted to recover the metals. This technique is known as *phytoremediation* or *green remediation*.

Based on traditional terminology, *phytoremediation* technologies use green or higher terrestrial plants for treating chemically or radioactively polluted soils (Baker et al., 1991, 1992; Raskin et al., 1994). Essentially, there are two process categories by which plants can be used to remediate contaminated soils, sediments, and waters (Cunningham et al., 1995a; Salt et al., 1995). Depending on cleanup strategy, these processes result in the plant-based containment or removal of soil contaminants (Wenzel et al., 1999). The two strategies are:

1. Containment strategy, which includes

Phytostabilization uses metal-tolerant plants to mechanically stabilize contaminated soils to prevent erosion and airborne transport to other ecosystems. In addition, leachability of contaminants may be reduced due to higher evapotranspiration rates relative to bare soils.

Phytoimmobilization processes prevents the movement and transport of dissolved constituents, using plants to minimize contaminant mobility in soils.

2. Removal strategy, which includes

Phytoextraction can takeup both metallic and organic constituents from soil by direct root absorption and translocation to above-ground biomass (Salt et al., 1995).

Phytodegradation involves the uptake and internal plant degradation and/or plant-assisted microbial degradation of organic contaminants in the rhizosphere (Anderson et al., 1993; Cunningham et al., 1995a; Walton and Anderson, 1992).

Phytovolatilization involves specialized enzymes that can transform, degrade, and eventually volatilize contaminants in the plant–microbe–soil system (Meagher and Rugh, 1996; Schnoor et al., 1995; Terry et al., 1992).

Among the five processes, cleanup may be achieved with phytoextraction, phytodegradation, and phytovolatilization, the extent of which depends on the type of contaminants and soil subsurface properties. These are processes that result in the removal of contaminants in soil, sediments, and waste streams. In contrast, phytostabilization and phytoimmobilization are containment processes.

Because of involvement of the plant(root)–microbe interactions, such as the association of fungi and bacteria with plant roots, i.e., in the rhizosphere, plant-based cleanup technology is also referred to as the plant–microbe treatment system (Wenzel et al., 1999). Because of its infancy, only few full-scale implementations or demonstrations of phytoremediation technologies have been carried out (Cunningham et al., 1995a; Raskin et al., 1994; Salt et al., 1995; Stomp et al., 1994). The rhizosphere plays an important role in biologically catalyzed redox reactions that lead to immobilization of redox-sensitive metals. The source of electrons is the organic compounds that are secreted as exudates from roots. This may result in the transformation of oxidized, soluble, and mobile species of $Cr(VI)$, $U(VI)$, and $Tc(VII)$ to reduced, insoluble, and immobile species of $Cr(III)$, $U(IV)$, and $Tc(IV)$ as shown:

$$Cr(VI) \rightarrow Cr(III); U(VI) \rightarrow U(IV); \text{ and } Tc(VII) \rightarrow Tc(IV)$$

Plants have evolved the ability to acquire nutrient ions and organic compounds in the soil and accumulate them in roots and shoots. Phytoextraction depends on this ability to remove organic and inorganic soil contaminants using plants as solar-driven pumps (Chaney, 1983; Chaney et al., 1997; Cunningham et al., 1995a, 1996; Raskin et al., 1994, 1997; Salt et al., 1995). The process involves the use of selected high-biomass plants to uptake contaminants from the soil into aboveground tissues where inorganic and organic constituents are accumulated for subsequent harvest. Although harvesting roots may be a viable option for removing soil contaminants (Entry et al., 1996), bioaccumulation in harvestable shoots is considered as the key process of phytoextraction.

The basic steps involved in phytoextraction are root absorption and subsequent translocation to and accumulation of contaminants in harvestable tissues. The plant's root system may enhance uptake by solubilizing contaminants through exudation of chelating agents and by influencing soil chemical properties that affect the contaminant solubility. In addition, microbial and mycorrhizal processes may be enhanced to assist in the mobilization and root uptake of contaminants (Salt et al., 1995; Stomp et al., 1995).

Bioaccumulation of metals by plants has been studied extensively, especially in species endemic to metalliferous soils (Brooks, 1984; Brooks and Radford, 1978; Brooks et al., 1980, 1992; Reeves et al., 1985). Generally, metals, such as Fe, Cu, Mn, Mo, Co, Ni, and Zn occur at low concentrations in plant tissues. However, unique plants, known as hyperaccumulators, have the natural ability to accumulate and detoxify metals such as Ni, Zn, Cu, and Mn at very high concentrations in their shoots (0.1 to 5% DW). The geobotanical literature has a good inventory of these species.

The value of metal-hyperaccumulating plant species for remediation of metal-contaminated soils has been demonstrated both in the greenhouse and in the field (Baker et al., 1991, 1992; Brown et al., 1994, 1995). These trials have clearly shown that metal-hyperaccumulating species can concentrate significantly more metals from contaminated soils than nonaccumulating species. However, using even the best metal accumulator species, *Thlaspi caerulescens,* would take a minimum of 13 to 14 yr of continuous cultivation to clean a site (Baker et al., 1994b). This exemplifies the problem associated with the application of hyperaccumulator species for remediation purposes—that these plants are relatively small, have low biomass production, and lack any protocol with respect to cultivation, pest management, and harvesting practices. To overcome these limitations, high-biomass metal-hyperaccumulator plants need to be developed, perhaps by modifying traditional agronomic species (Chaney et al., 1997; Cunningham et al., 1995a; Salt et al., 1995). This may be achievable by screening existing genotypes and mutant lines for metal accumulation in shoots (Kumar et al., 1995), or by transferring genetic information to high-biomass crop plants from hyperaccumulator species via somatic hybridization or genetic engineering (Fahleson et al., 1994).

It can be demonstrated that yield is a less important trait than hyperaccumulation (McGrath, 1998). Table 6.6 was constructed with Zn in mind, but the same kinds of considerations are relevant to all other elements. From this, it can be deduced that 25 tonnes ha^{-1}, the maximum potential yield of an annual crop (or biomass that can be cut from a long-lived perennial), still only results in 12.5 kg ha^{-1} yr^{-1} removal, even if the nontolerant nonhyperaccumulator has 500 μg g^{-1} of Zn, which is the usual threshold for toxicity of this element in many species. In contrast, growing a hyperaccumulator with a biomass yield of 5 tonnes ha^{-1} and a maximum concentration of 20,000 μg g^{-1} (Brown et al., 1994, 1995) results in a more

TABLE 6.6. Annual metal removal by plant shoot harvest in relation to biomass yield and metal concentration in the plant.

Type of plant	Concentration (μg g^{-1} DW)	Yield (tonnes ha^{-1})	Removal (kg ha^{-1})
Nonhyperaccumulator	50	5	0.25
	50	10	0.50
	50	15	0.75
	50	20	1.00
	50	25	1.25
	500	5	2.50
	500	10	5.00
	500	15	7.50
	500	20	10.00
	500	25	12.00
Hyperaccumulator	1000	5	5.00
	10,000 (or 1%)	5	50.00
	20,000 (or 2%)	5	100.00

Source: Modified from McGrath (1998).

effective 100 kg ha^{-1} removal. Future developments in phytoextraction must strive for hyperaccumulation, rather than simply increased biomass without substantial accumulation.

Phytoextraction has become more appealing as a remediation option for Pb. In the United States, remediation may be required when total soil Pb exceeds 300 to 500 mg kg^{-1} in residential soils, or 2000 mg kg^{-1} in soils used for industrial purposes (U.S. EPA, 1996). Remediation methods for Pb-contaminated soils are often expensive and disruptive to the site. *In situ* remediation techniques usually involve physical stabilization (e.g., mixing soils with cement of other solidifying agent or instillation of soil or asphalt caps) and/or chemical amendment with lime or phosphate fertilizer. None of these techniques actually removes Pb from the soil, which usually requires soil washing with strong acids that may have drastic and undesirable effects on soil properties.

Recent research has shown that chemical amendments, such as synthetic organic chelates may enhance phytoextraction by increasing soil Pb solubility, plant uptake, and translocation of Pb. Huang et al. (1997) found EDTA to be the most effective chelate (also HEDTA, CDTA, DTPA, and EDDHA) for enhancing Pb uptake and Pb solubility in soils, resulting in Pb concentrations in peas of $\sim$11,000 mg kg^{-1} and in corn of 3500 mg kg^{-1}, and increasing soil solution Pb concentrations from 3.4 to 1100 mg L^{-1}. They also found that Pb translocation and concentration in xylem sap of corn was increased to 6.9 mg L^{-1} by addition of 0.5 g EDTA kg soil^{-1} and to 21.15 mg L^{-1} by addition of 1.0 g EDTA kg^{-1}soil, compared with the control (0.15 mg L^{-1}). Similarly, Blaylock et al. (1997) found EDTA added to soil at 10 mmol kg^{-1} increased plant (*Brassica juncea*) Pb from <100 mg kg^{-1} in the control to 1.6% of dry shoot weight and increased water-soluble soil Pb from <10 to ~470 mg kg^{-1}. Similar although less dramatic increases occurred with CDTA and DTPA. In evaluating the efficacy of seven chelates to desorb from contaminated soils (i.e., 1278 to 14,349 mg Pb kg^{-1} soil), Cooper et al. (1999) obtained the following rank: HEDTA > CDTA > DTPA > EGTA > HEIDA > EDDHA~NTA. Plant uptake of Pb from the contaminated soil was enhanced by CDTA, DTPA, and HEDTA, but with even the most effective treatment (corn, high CDTA rate) the amount of Pb extracted by plants was rather low (0.4 kg Pb ha^{-1}).

5.4 Selection of Remediation Technology

Selection of remediation technology for treatment of metal-contaminated soils depends largely, among other things, on the treatment objectives or on the acceptable level of metal content after treatment. During the site assessment and site characterization process, data are collected on the forms, amounts, and spatial distributions of metal concentrations. If groundwater is contaminated with metals, the target cleanup level is usually set at drinking water standards. Initially, when CERCLA was enacted, government regulators expressed a preference for achieving the maximum soil cleanup to background metal concentrations. Because it is not always possible or cost effective to achieve such low levels of metal concentrations, cleanup levels are determined by site-specific assessment of the potential for migration and exposure. In the absence of groundwater contamination or other indications of migration, soil metal concentrations that meet the criteria for hazardous waste by the TCLP test would require remedial action or a permit as a hazardous waste disposal site. Regulators also indicated that the TCLP toxicity limits would typically represent the minimum soil cleanup requirement for a site.

The U.S. EPA has summarized the evaluation criteria of remedial technologies (Smith et al., 1995). They include the following: protection of human health and the environment; long-term effectiveness and permanence; reduction of mobility, toxicity, or volume; short-term effectiveness; implementability; cost; and general acceptance. The remedial technologies more widely employed by the industry include barriers, solidification, stabilization, and chemical treatment. Electrokinetics is acceptable but rarely chosen. Phytoremediation of metal-contaminated media still has to be demonstrated successfully under field conditions.

References

Acton, D.F., and L.J. Gregorich. 1995. In *The Health of Our Soils: Toward Sustainable Agriculture in Canada.* Cent Exp Sta, Ottawa, Canada.

Adriano, D.C. 1986. *Trace Elements in the Terrestrial Environment.* Springer-Verlag, New York.

Adriano, D.C., A. Chlopecka, D.I. Kaplan, H. Clijsters, and J. Vangronsveld. 1997a. In R. Prost, ed. *Contaminated Soils*. INRA 85, INRA, Paris.

Adriano, D.C., J. Albright, F.W. Whicker, and I.K. Iskandar. 1997b. In I.K. Iskandar and D.C. Adriano, eds. *Remediation of Metal-contaminated Soils*. Science Reviews, Northwood, United Kingdom.

Allen, H.E. 1993. *Sci Total Environ*. Suppl Part 1:23–45.

Alloway, B.J., ed. 1995. *Heavy Metals in Soils*. Chapman & Hall, London.

Anderson, T.A., E.A. Guthrie, and B.T. Walton. 1993. *Environ Sci Technol* 27:2630–2636.

Asami, T. 1997. In D.C. Adriano, Z.S. Chen, and S.S. Yang, eds. *Biogeochemistry of Trace Elements*. Applied Science Publ, Northwood, United Kingdom.

Baker, A.J.M., R.D. Reeves, and S.P. McGrath. 1991. In R.E. Hinchee and R.F. Olfenbuttel, eds. *In Situ Bioreclamation*. Butterworth-Heinemann, Stoneham, M.A.

Baker, A.J.M., S.P. McGrath, C. Sidoli, and R.D. Reeves. 1992. In *Workshop 1992: Soil Remediation*. Assoc. Francaise Interprof. du Cadmium, Paris.

Baker, A.J.M., R.D. Reeves, and A.S.M. Hajar. 1994a. *New Phytol* 127:61–68.

Baker, A.J.M., S.P. McGrath, C.M.D. Sidoli, and R.D. Reeves. 1994b. *Resources Conserv Recyc* 11:41–49.

Barnes, D.G., and M. Dourson. 1988. *Regul Toxicol Pharmacol* 8:471–486.

Berti, W.R., and S.D. Cunningham. 1997. *Environ Sci Technol* 31:1359–1364.

BioTrol, Inc. 1993. *Personal communication,* BioTrol, Inc., Chaska, MN.

Blaylock, M., D.E. Salt, S. Dushenkov, O. Zakharova, and I. Raskin. 1997. *Environ Sci Technol* 31:860–865.

Blum, W.E.H., and A.A. Santelises. 1994. In D.J. Greenland and I. Szabolcs, eds. *Soil Resilience and Sustainable Land Use*. CAB, Wallingford, England.

Breck, D.W. 1974. *Zolite Molecular Sieves*. Krieger Publ, Malabar, FL.

Brenner, W., and B. Rugg. 1982. In *Proc 8th Ann Res Symp Land Disposal of Hazardous Waste,* Cincinnati, Ohio.

Bricka, R.M., C.W. Williford, and L.W. Jones. 1994. In *Heavy Metal Soil Contaminant at U.S. Army Installations: Proposed Research Strategy for Technology Development*. U.S. Army Eng Waterway Exp Stat Tech Rep IRRP-94, Vicksburg, MS.

Brisbin, I.L., and M.C. Newman. 1991. *Water Air Soil Pollut* 57–58:691–696.

Brooks, R.R. 1994. In M.F. Farago, ed. *Plants and the Chemical Elements: Biogeochemistry, Uptake, Tolerance and Toxicity*. VCH Publ. Weinheim, Germany.

Brooks, R.R., and C.C. Radford. 1978. *Proc R Soc Lond B* 200:217–224.

Brooks, R.R., R.D. Reeves, R.S. Morrison, and F. Malaisse. 1980. *Bull Soc Roy Bot Belg* 113:166–172.

Brooks, R.R., A.J.M. Baker, and F. Malaisse. 1992. *Nat Geog Res Explor* 8(3):338–351.

Brown, S.L., R.L. Chaney, J.S. Angle, and A.J.M. Baker. 1994. *J Environ Qual* 23:1151–1157.

Brown, S.L., R.L. Chaney, J.S. Angle, and A.J.M. Baker. 1995. *Soil Sci Soc Am J* 59:125–132.

Bryda, L.D., and J.A. Sellman. 1994. *Remediation* 5:137–148.

Chaney, R.L. 1983. In J.F. Parr et al. eds. *Land Treatment of Hazardous Wastes*. Noyes Data Corp, Park Ridge, NJ.

Chaney, R.L., and J.A. Ryan. 1986. *Basis for Risk Reference Dose for Dietary Cadmium Intake*. Cadmium RfD Verification. U.S. Environmental Protection Agency, Washington, DC.

Chaney, R.L., and J.A. Ryan. 1993. In H.A.J. Hoitink and H.M. Keener, eds. *Science and Engineering of Composting: Design, Environmental, Microbiological and Ultilization Aspects.* Ohio State Univ, Columbus, OH.

Chaney, R.L., and J.A. Ryan. 1994. Risk base standards for arsenic, lead, and cadmium, In DECNMA, ISBN 3-926959-63-0.

Chaney, R.L., M. Malik, Y.M. Li, S.L. Brown, J.S. Angle, and A.J.M. Baker. 1997. *Curr Opin Biotechnol* 8:279–284.

Chappell, C.L., and S.L. Willitts. 1980. *J Haz Materials* 3:285–291.

Chen, X., J.V. Wright, J.L. Conca, and L.M. Peurrung. 1997. *Environ Sci Technol* 31:624–631.

Chlopecka, A., and D.C. Adriano. 1996. *Environ Sci Technol* 30:3294–3303.

Chlopecka, A., and D.C. Adriano. 1997a. *Sci Total Environ* 207:195–206.

Chlopecka, A., and D.C. Adriano. 1997b. *Soil Sci Plant Nutr* 43:1031–1036.

Cooper, E.M., J.T. Sims, S.D. Cunningham, J.W. Huang, and W.R. Berti. 1999. *J Environ Qual* 28:1709–1719.

Cullinane, M.J., et al. 1986. In *Handbook for Stabilization/Solidification of Hazardous Wastes.* U.S. EPA Rep 540/2-86-001, U.S. Environmental Protection Agency, Washington, DC.

Cunningham, S.D., W.R. Berti, and J.W. Huang. 1995a. *Trends Biotechnol* 13:393–397.

Cunningham, S.D., W.R. Berti, and J.W. Huang. 1995b. In R.E. Hinchee, J.L. Means, and D.R. Burris, eds. *Bioremediation of Inorganics.* Batelle Pr, Columbus, OH.

Denneman, C.A.J., and J.G. Robberse. 1990. In F. Arendt, M. Hinselveld, and W.J. van den Brink, eds. *Contaminated Soil '90.* Kluwer, Dordrecht, Netherlands.

[Dept. Environ.] Department of the Environment. 1987. Guidance 59/83. Department of the Environment, London.

Derelanko, M.J., and M.A. Hollinger. 1995. *CRC Handbook of Toxicology.* CRC Pr, Boca Raton, FL.

Eikmann, T., and A. Kloke. 1993. In D. Rosenkranz, G. Einsele, and H.M. Harress, eds. *Bodenschutz. Ergänzbares Handbuch der Massnahmen und Empfehlungen für Schutz, Pflege und Sanierung von Böden, Landschaft und Grundwasser, vol. 1: Grundlagen, Informationen, Bodenbelastung.* 9305. 1–7. E. Schmidt, Berlin.

Entry, J.A., N.C. Vance, M.A. Hamilton, D. Zabowski, L.S. Watrud, and D.C. Adriano. 1996. *Water Air Soil Pollut* 88:167–176.

Fahleson, J., I. Eriksson, M. Landgren, S. Stymne, and K. Glimelius. 1994. *Theor Appl Genet* 87:795–804.

Fan, A.M. 1996. In L.W. Chang, ed. *Toxicology of Metals.* CRC Pr, Boca Raton, FL.

Florence, T.M., B.G. Lumsden, and J.J. Fardy. 1984. In C.J. Kramer and J.C. Dunker, eds. *Complexation of Trace Metals in Natural Waters.* Martinus Publ, The Hague, Netherlands.

Hallenbeck, W.H., and K.M. Cunningham. 1986. *Quantitative Risk Assessment for Environmental and Occupational Health.* Lewis Publ, Chelsea, MI.

Hirose, K., and Y. Sugimura. 1985. *Marine Chem* 16:239–247.

Huang, J.W., J. Chen, W.R. Berti, and S.D. Cunningham. 1997. *Environ Sci Technol* 31:800–805.

Iskandar, I.K., and K. Mobley. 1988. *Potential Use of Chelating Agents for Decontamination of Soils.* Unpublished data. U.S. Army Cold Reg Res Eng Lab, Hanover, NH.

Iskandar, I.K., and D.C. Adriano, eds. 1997. *Remediation of Metal-contaminated Soils*. Science Reviews, Northwood, England.

Jones, K.C., T. Keating, P. Diage, and A.C. Chang. 1991. *J Environ Qual* 20: 317–329.

Kabata-Pendias, A., and H. Pendias. 1992. *Trace Elements in Soils and Plants*. CRC Pr, Boca Raton, FL.

Kabata-Pendias, A., and D.C. Adriano. 1995. In J.E. Rechcigl, ed. *Soil Amendments and Environmental Quality*. Lewis Publ, Boca Raton, FL.

Karlen, D.L., M.J. Mausbach, J.W. Doran, R.G. Cline, R.F. Harris, and G.E. Schuman. 1997. *Soil Sci Soc Am J* 61:4–10.

Kelly, M. 1988. *Mining and the Freshwater Environment*. Elsevier, Essex, United Kingdom.

Kesari, J., P.S. Puglionese, S. Popp, and M.H. Corbin. 1987. In *Interim Technical Report: Heavy Metal Contaminated Soil Treatment: Conceptual Development*. U.S. Army Environ Center Rep AMXTH-TE-CR-86101, Aberdeen, MD.

Kjellstrom T., and G.F. Nordberg. 1978. *Environ Res* 248–296.

Kumar, P., S. Dushenkov, H. Motto, and I. Raskin. 1995. *Environ Sci Technol* 29:1232–1238.

Lafornara, P. 1991. In *Remedial Action, Treatment, and Disposal of Hazardous Waste*. U.S. Environmental Protection Agency (EPA) Rep 600/9-91/002, Cincinatti, OH.

Lal, R., and F.J. Pierce. 1991. In R. Lal and F.J. Pierce, eds. *Soil Management for Sustainability*. Soil Conserv Soc Am, Ankeny, IA.

Langley, A. 1996. In R. Naidu et al., eds. *Contaminants and the Soil Environment in the Australasia-Pacific Region*. Kluwer, Dordrecht, Netherlands.

Laperche, V., T.J. Logan, P. Gaddan, and S.J. Traina. 1997. *Environ Sci Technol* 31:2745–2753.

Larson, W.E., and F.J. Pierce. 1991. In *Evaluation for Sustainable Land Management in Developing World*. IBSRAM, Bangkok, Thailand.

Leppert, D. 1990. *Min Eng* 42:604–608.

Lindberg, S.E., A.L. Page, and S.A. Norton, eds. 1990. *Acidic Precipitation, vol. 3: Sources, Deposition, and Canopy Interactions. Adv Environ Sci*. Springer-Verlag, New York.

Linz, D.G., and D.V. Nakles, eds. 1997. *Environmentally Acceptable Endpoints in Soil*. Am Acad Environ Engr, Annapolis, MD.

Logan, T.J., and R.L. Chaney. 1983. In A.L. Page et al., eds. *Utilization of Municipal Wastewater and Sludge on Land*. Univ California, Riverside, CA.

Lu, F.C. 1983. *Reg Toxicol Pharmacol* 3:121–132.

Ma, L.Q., S.J. Traina, and T.J. Logan. 1993. *Environ Sci Technol* 27:1803–1810.

Ma, L.Q., T.J. Logan, S.J. Traina, and J.A. Ryan. 1994. *Environ Sci Technol* 28:408–418.

Ma, L.Q., T.J. Logan, and S.J. Traina. 1995. *Environ Sci Technol* 29:1118–1126.

Marmo, L. 1999. EC DG XI, Brussels, personal communication.

McBride, M.B. 1994. *Environmental Chemistry of Soils*. Oxford Univ Pr, New York.

McGrath, S.P. 1998. In R.R. Brooks, ed. *Plants That Hyperaccumulate Heavy Metals*. CAB, Wallingford, United Kingdom.

McLaughlin, M.J., L.T. Palmer, K.G. Tiller, T.A. Beech, and M.K. Smart. 1994. *J Environ Qual* 23:1013–1018.

Meagher, R.B., C. Rugh, D. Wilde, M. Wallace, S. Merkle, and A.O. Summers. 1995. In *Proc 14th Ann Symp Current Topics in Plant Biochemistry, Physiology, and Molecular Biology*. Univ Missouri, Columbia, MO.

Meagher, R.B., and C. Rugh. 1996. In *IBC Symp. on Phytoremediation*, May 8–10, 1996. Arlington, VA.

Mench, M., J. Vangronsveld, V. Didier, and H. Clijsters. *Environ Pollut* 86:279–296.

Merian, E., ed. 1991. In *Metals and Their Compounds in the Environment*. VCH Publ, Weinheim, Germany.

Mineyev, V.G., A.V. Kochetavkin, and N. Van Bo. 1990. *Soviet Soil Sci* 22:72–79.

Misra, D.N., and R.L. Bowen, 1981. In P.H. Tewari, ed. *Adsorption from Aqueous Solutions*. Plenum, New York.

Mumpton, F.A. 1984. In W.G. Pond, and F.A. Mumpton, eds. *Zeo-Agriculture: Use of Natural Zeolites in Agriculture and Aquaculture*. Westview Pr, Boulder, CO.

[NAS] National Academy of Sciences. 1977. *Drinking Water and Health*. National Research Council (NRC)-NAS, Washington, DC.

[NRC] National Research Council. 1993. *Soil and Water Quality: An Agenda for Agriculture*. National Academy Pr, Washington, DC.

Newman, M.C., ed. 1995. *Quantitative Methods in Aquatic Ecotoxicology*. Lewis Publ, Boca Raton, FL.

Nogawa, K., A. Ishizaki, and S. Kawano. 1978. *Environ Res* 18:397–409.

Norton, S.B. et al. 1995. In D.J. Hoffman et al., eds. *Handbook of Ecotoxicology*. Lewis Publ, Boca Raton, FL.

Page, A.L., T.L. Gleason, I. Iskandar, and L.E. Sommers, eds. 1983. *Utilization of Municipal Wastewater and Sludges on Land*. Univ California, Riverside, CA.

Page, A.L., T.J. Logan, and J.A. Ryan, eds. 1987. *Land Application of Sludge: Food Chain Implications*. Lewis Publ, Chelsea, MI.

Palazzo, A.J., and C.M. Reynolds. 1991. *Water Air Soil Pollut* 57–58:839–848.

Parr, J.F., R.I. Papendick, S.B. Hornick, and R.E. Meyer. 1992. *Am J Altern Agric* 7:5–11.

Pierzynski, G.M., and A.P. Schwab. 1993. *J Environ Qual* 22:247–254.

Pojasek, R.B., ed. 1979. *Toxic and Hazardous Waste Disposal, vol 1: Processes for Stabilization Solidification*. Ann Arbor Sci Publ, Ann Arbor, MI.

Prost, R., V. Laperche, and D. Tinet. 1997. In I.K. Iskandar, and D.C. Adriano, eds. *Remediation of Metal-contaminated Soils*. Applied Science Publ, Northwood, United Kingdom.

Raskin, I., N.P. Kumar, S. Dushenkov, and D.E. Salt. 1994. *Curr Opin Biotechnol* 5:285–290.

Raskin, I., R.D. Smith, and D.E. Salt. 1997. *Curr Opin Biotechnol* 8:221–226.

Rebedea, I., and N.W. Lepp. 1994. In D.C. Adriano et al., eds. *Biogeochemistry of Trace Elements*. Sci Technol Letters, Northwood, United Kingdom.

Reeves, R.D., R.R. Brooks, and T.R. Dudley. 1985. *Taxon* 32(2):184–192.

Reiniger, P. 1997. In I.K. Iskander, ed. *Proc 4th Intl Conf Biogeochemistry of Trace Elements, June 1997*, Berkeley, CA. Univ California, Berkeley, CA.

Rubenstein, R., and S.A. Segal. 1993. In H.E. Allen et al., eds. *Metals in Groundwater*. Lewis Publ, Chelsea, MI.

Ruby, M.V., A. Davis, and A. Nicholson. 1994. *Environ Sci Technol* 28:646–654.

Ryan, J.A., and R.L. Chaney. 1993. In H.A.J. Hoitnik and H.M. Keener, eds. *Science and Engineering of Composting: Design, Environmental, Microbiological, Utilization Aspects*. Ohio State Univ, Columbus, OH.

Ryan J.A., and R.L. Chaney. 1994. *Trans 15th World Congress Soil Science, Acapulco, Mexico*. 3A:534–553.

Ryan, J.A., and R.L. Chaney. 1997. In R. Prost, ed. *Contaminated Soils*. INRA 85, INRA, Paris.

Ryan, J.A., H.R. Pahren, and J.B. Lucas. 1982. *Environ Res* 28:251–302.

Salt, D.E., M. Blaylock, N.P. Kumar, V. Dushenkov, B.D. Ensley, I. Chet, and I. Raskin. 1995. *Biotechnology* 13:468–474.

Schnoor, J.L.L.A. Licht, S.C. McCutcheon, N.L. Wolfe, and L.H. Carreira. 1995. *Environ Sci Technol* 29:318–323.

Siller, G. 1993. *Remediation of Oxidation Lagoons*. Unpublished data from Sacramento Army Depot, Sacramento, CA.

Smith, L.A. et al. 1995. In *Remedial Options for Metals-contaminated Sites*. CRC Pr, Boca Raton, FL.

Sposito, G. 1989. *The Chemistry of Soils*. Oxford Univ Pr, New York.

Stomp, A.M., K.H. Han, S. Wilbert, M.P. Gordon, and S.D. Cunningham. 1994. *Ann NY Acad Sci* 721:481–492.

Stringer D.A. 1990. *Hazard Assessment of Chemical Contaminants in Soil*. ECETOC Tech Rep 40, Brussels, Belgium.

Suter, G.W. 1993. *Ecological Risk Assessment*. Lewis Publ, Boca Raton, FL.

Suter, G.W., R.A. Efroymson, B.E. Sample, and D.S. Jones. 2000. *Ecological Risk Assessment for Contaminated Soils*. Lewis Publ, Boca Ration, FL.

Tack, F.M., K. Martens, M.G. Verloo, L. Van Mechlen, and E. Van Ranst. 1995. In R.D. Wilken, U. Förstner, and A. Knechel, eds. *Heavy Metals in the Environment*. CEP Consultants, Edinburgh, Scotland.

Terry N., C. Carlson, T.K. Raab, and A.M. Zayed. 1992. *J Environ Qual* 21: 341–344.

Theelen, R.M.C., A.G. Nijhof, and H. Bomer. 1995. In R. Prost, ed. *Contaminated Soils*. INRA 85, INRA, Paris.

Tiller, K.G., D.P. Oliver, M.J. McLaughlin, R.H. Merry, and R. Naidu. 1997. In I.K. Iskandar and D.C. Adriano, eds. *Remediation of Metal-contaminated Soils*. Applied Science Publ, Northwood, United Kingdom.

U.S. EPA. United States Environmental Protection Agency. 1985. *Summary of Environmental Profiles and Hazard Indices for Constituents of Municipal Sludge*. Office of Water Regulations and Standards, Wastewater Criteria Branch, Washington, DC.

U.S. EPA. 1986a. *Fed Register* 51:33992–34003.

U.S. EPA. 1986b. *Fed Register* 51:34042–34054.

U.S. EPA. 1987. *Superfund: Looking Back, Looking Ahead*. Office of Public Affairs, U.S. EPA Rep OPA-87-007, U.S. EPA, Washington, DC.

U.S. EPA. 1988. *Reference Dose (RfD): Description and Use in Health Risk Assessment. Integrated Risk Information Systems (IRIS). Intra Agency Reference Dose (RfD) Work Group*. Office of Health and Environ Assess, Cincinnati, OH.

U.S. EPA. 1989b. *Fed Register* 54:5746–5902.

U.S. EPA. 1989a. *Technology Demonstration Summary, Technology Evaluation Report, Site Program Demonstration Test, HAZCON Solidification*. Douglassville, PA.

U.S. EPA. 1990a. *Technical Support Documentation for Part I the National Sludge Survey Notice of Availability*. U.S. EPA, Analysis and Evaluation Division, Washington, DC.

U.S. EPA. 1990b. *Exposure Factors Handbook*. EPA/60-89/043. Exposure Assessment Group, Office of Health and Environ Assess, Washington, DC.

U.S. EPA. 1991. *Guidelines for Exposure Assessment*. Draft for Risk Assessment Forum. Washington, DC.

U.S. EPA. 1992. *Framework for Ecological Risk Assessment*. EPA/630/R-92/001 Washington, DC.

U.S. EPA. 1993. *Fed Register* 58:9248–9415.

U.S. EPA. 1996. *Soil Screening Guidance*. Rep 540/R-95/128. U.S. EPA, Washington, DC.

U.S. EPA. 1997a. *Recent Developments to In Situ Treatment of Metal Contaminated Soils*. EPA-542-R-97-004. Office of Solid Waste and Emergency Response, Washington, DC.

U.S. EPA. 1997b. *Technology Alternatives for the Remediation of Soils Contaminated with Arsenic, Cadmium, Chromium, Mercury, and Lead*. EPA/540/S-97/500. Cincinnati, OH.

U.S. EPA. 1998. *Fed Register* 63:26846–26924.

Vaniman, D.T., and D.L. Bish. 1995. In D.W. Ming and F.A. Mumpton, eds. *Natural Zeolites '93: Occurrence, Properties, Use*. Intl Comm Natural Zeolites, Brockport, NY.

Van den Berg, R., C.A.J. Denneman, and J.M. Roels. 1993. In F. Arendt et al., eds. *Contaminated Soil '93*. Kluwer, Dordrecht, Netherlands.

Van Straalen, N.M., and C.A.J. Denneman. 1989. *Ecotox Environ Safety* 18:241–251.

Vegter, J.J. 1995. In W. Salomons, U. Förstner, and P. Mader, eds. *Heavy Metals: Problems and Solutions*. Springer-Verlag, Berlin.

Vegter, J.J. 1997. In R. Prost, ed. *Contaminated Soils*. INRA 85, INRA, Paris.

Vollmer, M.K., S.K. Gupta, and R. Krebs. 1997. In R. Prost, ed. *Contaminated Soils*. INRA 85, INRA, Paris.

Walton, B.T., and T.A. Anderson. 1992. *Curr Opin Biotechnol* 3:267–270.

Wenzel, W.W., D.C. Adriano, D. Salt and R. Smith. 1999. In D.C. Adriano et al., eds. *Bioremediation of Contaminated Soils*. ASA 37, American Society of Agronomy, Madison, WI.

Wright, J.V., et al. 1995. *In Situ Immobilization of Heavy Metals in Apatite Mineral Formulation*. Spec Rep Apatite mineral formulations and emplacement options, September 1995. Pac North Lab, Richland, WA.

World Commission on Environment and Development. 1987. *Our Common Future*. Oxford Univ Pr, Oxford.

[WHO] World Health Organization. 1984. *Guidelines for Drinking Water Quality*. Geneva.

WHO. 1987. *Principles for the Safety Assessments of Food Additions and Contaminants in Food*. Environ Health Criteria, 70. WHO/IPCS, Geneva.

Zhang, P., J.A. Ryan, and L.T. Bryndzia. 1997. *Environ Sci Technol* 207:2673–2678.

7
Arsenic

1 General Properties of Arsenic

Arsenic (atom. no. 33) is a steel-gray, brittle, crystalline metalloid with three allotropic forms that are yellow, black, and gray. It tarnishes in air and when heated is rapidly oxidized to arsenous oxide (As_2O_3) with the odor of garlic. It belongs to Group V-A of the periodic table, has an atom. wt. of 74.92, and closely resembles phosphorus chemically. Gray As, the ordinary stable form, has a density of 5.73 g cm^{-3}, a melting pt. of 817 °C, and sublimes at 613 °C. The more common oxidation states for As are –III, 0, III, and V. Arsines and methylarsines, which are –III species, are generally unstable in air. Elemental As (As^0) is formed by the reduction of arsenic oxides. Arsenic trioxide (As^{3+}) is a product of smelting operation and is the raw material used in forming most arsenicals. It is oxidized catalytically or by bacteria to arsenic pentoxide (As^{5+}) or orthoarsenic acid (H_3AsO_4). Arsenic compounds compete with their phosphorus analogs for chemical binding sites. Arsenic covalently bonds with most nonmetals and metals, and forms stable organic compounds in both its trivalent and pentavalent states. The most important compounds are white As (As_2O_3), the sulfide, Paris green (copper acetoarsenite) [$3Cu(AsO_2)_2 \cdot Cu(CH_3COO)_2$], calcium arsenate, and lead arsenate, the last three being used as pesticides and poisons.

2 Production and Uses of Arsenic

World production of As is estimated to be 75 to 100×10^3 tonnes annually, of which the United States uses about half. Sweden is the world's leading producer. Arsenic trioxide (As_2O_3), also known as white arsenic, constitutes 97% of As produced that enters end-product manufacturing. The other form needed by end-product manufacturers is the metal form, which is used as an additive in special lead and copper alloys. The primary commercial sources of As are copper and lead ores. Arsenic is recovered as a by-product during the smelting process.

Arsenic compounds are used mainly in agriculture and forestry as pesticides, herbicides, and silvicides. In the United States, agricultural products accounted for about 81% of As use during the 1970s, ceramics and glass for 8%, chemicals for 5%, and the balance for other uses. By the 1980s, agricultural application accounted for only 46% of As use in the United States. Arsenic trioxide is the raw material for arsenical pesticides, including lead arsenate, calcium arsenate, sodium arsenite, and organic arsenicals. These compounds are used in insecticides, herbicides, fungicides, algicides, sheep dips, wood preservatives, and dyestuffs, and for the eradication of tapeworm in sheep and cattle.

Arsenic is used primarily for its toxic properties. Inorganic arsenicals have been used in agriculture as pesticides or plant defoliants for many years (Table 7.1). Recently, the major use for inorganic As is in wood preserva-

TABLE 7.1. Names and uses of some important arsenical pesticides.

Pesticide	Application rates and methods	Commercial uses
Arsenic acid	0.0035 m^3 ha^{-1} of the 75% concentrate	Cotton desiccant to facilitate mechanical harvesting
CA (cacodylic acid)	3.4–11.2 kg ha^{-1}	Lawn renovation and general weed control in noncrop areas
DSMA (disodium methanearsonate)	Directed after emergence on cotton at 2.52 kg ha^{-1}; 2.24–4.26 kg ha^{-1} for lawn and ornamental uses	Cotton and noncrop areas; crabgrass
MSMA (monosodium methanearsonate)	Directed after emergence on cotton at 2.52 kg ha^{-1}; 2.24–4.26 kg ha^{-1} for lawn and ornamental uses	Cotton and noncrop areas; crabgrass
Calcium arsenate	1.68–3.36 kg ha^{-1} in 100 gal (0.38 m^3) of water or dust at 2.24–28 kg ha^{-1}	Cotton insecticide; fruits, vegetables, and potatoes
Lead arsenate	3.3–67.3 kg ha^{-1} or 1.2–72 kg m^{-3} of water	Fruits, vegetables, nuts, turf, and ornamentals
Paris green	1.12–17.9 kg ha^{-1}	Baits and mosquito larvicide
Sodium arsenite	1.12–22.4 kg ha^{-1} in dry baits	Baits and a livestock dip; a nonselective herbicide, rodenticide, desiccant, and aquatic weed killer

Source: Modified from NAS (1977).

tion; arsine (AsH) is used in the microelectronics industry and in semiconductor manufacture. Until the 1940s, inorganic As solutions were widely used in the treatment of various diseases, such as syphilis and psoriasis. Inorganic As is still used as an antiparasitic agent in veterinary medicine and in homeopathic and folk remedies in the United States and other countries. Calcium arsenate and lead arsenate were the backbone of the insecticide industry from the early 1900s until the advent of the organic pesticides in the 1940s. Because they were less phytotoxic, they replaced the earlier arsenicals, Paris green and London purple, which had been in use since the 1860s.

Paris green (copper acetoarsenite) was successfully used to control the Colorado potato beetle (*Leptinotarsa decemlineata*) in the eastern United States by around 1870. Calcium arsenate has been applied to cotton and tobacco fields for boll weevil (*Anthonomus grandis*), beetle, and other insect control. Lead arsenate has been used for insect control on a variety of fruit trees and was particularly effective in the control of codling moth (*Cydia pomonella*) in apple orchards and hornworm (larvae of *Manduca quinquemaculata*) on tobacco. Arsenic trioxide has been widely used as a soil sterilant, while sodium arsenite has been used for aquatic weed control and as a defoliant for potato prior to tuber harvest.

Organic arsenicals have largely replaced inorganic As compounds as herbicides. The use of sodium arsenite has been limited, and the compound can no longer be used as a defoliant or vine-killer (Walsh and Keeney,

1975). Lead arsenate is no longer used in orchards, since fruit growers rely primarily on carbamates and organic phosphates for insect control. Calcium arsenate and Paris green are also seldom used nowadays.

Since the mid-1970s, the use of organoarsenical herbicides, e.g., MSMA (monosodium methanearsonate), DSMA (disodium methanearsonate), and cacodylic acid (CA), has grown rapidly (see Table 7.1). Both MSMA and DSMA have been used as selective herbicides for postemergence control of weeds in turf. They are also extensively used as selective postemergence herbicides in citrus, cotton, and noncrop areas for control of a number of weeds. MSMA is also used for chemical mowing along highway rights-of-way; CA is used for general weed control and is an excellent herbicide for monocotyledonous weeds.

In cotton production, As chemicals being used as herbicides and harvest aids are MSMA, DSMA, As acid, and CA. Herbicidal control of weeds is obtained with MSMA, DSMA, and CA, while As acid and CA are used as desiccants and defoliants in cotton. The use of MSMA and DSMA herbicides now accounts for more than 90% of the As used in agriculture and it is one of the largest volume pesticides (Wauchope and McDowell, 1984). However, because of economic and environmental considerations, glyphosate [N-(phosphonomethyl)glycine] has become a viable alternative to about half of the MSMA and DSMA use.

Arsenic is also used as a feed additive, although lesser amounts are used than in pesticides, defoliants, or herbicides. Arsenicals are used in poultry feeds to control coccidiosis and to promote chick growth (Calvert, 1975). They are added to poultry rations at rates of about 100 mg kg^{-1} for arsanilic acid and 50 mg kg^{-1} for 3-nitro-4-hydroxyphenylarsonic acid.

3 Arsenic in Nature

Arsenic is ubiquitous in nature and is found in detectable concentrations in all soils and nearly all other environmental media (Table 7.2). The occurrence of As in the continental crust of the earth is generally given as 1.5 to 2 ppm. Arsenic ranks 52nd in crustal abundance, ahead of Mo. It is a major constituent of more than 245 minerals and is found in high concentrations in sulfide deposits: arsenides (27 minerals), sulfides (13 minerals), and sulfosalts (65 minerals).

The geologic background of a particular soil determines its native As content. Soils overlying sulfide ore deposits usually contain As at several hundred ppm, the reported average being 126 ppm and the range being from 2 to 8000 ppm As (NRCC, 1978). Background soil concentration of As generally does not exceed 15 ppm, although concentrations ranging from 0.2 to 40 ppm have been reported (Walsh et al., 1977). A natural background level of 6.3 ppm total As was reported for agricultural soils in Ontario (Frank et al., 1976). Similarly, an average of 5 ppm As was reported for chernozems and luvisols in Alberta (Dudas and Pawluk, 1980). However, much higher As concentrations (8 to 40 ppm) have been found for acid sulfate soils attributed to the weathering of pyrite in the parent material.

The median As levels in South Australian and Tasmanian soils were 3.9 ppm and 0.6 ppm As, respectively (Merry et al., 1983). In contrast, As background concentrations for urban soils in Australia ranged from <0.2 to 45 ppm (Tiller, 1992).

The As concentrations of rocks depend on the rock type, with sedimentary rocks containing much higher levels than igneous rocks (see Table 7.2). Generally, the mean As concentrations in igneous rocks range from 1.5 to 3.0 ppm, whereas the mean As concentrations in sedimentary rocks range from 0.3 to 500 ppm. Limited data indicate that shale and clays usually exhibit rather high As concentrations. Coal and its by-product fly ash also contain significant quantities. Thus, combustion of coal for generating power, as well as the disposal of fly ash, may contribute to As input in the environment.

TABLE 7.2. Commonly observed arsenic concentrations (ppm) in various environmental media.

Material	Average concentration	Range
Igneous rocks[a]	1.5–3	0.06 – 113
Limestone[a]	1.7	0.1 – 20
Sandstone[a]	4.1	0.6 – 120
Shale[a]	14.5	0.3 – 500
Phosphate	22.6	0.4 – 188
Petroleum[a]	0.18	<0.003– 1.11
Coal[a,b]	13	trace –2000
Coal ash[b]		
Fly ash	156	(8 –1385)
Bottom ash	8	<5 – 36
FGD sludge	25	<5 – 53
Oil ash[b]	112	75 – 174
Sewage sludge[d]	14.3	3 – 30
Soils (world, normal)[a]	7.2	0.1 – 55
Soils (USA, noncontaminated)	7.4	
Forest soils (Norway)[a]	—	0.59 – 5.70
Common crops[c]	—	0.03 – 3.50
Drinking water (national USA) (μg L^{-1})	2.4	0.5 – 2.4
River water (μg L^{-1})[a]		≤5
Lake Superior water (μg L^{-1})[a]		0.1 – 1.6
Japan, several lakes (μg L^{-1})[a]		0.2 – 1.9
Groundwater (USA) (μg L^{-1})[a]		≤10
Great Lakes sediments	—	0.50 – 14.00
Ocean sediments	33.7	<0.40 – 455

Sources:
[a]Eisler (1994).
[b]Ainsworth and Rai (1987).
[c]Liebig (1965).
[d]Furr et al. (1976).

4 Arsenic in Soils/Sediments

4.1 Total Arsenic in Soils

The main origin of As in soils is the parent material from which the soil is derived. Arsenic levels in uncontaminated, nontreated soils seldom exceed 10 ppm. However, anthropogenic sources of As have elevated the background levels of As. Consequently, As residues can accumulate to very high levels in agricultural areas where As pesticides or defoliants were repeatedly used (Table 7.3). Thus, agricultural use of arsenicals has caused surface soil accumulation of 600 ppm or more. Survey of soils in the United States indicates that As levels for normal soils ranged from 0.2 to 40 ppm (Olson et al., 1940). More recently, an As average of 7.2 ppm (range of <0.1 to 97 ppm) for U.S. surface soils was reported.

A mean of 8.7 ppm As in soil samples ($n = 1140$) collected from Missouri has been reported (Selby et al., 1974). Thus, it appears that uncontaminated soils seldom contain >10 ppm As. The As concentrations of soils in the Russian Plain were estimated to be generally uniform at 1 to 10 ppm, with an average of 3.6 ppm (Vinogradov, 1959). The highest As content is found in the chernozem and gray forest soils, and the lowest in the northern tundra soils and podzolic soils. World soil As content averaged 10 ppm (Berrow and Reaves, 1984). In areas near As mineral deposits, soil levels may average 400 to 900 ppm As.

TABLE 7.3. Total arsenic concentration (ppm) in field soils in North America repeatedly treated with contaminated arsenical pesticides or defoliants and in non-treated soils.

Location	Control soil	Contaminated soil	Crop
Colorado	1.3– 2.3	13– 69	Orchard
Florida	8	18– 28	Potato
Idaho	0 –10	138– 204	Orchard
Indiana	2 – 4	56– 250	Orchard
Maine	9	10– 40	Blueberry
Maryland	19 –41	21– 238	Orchard
New Jersey	10	92– 270	Orchard
New York	3 –12	90– 625	Orchard
North Carolina	4	1– 5	Tobacco
Nova Scotia	0 – 7.9	10– 124	Orchard
Ontario	1.1– 8.6	10– 121	Orchard
Oregon	2.9–14.0	17– 439	Orchard
	3 –32	4– 103	Orchard
Washington	6 –13	106– 830	Orchard
	8 –80	106–2553	Orchard
	4 –13	48	Orchard
Wisconsin	2.2	6– 26	Potato

Source: Adapted from Walsh and Keeney (1975).

4.2 Bioavailable Arsenic in Soils

The extractable fraction of As may give a better indication of its bioavailability and mobility in soils. It is a general consensus that total As in soil does not reliably reflect its phytoavailability. Indeed it has been reported that correlation is better between extractable As and plant growth than between total As and plant growth (Woolson et al., 1971). The following are some extractants used for soil As: distilled water, 0.1 N NH$_4$OAc, Bray P-1 solution (0.03 N NH$_4$F + 0.025 N HCl), 0.05 N HCl, mixed acid (0.05 N HCl + 0.025 N H$_2$SO$_4$), 0.5 N NaHCO$_3$, 0.05 M KH$_2$PO$_4$, 0.1 N NH$_4$NO$_3$, and 0.1 N KNO$_3$. Arsenic concentrations in apricot and apple fruits growing on highly contaminated orchards were significantly correlated with concentrated HCl-extractable soil As (Fig. 7.1).

In soils under reducing conditions, water-soluble As is well correlated to total As (Deuel and Swoboda, 1972b). In addition, water-soluble As increased with increased level of As added; <5% of the added As was water soluble in samples under room temperature. On the average, only <5% of the total As can be extracted with 1 N NH$_4$OAc (pH 7.0), while Bray P-1 can remove up to 38% of the As in the soil (Jacobs et al., 1970a). The efficiency of some basic extractants in removing soil As is as follows: 0.5 M NH$_4$F ~ 0.5 M NaHCO$_3$ < 0.5 M (NH$_4$)$_2$CO$_3$ < 0.5 M Na$_2$CO$_3$ < 0.1 M NaOH, with the effectiveness increasing with increasing pH (Johnston and Barnard, 1979). For acid extractants, the order is (0.05 N HCl + 0.025 N H$_2$SO$_4$) ~ 0.5 N HCl < 0.5 M KH$_2$PO$_4$ < 0.5 N H$_2$SO$_4$.

The water-, 1 N NH$_4$Cl-, and 0.5 N HCl-extractable values were significantly correlated with plant growth (Deuel and Swoboda, 1972a). Total As,

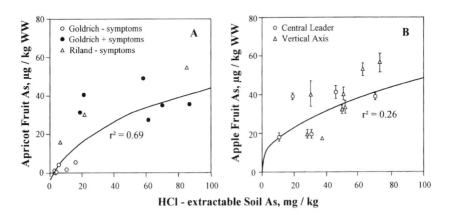

FIGURE 7.1. Arsenic concentrations in apricot (A) and apple (B) fruits as a function of HCl-extractable soil arsenic. The apricots were sampled from trees with and without As phytotoxicity symptoms. Error bars indicate 95% CI for mean values. (Modified from Creger and Peryea, 1994.)

NH$_4$OAc-, and Bray P-1-extractable As were equally effective in predicting reduced yields (Jacobs et al., 1970a). Better correlation between plant growth and 0.05 N HCl-extractable As than with water-soluble As also has been observed (Reed and Sturgis, 1936). Both mixed acid– (0.025 N H$_2$SO$_4$ and 0.05 N HCl) and 0.5 N NaHCO$_3$-extractable As had high and nearly equal correlations with plant growth ($r = 0.81$ and 0.82, respectively). The mixed acid method is routinely used to test for available P in acid soils of the eastern United States; the bicarbonate procedure is used to test for available P in alkaline soils in the western states. These P extractants, along with Bray P-1 solution, may be more favorable over the other extractants in determining bioavailable As because they are less laborious to perform, are routinely used for P analysis, and extract relatively higher As levels, favoring analytical detectability (Jacobs et al., 1970a; Woolson et al., 1971).

4.3 Arsenic in Soil Profile

Appreciable As can move downward in the profile with leaching water, especially in coarse-textured soils. Losses of As applied as sodium arsenite to sandy soils in the Netherlands were directly related to the amount of As in soil, with an average half-life in surface soil of 6.5 ± 0.4 yr (Tammes and deLint, 1969). Most of the As lost from the upper 20 cm of soil was found in the 20- to 40-cm depth, although net loss of As from the soil occurred continuously.

The profile distribution of Pb, As, and Zn in an abandoned apple orchard indicates insignificant leaching of the metals below the 20-cm depth (Veneman et al., 1983). The site, used for fruit production since the late 1800s, had received large applications of lead arsenate pesticides containing Pb and As, and trace amounts of Zn. Maximum As and Pb concentrations in surface soils were 120 and 870 ppm, respectively.

In As-contaminated region of Bangladesh, annual irrigation of a paddy soil with 500 mm of water containing 35 µg As L^{-1} (well 1) is equivalent to an input of 150 g As ha^{-1} yr^{-1} and would increase As concentration in the top 15 cm of soil by 0.1 mg kg^{-1} yr^{-1}. Utilizing the same amount of water containing 1100 µg L^{-1} (well 2) would increase As levels (through an input of 4.5 kg ha^{-1} yr^{-1}) for about 3 mg kg^{-1} ha^{-1}. On a particular site, As concentration in the surface soil under rice cultivation is much larger than in soil under vegetable production (Fig. 7.2) because rice production is more intensively irrigated (Brandstetter et al., 2000).

The profile distribution of As in plots of a Plainfield sand that had received varying amounts of As as sodium arsenite in 1967, receiving about 75 cm of precipitation and 50 cm of irrigation water per year until 1970, was monitored for several years (Steevens et al., 1972). Irrigation was stopped after 1970. Phytotoxicity persisted at the 180- and 720-kg ha^{-1} plots from 1967 to 1970. The 720-kg ha^{-1} plots were still barren in 1974, 7 yr after As application. With time however, significant declines in the amount of total

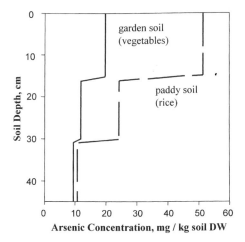

FIGURE 7.2. Total As concentration in soil profiles from an irrigated vegetable garden and periodically flooded paddy site in an arsenic-contaminated area (Faridpur) in Bangladesh. (Reprinted from Brandstetter et al., 2000, p. 721, by courtesy of Marcel Dekker, Inc.)

As in surface soil of all treated plots were observed. In the 720-kg ha^{-1} plots, total As in surface soil had declined to 40% of the original value in 3 yr and to 33% by 7 yr.

It has been estimated that As applied at recommended rates can accumulate in soil gradually and can basically reach equilibrium levels after 25 to 30 yr of application (Woolson and Isensee, 1981). Losses in surface soils due largely to leaching caused increases in As at lower depths. Seven years after As application on the soil surface, As had leached to 83 cm, indicating some mobility of the element. When applied as methanearsonates to turf over a 4-yr period on a sandy loam, As was highest in concentration in the upper 5 cm of soil and decreased at the 30-cm depth (Johnson and Hiltbold, 1969). Although most of the As residue from MSMA application accumulated in the top 15 cm of soil, the amount of As in the 15- to 30-cm depth was 2.5 times greater after 5 yr than after 3 yr of application, indicating As buildup over time (Robinson, 1975).

4.4 Sorption of Arsenic in Soils/Sediments

In soils and sediments, As has been known to have a high affinity for oxide surfaces, the extent of which depends on several biogeochemical factors. Particle size, OM, nature of constituent minerals, pH, redox potential, and competing ions have all been shown to influence adsorption processes (Chiu and Hering, 2000; Jones et al., 2000; Smith et al., 1999).

Formation of inner sphere complexes of As(V) with soil constituents has been directly demonstrated using EXAFS (extended X-ray absorption fine structure spectroscopy) and wide-angle X-ray scattering on ferrihydrite (Waychunas et al., 1993, 1996), and infrared (IR) spectroscopy on goethite (Lumsdon et al., 1984). Formation of binuclear, inner-sphere complexes was postulated as a main mechanism for the sorption (i.e., specific adsorption) of As(V) onto ferrihydrite. However, monodentate complexes were also observed that accounted for about 30% of all As–Fe complexation. In addition, Fendorf et al. (1997) indicated that As(V) formed three different complexes on goethite using EXAFS technique; monodentate complexation was favored at low As concentrations, whereas bidentate complexation was favored at higher concentrations.

The As sorbing inorganic constituents of soils include Fe and Al oxides, clay minerals, and calcite. The sorption behavior of As is largely dependent on its oxidation state, and the pH and redox potential of the medium. Arsenite sorption rate was significantly correlated with Fe oxide content of five West Virginia soils (Elkhatib et al., 1984a,b) and with dithionite-extractable Fe content of 15 Japanese soils (Sakata, 1987). Sorption of arsenate in soils was significantly influenced by oxalate-extractable Al and Fe and with clay (Livesey and Huang, 1981; Wauchope, 1975). Sorption of arsenate on amorphous Fe hydroxide, goethite, gibbsite, amorphous Al hydroxide, and activated alumina exhibited maxima in the pH range of 3 to 7, followed by a decline with increasing pH. Sorption of arsenite on amorphous Fe hydroxide and activated alumina increased at low pH, peaked between pH 7 and 8, and decreased at high pH. Arsenite sorption was less than arsenate sorption on activated alumina and amorphous Fe hydroxide. The retention of As on clay is dependent on the type and quantity of clay. Soils having higher clay content are expected to retain more As than soils with lesser amounts. Kaolinite sorbs more As from solution than vermiculite, which sorbs more As than montmorillonite (Dickens and Hiltbold, 1967).

Arsenate sorption on calcareous, montmorillonitic soils increased with increasing pH, attained a maximum near pH 10.5, and decreased at higher pH (Goldberg and Glaubig, 1988). Sorption of arsenate on montmorillonitic and kaolinitic reference minerals increased at low pH, exhibited a peak near pH 5, and decreased at higher pH; whereas arsenate sorption on calcite increased from pH 6 to 10, peaked between pH 10 and 12, and decreased above pH 12. Carbonates can play a major role in arsenate sorption above pH 9.

The adsorption maxima of soils for arsenate were found not to be related to pH and inorganic C (carbonates) content but were related to ammonium oxalate–extractable Al and, to a lesser extent, to the clay content and ammonium oxalate–extractable Fe (Livesey and Huang, 1981). Among the competing anions (i.e., Cl^-, NO_3^-, SO_4^{2-}, and HPO_4^{2-}), the phosphate substantially suppressed As sorption by the soil. In evaluating the sorption of three As species relative to phosphate in aerobic soils, the order of sorption was phosphate < cacodylate < MSMA = arsenate (Wauchope, 1975).

In anaerobic river sediments, the order of sorption was cacodylate < MSMA < arsenate (Holm et al., 1980).

Because of their geochemical similarity, As has been assumed to react similarly to P in soil, forming insoluble compounds with Al, Fe, and Ca. Indeed the sorption capacities for P and As were found similar ($r = 0.98$) in 19 German acidic forest soils and that the sorbed As and P were primarily bound to Fe and Al (Fassbender, 1974). This is expected because both As and P belong to the same chemical group, having comparable dissociation constants for their acids and solubility products for their salts.

In general, As(III) is less strongly sorbed than As(V) to a variety of sorbents; is generally more mobile in the +III oxidation state than in the +V oxidation state, and immobilization of As is enhanced by the oxidation of As(III) to As(V) (Chiu and Hering, 2000; Masscheleyn et al., 1991).

The Langmuir and Freundlich adsorption isotherms have been employed for characterizing As sorption by soils and clay minerals. As a result, similar reactions of As and P in soils have been advanced. Jacobs et al. (1970b) found that the sorption of As against extraction by NH_4OAc and Bray P-1 extractants from Wisconsin soils equilibrated with As (80 and 320 µg As g^{-1} added as Na_2HAsO_4) increased as the sesquioxide content increased. More specifically, the amount of As sorbed increased as the free Fe_2O_3 content of the soil increased. Furthermore, they found that removal of amorphous Fe and Al components with oxalate treatment minimized As sorption by the soil.

4.5 Forms and Chemical Speciation of Arsenic

In evaluating As forms in soil, the traditional Chang and Jackson (1957) procedure used to fractionate soil P has been adopted. This fractionation scheme approximates water-soluble or adsorbed As (NH_4Cl-soluble), Al arsenate (NH_4F-soluble), Fe arsenate (NaOH-soluble), and Ca arsenate (H_2SO_4-soluble).

In an extensive soil survey encompassing 12 states, Woolson et al. (1971) found that most of the As residues in surface soils with a history of inorganic As application was in the Fe arsenate form. However, when the amount of reactive Al or Ca was high and reactive Fe was low, Al arsenate and Ca arsenate predominated. Water-soluble As was detected in soils from only two states. The distribution of As forms in soils and the influence of soil chemical properties on these forms are shown in Table 7.4.

In another fractionation study of As, added to soils as DSMA, Fe arsenate was the predominant form present, accounting for 31 to 54% the total As present, with an average of nearly 44% (Akins and Lewis, 1976). This was followed by Al arsenate with over 27% (11 to 41% of added As), followed by Ca arsenate with 16%, water-soluble arsenate with 6%, and nonextractable fraction, 7%. Thus, it can be generalized that in acid soils, As retention follows the dominance of Fe and Al ions over those of the Ca

TABLE 7.4. Total arsenic concentrations and distribution of forms of arsenic in contaminated (treated) and control U.S. soils.

Location and soil type	Treatment	pH	Available cations, meq/100 g			Total, ppm	Arsenic[a]					Corn growth reduction, %
			Ca	Fe	Al		Easily soluble, %	Al–As, %	Fe–As, %	Ca–As, %		
Mississippi Memphis sil	Control	5.40	3.2	1057	6.3	21	0	0	91	9		—
	Treated	5.62	4.0	934	6.6	96	0	10	78	12		-13
Alabama Chesterfield sil	Control	—	—	—	—	8	0	33	67	0		—
	Treated	5.61	3.4	369	4.0	16	0	31	62	7		-28
Idaho Greenleaf sil	Control	7.11	8.6	925	3.3	28	0	0	0	0		—
	Treated	7.01	8.2	906	4.0	170	7	34	46	13		69
Oregon Agate grl	Control	—	—	—	—	10	—	—	—	—		—
	Treated	5.78	8.5	1155	2.7	67	0	22	56	22		-5
New York Newfane sl	Control	—	—	—	—	4	—	—	—	—		—
	Treated	4.40	1.3	745	4.8	319	0	24	63	13		42
Florida Lakeland fs	Control	—	—	—	—	8	—	—	—	—		—
	Treated	7.14	6.9	35	1.5	28	10	30	0	60		15
Indiana (loess soil)	Control	—	—	—	—	2	—	—	—	—		—
	Treated	4.40	1.4	572	3.6	250	5	26	53	16		67

[a] Arsenic forms estimated by a modified Chang and Jackson procedure for P (1957).

Source: Extracted from Woolson et al. (1971).

ions, whereas in alkaline, calcareous soils, the sorption follows the dominance of Ca over the other two ions.

Analysis of surface soils (0 to 15 cm) collected in Japan from various agricultural lands contaminated with As revealed that arsenate(V) was the dominant form of As in these soils, with lower levels of dimethylarsinate $[(CH_3)_2AsOO^-]$, and detectable levels of monomethylarsenate $(CH_3-AsO_2OH^-)$ in most of the samples (Takamatsu et al., 1982). This study also found seasonal variation of As forms depending on the water status of the paddy field, i.e., under flooded conditions, there was an increase in pH and the amounts of arsenite(III) and dimethylarsinate, whereas under upland conditions, the pH decreased, but the amounts of arsenate and monomethylarsonate increased. Sequential extraction of As by NH_4Cl (water-soluble and exchangeable), NH_4F (Al-bound), NaOH (Fe- and organic-bound) and H_2SO_4 (Ca-bound) from a sandy soil (pH 4.5) where sodium arsenate or sodium arsenite was added under saturated or partially saturated conditions indicates that the added As became more recalcitrant in soils with time (Onken and Adriano, 1997). The rate of conversion to the more recalcitrant forms was as rapid in saturated soils as in partially saturated soils.

Sequential extraction of As-contaminated soils from the Austrian Alps shows that As was most abundant in the two oxalate fractions, indicating that As is primarily associated with amorphous and crystalline Fe oxides (Table 7.5). The fraction of As extracted by $(NH_4)_2HPO_4$ represents only about 10% of total As. This fraction may provide a relative measure of specifically sorbed As in soils. The amount of readily mobile As extracted by $(NH_4)_2SO_4$ is generally small, but represents the most important fraction related to risk assessment. The residual fraction is typically small, since most As is bound to amorphous and crystalline Fe oxides.

TABLE 7.5. Distribution of arsenic forms for As-contaminated soils (mg As kg^{-1}) ($n = 24$) of the Austrian Alps.

	$(NH_4)_2SO_4$[a]	$NH_4H_2PO_4$[b]	Oxalate[c]	Oxalate/Asc.[d]	Residual[e]	Sum
Mean	1.81	57.5	190.2	193	84.9	527.4
Median	0.71	28.7	118.8	63.5	30.2	222.3
Max.	11.13	179.9	695.6	763.4	343.8	—
Min.	0.02	3.1	17.7	8.7	2.3	—
% of Total	0.34	10.9	36.1	36.6	16.1	100.0

[a]Exchangeable fraction extracted with 0.05 M $(NH_4)_2SO_4$, 4 hr shaking at 20 °C.
[b]Specifically adsorbed fraction extracted with 0.05 M $(NH_4)_2HPO_4$, 16 hr shaking at 20 °C.
[c]Amorphous-Fe bound fraction extracted with 0.2 M NH_4-oxalate pH 3.25, 4 hr shaking in the dark.
[d]Crystalline-Fe bound fraction extracted with 0.2 M NH_4-oxalate + 0.1 M ascorbic acid pH 3.25, 30 min shaking in the water bath at 96 °C in the light.
[e]Residual fraction extracted with acid digestion: HNO_3 + H_2O_2, closed vessel in microwave digestion.
Source: Brandstetter et al. (2000).

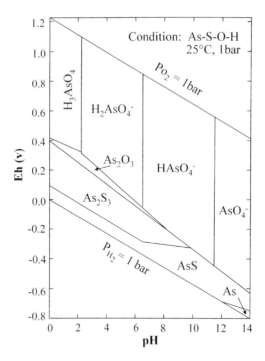

FIGURE 7.3. Predicted Eh–pH stability field for arsenic under the system As–S–O–H, with assumed activities of As $= 10^{-6}$, S $= 10^{-3}$. (From Brookins, 1988.)

Arsenic is predominantly bound to sediments in most aquatic systems. In river sediments, the main As forms are: Al arsenate, Fe arsenate, Ca arsenate, and Fe-occluded As (Chunguo and Zihui, 1988). Oxidation of As(III) to As(V) by freshwater sediments has been attributed to the reaction of As(III) with Mn oxides (Chiu and Hering, 2000).

The predicted species of As in a Eh–pH stability field is displayed on Figure 7.3.

4.6 Transformation of Arsenic in Soils/Sediments

Arsenic in soils and sediments can undergo transformations via abiotic or biotic processes. Sediments from five lakes in Saskatchewan (Canada) oxidized As(III) to As(V), plausibly through the oxidation of As(III) by Mn(IV) and Fe(III) as shown below (Oscarson et al., 1980):

$$MnO_2 + HAsO_2 + 2H^+ \rightarrow Mn^{2+} + H_3AsO_4$$

$$2Fe^{3+} + HAsO_2 + 2H_2O \rightarrow 2Fe^{2+} + H_3AsO_4 + 2H^+$$

Also, a synthetic, amorphous dioxide mineral (birnessite, δ-MnO$_{2(s)}$) was found to be very effective in oxidizing As(III). Thus, the Mn(IV) and Fe(III) oxides present on the surface of mineral and organic compounds and existing as discrete particles may play a significant role in catalyzing the oxidation of As(III) through an electron transfer mechanism.

Similar abiotic processes may be responsible for the release of As from arsenopyrite through its oxidation by Fe(III) (Rimstidt et al., 1994):

$$FeAsS + 13Fe^{3+} + H_2O \rightarrow 14Fe^{2+} + SO_4^{2-} + 13H^+ + H_3AsO_4$$

This reaction is believed to be a predominating process inducing the release of As into the groundwater in areas where well waters are highly contaminated with As.

As(V) can become immobilized by coprecipitation with hydrous iron oxides, with subsequent formation of the mineral scodorite, $FeAsO_4 \cdot 2H_2O$ (Mok et al., 1988):

$$Fe(OH)_3 + H_3AsO_4 \rightarrow FeAsO_4 \cdot 2H_2O + H_2O$$

Similarly, As(V) can be immobilized in sediments by coprecipitation with hydrous Mn oxides (Masscheleyn et al., 1991):

$$3MnOOH + 2HAsO_4^{2-} + 7H^+ + 3e^- \rightarrow Mn_3(AsO_4)_2 + 6H_2O$$

or by Mn(III) oxide manganite (γ-MnOOH) (Chiu and Hering, 2000):

$$2\gamma\text{-}MnOOH_{(s)} + H_3AsO_3 + 3H^+ \rightarrow 2Mn^{2+} + H_2AsO_4^- + 3H_2O$$

The formation of sulfides in reducing environment accompanies the reduction of As(V) to As(III), with the latter species dominating in the porewater (Moore et al., 1988):

$$H_2AsO_4^- + 3H^+ + 2e^- \rightarrow H_3AsO_3 + H_2O$$
$$2H_3AsO_3 + 6H^+ + 3S^{2-} \rightarrow As_2S_3 + 6H_2O$$
$$2As_2S_3 + 4e^- \rightarrow 4AsS + 2S^{2-}$$

However, the sulfides, including the arsenopyrite FeAsS, can undergo oxidation releasing As into the water or surface sediment layers (Hallacher et al., 1985):

$$As_2S_3 + 7O_2 + 6H_2O \rightarrow 3H_2SO_4 + 2H_3AsO_4$$
$$4AsS + 11O_2 + 10H_2O \rightarrow 4H_2SO_4 + 4H_3AsO_4$$
$$4FeAsS + 13O_2 + 6H_2O \rightarrow 4FeSO_4 + 4H_3AsO_4$$

Arsenic undergoes a series of biological transformations in aquatic systems, yielding organoarsenicals and other compounds (see Fig. 7.4).

Inorganic As can undergo biochemical transformations, i.e., the hydroxyl group of arsenic acid [$AsO(OH)_3$], is replaced by the CH_3 group to form MMAA, DMAA, and $(CH_3)_3AsO$ (Maeda, 1994). These transformations are usually biologically mediated in aquatic systems (Fig. 7.4). Initially,

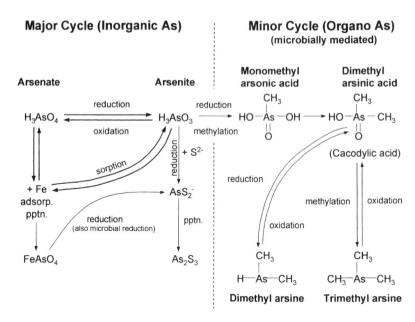

FIGURE 7.4. Simplified transformation pathways of arsenic in the environment. Interconversion of inorganic As species predominates the major cycle in typical arable and orchard soils (left half). (With input from Dr. E. Smith, CSIRO, Adelaide, Australia.)

inorganic As is incorporated by autotrophic organisms, such as algae, and then transported through the food chain. During this transfer, the methylation of As progresses in the organisms. Methylation of As is considered a major detoxifying process for these organisms. In general, organoarsenical species are found as dimethylarsenic compounds in aquatic plants such as algae and as trimethylarsenic compounds in fish, mollusks, and crustaceans in aquatic systems (Maeda et al., 1990, 1992). The methylated As species may also be degraded photochemically; the effect of light is likely to be more significant in freshwater than in saltwater systems (Brockbank et al., 1988).

As(III) in riverine water can be oxidized to As(V) by bacteria (Wakao et al., 1988). Heterotrophic As(III)-oxidizing bacteria including the *Bacillus* and *Pseudomonas* spp., have been identified. For example, microbial As(III) oxidation occurred in acid mine waters of the Matsuo sulfur pyrite mine in Japan and the stream of a nearby Akagawa river that received mine drift waters. Similarly, various bacteria, fungi, and algae that can reduce As compounds have been identified. The reduction of As(V) to As(III) has been reported to be mediated by *Pseudomonas fluorescens* (under reducing conditions), wine yeast, rumen bacteria, and cyanobacteria (Cullen and Reimer, 1989).

Monomethylarsenate ($CH_3AsO_2OH^-$) and dimethylarsinate [($CH_3)_2As$-OO^-] are the common organoarsenicals in river water (Maher, 1988). Methylated As species can result from direct excretion from algae or microbes, or from degradation of excreted arsenicals or more complex cellular organoarsenicals (Mok and Wai, 1994). In a river severely contaminated from the disposal of As waste effluent, both methylated (MMAA and DMAA) and inorganic As(V) and As(III) species were abundant in the porewater (Mok and Wai, 1994).

Biological transformations are important in the cycling of As in soils. In strongly reducing environments, elemental As and arsine($-$III) may be the predominant species, but arsenate(V) is the stable oxidation state in aerobic environments (see Fig. 7.4). Under moderately reducing conditions, such as in flooded soils, arsenite(III) may be the dominant species. As depicted in Figure 7.4, the major cycle of As involves the transformation of inorganic As in typical agricultural and orchard soils (left side of the diagram).

Arsenite is a common commercial form of As and one of the most toxic As compounds, being 25 times more potent than dimethylarsinic acid (Braman and Foreback, 1973). The trivalent state of As is known to be 4 to 10 times more soluble in soils than the pentavalent state. Trivalent As compounds have proven to be more toxic to rice plants grown under flooded conditions (Epps and Sturgis, 1939; Reed and Sturgis, 1936). Furthermore, arsine (AsH_3) could have been lost from these flooded soils.

An increase in As(III) concentrations in soil solution over time under flooded conditions has been attributed to the release of As during dissolution of iron oxyhydroxide minerals that have a strong affinity for As(V) under aerobic conditions (Deuel and Swoboda, 1972a,b). Similarly, at higher soil redox potential (500 to 200 mV), As solubility was low and As(V) was the predominant species in solution of contaminated soils (Masscheleyn et al., 1991). Increasing the pH or reducing As(V) to As(III) increased the concentration of As species in the solution. The predominance of As(III) species under reducing conditions is consistent with the thermodynamic stability of this species over that of As(V) under this condition. However, complete reduction to As(III) under reducing conditions cannot be expected (Onken and Hossner, 1995). Thus, minerals such as $FeAsO_4$ and other Fe(III) phases are reduced to soluble Fe(II) and the sorbed As(V) is released into the solution (Takamatsu et al., 1982).

The $-$III, III, and V oxidation states of As can form compounds containing the C–As bond and are readily interconverted by microorganisms. In reducing environments such as sediments, methanogenic bacteria can reduce arsenate(V) to arsenite(III) and methylate it to methylarsonic acid(III), or dimethylarsinic acid (Wood, 1974). These compounds may further methylate (to trimethylarsine, $-$III), or reduce (to dimethylarsine, $-$III), and may volatilize to the atmosphere with the formation of cacodylic acid via oxidation reactions (see Fig. 7.4).

Both oxidative and reductive transformations of methanearsonates could occur in soils. Cheng and Focht (1979) demonstrated that microorganisms in soils amended with inorganic and methylated forms of As could produce volatile arsenicals by a reductive and/or reductive and demethylative pathway. In addition, soil isolates of *Alcaligenes* and *Pseudomonas* were also found to produce arsine from the reduction of arsenate and arsenite.

It has been postulated that cacodylic acid can be metabolized by two pathways: an oxidative pathway leading to C–As bond cleavage and a reductive pathway leading to alkyl arsine production (Woolson and Kearney, 1973). Oxidation of the methyl substituent to CO_2 occurs in association with microbial oxidation of soil OM, producing arsenate (Hiltbold, 1975). Carbon dioxide was evolved from soils, i.e., ranging from 0.7 to 5.5% of the added compound receiving methanearsonate during a 1-mo incubation (Dickens and Hiltbold, 1967). Similar values (1.7 to 10%) were reported on the oxidation of methyl carbon of MSMA in soils during a 3-wk period (Von Endt et al., 1968). As much as 14 to 15% of the As applied in soil can be lost through volatilization of alkyl arsines each year (Woolson and Isensee, 1981).

5 Arsenic in Plants

5.1 Arsenic in Mineral Nutrition in Plants

No evidence exists that As is essential for plant nutrition, although stimulation of root growth with small amounts of As added in solution culture has been reported (Liebig et al., 1959). Root growth of lemon plants in solution culture was enhanced by the presence of 1 ppm As as arsenate or arsenite. However, at 5 ppm of either form of As, both top and root growth were reduced. In addition, small yield increases have been observed at low levels of As especially for tolerant crops such as corn, potatoes, rye, and wheat (Jacobs et al., 1970a; Woolson et al., 1971). Stewart (1922) noted that 75 ppm As in a calcareous loam soil caused only slight detrimental effects to the more sensitive plants but that 25 ppm As in the soil appeared to have stimulated them. Liebig (1965) noted that stimulation does not always occur, is sometimes temporary, and may result in the reduction of top growth. Two possibilities may explain the growth stimulation by As: first, stimulation of plants by small amount of As since other pesticides, like 2,4-D, stimulate plant growth at sublethal levels (Woolson et al., 1971); second, displacement of phosphate ions from the soil by arsenate ions with a resultant increase in phosphate availability (Jacobs et al., 1970b).

The uptake and translocation of As by plants are also influenced by the source of As. For example, As added as As_2O_3 in solution culture was readily taken up by the cotton roots, but not translocated to the shoot (Marcus-Wyner and Rains, 1982). However, when cacodylic acid was applied, As was translocated to the shoot as well as reproductive tissues. Arsenic also accumulated in the roots and shoot with MSMA and DSMA as the source.

5.2 Bioaccumulation and Phytotoxicity of Arsenic

Arsenic is naturally present in plants, but concentrations in plant tissues rarely exceed 1 mg kg^{-1}. Few higher plants are known to accumulate As in their tissues. Its distribution, in general, decreases from root to stem and leaf to edible parts. Arsenic uptake by some species, such as *Taraxacum officinalis,* is known to correlate with As concentration in the soil and, thus, may provide a tool for biomonitoring As concentration.

In a survey of As-contaminated locations in the Austrian Alps, a total of 128 plant samples were collected from contaminated sites. Sampling included 31 different plant species, forest litter, and 14 bulk samples of grass and herbaceous species at grassland sites. All sites had total As concentrations in soils greater than 90 mg kg^{-1}. Arsenic concentrations in plant tissues ranged from 0.03 to 34.5 mg kg^{-1} DW, with an average of 1.9 mg kg^{-1} ($n = 128$) and a median of 0.8 mg kg^{-1} (Fig. 7.5). About 30% of plant samples exceeded the regulatory limit of 2 mg As kg^{-1} according to the "Austrian legislation on fodder quality" (Brandstetter et al., 2000). No correlations between total or $(NH_4)_2SO_4$-extractable As in soils and As in plants ($r^2 = 0.20$ for total As; $r^2 = 0.24$ for $(NH_4)_2SO_4$-extractable As; $n = 88$) were found. Some species as well as the mixed fodder displayed large

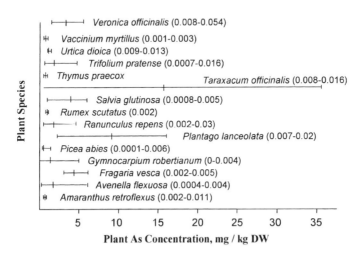

FIGURE 7.5. Arsenic concentration (minimum, mean, maximum) in different plant species ($n = 128$) from 10 contaminated sites in the Austrian Alps. The range values in parentheses show the soil-plant transfer coefficients. (Reprinted from Brandstetter et al., 2000, p. 729, by courtesy of Marcel Dekker, Inc.)

variations of As concentrations among specimens collected at the same time from the same site. Soil–plant transfer coefficients typically were <0.01, the maximum being 0.05.

Crops have variable tolerance to soil As (Table 7.6). Members of the bean family, rice, and most of the legumes are fairly sensitive to As. In pot experiments, the yield-limiting As concentrations in plant tissues were 4.4 ppm in cotton and 1 ppm in soybeans (Deuel and Swoboda, 1972a). A series of crop tolerance tests to As reveal the following: (1) snap beans < sweet corn < peas < potatoes (Jacobs et al., 1970a); (2) green beans < lima beans < spinach < radish < tomato < cabbage (Woolson, 1973); and (3) snap beans > rice > soybeans > potatoes > cotton (Baker et al., 1976). The critical level for barley grown in sand culture was 20 ppm As in the leaves and shoots (11 to 26 ppm range) (Davis et al., 1978). In rice, the critical level in tops ranged from 20 to 100 ppm As, and 1000 ppm in roots (Chino, 1981). Normal leaves from fruit trees contained 0.9 to 1.7 ppm

TABLE 7.6. Comparative sensitivity of plants to arsenic.

Tolerant	Moderately tolerant	Low tolerant
	Fruit Crops	
Apple	Cherries	Peach
Pears	Strawberries	Apricot
Grapes		
Raspberries		
Dewberries		
	Vegetables and Field Crops	
Rye	Beets	Peas
Mint	Corn	Onion
Asparagus	Squash	Cucumber
Cabbage	Turnips	Snap beans
Carrots	Radish	Lima beans
Parsnips		Soybeans
Tomato		Rice
Potato		Spinach
Swiss chard		
Wheat		
Oats		
Cotton		
Peanuts		
Tobacco		
	Forage Crops	
Sudangrass	Crested wheat grass	Alfalfa
Bluegrass	Timothy	Bromegrass
Italian ryegrass		Clover
Kentucky bluegrass		Vetch
Meadow fescue		Other legumes
Red top		

Source: Benson and Reisenauer (1951); Liebig (1965).

As, but leaves from trees suffering from As excess contained 2.1 to 8.2 ppm (NAS, 1977). Paddy rice is known to be very susceptible to As toxicity as compared with upland rice because As(III) is more prevalent under reducing conditions and because As toxicity to paddy rice can be exacerbated by the toxic effect of Fe(II) to the rice plant (Tsutsumi, 1980).

The primary mode of action of As in plants is inhibition of light activation, plausibly through interference with the pentose phosphate pathway (Marques and Anderson, 1986). Arsenites penetrate the plant cuticle to a greater degree than the arsenates. One of the first signals of plant injury from arsenite phytotoxicity is wilting caused by the loss of turgor, whereas stress due to arsenate does not involve rapid loss of turgor. Organoarsenicals, such as cacodylic acid, enter plants by absorption of sprays with only minor contribution from root uptake. Phytotoxicity symptoms caused by organoarsenical herbicides include chlorosis, gradual necrosis, dehydration, cessation of growth, then eventual death (NAS, 1977). In addition to wilting of new-cycle leaves, severely affected plants exhibit retarded root and top growth (Liebig, 1965). This is often accompanied by root discoloration and necrosis of leaf tips and margins. In rice plants, tillering is severely depressed as in the case of P deficiency (Chino, 1981). These symptoms indicate a restriction in the movement of water into the plant, which may result in death.

In deciduous tree fruit orchards where soils were contaminated with as much as 80 kg As ha^{-1} yr^{-1} till the mid-1900s from the applications of lead arsenates, fruit trees often exhibit retarded early growth, a phenomenon called as the *replant problem* (Davenport and Peryea, 1991). Phytotoxicity to As is a causal factor to this replant problem. Using the GR_{50} (50% reduction in growth) technique, Woolson (1973) found that the GR_{50} values correspond with 76 ppm (8 ppm WW) for radish tuber and 10 ppm (1 ppm WW) for spinach, exceeding the tolerance limit of 2.6 ppm of As for vegetables treated with calcium arsenate. The soil used for the GR_{50} test for radish contained about 19 ppm available As.

In general, total As in soil does not accurately reflect phytotoxicity. Correlation is better between plant growth and bioavailable As than between plant growth and total As (Woolson et al., 1971). Others have also reported a direct relationship between phytotoxicity and soluble As in soils (Deuel and Swoboda, 1972a; Reed and Sturgis, 1936; Vandecaveye et al., 1936).

In practice, ordinary crop plants do not accumulate enough As to be toxic to humans. Instead, growth reductions and crop failures are the main consequences from As pollution, and only small increases in the total As content of crops are noted in contaminated as compared with noncontaminated soils. Edible portions of crops usually contain less As than other plant parts. There is very little danger of As residues accumulating to phytotoxic levels under normal application rates of arsenicals.

6 Ecological and Health Effects of Arsenic

Historically, the toxicological importance of As as a suicidal and homicidal agent is well known since the Middle Ages. The poisonous nature of this element has been exploited by society into the modern era as arsenicals that are widely used in agriculture and forestry to eradicate pests and obnoxious weeds, and as defoliants. Today, numerous terrestrial and aquatic systems as well as groundwater aquifers contaminated with As have been reported in many parts of the world. It is therefore pertinent to highlight the ecological and health effects of As.

6.1 Ecological Effects of Arsenic

In general, As phytotoxicity is influenced by the form and nature of As itself, soil properties, and environmental conditions. Its trivalent form is considered to be much more toxic than its pentavalent form, the former being more predominant in reducing conditions. Additionally, As(III) as compared with As(V) is more mobile in sediments and groundwater systems (Harrington et al., 1998). Thus, rice plants growing on paddies are more susceptible to As(III) toxicity than similar rice plants growing on dryland areas. Arsenic in coarse-textured soils (e.g., sandy soils) is more mobile and bioavailable than in soils having higher clay content; the clay being characterized with greater contents of oxide minerals that are effective sorbers for As.

Both As(III) and As(V) readily accumulate in living tissues due to their affinity for proteins, lipids, and other cellular components (Fergusson and Gavis, 1972).

6.1.1 Effects of Arsenic in Aquatic Invertebrates and Fish

In general, As concentrations in freshwater organisms are lower than in marine organisms (Lunde, 1977). However, bioaccumulation patterns are similar in both ecosystems in that total As decreases and biomethylation of As increases with ascent in trophic level (Maeda et al., 1990, 1992). Arsenic concentrations in invertebrates can be expected to be directly proportional to substrate As concentrations. For example, As uptake by mussels was positively correlated with As concentrations, with the highest levels recorded in small mussels (U.S. Dept Int, 1988). Tamaki and Frankenberger (1992) reported that As concentrations were generally higher in crustaceans and mollusks than in fish.

Several reports on the biogeochemical cycling of As on Lake Rotoroa (New Zealand) in the 1950s suggest that the levels of As(III) applied on the lake would have been toxic to many of the lake's benthic invertebrates (Tanner and Clayton, 1990). Persistence and biotoxicity of As in the sediments of this lake could have been one of the factors responsible for their low abundance and diversity in this ecosystem. For example, common New Zealand mollusk species, such as the mussel *Hydridella menziesi* (Gray) and the snail *Potamopyrgus antipodarum* (Gray), are absent from the lake.

In contrast, As does not appear to bioaccumulate in fish. After 24 yr of As herbicide application on Lake Rotoroa, the concentrations of As on rudd fish were <0.20 mg kg^{-1} (flesh, WW) or <1 mg kg^{-1} DW (Tanner and Clayton, 1990) despite their diet consisting primarily of aquatic plants with high As content (200 to 300 mg kg^{-1} for *Egeria densa* and *Lagarosiphon major*). Tissue As levels for perch (<60 mg kg^{-1}) and eel (0.40 mg kg^{-1}) were also below the permissible level for human consumption (i.e., 2 mg kg^{-1}). Similar results obtained in 1992 indicate that As is not accumulating in the common fish (rudd, perch, eel, and catfish) in the lake.

The ability of aquatic organisms to transform inorganic As into complex organoarsenicals appears to be retained along the food chain, with consumers such as fish retaining as much as 99% of the As in an organic form (Philips, 1994; Tamaki and Frankenberger, 1992); in contrast, inorganic As rarely exceeds 1 mg kg^{-1} in tissues. In general, As does not biomagnify in aquatic food chain. Bioconcentration factors experimentally determined for As in aquatic organisms are generally relatively low, except for algae (Eisler, 1994).

6.2 Effects of Arsenic on Human Health

Some studies have suggested that inorganic As at trace levels is an essential nutrient in goats, chicks, mini pigs, and rats (U.S. EPA, 1993a,b). However, no comparable data are available for humans. The consensus in the scientific community regarding the role of As in environmental health can be summarized as follows (Eisler, 1994; Hindmarsh and McCurdy, 1986):

1. As may be absorbed by ingestion, inhalation, or through permeation of the skin or mucous membrane.
2. Cells accumulate As by using an active transport system normally used in phosphate transport.
3. Arsenicals are readily absorbed after ingestion, most being rapidly excreted in the urine during the first few days, or at most a week.
4. The toxicity of arsenicals conforms to the following order, from highest to lowest toxicity: arsines > inorganic arsenites > organic trivalent compounds (arsenoxides) > inorganic arsenates > organic pentavalent compounds > arsonium compounds > elemental arsenic.
5. Solubility in water and body fluids appears to be directly related to toxicity (e.g., the low toxicity of As^0 is due to its virtual insolubility in water and body fluids, whereas the highly toxic arsenic trioxide is fairly soluble even at room temperature).
6. The mechanisms of As toxicity differ considerably among As species, through signs of poisoning appear to be similar for all arsenicals.

The U.S. EPA (1993a,b) has classified inorganic As as a human carcin-ogen, but has not classified arsine (AsH_3) for carcinogenicity. Inorganic As, especially As(III), is readily absorbed from the GI tract to the blood. The mechanisms through which As causes cancer are not well understood, but data suggest that As probably causes chromosomal abnormalities that lead to cancer. Chronic *arsenicosis,* when treated early enough, is reversible. But once cancer begins to develop after 8 to 10 yr, or the liver and kidneys begin to deteriorate, the victims are most likely beyond help. The U.S. EPA (1993a,b) has not established a reference concentration (RfC) for inorganic As; the RfC for arsine is 0.00005 mg m^{-3} based on increased hemolysis, abnormal red blood cell morphology, and increased spleen weight in rats, mice, and hamsters. (Note: The EPA estimates that inhalation of this concentration or less over a lifetime would not likely result in the occurrence of chronic noncancer effects.) The reference dose (RfD) for inorganic As is 0.0003 mg kg^{-1} per day based on hyperpigmentation, keratosis, and possible vascular complications in humans. (Note: The EPA estimates that con-sumption of this dose or less over a lifetime would not likely result in the occurrence of chronic noncancer effects.) No RfD for arsine has been established.

6.2.1 Exposure to Drinking Water

The two most common exposure routes of As to humans are ingestion and inhalation; drinking of As-contaminated groundwater has been suggested as the main risk for humans (Borum and Abernathy, 1994). However, for most people, food is the largest source of inorganic As exposure (about 25 to 50 µg day^{-1}), with lower amounts coming from drinking water and air (ATSDR, 1989). If the route is by ingestion, then symptoms caused by GI irritation will dominate the early reaction (Morton and Dunnette, 1994). Ingested As has a shorter half-life than inhaled As due to a more rapid biotransformation in the liver.

Humans may ingest As in water from wells drilled into As-rich geologic strata, or in water contaminated by industrial or agrochemical wastes. Although As in drinking water is generally not a concern to the general population, there are areas and regions where consumption of well water or water from public water supplies has resulted in signs and symptoms of As poisoning. Such incidents (Table 7.7) have been reported from Taiwan (Tseng et al., 1968), the United States (Feinglass, 1973), Thailand (Chop-rapawon and Rodcline, 1997), Canada (Grantham and Jones, 1977), Argentina (Astolfi, 1971), Chile (Zaldivar, 1974), China (Wang and Huang, 1994), Mongolia (Luo et al., 1997), and Mexico (Cebrian et al., 1994). More recently, allegedly the most extensive metal poisoning with catastrophic proportion in the history of environmental toxicology, exposing millions of inhabitants, has been reported for West Bengal (India) and Bangladesh (Chowdhury et al., 1997; Das et al., 1996). In Bangladesh alone, estimates

TABLE 7.7. Major groundwater contamination by arsenic in the world.

Location	Year	Number of people exposed	People showing arsenical skin lesions (%)
West Bengal, India and Bangladesh	1978–1998	20,000,000+	20+
China[a]	1980s to present	250,000	
Taiwan	1961–1985	103,154	19
Antofagasta, Chile	1958–1970	130,000	16
Monte quemado Argentina	1938–1981	10,000	Many
Lagunera region Mexico	1963–1983	200,000	21
Ronpiboon, Thailand	1987–1988	[b]	18

[a]Shanxi province and inner Mongolia.
[b]Total population of Ronpiboon subdistrict of Thailand was 14,085 and 5.9% of the total population was suffering from chronic arsenic poisoning and in the three main arsenic affected villages, 18% of the total population showing arsenical lesions.
Source: Modified from Chowdhury et al. (1997).

of over 20 million of the country's 120 million population are potentially at risk from drinking As-contaminated tubewell water (World Bank Group, 1999).

In China, the first case of chronic As poisoning from drinking water was recognized in the early 1980s in Xinjiang Province, where As levels in deep well water were over 0.6 mg L^{-1} (Niu et al., 1997). Similar As poisonings were found in other places in China (Shanxi Province and Inner Mongolia) in the early 1900s where waters from mostly deep wells were contaminated with As in the range 0.50 to 1.2 mg L^{-1}. The chronic poisoning symptoms among the patients include peripheral neuritis, gastritis, enteritis, liver and kidney diseases, and abnormal cardiograms. The most typical symptoms are hyperpigmentation of skin and symmetric hyperkeratosis on palms and soles; their prevalence is as high as 70 to 80% when As levels in water approach 1.2 mg L^{-1}. Approximately 250,000 people are exposed to these As levels.

Groundwater aquifers in the Lagunera region of northern Mexico are heavily contaminated with As (del Razo et al., 1990). Total As in water samples ($n = 128$) ranged from 0.008 to 0.624 mg L^{-1}, and concentrations exceeding 0.050 mg L^{-1} were found in half of the samples. Approximately 400,000 people living in rural areas are exposed to high As levels. Most of the As is in inorganic form, predominantly As(V), followed by As(III), with only trace amounts of organoarsenicals.

6.2.2 The West Bengal (India)–Bangladesh Case Study

In six districts of West Bengal, groundwater contains As above the MCL recommended by WHO (Chowdhury et al., 1997). The current WHO provisional As level in drinking water is 0.010 mg L^{-1} (WHO, 1993); that for the U.S. EPA is at 0.050 mg As L^{-1} (new proposal at 0.005 mg L^{-1}); the German standard is at 0.010 mg L^{-1}, which is also the target for the EU

(Jekel, 1994). These waters are used by the inhabitants for drinking, cooking, household, and other purposes. The total population in affected districts is $\sim$30 million, scattered in about 34,000 km^2. West Bengal is about 89,000 km^2 area with a total population of over 68 million people. At least 1 million people have been drinking the high As-contaminated well waters and about 200,000 people already have As skin lesions. Virtually, the general population in the region is at risk, as more districts in the region with As-contaminated well waters are being discovered.

In Bangladesh, a survey report for 26 districts indicates that groundwater in 21 districts contains >0.050 mg As L^{-1}, and in 18 districts (covering 40,000 km^2, 35 million population) many people suffering from arsenical skin lesions, including skin cancer, have already been identified. Arsenic poisoning affects primarily the rural villagers in Bangladesh who use the untreated water and have protein-poor diets. Moreover, their water uptake is larger than the rest of the population, because they are subsistent farmers who toil under tropical climate. Chakraborti (1999) collected about 2550 water samples from Bangladesh and 35,000 from West Bengal, and some hair and nail samples from the As victims of the affected districts. In hair collected from the affected West Bengal population, 49% of all samples have As content above the toxic level of 0.08 to 0.20 mg kg^{-1}; in Bangladesh, 92% were above the toxic level. The nail As concentrations of 74% of the tested persons in West Bengal and 98% in Bangladesh were higher than the normal range of 0.40 to 1.1 mg g^{-1}. More recently, the exposure scenario in Bangladesh has been upgraded to as many as 70 million people who are potentially at risk from drinking As-contaminated water drawn from millions of wells (Lepkowski, 1998).

In India as well as in Bangladesh, the source of As is geological in nature. The As-affected area is in the region where sediment deposition took place during the Quaternary Period, i.e., 25,000 to 80,000 years ago, commonly known as the Younger Deltaic deposition. This sediment contains As-rich pyrite and covers almost the entire alluvial region of the river Ganges (Das et al., 1996). In this region, millions of cubic meters of groundwater are extracted for irrigation by shallow and deep tubewells. Due to such high groundwater withdrawal rates, air enters the aquifer, and due to available oxygen, As-rich pyrite is oxidized and mobilized from the vadoze zone (i.e., "pyrite oxidation" hypothesis) (Chakraborti, 1999). However, an alternative mechanism of As release has been offered by Nickson et al. (1998, 2000), whereby the OM comingled with the As-rich Fe oxyhydroxides deposited millennia ago within the Ganges Basin has been reduced to the soluble state by the anoxic nature (primarily due to the OM) of the geologic strata (i.e., "oxyhydroxide reduction" hypothesis) and leaches upon exposure to aquifer water, thereby enriching the water.

The pumped water is used by local residents for irrigation of crops, drinking, cooking, and other uses as shown on Figure 7.6, which indicates that the population is directly exposed via consumption of groundwater, food crops, livestock tissues, and invertebrates and fish from affected

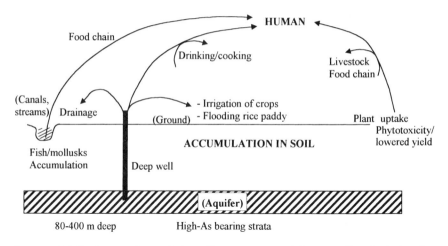

FIGURE 7.6. Pathways of arsenic in well waters pumped from aquifers depicting the arsenic poisoning scenario in West Bengal and Bangladesh. (Reprinted from Brandstetter et al., 2000, p. 727, by courtesy of Marcel Dekker, Inc.)

waterways. The frequency of As content in groundwater samples from 16 heavily contaminated districts in Bangladesh is displayed on Figure 7.7.

The concentrations of As in tubewell water in the impacted areas (6 districts) averaged 0.32 mg As L^{-1} (range of 0.050 to 3.7 mg L^{-1}, $n = 427$) (Das et al., 1996). Only As(V) and As(III) were detected in the groundwater, having an average ratio of 1:1. No organoarsenical species were detected. However, urine assays of patients have revealed the presence organoarsenical species at trace levels in the urine in the order DMAA > MMAA > As(III) > As(V). As concentrations in liver tissue of patients are in the ppm range (i.e., >1 mg kg^{-1}) compared with the common liver tissue content of 5 to 15 µg kg^{-1}. The concentrations of As in hair, nails, and skin scales of these patients are also very high. In the impacted areas of West Bengal, affected residents manifest most of the three stages of clinical symptoms of As poisoning (Das et al., 1996):

Initial stage: dermatitis, keratosis, conjunctivitis, bronchitis, and gastroenteritis
Second stage: peripheral neuropathy, hepatopathy, melanosis, depigmentation and hyperkeratosis
Last stage: gangrene in the limbs and malignant neoplasm

Malnutrition, low socioeconomic status, illiteracy, bad food habits, and long-term intake of As-contaminated water have exacerbated As toxicity in the region. In addition, major water demands for irrigation and household uses have caused high withdrawal rates of groundwater from the aquifers thereby inducing rapid oxidation, dissolution, and subsequent leaching of As from the enriched strata. In most cases of chronic

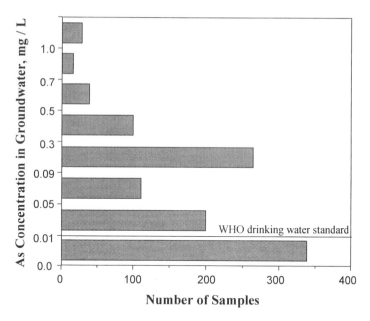

FIGURE 7.7. Frequency of arsenic in groundwater samples ($n = 1088$) from 16 districts in Bangladesh where arsenic poisoning is prevalent. (Reprinted from Brandstetter et al., 2000, p. 722, by courtesy of Marcel Dekker, Inc.)

poisoning (except in children), melanosis and hyperkeratosis are common symptoms. Postmortem analyses show high concentrations of As in urine of deceased victims.

In view of new information on As exposure and cancer, there is a concern in the scientific community that the U.S. EPA's current MCL for As in drinking water (i.e., 50 ppb) does not sufficiently protect public health. While the WHO and the EU have agreed on the 10-ppb standard, that in the United States is still uncertain, although a new proposal of 5 ppb As is under consideration (Aberrathy, 2000).

6.2.3 Dietary Exposure

Contamination of vegetable crops and fruits occurs through spraying application of arsenicals or from root uptake of As from soils. For many years, the main source of As in food was the spray residue in vegetables and fruits (Bishop and Chisholm, 1966). For example, it was reported that nearly 45% of the 335 samples of German wine in 1938 contained over 3 mg As L^{-1}, with 14 mg L^{-1} the highest concentration observed (Nriagu and Azcue, 1990). The accumulation of elevated levels of As in grapes (for consumption as table grapes, wine, and grape juice) grown in contaminated vineyards remained a problem. In 1976, over 50% of samples of table wines sold in the United States contained over 0.05 mg As L^{-1}. Thus, it would seem prudent to postulate that wine consumption is a potentially major route of As exposure to the general population. However, the banning of arsenical pesticides in many countries has rendered this route of food contamination, except possibly grapes, no longer relevant.

Consumption of seafoods from various sources may supplement dietary intake of As by humans. Edible seaweed, a common food item in Japan and Pacific Rim countries, has been reported to contain 112 μg As g^{-1} DW on the average (range of 19 to 172 μg g^{-1} DW) (WHO, 1981). Similar As levels for marine algae collected from the coast of Norway have been reported. In general, organoarsenical compounds account for 60 to 99% of the As in marine algae. Lobsters, shrimps, canned clams, and smoked oysters sold in Canada had As in the 5 to 45 μg g^{-1} level; shellfish As levels typically fall within the 5 to 20 μg g^{-1} DW range (Nriagu and Azcue, 1990).

The distribution of As in fish varies with fish species and location. Arsenic levels in pelagic fish (in the order of 0.30 to 3 ppm) are substantially lower than those of bottom feeders and shellfish (in the order of 1 to 55 ppm) (Cappon, 1990). Concentrations of As in Canadian freshwater fish of only 0.010 to 0.24 μg g^{-1} WW are much lower than those in marine fish (Nriagu, 1994). Even fish from a relatively more contaminated lake in New Zealand contained low As concentrations. The higher concentrations of As in marine fish and shellfish is due to lower As concentrations in freshwater and to the difference in the P/As ratios in the two aquatic systems. In general, higher levels of total As are found in commercially available marine seafood products compared with terrestrial animals. A typical U.S. adult diet supplies about 48 μg As day^{-1}, with most of the sources being dairy products, meat, fish, and poultry (Nriagu and Azcue, 1990). Compilation of national data suggests large differences in calculated As intakes among countries, with the lowest values of 12 μg As day^{-1} for the Netherlands. The daily intake for adults in Canada was calculated as 19 μg As day^{-1}, with about one-third of the total coming from meat, fish, and poultry (Table 7.8).

In summary, the toxicity of As to humans largely depends on the chemical form and physical state of the compound involved. Inorganic As(III) is generally regarded as being more acutely toxic than inorganic As(V), which in turn is more toxic than the methylated forms. Absorbed inorganic As is widely distributed among the organs, converted in the liver to less-toxic methylated forms, and efficiently excreted in the urine. Several organoarsenicals, mainly arsenobetaine and arsenocholine, are found to accumulate in fish and shellfish, but are generally regarded as nontoxic. Elemental As is nontoxic, even if ingested in relatively substantial amounts. Although organoarsenicals are usually regarded as less toxic than inorganic arsenicals, several methyl and phenyl derivatives widely used in agriculture and forestry are potentially toxic to humans. Chief among these are MMAA, MSMA, DSMA, and CA (Garcia-Vargas and Cebrian, 1996).

TABLE 7.8. Description of food categories and average arsenic concentrations in food items and daily intake in Canada.

Category	Food description	Mean food intake (g)	Average As concentration in food item (ng g^{-1})	Average As intake (µg day^{-1})	Percentage of total intake
I	Cereals, porridge, bread, toasted buns, doughnuts, including bread from sandwiches	256	8.6	2.2	18.1
II	Water consumed directly, including duplicates from drinking fountains, etc.	319	0.46	0.15	1.4
III	Coffee, tea, beer, liquor, soft drinks (as consumed with or without milk, sugar, mixes, etc.)	780	1.0	0.78	8.6
IV	Fruit juices, fruits (canned or fresh)	297	3.8	1.13	6.4
V	Dairy products and eggs consumed directly; milk, eggs, omelets, yogurt, ice cream, milk shakes, cottage cheese, etc.	293	2.58	0.76	6.6
VI	Starch vegetable: rice, potatoes, yams, etc.	164	13.69	2.25	14.9
VII	Other vegetables: mushrooms, salads, vegetable-base soups, etc., including tomatoes and tomato juice	203	2.6	0.53	4.0
VIII	Meat, fish, poultry, meat-base soups, etc.	170	60.1	10.2	32.1
IX	Miscellaneous: pies, puddings other than rice, tarts, nuts, potato chips, chocolate bars, etc.	70	12.19	0.85	6.5
X	Cheese (other than cottage cheese)	17	8.46	0.14	1.4

Source: Nriagu (1994).

Human data indicate that As doses of 1 to 3 mg kg^{-1} per day are usually fatal. Chronic oral exposure to inorganic As at doses of 0.05 to 0.1 mg kg^{-1} per day is frequently associated with neurological or hematological signs of toxicity (Garcia-Vargas and Cebrian, 1996). Recent studies in Taiwan have shown significant dose–response relationships between long-term exposure to inorganic As in drinking water and risk of malignant neoplasms of the liver, nasal cavity, lung, skin, bladder, kidney, and prostate (Chen and Wang, 1990; Chiou et al., 1995).

7 Factors Affecting Mobility and Bioavailability of Arsenic

7.1 Chemical Species of Arsenic

Arsenate(V) and arsenite(III) are the primary species of As in soils and natural waters. Of these forms, As(III) is the most soluble and mobile as well as the most toxic species.

Chemical species also influences the sorption affinity for As. For example, the adsorption of As on alumina at pH below 6 decreased in the order: As(V) > MMAA = DMAA > As(III), whereas at pH above 6, the order was As(V) > As(III) > MMAA = DMAA (Xu et al., 1991). In general, the affinity of As for soils and sediments is greater for As(V) than for As(III). Arsenic is present as $H_2AsO_4^-$ in well-drained acidic soils, or as $HAsO_4^{2-}$ in well-drained alkaline soils.

Phytotoxicity of As in soils depends on its chemical form. The most phytotoxic is As(III), followed by As(V), MSMA, and cacodylic acid (CA); however, when sprayed onto leaves, CA is the most phytotoxic (Sachs and Michael, 1971).

In most oxidized riverine systems (200 to 500 mV), both $H_2AsO_4^-$ and $HAsO_4^{2-}$ are expected to occur at pH 5 to 8, while H_3AsO_3 is the dominant species under reducing (0 to 100 mV) conditions (Cherry et al., 1979).

In aquatic systems in general, the formation of As(V) species, the most common species in water, is favored under conditions of high dissolved oxygen, alkaline pH, high redox, and reduced OM content, and As solubility is controlled by the formation of $Mn_3(AsO_4)_2$, $FeAsO_4$, and $Ca(AsO_4)$. In contrast, opposite conditions usually favor the predominance of As(III) species and arsenic sulfides.

7.2 pH

The effect of pH on As adsorption varies considerably among soils. The effect of pH on As(III) sorption has been shown to be dependent on the nature of the mineral surface. In soils with low oxide content, increasing the pH had little effect on the amount of As(V) adsorbed, while in highly oxidic soils, adsorption of As(V) decreased with increasing pH (Smith et al., 1999). This decrease was attributed to two interacting factors—the increasing negative surface potential on the plane of adsorption and increasing amount of negatively charged As(V) species present in the soil solution. In general, adsorption of As(V) decreases with increasing pH. In contrast to As(V), adsorption of As(III) increases with increasing pH. As discussed earlier, sorption maxima occur at different pH values depending on the dominant soil component. In adsorption studies of $H_2AsO_4^-$ from landfill leachates by various clay minerals, the adsorption maxima for As were at approximately pH 5 (Frost and Griffin, 1977).

The effect of liming the soil on As mobility has been inconsistent. Because agricultural lime has Ca^{2+} ions as a principal constituent, it may affect the form or solubility of As. Because the solubility product constant of $Ca_3(AsO_4)_2$ is greater than that of Fe_x and $Al_x(AsO_4)_x$, the role of Ca in the sorption process of As is not as pronounced as the role of Fe and Al. As previously mentioned, Al–As and Fe–As are the dominant forms of As in acid soils. Therefore, applying lime to acid soils to raise pH is very likely to shift the As form to the Ca–As species. Thus, it is not surprising to find lime as an ineffective soil amendment to alleviate As phytotoxicity on soils containing toxic levels of As (MacPhee et al., 1960). However, a positive response in barley with the addition of lime to acid soils had been reported (Vandecaveye et al., 1936). Similarly, improved cotton growth in contaminated soil with lime addition was found (Paden and Albert, 1930). Apparently in these latter cases, the added lime reduced the level of water-soluble As, effectively reducing As phytotoxicity.

7.3 Oxides of Manganese and Iron

Sorption of As(V) and As(III) varies among soils and appears to be related to the oxide content of the soil (Jain et al., 1999; Roussel et al., 2000; Smith et al., 1999). In addition, amorphous Al hydroxide and phyllosilicate mineral surfaces have been shown to play an important role in the adsorption of As(III) by soils (Manning and Goldberg, 1997). Like phosphate, As is strongly adsorbed by amorphous Fe oxide but shows less affinity for Al oxides than does the phosphate. Highly oxidic Austrian soils adsorbed 3 times more As(V) than soils containing lesser amounts of oxide minerals (Smith et al., 1999). Similarly, dissolved As in submerged soils was inversely correlated with less crystalline and crystalline Fe (McGeehan et al., 1998). Spectroscopic analyses [electron microprobe analysis (EDAX) and backscatter electron detector] of particle-size fractions of As-contaminated soils from the Austrian Alps reveal that Fe oxides represent the major sink for As in those soils (Fig. 7.8). These spectroscopic results corroborate the chemical extraction findings. More specifically sequential extraction analysis of drainage effluents from a mine tailing indicates that most of the As is bound to lepidocrite (γ-FeOOH) (Roussel et al., 2000). Similarly, the bulk of the contaminant As in soils at sites impacted by spills from cattle dip has been associated with soil amorphous Fe and Al minerals (McLaren et al., 1998). Furthermore, ammonium oxalate–extractable Fe and Mn

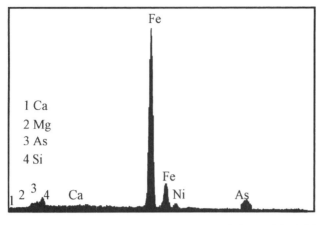

FIGURE 7.8. Backscatter image of an As-enriched soil from the B horizon of an Austrian eutric cambisol derived from schist and an electron diffraction spectra confirming the association of As with Fe–oxides. (Reprinted from Brandstetter et al., 2000, p. 720, by courtesy of Marcel Dekker, Inc.)

concentrations were found to be positively correlated with K_d values for As ($r^2 = 0.92$) (Jones et al., 1999).

The redox-sensitive Mn and Fe oxides constitute a significant sink of As in aquatic systems (Mok and Wai, 1994). These hydrous oxides in sediments occur as partial coatings on the silicate minerals rather than as discrete, crystalline minerals. Under reducing conditions, the As sorbed onto hydrous Fe and/or Mn oxides are solubilized (Cummings et al., 1999). In highly reducing conditions, As will coprecipitate with iron sulfides as arsenopyrite (FeAsS) or form As sulfides (e.g., AsS, As_2S_3) (Ferguson and Gavis, 1972). However, under aerobic conditions, these sulfides are easily oxidized, releasing As into the environment.

In the pH range of 2 to 11 the release of As(V) and As(III) from sediments to the overlying water generally increases with decreasing pH (Mok and Wai, 1994). The release of As also rises sharply at high pH. At the lower pH values, the hydrous oxides of Fe and Mn are solubilized from the sediments releasing As species. At high pH levels, the increased hydroxide concentrations caused displacement of these As species from their binding sites in a ligand exchange reaction.

Thus, the role of hydrous Fe, Mn, and Al oxides and sulfides as an important sink influencing the mode of transport for As in the environment is of great relevance, since changes in redox potential and pH can induce the formation and solubilization of these solid phases, hence directly controlling the mobilization of As in aquatic systems. In general, As concentrations are much higher in sediments (in the ppm range) than in the overlying water (in ppb range) in contaminated aquatic ecosystems.

7.4 Soil Texture and Clay Minerals

In general, As mobility and bioavailability are greater in sandy than in clayey soils. Woolson (1973) reported that As phytotoxicity to horticultural crops was highest on a loamy sand soil and lowest on silty clay loam. Similarly, As was more phytotoxic to corn when grown on a sandy than on a silt loam soil (Jacobs and Keeney, 1970). With rice, plants grown on sandy soils were more susceptible to As phytotoxicity than those grown on clayey soils (Reed and Sturgis, 1936). Others have also noted this inverse relationship between clay content and trend in bioaccumulation (Crafts and Rosenfels, 1939). The main reason for this phenomenon is that both hydrous Fe and Al oxides vary directly with the clay content of the soil. Consequently, the water-soluble fraction of As is higher in soils with low clay content than in soils with higher clay content. In addition, about 90% of the soil As, applied repeatedly over a 4-year period as either MSMA, MAMA (monoammonium methanearsonate), or DSMA was retained in the clay fraction (Johnson and Hiltbold, 1969). Likewise, leaching with water removed 52% of surface-applied DSMA from columns of a loamy sand soil, while none was leached from a clay loam soil (Dickens and Hiltbold, 1967).

Frost and Griffin (1977) found that montmorillonite adsorbed greater amounts of As(V) and As(III) than kaolinite. The higher adsorption by montmorillonite was attributed to its higher edge surface area (2.5 times greater) than kaolinite. Another possibility for the higher retention by montmorillonite is its content of interlayer hydroxy aluminum polymers.

7.5 Redox Potential

Anaerobic conditions activate a suite of chemical and biological processes that can change the physicochemical properties of soils and sediments for As sorption. Reducing conditions result in the dissolution of Fe(III) oxides and conversion of their semicrystalline Fe minerals to amorphous Fe(II) hydroxides, Fe_3O_4 (magnetite), Fe(II)-Fe(III) hydroxides (green rust), $FeCO_3$ (siderite), and Fe sulfides. Indeed, prolonged soil flooding resulted in a decrease in soil redox potential and an increase in dissolved Al, Fe, Mn, and Si (McGeehan et al.,1998). Less-crystalline Fe(III) phases increased as the length of flooding period increased. In contrast, the crystalline Fe(III) oxide decreased as the flooding time increased.

As(III) species that predominate in reducing conditions have been reported to be 4 to 10 times more soluble in soils than As(V). Adding sucrose to soils equilibrated in an oxygen-free environment resulted in greater solubilization of As (Deuel and Swoboda, 1972b). The higher As solubility was attributed to the reduction of iron from Fe(III) to Fe(II) with subsequent dissolution of ferric arsenate. Thus, in flooded soils such as rice paddies, reducing conditions can result in the dissolution of Fe-As solid phase, with the subsequent reduction of dissolved arsenate to arsenite. The concentrations of As(III) in the soil solution can become very high, resulting in phytotoxicity. In addition, Ferguson and Gavis (1972) predicted that at high Eh values, the pentavalent species (H_3AsO_4, $H_2AsO_4^-$, $HAsO_4^{2-}$, and AsO_4^{3-}) are more stable. Under mildly reducing conditions, the trivalent species (AsO_3^{3-}, $H_2AsO_3^-$, and $HAsO_3^{2-}$) become more stable. At very low Eh, arsine (AsH_3) may be formed and volatilized.

7.6 Competing Ions

Arsenate ($H_2AsO_4^-$), arsenite ($H_2AsO_3^-$), and phosphate ($H_2PO_4^-$) are oxyanions that similarly behave in aerated soils (Reynolds et al., 1999). Phosphate and, particularly, As(V) appear to compete for similar sorption sites, though some sites were preferentially available for the sorption of either P or As(V). Because of the competitive nature of phosphate with arsenate, application of P fertilizer on the surface of a lead-arsenate contaminated (orchard) soil column significantly increased the amount of As leached or solubilized from the soil (Creger and Peryea, 1994; Davenport and Peryea, 1991). Heavy additions of P to As-contaminated soils have displaced substantial amounts ($\sim 77\%$) of total soil As, with the more-soluble fractions being redistributed to lower depths in the soil profile (Woolson et al., 1973). About 60% of the sorbed As(V) and 70% of the sorbed As(III) were displaced by $H_2PO_4^-$ in a solution of 10^{-6} M phosphate (Thanabalasingam and Pickering, 1986). Other competing anions were less effective in displacing As from the exchange complex, the pattern for As(V) being $H_2PO_4^- > H_2AsO_4^- > SO_4^{2-} > CO_3^{2-}$, and for As(III) being $H_2PO_4^- > H_3AsO_3 > F^- > SO_4^{2-} > CO_3^{2-}$.

The relative affinity of As(V), P, and Mo for goethite and gibbsite surfaces is pH dependent and tended to be P $\sim$ As(V) $>$ Mo at neutral pH (Manning and Goldberg, 1996). However, in allophanic (volcanic) soils, alteration of

As contamination in soil solution by P addition may have an insignificant influence on the behavior of As in soil. Thus, soil type is very important in the behavior of As in soil, with allophanic soils being not so sensitive to competing ions.

The effect of P on plant nutrition of As has been unreliable. Although most reports indicate the favorable influence of P on As bioaccumulation in plants, some were not as positive. Because of its physicochemical similarity to As, P competes for As sorption sites on soil and thus may affect As bioavailability. Nutrient culture studies have demonstrated that the amount of P relative to As (P/As ratio) may influence plant growth. At P/As ratios of 4:1 or greater, phytotoxicity on wheat was markedly reduced (Hurd-Karrer, 1939). However, at a lower ratio of 1:1, stunting occurred at concentrations of 10 ppm As and higher.

Enhanced reductions in As phytotoxicity on a wide variety of crops occurred as P levels in soil increased. Kardos et al. (1941) noted that at P/As ratios above 1.3:1, As toxicity was minimized, whereas below 1:1 toxicity increased. Tsutsumi (1980) noticed that As was harmless to rice plants when the P/As ratio was 5:0.4 to 1.3 but was almost lethal when the ratio was in the range 1:0.8 to 2.4. Reductions in As phytotoxicity at ratios of P/As of 0.7:1 to 42.5:1 occurred, but in another instance, no influence at a ratio of 6.8:1 was obtained (Woolson et al., 1973). Addition of P to soil had only a minimal effect on As toxicity to turf grass under conditions simulating normal turf management practices (Carrow et al., 1975). Nevertheless, a common practice when using tricalcium arsenate for control of the weed *Poa annua* L. in managed turfs is to minimize or eliminate P fertilization in order to increase the effectiveness of As.

7.7 Plant Factors

Numerous reports have indicated the differential tolerance to As among plant species. Arsenic concentrations among plant parts also vary and occur at only trace levels (<0.020 ppm) in corn kernels and shelled peas (Jacobs et al., 1970a). In potatoes, most of the As in tubers was found in the peelings and was only slightly above the trace levels (<0.10 ppm) in potato flesh from highly polluted plots.

8 Sources of Arsenic in the Environment

Arsenic compounds are released into the environment by mining and smelting operations, agriculture, forestry, and manufacturing. The largest anthropogenic sources of As in 1988 onto land originated from the manufacture of As-based commercial goods (~40%), followed by enrichment of coal ash (~25%) (Nriagu and Pacyna, 1988).

8.1 Pesticide Manufacture and Use

Inorganic arsenicals in the form of Pb, Ca, Mg, and Zn arsenate, Zn arsenite, or Paris green were widely used as pesticides in orchards from the late 1800s to mid-1900s. There have been concerns about the contamination of the food chain from residual As from old orchard soils that have received large amounts of As from past uses. Thus, it is rather common to observe *replant syndrome* in young fruit trees as a result of phytotoxic levels of As in these soils. Many metal-based arsenical pesticides have already been banned by the EPA in the United States, and the use of other arsenicals, including organoarsenicals, is expected to decline.

Dumping on the ground of industrial effluents from the production of Paris green has contaminated groundwater in a residential area of Calcutta, India (Chatterjee et al., 1993). The groundwater contained very high levels of Cu and As [both As(V) and As(III)], with some samples showing total As in the 30 to 50 ppm range. The tubewell water is used by residents for drinking and other household purposes. Sodium arsenite was used in the United States for the control of aquatic weeds from the 1970s (Tanner and Clayton, 1990). This herbicidal use has contaminated small fish ponds and lakes in several areas of the world.

8.2 Mining and Smelting

Arsenic is a natural constituent in lead, zinc, gold, and copper ores and can be released during the smelting process. The flue gasses and particulates from smelters can contaminate nearby ecosystems downwind from the operation with As.

In Japan, the main thrust of the environmental As problem has been its toxicity to rice plants (Takamatsu et al., 1982). Irrigation of paddy fields with water contaminated by mining wastes or wastewater from geothermal electric power stations has frequently produced yield depression of rice. Toxicity can also occur to rice being grown in paddy fields converted from old apple orchards, where inorganic arsenicals were used as pesticides.

Arsenic oxide, a common waste product from the refining of metal ores, has led to disposal problems with an estimated 100,000 tonnes per year deposited as slag in landfills (Tamaki and Frankenberger, 1992). Contamination of groundwater and surface water can arise from such disposal of As wastes. For example, disposal of mill tailings at a site in Ontario, Canada has resulted in substantial leaching of As, contaminating the groundwater and Lake Moira (Bernard, 1983).

Li and Thornton (1993) reported that As contamination of soil from three ore smelting areas in England has resulted in the surface soil (0 to 15 cm) having much greater As concentrations (16 to 975 mg kg^{-1}) than the background samples (7.7 to 9.0 mg kg^{-1}). Despite the cessation of most mining and smelting in these areas (Derbyshire, Cornwall, and Somerset),

As contamination in some areas is still very high. Asami (1988) has enumerated the impacts of mining and smelting on ecosystems worldwide with regard to metal contamination.

8.3 Coal Combustion and Its By-Products

Fly ash and other coal combustion residues contain variable amounts of As in the ppm range, with some sample having As concentrations exceeding 100 ppm up to over 1000 ppm (Adriano et al., 1980; Adriano and Weber, 1998; Carlson and Adriano, 1993). Indeed, bioaccumulation of As in plants and eventual leaching of As from ash are great concerns when this by-product is applied on land at large amounts (Adriano and Weber, 1998; Qafoku et al., 1999).

An unusual source of As that can directly expose a significant segment of the population in certain places is the domestic combustion of coal containing high levels of As (90 to 1200 mg As kg^{-1}) (Niu et al., 1997). This is a common practice in southern China, where coal is burned inside the home in open pits for daily cooking and drying corn. This traditional practice results in high As concentrations in indoor air as well as As accumulation on the corn, which is hung under the ceiling to dry. The air As levels can be as high as 0.04 to 0.13 mg m^{-3}, which is 10 to 40 times higher than the standard. The As in exposed corn can range from 1.5 to 11 mg kg^{-1}, which is 2 to 15 times higher than the proposed limit of 0.7 mg kg^{-1}. Approximately 100,000 people are at great risk from this exposure in this part of China.

References

Abernathy, C.O. 2000. U.S. Environmental Protection Agency, Washington, DC, pertoral communication.

Adriano, D.C., and J. Weber. 1998. *Coal Ash Utilization for Soil Amendment to Enhance Water Relations and Turf Growth.* EPRI Rep TR-111318. Electric Power Res Inst, Palo Alto, CA.

Adriano, D.C., A.L. Page, A.A. Elseewi, A.C. Chang, and I. Straughan. 1980. *J Environ Qual* 9:333–344.

Ainsworth, C.C., and D. Rai. 1987. *Chemical Characterization of Fossil Fuel Combustion Wastes.* EPRI Rep EA-5321. Electric Power Res Inst, Palo Alto, CA.

Akins, M.B., and R.J. Lewis. 1976. *Soil Sci Soc Am J* 40:655–658.

Asami, T. 1988. In W. Salomons and U. Forstner, eds. *Chemistry and Biology of Solid Wastes.* Springer-Verlag, Berlin.

Astolfi, E., A. Maccagno et al. 1981. *Biol Trace Elem Res* 3:133–143.

[ATSDR] Agency for Toxic Substances and Disease Registry. 1989. *Toxicological Profiles for Arsenic.* U.S. Dept of Health and Human Services, Atlanta, GA.

Baker, R.S., W.L. Barrentine, D.H. Bowman, W.L. Hawthorne, and J.V. Pettiet. 1976. *Weed Sci* 24:322–326.

Benson, N.R., and H.M. Reisenauer. 1951. *Washington Agric Exp Sta Circ* 175:1–3.

Bernard, D.W. 1983. In *Proc Intl Conf Groundwaters and Man*. Aust Gov Public Serv, Canberra, Australia.

Berrow, M.L., and G.A. Reaves. 1984. In *Proc Intl Conf Environmental Contamination*. CEP Consultants, Edinburgh.

Bhumbla, D.K., and R.F. Keefer. 1994. In J.O. Nriagu, ed. *Arsenic in the Environment, part 1: Cycling and Characterization*. Wiley, New York.

Bishop, R.F., and D. Chisholm. 1966. *Can J Plant Sci* 46:225–231.

Borum, D.R., and C.O. Abernathy. 1994. In W.R. Chapell, C.O. Abernathy, and C.R. Cothern, eds. *Arsenic Exposure and Health*. Sci Technol Letters, Northwood, United Kingdom.

Braman, R.S., and C.C. Foreback. 1973. *Science* 182:1247–1249.

Brandstetter, A., E. Lombi, W.W. Wenzel, and D.C. Adriano. 2000. In D.L. Wise et al., eds. *Remediation Engineering of Contaminated Soils,* Marcel Dekker, New York.

Brockbank, C.I., G.E. Batley, and G.K. Low. 1988. *Environ Technol Lett* 9:1361–1366.

Brookins, D.G. 1988. *Eh–pH Diagrams for Geochemistry*. Springer-Verlag, New York.

Calvert, C.C. 1975. In E. A. Woolson, ed. *Arsenical Pesticides*. American Chemical Society, Washington, DC.

Cappon, C.J. 1990. In J. O. Nriagu and M. Simmons, eds. *Food Contamination from Environmental Sources*. Wiley, New York.

Carlson, C., and D.C. Adriano. 1993. *J Environ Qual* 22:227–247.

Carrow, R.N., P.E. Rieke, and B.G. Ellis. 1975. *Soil Sci Soc Am Proc* 39:1121–1124.

Cebrian, M.E., A. Albores, G. Garcia-Vargas, L.M. del Razo, and P. Ostrosky-Wegman. 1994. In J.O. Nriagu, ed. *Arsenic in the Environment, part 1: Cycling and Characterization*. Wiley, New York.

Chakraborti, D. 1999. School of Environ Studies, Jadavpur Univ, Calcutta, India, personal communication.

Chang, S.C., and M.L. Jackson. 1957. *Soil Sci* 84:133–144.

Chatterjee, A., D. Das, and D. Chakraborti. 1993. *Environ Pollut* 80:57–65.

Chen, C.J., and C.J. Wang. 1990. *Cancer Res* 50:5470–5474.

Cheng, C.N., and D.D. Focht. 1979. *Appl Environ Microbiol* 38:494–498.

Cherry, J.A., A.U. Shaikh, D.E. Tallman, and R.V. Nicholson. 1979. *J Hydrol* 43:373–392.

Chino, M. 1981. In K. Kitagishi and L. Yamane, eds. *Heavy Metal Pollution in Soils of Japan*. Japan Sci Soc Pr, Tokyo.

Chiou, H.Y., Y.M. Hsueh, and Y.M. Liaw. 1995. *Cancer Res* 55:1296–1300.

Chiu, V.Q., and J.G. Hering. 2000. *Environ Sci Technol* 34:2029–2034.

Choprapawon C., and A. Rodcline. 1997. In C.O. Abernathy, R.L. Calderon, and W.R. Chappell, eds. *Arsenic Exposure and Health Effects*. Chapman & Hall, London.

Chowdhury, T.R., B.K. Mandal, D. Chakraborti, et al. 1997. In C.O. Abernathy, R.L. Calderon, and W.R. Chappell, eds. *Arsenic Exposure and Health Effects*. Chapman & Hall, London.

Chunguo, C., and L. Zihui. 1988. *Sci Total Environ* 77:69–82.

Crafts, A.S., and R.S. Rosenfels. 1939. *Hilgardia* 12:177–200.

Creger, T.L., and F.J. Peryea, 1994. *Hort Science* 29:88–92.

Cullen, W.R., and K.J. Reimer. 1989. *Chem Nov* 89:713–764.

Cummings, D.E., F. Caccavo, S. Fendorf, and R.F. Rosenzweig. 1999. *Environ Sci Technol* 33:723–729.

Das, D., G. Sawanta, and D. Chakraborti. 1996. *Environ Geochem Health* 18:5–15.

Davenport, J.R., and F.J. Peryea. 1991. *Water Air Soil Pollut* 57–58:101–110.

Davis, R.D., P.H.T. Beckett, and E. Wollan. 1978. *Plant Soil* 49:395–408.

del Razo, L.M., M.A. Arellano, and M.E. Cebrian. 1990. *Environ Pollut* 64:143–153.

Deuel, L.E., and A.R. Swoboda. 1972a. *J Environ Qual* 1:317–320.

Deuel, L.E., and A.R. Swoboda. 1972b. *Soil Sci Soc Am Proc* 36:276–278.

Dickens, R., and A.E. Hiltbold. 1967. *Weeds* 15:299–304.

Dudas, M.J., and S. Pawluk. 1980. *Can J Soil Sci* 60:763–777.

Eisler, R. 1994. In J.O. Nriagu, ed. *Arsenic in the Environment, part 1: Cycling and Characterization*. Wiley, New York.

Elkhatib, E.A., O.L. Bennett, and R.J. Wright. 1984a. *Soil Sci Soc Am J* 48:758–762.

Elkhatib, E.A., O.L. Bennett, and R.J. Wright. 1984b. *Soil Sci Soc Am J* 48:1025–1030.

Epps, E.A., and M.B. Sturgis. 1939. *Soil Sci Soc Am Proc* 4:215–218.

Fassbender, H.W. 1974. *Zeitschrift für Pflanzenernährung und Bodenkunde* 137:188–203.

Feinglass, E.J. 1973. *N Engl J Med* 228:828–830.

Fendorf, S.E., M.J. Eick, P. Grossl, and D.L. Sparks. 1997. *Environ Sci Technol* 31:315–320.

Ferguson, J.F., and J. Gavis. 1972. *Water Res* 6:1259–1274.

Frank, R., K. Ishida, and P. Suda. 1976. *Can J Soil Sci* 56:181–196.

Frost, R.R., and R.A. Griffin. 1977. *Soil Sci Soc Am J* 41:53–57.

Furr, K.A., A.W. Lawrence, S.S.C. Tong, M.G. Grandolfo, R.A. Hofstader, C.A. Bache et al. 1976. *Environ Sci Technol* 7:683–687.

Garcia-Vargas, G., and M.E. Cebrian. 1996. In L. Chang, ed. *Toxicology of Metals*, CRC Pr, Boca Raton, FL.

Goldberg, S., and R.A. Glaubig. 1988. *Soil Sci Soc Am J* 52:1297–1300.

Grantham, S.A., and J.P. Jones. 1977. *J Am Water Works Assoc* 69:653–657.

Hallacher, L.E., E.B. Kho, N.D. Bernard, A.M. Orcutt, W.C. Dudley, and T.M. Hammond. 1985. *Pac Sci* 39:266–273.

Harrington, J.M., S.E. Fendorf, and R.F. Rosenzweig. 1998. *Environ Sci Technol* 32:2425–2430.

Hiltbold, A.E. 1975. In E.A. Woolson, ed. *Arsenical Pesticides*. American Chemical Society, Washington, DC.

Hindmarsh, J.T., and R.F. McCurdy. 1986. *CRC Crit Rev Clin Lab Sci* 23:315–347.

Holm, T.R.I., M.A. Anderson, R.A. Stanforth, and D.G. Iverson. 1980. *Limnol Oceanogr* 25:23–30.

Hurd-Karrer, A.M. 1939. *Plant Physiol* 14:9–29.

Jacobs, L.W., and D.R. Keeney. 1970. *Commun Soil Sci Plant Anal* 1:8593.

Jacobs, L.W., D.R. Keeney, and L.M. Walsh. 1970a. *Agron J* 62:588–591.

Jacobs, L.W., J.K. Syers, and D.R. Keeney. 1970b. *Soil Sci Soc Am Proc* 34:750–754.

Jain, A., K.P. Raven, and R.H. Loeppert. 1999. *Environ Sci Technol* 33:1179–1184.

Jekel, M.R. 1994. In J.O. Nriagu, ed. *Arsenic in the Environment, part 1: Cycling and Characterization*. Wiley, New York.

Johnson, L.R., and A.E. Hiltbold. 1969. *Soil Sci Soc Am Proc* 33:279–282.

Johnston, S.E., and W.M. Barnard. 1979. *Soil Sci Soc Am J* 43:304–308.

Jones, C.A., W.P. Inskeep, J.W. Bauder, and K.E. Keith. 1999. *J Environ Qual* 28:1314–1320.

Jones, C.A., H.W. Langer, K. Anderson, T.R. McDermott, and W.P. Inskeep. 2000. *Soil Sci Soc Am J* 64:600–608.

Kardos, L.T., S.C. Vandecaveye, and N. Benson. 1941. *Washington Agric Exp Sta Bull* 410:25.

Lepkowski, W. 1998. C & EN (November Issue).

Li, X., and I. Thornton. 1993. *Environ Health Perspect* 15:135–144.

Liebig, G.F. 1965. In H.D. Chapman, ed. *Diagnostic Criteria for Soils and Plants*. Quality Printing, Abilene, TX.

Liebig, G.F., G.R. Bradford, and A.P. Vanselow. 1959. *Soil Sci* 88:342–348.

Livesey, N.T., and P.M. Huang. 1981. *Soil Sci* 131:88–94.

Lombi, E., R.S. Sletten, and W.W. Wenzel. 2000. *Water Air Soil Pollut* (in press).

Lumsdon, D.G., A.R. Fraser, J.D. Russel, and N.T. Livesey. 1984. *J Soil Sci* 35: 381–386.

Lunde, G. 1977. *Environ Health Perspect* 19:47–52.

Luo, Z.D., Y.M. Zhang, K. Grumski, and S.H. Lam. 1997. In C.O. Abernathy, R.L. Calderon, and W.R. Chappell, eds. *Arsenic Exposure and Health Effects*. Chapman & Hall, London.

MacPhee, A.W., D. Chisolm, and C.R. MacEachern. 1960. *Can J Soil Sci* 40: 59–62.

Maeda, S. 1994. In J.O. Nriagu, ed. *Arsenic in the Environment, part 1: Cycling and Characterization*. Wiley, New York.

Maeda, S., A. Ohki, T. Tokuda, and M. Ohmine. 1990. *Appl Organomet Chem* 4:251–254.

Maeda, S., A. Ohki, K. Kusadome, T. Kuroiwa, I. Yoshifuku, and K. Naka. 1992. *Appl Organomet Chem* 6:213–219.

Maker, W.A. 1988. *The Biological Alkylation of Heavy Metals*. Royal Soc Chem, London.

Manning, B.A., and S. Goldberg. 1996. *Soil Sci Soc Am J* 60:121–131.

Marcus-Wyner, L., and D.W. Rains. 1982. *J Environ Qual* 11:715–719.

Marques, I.A., and L.E. Anderson. 1986. *Plant Physiol* 82:488–493.

Masscheleyn, P.H., R.D. Delaune, and W.H. Patrick. 1991. *Environ Sci Technol* 25:1414–1418.

McGeehan, S.L., S.E. Fendorf, and D.Y. Naylor. 1998. *Soil Sci Soc Am J* 62:828–833.

McLaren, R.G., R. Naidu, J. Smith, and K.S. Tiller. 1998. *J Environ Qual* 27:348–354.

Merry, R.H., K.G. Tiller, and A.M. Alston. 1983. *Aust J Soil Res* 21:549–561.

Mok, W.M., and C.M. Wai. 1994. In J.O. Nriagu, ed. *Arsenic in the Environment, part 1: Cycling and Characterization*. Wiley, New York.

Mok, W.M., J.A. Riley, and C.M. Wai. 1988. *Water Res* 22:769–774.

Moore, J.N., W.H. Picklin, and C. Johns. 1988. *Environ Sci Technol* 22:432–437.

Morton, W.E., and D.A. Dunnette. 1994. In J.O. Nriagu, ed. *Arsenic in the Environment, part 2. Human Health and Ecosystem Effects*. Wiley, New York.

[NAS] National Academy of Sciences. 1977. *Arsenic*. NAS, Washington, DC.

Nickson, R.T., et al. 1998. *Nature* 395:338.

Nickson, R.T., J.M. McArthur, P. Ravenscroft, W.G. Burgess, and K.M. Ahmed. 2000. *Applied Geochem* 15:403–413.

Niu, S., S. Cao, and E. Shen. 1997. In C.O. Abernathy, R.L. Calderon, and W.R. Chappell, eds. *Arsenic Exposure and Health Effects*. Chapman & Hall, London.

[NRCC] (National Research Council of Canada). 1978. *Effects of Arsenic in the Canadian Environment*. NRCC N391, Ottawa.

Nriagu, J.O. 1994. *Arsenic in the Environment, part 2. Human Health and Ecosystems Effects*. Wiley, New York.

Nriagu, J.O., and J. Pacyna. 1988. *Nature* 333:134–139.

Nriagu, J.O., and J.M. Azcue. 1990. In J.O. Nriagu and M. Simmons, eds. *Food Contamination from Environmental Sources*. Wiley, New York.

Olson, O.E., L.L. Sisson, and A.L. Moxon. 1940. *Soil Sci* 50:115–118.

Onken, B.M., and L.R. Hossner. 1995. *J Environ Qual* 27:373–381.

Onken, B.M., and D.C. Adriano. 1997. *Soil Sci Soc Am J* 61:746–752.

Oscarson, B.M., P.M. Huang, and W.K. Liaw. 1980. *J Environ Qual* 9:700–703.

Paden, W.R., and W.B. Albert. 1930. *SC Agric Exp Sta Ann Rep* 43:129. Clemson, SC.

Phillips, D.J.H. 1994. In J.O. Nriagu, ed. *Arsenic in the Environment, part 1: Cycling and Characterization*. Wiley, New York.

Qafoku, N.P., U. Kukier, M.E. Sumner, and D.E. Radcliffe. 1999. *Water Air Soil Pollut* 114:185–198.

Reed, J.F., and M.B. Sturgis. 1936. *J Am Soc Agron* 28:432–436.

Reynolds, J.G., D.V. Naylor, and S.E. Fendorf. 1999. *Soil Sci Soc Am J* 63:1149–1156.

Rimstidt, J.D., J.A. Chermack, and P.M. Gagen. 1994. In C.N. Alpers and D.W. Blowes, eds. *Environmental Geochemistry of Sulfide Oxidation*. American Chemical Society, Washington, DC.

Robinson, E.L. 1975. *Weed Sci* 23:341–343.

Roussel, C., H. Bril, and A. Fernandez. 2000. *J Environ Qual* 29:182–188.

Sachs, R.M., and T.L. Michael. 1971. *Weed Sci* 19:558–564.

Sakata, M. 1987. *Environ Sci Technol* 21:1126–1130.

Selby, L.A., A.A. Case, C.R. Dorn, and D.J. Wagstaff. 1974. *J Am Vet Med Assoc* 165:1010–1014.

Smith, E., R. Naidu, and A.M. Alston. 1999. *J Environ Qual* 28:1719–1726.

Steevens, D.R., L.M. Walsh, and D.R. Keeney. 1972. *J Environ Qual* 301–303.

Stewart, J. 1922. *Soil Sci* 14:111–118.

Takamatsu, T., H. Aoki, and T. Yoshida. 1982. *Soil Sci* 133:239–246.

Tamaki, S., and W. Frankenberger. 1992. *Rev Environ Contam Toxicol* 124:79–110.

Tammes, P.M., and M.M. deLint. 1969. *Netherlands J Agric Sci* 17:128–132.

Tanner, C.C., and J.S. Clayton. 1990. *NZ J Mar Freshwater Res* 24:173–179.

Thanabalasingam, P., and W.F. Pickering. 1986. *Environ Pollut* 12:233–246.

Tiller, K.G. 1992. *Aust J Soil Res* 30:937–957.

Tseng, W.P., H.M. Chu, S.W. How, J.J. Fong, C.S. Lin, and S. Yeh. 1968. *J Natl Cancer Inst* 40:453–463.

Tsutsumi, M. 1980. *Soil Sci Plant Nutr* 26:561–569.

Tsutsumi, M. 1983. *Soil Sci Plant Nutr* 29:63–69.

US Bureau Mines. 1998. *Minerals Yearbook*. U.S. Dept Interior, Washington, DC.

[U.S. EPA] United States Environmental Protection Agency. 1993a. *Integrated Risk Information System on Arsenic*. Office Health Environ Assess, Cincinnati, OH.

Vandecaveye, S.C., G.M. Horner, and C.M. Keaton. 1936. *Soil Sci* 42:203–215.

Veneman, P.L.M., J.R. Murray, and J.H. Baker. 1983. *J Environ Qual* 12:101–104.

Vinogradov, A.P. 1959. *The Geochemistry of Rare and Dispersed Chemical Elements in Soils,* 2nd ed. Consultants Bureau, New York.

Von Endt, D.W., P.C. Kearney, and D.D. Kaufman. 1968. *J Agric Food Chem* 16:17–20.

Wakao, N., H. Koyatsu, Y. Komai, and H. Shiota. 1988. *Geomicrobiol J* 6:11–24.

Walsh, L.M., and D.R. Keeney. 1975. In E.A. Woolson, ed. *Arsenical Pesticides.* American Chemical Society, Washington, DC.

Walsh, L.M., M.E. Sumner, and D.R. Keeney. 1977. *Environ Health Perspect* 19:67–71.

Wang, L.F., and J.Z. Huang. 1994. In *Adv Environ Sci Technol* vol. 27. Wiley, New York.

Wauchope, R.D. 1975. *J Environ Qual* 4:355–358.

Wauchope, R.D., and L.L. McDowell. 1984. *J Environ Qual* 13:499–504.

Waychuras, G.A., B.A. Rea, C.C. Fuller, and J.A. Davis. 1993. *Geochim Cosmochim Acta* 57:2251–2269.

Waychuras, G.A., C.C. Fuller, B.A. Rea, and J.A. Davis. 1996. *Geochim Cosmochim Acta* 60:1765–1781.

Wood, J.M. 1974. *Science* 183:1049–1052.

Woolson, E.A. 1973. *Weed Sci* 21:524–527.

Woolson, E.A., and P.C. Kearney. 1973. *Environ Sci Technol* 7:47–50.

Woolson, E.A., and A.R. Isensee. 1981. *Weed Sci* 29:17–21.

Woolson, E.A., J.H. Axley, and P.C. Kearney. 1971. *Soil Sci Soc Am Proc* 35:101–105.

Woolson, E.A., J.H. Axley, and P.C. Kearney. 1973. *Soil Sci Soc Am Proc* 37:254–259.

World Bank Group. 1999. The Bangladesh Arsenic Mitigation Water Support Project: Addressig a Massive Public Health Crisis. World Bank Bangladesh Office, Dhacca.

[WHO] World Health Organization. 1981. *Environmental Health Criteria 18: Arsenic.* WHO, Geneva.

WHO. 1993. *Guidelines for Drinking-Water Quality, 2nd ed, vol. 1 Recommendations,* WHO, Geneva.

Xu, H., B. Allard, and A. Grimvall. 1991. *Water Air Soil Pollut* 57–58:269–278.

8
Cadmium

1 General Properties of Cadmium

Cadmium is a soft, ductile, silver-white, lustrous, electropositive metal with an atom. wt. of 112.4, density of 8.64 g cm^{-3}, and melting pt. of 321 °C. It has eight stable isotopes with the following percentages of abundance: ^{106}Cd (1.22%), ^{108}Cd (0.88%), ^{110}Cd (12.39%), ^{111}Cd (12.75%), ^{112}Cd (24.07%), ^{113}Cd (12.26%), ^{114}Cd (28.86%), and ^{116}Cd (7.58%). Like Zn and Hg, Cd is a transition metal in Group II-B of the periodic table. Cadmium and Zn however, differ from Hg in that the Hg forms particularly strong Hg–C bonds. Like Zn, Cd is almost always divalent in all stable compounds, and its ion is colorless. Its most common compound in nature is CdS. It forms hydroxides and complex ions with ammonia and cyanide, e.g., $Cd(NH_3)_6^{4-}$ and $Cd(CN)_4^{2-}$. It also forms a variety of complex organic amines, sulfur complexes, chloro-complexes, and chelates. Cadmium ions form insoluble usually hydrated white compounds, with carbonates, arsenates, phosphates, oxalates, and ferrocyanides.

Cadmium is readily soluble in nitric acid but only slowly soluble in hydrochloric and sulfuric acids. Its low melting point is a valuable property to form important low-melting alloys. Although the metal surface oxidizes readily, it is very resistant to rusting.

2 Production and Uses of Cadmium

Cadmium is produced commercially as a by-product of the Zn industry. The most important uses of Cd are as alloys, in electroplating (auto industry), in pigments (cadmium sulfide, cadmium selenide), as stabilizers for polyvinyl plastics, and in batteries (rechargeable Ni–Cd batteries). Electrodeposited Cd is widely used for protecting iron and steel against corrosion—even a thickness of 0.008 mm is sufficient for protection. As an impurity in Zn, significant amounts of Cd are also present in galvanized metals. Consequently, Cd can be found in a wide variety of consumer goods, and virtually all households and industries have products that contain some Cd. Cadmium is also used in photography, lithography, process engraving, rubber curing, and as a fungicide primarily for golf course greens.

Cadmium is basically recovered as a by-product from the smelting and refining of Zn concentrates at a rate of about 3.0 to 3.5 kg tonne^{-1} of primary Zn. Different yields are due to different Cd contents of the ores used and/or fluctuations in the addition of residues (recycling). No ores are mined and processed exclusively to provide Cd. Therefore, Cd production can be expected to more or less parallel Zn production. World production in 1994 was 18,100 tonnes, with the United States accounting for about 5.6% of the production (1010 tonnes). This is in contrast to the 1940s when the United States produced approximately 70% of the world's supply. Starting in the 1950s, the U.S. dominance of the world production of Cd gradually diminished partly due to large increases in production in Japan and the former Soviet Union. In 1969, the United States produced about 34% of the world's supply, while Japan and the former Soviet Union produced approximately 16 and 14%, respectively. While the rest of the world has practically maintained production levels of the 1970s, U.S. production has steadily declined to its lowest of 5.6% of the world's output in 1994. The majority of Cd-producing countries, particularly Japan and western Europe, recorded significant increases in production in 1977 caused by the expansion of Zn output. As of 1977, Belgium, Germany, the former Soviet Union, Canada, the United States, and Japan were the major Cd producers; the major consumers were: Belgium, France, Germany, the United Kingdom, the former Soviet Union, the United States, and Japan. During the last two decades, pigments (25%), plastic stabilizers (15%), plating (35%), and Ni–Cd batteries (15%) represented the major end uses for Cd in the Western world.

Until recently, the price index has been a major factor in Cd consumption, but environmental considerations have now assumed greater importance. For example, in Japan, Cd use dramatically declined following the incidence of *itai-itai* disease in the Jinzu River basin in the late 1960s; in Sweden, the use of Cd was restricted, based on rising levels of Cd in human tissues and increases of its use in electroplating, as a stabilizer, and as a coloring agent. The Swedish government also ordered reduction in Cd levels in phosphate fertilizers. In addition, importation of Cd-containing goods, on a commercial basis, was temporarily curtailed by the Swedish government. The level of Cd in phosphate fertilizers has also been restricted in several other countries, including Australia because of the concern of Cd transfer in the crop food chain from soil fertilization with phosphates (McLaughlin et al., 1996).

3 Cadmium in Nature

Cadmium is a naturally occurring metallic element found in soils, waters, plants, and other environmental samples. Cadmium exists in the II oxidation state in nature (Table 8.1). There is little difference in Cd content

TABLE 8.1. Normal cadmium concentrations (ppm) in various environmental media.

Material	Average concentration	Range
Igneous rocks[c]	0.082	0.001–0.60
Metamorphic rocks[c]	0.06	0.005–0.87
Sedimentary rocks[c]	3.42	0.05–500
Recent sediments[c]	0.53	0.02–6.2
Crude oil[d]	0.008	0.0003–0.027
Coal[d]	0.10	0.07–0.18
Fly ash[d]	11.7	6.5–17
Phosphate rocks[h]	25	0.2–340
Phosphated fertilizers (mixed)[e]	4.3	1.5–9.7
Sewage sludges[b]	74	2–1100
Soils[a] (world nonpolluted)[f]	0.35	0.01–2.0
Australia[h]		
Fertilized rural soils	0.42	0.01–13.9
Unfertilized soils	0.37	0.01–12.02
U.S. soils (national arable, $n = 3202$)[i]	0.255	0.03–0.94
Fruits (USA, $n = 190$)[j]	0.005	0.0043–0.012
Vegetables (USA, $n = 1891$)[j]	0.028	0.016–0.13
Crop grains (field-USA, $n = 1302$)[j]	0.047	0.014–0.21
Field crops (USA, $n = 2858$)[j]	0.21	<0.001–3.80
River sediments (polluted)[k]	—	30–>800
River sediments (unpolluted)[k]	—	0.04–0.8
Grasses[g]	—	0.03–0.3
Ferns[f]	0.13	—
Mosses[g]	—	0.7–1.2
Lichens[g]	—	0.1–0.4
Trees, deciduous[g]		
Leaves	—	0.1–2.4
Branches	—	0.1–1.3

among the igneous rocks. Among the sedimentary rocks, the carbonaceous shales, formed under reducing conditions, contain the most Cd. The Zn/Cd ratio in rocks is about 250. The most often-quoted average concentration of Cd in the earth's crust is 0.15 to 0.20 ppm. Krauskopf (1979) ranks Cd 64th in crustal abundance among the elements. Since Cd is closely related geochemically to Zn, it is found mainly in Zn, Pb–Zn, and Pb–Cu–Zn ores. The amount in the principal Zn ore, sphalerite (ZnS) varies markedly from a low of about 0.1% to a high of 5%, and sometimes even higher. The Cd content of the majority of Cu–Zn deposits is 0.3 part of Cd/100 parts of Zn, and in Pb–Zn deposits it is 0.4 part of Cd/100 parts of Zn. Cadmium is also found in wurtzite, another ZnS, and in trace amounts in galena, tetrahedrite, and a variety of other sulfides and sulfosalts.

In the Malibu Canyon area near Los Angeles, California, Cd in soils ranged up to 22 ppm. The soil levels of Cd were reflected by the Cd concentrations in native vegetation, wild oats, and mustard. Additional surveys indicated that residual soils developed from shale parent materials

TABLE 8.1. (*Continued*)

Trees, coniferous[g]		
Leaves	—	0.1–0.9
Freshwater[f]	0.10[a]	0.01–3[a]
Seawater[f]	0.11[a]	<0.01–9.4[a]
Oysters[k]	—	<1–12 (DW)
Fish muscle (fillet)[k]	—	<0.01 (DW)
Fish organs[k]	—	2–20 (DW)
Mussels (polluted freshwater)[k]	—	≤3 (DW)
Cigarettes[k]	—	<0.5–3 (DW)
Kidneys (exposed persons)[k]	—	up to 500 (WW)
Kidneys (smokers)[k]	—	≤6 (WW)
Kidneys (nonsmokers)[k]	—	≤3 (WW)
Liver (human)[k]	—	0.1–3 (WW)
Lung tissue (unexposed)[k]	—	≤0.1 (WW)
Bones (human)[k]	—	<0.01–0.3 (WW)
Whole blood (exposed persons)[k]	—	up to 0.2 (WW)
Whole blood (smokers)[k]	—	≤0.0002–0.006 (WW)
Whole blood (nonsmokers)[k]	—	<0.0002–0.002 (WW)
Urine (exposed persons)[k]	—	up to 0.2 (WW)
Urine (unexposed persons)[k]	—	<0.0001–0.003 (WW)

[a] $\mu g\ L^{-1}$.
[b] For 57 sludges from Michigan.
Sources:
[c] Page et al. (1987).
[d] Ainsworth and Rai (1987).
[e] Lee and Keeney (1975).
[f] Bowen (1979).
[g] Shacklette (1972).
[h] McLaughlin et al. (1996).
[i] Holmgren et al. (1993).
[j] Wolnik et al. (1985).
[k] Stoeppler (1992).

had the greatest Cd concentrations, with a mean of 7.5 ppm, whereas soils originating from sandstone and basalt had the lowest concentrations, with a mean of 0.84 ppm (Lund et al., 1981). Alluvial soils with parent materials from mixed sources had an intermediate Cd concentration of 1.5 ppm. Even though high-Cd soils are widespread in this area, the unfavorable terrain for farming would preclude Cd transfer into the agricultural food chain.

Although the fate and behavior of Cd in agroecosystems—primarily Cd from sewage sludge and phosphatic fertilizers—have been extensively studied, relatively limited information is available concerning the fate of Cd in forest ecosystems. Cadmium concentrations in deciduous foliage collected from polluted areas were in the range of 4 to 17 ppm, compared with Cd levels of 0.1 to 2.4 ppm in similar vegetation from normal environments (Shacklette, 1972). Cadmium concentrations in coniferous foliage were 0.05 to 1.0 ppm in polluted areas and 0.1 to 0.9 ppm in normal areas. Foliage of sugar maple growing in remote sections of New England, in the United States, had Cd concentrations in the range of 0 to 5 ppm (Smith, 1973). In spruce forest sites in central Sweden that were subjected to industrial

TABLE 8.2. Concentrations or inventory of cadmium in components of forests affected by anthropogenic input of cadmium.

| Component | Spruce forest, central Sweden[b] | | Oak woodland/Hallen Wood, Avonmouth, United Kingdom[c] | |
	ppm Cd	ER[a]	Component	mg Cd m^{-2}
Roots			Trees	
<5 mm diam	2.7	6.8	leaves	0.456
≥5 mm diam	1.5	7.5	branches	31.92
Bole	<0.1		bark	21.19
Bark	2.5	13	wood	8.01
Branch			total	61.57
1st yr	5.4	14	Shrubs	
2nd yr	4.6	12	leaves	0.186
3rd yr	4.2	10	branches	0.610
4th yr	3.3	7.5	bark	0.320
5th–7th yr	2.7	6.0	wood	0.920
Needles			total	2.036
1st yr	0.6	3.0	Ground flora	1.611
2nd yr	0.4	1.5	Organic litter	531.6
3rd yr	0.5	1.4	Mineral soil[b]	4348
4th yr	0.5	1.4		
5th–7th yr	1.0	2.4	Total	4945
Litter				
O1	24	63		
O2	44	40		
Moss	30	33		
Epiphytic lichens	12	30		

[a]Enrichment ratio = ratio between the metal concentration in this site and in a similar site with no local deposition.
Sources:
[b]Tyler (1972).
[c]Martin and Coughtrey (1975).

pollution, Cd levels were 0.4 to 1.0 ppm in spruce needles, compared with 0.2 to 0.4 ppm in spruce foliage from nonpolluted areas (Table 8.2). Essentially all the components of this forest in Sweden showed high enrichment ratios, particularly in the litter component. In the United States, Cd levels for pine trees in a mixed deciduous forest at the Walker Branch Watershed in eastern Tennessee, considered as a relatively unpolluted ecosystem, are generally lower than those in central Sweden (Van Hook et al., 1977).

In normal forest ecosystems, the concentration of Cd in vegetation followed the pattern: roots > foliage > branch > bole. However, this may not be the case in polluted ecosystems, which show branches (see Table 8.2) having relatively high Cd levels (Tyler, 1972). The major sites of Cd accumulation in a forest ecosystem are in the litter component, partly because of interception of the element by the litter from throughfall and by the immobilization of the element by the OM of the litter. However, the soil should still be expected as the major sink for metals.

In another relatively unpolluted forest ecosystem in the Bärhalde Watershed, southern Black Forest, Germany the retention of Cd is highest in the surface soil, where it increases with increasing pH and OM content (Stahr et al., 1980). The watershed has Cd input of 400 μg m^{-2} yr^{-1}. The increases in atmospheric input of Cd were reflected by the enrichment and higher turnover of Cd, Cu, and Pb throughout this watershed.

Other studies also have indicated substantial enrichment of ecosystem components by atmospheric deposition from industrial activities. Forest components had elevated Cd levels due to smelting operations (Martin and Coughtrey, 1975; Wixson et al., 1977). In a mixed oak woodland close to a primary Pb–Zn smelter in Avonmouth, England Cd concentrations vary among species and among plant parts of certain species. Some of the highest concentrations were observed in bryophytes, ferns, and on small woody twigs and bark of shrubs and trees (see Table 8.2). Similar data have been obtained for other impacted ecosystems in Belgium, the United States, Germany, and Poland. However, the input of Cd into Hallen Wood (Avonmouth) (i.e., 9.2 mg Cd m^{-2} yr^{-1}) resulted in a 4.2 times contamination of aboveground biomass compared with an oak–hornbeam forest in Poland (Martin and Bullock, 1994).

In comparing undisturbed forest ecosystems in urban and rural northwestern Indiana (United States), much higher levels of Cd in soils and vegetation were found in the urban site compared with a similar system in a rural setting 67 km away (Parker et al., 1978). The urban area, which has been exposed for about 100 years to contamination from industrial and other sources, has about 10 ppm Cd in the top 2.5 cm of the soil vs. 0.2 ppm in the rural site. These higher soil levels were reflected in higher concentrations in most plant species. Contrary to these findings, the foliage of woody plants in New Haven, Connecticut had Cd contents within the "normal" range of values (0 to 5 ppm) found for foliage of red maples collected from remote areas in Maine and New Hampshire (Smith, 1973).

Forest ecosystems are also potential recipients of municipal sewage sludge and wastewater because of their fertilizer value. Furthermore, contamination of the food chain is not much of a concern since the vegetation is not harvested for human consumption, except for the potential contamination of wildlife feeding on vegetation and contamination of groundwater. Cadmium levels were not significantly elevated in vegetation and soils of sites receiving either sludge or wastewater effluent at loadings of <1 kg Cd ha^{-1} (Sidle and Sopper, 1976). However, at higher loading rates of Cd, the amount of Cd moving past the root zone and eventually to the groundwater could be of some concern.

Of the Cd added to the soil in a *Filipendula ulmaria* meadow ecosystem, <10% of the total Cd in the ecosystem was found in plant biomass

(Balsberg, 1982). Root Cd concentration exceeded that in the soil and was several times the concentration in aboveground tissues. The Cd concentrations in various plant parts decreased in the order: new roots > old roots > rhizomes > stem > leaf-stalks > stems ~ stem leaf-blades > reproductive organs.

4 Cadmium in Soils/Sediments

4.1 Total Cadmium in Soils

In natural soils, Cd concentration is largely influenced by the amount of Cd in the parent rock. Based on the concentration reported for common rocks, one can expect on the average, soils derived from igneous rocks would contain the lowest Cd (< 0.10 to 0.30 ppm), soils derived from metamorphic rocks would be intermediate (0.10 to 1.0 ppm), and soils derived from sedimentary rocks would contain the largest amount of Cd (0.30 to 11 ppm). The Malibu Canyon study demonstrates the dramatic effect of parent material on the Cd content of soils (Lund et al., 1981).

The abundance of Cd in soil is on the order of 0.30 ppm, as indicated by numerous data. Normal Canadian soils were reported to contain from 0.01 to 0.10 ppm total Cd (mean = 0.07 ppm); normal glacial tills and other glacial materials had 0.01 to 0.70 ppm total Cd (mean = 0.07 ppm); whereas, soils and glacial tills near Cd-bearing deposits contained up to 40 ppm Cd. For Arctic soils, an average of 0.67 ppm Cd, with much higher values (5.4 ppm) for areas containing Cd-mineral deposits was reported. An average of 1.8 ppm Cd for northern Canadian surface soils was also reported (Hutchinson, 1979). However, areas affected by smelting operations showed Cd concentrations ranging from 0.20 to 350 ppm in the surface soil.

In general, background Cd level in agricultural soils is about 1 ppm (or less). Surface agricultural soils in Ontario, Canada had 0.56 ppm Cd (0.10 to 8.10 ppm range), with only 5% of all samples exceeding the 1.25 ppm level (Frank et al., 1976). Similar background levels were reported for rural nonmineralized areas of Wales (Bradley, 1980). A more extensive survey of the soils of England and Wales revealed a median value of < 1.0 ppm Cd (0.08 to 10 ppm range, $n = 689$) (Archer, 1980).

Cadmium concentrations of arable soils in the United States, by region and soil type, are given in Table 8.3. In Denmark, an average Cd concentration of 0.22 ppm (range of 0.03 to 0.90 ppm) was found for agricultural soils, projected to increase at the rate of 0.6% per year due to atmospheric deposition and the use of phosphatic fertilizers (Tjell et al., 1980). In an extensive survey of both cultivated and noncultivated soils of Sweden, an average of 0.22 ppm Cd (0.03 to 2.3 ppm range) for all soils was

TABLE 8.3. Cadmium concentrations of arable soils in the United States by region and soil type.

| | No. of samples | Cd concentration (mg kg^{-1}) | |
		Range[a]	Mean
Region			
Western	742	0.20–0.49	0.33
North Central	937	0.20–0.94	0.37
Northeast	293	0.08–0.21	0.17
Southern	1230	0.03–0.44	0.15
Soil type			
Histosol	288	0.36–1.44	0.72
Aridisol	152	0.17–0.71	0.35
Entisol	256	0.07–0.70	0.28
Mollisol	1076	0.08–0.68	0.28
Inceptisol	256	0.05–0.71	0.27
Vertisol	91	0.13–0.55	0.27
Spodosol	43	0.08–0.47	0.21
Alfisol	570	0.03–0.42	0.16
Ultisol	571	<0.01–0.30	0.08

[a]Range was estimated from the standard deviation (95% confidence interval).
Source: Extracted from Holmgren et al. (1993).

found (Andersson, 1977). The same value of 0.22 ppm for both cultivated and noncultivated soils was established. In Japan, a nationwide survey on Cd content of noncontaminated paddy field soils revealed an average Cd concentration of 0.45 ± 0.23 ppm (Iimura, 1981). For world soils, a mean of 0.40 ppm Cd as a background level, was reported (Berrow and Reaves, 1984).

It has become apparent that although Cd concentrations in noncontaminated soils are typically < 1 ppm, its level in soils may be substantially elevated due to human activities, or due to weathering of parent rocks having high Cd contents.

4.2 Bioavailable Cadmium in Soils

Several chemical extractants have been tested to provide an index of Cd phytoavailability or Cd recovery from soils. They include weak acids, neutral salts, and chelating agents. Because the efficiency and predictability of a given extractant can be strongly influenced by soil and plant factors, there is no universal extractant for Cd. In using $CdCl_2$-treated soils, a significant relationship between plant Cd (radish and lettuce) and Cd extracted from soil by neutral 1 N NH_4OAc, but not by 1 N HCl and 1 N HNO_3 solutions, was reported (John et al., 1972). The Cd extracted by the NH_4OAc was regarded as in the exchangeable form, whereas the HCl and

HNO$_3$ extractions closely estimated the total amount. Similar results were obtained by others with a NH$_4$OAc solution using oats as the test plant (Andersson and Nilsson, 1974; Haghiri, 1974). Although significant relationships were obtained for various crops (potatoes, cabbage, barley, parsley, and lettuce) using this extractant, some had reservations about this extractant since it extracted less than 20% of total Cd, potentially causing detectability problems (Lag and Elsokkary, 1978).

In evaluating nine different extractants, 0.5 N HOAc was the best extractant for predicting plant-available Cd when soil pH was included in the regression equation (Haq et al., 1980). Extractable Cd and soil pH accounted for 81% of the variability of plant Cd concentration. Using seven various extractants, the highest significant relationship ($r = 0.97$, $p < 0.001$) between Cd concentration in radish and extractable Cd was from unbuffered 1 N NH$_4$NO$_3$ solution (Symeonides and McRae, 1977). Significant relationships were also found using neutral 1 N NH$_4$OAc ($r = 0.50$, $p < 0.001$) and 5% HOAc (pH 2.5) ($r = 0.36$, $p < 0.01$). The greater efficiency of the 1 N NH$_4$NO$_3$ solution was due to its ability in maintaining the natural pH of the soil, which has a marked influence on Cd phytoavailability. In testing industrially contaminated soils in Norway, a 1.0 N HNO$_3$ solution was recommended as the extractant of choice (Lag and Elsokkary, 1978).

In metal-contaminated soils from Ontario, Canada the exchangeable, complexed, or HNO$_3$-soluble fractions of Cd, or the total of these three fractions, predicted Cd uptake by plants (Soon and Bates, 1982). The fractions were extracted with 1 M NH$_4$OAc (pH 7), 0.125 M Cu(OAc)$_2$, and 1 M HNO$_3$, respectively. Others found 0.1 N HCl as a favorable extractant for predicting plant Cd (Jones et al., 1973; Takijima et al., 1973a); others favor the use of DTPA extracting solution in predicting Cd uptake and yield by crops (Bidwell and Dowdy, 1987; Bingham et al., 1976a; Street et al., 1978). Similarly, a DTPA extractant proved to be more predictive of Cd concentrations in tobacco and peanut plants than the Mehlich 3 extracting solution, which is widely used in the southeastern United States (King and Hajjar, 1990). Furthermore, in evaluating four soil tests, the 0.005 M DTPA proved superior to 0.05 M CaCl$_2$, 1 M NH$_4$NO$_3$, and 0.05 M EDTA-(Na)$_2$ in predicting crop Cd concentrations (Jackson and Alloway, 1991). In another instance, highly significant linear relationships were obtained between soil Cd in the surface soil (0 to 20 cm) with 0.1 M CaCl$_2$ extraction and seed Cd content in field-grown soybeans (Bell et al., 1997). However, there was no significant relationship between kernel Cd in peanuts and 0.1 M CaCl$_2$-extracted soil Cd. This indicates interspecies variation in Cd accumulation, with soybean being more responsive.

It has become apparent that the bioavailability of Cd to crop plants can be more accurately predicted using *bioavailable* extractants rather than the soil total Cd content.

4.3 Cadmium in Soil Profile

Cadmium is generally fairly immobile in the soil profile. Based on a Scottish soil profile examination, Cd like Zn was somewhat higher in soils formed on basic igneous rocks than soils formed on other rock types (Berrow and Mitchell, 1980). The Cd (≤1 ppm) was fairly uniformly distributed throughout the profile. Higher accumulation of Cd occurred in the upper part of a Swedish profile, paralleling that of humus distribution (Andersson, 1977). The levels of Cd in a profile of virgin muck soils in Canada (> 2 ppm) were much higher than the levels reported in Scotland and Sweden, but nevertheless, they showed increasing trends in the top layers (Hutchinson et al., 1974). Long-term cultivation tended to decrease Cd levels in the muck soil profile.

Contaminated soils also show the general immobility of Cd in the soil profile. Soils contaminated by smelting operations showed Cd concentrations close to background levels at a depth of about 30 to 40 cm (John et al., 1972; Kobayashi, 1979). In sludged soils, practically all of the Cd remained in the surface 20 cm of soil following application of 84 tonnes ha^{-1} of sewage sludge over a 12-yr period (Andersson and Nilsson, 1972). Likewise, little Cd moved beyond 15 cm in an acid soil receiving surface application of 17 tonnes ha^{-1} of sewage sludge (Boswell, 1975). However, evidence of Cd movement below 15 cm in soil after application of 88 to 166 tonnes ha^{-1} of sludge over a 3-yr period was observed in another situation (Hinesly et al., 1972). In a silt loam soil that was furrow irrigated with anaerobically digested sewage sludge for 14 yr, total soil Cd has increased to a depth of 75 cm (Baveye et al., 1999). Approximately 90% of the total Cd applied on a soil after 6 yr of continuous annual sewage sludge applications was found in the 0 to 15 cm soil depth (Chang et al., 1984a). This zone of accumulation corresponded to the depth where the sludge was incorporated. In another sludged-soil instance, Cd movement in the profile was limited to a depth of 5 cm below the zone of sludge incorporation (Williams et al., 1984). Recently, no evidence was found to indicate offsite movement of Cd that had been applied on plots since 1948 via repeated fertilization with superphosphate, either through horizontal transfer or via leaching in the soil profile (Hamon et al., 1997).

Cadmium arising from phosphate fertilization of citrus groves over a 36-yr period proved to be less mobile than the P carrier itself in the soil profile (Mulla et al., 1980). About 71% of the accumulated Cd resided in the surface soil (0 to 15 cm) compared with only 45% for P.

More recent studies confirm the inconsequential movement of Cd below the application zone of sludge in agricultural soils over the long term (Dowdy et al., 1991; McGrath and Lane, 1989; Yingming and Corey, 1993). Whether this is due to the sequestration effect of sludge OM and other constituents on Cd is under contention. In contrast however, heavily polluted soils, such as the soil impacted by a copper refinery (Fig. 8.1)

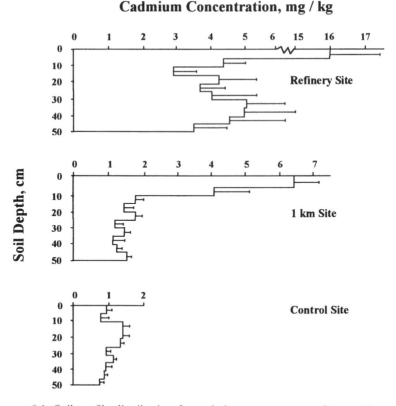

Cadmium Concentration, mg / kg

FIGURE 8.1. Soil profile distribution for cadmium at a copper refinery, 1 km, and control sites. (From Hunter et al., 1987.)

indicate substantial movement of the Cd beyond the 50-cm depth (Hunter et al., 1987a,b,c).

4.4 Forms and Chemical Speciation of Cadmium in Soils/Sediments

The mobility and bioavailability of Cd largely depends on its form or chemical species.

4.4.1 Forms in Soils/Sediments

In the analyses of three heavily polluted soils and a sediment, the greatest amount of Cd (i.e., 37%) was in the exchangeable fraction (Table 8.4) (Hickey and Kittrick, 1984). The Fe-Mn oxide and the residual fractions accounted for 23 and 15% of total Cd, respectively. The rest of the Cd was detected in the carbonate fraction, with the least and insignificant amount in

TABLE 8.4. Forms of cadmium (in ppm) in heavily contaminated soils and sediments.

Form[a]	Samples[b]			
	Soil A	Soil B	Soil C	Sediment
Exchangeable	6.56 ± 0.42[c]	14.0 ± 0.21	6.18 ± 0.21	16.8 ± 0.4
Carbonate	2.92 ± 0.20	14.7 ± 0.1	1.19 ± 0.10	17.9 ± 0.7
Fe–Mn oxide	5.70 ± 0.37	9.16 ± 0.23	2.80 ± 0.08	10.8 ± 0.8
Organic	$0.37 \pm NS$[d]	0.73 ± 0.07	0.39 ± 0.06	0.25 ± 0.01
Residual	3.85 ± 0.20	4.50 ± 0.02	2.93 ± 0.10	3.64 ± 0.04

[a]The sequential extractants used were 1 M MgCl$_2$ (pH 7), 1 M NaOAc (pH 5), 0.04 M NH$_2$OH · HCl in 25% HOAc, (0.02 M HNO$_3$ + 30% H$_2$O$_2$ + 3.2 M NH$_4$OAc in 20% HNO$_3$), and finally (conc HF + conc HClO$_4$) for the following fractions: exchangeable, carbonate, Fe-Mn oxide, organic, and residual, respectively.
[b]Soil A was a surface soil taken within 1 km of a Cu smelter; soils B and C were surface soils from sewage sludge disposal sites; and the sediment was a contaminated sample.
[c]Values are mean ± SD ($n = 4$).
[d]Not significant, but not 0.
Source: Hickey and Kittrick (1984).

the organic fraction. The amounts of Cd in the residual fractions of these materials were about 50 times higher than typical for unpolluted soils. Soils in southwest Poland that are heavily contaminated by the metallurgical industry were sequentially extracted after grouping into background level (group 1), intermediate (i.e., background to maximum tolerance level) (group 2), and in excess of maximum tolerable level (group 3) (Chlopecka et al., 1996). Of the soil parameters investigated to influence the fractional distribution of Cd, the most significant effect was from the extent of contamination, as modified by soil pH. Cadmium became increasingly associated with the exchangeable fraction in the contaminated soils in groups 2 and 3. Where the metals are present at background levels, the residual form is usually dominant, indicating that the metals are mainly associated with primary minerals. In the most highly contaminated soils (group 3), the residual fraction only accounts for ~12% of total Cd, and the oxide and carbonate forms increased proportionally compared with the background soils. A similar fractional trend was observed in Doñana National Park in Spain, which has been impacted by mining, where the bioavailable fraction of Cd represents >50% of the total amount of Cd in soils (Ramos et al., 1994). Similarly, soils contaminated by a zinc smelter in Japan contained ~55% of the total Cd in the exchangeable fraction (Xian, 1989). Cadmium uptake by cabbage plants in this case was strongly controlled by the amount in the exchangeable and carbonate forms rather than the total Cd content in the soil.

Based on several studies, it has become apparent that Cd from metallurgical activities (i.e., mining/smelting) is likely to be more bioavailable to organisms than Cd from relatively unimpacted soils, as indicated by the increasingly larger fraction of Cd in labile fractions with increasing intensity of contamination (Asami et al., 1995; Chlopecka et al., 1996; Ramos et al.,

1994; Xian, 1989). This infers that metals in soils from anthropogenic sources are potentially more bioavailable. In soils treated with sewage sludge, sequential extractions revealed increases of Cd in every fraction (as compared with nontreated soils), with the most significant increases occurring in the carbonate and organic fractions (Chang et al., 1984b).

The following distribution of added Cd in river sediments was observed (Khalid et al., 1978): at pH 5.0—water-soluble > exchangeable ≫ DTPA-extractable ~ reducible ~ residual organic-bound; and at pH 6.5— exchangeable > DTPA-extractable > residual organic-bound ~ water-soluble > reducible. In an alkaline (pH 8) sediment, essentially all the Cd extracted was associated with the DTPA-extractable, residual organic-bound, and reducible (oxides of Fe and Mn) fractions (Khalid et al., 1981).

4.4.2 Chemical Speciation of Cadmium in Soils

In typical soil solutions, there may be 10 to 20 different metal cations (including trace metals) that can react with as many different inorganic and organic ligands to form 300 to 400 soluble complexes and up to 80 solid phases. Chemical speciation modeling can be used to predict these soluble complexes. Cadmium uptake by sweet corn, tomato, or Swiss chard correlated equally well with either total Cd concentration in the saturation extract, free Cd^{2+} ions, or Cd^{2+} activity (Mahler et al., 1980). However, uptake was much lower in calcareous soils than in acid soils at equal Cd concentrations in the soil solutions. Using the GEOCHEM program, similar free Cd percentages (64 to 73%) in solutions were found for both acid and calcareous soils. In solutions amended with Cd, most of the Cd was identified as free Cd^{2+} in a $CaCl_2$ solution (Table 8.5). Virtually all Cd in the EDTA solution was present as stable complexes; the humic solution contained a substantial fraction of Cd as labile (chloro-complexes) and slowly labile complexes. In the soil solution, most of the Cd was present as free Cd^{2+}, although the fraction of free ion in the sample from rhizosphere soil was less than in the sample from the nonrhizosphere soil (see Table 8.5). In another instance, much of the Cd added to calcareous soils (pH 7.5 to 8.5) at low input (0.1 mg Cd kg^{-1} soil) was rapidly adsorbed or precipitated

TABLE 8.5. Cadmium speciation in five solutions, including one from a rhizosphere.

	Total Cd µg L^{-1}	Free Cd ions[a]	Cd complexes[a]		
			Labile	Slowly labile	Stable
Chloride solution (10^{-2} M $CaCl_2$, pH 6.0)	4.2	65	35	0	0
EDTA solution (10^{-4} M EDTA, pH 5.7)	4.7	1	0	0	99
Humic solution (30 mg C L^{-1}, pH 7.0)	5.0	22	52	24	2
Nonrhizosphere soil solution (pH 5.6)[b]	6.8	90	10	0	0
Rhizosphere soil solution (pH 6.7)[c]	2.1	80	20	0	0

[a]% Metal of total in solution.
[b]Soil solution from unplanted potted soil.
[c]Soil solution from planted soil after 21 days of radish growth.
Source: Extracted from Holm et al. (1995).

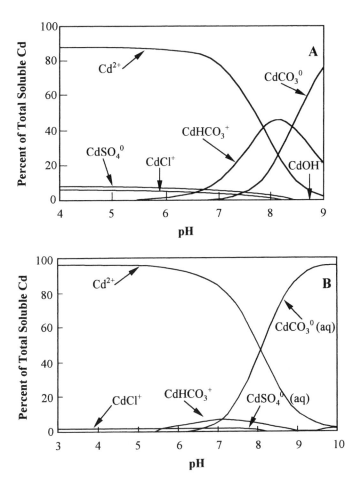

FIGURE 8.2. (A) Calculated distribution of cadmium species between pH 4 and 9 in the solution of a typical calcareous soil (soil at $P_{CO_2} = 0.01$ atm); (B) calculated distribution of cadmium aqueous species as a function of pH, i.e., based on mean composition of river water of the world from Hem (1985). (From [A] Hirsch and Banin, 1990; [B] Krupka et al., 1999.)

in the solid phase, and the Cd concentrations in the soil solution were low (Hirsch and Banin, 1990). The dominant Cd species in the soil solution were Cd^{2+} and $CdHCO_3^+$ (35 and 45%, respectively) while the organo-Cd complexes were only minimal (Fig. 8.2A).

Soil solution speciation of Cd in contaminated soils containing between 0.10 and 38 mg Cd kg^{-1} indicates that free Cd^{2+} species varied between 0 and 60% (as percentage of total dissolved Cd in solution) and averaged about 20% (Sauve et al., 2000). The soils ($n = 64$) were collected from diverse land uses: orchards, forest, agricultural, and urban. The predictability of free Cd^{2+} activity is largely controlled by the level of soil contamination (total Cd) and soil pH; the effect of dissolved OM on the predictability was insignificant, i.e., <2% ($r^2 = 0.70$, $p < 0.001$).

In soils where very high Cl^- concentrations occur in solution, Garcia-Miragaya and Page (1976) predicted that Cd would be primarily complexed with chloride (e.g., $CdCl_2$, $CdCl_3^-$, and $CdCl_4^{2-}$) rather than as a free cation (Cd^{2+}). Indeed in arid–saline soils, complexes of Cd with Cl^- and SO_4^{2-} are important (McLaughlin et al., 1996). For example, Cd chloro-complexes constituted the dominant species of inorganic Cd in soil solutions from irrigated agricultural soils in South Australia (McLaughlin and Tiller, 1994). Up to 75% for chloro- and 20% for sulfato-complexes of Cd were found in these soils. These soluble Cd complexes are also likely to predominate in the soil solutions of saline soils in the United States whose composition favors such complexation (Jurinak and Suarez, 1990). This is expected, because Cd readily forms chloro-complexes (Holm et al., 1995).

In contrast to other metals, such as Cu, Pb, Hg, or Zn, it is plausible that organic ligands do not have great relevance in the complexation of Cd in soil solutions. Indeed, a number studies have indicated that either free Cd^{2+} ions (see Table 8.5) or Cd complexed by inorganic ligands are the dominant Cd species present in soil solution of most sludged and arable soils (Emmerich et al., 1982; Hirsch and Banin, 1990; Holm et al., 1993).

4.4.3 Chemical Speciation of Cadmium in Groundwater

The distribution of Cd aqueous species was predicted based on a generalized river water composition (Hem, 1985; Krupka et al., 1999) and using a concentration of 1 µg L^{-1} total dissolved Cd. The MINTEQ A2 calculations indicate that in water systems with pH <6, essentially all of the dissolved Cd is expected to exist as the free Cd^{2+} (see Fig. 8.2B). At pH between 6 and 8.2, Cd carbonate species ($CdHCO_3^+$ and $CdCO_3^0$) become increasingly important. At pH between 8.2 and 10, virtually all of the Cd in solution is predicted to exist as a neutral carbonate complex $CdCO_3^0$. The species $CdSO_4^0$, $CdHCO_3^+$, $CdCl^+$, and $CdOH^-$ are also present but at much lower concentrations.

In view of the widely held belief that complexations of Cd with naturally occurring organic matter (DOC) are weak because of the competition for binding sites with Ca, which is generally present in much higher concentrations, it is not surprising to see generally small amounts of Cd, if any, complexed with organic ligands (Giesy, 1980; Hem, 1972).

4.5 Sorption of Cadmium in Soils

Adsorption is the operating mechanism of the reaction of Cd at low concentrations with soils. In most sorption studies, the results conform to either the Langmuir or the Freundlich isotherm. The sorption of Cd by hydrous iron oxides conforms to the Langmuir isotherm (Kuo and McNeal, 1984).

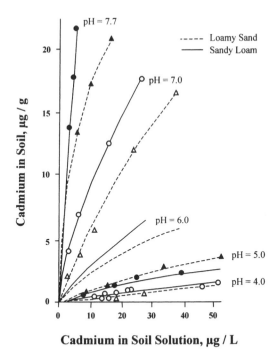

FIGURE 8.3. Cadmium sorption isotherms for two soils as influenced by soil texture and pH. (From Christensen, 1984a.)

In addition to adsorption, precipitation can play a key role in controlling Cd levels in soils. Using equilibrium batch studies of three soils having different chemical properties, Cd solubility in soils decreased as pH increased (Santillan-Medrano and Jurinak, 1975). The lowest values were obtained in calcareous soils (pH 8.4). At high Cd concentrations, the precipitation of $Cd_3(PO_4)_2$ and/or $CdCO_3$ controlled Cd solubility. Precipitation of $CdCO_3$ also occurred in sandy soils having low CEC (<6 meq/100 g), low OM (<1%), and alkaline pH (pH >7.0) (Street et al., 1977).

Using loamy sand and sandy loam soils, sorption of Cd was demonstrated as a fast process where >95% of the sorption took place within the first 10 min and equilibrium was attained within an hour (Christensen, 1984a). In addition, the sorption capacity of the soil was observed to increase approximately three times per unit increase in pH (Fig. 8.3).

In general, precipitation occurs at higher Cd^{2+} activities and ion exchange predominates at lower Cd^{2+} activities. McBride (1980) found that initial chemisorption of Cd^{2+} on $CaCO_3$ was very rapid, while $CdCO_3$ precipitation at higher Cd concentrations was slow. This indicates that chemisorption may regulate Cd^{2+} activity in some calcareous soils, producing solubilities much lower than predicted by the solubility product for $CdCO_3$.

Several factors may influence the degree by which Cd is adsorbed on soil surfaces. In addition to pH, ionic strength and exchangeable cations also influence Cd adsorption. Ionic strength of various salt solutions ($NaClO_4$, NaCl, and Na_2SO_4) affected the amount of Cd sorbed on montmorillonite surfaces (Garcia-Miragaya and Page, 1976). The adsorption of Cd^{2+} from Cl^- solutions diminished with increasing salt concentration, concomitant with the formation of uncharged ($CdCl_2^0$) and negatively charged complexes of Cd with Cl^- ligands (e.g., $CdCl_3^-$, $CdCl_4^{2-}$, etc.). The chloro species of Cd are less strongly adsorbed than the Cd^{2+}. Organic ligands can also influence the sorption of Cd by soils. Among several organic compounds tested, the ability to reduce Cd retention by soil followed the trend: EDTA > NTA > oxalate $\approx$ acetate (Elliott and Denneny, 1982). Thus, the presence of "precipitation inhibitors," such as dissolved organic C and/or chelates, could prevent metal coprecipitation with $CdCO_3$ or minimize adsorption of metals onto solid phases (Holm et al., 1996).

The presence of competing cations can also greatly affect Cd sorption by soils. Divalent Ca^{2+} and Zn^{2+}, or H^+ present in the soil solution can compete effectively with Cd for sorption sites in soils or can desorb Cd from the soil (Christensen, 1984a,b). Divalent ions greatly reduce the efficiency of Cd sorption by permanent charge clays (Bittell and Miller, 1974). The selectivity coefficient for Cd^{2+}–Ca^{2+} exchange on mineralogically pure clays is near unity, i.e., Cd^{2+} would compete on equal basis with Ca^{2+} for clay adsorption sites. However, with field soils, Cd^{2+} was preferentially adsorbed over Ca^{2+}, with the selectivity greatest in montmorillonitic and least in kaolinitic clay (Milberg et al., 1978). Apparently, the difference between the model clay and field soils lies in the fact that soil colloids carry a number of specific exchange sites with higher bonding energy for Cd than are present on pure clays. For example, Cd adsorption diminished when Ca was present at higher Cd loadings on hydrous manganese oxide (δ-MnO_2) but competition from Ca^{2+} became negligible at low Cd loadings (Zasoski and Burau, 1988). In addition, simultaneous additions of Cd and Zn reduced adsorption of both metals. Competition among Cd, Zn, and Ca was observed at pH 4, 6, and 8, with the higher pHs favoring the adsorption of Zn over Cd. However, at typical environmental concentrations, the presence of alkaline-earth elements, such as Mg, Sr, and Ba, would have little effect, if any, on the adsorption of Cd on amorphous iron oxyhydroxide ($Fe_2O_3 \cdot H_2O_{am}$) (Cowan et al., 1991).

5 Cadmium in Plants

5.1 Uptake and Translocation of Cadmium in Plants

Cadmium is a nonessential element in plant nutrition. Under normal conditions, plants take up only small quantities of Cd from soils. In an extensive survey in the United States in which samples of wheat and perennial grasses were collected from major agricultural soils with no known input of Cd except from fertilizers, the levels of Cd found were generally below 0.30 ppm (wheat, 0.20 ppm; grasses, 0.17 ppm) (Huffman and Hodgson, 1973). No distinct regional pattern of Cd concentration was observed in this survey. Additionally, the levels of Cd in the shoots of 10 plant species grown without added Cd in a neutral soil ranged from 0.21 ppm in timothy to 0.71 ppm in soybean leaves (MacLean, 1976).

When present in growth medium in higher than background concentrations, Cd is readily taken up by the roots and distributed throughout the plant. The amount of uptake is tempered by soil factors such as pH, CEC, redox potential, OM, other metals, fertilization, and other factors. Plant factors, such as species and genotype, also influence the total uptake. In addition to these factors, the level of Cd in the growth medium generally influences uptake. Hence a consensus exists that over a certain range there is a positive, almost linear correlation between the levels of added Cd in the medium and the resulting Cd concentrations in plant tissues. In addition, the Cd uptake–medium relationship suggests that yields of plants may be predicted on a basis of plant tissue analysis. Using this technique, one can predict the yield of a given crop based on either the plant tissue content of Cd or soil Cd content.

Numerous studies lend support to plants' tendencies to accumulate Cd. Radish shoots can accumulate 5 ppm of Cd when grown on soils containing 0.6 ppm Cd (Lagerwerff, 1971). Leafy plants, such as lettuce, spinach, and turnip greens, accumulated 175 to 354 ppm Cd when grown on soils pretreated with sewage sludge enriched with Cd of up to 640 ppm (Bingham et al., 1975). Fruits and seeds of other plants tested usually accumulated no more than 10 to 15 ppm. Uptake by several crop species (lettuce, spinach, broccoli, cauliflower, and oats) grown on soil pretreated with Cd in variable amounts (from 40 to 200 ppm) ranged from 100 to 600 ppm in leaf tissues. In solution culture, Cd accumulation by plants can even be greater. The leaves of tomato, cabbage, pepper, barley, corn, lettuce, red beet, and turnips grown on complete nutrient solutions with Cd concentrations of 0.1 to 0.5 ppm contained up to 200 to 300 ppm (Page et al., 1972). Excessive uptake of up to 158 ppm Cd in leaves by tomatoes grown in solution culture containing Cd as low as only 0.10 ppm has occurred (Turner, 1973).

Cadmium is rather readily translocated throughout the plant following its uptake by the roots. Distribution between roots and shoots differs with plant species, rooting medium, and time of treatment. In rice, about 99% of the total Cd taken up by the plants was found in the shoot in a wide range of redox potentials and pHs (Reddy and Patrick, 1977). Some environmental factors, such as Cd concentration in the medium, ambient temperature, and light intensity, can affect the distribution of the metal between the shoots and roots of the rice plant (Chino and Baba, 1981). In general, roots contain at least twice the Cd concentrations found in shoots (Koeppe, 1977). Similarly, in field-grown fruit trees (apple, peach, plum, and pear), much lower Cd concentrations were found in fruits than in the roots, especially under low soil pH (Korcak, 1989). Among aerial tissues of corn plants grown in the field, stems and leaves generally concentrated more Cd than did the husks, cobs, kernels, silk, or tassels (Peel et al., 1978). In tobacco, Cd concentrations declined markedly with higher leaf position on the stalk (King and Hajjar, 1990). This general decline with increasing leaf height in the tobacco plant suggests that Cd has limited mobility within the plant and its concentration is related to the age of the leaf.

In spite of the high bioavailability of Cd to plants, only a small fraction of the Cd pool in soils is recovered by plants. One reason is that Cd is phytotoxic up to certain levels, which can drastically reduce plant yield. Under field conditions, the recovery of Cd by crops from soils treated with sewage sludge was usually only <1% of the total added (Kelling et al., 1977; Stenström and Lönsjö, 1974). However, under controlled experiments, total uptake from soil can be >1% of the total added (Andersson and Nilsson, 1974; Williams and David, 1973), with up to 6% recovered by tomatoes from added superphosphate. The recovery of Cd added to a podzolic soil using pot culture ranged from 0.9% for beans to 13.6% for tobacco (Williams, 1977). This is expected because the medium in small containers can be fully exploited by the roots.

5.2 Cadmium–Zinc Interactions in Plants

The close association of Cd and Zn in geological deposits and the chemical similarity of the two elements carry over into biological systems. Cadmium has no known biological function, but Zn is an essential element. Cadmium and Zn appear to compete for certain organic ligands *in vivo,* accounting in part for the toxic effects of Cd and the ameliorative effects of Zn on Cd toxicity. Because Cd competes with Zn in forming protein complexes, a negative association between the two can be expected (Vallee and Ulmer, 1972). For this reason, the ratio between these two elements in plant tissues is thought to be biologically important.

Indeed, one way to ensure that the Cd contents of food crops grown in sludged soil would not be hazardous in the food chain is to reduce the Cd content of sludge to 1.0% or less of the Zn content (Chaney, 1974). The rationale behind this recommendation is that Zn toxicity to the crop would occur first, before the Cd content of the crop would reach levels that could constitute a health hazard. With lower Cd/Zn ratios in the sludge, marked reductions of this ratio in grain, fruit, and root crops can be expected relative to the ratio in the sludge (Chaney, 1973). In most soils treated with sludge, the Cd/Zn ratio in tissues of plants grown thereon usually is 0.01 or less (Dowdy and Larson, 1975; Kelling et al., 1977), although in some instances, a higher ratio of 0.03 for edible tissues was obtained (Schauer et al., 1980). Under agricultural conditions, crops could be expected to have much lower Cd/Zn ratios, based on a survey involving immature wheat and grass samples in 19 states in the United States (Huffman and Hodgson, 1973). An average Cd/Zn ratio of 0.008 was obtained with the ratio for wheat slightly higher. It appears that the ratio tends to be lower in the edible portions of crops (radish, potato, pea, and corn) grown on sludge-treated soils, in which the ratio did not exceed that of the applied sludge (Dowdy and Larson, 1975). There is a possibility of the exclusion of Cd relative to Zn during the filling of the grain causing a drop in the Cd/Zn ratio (Chaney, 1973; Hinesly et al., 1972). The net result is that the Cd/Zn ratio in grain is one-tenth to one-half the value found for leaves and stems.

The interaction between Cd and Zn is either antagonistic or synergistic. The additions of Cd-treated sewage sludge to a calcareous soil (pH 7.4) decreased the concentrations of Zn in shoots of many but not all crops studied (Bingham et al., 1975). Increasing concentrations of Zn suppressed Cd uptake at low Cd concentrations (2 and 20 ppb Cd) in the culture solution (Lagerwerff and Biersdorf, 1971). However, at the 100 ppb level of Cd, increasing amounts of Zn increased Cd uptake. On the other hand, Japanese scientists reported that addition of Zn to the soil could increase Cd uptake by the rice plants (Chino, 1981b). Indeed, additions of Zn from 5 to 50 ppm to soil significantly increased Cd concentrations of soybean shoots, but at rates > 50 ppm the Cd uptake was decreased (Haghiri, 1974). One possible explanation for this synergistic effect is that Zn-induced dissociation of Cd sorbed onto the binding sites as a result of competition for these sites and increased Cd in soil solution. And vice versa, low levels of Cd in culture solution increased Zn concentration in the foliage of soybean but decreased it at high Cd treatment (Cunningham, 1977). In another soil, increasing Zn levels appeared to increase Cd uptake by crops at high soil solution concentrations of Cd and to decrease uptake at low solution concentrations (Gerritse et al., 1983).

Under field conditions, applications of low rates of Zn fertilizer as zinc sulfate (up to 5.0 kg Zn ha^{-1}), were found to markedly decrease the Cd concentration in wheat grown in Australian soils that have marginal to severe Zn deficiency (Oliver et al., 1994). This decreasing pattern of Cd uptake by wheat effected by Zn addition is displayed on Figure 8.4A. However, in this situation, no further significant decreases in Cd concentration occurred at higher rates of Zn application. A similar pattern for the depressive effect of bioavailable Zn on Cd uptake by potato tuber is exhibited on Figure 8.4B. Similarly, addition of ZnSO$_4$ at planting (up to 100 kg Zn ha^{-1}) significantly reduced potato tuber levels at four of the five sites studied in Australia (McLaughlin et al., 1995).

Zinc application decreased the Cd concentrations in rice grain grown under Zn-deficient conditions in solution culture (Honma and Hirata, 1978). Similarly, the addition of Zn as ZnSO$_4$ at 20 mg kg^{-1} to potted soils reduced the shoot Cd concentrations of durum wheat (Choudhary et al., 1994). Thus, it is likely that when Zn is added to soils in which this element is marginally to severely deficient as determined by one of the extraction solutions, the antagonistic effect on Cd uptake by plants can be more pronounced.

5.3 Interactions of Cadmium with Other Ions in Plants

In addition to Cd–Zn interaction, Cd is known to interact in plants with other ions. Cadmium has depressed the Mn uptake by plants (Iwai et al.,

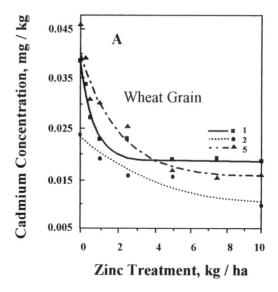

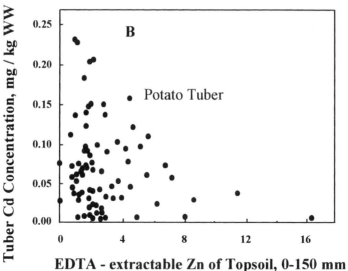

FIGURE 8.4. (A) Variation in mean wheat grain Cd concentration (mg kg⁻¹) with Zn treatment (ka ha⁻¹) for three field sites: site 1 is loamy sand (clay ~12%) with pH 6.9–7.8, site 2 is sandy clay (38% clay) with pH 7.7–7.8, site 5 is sand (~5% clay) with pH 5.7–6.7; background Cd and Zn were 0.48 to 0.66 and 0.05 to 0.07, respectively. (B) Relationship between potato tuber Cd concentration and EDTA-extractable Zn (mg kg⁻¹) in the surface soil (0 to 15 cm). (Modified from Oliver et al., 1994.)

1975; Patel et al., 1976; Root et al., 1975). Cadmium causes a type of Fe deficiency in plants (Root et al., 1975) and could also depress the uptake of other ions (Ca, Mg, and N) (Cunningham, 1977; Iwai et al., 1975). Added Fe, as FeEDDHA, reduced Cd uptake by about 17% in flax grown on potted calcareous and neutral soils (Moraghan, 1993). On the other hand, Se (Francis and Rush, 1983) and Ca (Tyler and McBride, 1982) have been shown to depress Cd uptake.

Often, Cd and Pb are closely scrutinized together because they originate from the same emission sources, such as mining and smelting operations, sewage, sludge, combustion of energy materials. The two elements have been shown to act synergistically on plant growth. Cadmium content of ryegrass and red fescue treated with Pb and Cd was greater than that of plants treated with Cd alone, with the effect being greater for fescue than for ryegrass (Carlson and Rolfe, 1979). A synergistic response to Pb and Cd was found for root and shoot growth of American sycamore (Carlson and Bazzaz, 1977) and for root growth of maize (Hassett et al., 1976). Furthermore, while treatment with Cd or Pb alone caused a reduction in photosynthesis and transpiration of sycamore seedlings, the addition of Cd to Pb-stressed plants did not reduce rates of photosynthesis and transpiration below those observed for plants treated with Pb alone. A tendency for soil Pb to increase both the plant Cd concentration and the total Cd uptake of the corn shoots was also observed (Miller et al., 1977). Both Cd (at 2.5- and 5-ppm levels in soil) and Pb (at 125- and 250-ppm levels in soil) reduced corn shoot yield, with a concomitant positive interaction of the two metals on growth. The enhancement of Cd uptake by soil Pb was attributed to the ability of Pb to more effectively compete for exchange sites on colloidal surfaces than Cd, releasing Cd into the soil solution.

6 Ecological and Health Effects of Cadmium

6.1 Cadmium Toxicity in Agronomic–Horticultural Crops

Cadmium is not only bioavailable to plants from soil and other media but is also known to be toxic to them at much lower concentrations than other metals, such as Zn, Pb, Cu, etc. Phytotoxicity has been observed to be dependent upon plant species as well as concentration of Cd in the medium. For example, using nutrient culture, the following Cd concentrations (in ppm) in solution were found to be associated with 50% yield decrements: beet, bean, and turnip, 0.2; corn and lettuce, 1.0; tomato and barley, 5; and cabbage, 9 (Bingham and Page, 1975). In contrast, there is a wider range of Cd phytotoxicity in plants grown on soils. For 25% yield decrement, the total soil Cd level ranged from 4 ppm for spinach to > 640 ppm for paddy rice. However, these values correspond to 75 and 3 ppm in the diagnostic leaf tissues of spinach and rice, respectively. Roots of subterranean clover were depressed by the addition of 1 ppm Cd to soil, and at 5 ppm or higher, toxicity symptoms began to appear (Williams and David, 1977).

Because of the highly toxic nature of Cd to plants and animals, it is necessary to obtain information on Cd contents of edible tissues of field and vegetable crops. The following Cd concentrations in edible tissues of potted soil–grown plants were associated with 50% yield decrement (Bingham and Page, 1975): field bean, paddy rice, upland rice, and sweet corn, ≤ 3 ppm;

zucchini squash, wheat, tomato, cabbage, beet, and soybean, 10 to 20 ppm; turnip, radish, and carrot 20 to 30 ppm; and romaine lettuce, Swiss chard, curlycress, and spinach, >80 ppm. These levels, although much greater than what can be expected for field-grown crops, are well above the normal levels of Cd found in foodstuff, which is around 0.05 ppm (Friberg et al., 1971). In young spring barley, the critical level of Cd in plant tissues that affected their growth was about 15 ppm (Davis et al., 1978).

In solution culture, Cd concentration as low as 0.10 ppm in solution produced Cd accumulations varying between 9 ppm in bean leaves and 90 ppm in corn leaves (Page et al., 1972). At solution concentrations of 1 ppm Cd, the range was 35 ppm in bean leaves to 469 ppm in turnip leaves. Reduced yield and chlorosis occurred in *Brassica chinensis* also at 1 ppm Cd level in the nutrient solution (Wong et al., 1984). In ryegrass cultured in nutrient solution spiked with Cd up to 82 ppm, the shoots accumulated 100 to 500 ppm Cd, resulting in chlorotic plants and subsequent death (Dijkshoorn et al., 1974).

In rice, the critical Cd content of plant tissues above which growth is retarded has been reported at about 10 ppm (Iimura and Ito, 1971). This level was later modified (Chino, 1981a) to be 5 to 10 ppm for rice shoots and 100 to 600 ppm for rice roots. Concentrations of about 15.5 ppm Cd extracted with 0.1 N HCl from soils can be associated with excessive levels of Cd in the rice grain (Takijima et al., 1973b).

Cadmium toxicity symptoms in crops, in general, resemble Fe chlorosis. In addition to chlorosis, plants may exhibit necrosis, wilting, red–orange coloration of leaves, and general reduction in growth. In rice, the number of tillers is usually reduced, accompanied by severely depressed root growth (Chino, 1981a). In radish, Cd toxicity was associated with the reduction of Zn levels in plant tissues to near the deficiency threshold value (Khan and Frankland, 1983).

Phytotoxicity is due to interference with metabolic processes in plants. Obvious symptoms induced by the absorption of plants of elevated levels of Cd include: root growth retardation, suberization, damage to internal and external structures of the root, decreased root hydraulic water conductivity, interference with absorption and translocation of nutrients leading to nutrient imbalance, reduction of chlorophyll content, interference with enzymatic activities related to photosynthesis, and decrease in stomatal opening and conductance (Breckle, 1989; Marchiol et al., 1996; Van Assche and Clijsters, 1990). One of the main toxic effects of trace metals particularly Cd and Cu, is due to lipid peroxidation of cellular membranes and is termed oxidative stress (deVos et al., 1991; Ernst et al., 1992). The damage is the net effect caused by the active forms of oxygen, such as superoxide ($O_2^{\bullet-}$), hydrogen peroxide (H_2O_2), hydroxyl radical ($OH^{\bullet-}$), and single oxygen ($^1O_2$). Measurement of certain enzyme activities in plant tissues (e.g., peroxidase, dehydrogenase, malic, etc.) may serve as biosensors to assay the extent of physiological damage of Cd (Lagriffoul et al., 1998).

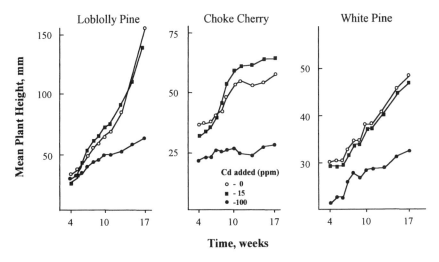

FIGURE 8.5. Effect of soil cadmium on the growth of woody species. (From Kelly et al., 1979.)

6.2 Cadmium Toxicity in Woody Plants

Tree species are also sensitive to soil Cd, but apparently at much higher concentrations than levels phytotoxic to agronomic and horticultural crops. In testing several tree species, increased Cd content in roots and shoots in response to soil (pH 4.8) Cd levels was observed but the growth was drastically retarded only by the 100-ppm treatment (Fig. 8.5) (Kelly et al., 1979). Using sand culture, excessive Cd accumulation in white pine, red maple, and Norway spruce was induced, resulting in reduced root initiation, poor development of root laterals, chlorosis, dwarfism, early leaf drop, wilting, and necrosis of current season's growth (Mitchell and Fretz, 1977). A severe reduction in the growth of silver maple with $CdCl_2$ application to sand culture was partly attributed to reduced relative conductivity of the stem, which in turn, was caused by deterioration of xylem tissues (Lamoreaux and Chaney, 1977).

Woody plants are a key component in polluted ecosystems. Tree dieback has been related to metal contamination of soils, though other constituents of pollution may have stronger effects (Breckle, 1991). It has been shown that transpiration in young *Fagus sylvatica* trees is affected by Cd stress, which can be depressed even before the onset of toxicity (Hagemeyer et al., 1986). Root architecture is affected by metal stress in general. Even though the number of laterals might be slightly increasing, as well as the number of short roots, the main root becomes shorter, producing a more compact root system (Arduini et al., 1994). Similar results were obtained for *Pinus sylvestris,* where the length of 1-yr-old needles was highly correlated with trace metal concentration in the growth medium (Breckle, 1991).

6.3 Cadmium Toxicity in Terrestrial Fauna and Wildlife

The fauna inhabiting the polluted soil/litter in Hallen Wood, Avonmouth, England, showed marked differences in the taxa of invertebrates compared with uncontaminated woodlands (Martin and Bullock, 1994) (see Chapter 5). The millipedes, earthworms, and wood lice were severely reduced in number per unit area in the polluted site. This reduction in populations of these invertebrates explains the reduced decomposition of OM often observed in impacted areas since they are largely responsible for the initial breakdown and subsequent decomposition of this material. In this polluted ecosystem, the soil and humus had 11.6 and 15.0 ppm Cd DW, respectively. The Cd accumulator invertebrates include: earthworm (Annelida —*Lumbricus terrestris*, 56.6 ppm; *Aporrectodea longa*, 23.7 ppm), wood lice (Isopoda—*Oniscus asellus*, 125.7 ppm), and snail and slug (Mollusca— *Cepaea nemoralis*, 26.9 ppm; *Discus rotundatus*, 31.1 ppm). In contrast, Cd concentrations were usually below 1 ppm in invertebrates in unpolluted spruce stands in southern Germany (Roth, 1992). Similarly, invertebrates from contaminated grasslands in the vicinity of a major refinery, housing a copper/cadmium alloying plant, showed significant elevation of the total body burden of Cu and Cd concentrations relative to control values (Hunter et al., 1987a,b,c). Of the taxa examined, Cd concentrations were highest in the Isopoda, Oligochaeta, and Lycosidae.

Cadmium-induced injury to wildlife is virtually unheard of. When it happens, the main clinical signs of Cd toxicity in animals include anemia, retarded gonad development, enlarged joints, scaly skin, liver and kidney damage, and reduced growth. In the U.S., high metal concentrations in tissues of white-tailed deer shot near a zinc smelter in Palmerton, Pennsylvania were attributed to metal poisoning of the animal (Sileo and Beyer, 1985). The kidneys of deer collected less than 8 km from the smelter had an average Cd level of 310 μg g^{-1} DW compared with only 15.5 μg g^{-1} DW in deer collected more than 100 km from the smelter. One deer with a renal Cd level of 600 μg g^{-1} DW had lesions in joints, similar to those observed in nearby horses diagnosed to be suffering from Zn poisoning. Osteoporosis and nephrocalcinosis were also observed in foals and were attributed to Cd toxicity (Gunson et al., 1982). Deer living near the smelter may also be suffering from chronic Cd poisoning. Metal concentrations in tissues decreased in the order shrews > birds > mice in the Palmerton area (Beyer et al., 1984). More recent studies at this site indicate that levels of Cd in kidneys and liver of white-tailed deer to be about 5 times higher than those for control areas (180 km away from the smelter) (Storm et al., 1994). These studies, conducted after the cessation of the smelting operation in 1980, indicate that the abnormal amounts of metals in the tissues of terrestrial vertebrates and the absence or low abundance of wildlife at Palmerton caused an alteration in ecological processes within 51 km of the smelter even 6 yr after its termination.

Thus, it has been suggested that Cd concentrations in the livers and kidneys of vertebrates exceeding 10 μg g^{-1} WW be viewed as evidence of Cd contamination and residues above 200 μg g^{-1} in these tissues be viewed as life threatening (Eisler, 1985). This should be viewed with caution however, as wildlife, particularly birds, have been demonstrated to accumulate metals even in remote areas far from sources of contamination.

6.4 Cadmium Toxicity in Freshwater Invertebrates and Fish

The average total Cd concentrations in rivers, streams, and lakes are expected to be in the range of <0.010 to 0.070 μg L^{-1} (Wren et al., 1995). Total background Cd concentrations in the Great Lakes ranged from <0.020 to 0.10 μg L^{-1}, with filtered total Cd concentrations in the range 0.001 to 0.068 μg L^{-1}. In contrast, contaminated lakes in the Sudbury (Ontario, Canada) area ranged from 0.08 to 0.22 μg L^{-1} (Stephenson and Mackie, 1988a,b).

In general, Cd bioaccumulates in freshwater organisms, but it does not biomagnify in the food chain as does Hg. Bioconcentration (whole body) factors for rainbow trout and whitefish are usually <1 (Harrison and Klaverkamp, 1989). However, bioconcentration factors of 2000 to 4000 have been reported for mollusks (Graney et al., 1983; Pesch and Stewart, 1980). Once metabolized, Cd is sequestered as metallothionein (MT) complexes (Klaverkamp and Duncan, 1987; Olson and Hogstrand, 1987; Taylor, 1983). The MT proteins are primarily S-containing amino acids that bind Cd, rendering it unavailable for interaction with intracellular sites. The bioaccumulation of Cd is dependent on exposure route (dietary or aqueous vector), with a large proportion of the body burden of Cd in fish accumulating in the gill, liver, and kidney (Harrison and Klaverkamp, 1989). Cadmium levels in various freshwater fish species in Canada were generally <0.40 μg g^{-1} DW (Suns et al., 1987; Wren et al., 1983); in edible salmonid tissue from Lake Ontario, Cd concentrations ranged from 0.040 to 0.21 μg g^{-1} DW (Cappon, 1987).

Cadmium bioaccumulation in invertebrates is also a function of species and tissues. Levels of Cd from 3.8 to 7.3 μg g^{-1} DW in cladoceran zooplankton from Lake Huron were reported (Wilson, 1982); mollusks in Ontario had 1.0 to 51.3 μg g^{-1} DW; the amphipod *Hyalella azteca* ranged from 0.13 to 56.6 μg g^{-1} DW in Ontario lakes (Stephenson and Mackie, 1988a,b). In general, for all species of fish, tissue Cd levels were positively correlated with aqueous Cd levels and inversely related to water hardness (i.e., CaCO$_3$ concentrations) and DOC concentration. Young life stages are more sensitive to Cd than adult. Nonsalmonid species are less sensitive to Cd than salmonid species. The salmonid species are 10 times more sensitive than the nonsalmonid species, and an increase in water hardness from 20 to 200 mg L^{-1} (as CaCO$_3$) increases the results of acute toxicity tests approximately 10 times for nonsalmonid species (Mance, 1987). Extensive toxicity data indicate the following (Mance, 1987); *Daphnia magna* is the most sensitive organism tested with a 4-day LC_{50} of 0.005 mg L^{-1} and a 20-day LC_{50} of 0.00067 mg L^{-1}; the crustacea are the most sensitive taxa and the insect larvae are the least sensitive. Adverse effects to aquatic invertebrates have been reported at relatively low Cd concentrations (0.28 to 3.0 µg L^{-1}) (Chapman et al., 1980; Lawrence and Holoka, 1987; Marshall, 1978; Winner, 1988).

6.5 Exposure of Humans to Cadmium

As discussed earlier, the entry and bioaccumulation of soil Cd in humans is influenced by edaphic factors (soil type, pH, CEC, salinity, etc.), plant factors (species, cultivars, tissues, etc.), form and species of Cd, livestock species, Cd–other element interactions once absorbed (in plants, livestock, human), and agronomic management practices. Transfer of Cd to food crops is of major concern. The consumption of agronomic–horticultural food crops represents the most critical exposure pathway for Cd in the general population. The current guideline limit for Cd in cereal grain is 0.12 mg kg^{-1} DW in Germany (Bundesgeshundheitsamt, 1986) and adoption of the same limit in the European Union is being discussed. Drinking contaminated water, smoking, and consumption of animal organs, especially kidney and liver, also contribute to the exposure. For drinking water, the WHO (1984) has a limit of 5 ppb Cd. A survey of 969 community water supply systems in the United States showed the average Cd concentration to be 1.3 ppb (Craun and McCabe, 1975). Bottled water ($n = 172$) offered for sale in Canada had a mean Cd concentration of 0.18 ppb (range < 0.10 to 0.40 µg L^{-1}) (Dabeka et al., 1992). Thus, drinking water contributes very little to the daily intake.

There are basically four sources of Cd that may contaminate food: agricultural technology (e.g., pesticides, phosphate fertilizer, sewage sludge, etc.), industrial pollution, geological sources, and food processing (e.g., food additives, physical and chemical contact with equipment and vessels). Dietary intake of Cd varies from country to country, due to differences in eating habits, amounts and types of food consumed, and levels of Cd residues (Table 8.6). Estimation of the daily intake of Cd for several countries ranged from 25 to 60 µg day^{-1} per person (Friberg et al., 1971). This range may be compared with a WHO/FAO provisional tolerable weekly intake of 400 to 500 µg Cd, or about 57 to 72 µg day^{-1} per person. The most recent estimate for the United States is about 39 µg day^{-1} (see Table 8.6). There are no official limitations for Cd in foods in the United States for adults. The intake for Canadians, based on a 1970 to 1971 survey, was about 67 µg day^{-1} (Somers, 1974). The Total Diet Survey of the U.S. FDA between 1968 and 1974 gave a range of 26 µg day^{-1} in 1968 to 61 µg day^{-1} in 1974. In this survey, which involved "market baskets" of typical foods and beverages, of the six metals (Pb, Cd, Hg, Zn, As, and Se), Cd has the most widespread distribution and Hg the most limited. These results show that the levels of these elements in foods do not vary significantly from one year to the next. Cadmium was most frequently detected in potatoes, leafy

TABLE 8.6. Cadmium dietary (total) intake in various countries.

Country	Period	Sampling procedure	Mean Cd intake (μg/person/day)
United States	1975–1976	Market basket	39
United States	1975–1976	Market basket	33
United Kingdom	1971–1972	Market basket	<35
Canada	1978	Total diet, students	73
Netherlands	1976–1978	Market basket	20
			35 (max.)
	1976–1978	Duplicate meals	<15–80
			<20 (median)
Japan[a]	1991–1992	Duplicate meals	27–37
		Market basket	32–35

[a]For women in Shiga Prefecture (Tsuda et al., 1995).
Source: Modified from Copius Peereboom-Stegeman and Copius Peereboom (1989).

vegetables, grain and cereal products, and oils and fats (Appendix Table A.28); and least detected in dairy products, fruits, and beverages. The intake from consumption of fish and seafoods in several countries are presented on Appendix Table A.29.

Traditionally, the mean concentrations for dairy products, red meats, and poultry are generally low (Mitchell and Aldous, 1974). However, concentrations in the kidneys of animals are generally much higher than in other tissues, averaging about 50 ppb WW (Neathery and Miller, 1975). Aquatic foods were reported to possess high mean concentrations of about 20 ppb WW, but some oysters were known to have concentrations as high as 2.0 ppm (Childs and Gaffke, 1974).

Because of the incidence of *itai-itai* disease in some areas of Japan, partly caused by consumption of unpolished rice with Cd levels as high as 3.4 ppm (Yamagata and Shigematsu, 1970), the Japanese government has set 1 ppm of Cd in rice grain as the maximum allowable limit (Asami, 1988). In a survey of Cd content of rice involving 22 countries, polished and unpolished rice samples had similar contents of 29 ppb Cd (Masironi et al., 1977). In this survey, Japanese rice was found to contain high Cd levels, averaging about 65 ppb. Daily intakes of Cd by women in Shiga Prefecture, Japan, determined by the market basket method, were 32 and 35 μg Cd day^{-1} for 1991 and 1992, respectively (Tsuda et al., 1995).

In the Netherlands, estimates for the rural daily intake for Cd in the general population range from 20 μg day^{-1}, to a maximum intake of 32 to 35 μg day^{-1} (Copius Peereboom-Stegeman and Peereboom, 1989). This corresponds to approximately 184 μg Cd wk^{-1} for the general population of the Netherlands. Intake can drastically change according to the extent of soil contamination with Cd. In a worst-case scenario at the Kempen area (Belgium–Netherlands border) where Cd levels in soil exceed 1.0 ppm (pH <5), the WHO provisional acceptable daily intake (400 to 500 μg wk^{-1}) is surpassed, even in the case of nonsmokers. In the highly polluted area where soil Cd exceeds 2.4 ppm, the intake by smokers is estimated at 12 μg Cd day^{-1}, 3 times the acceptable amount. The intake through food is at 1235 μg wk^{-1}, which is almost 7 times the normal intake of 184 μg day^{-1}.

The increases in the Cd intake in the polluted areas in Kempen are due to increased proportion of Cd from the consumption of vegetables and potatoes.

In central Italy, blood Cd concentrations were more than double in cigarette smokers than in nonsmokers; blood Cd level correlated significantly with the number of cigarettes smoked daily (dell'Omo et al., 1999). In Catalonia (in northeastern Spain), which is an industrial area, the Cd intake was estimated at 56 μg day^{-1}, with about 27% of total intake coming from green vegetables (Schuhmacher et al., 1991).

In the Upper Silesia region of Poland, which is the most industrialized and extensively contaminated area in that country, only 17 out of 756 vegetable samples met the tolerable concentration of 0.030 mg Cd kg^{-1} WW (Gzyl, 1990). Of the vegetables (parsley, carrot, celery, and red beet), the highest Cd concentrations were found in celery (19 mg kg^{-1} DW, soil pH 7.5). Furthermore, in the most contaminated area of Tarnowskie Gory in this region of Poland, about 95% of the cereal and all potato samples had Cd contents exceeding the limit value in feed for that country (0.10 Cd mg kg^{-1} DW) (Dudka et al., 1995a). In this region, the Cd concentration of soils is the most serious issue affecting the use of such soils for agricultural production.

As indicated on Appendix Table A.28, consumption of grain and cereal, potatoes, and leafy vegetables may constitute over 50% of the total Cd intake in most populations. Cadmium concentrations in potatoes can vary substantially among countries (Table 8.7), with some of the highest reported for Australia. However, the estimated daily dietary Cd intake in the United Kingdom through the consumption of wheat-based products was only about 8 μg per person or about 11% the WHO limit of 70 μg day^{-1} (Chaudri et al., 1995). The Cd content in the wheat grain decreased by about 20% from 0.052 ppm in 1982 to 0.042 and 0.038 ppm in 1992 and 1993, respectively. The proportion of samples in this survey exceeding the WHO limit of 0.12 ppm Cd DW was 4% in 1982, decreasing to 2% 10 yr later.

TABLE 8.7. Concentrations of cadmium reported in potato tubers from various countries.

Country	No. sites	Cd concentration in potato tubers (mg kg^{-1} WW)			
		Median	Mean	Min.	Max.
Australia	359	0.033	0.041	0.004	0.232
Aust.-WA	116	0.030	0.035	0.005	0.120
Germany	223	0.030	0.033	<0.013	0.200
Netherlands	94	0.030	0.030	0.002	0.090
Poland 1986–1988	254	0.010	0.014	0.001	0.090
Spain	8	—	0.013	0.008	0.017
Sweden	62	—	0.016	0.005	0.055
Switzerland	101	0.013	0.015	0.003	0.044
United Kingdom	39	—	0.080	<0.010	0.170
United States	297	0.028	0.031	0.002	0.182

Source: Modified from McLaughlin et al. (1996).

The decrease in Cd content in grain can be attributed to a decrease in Cd input from atmospheric deposition, change in phosphate fertilization management, and a dilution effect due to increased yields (15% increase in 10 yr).

Cigarette tobacco contains about 1 ppm Cd. Data indicate that 0.10 to 0.20 μg of Cd is inhaled for each cigarette smoked (Friberg et al., 1971), giving a total intake of about 3 μg per pack of cigarettes smoked.

Exposure to Cd may promote several disorders through Cd interference with the metabolism of Ca, vitamin D, and collagen, and bone degeneration, such as osteomalacia, or osteoporosis, which are late manifestations of severe Cd poisoning. Even chronic low-level exposure to Cd may promote Ca loss through urinary excretion. For example, a study (1985 to 1989) conducted in northeastern Belgium to assess the long-term effect of Cd exposure from zinc smelter emissions indicates that even at a low degree of environmental exposure, Cd may promote skeletal demineralization, which may lead to increased bone fragility and elevated risk of fractures in exposed individuals (Jarup et al., 1998; Kjellstrom et al., 1985).

Chronic (long-term) inhalation and oral exposure to Cd in humans affects the kidneys, with proteinuria, decreased glomerular filtration rate, and increased frequency of kidney stone formation noted (ATSDR, 1992). Other effects noted from chronic exposure to Cd in air in humans are effects on the lungs, including bronchiolitis and emphysema (see also Chapter 5). The U.S. EPA (1993) has not yet established a reference concentration (RfC) for Cd; the reference dose (RfD) for Cd in drinking water is 0.0005 mg kg^{-1} per day and the RfD for Cd in food is 0.001 mg kg^{-1} per day, based on significant occurrence of proteinuria in humans. The U.S. EPA estimates that consumption of these levels or less, over a lifetime, would not likely result in the occurrence of chronic noncancer effects.

7 Factors Affecting Mobility and Bioavailability of Cadmium

The primary factors that affect mobility and bioavailability of Cd in soils are pH, texture, OM, Cd concentration, Cd species, Zn status, and salinity.

7.1 pH

Soil pH is the most important single soil property that determines Cd bioavailability to plants. Consequently, it is recommended that soil pH be maintained at pH 6.5 or greater in land receiving biosolids containing Cd. In general, Cd uptake by plants almost always increases with decreasing pH. Cadmium uptake by plants is usually higher in acidic than in alkaline or calcareous soils. Higher Cd concentrations were obtained for lettuce and Swiss chard grown on acid soils (pH 4.8 to 5.7) than on calcareous soils (pH 7.4 to 7.8) (Mahler et al., 1978). In wheat, liming reduced the grain Cd concentration by approximately 50% (Bingham et al., 1979). The liming potential of fly ash also significantly reduced Cd absorption by sudan grass grown on soils treated with composted sewage sludge having 31 ppm Cd (Adriano et al., 1982). Increasing the soil pH with the addition of CaO decreased the Cd content of fodder rape (Andersson and Nilsson, 1974). This pH effect on Cd uptake was attributed, in part, to competition between Ca^{2+} and Cd^{2+} ions at the root surface. Similarly, increasing the pH decreased the Cd uptake of radish plants grown on Cd-contaminated soils due to a decrease in mobility of Cd in the soil (Lagerwerff, 1971).

The retention capacity of northeastern United States soils was strongly dependent on the exchangeable Ca^{2+} concentration of the soil, i.e., the retention capacity increases with increased level of Ca^{2+} in the soil (McBride et al., 1981). The best indicators of Cd bioavailability to corn plants include the retention capacity and exchangeable base (mainly Ca^{2+}) content of the soil. In general, beneficial effects of elevated soil pH on Cd uptake by plants can be expected. However, there may be exceptions. Liming soils that were treated with anaerobically digested sewage sludge to raise soil pH to 6.5 did not reduce Cd uptake by silage corn (Pepper et al., 1983). While lime addition to soil reduced the mean concentration of sludge-borne Cd in cabbage and lettuce, it did not affect Cd levels in potato tubers (Jackson and Alloway, 1991).

Soil pH was also negatively correlated with Cd content of rice (Bingham et al., 1980; Chino, 1981b; Takijima et al., 1973a). Total Cd uptake and shoot uptake increased with an increase in redox potential and a decrease in pH. They suggested that increased uptake by the rice plant under low soil pH conditions may also be due to the increased solubility of solid phases of Cd such as the carbonates, hydroxides, and phosphates, and increased concentrations of Fe^{2+}, Mn^{2+}, Zn^{2+}, Cu^{2+}, and H^+ which may then compete for exchange sites.

In addition to decreased solubility of Cd in soils associated with the formation of carbonates and phosphates under increasing pH, raising the pH of soil solution can lead to the formation of hydrolysis products that have a different affinity for sorption sites. The hydrolysis of Cd^{2+} can become important above pH 8. Elevated pH can also change the nature of the exchange sites by hydrolyzing or precipitating Al^{3+} ions that occupy the exchange sites, thus creating more exchange sites. Thus, reduction in plant uptake of Cd with increasing pH is due to reduced solubility of Cd, physiological changes in ion uptake, and transport in the plant. In the case of soils containing greater amounts of Ca^{2+}, the difference is considered to be due to ion competition between Ca^{2+} and Cd^{2+} ions at the root surface or interaction within the plant. In solution culture, however, the changes in the pH of the solution affect the membrane permeability of the metals and/or the competition between H^+ and Cd^{2+} ions.

In general, with the exception of Se and Mo, metals are more soluble in soils at low pH due to the dissolution of the carbonates, phosphates, and other solid phases. Low pH also weakens the sorption of metals to specific adsorption sites on mineral surfaces and lowers the CEC of OM. There are exceptions, however. At pH above 7, solubility or uptake of Cd can be enhanced due to facilitated complexation of Cd with humic or organic acids (Naidu and Harter, 1998).

7.2 Cation Exchange Capacity

Cation exchange capacity is a net expression of the amount and type of OM, and soil texture and minerals. The higher the OM content and the finer the texture, the higher the CEC. Thus, recommended limits on total cumulative added amounts of sludge-borne Cd vary according to the CEC and pH of the soil.

Results assessing the effect of soil CEC on the phytoavailability of Cd are variable. In an experiment in which an increase in CEC was induced by OM addition to soil, ryegrass Cd uptake was highly but negatively correlated with soil CEC (He and Singh, 1993). Haghiri (1974) found that Cd concentrations in the oat shoot decreased with increasing CEC of the soil. The soil CEC was elevated by adding muck up to 7% by weight, producing a CEC of about 30.5 meq/100 g at this rate. The increase in CEC by the addition of OM also resulted in the increased growth of the oat shoots. This probably was due to suppression of Cd phytoavailability, which, in turn, diminished the deleterious effects of Cd on the growth. In general, a negative correlation between leaf Cd values and CEC can be expected.

In a pot culture study, Cd uptake and yields of corn were inversely related to the soil CEC (range 5.3 to 15.9 meq/100 g) on $CdCl_2$-treated soils but were not affected by CEC on soils treated with sewage sludge (Hinesly et al., 1982). It was concluded that Cd source (soluble salt vs. sludge) exerted a far greater effect on Cd uptake than did the soil CEC.

Thus, the CEC can play an important role in the sorption of Cd by soils. The CEC is more important than OM in the sorption of Cd because the pH buffering capacity of soils increases with the CEC. Also, the potential for increased Cd uptake by plants in acidic soils is less in soils with high CEC than in those with low CEC.

7.3 Organic Matter

The presence of OM in soils and sediments affects the biogeochemical processes in these systems. In fact, humic substances are considered the major adsorbent for Cd in oxic sediments (Fu et al., 1992).

One way that OM adsorbs metal ions is by the ion exchange mechanism rendering them less mobile, therefore less bioavailable. The sorption ability of OM for Cd is predominantly through its CEC rather than chelating ability. In addition to CEC, soil humus has chelating ability, and certain metals have a tendency to combine with certain chelating groups. Organic complexes of Cu^{2+} and Pb^{2+} were considerably more stable than those for

Cd^{2+} (Stevenson, 1976). The stability constants with humic acids were in the order Cu > Pb > Cd (Bondietti and Sweeton, 1973).

Cadmium added to soil was retained by alkali-soluble humic substances so strongly that it was not taken up by wheat seedlings (Petruzelli et al., 1977). The soil used in this study was a Histosol with 24% OM content, pH of 5.4, and CEC of 318 meq/100 g. Soil OM appears to be more important than hydrous oxides of Fe and Mn in limiting the uptake of Cd by soybean plants (White and Chaney, 1980). Comparing two soils, they obtained higher binding with an OM-rich soil (3.8% OM; CEC of 16 meq/100 g) than with a soil having lower OM content (1.2% OM; CEC of 5.4 meq/100 g). In general, when Cd is added with sewage sludge, it is taken up less by plants than when added as inorganic Cd alone. Somers (1978) found that both fresh and partially decomposed residues of plant material from an old field, from a deciduous hardwood forest, and from under pine trees exhibited high affinity for Cd^{2+}. This explains the typical tendency for metals to accumulate in the humus layer in forest ecosystems.

Because of the importance of organic acids from OM decomposition and root exudates on the solubility of micronutrients and metals in the rhizosphere (Mench and Martin, 1991), it was recently demonstrated that low-molecular-weight organic acids (e.g., oxalic, citric, etc.) have some ability to desorb Cd from soils, with malate, fumarate, and succinate being more effective than the others (Naidu and Harter, 1998).

7.4 Redox Potential

In the Jinzu region of Japan, the cradle of *itai-itai* disease, it has been observed that the greater the number of days that rice paddies were drained prior to harvest, the greater the uptake and accumulation of Cd in rice grain (Takijima et al., 1973a,b). Marked reduction in the bioavailability of Cd to the rice plants also occurred upon flooding the soil (Bingham et al., 1976b). Other results (Chino, 1981b; Chino and Baba, 1981) also indicate greatly reduced solubility of Cd under flooded regimes. This behavior of Cd is due to the formation of CdS solid phase at low redox potential. Under low Eh–high pH conditions, the adsorption of Cd by amorphous oxyhydroxides of Fe and Mn may also be important.

Thus, it is common to observe that flood-managed rice has extreme tolerance to excess soil Cd compared with nonflooded rice. For example, increasing (potted) soil Cd to a maximum rate of 640 ppm Cd resulted in a yield decrement of only 30% for flooded rice; the nonflooded rice did not survive treatments with ≥ 160 ppm Cd (Bingham et al., 1976b). Chino (1981b) indicated that the activity of the rice roots may be involved in Cd absorption and translocation from flooded soil. Cadmium placed in the subsurface layer was absorbed by the roots in that layer but was not transported to the shoots, whereas Cd placed in the surface layer was transported efficiently. This implies that roots in the reductive subsurface layer may be too inactive to transport Cd.

7.5 Chemical Species of Cadmium

In soil, the bioavailability of Cd to plants can be affected by the solid–solution equilibrium of Cd^{2+} and the extent of complexation with organic and inorganic ligands in soil solution. It has been suggested that, typically, strongly complexed metals are less toxic to organisms than weakly complexed forms, which in turn, are less toxic than the free ions (Allen et al., 1980; Cabrera et al., 1988; Petersen, 1982). Indeed, using nutrient solution, barley plants preferred Cd^{2+} over $CdCl^{+}$ for uptake, while Cd complexed by humic acid was not adsorbed (Cabrera et al., 1988). In the absence of humic acid, Cl^{-} suppressed Cd uptake. In contrast, a recent survey of Cd contents in potato tubers in South Australia indicates that water-extractable Cl^{-} concentrations in soils were found to account for more of the variations in Cd concentrations in tubers than any other soil factor (McLaughlin et al., 1994). Elevated tuber Cd contents were associated with high Cl^{-} concentrations in soils (up to 1500 mg kg^{-1}), caused by the salinity of irrigation water used. Similarly, elevated Cd concentrations in sunflower kernels were correlated with high Cl^{-} concentrations in soil (Li et al., 1994). The promotive effect of the presence of Cl^{-} in soil on Cd uptake by potato tuber is indicated on Figure 8.6. Likewise, the bioaccumulation of Cd in durum wheat is enhanced by the presence of Cl^{-} in soil in northeastern North Dakota (Norvell et al., 2001).

Possible mechanisms offered to explain the synergistic effect of Cl^{-} on Cd uptake include (1) ion exchange of Na^{+} or Ca^{2+} for Cd^{2+} displacing Cd^{2+} from the sorption site, (2) chloro-complexation increasing the diffusion of Cd through soil to plant roots, and (3) increasing Cd concentrations in soil solutions and direct uptake of $CdCl_{2}^{2-n}$ species by plants (Smolders and McLaughlin, 1996). Using Swiss chard grown on resin-buffered solution culture (to stabilize solution Cd^{2+} activities), $CdCl_{2}^{2-n}$ species in solution are also available for plant uptake, in addition to Cd^{2+} ions (Smolders and McLaughlin, 1996). The predicted speciation of Cd in soil solution in potato-growing areas in South Australia is as follows: Cd^{2+} (56%), $CdSO_{4}^{0}$ (23%), and $CdCl_{2}^{2-n}$ (21%) in low OM soils. Presence of DOC lowers the distribution of Cd to inorganic species to 12% due to Cd-DOC complexations (McLaughlin et al., 1997a,b). The soil solution activities of the various Cd inorganic species are shown on Table 8.8.

Because sulfate is commonly found in high concentrations in soil solutions and irrigation waters, it is intuitive to wonder if such inorganic ligand would have a similar beneficial effect on the phytoavailability of Cd as does Cl^{-}. Several studies indicate however, that SO_{4}^{2-} does not have the same effect as does Cl^{-} in enhancing Cd uptake by crops (McLaughlin et al., 1998a,b).

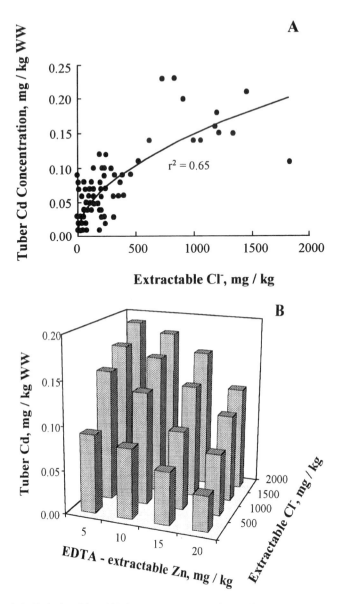

FIGURE 8.6. Relationships (A) between potato tuber Cd concentration and water-extractable Cl, and (B) in surface soils (0 to 15 cm), among tuber Cd concentration, water-extractable Cl, and EDTA-extractable Zn in commercial potato-growing areas in South Australia. (Extracted from McLaughlin et al., 1994a,b,c.)

In traditional environmental toxicology, the general hypothesis is that the toxicity or bioavailability of metal ions is directly related to the activity in solution of the uncomplexed aquo species. Therefore, the uptake of chloro-

TABLE 8.8. Activities (nM) of inorganic cadmium species in soil solutions from soils used to grow commercial potatoes in South Australia ($n = 50$).

Cd species	Range	Mean	Median
Cd^{2+}	0.4–24.6	7.3	6.4
$CdCl^+$	0.8–96.9	17.9	10.0
$CdCl_2$	0–54.3	4.3	0.6
$CdSO_4$	0.3–45.3	5.8	3.3

Source: McLaughlin et al. (1997a,b).

Cd complexes by plants run counter to this hypothesis. These investigators pointed out, however, that in spite of a large number of solution culture studies pointing to the uptake of uncomplexed ions, exceptions to this rule have been found (Smolders and McLaughlin, 1996a,b).

7.6 Plant Species, Cultivars, and Parts

Plant species exert marked differences with regard to Cd uptake, accumulation, and tolerance by plants. Sensitive crops grown in the greenhouse, such as spinach, curlycress, romaine lettuce, soybean, and field bean, have their yields reduced 25% by soil Cd additions of as low as 5 to 15 ppm (Bingham et al., 1975). More tolerant crops, such as tomato, zucchini squash, and cabbage, required 10-fold additions of Cd to produce similar yield reductions. John (1973) found that among edible parts, the highest Cd levels were found in lettuce and spinach leaves, followed by *Brassica* tops, radish and carrot tubers, pea seeds, and oat grains. Similarly, Cd concentrations for various crops decreased in the order: lettuce > radish tops > celery stalks > celery leaves > green pepper > radish roots (Haghiri, 1973). Using solution culture, Cd concentrations in the shoot decreased in the order: lettuce > sorrel, carrot, tomato > rape, kale, radish > mustard, corn > cucumber, sunflower, peas, bean > wheat and oats (Pettersson, 1977). Among forage species grown in potted soil, the sensitivity increased in the order: bermuda grass < fescue < clover < alfalfa < sudan grass (Bingham et al., 1976a). Among five plant families (Cruciferae, Cucurbitaceae, Gramineae, Leguminosae, and Solanaceae) examined, based on the critical levels of Cd in the tops, the Cruciferae and Leguminosae species were the most and least tolerant to Cd, respectively (Kuboi et al., 1987).

Proper selection of cultivar, in addition to species selection, is very important in managing Cd levels in crops. For example, potato tuber Cd concentrations can be reduced by up to 50% by proper choice of cultivar (McLaughlin et al., 1994a,b,c). Additionally, Cd accumulation in wheat grain has been observed to vary by a factor of up to 2.5 among cultivars (Wenzel et al., 1996). Furthermore, intracultivar variation in the uptake of Cd by wheat could vary by as much as a factor of 10 (Wenzel, 2000).

TABLE 8.9. Variation in mean cadmium concentrations (mg kg^{-1}) in seeds of peanut, navy bean, and lima bean varieties field-grown in northeastern Australia.

Cultivar	Seed or kernel Cd
Peanut	
RMP 91	0.667
A140 L31	0.471
Florunner	0.460
New Mexico (Valencia C)	0.446
NC7	0.441
Southern runner	0.404
B57 P5 1	0.351
55-437	0.334
Q24168	0.333
Streeton	0.326
A166 L17	0.307
B57 P4 1	0.289
Tifrust	0.284
Peanut l.s.d. ($p = 0.05$)	0.086
Navy bean	
Kerman	0.019
Rainbird	0.012
Spearfelt	0.012
Actolac	0.007
Sirrius	0.005
Lima bean	
Bridgeton	0.006
Phaseolus spp. l.s.d. ($p = 0.05$)	n.s.

Source: Bell et al. (1997).

In studying grain legumes grown on acidic, light-textured soils in the coastal areas of northeastern Australia, the relative risk of exceeding the allowable Cd concentrations in foodstuff (~0.05 mg Cd kg^{-1}) was in the order: peanut > soybean > navy bean (Bell et al., 1997). Cadmium concentrations in plant tops always exceeded that in seeds or kernels, with the testa in peanut kernels showing Cd levels that were 50 times greater than that in embryonix axis and cotyledons (Table 8.9). Significant variation in Cd content (at least 2 times) was recorded among peanut varieties, with lesser variation evident among the commercial navy bean varieties.

Because of genetic variation among cultivars within a plant species, differences in Cd uptake and accumulation in various plant parts have been observed for several crops, including soybean, wheat, barley, lettuce, rice, and corn. In addition to the influence of genotype on Cd uptake, translocation and accumulation among different plant parts vary (Table 8.10). In general, Cd concentrations are lower in seed, tuber, and fruit tissues relative to the concentrations in other parts, such as roots and other leafy tissues as

TABLE 8.10. Accumulation of cadmium (mg kg^{-1}) and distribution between vegetative and reproductive plant parts in two varieties of greenhouse-grown navy beans and peanuts.

Cultivar	Plant tops at: Flowering	Maturity	Pods	Mature kernels or seeds
		Peanut[a]		
Streeton	0.349 (0.026)	0.223 (0.082)	nd[c]	0.061 (0.011)
NC7	0.217 (0.059)	0.180 (0.019)	nd	0.057 (0.014)
		Navy bean[b]		
Kerman	0.072 (0.011)	0.058 (0.027)	0.046 (0.011)	0.032 (0.008)
Spearfelt	0.083 (0.009)	0.111 (0.062)	0.044 (0.014)	0.036 (0.019)

[a]Values are means ± SE; $n = 3$.
[b]Values are means ± SE; $n = 2$.
[c]nd, Not determined.
Source: Extracted from Bell et al. (1997).

those in lettuce, spinach, and tobacco. The gradient of Cd concentration within a plant can be generalized as follows: roots ≫ stalk base ≫ shoots (stems/sheaths > leaves) > grain (or fruit).

7.7 Other Factors

Increases in soil temperature are also known to enhance Cd uptake by plants. Soil type and the mineralogy of the clay fraction may also influence the sorption of Cd by soil. Higher sorption capacity is favored by fine-textured soils, or soils high in OM or clay content. When the clay fraction is dominated by 2:1 layer silicates, such as montmorillonite, the adsorptive capacity may be expected to be higher (Farrah and Pickering, 1977) than when the dominant clay mineral is a 1:1 type, such as kaolinite. The exception however, is when soils are high in OM and oxides of Fe and Mn; then, higher affinity for Cd can be expected even in the presence of 1:1 layer silicates. Sorption can be expected to be greater by soils dominated by Mn oxide than by Fe oxide (Backes et al., 1995).

8 Sources of Cadmium in the Environment

The natural level of Cd in soil is usually <1 ppm, including most arable soils. It can occur naturally in higher concentrations when associated with Zn ores, or in areas near Cd-bearing deposits. Because of its many uses in industry, significant amounts of Cd may be released to the environment.

In a UNESCO Man and the Biosphere Programme in Japan, Cd concentrations of domestic wastewater from lavatory, kitchen, laundry, and bath were 38, 0.4, 0.6, and 0.5 ppb Cd, respectively (Chino and Mori, 1982). The corresponding total amount of Cd discharged with such wastewaters was 128 µg day^{-1} per person: 18 µg from the kitchen, 42 µg from the laundry, 18 µg from the bath, 46 µg from the lavatory, and 4 µg from other sources.

Atmospheric fallout and the use of phosphate fertilizers are the major sources of Cd to agricultural soils. Although significant improvement in controlling air emission of metals has occurred during the last two decades in developed countries, further significant increases in atmospheric emission of Cd from anthropogenic activities are predicted beyond the year 2000.

8.1 Phosphate Fertilizers

Cadmium occurs in ores used in the production of phosphate fertilizers. The concentration of Cd in the ores can be as great as 340 ppm (McLaughlin et al., 1996). Ores from the western United States contain considerably higher concentrations of Cd than ores from the southeastern states. Cadmium concentrations in concentrated superphosphate fertilizers (0-45-0) originating from western sources range from 50 to 200 ppm, whereas concentrations from southeastern sources range from 10 to 20 ppm. Only about 20% of all ores used for production of P fertilizers are derived from western sources. In diammonium phosphates (18-45-0), mostly of western United States origin, the Cd content ranged from 7.4 to 156 ppm, whereas concentrated superphosphates contained from 86 to 114 ppm Cd.

The range of Cd concentrations of several Australian commercial fertilizers was 18 to 91 ppm, with the superphosphates containing 38 to 48 ppm (Williams and David, 1973), whereas Swedish fertilizers were shown to contain <0.1 to 30 ppm Cd (Stenström and Vahter, 1974). Almost pure phosphate ores exist in the Kola Peninsula of the former Soviet Union (0.10 to 0.40 ppm), whereas ores in Senegal are known to contain 70 to 90 ppm Cd, and those in Togo, about 50 ppm.

With the widespread use of phosphate fertilizers in agriculture, this source can potentially contribute to the contamination of arable soils with Cd. Estimates indicate that this source can account for a substantial input of Cd to agricultural soils. A specific example is that of the soils in the Rhine Basin (see Chapter 4), where Cd in soils would level off if application of Cd-enriched fertilizers were stopped. Also, it was estimated that with current fertilizer practice, extractable-Cd concentrations in surface soils in South Australia would approximately double within about 20 years (Merry and Tiller, 1991).

Although long-term soil P fertility studies indicate that addition of phosphate fertilizers to soil can increase the Cd burden of the soil, such increases have not substantially enhanced Cd uptake by plants. For example, in nine long-term (>50 years) soil fertility studies in the United States, where Cd from the phosphate fertilizers was added at the rate of 0.3 to 1.2 g ha^{-1}yr^{-1}, no significant increases in Cd uptake by plants were observed (Mortvedt, 1987). It was suggested that plant uptake of Cd from fertilizers containing <10 mg Cd kg^{-1} would be negligible. Likewise, long-term fertility experiments (>70 years) in Norway show insignificant buildup of Cd in soils producing inconsequential plant uptake (Jeng and Singh, 1995). In these studies, the annual Cd input (atmospheric and fertilizer) ranged from 1.20 to 2.56 g Cd ha^{-1} yr^{-1}, while the Cd removal (crop harvest and leaching) varied from 1.16 to 1.68 g Cd ha^{-1} yr^{-1} among the treatments presented (Table 8.11), resulting in hardly noticeable changes in soil Cd. In contrast, long-term (1861 to 1992) studies at Rothamsted, England, indicate that superphosphate addition to soil can increase the Cd content of herbage grown on acid soils (pH 4.8 to 5.7) (Nicholson and Jones, 1994). The increased uptake by the herbage on acid soils was easily negated however, by the addition of lime.

Because of the large amounts of Cd that originate from these fertilizers, concern arises about food chain contamination from long-term accumulation of this element in soil. There is a direct correlation between phosphate

TABLE 8.11. Cadmium fluxes in long-term fertilizer experiments in Norway (1922–1992).

| | Control plot | Treatment plots[*] | |
		E2	E6
Cd input, g ha^{-1} yr^{-1}			
P fertilizers	0.00	1.36	0.00
Farm yard manure	0.00	0.00	0.72
Atmospheric deposition	1.20	1.20	1.20
Total (a)	1.20	2.56	1.92
Cd removal, g ha^{-1} yr^{-1}			
Crops	0.76	1.28	1.13
Leaching	0.40	0.40	0.40
Total (b)	1.16	1.68	1.53
Accumulated (a − b)	0.04	0.88	0.39
[a]Total Cd, g ha^{-1}	820	813	926
[b]Available Cd, g ha^{-1}	30	40	22
Change in total Cd, % yr^{-1}	0.004	0.11	0.04
Change in avail. Cd, % yr^{-1}	0.1	2.2	1.8
Accumulated, % of added as fertilizer/manure	0	65	54
Accumulated, % of added	3	34	20

[*]E2 (15 kg P ha^{-1} yr^{-1} from 1922–1982, then 44 kg P ha^{-1} yr^{-1} from 1983–1992).
E6 (1.5 t DW FYM ha^{-1} yr^{-1} from 1922–1982, then 3 t DW FYM ha^{-1} yr^{-1} from 1983–1992).
Source: Extracted from Jeng and Singh (1995).

and Cd accumulation in soil and fertilizer. The concentrations of Cd in several vegetable crops were increased by heavy applications of superphosphates (Schroeder and Balassa, 1963). Similarly, applications of superphosphates increased the Cd content of soils and of cereal and fodder crops (Williams and David, 1973). In addition, the uptake by horticultural crops under greenhouse conditions was observed to be a linear function of the Cd content of the fertilizer (Reuss et al., 1978). Slight increases in uptake of Cd by forage from applications of commercial fertilizers were obtained, with the highest uptake resulting from western United States diammonium phosphate (18 μg/pot) and 10-15-0 fluid fertilizer (16 μg/pot) (Mortvedt and Giordano, 1977). The nontreated soil had only 8 μg Cd/pot uptake. Cadmium concentrations in (field) wheat grain and straw were significantly increased with application of high-Cd diammonium phosphate to an acidic (pH 5.1) soil (Mortvedt et al., 1981). If a customary annual application rate of 120 kg P ha^{-1} for U.S. arable soils is assumed, the annual input of Cd from concentrated superphosphates (0-45-0) containing 175 μg Cd g^{-1} of fertilizer would be 0.10 kg Cd ha^{-1}.

8.2 Sewage Sludge

Land application of municipal sewage sludge is becoming more popular because of constraints placed on alternative disposal methods, such as restrictions on ocean disposal, and air pollution problems and energy requirements associated with incineration. In addition, the material can be used as a source of N and P, and as a soil structure builder. Thus, when the constraints on the alternative methods become more stringent, a greater proportion of sludge would have to be applied on land. Increasing applications of sewage sludge on arable lands increase the probability of Cd in foodstuff to exceed a given allowable limit, as reflected by the increases in Cd levels in plant tissues concomitant with increases in Cd input (Table 8.12).

Utilization of sewage sludge on arable lands is often associated with a potential risk of contaminating the food chain, with Cd posing the greatest concern (see Chapter 6). Because of variable Cd contents in sewage sludge, the Cd concern appears to be more local than national, with big cities generating the greatest amount and also, most likely, the sludge with the highest Cd concentrations (Appendix Table A.9). Contamination of crops by sludge-borne Cd is well documented.

When sludge is used as a source of plant nutrients, the amount applied per year can be based on (1) an annual limit on Cd, (2) the amount of N or P required by the crop grown, or (3) a combination of both criteria. The third option is typically used for privately owned farm land on which food or fuel crops are grown and the farmer uses a conventional soil testing program to monitor the soil after sludge application. The rationale for limiting sludge application on the basis of crop needs for N is that nitrate (NO$_3^-$) leaching

TABLE 8.12. Cadmium concentrations of vegetative and reproductive tissues from crops grown in sewage sludge–amended soils in relation to species, application rate, and soil pH.

Cd added from sludge (kg ha^{-1})	Soil pH	Cadmium concentration (mg kg^{-1} DW)					
		Peas		Wheat		Corn	
		Leaves	Seeds	Leaves	Grain	Leaves	Grain
0	4.4	0.94	0.20	0.19	0.06	0.11	0.03
2.2	4.6	4.40	0.64	0.82	0.25	1.44	0.06
4.5	5.0	3.80	1.11	1.10	0.32	1.83	0.08
9.0	5.2	6.05	1.45	0.89	0.58	3.37	0.09
0	7.8	0.43	0.14	0.14	0.04	0.07	0.01
2.2	7.6	0.43	0.17	0.18	0.12	0.14	0.02
4.5	7.4	0.42	0.18	0.25	0.23	0.41	0.03
9.0	7.3	0.48	0.30	0.33	0.34	0.92	0.03

Source: Extracted from Page et al. (1987).

and its subsequent contamination of the groundwater would not be greater than that caused by the use of commercial fertilizers. The U.S. EPA and the European Union (EU) in response to concern over Cd entering the food chain, established limits on both the annual and cumulative amounts of Cd that may be added to agricultural soils in the form of sewage sludge (see Chapter 4). The U.S. EPA (1992) annual limit on Cd addition to soil is 1.9 kg Cd ha^{-1} yr^{-1} and a cumulative loading limit of 39 kg ha^{-1}.

8.3 Atmospheric Deposition

Primary sources of air pollution emitting Cd rank as follows: smelter > incineration of plastics and Cd pigments > fossil fuel (including coking) > steel mills > metallurgical. Numerous studies have demonstrated significant Cd contamination of soils, plants, and animals in the vicinity of lead-, zinc-, and battery-smelting facilities. For example, in the United States, increased concentrations of Cd and other metals in the forest near the AMAX lead smelter in southeastern Missouri were observed (Wixson et al., 1977). Aerial deposition of Cd from smelting caused decomposing leaf litter to contain Cd from 10 ppm at the 4.4 km location to 200 ppm at the 0.4 km location along forest ridges. In Annaka City, Japan which has the country's largest Zn refinery, Cd and other metal contents in mulberry leaves, wheat, barley, and vegetables (especially in Chinese cabbage—40 ppm) were enriched in areas near the refinery (Kobayashi, 1971). The accumulation of metals in these plants was attributed to root uptake from the polluted soils rather than to direct deposition of the metals on the plant surfaces. Near the Avonmouth, England industrial complex, elm leaves collected close to the complex had 50 ppm Cd; this decreased to background level (0.25 ppm) at 10 to 15 km away (Little and Martin, 1972).

In the United States, in Deer Lodge Valley, Montana average Cd concentrations in contaminated (from smelting) vs. noncontaminated areas were: grasses, 1.72 and 0.07; alfalfa, 0.83 and 0.06; and barley grain, 0.65 and 0.08 ppm, respectively (Munshower, 1977). Average levels in animal tissues were: cattle liver, 0.34 and 0.06; cattle kidney, 1.67 and 0.22; swine liver, 0.24 and 0.14; and swine kidney, 0.99 and 0.39 ppm, from contaminated and noncontaminated areas, respectively. Much higher enrichments of surface soils and grasses in the vicinity of the Pb smelting complex in Kellogg, Idaho were also observed (Ragaini et al., 1977).

Other sources of atmospheric Cd originate from the combustion of coal, oil, paper, and urban organic trash. Near roads, motor oils and tread wear from vehicular tires are sources of Cd and other metals. Leafy vegetables grown in gardens in the Boston area had 0.8 to 9.1 ppm Cd DW, indicating the influence of automobile emissions on the soil contamination in urban areas (Preer and Rosen, 1977).

8.4 Mining and Smelting

Although usually localized, contamination of rivers and streams by Cd in wastewater from mining operations can affect broad areas and can afflict hundreds of its inhabitants. Such is the case in the Jinzu River basin in Japan, where the *itai-itai* disease occurred (Asami, 1988). The residents had been ingesting Cd over a 30-yr period in both their drinking water and in their rice, in which Cd had accumulated through the river water used for irrigation. Similarly, in areas in Fukui Prefecture in Japan, unpolished rice harvested along the Kuzuryu River had Cd concentrations ranging from 0.02 to 1.82 ppm, with some rice containing >1 ppm found near a Zn mining station (Takijima and Katsumi, 1973). Cadmium in soils ranged from 0.20 to 10.4 ppm.

In Ontario, Canada soils near the Sudbury mining and smelting complex are also heavily contaminated with metals including Cd from years of operation (Dudka et al., 1995b). In this case, the soils were also acidified (pH 2 to 4.5) by deposition of SO_2 from the smelters, a condition that promotes enhanced mobility and bioavailability of metals. Similar results on the effect of mining/smelting on metal loads in soils and their consequential effects on the quality of food crops were observed in the Upper Silesia region of Poland (Dudka et al., 1995a; Gzyl, 1990).

References

Adriano, D.C., A.L. Page, A.A. Elseewi, and A.C. Chang. 1982. *J Environ Qual* 11:197–203.

Ainsworth, C.C., and D. Rai. 1987. *Chemical Characterization of Fossil Fuel Combustion Wastes*. EPRI Rep EA-5321. Electric Power Res Inst, Palo Alto, CA.

Allen, H.E., R.H. Hall, and T.D. Brisbin. 1980. *Environ Sci Technol* 14:441–443.

Andersson, A. 1977. *Swedish J Agric Res* 7:7–20.

Andersson, A., and K.O. Nilsson. 1972. *Ambio* 1:176–179.

Andersson, A., and K.O. Nilsson. 1974. *Ambio* 3:198–200.

Archer, F.C. 1980. *Minist Agric Fish Food* (Great Britain) 326:184–190.

Arduini, I., D.L. Godbold, and A. Onnis. 1994. *Physiol Plant* 92:675–680.

Asami, T. 1988. In W. Salomons and U. Forstner, eds. *Chemistry and Biology of Solid Waste.* Springer-Verlag, Berlin.

Asami, T., M. Kubota, and K. Orikasa. 1995. *Water Air Soil Pollut* 83/84:187–194.

[ATSDR] Agency for Toxic Substance and Disease Registry. 1992. *Toxicological Profile for Cadmium.* U.S. Dept Health and Human Services, Atlanta, GA.

Backes, C.A., R.G. McLaren, A.W. Rate, and R.S. Swift. 1995. *Soil Sci Soc Am J* 59:778–785.

Balsberg, A.M. 1982. *Oikos* 38:91–98.

Baveye, P., M.B. McBride, D. Bouldin, T.D. Hinesly, et al. 1999. *Sci Total Environ* 227:13–28.

Bell, M.J., M.J. McLaughlin, G.C. Wright, and A. Cruickshank. 1997. *Aust J Agric Res* 48:1151–1160.

Berrow, M.L, and R.L. Mitchell. 1980. *Trans Royal Soc Edinburgh Earth Sci* 71:103–121.

Berrow, M.L., and G.A. Reaves. 1984. In *Proc Intl Conf Environmental Contamination.* CEP Consultants, Edinburgh.

Beyer, W.N., G.W. Miller, and E.J. Cromartie. 1984. *J Environ Qual* 13:247–251.

Beyer, W.N., O.H. Pattee, L. Sileo, O.J. Hoffman, and B.M. Mulhern. 1995. *Environ Pollut* 38:63–86.

Bidwell, A.M., and R.H. Dowdy. 1987. *J Environ Qual* 16:438–442.

Bingham, F.T., and A.L. Page. 1975. In *Proc Intl Conf Heavy Metals in the Environment.* Univ. of Toronto, Ontario, Canada.

Bingham, F.T., A.L. Page, R.J. Mahler, and T.J. Ganje. 1975. *J Environ Qual* 4:207–211.

Bingham, F.T., A.L. Page, R.J. Mahler, and T.J. Ganje. 1976a. *J Environ Qual* 5:57–60.

Bingham, F.T., A.L. Page, R.J. Mahler, and T.J. Ganje. 1976b. *Soil Sci Soc Am J* 40:715–719.

Bingham, F.T., A.L. Page, G.A. Mitchell, and J.E. Strong. 1979. *J Environ Qual* 8:202–207.

Bingham, F.T., A.L. Page, and J.E. Strong. 1980. *Soil Sci* 130:32–38.

Bittell, J.E., and R.J. Miller. 1974. *J Environ Qual* 3:250–253.

Bondietti, E.A., and F.H. Sweeton. 1973. *Agron Abst* 89–90.

Boswell, F.C. 1975. *J Environ Qual* 4:267–272.

Bowen, H.J.M. 1979. *Environmental Chemistry of the Elements.* Academic Pr, New York.

Bradley, R.I. 1980. *Geoderma* 24:17–23.

Breckle, S.W. 1989. In Y. Waisel, ed. *The Root System: The Hidden Half.* Marcel Dekker, New York.

Breckle, S.W. 1991. In Y. Waisel, U. Kafkafi, and A. Eshel, eds. *Plant Roots: The Hidden Half.* Marcel Dekker, New York.

Bundesgeshundheitsamt. 1986. *Bundesgesundhbl* 29:22–23.

Cabrera, D., S.D. Young, and D.L. Rowell. 1988. *Plant Soil* 105:195–204.

Cappon, C.J. 1987. *Bull Environ Contam Toxicol* 38:695–699.

Carlson, R.W., and F.A. Bazzaz. 1977. *Environ Pollut* 12:243–253.

Carlson, R.W., and G.L. Rolfe. 1979. *J Environ Qual* 8:348–152.

Chaney, R.L. 1973. *Recycling Municipal Sludge and Effluents on Land.* National Assoc. State Univ. Land-Grant Colleges, Washington, DC.

Chaney, R.L. 1974. *Factors Involved in Land Application of Agricultural and Municipal Wastes.* Rep 67-120. U.S. Dept Agriculture, Ag Res Service, Beltsville, MD.

Chang, A.C., J.E. Warneke, A.L. Page, and L.J. Lund. 1984a. *J Environ Qual* 13: 87–91.

Chang, A.C., A.L. Page, J.E. Warneke, and E. Grgurevic. 1984b. *J Environ Qual* 13:33–38.

Chapman, G.A., S. Ota, and F. Recht. 1980. *Status Report.* U.S. Environmental Protection Agency, Corvallis, OR.

Chaudri, A.M.M., F.J. Zhao, S.P. McGrath, and A.R. Crosland. 1995. *J Environ Qual* 24:850–855.

Childs, E.A., and J.N. Gaffke. 1974. *J Food Sci* 39:453–454.

Chino, M. 1981a. In K. Kitagishi and I. Yamane, eds. *Heavy Metal Pollution in Soils of Japan.* Japan Sci Soc Pr, Tokyo.

Chino, M. 1981b. In K. Kitagishi and I. Yamane, eds. *Heavy Metal Pollution in Soils of Japan.* Japan Sci Soc Pr, Tokyo.

Chino, M., and A. Baba. 1981. *J Plant Nutr* 3:203–214.

Chino, M., and T. Mori. 1982. In *Research Related to the UNESCO's Man and Biosphere Programme in Japan.* Univ of Tokyo, Tokyo.

Chlopecka, A., J.R. Bacon, M.J. Wilson, and J. Kay. 1996. *J Environ Qual* 25:69–79.

Choudhary, M., L.D. Bailey, and C.A. Grant. 1994. *Can J Plant Sci* 74:549–552.

Christensen, T.H. 1984a. *Water Air Soil Pollut* 21:105–114.

Christensen, T.H. 1984b. *Water Air Soil Pollut* 21:115–125.

Copius Peereboom-Stegeman, J.H.J., and J.W. Copius Peereboom. 1989. *Ecotoxicol Environ Saf* 18:93–108.

Cowan, C.E., J.M. Zachara, and C.T. Resch. 1991. *Environ Sci Technol* 25:437–446.

Craun, G.F., and L.J. McCabe. 1975. *J Am Water Works Assoc* 67:593–599.

Cunningham, L.M. 1977. *Trace Subs Environ Health* 11:135–145.

Dabeka, R.W., H.B.S. Conacher, J. Salminen, et al. 1992. *J AOAC Int* 75(6):949–953.

Davis, R.D., P.H.T. Beckett, and E. Wollan. 1978. *Plant Soil* 49:395–408.

dell'Omo, M., G. Muzi, and R. Piccinini, et al. 1999. *Sci Total Environ* 226:57–64.

DeVos, C.H.R., H. Schat, M. deWaal, R. Youijs, and W.H.O. Ernst. 1991. *Physiol Plant* 82:523–528.

Dijkshoorn, W., J.E.M. Lampe, and A.R. Kowsoleea. 1974. *Netherlands J Agric Sci* 22:66–71.

Dowdy, R.H., and W.E. Larson. 1975. *J Environ Qual* 4:278–282.

Dowdy, R.H., J.J. Latterell, T.D. Hinesly, R.B. Grossman, and D.L. Sullivan. 1991. *J Environ Qual* 20:119–123.

Dudka, S., M. Piotrowska, A. Chlopecka, and T. Witek. 1995a. *J Geochem Explor* 52:237–250.

Dudka, S., R. Ponce-Hernandez, and T.C. Hutchinson. 1995b. *Sci Total Environ* 162:161–171.

Eisler, R. 1985. *U.S. Fish and Wildlife Serv Biol Rep* 85. U.S. Fish and Wildlife Serv, Washington, DC.

Elliott, H.A., and C.M. Denneny. 1982. *J Environ Qual* 11:658–662.

Emmerich, W.E., L.J. Lund, A.L. Page, and A.C. Chang. 1982. *J Environ Qual* 11:182–186.

Ernst, W.H.O., J.A.C. Verkleij, and H. Schat. 1992. *Acta Bot Neer* 41:229–248.

Farrah, H., and W.F. Pickering. 1977. *Aust J Chem* 30:1417–1422.

Francis, C.W., and S.G. Rush. 1973. *Trace Subs Environ Health* 7:75–81.

Frank, R., K. Ishida, and P. Suda. 1976. *Can J Soil Sci* 56:181–196.

Friberg, L., M. Piscator, and G. Nordberg. 1971. *Cadmium in the Environment*. CRC Pr, Cleveland, OH.

Fu, G., H.E. Allen, and Y. Cao. 1992. *Environ Toxicol Chem* 11:1363–1372.

Garcia-Miragaya, J., and A.L. Page. 1976. *Soil Sci Soc Am J* 40:658–663.

Gerritse, R.G., W. Van Driel, K.W. Smilde, and B. Van Luit. 1983. *Plant Soil* 75:393–404.

Giesy, J. 1980. In J.O. Nriagu, ed. *Cadmium in the Environment, part 1: Ecological Cycling*. Wiley, New York.

Godbeer, W.C., and D.J. Swaine. 1979. *Trace Subs Environ Health* 13:254–260.

Graney, R.L., D.S. Cherry, and J. Cairns. 1983. *Hydrobiologia* 102:81–88.

Gunson, D., D. Kowalczck, R. Shoop, and C. Ramberg. 1982. *J Am Vet Med Assoc* 180:295–299.

Gzyl, J. 1990. *Sci Tot Environ* 96:199–209.

Hagemeyer, J., H. Kahle, S.W. Breckle, and Y. Waisel. 1986. *Water Air Soil Pollut* 29:347–359.

Haghiri, F. 1973. *J Environ Qual* 2:93–96.

Haghiri, F. 1974. *J Environ Qual* 3:180–183.

Hamon, R.E., J. Wundke, M. McLaughlin, and R. Naidu. 1997. *Aust J Soil Res* 35:1267–1277.

Haq, A.U., T.E. Bates, and Y.K. Soon. 1980. *Soil Sci Soc Am J* 44:772–777.

Harrison, S.E., and J.F. Klaverkamp. 1989. *Environ Toxicol Chem* 8:87–97.

Hassett, J.J., J.E. Miller, and D.E. Koeppe. 1976. *Environ Pollut* 11:297–302.

He, Q.B., and B.R. Singh. 1993. *J Soil Sci* 44:641–650.

Hem, J.D. 1972. *Water Resources Res* 8:661–679.

Hem, J.D. 1985. *USGS Water Supply Rep 2254*. U.S. Geological Survey (USGS), Alexandria, VA.

Hickey, M.G., and J.A. Kittrick. 1984. *J Environ Qual* 13:372–376.

Hinesly, T.D., R.L. Jones, and E.L. Ziegler. 1972. *Compost Sci* 13:26–30.

Hinesly, T.D., K.E. Redborg, E.L. Ziegler, and J.D. Alexander. 1982. *Soil Sci Soc Am J* 46:490–497.

Hirsch, D., and A. Banin. 1990. *J Environ Qual* 19:366–372.

Holm, P.E., J. Futtrup, T.H. Christensen, and J.C. Tjell. 1993. In J.P. Vernet, ed. *Heavy Metals in the Environment*, vol. 1. Elsevier, Amsterdam.

Holm, P.E., T.H. Christensen, J.C. Tjell, and S.P. McGrath. 1995. *J Environ Qual* 24:183–190.

Holm, P.E., B.B.H. Andersen, and T.H. Christensen. 1996. *Soil Sci Soc Am J* 60:775–780.

Holmgren, G.G.S., M.W. Meyer, R.L. Chaney, and R.B. Daniels. 1993. *J Environ Qual* 22:335–348.

Honma, Y., and H. Hirata. 1978. *Soil Sci Plant Nutr* 24295–297.

Huffman, E.W.D., and J.F. Hodgson. 1973. *J Environ Qual* 2:289–291.

Hunter, B.A., M.S. Johnson, and D.J. Thompson. 1987a. *J Appl Ecol* 24:573–586.

Hunter, B.A., M.S. Johnson, and D.J. Thompson. 1987b. *J Appl Ecol* 24:587–599.

Hunter, B.A., M.S. Johnson, and D.J. Thompson. 1987c. *J Appl Ecol* 24:601–614.

Hutchinson, T.C. 1979. *Effects of Cadmium in the Canadian Environment*. NRCC 16743. National Research Council of Canada (NRCC), Ottawa.

Hutchinson, T.C., M. Czuba, and L. Cunningham. 1974. *Trace Subs Environ Health* 8:81–91.

Iimura, K. 1981. In K. Kitagishi and I. Yamane, eds. *Heavy Metal Pollution in Soils of Japan*. Japan Sci Soc Pr, Tokyo.

Iimura, K., and H. Ito. 1971. *Rev Bull Science Soil Manure of Chuba Branch, Japan* 34:1–15.

Iwai, I., T. Hara, and Y. Sonoda. 1975. *Soil Sci Plant Nutr* 21:37–46.

Jackson, A.P., and B.J. Alloway. 1991. *Water Air Soil Pollut* 57–58:873–881.

Jarup, L., M. Bergland, C.G. Elinder, G. Nordberg, and M. Vahter. 1998. *Scand J Work Environ Health* 24:1–51.

Jeng, A.S., and B.R. Singh. 1995. *Plant Soil* 175:67–74.

John, M.K. 1973. *Environ Pollut* 4:7–15.

John, M.K., C.J. Van Laerhoven, and H.H. Chuah. 1972. *Environ Sci Technol* 6:1005–1009.

Jones, R.L., T.D. Hinesly, and E.L. Ziegler. 1973. *J Environ Qual* 2:351–353.

Jurinak, J.J., and D.L. Suarez. 1990. In K.K. Tanji, ed. *Agricultural Salinity Assessment and Management*. Am Soc Civil Eng, New York.

Kelling, K.A., D.R. Keeney, L.M. Walsh, and J.A. Ryan. 1977. *J Environ Qual* 6:352–358.

Kelly, J.M., G.R. Parker, and W.W. McFee. 1979. *J Environ Qual* 8:361–364.

Khalid, R.A., R.P. Gambrell, and W.H. Patrick, Jr. 1978. In D.C. Adriano and I.L. Brisbin, Jr., eds. *Environmental Chemistry and Cycling Processes*. CONF-760429. NTIS, Springfield, VA.

Khalid, R.A., R.P. Gambrell, and W.H. Patrick, Jr. 1981. *J Environ Qual* 10:523–528.

Khan, D.H., and B. Frankland. 1983. *Plant Soil* 70:335–345.

King, L.D., and L.M. Hajjar. 1990. *J Environ Qual* 19:738–748.

Kjellstrom, T. 1985. In L. Friberg et al., eds. *Cadmium and Health: A Toxicological and Epidemiological Appraisal*, vol 2. CRC Pr, Boca Raton, FL.

Klaverkamp, J.F., and D.A. Duncan. 1987. *Environ Toxicol Chem* 6:225–289.

Kobayashi, J. 1971. *Trace Subs Environ Health* 5:117–128.

Kobayashi, J. 1979. In F.W. Oehme, ed. *Toxicity of Heavy Metals in the Environment*, part 1. Marcel Dekker, New York.

Koeppe, D.E. 1977. *Sci Total Environ* 7:197–206.

Korcak, R.F. 1989. *J Environ Qual* 18:519–523.

Krauskopf, K.B. 1979. *Introduction to Geochemistry*, 2nd ed. McGraw-Hill, New York.

Krupka, K.M., D.I. Kaplan, G. Whelan, S.V. Martigod, and R.J. Serne. 1999. *Understanding Variation in Partition Coefficient, K_d, Values*. U.S. EPA 402-R-99-004A,B. U.S. Environmental Protection Agency, Washington, DC.

Kuboi, T., A. Noguchi, and J. Yazaki. 1987. *Plant Soil* 104:275–280.

Kuo, S., and B.L. McNeal. 1984. *Soil Sci Soc Am J* 48:1040–1044.

Lag, J., and I.H. Elsokkary. 1978. *Acta Agric Scan* 28:76–80.

Lagerwerff, J.V. 1971. *Soil Sci* 111:129–133.

Lagerwerff, J.V., and G.T. Biersdorf. 1971. *Trace Subs Environ Health* 5:515–522.

Lagriffoul, A., B. Macquot, M. Mench, and J. Vangronsveld. 1998. *Plant Soil* 200:241–250.

Lamoreaux, R.J., and W.R. Chaney. 1977. *J Environ Qual* 6:201–205.

Lawrence, S.G., and M.H. Holoka. 1987. *Can J Fish Aquat Sci* 44:163–172.

Lee, K.W., and D.R. Keeney. 1975. *Water Air Soil Pollut* 5:109–112.

Li, Y.-M., R.L. Chaney, and A.A. Schneiter. 1994. *Plant Soil* 167:275–280.

Little, P., and M.H. Martin. 1972. *Environ Pollut* 3:241–254.

Lund, L.J., E.E. Betty, A.L. Page, and R.A. Elliott. 1981. *J Environ Qual* 10:551–556.

MacLean, A.J. 1976. *Can J Soil Sci* 56:129–138.

Mahler, R.J., F.T. Bingham, and A.L. Page. 1978. *J Environ Qual* 7:274–281.

Mahler, R.J., F.T. Bingham, G. Sposito, and A.L. Page. 1980. *J Environ Qual* 9:359–363.

Mance, G. 1987. *Pollution Threat of Heavy Metals in Aquatic Environments.* Elsevier, London.

[MARC] Monitoring and Assessment Research Center. 1981. *Cadmium in the European Environment* MARC Rep 28. MARC, London.

MARC (Monitoring and Assess. Research Center). 1982. *Cadmium Exposure and Indicators of Kidney Function.* MARC Rep 29. MARC, London.

Marchiol, L., L. Leita, M. Martin, A. Peressotti, and G. Zerbi. 1996. *J Environ Qual* 25:562–566.

Marshall, J.S. 1978. *Bull Environ Contam Toxicol* 20:387–393.

Martin, M.H., and P.J. Coughtrey. 1975. *Chemosphere* 3:155–160.

Martin, M.H., and R.J. Bullock. 1994. In S.M. Ross, ed. *Toxic Metals in Soil-Plant System.* Wiley, New York.

Masironi, R., S.R. Koirtyohann, and J.O. Pierce. 1977. *Sci Total Environ* 7:27–43.

McBride, M.B. 1980. *Soil Sci Soc Am J* 44:26–28.

McBride, M.B., L.D. Tyler, and D.A. Hovde. 1981. *Soil Sci Soc Am J* 45:739–744.

McGrath, S.P., and P.W. Lane 1989. *Environ Pollut* 60:235–256.

McLaughlin, M.J., and K.G. Tiller. 1994. *Trans 15th World Congress Soil Science* 36:195–196.

McLaughlin, M.J., L.T. Palmer, et al. 1994a. *J Environ Qual* 23:1013–1018.

McLaughlin, M.J., K.G. Tiller, T.A. Beech, and M.K. Smart. 1994b. *J Environ Qual* 23:1013–1018.

McLaughlin, M.J., C.M.J. Williams, A. McKay, and K.G. Tiller. 1994c. *Aust J Agric Res* 45:1483–1495.

McLaughlin, M.J., N.A. Maier, K. Freeman, K.G. Tiller, C.M.J. Williams, and M.K. Smart. 1995. *Fertilizer Res* 40:63–70.

McLaughlin, M.J., K.G. Tiller, et al. 1996. *Aust J Soil Res* 34(1):1–54.

McLaughlin, M.J., N.A. Maier, G.E. Rayment, et al. 1997a. *J Environ Qual* 26:1644–1649.

McLaughlin, M.J., K.G. Tiller, and M.K. Smart. 1997b. *Aust J Soil Res* 35:183–198.

McLaughlin, M.J., R.M. Lambrechts, E. Smolders, and M.K. Smart. 1998a. *Plant Soil* 202:217–222.

McLaughlin, M.J., R.M. Lambrechts, E. Smolders, and M.K. Smart. 1998b. *Plant Soil* 202:211–216.

Mench, M., and E. Martin. 1991. *Plant Soil* 132:187–196.

Merry, R.H., and K.G. Tiller. 1991. *Water Air Soil Pollut* 57–58:171–180.

Milberg, R.P., D.L. Brower, and J.V. Lagerwerff. 1978. *Soil Sci Soc Am J* 42:892–894.

Miller, J.E., J.J. Hassett, and D.E. Koeppe. 1977. *J Environ Qual* 6:18–20.

Mitchell, C.D., and T.A. Fretz. 1977. *J Am Soc Hort Sci* 102:81–84.

Mitchell, D.G., and K.M. Aldous. 1974. *Environ Health Perspect* 7:59–64.

Moragan, J.T. 1993. *Plant Soil* 150:61–68.

Mortvedt, J.J. 1987. *J Environ Qual* 16:137–142.

Mortvedt, J.J., and P.M. Giordano. 1977. In H. Drucker and R.A. Wildung, eds. *Biological Implications of Heavy Metals in the Environment*. CONF-750929. NTIS, Springfield, VA.

Mortvedt, J.J., D.A. Mays, and G. Osborn. 1981. *J Environ Qual* 10:193–197.

Mulla, D.J., A.L. Page, and T.J. Ganje. 1980. *J Environ Qual* 9:408–412.

Munshower, F.F. 1977. *J Environ Qual* 6:411–413.

Naidu, R., and R.D. Harter. 1998. *Soil Sci Soc Am J* 62:644–650.

Neathery, M.W., and W.J. Miller. 1975. *J Dairy Sci* 58:1767–1781.

Norvell, W.A., J. Wu, D.G. Hopkins, and R.M. Welch. 2001. *Soil Sci Soc Am J* (in press).

Nicholson, F.A., and K.C. Jones. 1994. *Environ Sci Technol* 28:2170–2175.

Nriagu, J.O., and M.S. Simmons, eds. 1990. *Food Contamination from Environmental Sources*. Wiley, New York.

Oliver, D.P., R. Hannam, K.J. Tiller, N.S. Wilhem, R.H. Merry, and G.D. Cozens. 1994. *J Environ Qual* 23:705–711.

Olsson, P.E., and C. Hogstrand. 1987. *Environ Toxicol Chem* 6:867–874.

Page, A.L., F.T. Bingham, and C. Nelson. 1972. *J Environ Qual* 1:288–291.

Page, A.L., A.C. Chang, and M. El-Amamy. 1987. In *Lead, Mercury, Cadmium, and Arsenic in the Environment*. SCOPE 31. Wiley, New York.

Pahren, H.R., J.B. Lucas, J.A. Ryan, and G.K. Dotson. 1979. *J Water Pollut Cont Fed* 51:2588–2601.

Parker, G.R., W.W. McFee, and J.M. Kelly. 1978. *J Environ Qual* 7:117–142.

Patel, P.M., A. Wallace, and R.T. Mueller. 1976. *J Am Soc Hort Sci* 101:553–556.

Peel, J.W., R.J. Vetter, J.E. Christian, W.V. Kessler, and W.W. McFee. 1978. In D.C. Adriano and I.L. Brisbin, eds. *Environmental Chemistry and Cycling Processes*. CONF-760429. NTIS, Springfield, VA.

Pepper, I.L., D.F. Bezdicek, A.S. Baker, and J.M. Sims. 1983. *J Environ Qual* 12:270–275.

Pesch, G.C., and N.E. Stewart. 1980. *Marine Environ Res* 3:145–154.

Petersen, R. 1982. *Environ Sci Technol* 16:443–447.

Petruzzelli, G., G. Guidi, and L. Lubrano. 1977. *Water Air Soil Pollut* 8:393–399.

Pettersson, O. 1977. *Swedish J Agric Res* 7:21–24.

Preer, J.R., and W.G. Rosen. 1977. *Trace Subs Environ Health* 11:399–404.

Ragaini, R.C., H.R. Ralston, and N. Roberts. 1977. *Environ Sci Technol* 11:773–781.

Ramos, L., L.M. Hernandez, and M.J. Gonzalez. 1994. *J Environ Qual* 23:50–57.

Reddy, C.N., and W.H. Patrick, Jr. 1977. *J Environ Qual* 6:259–262.

Reuss, J.O., H.L. Dooley, and W. Grims. 1978. *J Environ Qual* 7:128–133.

Root, R.A., R.J. Miller, and D.E. Koeppe. 1975. *J Environ Qual* 4:473–476.

Roth, M. 1992. In D.C. Adriano, ed. *Biogeochemistry of Trace Metals*. Lewis Publ, Boca Raton, FL.

Santillan-Medrano, J., and J.J. Jurinak. 1975. *Soil Sci Soc Am Proc* 39:851–856.

Sauve, S., W.A. Norvell, M. McBride, and W. Hendershot. 2000. *Environ Sci Technol* 34:291–296.

Schauer, P.S., W.R. Wright, and J. Pelchat. 1980. *J Environ Qual* 9:69–71.

Schroeder, H.A., and J.J. Balassa. 1963. *Science* 140:810–820.

Schuhmacher, M., M.A. Bosque, J.L. Domingo, and J. Corbella. 1991. *Bull Environ Contam Toxicol* 46:320–328.

Shacklette, H. 1972. *Cadmium in Plants.* USGS Bull 1314-G. U.S. Geological Survey (USGS), Washington, DC.

Sidle, R.C., and W.E. Sopper. 1976. *J Environ Qual* 5:419–422.

Sileo, L., and W.N. Beyer. 1985. *J Wildlife Dis* 21:289–296.

Smith, W.H. 1973. *Environ Sci Technol* 7:631–636.

Smolders, E., and M.J. McLaughlin. 1996a. *Plant Soil* 179:57–64.

Smolders, E., and M.J. McLaughlin. 1996b. *Soil Sci Soc Am J* 60:1443–1447.

Somers, E. 1974. *J Food Sci* 39:215–217.

Somers, G.F. 1978. *Environ Pollut* 17:287–295.

Soon, Y.K., and T.E. Bates. 1982. *J Soil Sci* 33:477–488.

Stahr, K., H.W. Zöttl, and Fr. Hädrich. 1980. *Soil Sci* 130:217–224.

Stenström, T., and H. Lönsjö. 1974. *Ambio* 3:87–90.

Stenström, T., and M. Vahter. 1974. *Ambio* 3:91–92.

Stevenson, F.J. 1976. *Soil Sci Soc Am J* 40:665–672.

Stephenson, M., and G.L. Mackie. 1988a. *Can J Fish Aquat Sci* 45:1705–1710.

Stephenson, M., and G.L. Mackie. 1988b. *Water Air Soil Pollut* 38:121–136.

Stoeppler, M. 1992. *Hazardous Metals in the Environment.* Elsevier, London.

Storm, G.L., G.J. Fosmire, and E.D. Bellis. 1994. *J Environ Qual* 23:508–514.

Street, J.J., W.L. Lindsay, and B.R. Sabey. 1977. *J Environ Qual* 6:72–77.

Street, J.J., B.R. Sabey, and W.L. Lindsay. 1978. *J Environ Qual* 7:286–290.

Suns, K., G. Hitchin, B. Loescher, E. Pastorek, and R. Pearce. 1987. Ont Min Environ Tech Rep, Ontario, Canada.

Symeonides, C., and S.G. McRae. 1977. *J Environ Qual* 6:120–123.

Takijima, Y., and F. Katsumi. 1973. *Soil Sci Plant Nutr* 19:29–38.

Takijima, Y., F. Katsumi, and K. Takezawa. 1973a. *Soil Sci Plant Nutr* 19:173–182.

Takijima, Y., F. Katsumi, and K. Takezawa. 1973b. *Soil Sci Plant Nutr* 19:183–193.

Tanner, J.T., and M.H. Friedman. 1977. *J Radioanal Chem* 37:529–538.

Taylor, D. 1983. *Ecotox Environ Safety* 7:33–42.

Tjell, J.C., J.A. Hansen, T.H. Christensen, and M.F. Hovmand. 1980. In P. L'Hermite and H. Ott, eds. *Characterization, Treatment, and Use of Sewage Sludge.* Proc. 2nd Comm European Communities Workshop. Reidel, Dordrecht, Netherlands.

Tsuda, T., T. Inoue et al. 1995. *J AOAC Int* 78:1363–1368.

Turner, M.A. 1973. *J Environ Qual* 2:118–119.

Tyler, G. 1972. *Ambio* 1:52–59.

Tyler, L.D., and M.B. McBride. 1982. *Plant Soil* 64:259–262.

[U.S. EPA] United States Environmental Protection Agency. 1992. *The National Sewage Sludge Rule.* U.S. EPA, Washington, DC.

US-EPA. 1993. *Integrated Risk Information System (IRIS) on Cadmium.* Office of Health and Environ Assess, Cincinnati, OH.

Vallee, B.L., and D.D. Ulmer. 1972. *Ann Rev Biochem* 41:91–128.

Van Assche, and F.H. Clijsters. 1990. *Plant Cell Environ* 13:195–206.

Van Hook, R.I., W.F. Harris, and G.S. Henderson. 1977. *Ambio* 6:281–286.

Wenzel, W.W. 2000. Univ für Bodenkultur, Vienna, personal communication.

Wenzel, W.W., W.E.H. Blum, A. Brandstetter, and F. Jockwer. 1996. *Z Pflanzenernähr Bodenk* 159:609–614.

White, M.C., and R.L. Chaney. 1980. *Soil Sci Soc Am J* 44:308–313.

Williams, C.H. 1977. *J Aust Inst Agric Sci* (Sept–Dec):99–109.

Williams, C.H., and D.J. David. 1973. *Aust J Soil Res* 11:43–56.

Williams, C.H., and D.J. David. 1977. *Aust J Soil Res* 15:59–68.

Williams, D.E., J. Vlamis, A.H. Pukite, and J.E. Corey. 1984. *Soil Sci* 137:351–359.

Wilson, J.B. 1982. PhD Thesis, Univ. of Guelph, Guelph, Ontario, Canada.

Winner, R.W. 1988. *Environ Toxicol Chem* 7:153–159.

Wixson, B.G., N.L. Gale, and K. Downey. 1977. *Trace Subs Environ Health* 11:455–461.

Wolnik, K.A., F.L. Fricke, S.G. Capar, et al. 1983. *J Agric Food Chem* 31:1240–1244.

Wolnik, K.A., F.L. Fricke, et al. 1985. *J Agric Food Chem* 33:807–811.

Wong, M.K., G.K. Chuah, L.L. Koh, K.P. Ang, and C.S. Hew. 1984. *Environ Exp Bot* 24:189–195.

[WHO] World Health Organization. 1984. *Guidelines for Drinking-Water Quality, vol 1: Recomendations.* WHO, Geneva.

Wren, C.D., H.R. MacCrimmon, and B.R. Loescher. 1983. *Water Air Soil Pollut* 19:277–291.

Wren, C.D., S. Harris, and N. Harttrup. 1995. In D.J. Hoffman et al., eds. *Handbook of Ecotoxicology.* Lewis Publ, Boca Raton, FL.

Xian, X. 1989. *Plant Soil* 113:257–264.

Yamagata, N., and I. Shigematsu. 1970. *Inst Public Health* (Tokyo) 19(1):1–27.

Yingming, L., and R.B. Corey. 1993. *J Environ Qual* 22:1–8.

Zasoski, R.J., and R.G. Burau. 1988. *Soil Sci Soc Am J* 52:81–87.

9
Chromium

1 General Properties of Chromium

Chromium, a member of Group VI-B of the periodic table has atom. no. 24, atom. wt. of 52.0, specific gravity of 7.2 at 20 °C, melting pt. of 1857 °C, and four stable isotopes with the following percentages of abundance: ^{50}Cr (4.31%), ^{52}Cr (83.76%), ^{53}Cr (9.55%), and ^{54}Cr (2.38%). It has five radioactive isotopes but only ^{51}Cr with a half-life of 27.8 days is commonly used for tracer studies. Chromium is a silvery, lustrous, malleable metal that takes a high polish. It dissolves readily in nonoxidizing mineral acids but not in cold aqua regia or HNO_3. Thus, it is resistant to attack by oxidizing acids and a range of other chemicals, hence its use in corrosion-resistant alloys.

Chromium may occur in any of the oxidation states from $-II$ to VI, but it is not commonly found in oxidation states other than 0, III, and VI, with III being the most stable. Within the range of pH and redox potentials found in soils it can exist in four forms—two trivalent forms, the Cr^{3+} cation and the CrO_2^- anion, and two hexavalent anion forms, $Cr_2O_7^{2-}$ and CrO_4^{2-}. The trivalent form has a great tendency for coordination with oxygen- and nitrogen-containing ligands. Chromium compounds with oxidation states below III are reducing and those with greater than III are oxidizing. The chemistry of Cr of relevance to biology, ecology, and health is that of Cr(III) and Cr(VI) present in various environmental media. Of the two forms found in nature, the trivalent form is relatively benign while the hexavalent form is relatively toxic. Thus, the III and VI oxidation states are important in biology and environmental health.

2 Production and Uses of Chromium

World production of chromite has been in the order of over 9 million tonnes. The principal producers of chromite in decreasing order are: South Africa, the former Soviet Union, Albania, the Philippines, Zimbabwe, and Turkey. These countries account for over 90% of all the current chromite production. Mine production of chromite ceased in the U.S. in 1961 since it was not economically feasible to recover the chromite from the mine. While the U.S. is not a producer of chromite, it consumes about 12% of the world's output.

The most important Cr ore is chromite [(Fe, Mg)O(Cr, Al, Fe)$_2$O$_3$]. The end member $FeOCr_2O_3$ (also called chromite) contains 68% chromic oxide (Cr_2O_3) and 32% ferrous oxide (FeO). The highest grades of ore contain about 52 to 56% Cr_2O_3 and 10 to 26% FeO.

The big users of Cr are the metallurgical (ferroalloys and nonferrous alloys), refractory, and chemical industries. For example, of the total amount of chromite consumed in the U.S. the metallurgical industry uses about 60%; the refractory industry, 20%; and the chemical industry, 20%. Metallurgic-grade chromite is usually converted into one of several types of ferrochromium or chromium metal that are alloyed with Fe, Ni, or Co. The addition of chromium to steel or wrought iron enhances anticorrosion properties and improves its mechanical properties, including hardness, strength, and tempering resistance.

The most important industrial application for Cr is in the manufacture of stainless steel, which consumes about 75% of all the ferrochrome produced each year. The second most important use for the chrome ore is for refractory purposes because it has a high melting point and is virtually chemically inert. As such, it is used in the making of refractory bricks, mortars, castables, and other brick-related purposes. The chemical industry uses ore containing about 45% chromic oxide for preparation of sodium chromate and sodium dichromate from which most other Cr chemicals are produced. Chromium chemicals are used in leather tanning, catalysts, pigments, drilling muds, textiles, chemical manufacture, wood preservatives, and toner for copying machines.

3 Chromium in Nature

Chromium is omnipresent in the environment, found in varying concentrations in air, soil, water, and all biological matter. Chromium is abundant in the earth's crust, more so than Co, Cu, Zn, Mo, Pb, Ni, and Cd. It ranks 21st among the elements in crustal abundance, averages about 100 ppm in rocks, and ranks 4th among biologically important metals. Although about 40 Cr-containing minerals are known, chromite is the only one of commercial importance.

The levels of Cr in soil have been reported to vary from trace to as high as 5.23% (NAS, 1974). Soils derived from serpentine rocks usually contain high Cr. Serpentine, a type of ultramafic igneous rock, contains an average of 1800 ppm Cr (Cannon, 1978). In addition, shales and phosphorites usually contain high concentrations of Cr (Table 9.1). In rocks, Cr is often present as chromite ($FeOCr_2O_3$). World soils probably have an average Cr content of about 40 ppm, although a summary of the world literature yields 84 ppm (Appendix Table A.24) (Ure and Berrow, 1982). A recent survey of surface soils in England and Wales indicates a geometric mean of 34 ppm (McGrath, 1995). In the U.S. soils have a geometric mean of 37 ppm ($n = 863$) (Shacklette et al., 1971). Canadian soils have an arithmetic mean of 43 ppm ($n = 173$) and a range of 10 to 100 ppm (McKeague and Wolynetz, 1980). There are contrasting values reported for world soils: 200 ppm by Vinogradov (1959); 100 to 300 ppm by Aubert and Pinta (1977); and 40 ppm (10 to 150 ppm range) by the National Academy of Sciences (NAS, 1974). These discrepancies probably occurred from variations in sampling and analysis protocol and treatment of data.

4 Chromium in Soils/Sediments

4.1 Total Chromium in Soils

The Cr concentration of soil is largely determined by the parent material. Soils derived from ultramafic igneous rocks can be expected to contain high levels of Cr (see Table 9.1); so are soils derived from shales and clays. In Minnesota, 16 soil series were examined to establish baseline levels of several heavy metals (Pierce et al., 1982). The soils were representative of those formed from seven major parent materials in the state and comprised a broad range of soil properties. It was found that, in general, concentrations of Cr were highest in the parent materials of all soils (48 ppm in parent materials vs. 39 ppm in surface soil samples). The highest concentrations of total Cr (and Cu and Ni) were found in soils developed from the Rainy Lobe

TABLE 9.1. Commonly observed chromium concentrations (ppm) in various environmental media.

Material	Average concentration	Range
Continental crust	125	80–200
Ultramafic igneous[a]	1800	1000–3400
Basaltic igneous[b]	200	40–600
Granitic igneous	20	2–90
Limestone[a]	10	<1–120
Sandstone[b]	35	—
Shales and clays[a]	120	30–590
Black shales[b]	100	26–1000
Petroleum[c]	0.3	—
Coal[c]	20	10–1000
Coal ash[d]		
Fly ash	247	37–651
Bottom ash	585	<40–4710
FGD sludge	73	<40–168
Oil ash[d]	2411	364–4390
Phosphate fertilizers[a]	—	30–3000
Soils[c] (normal)	40	10–150
Herbaceous vegetables[b]	—	<0.05–14
Ferns[e]	1.9	—
Fungi[e]	2.6	—
Lichens[e]	—	0.6–7.3
Freshwater ($\mu g\ L^{-1}$)[e]	1	0.1–6
Seawater ($\mu g\ L^{-1}$)[e]	0.3	0.2–50

Sources:
[a]NRCC (1976).
[b]Cannon (1978).
[c]NAS (1974).
[d]Ainsworth and Rai (1987).
[e]Bowen (1979).

TABLE 9.2. Concentrations of chromium in surface soils of England and Wales, classified according to soil texture class.

Soil texture	No. of samples	Minimum	25th	50th	75th	Maximum
			\multicolumn{3}{c	}{Percentile}		
Clayey	479	18.7	50.3	59.0	69.6	837.8
Fine loamy	202	5.0	35.3	43.5	54.0	692.9
Fine silty	1063	4.9	38.4	48.0	57.6	285.4
Coarse silty	184	6.3	29.5	39.3	47.4	143.5
Coarse loamy	1143	0.2	21.1	27.4	36.0	256.3
Sandy	229	0.2	9.4	13.2	18.0	91.5
Peaty	557	0.2	6.2	12.2	24.8	153.7
All Soils	5692	0.2	26.5	39.3	52.6	837.8

Source: McGrath (1995).

till. In Belorussia where soil parent materials include various clay and sandy deposits mainly of glacial and fluvial origin, Cr was least in eolian deposits (11 ppm) and highest in the alluvial deposits (54 ppm) (Lukashev and Petukhova, 1975). Among the parent materials examined in the Nakhich-evan region of Russia, Cr was lowest in marl, sandstone, and granite; highest in serpentinite, gabbro, diorite, basalt, and diabase; and in between in shales and clays (Shakuri, 1978). In China, the geometric mean ($n = 3981$) for soils in the mainland is 54 ppm (Chen et al., 1991). In general, the contents of trace elements were highest in Lithosols, Inceptisols and Aridisols, and lowest in Ultisols and Oxisols.

In his synthesis of the literature, Mitchell (1964) stated that most soils can be expected to contain between 5 and 1000 ppm Cr, but soils containing <5 ppm and up to a few percent Cr are also known. Shacklette et al. (1971) however, indicated that the majority of soils in the U.S. contain between only 25 and 85 ppm Cr. A range of 20 to 125 ppm in Canadian soils was reported (Morley, 1975). Furthermore, Frank et al. (1976) reported that the average natural background level of Cr in agricultural soils of Ontario (Canada) was 14.3 ppm. A more recent survey of almost 5700 surface soils in England and Wales (Table 9.2) gives a geometric mean concentration of 34 ppm Cr (McGrath, 1995). The data indicate the profound effect of soil texture on the Cr content of British soils. For example, a value of 59 ppm (50th percentile) was found for clayey vs. only 13.2 ppm for sandy soils (see Table 2). Soils derived from serpentine rocks commonly have the highest Cr contents. (Note: Serpentine soils also have high Ni contents.) Values for British and Swedish serpentine soils were reported to range from 2500 to 4000 ppm Cr (Proctor, 1971).

A mean value of 50 ppm Cr for world soils has been indicated (Berrow and Reaves, 1984), whereas a mean value of 136 ppm (range of 46 to 467 ppm) has been reported for tropical Asian paddy soils (Domingo and Kyuma, 1983).

4.2 Bioavailable Chromium in Soils

Extractable Cr, just like any other extractable trace element, is the form most commonly used to indicate Cr bioavailability to plants. Often, the extractable form of a trace element is not directly related to the total content in the soil. In general, only a very small but variable fraction of the total Cr in soil is extractable with ordinary extractants. In extracting several metals from a contaminated soil by 1 N NH_4OAc–0.02 M EDTA, only 0.30% of the total Cr was extracted by this reagent, in contrast to 13% for Zn and 17% for Pb (Nakos, 1982). In Scottish peaty podzols, from 0.34 to 0.80% of the total Cr fraction was extractable by HOAc reagent (Mitchell, 1971). In another instance $< 2\%$ of the total (400 ppm) Cr added was extractable by 1 N NH_4OAc and $< 25\%$ of the total amount extractable by 0.1 N HCl (Schueneman, 1974). Extractions with 0.1 N HCl from two agricultural Nova Scotia soils contained 0.102 to 2.9 ppm Cr (MacLean and Langille, 1980). From < 0.03 to 2.5 ppm Cr was extractable by 2.5% HOAc from Scottish podzols (Swaine and Mitchell, 1960). No apparent relationships between EDTA- or NH_4OAc-extractable Cr and plant uptake of Cr were found (Mitchell et al., 1957). Less than 2 ppm Cr was extractable by 2.5% HOAc in Zimbabweian serpentine soils (Wild, 1974).

Chromium extracted from surface layers of arable Scottish soils is generally in the following ranges: < 0.01 to 1.0 ppm with 2.5% HOAc and 0.1 to 4.0 ppm with 0.05 M EDTA (Mitchell, 1964). Extraction of Scottish soils by 0.1 N HCl may yield up to 10-fold increases over the 2.5% HOAc extractant. Because of this low extraction rate by HOAc, this extractant can be regarded as unacceptable for determining bioavailable Cr in soils. Among four extractants tested (distilled water, 1 N NH_4OAc, 0.005 M DTPA, and 0.1 N HCl), phytotoxicity symptoms in several crop species became more pronounced as 1 N NH_4OAc-extractable Cr increased. In general, it is prudent to assume that the amount of Cr in soil that is bioavailable to the plant is relatively independent of the total soil concentration.

4.3 Chromium in Soil Profile

The patterns of profile distribution of Cr appear to be inconsistent. In some U.S. podzols, most of the metals examined including Cr, tended to be higher in the B or C horizon than in the A horizon (Connor et al., 1957). Chromium was about 2.5 to 4 times higher in the C than in the A horizon. In addition, substantial portions of the elements in the soil were associated with the clay fraction. Similar trends in the soil profile accumulation of Cr were observed in some British soils. For example, in six Lancashire gleyed soils supporting permanent pasture, the maximum concentration of Cr occurred in the lowest horizon (Butler, 1954). In this case, a somewhat consistent decrease of this element in the top horizon, which seemed to be unaffected by the redox conditions, occurred. Similarly, Mitchell (1971) observed an increasing pattern of Cr accumulation with increasing soil depth in some Scottish peaty podzols. As if corroborating, Berrow and Reaves (1986) found that, on the average, the organic-rich 0 to 6 cm of Scottish soils had smaller concentrations of Cr (and Ni) than the underlying mineral soils, but that below 6 cm there were only small changes in Cr level with depth. Conversely, some soil profiles display accumulations of Cr in the surface horizon and depletions with depth. This trend is particularly pronounced in some soil profiles in Papua, New Guinea (Bleeker and Austin, 1970). Chromium in these profiles decreased from 946 ppm in the A1 horizon to 294 ppm in the C horizon, and to 283 ppm in the parent material.

Some soils tend to exhibit a more uniform distribution in the soil profile. For example, some agricultural soils in Ontario (Ap = 53; B = 55; and C = 49 ppm) (Whitby et al., 1978) and serpentine soils in western Newfoundland (Roberts, 1980) display this particular pattern.

Mobility of Cr in the soil profile can occur in certain unnatural situations. For instance, Lund et al. (1976) found metal enrichment to depths of 3 m or more under sewage sludge disposal ponds. The metals apparently mobilized as soluble metal–organic complexes. Thus, in instances where Cr could be complexed or oxidized into hexavalent (VI) form, it can be mobile in the soil profile. In a silt loam soil furrow irrigated for 14 years with anaerobic sewage sludge, total Cr has moved to 60 cm (Baveye et al., 1999).

4.4 Sorption of Chromium in Soils

The fate of Cr in soils depends on several factors: redox potential, presence of electron donors or acceptors, oxidation state, pH, soil minerals, competing ions, complexing agents, and the like. These factors in turn control the various processes that determine the partitioning of Cr between the solid and aqueous phases in soils and geologic media. These processes include the hydrolysis of Cr(III), precipitation/dissolution of Cr(III) and Cr(VI), redox reactions of Cr(VI) and Cr(III), and adsorption/ desorption of Cr(VI). In this section, sorption connotes adsorption and precipitation.

Under acidic to circumneutral pH, Cr(VI) aqueous concentrations are controlled primarily by adsorption. Hexavalent Cr species are adsorbed by soil phases that have exposed hydroxyl groups on their surfaces, such as Fe, Mn, and Al oxides, kaolinite, and montmorillonite (Davis and Leckie, 1980; Griffin et al., 1977; Rai et al., 1986; Zachara et al., 1987, 1988). The adsorption of Cr(VI) on these surfaces increases with decreasing pH due to the protonation of the hydroxyl groups. This suggests that Cr(VI) adsorption is favored when the adsorbent surfaces become positively charged at low to neutral pH (i.e., that have high pH_{pzc} values). This adsorption can be

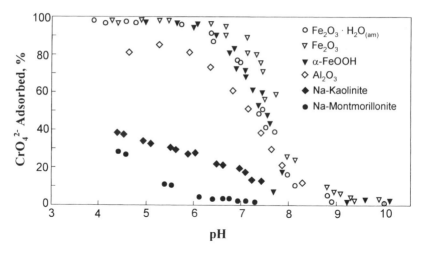

FIGURE 9.1. Sorption of Cr(VI) by various adsorbents for a fixed adsorption site concentration. (Reprinted from the *Science of the Total Environment*, Rai et al., copyright © 1989, with permission from Elsevier Science.)

described as a surface complexation reaction between the Cr(VI) species, such as CrO_4^{2-} and $HCrO_4^-$, and the surface hydroxyl sites:

$$S\text{–}OH + H^+ + CrO_4^{2-} \rightleftharpoons S\text{–}OH_2^+\text{–}CrO_4^{2-}$$

where S–OH is a mineral hydroxyl site (e.g., on Fe, Mn, or Al oxides, or edges of layer silicates), and $S\text{-}OH_2^+\text{–}CrO_4^{2-}$ is the surface complex (i.e., adsorbent–Cr complex). Adsorbents that possess strong affinity for Cr(VI) include Al and Fe oxides, kaolinite, and montmorillonite. Among these adsorbents, Fe oxides exhibit the strongest affinity for CrO_4^{2-}, followed by Al_2O_3, then kaolinite and montmorillonite (Fig. 9.1). More specifically, Cr(VI) adsorption was greatest in lower pH materials enriched with kaolinite and crystalline Fe oxides (Zachara et al., 1989).

Like other cationic metals, Cr(III) is rapidly and specifically adsorbed by Fe and Mn oxides and clay minerals (Bartlett and Kimble, 1976b; Griffin et al., 1977; Rai et al., 1986). The reaction is fairly rapid—about 90% of the added Cr is adsorbed by the oxides and clay minerals within 24 hr. The adsorption of Cr(III) increases with increasing pH and soil OM content (Griffin et al., 1977; Rai et al., 1986). It decreases in the presence of competing cations or dissolved organic ligands in the solution. Both the Langmuir and Freundlich isotherms have been used to describe the adsorption of Cr on solid phases (Davis and Olsen, 1995; Griffin et al., 1977; Palmer and Wittbrodt, 1991; Stollenwerk and Grove, 1985).

Under environmental conditions, Cr(III) polynuclear species [e.g., $Cr(OH)_4^{5+}$] do not play an important role in the aqueous chemistry of Cr and the solubility of $Cr(OH)_3$ keeps Cr concentrations less than the drinking water limit (i.e., 50 μg Cr L^{-1}) between pH 6 and 12 (Rai et al., 1987).

Another solid phase, $(Cr,Fe)(OH)_3$, has even lower solubility than $Cr(OH)_3$ and might also be an important solubility-controlling compound for Cr(III) in geologic environment. In contrast, most Cr(VI) salts, except possibly $BaCrO_4$, are expected to be relatively soluble under ordinary environmental conditions. Indeed, Rai et al. (1986) showed that $Ba(S,Cr)O_4$ is a probable solubility-controlling phase for Cr(VI) in oxidizing Cr-contaminated soils.

The effects of oxidation state, pH, and complexing agent (represented by landfill leachate) on the sorption of Cr by clay minerals have been elucidated (Griffin et al., 1977). The adsorption of Cr from either the landfill leachate or Cr salt solutions by montmorillonite or kaolinite was highly pH-dependent. Adsorption of Cr(VI) in the form of $HCrO_4^-$ decreased as pH increased, with no further sorption occurring above pH 8.5. More Cr(VI) was adsorbed from the leachate solution than from the K_2CrO_4 solution, which was attributed to the formation of polynuclear complexes in the leachate solution. No precipitation of Cr(VI) occurred in either solution over the pH range of 1.0 to 9.0.

In the case of trivalent Cr, Griffin et al. (1977) found that the adsorption increased as the pH of the suspension increased. Trivalent Cr is known to be extensively hydrolyzed in acid solutions to species such as $Cr(OH)^{2+}$, $Cr_2(OH)_4^{2+}$, or $Cr_6(OH_{12})^{6+}$. The increased adsorption of Cr(III) with increasing pH is due to cation exchange reactions of the hydrolyzed species. About 30 to 300 times more Cr(III) than Cr(VI) was adsorbed by either kaolinite or montmorillonite. In contrast to Cr(VI), the adsorption of Cr(III) is 3 to 14% lower from the landfill leachate than in $Cr(NO_3)_3$ solutions. This was caused by the presence of cations in the leachate competing with the cationic Cr(III) species for adsorption sites.

The presence of excess orthophosphate in equilibrating solution can depress the adsorption of Cr(VI) (Bartlett and Kimble, 1976b). This can be attributed to the phosphates competing with Cr(VI) for the same adsorption sites. Sorption of Cr(VI) by certain soils appears to be of specific type (Gebhardt and Coleman, 1974). This indicates that chromate ions are tightly bound compared with anions such as chloride, nitrate, or sulfate, but they can be desorbed by reaction of the soil with other specifically adsorbed anions, such as phosphates and sulfates. Amacher (1981) found that 0.01 M KH_2PO_4 solution was sufficient to extract Cr(VI) from soils, indicating the competitive nature of $H_2PO_4^-$ ions with $HCrO_4^-$ ions.

Adsorption by OM and iron oxides is the most likely explanation for the immobilization of Cr(III) in some soils, since humic acids have a high affinity for Cr(III) (Amacher, 1981). When Cr(III) was added to soil, a large fraction of added Cr was extracted with the iron oxides, confirming the high affinity of Cr for Fe oxides (Grove and Ellis, 1980a).

Trivalent Cr can be complexed by OM components, especially fulvic acid (Bartlett and Kimble, 1976a). In soil extraction by $Na_4P_2O_7$, considerably more Cr(III) was removed from soils collected from horizons where fulvic acid was more predominant than humic acid. The $Na_4P_2O_7$ extractant can remove up to about 80% of the OM, consisting largely of fulvic acid. For example, Na-citrate can complex about 90% of the Cr in solution. Citric acid, DTPA, fulvic acids, and water-soluble OM can keep most of Cr(III) in solution above pH 5.5 (James and Bartlett, 1983a,b). Animal manures have varying effects but less than the citrate in solubilizing Cr (Bartlett and Kimble, 1976a). Apparently, the effectiveness of manures depends on their degree of decomposition.

4.5 Transformation and Chemical Speciation of Chromium in Soils/Sediments

The two most environmentally important oxidation states of Cr, i.e., Cr(VI) and Cr(III), are redox- and pH-sensitive. Oxidation–reduction reactions can convert Cr(III) to Cr(VI) and vice versa. In the natural environment, the reduction reactions would be ordinarily favored, although both can occur simultaneously. Chromium (VI) is strongly oxidizing, as indicated by its stability under high redox potentials and in the absence of reductants (Fig. 9.2). The interconversion of trivalent and hexavalent Cr species also largely depends on the presence of electron acceptors (i.e., oxidants) and electron donors (i.e., reductants). In surface soils and shallow freshwater systems,

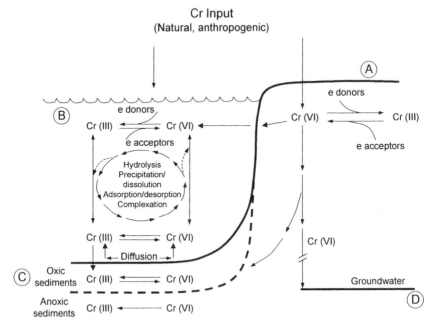

FIGURE 9.2. Biogeochemical cycling of Cr in the environment. The scheme shows the interconversion of Cr(III) and Cr(VI) in (A) soils, (B) freshwater ecosystem, (C) sediments, and (D) groundwater.

typical oxidants may include O_2 and high-valency Mn oxides; reductants may include OM, other organic compounds, Fe^{2+}, and sulfides.

The reduction of Cr(VI) by Fe^{2+} can be described by a generalized reaction:

$$Cr(VI) + 3Fe(II) \rightarrow Cr(III) + 3Fe(III)$$

or, when applied to remediation of Cr(VI) by the addition of zero-valent Fe (Ponder et al., 2000):

$$2Fe^0_{(s)} + 2H_2CrO_4 + 3H_2O \rightarrow 3(Cr_{.67}Fe_{.33})(OH)_{3(s)} + FeOOH_{(s)}$$

or, when there is a more discrete Fe source (e.g., biotite or hematite), the reaction follows (Rai et al., 1989):

$$3FeO + 6H^+ + Cr(VI) \rightarrow Cr(III) + 3Fe(III) + 3H_2O$$

Even small amounts of Fe(II) contained in biotite and hematite can reduce aqueous Cr(VI) species to Cr(III), resulting in the precipitation of $(Fe,Cr)(OH)_{3(am)}$ (Eary and Rai, 1989), or in the presence of soil organic C (SOC), [or just carbon (C^0) below], the overall reaction is:

$$2Cr_2O_7^{2-} + 3C^0 + 16H^+ \rightarrow 4Cr^{3+} + 3CO_2 + 8H_2O$$

or in the presence of free Fe^{2+} ions in aqueous systems:

$$3Fe^{2+} + HCrO_4^- + 3H_2O \rightarrow CrOH^{2+} + 3Fe(OH)_2^+$$

The presence of high-valency Mn oxides can result in the oxidation of Cr(III) to Cr(VI). A more generalized reaction follows:

$$Cr(III) + Mn(III,IV) \rightarrow Cr(VI) + Mn(II)$$

or more specifically:

$$Cr^{3+} + 1.5\delta\text{-}MnO_{2(s)} + H_2O \rightarrow HCrO_4^- + 1.5Mn^{2+} + H^+$$

or

$$Cr(OH)^{2+} + 3\beta\text{-}MnO_{2(s)} + 3H_2O \rightarrow HCrO_4^- + 3MnOOH_{(s)} + 3H^+$$

where $\delta\text{-}MnO_{2(s)}$ represents the birnessite mineral, and $\beta\text{-}MnO_{2(s)}$ represents the pyrolusite mineral.

As can be deduced from the preceding reactions, both oxidation and reduction can occur simultaneously in soils when OM, Fe(II) and Mn(III, IV) hydroxides and oxides are relatively abundant (James, 1996). The interconversion of Cr(VI) and Cr(III) is depicted in Figure 9.2. Such two-way reactions can be expected to occur in surface soils and in shallow aquatic systems but may not be expected to exist in subsurface environments (including groundwater) and anoxic sediments due to unfavorable conditions (e.g., too reducing in the case of sediment or the absence of

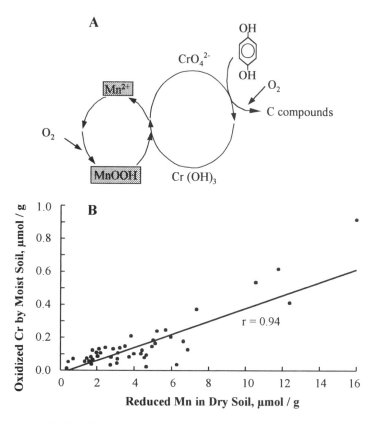

FIGURE 9.3. (A) Simplified scheme for the interconversion of Cr(III) and Cr(VI) in typical soils and sediments, highlighting the roles of manganese oxide (as oxidant) and organic compounds (as reductant) in the process. (Reprinted with permission from James, The challenge of remediating chromium-contaminated soil, *Environmental Science & Technology* 30:248A–251A. Copyright © 1996 American Chemical Society.) (B) Amounts of oxidized chromium formed in 50 fresh moist soils related to the amount of manganese reduced in the dried samples of the same soils. (Modified from James, 1996; Bartlett and James, 1979).

reductants in the case of groundwater). In soils and sediments, the inter-conversion of Cr(III) and Cr(VI) in the presence of oxidic and OM materials is depicted in Figure 9.3A. The source of reducing capacity in the presence of natural OM (NOM) originates from the hydroquinone groups in humic substances (Elovitz and Fish, 1995; Wittbrodt and Palmer, 1995). The reduction kinetics of Cr(VI) by tannic acid or gallic acid as representative low-molecular-weight DOC and humic substances representing the high-molecular-weight DOC indicate that the lower-molecular-weight compounds have smaller oxidation potential and therefore, are better facilitators in the reduction of Cr(VI) (Nakayasu et al., 1999).

The direction of conversion is also largely dictated by the pH (i.e., as a master variable). Reduction of Cr(VI) by OM and other reductants is favored by pH values <6, and both reduction and oxidation may be inhibited under more alkaline conditions (James, 1994). Surface-bound NOM (e.g., humic and fulvic acids) can effectively reduce Cr(VI) to Cr(III) in undisturbed highly acidic field soils (i.e., pH $\sim$ 4) (Jardine et al., 1999). The reduction reaction is catalyzed by the presence of soil mineral surfaces, and the reduced product Cr(III) is immobilized as tightly bound surface species [adsorption and Cr(OH)$_3$ precipitation products]. X-ray absorption near-edge structure spectroscopy (XANES) revealed that both Cr(VI) and Cr(III) resided on the solid mineral surface. The reduction of Cr(VI) to Cr(III) was dramatically more significant on soils with higher levels of surface-bound NOM. Cr(VI) could be reduced to Cr(III) in soils with the rate of conversion being slower in alkaline than in acid soils (Cary et al., 1977b). The insoluble forms of Cr had the properties of mixed hydrous oxides of Cr(III) and Fe(III), and there was somewhat similar chemistry between Cr(III) and Fe(III) in soils. Similarly, analogous behavior between Cr(III) and Al(III) and also between Cr$_4$O^{2-} and orthophosphate in soils can be expected (Bartlett and Kimble, 1976a,b). Cr(VI) was extensively reduced in soils of near neutral pH under anaerobic conditions, particularly when the soil contained OM (Bloomfield and Pruden, 1980). The Cr from organic sources may not oxidize in soil and may remain associated with the soil OM where it is resistant to oxidation.

Oxidation of Cr(III) in soils can occur under most field conditions. Bartlett and James (1979) found that for the oxidation of Cr to proceed, the presence in the soil of high-valency Mn oxides is essential because they serve as electron acceptors in the oxidation reaction (see Fig. 9.3A). A test was developed to determine the potential of a soil to oxidize Cr(III) by measurement of the amount of Mn reducible by hydroquinone (see Fig. 9.3B). In what may be rare circumstances, James and Bartlett (1983b) found that oxidation of Cr(III) in tannery wastes can proceed despite the high levels of reducing compounds in the sludge, effluent, and the soil used. However, the rate of oxidation of Cr(III) from this source was slower than the rates of pure chemical forms used, such as Cr(OH)$_3$. Cr(III) applied to alkaline soils can remain in the reduced form for a considerable length of time (Sheppard et al., 1984).

Furthermore, Bartlett and James (1993) and Zhang and Bartlett (1999) indicate that oxidation of aqueous Cr(III) can occur at pH levels of 3.2 to 4.4 in the presence of Fe(III) and in light. Likewise, it is possible to induce indirect photoreduction of Cr adsorbed on clay particles by photoproduced Fe(II) (Kieber and Helz, 1992). Sunlight acts on the particulate iron oxyhydroxides, which act as electron donors and release Fe(II) in the presence of organic ligands; Fe(II) then reduces Cr(VI) to Cr(III).

The basic requirements for light-induced oxidation of Cr(III) in the presence of Fe(III) are sunlight, a pH of about 2.5 to 4.5, and soluble Fe(III) species, which are present ubiquitously in polluted atmosphere as well as in soils and aquatic bodies (Zhang and Bartlett, 1999). Photolysis of FeOH$_2^+$, the major Fe(III) species at pH 2.5 to 5, has been considered as one important source of $\cdot$OH radicals in ambient environments. However, in light and in the presence of OM in surface soils, the Fe(III)-catalyzed photochemical reduction of Cr(VI) by organic acids will predominate in Cr cycling in most soils. But in atmospheric waters, which commonly have pH 2 to 5 and much less DOC, and in some acidic surface waters with minimal DOC, light-induced Cr(III) oxidation could be one potentially important pathway for Cr(III) oxidation to Cr(VI).

In typical surface soils, reduction may be favored due to higher accumulation of OM on the surface, while in the oxidized subsoil, oxidation may be more favored due to lower OM content but higher Mn oxide content. This can explain why once Cr is oxidized into the hexavalent form in the soil, it becomes more mobile and leachable, and hardly gets reduced because of low amounts (or lack) of reductant materials in the subsurface environment. Therefore, in the majority of soils, the relatively insoluble and immobile Cr(III) form predominates, occurring generally as insoluble hydroxides and oxides.

The kinetics of oxidation–reduction reactions of Cr apparently are different. Using isotopic exchange techniques, Cary et al. (1977b) found that most of the Cr(VI) added to soils was converted to Cr(III) within 1 wk, and that only <5% of the added ^{51}Cr(III) was oxidized to Cr(VI) within 90 days. Similarly, Ross et al. (1981) observed a rapid decrease of extractable Cr(VI) during a 3-wk incubation study, which indicates significant occurrence of reduction of Cr(VI). Thus, results indicate that although Cr(III) can be oxidized to Cr(VI) in soil, this is probably not all that common. On the other hand, when Cr(VI) is added to a typical soil, it can be expected to rapidly reduce to Cr(III).

In summary, the following general statements can be offered about the environmental chemistry of Cr:

1. In normal surface soils, reduction of Cr(VI) to Cr(III) should be favored due to the presence of relatively higher amounts of reducing materials.
2. Cr(III) can be oxidized to Cr(VI) in natural environment in the presence of strong oxidants, such as high-valency Mn oxides (e.g., birnessite, δ-MnO_2).
3. Reduction of Cr(VI) to Cr(III) can occur in natural systems in the presence of reductants (i.e., electron donors), such as OM, Fe(II), and sulfides.
4. Aqueous concentrations of Cr(III) are primarily controlled by precipitation–dissolution reactions, whereas aqueous concentrations of Cr(VI) are controlled by adsorption–desorption, and/or precipitation–dissolution reactions.
5. Due to cationic behavior of Cr(III) species and anionic behavior of Cr(VI) species, the adsorption of Cr(III) and Cr(VI) are favored by high and low pH, respectively.
6. Reduction of Cr(VI) in subsurface environment (e.g., in groundwater remediation) may not be expected to occur unless reductant compounds are introduced as a remedial agent.

Reduction of Cr(VI) to Cr(III) is a means of remediating contaminated soils and groundwater. Metallic Fe (steel wool) or Fe(II) readily reduces soil Cr(VI) (James, 1994). Reduction of Cr(VI) by naturally occurring Fe(II)-bearing minerals followed by sorption or precipitation has often been invoked to explain migration behavior in the field. In a similar manner, the use of solid-phase reduced-Fe sources has been proposed as reactive, flow-through barriers to Cr(VI) migration (Seaman et al., 1999). In another instance, cattle manure application to soils promoted reduction of Cr(VI) in irrigation water (Losi et al., 1994). The removal of Cr(VI) from enriched water ranged from 51 to 98%, and increased with OM loading. The effect of various reducing agents also largely depends on the pH, with lower pH favoring reduction.

In summary, Cr(VI) can be reduced to Cr(III) in soils by redox reactions with aqueous inorganic species, electron transfer at mineral surfaces, reactions with nonhumic organic substances such as carbohydrates and proteins, or reduction by soil humic substances. The latter, which constitute the majority of the organic fraction in most soils, represent a significant reservoir of electron donors for Cr(VI) reduction. In general, the oxidative capacity for soils is expected to be highest for clay soils and lowest for sandy soils (Kozuh et al., 2000). The degree of oxidation is proportional to the concentration of Mn(IV) oxides, but it also depends on the reductive capacity of soils. Soils high in Mn(IV) oxides and low in OM have a high oxidative capacity for the oxidation of soluble Cr(III). On the contrary, soils high in OM and low in Mn(IV) oxides have a high reductive capacity for the reduction of soluble Cr(VI). In ordinary soil, the reductive capacity for soluble Cr is typically higher than the oxidative capacity; the exception being soils with high content of Mn(IV) oxides and low content of OM. In general, the reductive capacity is expected to be highest in peat soils and lowest in low-OM soils.

5 Chromium in Plants

5.1 Chromium in Mineral Nutrition

Interest in Cr content of plants is spurred by the known essentiality of Cr in animal and human nutrition (Mertz, 1969, 1993). Stimulatory effects of trace amounts of Cr on plant growth, although they were generally small, inconsistent, and mostly inconclusive have been reported (Pratt, 1966). The stimulatory effect of Cr on plant growth has been interpreted as a limited substitution of Cr ions for Mo ions (Warington, 1946). Although there are numerous reports on the stimulation of growth or yield of plants grown either in the field, solution, or sand culture, it is difficult to relate responses in plant growth to Cr additions, since there is insufficient evidence to confirm Cr as an essential element or even as a beneficial element. This difficulty exists despite its experimental use to increase the yield of potatoes, enhance the growth of citrus and avocado trees, and increase the sugar content and ripening of grapes. Nevertheless, the nonessentiality of Cr for normal plant growth has been demonstrated conclusively (Huffman and Allaway, 1973a).

5.2 Uptake and Translocation of Chromium in Plants

Concentrations of Cr in the foliage of plants show little relationship with the total Cr in the soil. Background concentrations of Cr in plant tissue usually fall below the 1 ppm level across a wide range of Cr soil values.

Controlled studies have shown that Cr(III) and Cr(VI) are practically equally bioavailable to plants grown in culture solution (Breeze, 1973; Huffman and Allaway, 1973b). However, the effect of Cr in causing phytotoxicity in sugar beets grown on sand culture depended on oxidation state, with more detrimental effects observed for Cr(VI) (Hewitt, 1953). Similarly, Cr(VI) was found to be slightly more toxic than Cr(III) to bush beans (Wallace et al., 1977) and to corn grown on pot culture (Mortvedt and Giordano, 1975).

Absorbed Cr is apparently poorly translocated in plants. Indeed, Cr absorbed by plants grown in culture solution remained primarily in the roots and was poorly translocated to the shoot (Huffman and Allaway, 1973b). At maturity, bean plants contained about 55% and wheat about 81% of the added Cr. However, bean roots contained about 92% and wheat roots 95% of the total plant Cr. Similarly, >99% of the Cr(III) and Cr(VI) absorbed remained bound in the roots of rice grown in culture solution (Myttenaere and Mousny, 1974). There was greater translocation of Cr to the shoot when either Cr(III) or Cr(VI) was chelated with EDTA. Similar accumulations of absorbed Cr in roots were found by others (Cary et al., 1977a; Ramachandran et al., 1980; and Wallace et al., 1976). This barrier effect of the cell wall to translocation of Cr from roots to the shoot was not circumvented by supplying Cr in the form of organic acid complexes or by increasing the Cr concentrations in the nutrient solution (Cary et al., 1977a). Further, it was found that plant species that accumulate Fe also accumulate Cr. Even with a Cr accumulator species, *Leptospermum scoparium*, most of the absorbed Cr persisted in the roots (Lyon et al., 1969). Most of the Cr retained in the roots apparently is present in the soluble form in vacuoles of root cells (Shewry and Peterson, 1974), and more specifically in the protoplasmic fraction of the roots (Myttenaere and Mousny, 1974). However, leaves usually contain higher concentrations of Cr than the grain (Smith et al., 1989).

Despite the poor translocation of Cr to the shoot, its presence in soils can still increase the Cr content of these tissues. For example, adding $Cr(OH)_3$ to a neutral soil increased the Cr contents of bean shoots (James and Bartlett, 1984). However, bean shoots grown in soils amended with Cr-enriched tannery effluent and sewage sludge did not contain more Cr than the control but the roots did. Soluble Cr was not detected in soils amended with Cr(III) in tannery effluent or sewage sludge, indicating the lack of oxidization probably due to the presence of high OM in these wastes. Soluble Cr(VI) levels were higher in unplanted soil amended with Cr than in planted soil, but levels of soluble Cr(III) were higher in the rhizosphere soil. This indicates the influence of plant roots and their exudates on the form of Cr and its subsequent bioavailability to plants, i.e., plant root metabolites can influence the reduction of Cr(VI) to Cr(III) and the possible formation of Cr(III)–organic complexes facilitating their uptake by the roots.

5.3 Bioaccumulation of Chromium in Plants

In normal soils, Cr concentrations in plants are usually in the < 1 ppm range and seldom exceed 5 ppm (Pratt, 1966). However, some plant species (e.g., accumulators) can accumulate appreciable quantities of Cr (see also the later chapter on Nickel for accumulator species). Peterson (1975) reported that many plant species have adapted to serpentine soils, which have relatively high levels of total Cr but low bioavailable Cr. Some species however, can accumulate appreciable amounts of Cr despite the low solubility of Cr. For example, Peterson (1975) reported, in ppm ash weight: 48,000 ppm for *Sutera fodina;* 30,000 ppm for *Dicoma niccolifera*; and 2470 ppm for *Leptospermum scoparium*. Their plant-to-soil concentration ratios average about 0.30. Plant-to-soil concentration ratios can vary widely, with some values reported as low as 0.01. Plants growing on serpentine soils rarely contain Cr concentrations exceeding 100 ppm (Brooks and Yang, 1984). For example, serpentine plants from Zimbabwe have a maximum Cr level of 77 ppm, and in species of *Geissois* from New Caledonia concentrations ranged up to 45 ppm (Brooks and Yang, 1984; Jaffre et al., 1979).

5.4 Phytotoxicity of Chromium

Most of the reported cases of Cr phytotoxicity have been observed under controlled experimental conditions, either in solution culture or in soils spiked with high levels of Cr. Under field conditions however, Cr phytotoxicity is essentially nonexistent, except possibly in soils derived from ultrabasic or serpentinic rocks. These soils are characteristically high in Ni and Cr, and it is most likely that Ni rather than Cr causes the toxicity. Nevertheless, phytotoxicity associated with high levels of Cr has been well documented. It appears that levels of 1 to 5 ppm Cr present in the available form in soil solution, either as Cr(III) or Cr(VI), are the critical levels for a number of plant species (NRCC, 1976). When added to soil, the toxic threshold level for Cr(VI) varied from 5 ppm for tobacco grown in sandy soil (Soane and Saunder, 1959) to 500 ppm for *L. perenne* grown in potted compost (Breeze, 1973). Toxic limits for Cr(III) varied from 8 ppm for sugar beets grown in sand (Hewitt, 1953) to 5000 ppm for *L. perenne* grown in potted compost (Breeze, 1973). Apparently the high OM in the potting soil bound the Cr, or enhanced the reduction of Cr(VI) to Cr(III). It is apparent from Table 9.3 that Cr may prove to be toxic to plants at about 5 ppm or higher in nutrient solution and at about 100 ppm when added to a mineral soil.

TABLE 9.3. Effects of oxidation state and level of chromium in growth medium on phytotoxicity.

Plant species	Oxidation state of Cr	Cr concentration in growth media, ppm	Growth medium	Effects	Source
Corn	Cr(III)	0.5	Solution culture	Stimulation of growth	NRCC (1976)
Corn	Cr(III)	5	Solution culture	Moderate toxicity	NRCC (1976)
Corn	Cr(III)	50	Solution culture	Severely stunted growth	NRCC (1976)
Ryegrass (perennial)	Cr(III)	10	Solution culture	Increased plant mortality	NRCC (1976)
Ryegrass (perennial)	Cr(VI)	10	Solution culture	Increased plant mortality	NRCC (1976)
Barley	Cr(VI)	50	Soil	Severely stunted growth	NRCC (1976)
Barley	Cr(VI)	500	Soil	Death	NRCC (1976)
Oats	Cr(VI)	5	Sand	Iron chlorosis	NRCC (1976)
Sugar beet	Cr(VI)	8	Sand	Iron chlorosis	NRCC (1976)
Tobacco	Cr(VI)	5	Sand	Retarded stem development	NRCC (1976)
Corn	Cr(VI)	10	Sand	Stunted growth	NRCC (1976)
Sweet orange seedling	Cr(VI)	75	Soil	No toxicity	NRCC (1976)
Sweet orange seedling	Cr(VI)	150	Soil	Observed toxicity	NRCC (1976)
Soybean	Cr(VI)	5	Loam soil	Inhibition of uptake of Ca, K, Mg, P, B, Cu	NRCC (1976)
Barley	Cr(III)	8	Solution culture	Chlorotic leaves	Davis et al. (1978)
Bush beans	Cr(III)	0.5	Solution culture	Decreased yield	Wallace et al. (1976)
Corn, beans, and tomatoes	Cr(III)	100	Soil	Stunted growth	Schueneman (1974)

The range of Cr concentrations in plant tissues before toxicity symptoms can be manifested ranged from 5 ppm for barley, corn, oats, and citrus to 175 ppm for tobacco (Pratt, 1966). Concentrations in plant tissues definitely associated with phytotoxicity symptoms are usually in the several hundred ppm range. For rice plants, 35 to 177 ppm of Cr in stem and leaf can cause a 10% reduction in yield (Chino, 1981). Thus, while the toxic effects of most heavy metals (e.g., Cd, Ni, Co, Zn, Cu, etc.) are associated with high concentrations of the elements in the leaf tissue, this is not the case with Cr.

Several explanations pertinent to the cause of Cr phytotoxicity have been proposed, but the most plausible mechanism is not clear. Excess Cr may interfere with Fe, Mo, P, and N metabolisms in plants. Cr(III) and Fe(III) have acted analogously in causing the acute P deficiency symptoms observed in oats (Soane and Saunder, 1959). Grove and Ellis (1980b) explained Cr phytotoxicity in terms of Cr(III) effects on Mn and Fe chemistry in soil; severe Fe chlorosis associated with Cr(III) phytotoxicity can be a result of greater quantities of available Mn and lesser amounts of soluble Fe, indicating Cr(III)-induced Fe-Mn interactions in plants. Hence, at lower rates of application, Cr(III) may stimulate growth and yield of plants when Mn bioavailability to the plant is limiting. In addition, the uptake of several nutrient elements (K, Mg, P, Mn, Ca, and Fe) by plants was affected by high Cr levels (Turner and Rust, 1971). However, appreciable increases in available Fe concentration in soils were observed when applied Cr(VI) was reduced to Cr(III) resulting in decreased soil pH (Jaiswal and Misra, 1984).

In most soils, phytotoxicity if any can be expected from Cr(VI) due to much reduced bioavailability of Cr(III) above pH 5 (McGrath, 1995); Cr(VI) species become more bioavailable as pH increases for reasons already explained in the preceding section. Thus, the prevailing view is that phytotoxicity from Cr(III) is unlikely to occur, especially under agronomic conditions, except in highly acidic soils.

The visual symptoms of Cr phytotoxicity commonly observed include stunted growth, poorly developed root systems, and curled and discolored leaves. Some plants may exhibit brownish-red leaves containing small necrotic areas or purpling of basal tissues (Pratt, 1966).

6 Ecological and Health Effects of Chromium

The biological importance of Cr to animals and humans depends on its oxidation state; Cr(III) is required to maintain normal glucose metabolism. This essentiality was first demonstrated in rats in 1959 (Schwartz and Mertz, 1959). That demonstration was followed by documentation of improved glucose tolerance by Cr(III) supplementation in human subjects and by the demonstration of Cr deficiency and its reversal in patients receiving total parenteral nutrition (Mertz, 1993). Subsequently, the U.S. National Research Council and the WHO Expert Committee described trivalent Cr as an essential nutrient with typical dietary intakes of 50 to 200 µg Cr day^{-1}.

The most commonly observed symptoms of Cr deficiency in humans include impaired glucose tolerance, glycosuria, and elevations in serum insulin, cholesterol, and total triglycerides (Cohen et al., 1993). In animals, in addition to the above manifestations, there is also decreased longevity, impaired growth, altered immune function, disturbances in aortic plaque and size, corneal lesion formation, and an overall decrease in reproductive functions.

In contrast, Cr(VI) is not essential and is a toxin. As such, Cr is unique among regulated toxic elements in the environment in that the two most environmentally important forms, Cr(III) and Cr(VI), are regulated differently. All other toxic elements, such as Cd, Pb, As, and Hg, are regulated based on their total concentrations, irrespective of oxidation state. The U.S. EPA classifies materials as hazardous waste if they contain leachable Cr (40 CFR 261.4) but excludes those materials from classification if it can be shown that the leachable Cr is not Cr(VI). The U.S. EPA has also proposed a cleanup concentration for Cr(VI) in soil at much lower levels than for Cr(III) (U.S. EPA, 1993).

6.1 Ecological Effects of Chromium

Depending on the species, Cr can be less toxic to fish in warmer water, but appreciable decreases in toxicity occur with increasing pH or water hardness; changes in salinity have little effect if any on its toxicity (WHO, 1988). Chromium can increase the susceptibility of fish to infection; high concentrations can accumulate and/or damage various fish tissues. This can also occur in invertebrates, such as snails and worms.

Reproduction of *Daphnia magna* can be affected by exposure to 0.01 mg L^{-1} of Cr(VI). The 3-day LC_{50} for *Daphnia* is between 0.03 to 0.081 mg L^{-1} (Mance, 1987). Factors that can influence the bioavailability of Cr include the presence of other pollutants, organic compounds, and the temperature of the medium (WHO, 1988). Hexavalent Cr is bioaccumulated by aquatic species by passive diffusion (U.S. EPA, 1978). The physiological state and activity of the fish also affect bioaccumulation. In general, invertebrate species such as polychaete worms, insects, and crustaceans are more sensitive to Cr exposure than vertebrates such as some fish (U.S. EPA, 1980). The lethal Cr level for several aquatic and nonaquatic invertebrates has been reported to be 0.05 mg L^{-1} (NAS, 1972). The salmonid species are more sensitive than the nonsalmonids. For example, the 4-day LC_{50} for salmonids ranges from 3.3 to 65 mg L^{-1} in contrast to a range of 28 to 169 mg L^{-1} for nonsalmonids (Mance, 1987). Accordingly, tentative water quality criteria for freshwater habitats for salmonids have been proposed not to exceed 0.025 mg L^{-1} of soluble Cr, with the 95 percentile not exceeding 0.1 mg L^{-1}. However, more stringent values are necessary for soft acidic waters.

TABLE 9.4. Maximum safe concentrations of chromium for fish (extracted from the literature).

Species	Study type	Cr (mg L^{-1})	pH	Temperature (°C)	Hardness (mg L^{-1})
Chromium(VI)					
Rainbow trout	Egg-embryo	0.05–0.11	6.7–7.0	10	33.4
Brook trout	Full life cycle	0.02–0.35	7.0–8.0	7–15	45
Lake trout	Embryo-larval	0.11–0.19	6.8–7.1	10	33.4
Northern pike	Partial embryo-larval	0.54	6.7–7.0	17	37.8
Fathead minnow	Early life cycle	1.25–1.86	8.2	25	hard
Fathead minnow	Full life cycle	1.00–3.95	7.5–8.2	16–24	209
Bluegill sunfish	Embryo-larval	0.52–1.12	6.7–7.1	25	38.3
Channel catfish	Embryo-larval	0.15–0.31	7.0–7.4	22	36.2
White sucker	Embryo-larval	0.290–0.538	6.9–7.2	17	38.8
Chromium(III)					
Rainbow trout	Simultaneous gamete exposure during insemination, survival after 10 days postfertilization	0.000–0.005	9.0	10	—
Steelhead trout	Egg-embryo	0.030–0.048	6.6–7.4	12.5	25
Fathead minnow	Full life cycle	0.180–0.380	7.6	20–25	203

Source: Holdway (1998).

Based on a literature synthesis, maximum acceptable concentrations of Cr in various life stages for fish vary according to Cr species, fish species, and water temperature, pH, and hardness (Table 9.4).

6.2 Effects of Chromium on Human Health

Chromium has been recognized by the International Agency for Research on Cancer as a potent human carcinogen. Chromium carcinogenesis has been known since the late 19th century when the first nasal tumors were detected among Scottish chrome pigment workers. Subsequently, other chrome-related industrial activities (e.g., chrome plating, leather tanning, and stainless steel production) have been recognized as important exposure scenarios. Extensive epidemiological results have already been obtained for industrial workers in the United Kingdom, Germany, Japan, and the United States. Environmental contamination arising from the widespread use of Cr in industry serves as an additional source of human exposure to Cr.

The respiratory tract is the major target organ for both Cr(VI) and Cr(III) toxicity, for acute (short-term) and chronic (long-term) inhalation exposures. The effects seem to be similar, although Cr(III) is less toxic. Dypsnea, coughing, and wheezing were reported from a case of acute exposure to Cr(IV), while perforations and ulcerations of the septum, bronchitis, decreased pulmonary function, pneumonia, and other respiratory effects have been noted from chronic exposure (ATSDR, 1993; WHO, 1988).

Human studies have clearly established that inhaled Cr is a human carcinogen, resulting in an increased risk of lung cancer. Animal studies have shown Cr(IV) to cause lung tumors via inhalation exposure. The U.S. EPA has classified Cr(VI) as a human carcinogen of high carcinogenic hazard (U.S. EPA, 1993).

Chromium toxicity is dependent on the chemical species. In occupational settings, exposure to Cr is as important as exposure to Ni (see later chapter on Nickel). The following hexavalent compounds pose the most common risks in industrial environment: sodium chromate (Na_2CrO_4), sodium dichromate ($Na_2Cr_2O_7 \cdot 2H_2O$), calcium chromate ($CaCrO_7 \cdot 2H_2O$), zinc and lead pigment chromates, and chromic acid. The trivalent and elemental Cr species often encountered by industrial workers include ore ($FeOCr_2O_3$), calcium chromite ($CaCr_2O_3$), Cr oxide (Cr_2O_3), basic Cr sulfates, and Cr metal and alloys (Miller-Ihli, 1992).

Chromium is generally considered as being the second most common skin allergen in the general population, after Ni (Haines and Nieboer, 1988). Allergic contact *dermatitis* is the most prominent reaction from skin contact with Cr(VI); other effects from dermal exposure to Cr(VI) include increased sensitivity and ulceration of the skin (ATSDR, 1993).

The majority of Cr(VI) that enters the body via inhalation or ingestion is rapidly reduced to Cr(III) (Cohen et al., 1993). Inhaled Cr(VI) is acted upon by alveolar macrophages and epithelial-lining fluids within the bronchial system. These actions reduce the metal to its trivalent form thereby diminishing the amount of Cr(VI) that might enter the bloodstream after crossing the alveoli. The Cr(VI) that does not enter the blood is reduced to Cr(III) by redox reactions with several blood-borne constituents and within the red blood cells themselves. Similarly, oral intake of Cr(VI) results in its rapid reduction to poorly absorbed Cr(III) form by components of the saliva and gastric juice (Cohen et al., 1993).

Although Cr(VI) can be rapidly absorbed through the GI tract, any ingested Cr(VI) is quickly reduced in the stomach where the pH is $\sim$1 and numerous reducing agents are found. Also gastric juice is found to reduce Cr(VI) (WHO, 1988). Any nonreduced Cr(VI) that remains is absorbed from the intestines and enters erythrocytes during transport in the portal vein. Most of this Cr(VI) is then reduced to Cr(III) by intracellular reductants of the red (i.e., hemoglobin) or white blood cells, leaving only an insignificant amount of the original Cr(VI) dose, if any. Several agents that occur within the cells and as components of the blood may act as reductants. These include glutathione, lipoic acid, glucose, vitamin C, vitamin E (α-tocopherol), vitamin B_2 (riboflavin), nicotinamide (NADPH), and certain amino acids (Cohen et al., 1993).

The chronic effects of Cr species [both Cr(III) and Cr(VI)] include skin and mucous membrane irritation, bronchopulmonary effects, and systemic effects involving the kidneys, liver, GI tract, and circulatory system (WHO, 1988). Increased incidence of lung cancer has been reported in Europe and the United States for workers employed by the chromate-producing industry (Miller-Ihli, 1992).

In general, Cr(VI) species and their compounds are better absorbed through the intestinal mucosa than are the Cr(III) species. However, due to the action of stomach acids and other components within the GI tract, most of the ingested Cr(VI) is converted to Cr(III). Also if Cr(VI) is reduced to Cr(III) extracellularly, this form is not readily transported into cells and is therefore nontoxic. In this state, the uptake of Cr is low, i.e., in the range of 1 to 3% of any given dose (Cohen et al., 1993). Thus, Cr(III) and Cr(VI) have contrasting relevance in biological system; the former is an essential nutrient while the latter is a toxin.

6.2.1 Dietary Exposure of Humans to Chromium

Because microgram amounts of Cr [i.e., Cr(III)] are essential in human nutrition for normal glucose, protein, and fat metabolism, the amounts of Cr in drinking water and food are of interest. The general population is exposed to Cr [generally Cr(III)] by eating food, drinking water, and inhaling air that contains the chemical. The average daily intake from air, water, and food is estimated to be approximately 0.01 to 0.03, 2.0, and 60 μg Cr, respectively (ATSDR, 1993).

A study of more than 1500 surface waters showed a maximum Cr content of 110 μg L^{-1}, with a mean of 10 μg L^{-1} (NAS, 1977). About 96% of Canadian stream and river samples ($n = 4163$) had < 10 μg L^{-1} (NRCC, 1976). A 1962 survey of municipal drinking water in the U.S. indicates Cr levels ranging from below detection limits up to 35 μg L^{-1}, with a median of 0.43 μg L^{-1}. Waters from large rivers of North America have a median Cr content of 5.8 μg L^{-1} and a range of 0.72 to 84 μg L^{-1}. Maximum Cr content in cold water taps in homes in the eastern U.S. was only 11 μg L^{-1}, well below the U.S. EPA interim limit of 50 μg L^{-1} (Goldhaber and Vogt, 1989). The U.S. EPA has set a standard of 50 μg Cr L^{-1} for drinking water and 100 μg L^{-1} for freshwater aquatic life (U.S. EPA, 1976). Similarly, the WHO has set a 50 μg Cr L^{-1} standard for drinking water (WHO, 1984).

Chromium contents of various food groups in the United States, Canada, and Finland are presented in Table 9.5. Available food data indicate a range of 0.01 to 0.34 μg Cr g^{-1} WW. Low concentrations can be expected in most sugars and flour products, but high levels for pepper and yeasts (WHO, 1988). Chromium intake from diet and water varies considerably among regions (Table 9.6). The following intakes, in μg Cr day^{-1}, have been reported: USA, 52 to 78; Japan, 723 to 943; Canada, 189; UK 80 to 100;

TABLE 9.5. Chromium contents of various food groups ($\mu g \ g^{-1}$ WW).

Foodstuffs	Canada	United States	Finland
Milk and dairy products	0.05–0.11	0.23	0.01–0.13
Meat, fish, and poultry	0.06–0.18	0.23	0.01–0.06
Cereals	0.06–0.17	0.22	<0.02–0.05
Potatoes	0.04–0.26	0.24	—
Leafy vegetables	0.09–0.11	0.10	0.005–0.03
Legumes	0.06–0.16	0.04	—
Root vegetables	0.06–0.15	0.09	—
Garden fruits	0.05–0.23	—	—
Fruits	0.05–0.07	0.09	0.14[a]
Oils and fats	0.03–0.09	0.18	—
Sugars and candy	0.17–0.34	—	<0.02
Drinks	0.03–0.07	—	—
Condiments and spices	—	—	0.33

[a]Nuts.
Source: NRCC (1976).

Germany, 62; New Zealand, 81. In general, most important Cr intakes range from 50 to 200 $\mu g \ day^{-1}$. These variations are largely caused by dietary preference by peoples around the world. Diets that contain mostly processed foods may be Cr deficient (NAS, 1977). Dietary supplements of 200 μg Cr(III) day^{-1} are sometimes recommended for adults with diabetes or insulin resistance (Fisher, 1990).

The reference dose (RfD) for Cr(VI) is 0.005 mg kg^{-1} per day and the RfD for Cr(III) is 1 mg kg^{-1} per day (U.S. EPA, 1993). The U.S. EPA estimates that consumption of these doses or less over a lifetime would not likely result in the occurrence of chronic noncancer effects. Note: The RfD is

TABLE 9.6. Chromium intake from diet and water for various countries.

Country	Cr intake $\mu g \ day^{-1}$		Remarks
	Mean	Range	
Canada	189	136–282	—
Germany	62	11–195	DA[a], 4 subjects, 1 wk
Japan	723	202–1710	Urban adults
	943	>180–1190	Rural adults[b]
New Zealand	81 ± 32	39–190	DA[a], 11 women, self-selected diets
United Kingdom		80–100	
United States (1968)	52	5–115	DA[a]
United States (1979)	78 ± 42	25–224	DA[a], 28 diets completely (controlled by SRM)
Former Soviet Union		88–126	Children[c] (Tatar region)

[a]DA = direct analysis of composite diets as consumed.
[b]Analysis of composite of cooked servings for one complete day collected from 10 families in different localities.
[c]Analysis of diets in kindergartens.
Source: Modified from WHO (1988).

not a direct estimator of risk but rather a reference point to gauge the potential effects. Intake in excess of the RfD does not imply that an adverse health effect would necessarily occur. As the amount and frequency of exposures exceeding the RfD increase, the probability of adverse health effects also increases. The reference concentrations (RfCs) for Cr(III) and Cr(VI) are under review by the U.S. EPA.

7 Factors Affecting Mobility and Bioavailability of Chromium

Factors that will greatly influence the mobility, solubility, transformation, speciation, and uptake of Cr include pH, oxidation state of Cr, presence of reducing and oxidizing material, oxides of Fe and Mn, and redox potential.

7.1 pH

Perhaps as the most important factor, pH affects the solubility of the Cr forms and therefore its sorption by soils and its bioavailability to plants. Figure 9.4A indicates that the oxidation of Cr(III) as $CrCl_3$ equilibrated in a northern Vermont soil can be affected by pH. At pH 3.2, all of the Cr(III) present was oxidized to Cr(VI) after only 18 hr. On the other hand, Cary et al. (1977b) found that reduction is more rapid in acid than in alkaline soils. The same soil from Vermont and another soil from Rothamsted, England (Fig. 9.4B) both exhibit decreased adsorption with increasing pH.

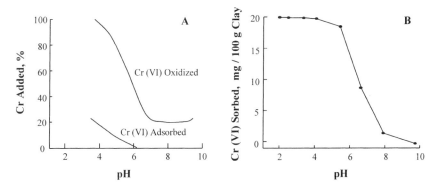

FIGURE 9.4. Influence of pH on oxidation of Cr(III) and soil sorption of Cr(VI). (A) Percentage of chromium in a 2000:1 (solution/soil ratio) equilibrium of 10^{-6} M $CrCl_3$ with soil that was oxidized or adsorbed as Cr(VI), as pH was varied by addition of HCl or $KHCO_3$. (B) Sorption of Cr(VI) on NaOH-extracted (to remove extractable organic matter) subsoil. (Modified from [A] Bartlett and James, 1979; [B] Bloomfield and Pruden, 1980.)

Desorption from the Rothamsted soil at pH 10 to 11 gave a quantitative recovery of Cr(VI) sorbed between pH 6 and 9; the small losses of sorbed Cr(VI) at pH <5 may be due to reduction by residual (i.e., not extracted by dilute NaOH) OM during the adsorption process.

Bloomfield and Pruden (1980) showed that reduction of Cr(VI) by water-soluble soil OM in soil equilibration studies occurred appreciably below pH 4, and that within experimental error there was no reduction in 40 days between pH 5 and 9. With surface soil, the extent of reduction after 3-wk aerobic incubation increased from about 5 mg Cr(VI) reduced per 100 g soil at pH 6.7 to 14 mg per 100 g soil at pH 4.1. It was concluded that the organic constituents of the soil became more effective in reducing Cr(VI) with decreasing pH. In raising the soil pH, the OM may solubilize and thus may increase the reduction of Cr(VI) during its determination with the acid S-diphenyl-carbazide reagent.

In clay mineral equilibration, the water-soluble Cr(VI) decreased as the pH of kaolinite and montmorillonite suspensions decreased (Griffin et al., 1977). This can be attributed to $HCrO_4^-$ adsorption by pH-dependent exchange sites on clay colloids. However, it is more likely that the anaerobic system maintained in that equilibration favored Cr(VI) reduction to Cr(III) and facilitated the cation Cr(III) sorption by the clay exchange sites. The effect of pH on the solubility-precipitation of Cr is also very important. The solubility of Cr(III) in pure Cr solutions decreased as the solution pH was raised above 4 with essentially complete precipitation occurring at pH 5.5. The precipitates were presumably composed of macromolecules with Cr ions in six coordination complexes with water and hydroxyl groups and can be likened to the formation of Al polymers. When soil was added, the solubility of Cr(III) decreased above pH 2.5 with virtually complete precipitation occurring at about pH 4.5. The formation of hydrolytic species of Cr(III) above pH 4 is indicated in Figure 9.5A.

7.2 Oxidation State of Chromium

The oxidation state of Cr is very important relative to its mobility and role in plant and human nutrition. The hexavalent form is more toxic in plant and human nutrition as explained earlier and also known to be more mobile in surface and subsurface environments than the trivalent form. A case in point is the landfill disposal of Cr(III)-bearing wastes which probably would not present a pollution problem, but when oxidation of Cr(III) to Cr(VI) occurs, it can present a serious problem because of the known mobility of Cr(VI) even at relatively low pH.

Cr(VI) was eluted prior to Cr(III) in anaerobic conditions (Artiole and Fuller, 1979). Again, this difference is due to the anionic nature of the Cr(VI) species, while Cr(III) exists as a cation and its hydrolytic forms are prone to precipitate starting at about pH 4.5 (see Fig. 9.5).

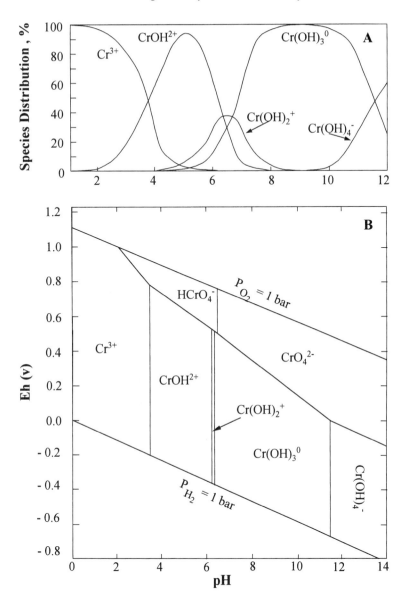

FIGURE 9.5. (A) Distribution of Cr(III) species as a function of pH where the solution is in equilibrium with $Cr(OH)_{3(s)}$. (From Rai et al., 1986, 1987.) (B) Predicted Eh–pH stability field for chromium species in aqueous systems. (Reprinted from the *Science of the Total Environment,* Rai et al., copyright © 1989, with permission from Elsevier Science.)

As discussed previously, Cr(III) is an essential element to humans and animals, but Cr(VI) is not and acts as a potent carcinogen, especially in occupational settings.

7.3 Presence of Reducing and Oxidizing Material

Reduction of Cr(VI) to Cr(III) may occur substantially under certain soil conditions. While the presence of Mn oxides appears to be the key to oxidation, serving as the electron acceptors in the reaction, electron donors are likewise necessary for reduction to proceed. Organic matter and Fe(II) can serve as electron donors in soils and waste disposal systems (Buerge and Hug, 1999). Thus, reduction may be insignificant in soils low in OM and/or soluble Fe (Bartlett and Kimble, 1976b; Korte et al., 1976).

Other organic compounds such as citric acid, gallic acid, acetic acid, etc. (i.e., low-molecular-weight organics) may serve a dual role as chelators for Cr(III) or electron donors for reducing Cr(VI) (James and Bartlett, 1983a–c). These compounds may exist in soils as degradation products of organic waste or as metabolites excreted by plant roots.

7.4 Oxides of Iron and Manganese

Bartlett and James (1979) proposed that adsorption of Cr(III) by Mn oxides is a first step in its oxidation by Mn. In aerobic soils, Mn oxides typically occur as surface coatings of clays and Fe oxides. Manganese oxides have high adsorptive capacities for heavy metals because of their large surface areas and high negative charges. Korte et al. (1976) observed that soils having high "free" Fe and Mn oxides would significantly retard migration of Cr(VI). Long (1983) found that Mn oxide and to a lesser extent, Fe oxide play very important roles in the adsorption of Cr by sediments. Under field conditions, the Mn oxides exist in reactive form and serve as the electron link between Cr(III) and atmospheric O_2 and that the amount of Cr(III) oxidized to Cr(VI) is proportional to the amount of Mn reduced (Bartlett and James, 1979). The oxidation of Cr(III) by Mn oxides is reported to be more rapid than by O_2.

The presence of Fe oxides (either amorphous or crystalline), especially in low-pH materials, facilitates the adsorption of Cr (Rai et al., 1989; Zachara et al., 1989). However, the presence of competing ions, such as SO_4^{2-}, and dissolved organic C which compete for adsorption sites, can depress Cr adsorption.

7.5 Redox Potential

Conversion of Cr(III) to Cr(VI) can be very slow or negligible in anaerobic soil environment of municipal landfills (Artiole and Fuller, 1979). It can be also due to the absence of electron acceptors in such conditions such as high-valency Mn and Fe. From the Eh–pH stability field (Fig. 9.5B), at low redox potential and low pH, Cr(III) species are dominant. At neutral pH, Cr(VI) could be a predominant form of Cr in some natural systems; at pH 7, the concentration of CrO_4^{2-} could be equal to the concentration of $Cr(OH)^{2+}$ in well-aerated soils.

8 Sources of Chromium in the Environment

Chromium can originate from the following major industries: paper products, organic chemicals including petrochemicals, inorganic chemicals, fertilizers, metal works and foundries, textile mill products, leather tanning and finishing, and power plants.

8.1 Atmospheric Emissions

Major portions of overall Cr emissions to the air come from the following industries: ferrochrome produced by electric furnaces, refractory brick (noncast), Cr steel production, and coal combustion. The metallurgical industry is the dominant source of anthropogenic Cr emission into the atmosphere (Miller-Ihli, 1992). These emissions are most often in the form of particulates; however, furnaces are now equipped with emission control devices, such as electrostatic precipitators, to limit releases. Refractory processing that produces bricks for electric furnaces constitutes the second most prevalent industrial use of Cr and, therefore, also releases substantial amounts of Cr into the atmosphere. Coal-fired power plants release substantially more Cr into the atmosphere than do commercial and residential furnaces. Other sources of Cr atmospheric emissions include cement production using kilns and incinerations.

8.2 Aqueous Emissions

Table 9.7 shows the various processes that release uncontrolled wastewater, which can contaminate surface water and groundwater. However, methods

TABLE 9.7. Aqueous releases of chromium from various industrial processes.

Process	Uncontrolled emissions, ppm	Oxidation state of Cr
Electroplating	trace–600	Cr(VI)
Metal pickling	600	Cr(VI)
Metal bright dip	10,000–50,000	Cr(VI)
Leather tanning	40	Cr(III)
Cooling tower blowdown	10–60	Cr(VI)
Animal glue manufacture	475–600	Cr(VI)
Textile dyeing	1	Cr(III)
Fur dressing and dyeing	20	Cr(III)
Laundry (commercial)	1.2	Cr(VI)

Source: NRCC (1976).

are available to reduce the levels of Cr in wastewater to trace amounts. Advanced electroplating technology now allows the use of Cr(III) in the plating process instead of the more toxic Cr(VI).

Wastes from Cr smelting can also be a source of water pollution from soluble chromates and other soluble salts. Chromium could leach from disposal pits and landfills and contaminate the groundwater.

8.3 Land Application and Disposal

Typical fertilizer materials contain from trace to sometimes several thousand ppm of Cr. The following values, in ppm, were reported: P fertilizers, 30 to 3000; superphosphates, 60 to 250; bone meal, <20 to 500; and limestone, <1 to 200 (NRCC, 1976). A range of 25 to 100 ppm Cr was reported for Australian superphosphates (Williams, 1977). Another report gave a range of 175 to 485 ppm Cr in P fertilizers (Mortvedt and Giordano, 1977). In general, with the exception of plant uptake of Cd (see previous chapter) from fertilizers, Cr is not of concern at the usual rates of P applied under field conditions.

Because of the increasing trend in the use of municipal sewage sludge in arable lands, heavy metal transfer from this source to the food chain may become a concern. Sommers (1977) reported a mean value of 2620 ppm Cr with a range of 10 to 99,000 ppm ($n = 180$) for the eight states in the north-central and eastern regions. In general, Cr in sewage sludge is extremely variable, ranging from trace amounts to several thousand ppm. Although repeated use or high application rates of sewage sludge can enrich the soil with metals, Cr can be expected to be relatively nonbioavailable to plants from this source.

Poor storage, leakage, and improper disposal practices at manufacturing and ore-processing facilities have released Cr into the environment, causing contamination of groundwater and surface water (Calder, 1988; Palmer and Wittbrodt, 1991). For example, 27 Superfund sites with RODs before 1987 report Cr as being a potential problem. Documented groundwater contamination with Cr has been reported for a Nassau County site on Long Island, New York (source: recharge basin for disposal of solution waste from an aircraft plant), Telluride, Colorado (source: tailings pond), United Chrome Products site in Vancouver, Washington (source: Cr plating facility), Palmetto Wood Preserving site in Dixiana, South Carolina (source: wood treatment facility), and Hudson County, New Jersey (source: waste mud from Cr ore processing). The New Jersey site used the waste mud as a "clean fill" for residential, commercial, and public buildings (James, 1996; Palmer and Wittbrodt, 1991). More than 2 million tonnes of the Cr-bearing fill were used from the early 1900s until the early 1970s in what is now a populous and industrial area just west of New York City.

References

Adriano, D.C., A.L. Page, A.A. Elseewi, A.C. Chang, and I. Straughan. 1980. *J Environ Qual* 9:333–344.

Ainsworth, C.C., and D. Rai. 1987. *Chemical Characterization of Fossil Fuel Combustion Wastes*. EPRI Rep EA-5321. Electric Power Res Inst, Palo Alto, CA.

Amacher, M.C. 1981. *Redox Reactions Involving Chromium*, Plutonium and Manganese in Soils. PhD dissertation. Pennsylvania State Univ, University Park, PA.

Artiole, J., and W.H. Fuller. 1979. *J Environ Qual* 8:503–510.

[ATSDR] Agency for Toxic Substances and Disease Registry. 1993. *Toxicological Profile for Chromium*. U.S. Dept Health and Human Services, Atlanta, GA.

Aubert, H., and M. Pinta. 1977. *Trace Elements in Soils*. Elsevier, New York.

Bartlett, R., and B. James. 1979. *J Environ Qual* 8:31–35.

Bartlett, R., and B. James. 1993. *Adv Agron* 50:151–208.

Bartlett, R.J., and J.M. Kimble. 1976a. *J Environ Qual* 5:379–383.

Bartlett, R.J., and J.M. Kimble. 1976b. *J Environ Qual* 5:383–386.

Baveye, P., M.B. McBride, D. Bouldin et al. 1999. *Sci Total Environ* 227:13–28.

Berrow, M.L., and G.A. Reaves. 1984. In *Proc Intl Conf Environmental Contamination*. CEP Consultants, Edinburgh.

Berrow, M.L., and G.A. Reaves. 1986. *Geoderma* 37:15–27.

Bleeker, P., and M.P. Austin. 1970. *Aust J Soil Res* 8:133–143.

Bloomfield, C., and G. Pruden. 1980. *Environ Pollut* A23:103–114.

Bowen, H.J. 1979. *Environmental Chemistry of the Elements*. Academic Pr, New York.

Breeze, V.G. 1973. *J Appl Ecol* 10:513–525.

Brooks, R.R., and X.H. Yang. 1984. *Taxon* 33:392–399.

Buerge, I.J., and S.J. Hug. 1999. *Environ Sci Technol* 33:4285–4291.

Butler, J.R. 1954. *J Soil Sci* 5:156–166.

Calder, L.M. 1988. In J.O. Nriagu and E. Nieboer, eds. *Chromium in the Natural and Human Environments*. Wiley, New York.

Cannon, H.L. 1978. *Geochem Environ* 3:17–31.

Cary, E.E., W.H. Allaway, and O.E. Olson. 1977a. *J Agric Food Chem* 25:300–304.

Cary, E.E., W.H. Allaway, and O.E. Olson. 1977b. *J Agric Food Chem* 25:305–309.

Chen, J., F. Wei, Y. Wu, and D.C. Adriano. 1991. *Water Air Soil Pollut* 57–58:699–721.

Chino, M. 1981. In K. Kitagishi and I. Yamane, eds. *Heavy Metal Pollution in Soils of Japan*, 1966–80. Japan Sci Soc Pr, Tokyo.

Cohen, M.D., B. Kargacin, C.B. Klein, and M. Costa. 1993. *Crit Rev Toxicol* 23:255–281.

Connor, J., N.F. Shimp, and J.C.F. Tedrow. 1957. *Soil Sci* 83:65–73.

Davis, A., and R.L. Olsen. 1995. *Groundwater* 33:759–768.

Davis, J.A., and J.O. Leckie. 1980. *J Colloid Interface Sci* 74:32–43.

Davis, R.D., P.H.T. Beckett, and E. Wollan. 1978. *Plant Soil* 49:395–408.

Domingo, L.E., and K. Kyuma. 1983. *Soil Sci Plant Nutr* 29:439–452.

Eary, L.E., and D. Rai. 1989. *Am J Sci* 289:180–213.

Elovitz, M.S., and W. Fish. 1995. *Environ Sci Technol* 29:1933–1943.

Fisher, J.A. 1990. *The Chromium Program*. Harper & Row, New York.

Frank, R., K. Ishida, and P. Suda. 1976. *Can J Soil Sci* 56:181–196.

Gebhardt, H., and N.T. Coleman. 1974. *Soil Sci Soc Am Proc* 38:263–266.

Goldhaber, S., and C. Vogt. 1989. *Sci Total Environ* 86:43–51.

Griffin, R.A., A.K. Au, and R.R. Prost. 1977. *J Environ Sci Health* A12(8):431–449.

Grove, J.H., and B.G. Ellis. 1980a. *Soil Sci Soc Am J* 44:238–242.

Grove, J.H., and B.G. Ellis. 1980b. *Soil Sci Soc Am J* 44:243–246.

Haines, A.T., and E. Nieboer. 1988. In J.O. Nriagu and E. Nieboer, eds. *Chromium in the Natural and Human Environments*. Wiley, New York.

Hewitt, E.J. 1953. *J Exp Bot* 4:59–64.

Holdway, D.A. 1988. In J.O. Nriagu and E. Nieboer, eds. *Chromium in the Natural and Human Environments*. Wiley, New York.

Huffman, E., Jr. 1973. *Chromium: Essentiality to Plants, Forms and Distribution in Plants and Availability of Plant Chromium to Rats*. PhD dissertation. Cornell Univ, Ithaca, NY.

Huffman, E.W.D., Jr., and W.H. Allaway. 1973a. *Plant Physiol* 52:72–75.

Huffman, E.W.D., Jr., and W.H. Allaway. 1973b. *J Agric Food Chem* 21:982–986.

Jaffre, T., R.R. Brooks, and J.M. Trow. 1979. *Plant Soil* 51:157–162.

Jaiswal, P.C., and S.G. Misra. 1984. *J Plant Nutr* 7:541–546.

James, B.R. 1994. *J Environ Qual* 23:227–233.

James, B.R. 1996. *Environ Sci Technol* 30:248A–251A.

James, B.R., and R.H. Bartlett. 1983a. *J Environ Qual* 12:169–172.

James, B.R., and R.H. Bartlett. 1983b. *J Environ Qual* 12:173–176.

James, B.R., and R.H. Bartlett. 1983c. *J Environ Qual* 12:177–181.

James, B.R., and R.H. Bartlett. 1984. *J Environ Qual* 13:67–70.

Jardine, P.M., S.E. Fendorf, M.A. Mayes, I.L. Larsen, S.C. Brookes, and W.B. Bailey. 1999. *Environ Sci Technol* 33:2939–2944.

Kieber, R.J., and G.R. Helz. 1992. *Environ Sci Technol* 26:307–312.

Korte, N.E., J. Skopp, W.H. Fuller, E.E. Niebla, and B.A. Alesii. 1976. *Soil Sci* 122:350–359.

Kozuh, N., J. Stupar, and B. Gorenc. 2000. *Environ Sci Technol* 34:112–119.

Long, D.T. 1983. *Geochemical Behavior of Chromium/Water-Sediment Systems*. Preliminary Report, EPA grant R808306010. Michigan State Univ, East Lansing, MI.

Losi, M.E., C. Amrhein, and W.T. Frankenberger. 1994. *J Environ Qual* 23:1141–1150.

Lukashev, K.I., and N.N. Petukhova. 1975. *Sov Soil Sci* 7:429–439.

Lund, L.J., A.L. Page, and C.O. Nelson. 1976. *J Environ Qual* 5:330–334.

Lyon, G.L., P.J. Peterson, and R.R. Brooks. 1969. *Planta (Berl)* 88:282–287.

MacLean, K.S., and W.M. Langille. 1980. *Commun Soil Sci Plant Anal* 11:1041–1049.

Mance, G. 1987. *Pollution Threat of Heavy Metals in Aquatic Environments*. Elsevier, London.

McGrath, S.P. 1995. In B.J. Alloway, ed. *Heavy Metals in Soils*. Blackie, London.

McKeague, J.A., and M.S. Wolynetz. 1980. *Geoderma* 24:299–307.

Mertz, W. 1969. *Physiol Rev* 49:163–239.

Mertz, W. 1993. *J Nutr* 123:626–633.

Miller-Ihli, N.J. 1992. In M. Stoeppler, ed. *Hazardous Metals in the Environment*. Elsevier, London.

Mitchell, R.L. 1964. In F.E. Bear, ed. *Chemistry of the Soil*. Reinhold, New York.

Mitchell, R.L. 1971. In Tech Bull 21. Her Majesty's Sta Office, London.

Mitchell, R.L., J.W.S. Reith, and I.M. Johnson. 1957. *J Sci Food Agric* 8:551–559.

Morley, H.V. 1975. NRCC 150017. National Research Council of Canada, Ottawa.

Mortvedt, J.J., and P.M. Giordano. 1975. *J Environ Qual* 4:170–174.

Mortvedt, J.J., and P.M. Giordano. 1977. In H. Drucker and R.E. Wildung, eds. *Biological Implications of Metals in the Environment*. CONF-750929. NTIS, Springfield, VA.

Myttenaere, C., and J.M. Mousny. 1974. *Plant Soil* 41:65–72.

Nakayasu, K., M. Fukushima, K. Sasaki, S. Taraka, and H. Nakamura. 1999. *Environ Toxicol Chem* 18:1085–1090.

Nakos, G. 1982. *Plant Soil* 66:271–277.

[NAS] National Academy of Sciences. 1972. *Water Quality Criteria. NAS*, Washington, DC.

National Academy of Sciences (NAS). 1974. *Chromium*. NAS, Washington, DC.

National Academy of Sciences (NAS). 1977. *Drinking Water and Health*. NAS, Washington, DC.

[NRCC] National Research Council of Canada. 1976. *Effects of Chromium in the Canadian Environment*. Publ. 15017. NRCC, Ottawa.

Palmer, C.D., and P.R. Wittbrodt. 1991. *Environ Health Perspect* 92:25–40.

Peterson, P.J. 1975. In *Proc Intl Conf Heavy Metals in the Environment*. Univ. of Toronto, Ontario, Canada.

Pierce, F.J., R.H. Dowdy, and D.F. Grigzl. 1982. *J Environ Qual* 11:416–422.

Ponder, S.M., J.G. Darab, and T.E. Mallouk. 2000. *Environ Sci Technol* 34:2564–2569.

Pratt, P.F. 1966. In H.D. Chapman, ed. *Diagnostic Criteria for Plants and Soils*. Quality Printing, Abilene, TX.

Proctor, J. 1971. *J Ecology* 59:827–842.

Rai, D., J.M. Zachara, L.E. Eary, et al. 1986. In *Geochemical Behavior of Chromium Species*. EPRI EA-4544. Electric Power Res Inst, Palo Alto, CA.

Rai, D., B.M. Sass, and D.A. Moore. 1987. *Inorg Chem* 26:345–349

Rai, D., L.E. Eary, and J.M. Zachara. 1989. *Sci Total Environ* 86:15–23.

Ramachandran, V., T.J. D'Souza, and K.B. Mistry. 1980. *J Nucl Agric Biol* 9:126–128.

Roberts, B.A. 1980. *Can J Soil Sci* 60:231–240.

Ross, D.S., R.E. Sjogren, and R.J. Bartlett. 1981. *J Environ Qual* 10:145–148.

Schueneman, T.J. 1974. *Plant Response to and Soil Immobilization of Increasing Levels of Zn^{2+} and Cr^{3+} Applied to a Catena of Sandy Soils*. PhD dissertation, Michigan State Univ, East Lansing, MI.

Schwartz, K., and W. Mertz. 1959. *Arch Biochem Biophys* 85:292–295.

Seaman, J.C., P.M. Bertsch, and L. Schwallie. 1999. *Environ Sci Technol* 33:938–944.

Shacklette, H.T., J.C. Hamilton, J.G. Boerngen, and J.M. Bowles. 1971. USGS Paper 574-D. U.S. Geological Survey (USGS), Washington, DC.

Shakuri, E.K. 1978. *Sov Soil Sci* 10:189–194.

Sheppard, M.I., S.C. Sheppard, and D.H. Thibault. 1984. *J Environ Qual* 13:357–361.

Shewry, P.R., and P.J. Peterson. 1974. *J Exp Bot* 25:785–797.

Smith, S., P.J. Peterson, and K.H.M. Kwan. 1989. *Environ Toxicol Chem* 24:241–251.

Soane, B.D., and D.H. Saunder. 1959. *Soil Sci* 88:322–330.

Sommers, L.E. 1977. *J Environ Qual* 6:225–232.

Stollenwerk, K.G., and D.B. Grove. 1985. *J Environ Qual* 14:150–155.

Swaine, D.J., and R.L. Mitchell. 1960. *J Soil Sci* 11:327–368.

Turner, M.A., and R.H. Rust. 1971. *Soil Sci Soc Am Proc* 35:755–758.

Ure, A.M., and M.L. Berrow. 1982. In *Environmental Chemistry*. Royal Soc Chem, London.

[U.S. EPA] United States Environmental Protection Agency. 1976. *Quality Criteria for Water*. U.S. EPA, Washington, DC.

U.S. EPA. 1978. *Reviews of the Environmental Effects of Pollutants. III Chromium*. U.S. EPA, Washington, DC.

U.S. EPA. 1980. *Ambient Water Quality Criteria for Chromium*. EPA-440/5-80-035. U.S. EPA, Washington, DC.

U.S. EPA. 1993. *Integrated Risk Information on Chromium*. Office of Health and Environ Assess, Cincinnati, OH.

Vinogradov, A.P. 1959. *The Geochemistry of Rare and Dispersed Chemical Elements in Soils,* 2nd ed. Consultants Bureau, New York.

Wallace, A., S.M. Soufi, J.W. Cha, and E.M. Romney. 1976. *Plant Soil* 44:471–473.

Wallace, A., G.V. Alexander, and F.M. Chaudry. 1977. *Commun Soil Sci Plant Anal* 8:751–756.

Warington, K. 1946. *Ann Appl Biol* 33:249–254.

Whitby, L.M., J. Gaynor, and A.J. MacLean. 1978. *Can J Soil Sci* 58:325–330.

Wild, H. 1974. *Kirkia* 9:233–242.

Williams, C.H. 1977. *J Aust Inst Agric Sci* (Sept-Dec):99–109.

Wittbrodt, P.R., and C.D. Palmer. 1995. *Environ Sci Technol* 29:255–263.

[WHO] World Health Organization. 1984. *Guidelines for Drinking-Water Quality. vol. 1: Recommendations*. WHO, Geneva.

World Health Organization (WHO). 1988. *Environmental Health Criteria 61: Chromium*. WHO, Geneva.

Zachara, J.M., D.C. Girvin, R.L. Schmidt, and C.T. Resch. 1987. *Environ Sci Technol* 21:589–594.

Zachara, J.M., C.E. Cowan, R.L. Schmidt, and C.C. Ainsworth. 1988. *Clays Clay Miner* 36:317–326.

Zachara, J.M., C.C. Ainsworth, C.E.G. Cowan, and C.T. Resch. 1989. *Soil Sci Soc Am J* 53:418–428.

Zhang, H., and R.J. Bartlett. 1999. *Environ Sci Technol* 33:588–594.

10
Lead

Remember the consumers' slogan "unsafe at any speed" regarding the fatalities associated with what many people considered to be an unsafe automobile design? With Pb, perhaps the public slogan should be "unsafe at any level in the blood" (i.e., above the background)—reminding us of the potential toxicity of this element to humans, especially the unborn.

1 General Properties of Lead

Lead (atom. no. 82) is a bluish-gray metal that is found in small quantities in the earth's crust. It is of bright luster, soft, highly malleable, ductile, and a poor conductor of electricity. It is very resistant to corrosion. It belongs to Group IV-A of the periodic table, has an atom. wt. of 207.2, melting pt. of 328 °C, and specific gravity of 11.4. It has oxidation states of II or IV. In most inorganic compounds, Pb is in the II oxidation state. Lead has four stable isotopes with the following percentages of abundance: ^{204}Pb (1.48%), ^{206}Pb (23.6%), ^{207}Pb (22.6%), and ^{208}Pb (52.3%). Two radioactive isotopes ^{210}Pb (half-life, $t_{1/2} = 22$ yr) and ^{212}Pb ($t_{1/2} = 10$ hr) are used in tracer experiments. Pure Pb is insoluble in water; however, its chloride and bromide salts are slightly soluble ($\sim$1%) in cold water, whereas carbonate and hydroxide salts are almost insoluble. Because of their use as antiknock agents in gasoline, tetramethyl and tetraethyl Pb are the most important organolead compounds.

2 Production and Uses of Lead

There is a long history of man's production and uses of Pb (Fig. 10.1). Ancient world technologies for smelting Pb–Ag alloys from sulfide (PbS) ores and cupeling Ag from the alloys were developed at least 5000 years ago. The need for Ag propelled Pb production in ancient times. In addition, the low melting point, the ease with which it can be worked, and its durability partially contributed to its use as a construction material during this era. World Pb production averaged about 145 tonnes yr^{-1} from 4000 years ago until about 2700 years ago, rose to about 9100 tonnes with the introduction of silver coinage, and rose again to about 73,000 tonnes during the era of the Roman Empire (2000 years ago). Lead production declined during medieval times but dramatically rose again during the onset of the industrial revolution—from 91,000 tonnes annually 300 years ago to about 1×10^6 tonnes at the beginning of the 20th century.

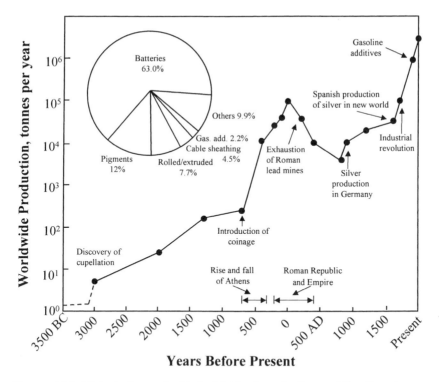

FIGURE 10.1. Historical record of industrial lead production and (inset) major uses of lead today. (Modified from Adriano, 1986; Alliance/EDF, 1994.)

Today, Pb mining and refining operations are found around the world. About 50 countries mine Pb in annual amounts ranging from a few hundred tonnes to >0.5 million tonnes (US Bureau of Mines, 1993). During the last three decades, the United States, Australia, and the former Soviet Union have been the major producers of ore concentrates (a *concentrate* is a partially processed ore containing 60 to 70% Pb and is the form in which the metal is shipped to the smelter).

In a few countries most notably the United States, production of Pb through recycling far outstrips ore production. The United States currently produces 30% of the world's secondary Pb. While most recycling occurs in North America, western Europe, and Japan, a growing number of secondary smelters are operating in developing countries to process locally available supplies, such as batteries.

Member countries of OECD (i.e., Europe, the United States, Canada, Australia, New Zealand, and Japan) accounted for 65% of the world Pb consumption in 1990 (OECD, 1993). Total Pb consumption has also increased significantly in eastern Europe, Latin America, Asia, and Africa. For example, Pb use in Asia has increased six fold in the last two decades.

In evaluating atmospheric Pb deposition in Swiss peat profiles, two enrichment spikes since the industrial revolution were noted (Weiss et al., 1999): between 1880 and 1920, and between 1960 and 1980, the latter coinciding with the widespread use of Pb additives to gasoline.

Lead is a very vital metal in any industrial economy. The only nonferrous metals that are used in greater amounts than Pb are Al, Cu, and Zn (Robinson, 1978). World consumption of Pb has followed a trend very similar to that of production, widespread globally over the last three decades. The predominant use of Pb is in large rechargeable batteries, which now accounts for nearly 65% of global Pb use (see Fig. 10.1, inset) (ILZSG, 1992). Other major uses of Pb include pigments and other compounds, rolled and extruded products, cable sheathing, alloys, shot and ammunition, and gasoline additives. Minor uses of Pb include radiation shielding, ceramic glazes, crystal (which can contain up to 36% Pb), fishing weights, as a heat stabilizer in PVC plastics, etc. While Pb-containing pesticides (e.g., lead arsenate) have been banned in the United States, Austria, Belgium, and Germany, they are still in use in other countries (OECD, 1993).

The United States and a few other countries began to reduce the use of Pb additives in gasoline in the early 1970s. Because Pb renders catalytic converters in cars inoperative, the U.S. EPA required that Pb-free gasoline be made available beginning in 1973. Although the use of Pb additives has not been totally phased out in some countries, the allowable levels of Pb in leaded gasoline range from 0.026 g Pb L^{-1} in Canada and the United States to 1.12 g Pb L^{-1} in the Virgin Islands (vs. >0.78 g Pb L^{-1} in premium gasoline prior to 1970 in the former countries). China uses leaded gasoline extensively (up to 0.78 g Pb L^{-1}), Russia allows up to 0.38 g Pb L^{-1}, while India and South Africa allow 0.56 and 0.40 g Pb L^{-1}, respectively (Alliance/ EDF, 1994). Most European countries allow 0.15 g Pb L^{-1} in premium gasoline although Belgium, Germany, and Luxembourg forbid Pb in regular grades; the United Kingdom, Denmark, and the Netherlands use tax incentives to sell unleaded gasoline cheaper than leaded gasoline.

3 Lead in Nature

Among the heavy metals with atomic numbers >60, Pb is the most abundant in the earth's crust. The average Pb content of the earth's crust is 13 to 16 ppm (Swaine, 1978). Probably, a more accurate estimate of the crustal abundance of Pb is that of Heinrichs et al. (1980) at $\sim$15 ppm.

Although there are more than 200 minerals of Pb, only a few are common, with galena (PbS, 87% Pb by weight), cerussite ($PbCO_3$), and anglesite ($PbSO_4$) as the most important economically. Lead occurs in rocks as a discrete mineral, or since it can replace K, Sr, Ba, and even Ca and Na, it can be fixed in the mineral lattice (Nriagu, 1978). Among the silicate minerals, potassium feldspars and pegmatites are notably enriched in Pb.

The most commonly reported values for Pb in various environmental media are given in Table 10.1. Lead content is higher in coal and shale, particularly organic shales, than in other types of rocks. It also accumulates with Zn and Cd in ore deposits. Lead contents of arable soils are reported to range from 2 to 300 ppm, with soils remote from human activity averaging about 10 to 30 ppm (Alloway, 1990). Much higher values can be expected in areas near densely populated centers, near industrial facilities, and in areas of geochemical deposits.

Background levels of Pb in 173 samples (53 soil types) in Canada averaged 20 ppm (McKeague and Wolynetz, 1980), with values ranging from 5 to 50 ppm. For Poland, a background value of 10.3 ppm Pb was reported for soils (from 62 profiles) derived from sedimentary rocks of glacial origin (Czarnowska and Gworek, 1991). This is comparable to the mean value of 19 ppm for U.S. soils (Appendix Table A.24), and 15 to 25 ppm (Aubert and Pinta, 1977), or 15 ppm (Berrow and Reaves, 1984) reported for world soils. For rural surface soils in Australia, background values of 2 to 160 ppm ($n = 160$) have been reported (Barzi et al., 1996). Similarly, for noncontaminated U.K. soils, 10 to 150 ppm has been reported (Thornton and Culbard, 1986).

Since its use as a gasoline additive, Pb concentration in the atmosphere has increased. Atmospheric deposition is generally the major input in the biogeochemical cycle of Pb. Thus, even remote forest ecosystems are subject to Pb enrichment. Studies in several parts of the world indicate considerable Pb enrichment of various forest ecosystem components from atmospheric input (Adriano, 1986). Consequently, Pb has been accumulating in the forest floor (i.e., organic horizon overlying the mineral soil).

The northeastern United States is a region of interest particularly with Pb because it receives large amounts of air pollution from upwind sources (Schwartz, 1989). Atmospheric deposition of Pb occurs primarily by rainfall at low elevations, with cloudwater interception playing a greater role at high elevations and dry deposition contributing up to 20% of the total Pb flux at some locations (Miller and Friedland, 1994). Both the amount of precipitation and the extent of cloudwater interception increase with increasing elevation, and thus Pb deposition is greater at higher elevations. Consequently, the amount and concentration of Pb in the organic horizon of forest soils increased with increasing elevation in northeastern ecosystems (Friedland et al., 1984; Johnson et al., 1982).

A number of studies have reported that Pb is immobile in organic soils and should have a long residence time in the forest floor (Friedland and Johnson, 1985; Heinrichs and Mayer, 1980; Smith and Siccama, 1981). Lead concentrations in the forest floor in the northeastern United States range from 90 to 225 ppm, while Pb concentrations in mineral soils range from 8 to 35 ppm (Friedland et al., 1992). Based on studies in the 1970s and 1980s, it is apparent that Pb in forest soils in North America and Europe accumulated at a rate of 50 to 700 g ha^{-1} yr^{-1} ($\sim$0.4 to 3.3% of the forest

TABLE 10.1. Common values for lead concentrations (ppm) in various environmental media.

Material	Average concentration	Range
Igneous rocks[a]	15	2–30
Limestone[a]	9	—
Sandstone[a]	7	<1–31
Shale[a]	20	16–50
Carbonates[b]	9	—
Coal[c]	16	up to 60
Coal ash[c]		
Fly ash	170	21–220
Bolton ash	47	5–843
FGD sludge	29	<4–181
Oil ash[c]	508	226–1029
Phosphate fertilizers[d]	—	4.4–488
Animal waste[d]	—	2.1–3.4
Sewage sludges[d]	1832	136–7627
Soils[b] (total)[g]		
Agricultural soils[f]		2–300
		10–30 (rural)
		30–100 (urban)
Remote areas		5–40
Rural areas		5–40
Urban		10–50
Near smelters		20–2000
Street dust[g]		
Remote areas		5–50
Rural areas		50–200
Urban areas		500–2000
Near smelters		1000–20,000
Drinking water[g]		
Normal (rural)		1–20
Normal (urban)		1–40
Plumbosolvent (rural)		100–1000
Plumbosolvent (urban)		100–1000
Plants[b]	—	0.1–30
Vegetables and fruits	1.5	<1.5–18
Forage legumes[a]	2.5	trace-3.6
Ferns[e]	2.3	—
Fungi[e]	—	0.2–40
Lichen[e]	—	1–78
Freshwater (μg L^{-1})[e]	3	0.06–120
Seawater (μg L^{-1})[e]	0.03	0.03–13

Sources:
[a]Cannon (1974).
[b]Swaine (1978).
[c]Ainsworth and Rai (1987).
[d]Appendix Tables A.9 and A.10.
[e]Bowen (1979).
[f]Alloway (1990).
[g]Merian (1991).

floor Pb pool per year), and the mean residence time has been estimated to range from 150 to 500 years. The distribution of Pb in major vegetative components and soils at the Walker Branch forest ecosystem in Tennessee indicates that most of the Pb in living vegetation is located in the bole (30%) and lateral roots (40%) (Van Hook et al., 1977). The average Pb concentrations (in ppm) in the biomass of 11 tree species were: foliage, 4.7; branch, 3.2; and bole, 1.5. Thus, the concentrations of Pb in vegetation followed the pattern: roots > foliage > branch > bole.

In Europe, heavy metal concentrations in the Solling forest ecosystems in central Germany often exceed the values reported from other regions that are not influenced by local pollution (Heinrichs and Mayer, 1977, 1980). For example, concentrations of Pb in the forest litter are higher in the Solling than in the mixed deciduous forest in eastern Tennessee (Van Hook et al., 1977) and in subalpine forests in New Hampshire (Reiners et al., 1975). It was also observed that Pb is the only element that shows a clear concentration increase with the aging of spruce needles and a further drastic increase in the freshly fallen litter and in the forest floor (Heinrichs and Mayer, 1980). This is consistent with Tyler's (1972) and Nilsson's (1972) observations for Swedish spruce needles. The pool of Pb in vegetation is about six times higher and that in the forest floor layers about 20 times higher in the Solling beech forest ecosystem than the ones reported for the Tennessee forest ecosystem (Van Hook et al., 1977).

In the northeastern United States, changes in the concentration and amount of Pb in the forest floor (organic horizon) from 11 forest stands on Camels Hump, Vermont between 1966 to 1980 and 1980 to 1990 have been monitored (Table 10.2). The 11 stands represent an elevational transect from the northern hardwood forest type (500 to 750 m) through the transition forest (750 to 850 m) to the subalpine spruce–fir forest (also referred to as the northern coniferous or boreal forest, 850 to 1200 m). The soils are Spodosols with well-developed surface horizons (forest floor). The OM content of the forest floor ranged from 50% of the total mass in the northern hardwood forest to >80% in the spruce–fir forest zone. Again the data indicate the rising Pb content of the forest floor with elevation. The accumulation of Pb in the forest floor peaked in 1980 and remained somewhat stable until 1990. This trend corresponds to the peak of Pb consumption as a gasoline additive in 1978 in the United States and declined rapidly to pre-1943 levels by 1988 (Table 10.3).

Lead concentrations in precipitation for the northeastern U.S. during the period 1966 to 1989 are summarized in Table 10.3. Lead concentrations were highest in 1966 and had declined in 1989 by 94%. The observed rate of decline from 1978 to 1982 (48%) was nearly twice the reduction observed at a rural site in Minnesota (26%) during roughly the same period but is consistent with the relative decrease in Pb consumption during this period. Estimated Pb concentrations in cloudwater and air are also given in

TABLE 10.2. Lead content of forest floor (organic soil horizon) at Camels Hump, Vermont.

Forest zone	Elevation	Pb amount, kg ha^{-1}					
		1966		1980		1990	
		Mean	SE	Mean	SE	Mean	SE
Subalpine	850–1200	12.70	(0.74)	19.98	(1.17)	19.38	(1.16)
Northern hardwood	500–750	5.98	(1.09)	8.44	(0.86)	5.89	(1.07)
Transition	750–850	4.41	(0.77)	12.72	(1.48)	11.30	(1.97)

Source: Modified from Miller and Friedland (1994).

Table 10.3. The concentration of Pb in air was linearly correlated with the record of Pb use as a gasoline additive ($r^2 = 0.63$, $p = 0.06$) during the period 1966 to 1989. Records of Pb accumulation in northeastern U.S. lake sediments show Pb accumulation rates that increase from < 0.10 mg cm^{-1} yr^{-2} in 1820 to approximately 1 mg cm^{-2} yr^{-1} in 1943.

Upon decomposition of the forest litter, Pb can be expected to mobilize upon its complexation with DOC to the mineral soil layer underneath. Lead can also be mobilized downward in the form of OM colloids. In essence, the mineral soil horizon can serve as the final sink for Pb because of its OM and Fe oxide contents. The results from four profiles sampled in the northern hardwood forest zones in New Hampshire are shown in Figure 10.2A. Ashed/acid-soluble Pb concentrations were highest in the Bh horizon ($\sim$3 cm depth) and decreased with depth (Miller and Friedland, 1994). The total ashed/acid-soluble Pb inventory averaged 15.2 ± 1.1 kg ha^{-1}. The integrated transport of Pb from forest floor to mineral soil estimated for the period 1800 to 1900 is 17 kg ha^{-1}. The small difference between these values may be accounted for by the pool of Pb stored in biomass ($\sim$0.9 to 1.2 kg ha^{-1}) and the cumulative loss of Pb from the forest in stream water. Measurements of total Pb in stream water draining from the forest at Camels Hump and similar forests elsewhere in the northeastern United

TABLE 10.3. Atmospheric Pb concentrations reported for the northeastern United States during the 1966–1989 period and Pb consumption as a gasoline additive.

Period	Precipitation (μg L^{-1})	Cloudwater (μg L^{-1})	Air (ng m^{-3})	Pb consumption (10^6 tonne yr^{-1})
1966–1967	32.7	98.1	160	0.231
1971–1972	31.2	93.6	152	0.252
1975–1976	25.2	75.6	123	0.210
1977–1978	15.6	46.8	76	0.201
1982	17.0	51.0	83	0.119
1988–1989	1.9	5.4	9	0.028

Source: Modified from Miller and Friedland (1994).

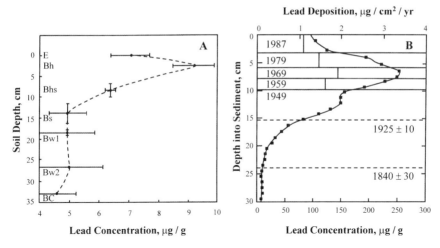

FIGURE 10.2. (A) Ashed/acid-soluble (pH = 1) Pb concentration in the mineral soil horizon of the northern hardwood forest zone on Camels Hump, Vermont. Average of four vertical profiles sampled at ~600-m elevation. The mean ±1 SE of the mean are plotted for seven field-distinguishable soil horizons. (Reprinted with permission from Miller and Friedland, Lead migration in forest soils, *Environmental Science & Technology*, 28, p. 667. Copyright © 1994 American Chemical Society.) (B) Stratigraphy of Pb concentrations and estimated Pb accumulation rates in a sediment core from Black Pond, New Hampshire. Annual laminations of the upper 10 cm provided a detailed recent chronology to supplement approximate pollen dates of 1925 (chestnut decline) and 1840 (increase in weeds) deeper in the core. (From Johnson et al., 1995.)

States indicate very stable and low-level (< 12 g ha^{-1} yr^{-1}) outputs of Pb from these ecosystems. Because stream water fluxes show limited annual variability in this forest type, an upper boundary for stream transport of Pb from this forest during 1940 to 1990 of ~0.6 kg ha^{-1} was estimated, assuming very low Pb losses prior to the dramatic rise in atmospheric Pb concentrations after 1940.

At the Hubbard Brook Experimental Forest (HBEF) in New Hampshire, the patterns in the concentration and flux of Pb in precipitation were remarkably similar to national emission trends (Johnson et al., 1995). The annual mean concentration of Pb in bulk precipitation at HBEF declined by 96% between 1975 and 1989 (from 23 μg L^{-1} in 1975 to 0.85 μg L^{-1} in 1989). The input of Pb in bulk deposition declined from 396 g ha^{-1} yr^{-1} in 1976 to 12.7 g ha^{-1} yr^{-1} in 1989. Early in the study period, the concentration of Pb in precipitation exceeded the U.S. drinking water standard (25 μg L^{-1}).

There was no apparent long-term trend in the concentration of Pb in stream water at HBEF (Johnson et al., 1995). Peak Pb concentrations in stream water were greater in the 1970s than in the 1980s, suggesting a decline. It is clear that precipitation concentrations have exceeded stream water concentrations throughout the study period. The Pb concentration of HBEF streams has been well below the U.S. drinking water standard throughout the record.

Between 1977 and 1987, the Pb content of the forest floor decreased by 29% (Johnson et al., 1995). These changes reflect decreased Pb concentrations since the mass and OM content of the forest flood did not change during this period. The timing of the decline in forest floor Pb content corresponds to the timings of the decreases in Pb consumption in gasoline and atmospheric deposition. This pattern is consistent with declines in forest floor Pb content and concentration observed in samples collected from across the northeastern United States (Friedland et al., 1992).

Sediments deposited in Black Pond, New Hampshire, showed elevated Pb concentrations as early as the beginning of the 20th century relative to early 19th century values. Peak Pb concentrations were found in sediments deposited during the mid-1960s (Fig. 10.2B). These observations are consistent with patterns established for other ponds in New England. The concentration of Pb in the most recently deposited sediments is similar to that of sediments deposited in the 1930s. The maximum accumulation rate calculated from the core was 190 g ha^{-1} yr^{-1}, which is also similar to maximum rates reported for other ponds in the northeastern United States, but considerably lower than their measurements of Pb deposition measured at HBEF in the mid-1970s.

Lead budgets for the periods 1975 to 1977 and 1985 to 1987 are shown in Table 10.4. In the earlier period, the exchangeable Pb pools in the mineral soil and soil solution Pb concentrations were not measured. Several important observations can be made from this study (Johnson et al., 1995):

The terrestrial ecosystem has served as a net sink for Pb. Though input of Pb in bulk precipitation in 1985 to 1987 declined to ~9% of the input in 1975 to 1977, Pb continued to accumulate in the forest ecosystem. The rate of ecosystem accumulation (input−output) of Pb decreased from 319 to 28 g ha^{-1} yr^{-1} (see Table 10.4).

Forest vegetation is not a major pool of the Pb budget at HBEF. The pool of Pb in forest vegetation is much smaller than the pools in forest floor and mineral soil (see Table 10.4). This pattern is expected since Pb is not a plant nutrient. Uptake of Pb into forest biomass was significant in the mid-1970s, when forest growth was rapid and Pb inputs were high. Small pools of Pb in forest vegetation have also been observed in other studies (e.g., Friedland and Johnson, 1985; Heinrichs and Mayer, 1977). There is evidence that foliar uptake of Pb from intercepted atmospheric deposition constitutes a significant fraction of Pb uptake in trees.

Lead leached from the forest floor is immobilized in the mineral soil, minimizing losses in drainage water. Soil solution fluxes of Pb decreased with depth within the mineral soil (see Table 10.4), indicating immobilization. It is likely that Pb is mobilized in the forest floor as Pb-OM complexes. Lead is known to form stable compounds with dissolved OM at low pH. Dissolved

TABLE 10.4. Ecosystem budgets of Pb at Hubbard Brook Experiment Forest, New Hampshire, for two time periods: high atmospheric deposition of Pb (1975–1977), and low Pb deposition (1985–1987).

Description	1975–1977 Pool (g ha^{-1})	1975–1977 Flux (g ha^{-1} yr^{-1})	1985–1987 Pool (g ha^{-1})	1985–1987 Flux (g ha^{-1} yr^{-1})
Bulk precipitation		325		29
Forest floor	10500		7500	
Oa horizon soil solution	ND	ND[a]		56
E + Bh exchangeable			2200	
Bh horizon soil solution		ND		48
Bsl exchangeable	ND		640	
Bs soil solution		ND		21
Bs2 exchangeable	ND		1500	
Stream water		6		4
Aboveground biomass	640		680	
Belowground biomass	620		660	
Net plant uptake		33		1
Total soil Pb	230,000		230,000	
Weathering release		7		7

[a]ND = not determined.
Source: Johnson et al. (1995).

OM has been immobilized in the Bh and Bs horizons of Spodosols at HBEF. Total concentrations of Pb are also highest in these horizons (Smith and Siccama, 1981). Thus, these horizons may serve as zones of Pb accumulation.

The forest floor has been a net source for Pb. In the mid-1970s, Pb accumulation in the forest floor at HBEF was reported to be ~305 g ha^{-1} yr^{-1}. However, in the mid-1980s, the input of Pb to the forest floor in bulk precipitation was less than the estimated output in forest floor leachate (see Table 10.4). Based on this difference (forest floor leachate−bulk precipitation), the average net loss of Pb from the forest floor in 1985 to 1987 was 27 g ha^{-1} yr^{-1}.

Patterns in Pb accumulation in aquatic sediments may not be consistent with changes in atmospheric inputs. The pattern of sharp decline in sediment Pb concentration is consistent with the atmospheric inputs of Pb at HBEF, which have decreased since at least 1975 (see Fig. 10.2B). However, a close examination of the core data illustrates the shortcomings of using sediment cores for the quantitative estimation of historical rates of atmospheric deposition, i.e., the peak in sediment Pb accumulation and concentration was observed in the mid-1960s, while the peak Pb input occurred in 1970 and 1972.

Complementary to the results of Johnson et al. (1995) for watershed 6 at HBEF is the study of Wang et al. (1995), who used clean techniques at all stages of sample collecting, processing, and analysis. Data on Pb indicate a concentration gradient starting with bulk precipitation and moving through the soil Oa, Bh, and Bs horizons to seeps and streams, indicating that a significant fraction of Pb is being transported out of the forest floor layer.

Estimation of Pb outflow via streams is a key component of the study of Wang ct al. The estimated value is the total output of both dissolved and suspended particulate Pb under base flow conditions. Compared with about 21 g ha^{-1} yr^{-1} Pb escaping from the forest floor layer, Pb outflow in streams is very low (<0.20 g ha^{-1} yr^{-1}), which indicates that stream-dissolved Pb derived from soil percolates is negligible. This reconfirms earlier findings that the ecosystem appears to be an excellent sink that effectively accumulates atmospheric Pb especially in the surface soil.

Just like in the northeastern United States, Scandinavia is subject to air pollution by long-distance transport of atmospheric particulates from upwind sources in western Europe. This regional pattern is most apparent within Norway, i.e., a southern Norway–central Norway heavy metal deposition gradient (Berthelsen et al., 1995). To evaluate temporal variations and plant species/parts effects, studies were conducted in 1982 and 1992 in Norway involving mixed deciduous–coniferous forest overlying podzols and organic surface soils. Of the metals examined, Pb was the only element showing a significant decrease in plant concentrations from 1982 to 1992 (see Table 10.5). These changes were evident both in southern and central Norway and confirm the important influence of atmospheric deposition on the biogeochemical cycling of Pb in nature. Rates of atmospheric Pb deposition also decreased distinctly during this 10-yr

TABLE 10.5. Lead concentration (ppm DW) in different plant species and plant parts collected from forest and ombrotrophic bogs in southern and central Norway in 1982 and 1992.

Species		1982	1992
	From forests		
Southern Norway			
Norway spruce	Twig	19.76 ± 9.83	12.63 ± 1.16
	Needle	0.34 ± 0.14	0.89 ± 0.50
Birch	Twig	12.04 ± 1.99	9.14 ± 2.28
	Leaf	1.30 ± 0.05	1.10 ± 0.23
Scotch pine	Twig	12.84 ± 1.50	6.02 ± 0.98
	Needle	4.00 ± 0.50	1.11 ± 0.47
Juniper	Twig	9.55 ± 4.07	7.38 ± 1.86
	Needle	7.07 ± 1.44	3.96 ± 1.18
Bilberry	Stem	1.86 ± 0.36	1.68 ± 0.56
	Leaf	0.34 ± 0.07	0.40 ± 0.08
Mountain cranberry	Stem	3.68 ± 0.72	4.42 ± 1.13
	Leaf	1.28 ± 0.22	0.63 ± 0.10
Heather		9.49 ± 1.10	2.82 ± 0.85
Crowberry		4.40 ± 1.90	2.82 ± 1.05
Feather moss		67.90 ± 24.00	22.20 ± 3.05
Bog whortleberry	Stem	2.64	2.06 ± 0.54
	Leaf	0.13	0.11 ± 0.06
Central Norway			
Norway spruce	Twig	5.49 ± 1.37	1.16 ± 0.59
	Needle	0.19 ± 0.04	0.03 ± 0.01

TABLE 10.5. (*Continued*)

Species		1982	1992
Birch	Twig	3.42 ± 0.77	0.87 ± 0.05
	Leaf	0.37 ± 0.12	0.28 ± 0.06
Scotch pine	Twig	1.94 ± 0.24	0.33 ± 0.15
	Needle	0.81 ± 0.10	0.21 ± 0.07
Juniper	Twig	2.71 ± 1.91	0.38 ± 0.30
	Needle	1.62 ± 0.38	0.26 ± 0.07
Bilberry	Stem	0.53 ± 0.13	0.08 ± 0.01
	Leaf	0.15 ± 0.02	0.08 ± 0.02
Mountain cranberry	Stem	1.26 ± 0.37	0.29 ± 0.03
	Leaf	0.35 ± 0.04	0.06 ± 0.02
Heather		1.65 ± 0.61	0.32 ± 0.03
Crowberry		0.75 ± 0.11	0.25 ± 0.01
Feather moss		5.10 ± 0.35	1.84 ± 0.00
Bog whortleberry	Stem	—	0.33 ± 0.09
	Leaf	—	0.06 ± 0.02
From ombrotrophic bogs			
Southern Norway			
Norway spruce	Twig	—	13.28 ± 3.17
	Needle	—	0.24 ± 0.05
Birch	Twig	15.76 ± 2.38	10.67 ± 1.55
	Leaf	1.27 ± 0.09	0.87 ± 0.12
Scotch pine	Twig	9.53 ± 0.98	7.67 ± 1.15
	Needle	4.42 ± 0.44	1.67 ± 0.08
Bilberry	Stem	3.61 ± 2.11	1.70 ± 0.16
	Leaf	0.49 ± 0.10	0.23 ± 0.02
Mountain cranberry	Stem	6.24	2.84 ± 0.33
	Leaf	1.38	0.57 ± 0.08
Heather		8.75 ± 2.43	5.19 ± 1.10
Crowberry		8.33 ± 1.09	3.24 ± 1.08
Bog whortleberry	Stem	8.25	2.11 ± 0.20
	Leaf	0.14	0.16 ± 0.05
Peat moss		46.23 ± 18.00	15.13 ± 1.81
Central Norway			
Norway spruce	Twig	10.07	1.49 ± 0.18
	Needle	0.30	0.04 ± 0.01
Birch	Twig	4.62	0.04 ± 0.01
	Leaf	0.49	0.16 ± 0.01
Scotch pine	Twig	1.43 ± 0.28	0.55 ± 0.22
	Needle	0.78 ± 0.02	0.19 ± 0.02
Bilberry	Stem	0.84	0.15 ± 0.01
	Leaf	0.18	0.12 ± 0.02
Mountain cranberry	Stem	1.60	0.44 ± 0.03
	Leaf	0.33	0.05 ± 0.02
Heather		2.14 ± 0.32	0.74 ± 0.30
Crowberry		0.68 ± 0.12	0.29 ± 0.17
Bog whortleberry	Stem	—	0.28 ± 0.07
	Leaf	—	0.04 ± 0.03
Peat moss		3.57 ± 0.88	1.39 ± 0.24

Values are arithmetic means ± SD.
Source: Modified from Berthelsen et al. (1995).

period, whereas Pb levels in surface soils and peat in southern and central Norway remained approximately unchanged from 1982 to 1992.

A recent study by Miller and Friedland (1994) shows that the residence time (response time) for Pb in organic surface soils is much lower even than previously estimated, which may indicate some availability of Pb for root uptake. In addition, the transport of Pb from roots to shoots is low. The significant decrease in plant Pb concentrations and decreased rates of atmospheric Pb deposition observed in the Norway study from 1982 to 1992 strongly suggest that direct Pb deposition on vegetation (both wet and dry) significantly influenced Pb concentration in vegetation. The strong influence of direct atmospheric Pb deposition on vegetation and the more than 15 times higher Pb deposition in southern Norway compared with central Norway caused the much higher Pb levels in southern Norway vegetation. The possibility that larger contributions from root uptake to Pb levels in vegetation in southern Norway causing the much higher Pb levels in southern Norway than in central Norway due to higher Pb levels in surface soils and peat cannot be ruled out.

No significant differences were observed between plant Pb concentrations in forests and on ombrotrophic bogs even in central Norway where contributions from atmospheric deposition to plant Pb levels were much smaller than in southern Norway (see Table 10.5). This indicates that contributions from current weathering of mineral material to Pb concentrations in forest plants in southern and central Norway are negligible.

4 Lead in Soils/Sediments

4.1 Total Lead in Soils

Lead is present in all soils at levels ranging from < 1.0 ppm in normal soils to well over 10% in ore materials. The following Pb contents (in ppm) of noncontaminated soils for several regions have been reported (Nriagu, 1978): United States, 18; former Soviet Union, 12; Canada, 12; Europe, from 10 for Spanish to 42 for Scottish soils; Australia and New Zealand, 15; Japan, 11; Africa, from 12 for South African to 21 for Egyptian soils; and Antarctica, 8. For major soil groups, organic soils (mean = 44 ppm) appear to have three times as much Pb as mineral soils. For soils in China ($n = 3989$), a mean value of 27 ppm Pb (geometric mean = 24) was reported (Chen et al., 1991), with the highest values obtained for Lithosols (mean = 30, $n = 205$) and the lowest values for Mollisols (mean = 18, $n = 240$). Reaves and Berrow (1984a) found a mean concentration of 14 ppm Pb ($n = 3944$) for soil samples collected from 8967 Scottish soil profiles, with a range of 2.5 to 85 ppm. Their results indicate that the average Pb content of organic soils is much higher than that of the mineral soil samples (i.e., 30 ppm vs. 13 ppm), but it is also much more variable. Davies (1983) reported that concentrations greater than 110 ppm total Pb should not be considered anomalous, with a high possibility that they may have originated from contamination. Swaine's (1955) 2 to 200 ppm range for total Pb is widely used and is regarded as a reasonable assessment of typical soil values.

Agricultural soils have a wide range of Pb content depending on a number of factors, such as parent material and anthropogenic input. In England and Wales, the median Pb content was 42 ppm, with range of 5 to 1200 ppm (Archer, 1980). These values are based on the analysis of 752 samples collected from 226 farms. In Ontario, Canada, Pb content of agricultural soils averaged 46 ppm ($n = 296$), with a range of 1.5 to 888 ppm (Frank et al., 1976). Soils from fruit orchards had the highest average contents of 123 ppm (range of 4.4 to 888 ppm) from the past use of lead arsenate pesticides. Other cropped soils had only 14 ppm Pb, with a range of 1.5 to 50 ppm. In addition, Chisholm and Bishop (1967) noted that the Pb content of orchard samples from Nova Scotia usually had > 50 ppm and nonorchard soils had concentrations below this level.

Frequent and heavy applications of lead arsenate (its use in the United States was banned in the 1960s) as a pesticide in deciduous fruit trees led to substantial accumulation of Pb and As in orchard surface soils, particularly in apple orchards throughout the world (Creger and Peryea, 1997). Air pollution can also enrich soils with Pb. In the Katowice district, the most industrialized and polluted part of Poland, Pb in soils averaged 221 ppm (17 to 1641 ppm range) (Kucharski et al., 1994), similar to the high values reported for garden soils in Upper Silesia, Poland (Gzyl, 1990). These values highly contrast to the reported values of 8 to 80 ppm for relatively noncontaminated soils for Poland (Kabata-Pendias, 2000).

The Pb values reported for normal agricultural soils in Canada are in the range of <1.0 to 12 ppm (Warren et al., 1969). In the Netherlands, field agricultural soils had a mean of 25 ppm ($n = 673$) vs. 58 ppm ($n = 147$) for greenhouse soils (Table 10.6), again indicating the influence of land use on soil content of metals. Other major anthropogenic sources of Pb that can enrich agricultural soils are automobile emissions, industrial deposition, and sewage sludge.

4.2 Bioavailable Lead in Soils

As with most trace elements, total Pb in soils usually is not a good indicator of Pb bioavailability. Therefore, extractable Pb is normally used as the indicator of amounts bioavailable for plant uptake. A number of extractants, e.g., HCl, HNO_3, NH_4OAc, $CaCl_2$, and organic acids have been used to extract soil Pb to predict its bioavailability to plants. The success of predictability seems to depend on several factors, such as type of extractant, soil properties especially pH, plant, and others.

In testing four extractants, it was found that soil Pb extractability was in the order: Grigg's reagent > 0.1 N HCl > 0.02 M EDTA > 1 N NH_4OAc (Misra and Pandey, 1976). Note: the Grigg's reagent is an acid ammonium

TABLE 10.6. Lead contents of soils (ppm DW) from several countries.

n	Min.	Max.	Mean	Country
708	0	460	32 (median=23)	Netherlands, soils used for agric and hortic
673	0	72	25	Netherlands, same soils, without the highest 5% (outdoors)
155	4	420	71 (median=45)	Netherlands, greenhouse soils
147	4	230	58	Netherlands, same soils, without the highest 5%
248			43	Netherlands, clay
63			31	Netherlands, sand
361	2	364	16	Sweden
51			24	Denmark
2223			20	Germany, Hessen
2742			40	Germany, Baden–Württemberg
472			50	Germany, Nordrhein–Westfalen
296	2	888	46	Canada, Ontario
3989			27 (median=24)	China

Source: Modified from Wiersma et al. (1986).

oxalate buffered solution. Both Grigg's reagent and NH_4OAc extractants gave significant correlations with Pb concentrations in wheat plants. Similarly, while HCl, DTPA, and Chelex resin extractions gave significant correlations with wheat plant Pb uptake, that for $CaCl_2$ did not (Lee and Zheng, 1994). The predictability of Pb uptake is also plant species–dependent. For example, among extractants evaluated (John, 1972), only 1 N HNO_3 gave significant correlations with Pb in lettuce, while 1 N NH_4OAc, 0.10 N HNO_3, and 0.01 N HNO_3 all gave significant correlations with Pb in oat shoots.

Soil Pb extracted by 0.5 M $BaCl_2$ or 0.05 M EDTA produced significant relationships (Fig. 10.3A) with Pb contents of ryegrass than either total soil Pb or Pb extracted by 2.5% HOAc. On the average, the solutions of HOAc, $BaCl_2$, and EDTA extracted 1.0, 16.3, and 32.7%, respectively, of the total Pb in soils. The higher efficiency of extraction by $BaCl_2$ over HOAc can be attributed to the greater degree of replacement of Pb^{2+} by Ba^{2+} than by H^+, due to the valence and similar radii of the ions (1.32 and 1.43 Å for Pb and Ba, respectively).

Additionally, the exchangeable concentrations of Pb in soil [extracted by widely used NH_4HCO_3–DTPA solution] correlates well ($r = 0.95$) with the total amount of Pb taken up by lettuce or spinach plants (Fig. 10.3B). The mine dump soils contained from 135 to 21,700 ppm total Pb, with corresponding spinach plants containing <5 to 45 ppm DW in their leaf tissues (Boon and Soltanpour, 1992).

The HOAc extract has been successfully used to estimate the exchangeable and dilute-acid–soluble fractions of trace elements such as Pb, Co, Ni, and Zn (Berrow and Mitchell, 1980). For Scottish soils, a mean HOAc-

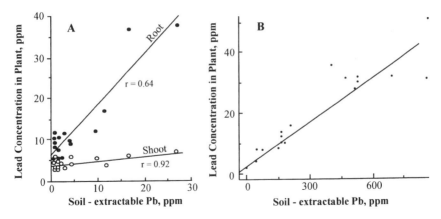

FIGURE 10.3. Relationships between exchangeable Pb in soils and Pb concentrations or amounts in (A) ryegrass and (B) lettuce or spinach plants. (Modified from [A] Jones et al., 1973a,b; [B] Boon and Soltanpour, 1992.)

extractable Pb content of 0.24 ppm, with a range of 0.016 to 3.4 ppm has been reported (Reaves and Berrow, 1984b). The extractable Pb concentration was positively and linearly correlated with that of total Pb. Extractable Pb was found to be generally higher in surface horizons of these soils than in underlying horizons and extractable levels were higher in poorly drained soils.

The extractability of Pb in soils is influenced by several soil properties. Soils limed to high pH yielded less extractable Pb (John, 1972; MacLean et al., 1969; Misra and Pandey, 1976). Also, soils with high phosphate (MacLean et al., 1969), OM, or clay contents (Karamanos et al., 1976; Scialdone et al., 1980) tend to reduce the extractability of Pb from soils.

4.3 Lead in Soil Profile

Pedogenic processes, climatic and topographic effects, and microbial activities may influence the distribution of Pb in soil profile. In general, Pb accumulates in the soil surface, usually within the top few centimeters and diminishes with depth. In Scotland, Swaine and Mitchell (1960) observed that the surface horizons of most soils have higher, often considerably higher, Pb than the subsurface horizons. This pattern of Pb accumulation can be attributed to OM accumulation on the surface due to plant dry matter recycling rather than to anthropogenic sources. The declining trend of Pb with soil depth has been observed also in farmed organic soils (>85% OM) in the Holland Marsh area near Toronto, Canada (Czuba and Hutchinson, 1980). Lead levels decreased from surface values of 22 to 10 ppm at a depth of 48 cm. In comparison, Pb did not decrease with depth in mineral soils with low Pb levels.

Lead of anthropogenic origin generally exhibits the same accumulation pattern on the surface layers, but also may migrate to deeper layers in some cases. In Cape Cod, Massachusetts, excess ^{210}Pb from atmospheric deposition has migrated to a depth of 40 cm in an undisturbed soil profile (Fisenne et al., 1978). This depth penetration of deposited material has been confirmed independently by ^{90}Sr measurements in the same soil profile (^{210}Pb and ^{90}Sr have some chemical similarity with comparable half-lives). Field plots at the University of Illinois amended with variable amounts of Pb (0 to 3200 kg Pb ha^{-1} as Pb acetate) had the Pb migrating down to at least the 90-cm depth, 6.5 yr after the soil treatment (Stevenson and Welch, 1979). The soil in the plots, a silty clay loam with a pH of 5.9 and CEC of 30.3 meq/100 g, is known to effectively bind Pb in nonexchangeable form with sorption capacity for Pb exceeding 20,000 μg g^{-1}. Downward movement of Pb was attributed to leaching as soluble chelate complexes with OM, transfer of soil particles by earthworms and other faunal organisms, translocation in plant roots, or a combination of these. Downward movement below the 30-cm depth could not have been caused by mechanical mixing of the soil during tillage operations. However, the retention of Pb in the surface layers was postulated due to reactions involving insoluble OM. In addition to vertical movement, significant horizontal movement to adjacent plots also occurred, primarily due to physical transfer of soil particles during tillage operations, or as windblown soil or plant particles.

Rains (1975) also showed the close association of Pb with OM in soil profiles of grassland ecosystems impacted by smelting operations. Similarly, Andersson (1977) found close associations between Pb and OM in profiles of forest ecosystems. Similarly, in contaminated soils in the Nile Delta of Egypt the surface soil was highly enriched with Pb relative to the subsurface soil (Elsokkary et al., 1995). In the alluvial soils, Pb ranged from 18 to 1850 ppm; likewise, the lacustrine soils had 39 to 1985 ppm total Pb (vs. the 14 ppm background level for Egypt). While the alkyl Pb also mostly accumulated in the surface soil, the enrichment peaked at the 15 to 30 cm depth; alkyl Pb was usually around 1% of the total Pb pool in the soil.

Because of the strong affinity of Pb for OM and its generally immobile nature, Pb can be expected to accumulate in the surface layers of soils. For example, essentially all of the Pb that has accumulated from automobile emissions has remained in the surface few centimeters of the soil profile (Al-Chalabi and Hawker, 2000; Milberg et al., 1980; Page and Ganje, 1970). Chang et al. (1984b) found that in cropped soils treated continuously for 6 yr with municipal sewage sludge, > 90% of the applied Pb was found in the surface 0 to 15 cm, the zone of application. No statistically significant increase in heavy metal contents was detected below the surface 30 cm of the soil profile. Similarly, Williams et al. (1984) found that metal movement within the profile was limited to a depth of 5 cm below the zone of sludge application (0 to 20 cm) for Pb, Cd, and Cu.

4.4 Forms and Chemical Speciation of Lead in Soils and Dusts

Sposito et al. (1982) fractionated Pb in arid field soils amended with sewage sludge using sequential extraction to estimate exchangeable, sorbed, organic, carbonate, and sulfide fractions of Pb. For the soil treated with 45 tonnes ha^{-1} yr^{-1} of composted sludge for 4 yr, the Pb distribution (in ppm) was as follows: total Pb, 70 ± 2; exchangeable + sorbed, 1.6 ± 0.6; organic, 3.9 ± 0.1; carbonate, 50.8 ± 0.3; and sulfide, 17.6 ± 0.0. Thus, the dominant fractions of Pb are: carbonate $\gg$ sulfide $>$ organic. These data indicate that the most dominant fraction is the carbonate regardless of the sludge application rate. The organic fraction did not change with application rate, whereas the sulfide fraction decreased slightly. The low percentages of Pb in the exchangeable and sorbed forms indicate low bioavailability of this metal to plants. Complementary to this, Chang et al. (1984a) observed that the forms of Pb in soils were not substantially affected by applications of sewage sludge.

In alkaline Nile Delta (Egypt) soils, most of the Pb accumulated in the OM-sulfide fraction especially in highly contaminated soils, confirming the high affinity of Pb for OM (Elsokkary et al., 1995). In contrast, most of the Pb in the mine-contaminated alkaline soils in the Doñana National Park, Spain were associated with the Fe–Mn oxide fraction and, to a lesser extent, with the carbonate fraction (Ramos et al., 1994). Similarly, contaminated acidic soils from southwest Poland had most of the Pb associated with the Fe–Mn oxide phase, followed by the OM and carbonate phases (Chlopecka et al., 1996).

For road dust, Pb was partitioned as follows (Thornton and Culbard, 1986): exchangeable (5%), carbonate (30%), Fe–Mn oxide (35%), organic (18%), and residual (12%). In general, the affinity of Pb for reactive soil phases largely depends on the soil type and source term of Pb.

Alkaline, calcareous soils of the arid and semiarid regions are an excellent sink for Pb (Jurinak and Santillan-Medrano, 1974). For example, the movement of Pb in the percolating water is essentially nil, even in the presence of large amounts of Pb salts. Cadmium however, is less effectively removed from percolating water by these soils. The solubility of Pb in soil is about 100 times less than that of Cd in the pH range of 5 to 9.

Lead undergoes hydrolysis at low pH values and displays multiple hydrolysis reactions at pH values encountered in the environment. Based on thermodynamic calculations, $Pb(OH)_2$ formation is important above pH 9, while $PbOH^+$ is predominant between pH 6 and 10 (Hahne and Kroontje, 1973; Schulthess and Huang, 1990). Computer calculations for aqueous systems indicate that at pH $\sim$7 to 8, the distribution of Pb^{2+} will be determined by the solubility of $PbOH^+$, with the latter predominating with increasing pH (Fig. 10.4).

In arid and semiarid regions, soils are not only alkaline but also high chloride content is often associated with high salinity. Furthermore, chloride is ubiquitous in the natural soil environment and can be regarded as a very mobile and persistent complexing agent for metal. In this scenario, chloride ions may influence the degree of complexation and mobility of Pb (Roy et al., 1993).

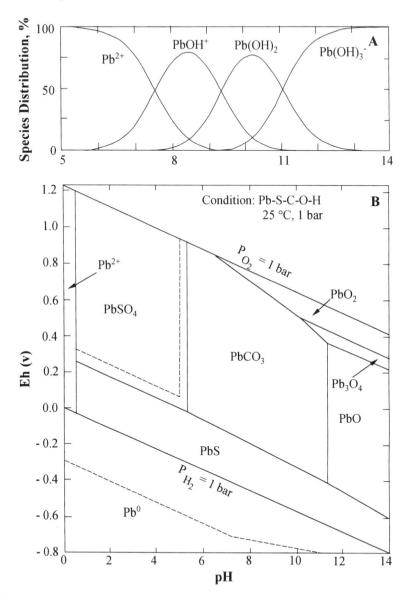

FIGURE 10.4. (A) Predicted aqueous monomeric chemical speciation of lead as a function of pH. (B) Predicted Eh–pH stability field for lead; the assumed activities of dissolved species are: $Pb = 10^{-6}$, $S = 10^{-3}$, $C = 10^{-3}$. (Extracted from [A] Schulthess and Huang, 1990; [B] Brookins, 1988.)

There is some evidence that certain microorganisms in soils and lake sediments can biotransform inorganic Pb to organolead (Pain, 1995). Organo-Pb compounds are more readily absorbed by organisms and potentially more toxic.

4.5 Sorption of Lead in Soils

The chemistry of Pb in soils can be qualitatively described as affected by (1) the specific adsorption to various solid phases, (2) the precipitation of sparingly soluble or highly stable compounds of which it is a constituent, and (3) the formation of relatively stable complexes or chelates that result from the interaction with OM.

4.5.1 Adsorption of Lead onto Soil Components

Adsorption of Pb by soils and clay minerals has generally conformed to either the Langmuir or Freundlich isotherm. Depending on the mass of clay used, Griffin and Au (1977) utilized the Langmuir equation over a concentration range of 0 to 1200 ppm Pb. Soldatini et al. (1976) found that Pb adsorption by soils conformed to either the Langmuir or Freundlich isotherm over a wide range of concentrations. Statistical analyses of their data indicate that OM and clay were the dominant constituents contributing to Pb adsorption, while the influence of other adsorbing phases, such as Mn oxides was negligible.

From equilibrium batch studies, Santillan-Medrano and Jurinak (1975) found that in noncalcareous soils the solubility of Pb is regulated by Pb-$(OH)_2$, $Pb_3(PO_4)_2$, $Pb_4O(PO_4)_2$, or $Pb_5(PO_4)_3OH$, depending on the pH. The $Pb(OH)_2$ appears to regulate Pb^{2+} activity when the solution pH is less than about 6.6. As the pH increases, the formation of Pb orthophosphate, Pb hydroxypyromorphite, and also tetraplumbite phosphate becomes a distinct possibility. In calcareous soils, $PbCO_3$ could assume importance (Elkhatib et al., 1991; Sposito et al., 1982).

In other situations, the oxides of Mn and Fe may exert a predominant role on Pb adsorption by soils. In measuring the adsorption of several trace elements by synthetic Mn and Fe oxides, McKenzie (1980) found that Pb adsorption by Mn oxides was up to 40 times greater than that by the Fe oxides, and that Pb was adsorbed more strongly than any other metal (Co, Cu, Mn, Ni, and Zn) studied. This is consistent with earlier findings on the accumulation of Pb in Mn oxides of soils (Norrish, 1975; Taylor and McKenzie, 1966). Three possible mechanisms may account for the binding of Pb by oxides of Mn: (1) strong specific adsorption, (2) a special affinity for oxides of Mn as found for Co (McKenzie, 1970, 1975) with the possibility of oxidation of the Pb, or (3) the formation of some specific Pb-Mn mineral such as coronadite. In aquatic systems (e.g., streams), Pb may be transported in particulate form due to its adsorption on hydrous Fe oxide (i.e., amorphous Fe oxide or ferrihydrite) (Erel et al., 1991).

Reactions involving insoluble OM may also play a predominant role in the adsorption of Pb by soils. In this case, precipitation by carbonate and sorption by hydrous oxides appear to be of secondary importance (Zimdahl and Skogerboe, 1977). While soil OM may immobilize Pb via specific adsorption reactions, mobilization of Pb can also be facilitated by its complexation with DOC or fulvic acids (Pinheiro et al., 1999; Saar and Weber, 1980; Stevenson and Welch, 1979).

Competitive adsorption data by Bittel and Miller (1974) using pure clay systems (montmorillonite, illite, and kaolinite) show that Pb^{2+} was preferentially adsorbed over Ca^{2+}. The importance of competing ions on Pb adsorption was reconfirmed subsequently when the adsorption of Pb by Ca-clay became dependent on the Pb/Ca ratio in solution (Griffin and Au, 1977; Riffaldi et al., 1976).

Lead sorption on α-Al_2O_3 (corundum) involves several mechanisms (Bargar et al., 1996; Strawn et al., 1998). (Note: Corundum is similar to Al hydroxide minerals commonly found in soils, such as gibbsite, bayerite, and boehmite.) It can be generalized that the adsorption kinetics of Pb are biphasic (see also Hg adsorption in the later chapter on mercury), i.e., an initial fast reaction is followed by slow reaction. The slow sorption reaction on α-Al_2O_3 is not due to surface precipitation of Pb (Strawn et al., 1998). Rather, it may due to (1) diffusion to internal sites; (2) adsorption onto sites that have slower reaction rates due to low affinity; and (3) formation of additional sorption sites due to slow transformation of α-Al_2O_3 into the less reactive solid phase. The initial fast reaction phase is most likely due to chemical reactions on readily accessible surface sites of the mineral (Strawn et al., 1998). For example, absorption of 76% of the total Pb adsorbed by α-Al_2O_3 occurred within the initial 15 min. Thereafter, the reaction continued for ~30 hr resulting in only small additional adsorption (~2.5% of the total).

Due to their differing physicochemical properties, metals vary in their affinity for sorption onto soil components. In general, Pb has the strongest affinity to silicate clays, peat, Fe oxides, and usual soils (Basta and Tabatabai, 1992a,b). For example, in two long-term cropped soils in Iowa, the following affinity sequence was observed: $Pb > Cu > Ni \geq Cd = Zn$. Likewise, Sauve et al. (2000) offered the following sequence of adsorption affinity of Pb for the following media: ferrihydrate > natural oxides > soils > OM.

4.5.2 Complexation of Lead with Phosphates

The major Pb species of interest relative to the transport and bioavailability of Pb are the Pb-bearing minerals, Pb-organic complexes, and adsorbed Pb since they represent the main sink and potential sources of Pb in the environment. In essence, they determine the free Pb^{2+} activity in aqueous solutions (see previous sections). It can be postulated that the complexed Pb species (or forms), as compared with free Pb^{2+} ions, are less bioavailable to organisms, including humans. Therefore, the source term of Pb is very important relative to its mobility and uptake.

Lead phosphates have been demonstrated to be a very stable environmental form of Pb. The relative solubilities of simple Pb compounds indicate that Pb phosphates are less soluble under equilibrium conditions than oxides, hydroxides, carbonates, and sulfates under soil-surface conditions (Ruby et al., 1994). The solubility products for common pyromorphite minerals (i.e., chloro-, bromo-, hydroxo-, and fluoropyromorphite) have been measured at 10^{-84} to 10^{-72}, yielding a thermodynamic stability sequence of

$$Pb_5(PO_4)_3Cl > Pb_5(PO_4)_3Br > Pb_5(PO_4)_3OH > Pb_5(PO_4)_3F$$

In contrast, the solubility products of $PbSO_4$ (anglesite), $PbCO_3$ (cerussite), PbS (galena) and PbO (litharge) are 10^{-8}, 10^{-13}, 10^{-28} and 10^{+13}, respectively (Ruby et al., 1994).

The solubility data indicate that Pb pyromorphites are substantially less soluble than the Pb forms generally found in ore minerals (e.g., galena, anglesite, cerussite), paint [e.g., $2\ PbCO_3 \cdot Pb(OH)_2$], litharge, and automobile emission-affected particulates. This invokes the necessity to formulate stable, insoluble Pb compounds to render Pb immobile or nontoxic to organisms. This also has immense implications in remediating Pb-contaminated soils. Accordingly, a rush of publications occurred in the 1990s where Pb-phosphate complexes, using phosphate rocks and other phosphate sources, were evaluated for detoxifying Pb (Chlopecka and Adriano, 1996; Laperche et al., 1996, 1997; Ma et al., 1993, 1994a,b, 1995; Rabinowitz, 1993; Ruby et al., 1994; Zhang et al., 1997).

The most attractive source of phosphate is hydroxyapatite [Ca_{10}-$(PO_4)_6(OH)_2$] because it is fairly abundant and relatively inexpensive, and most importantly, it dissolves to react with Pb:

$$Ca_{10}(PO_4)_6(OH)_{2(s)} + 14H^+_{(aq)} \overset{\text{dissol.}}{\rightleftharpoons} 10Ca^{2+}_{(aq)} + 6H_2PO^-_{4(aq)} + 2H_2O$$

$$10Pb^{2+}_{(aq)} + 6H_2PO^-_{4(aq)} + 2H_2O \overset{\text{precip.}}{\rightleftharpoons} Pb_{10}(PO_4)_6(OH)_{2(s)} + 14H^+_{(aq)}$$

The reactions above occur to completion, forming a fairly stable precipitate of hydroxy pyromorphite [$Pb_{10}(PO_4)_6(OH)_2$] in the absence of competing metals and ligands (Laperche et al., 1996). However, Ma et al. (1994a,b) indicated that the formation of hydroxopyromorphite could be perturbed by the presence of high concentrations of Al, Cu, Fe, Cd, and CO_3^{2-}.

The major aim of remediation of Pb-contaminated soils is to reduce risk relative to its mobility, bioavailability, and/or uptake by organisms. The pyromorphites can facilitate achievement of this goal. For example, dissolution of mine waste–related Pb phases, such as galena (PbS) and anglesite (PbSO$_4$), should be lower in simulated gastric conditions, compared to the more soluble Pb salts, such as Pb(OAc)$_2$ which is the form often used in toxicological studies (Ruby et al., 1992). As indicated earlier, the pyromorphites are much less insoluble than the mine-related minerals. From plant uptake studies, addition of hydroxyapatite to Pb-contaminated soils resulted in much lower Pb uptake by maize or oat plants compared with nontreated soils, or soils treated with zeolites (Knox and Adriano, 2000).

5 Lead in Plants

5.1 Root Uptake of Lead by Plants

The distribution of trace elements in plants can be categorized into three groups: (1) more uniformly distributed between roots and shoots, e.g., Zn, Mn, Ni, and B, (2) more in roots than in shoots, with moderate to sometimes large quantities in shoots, e.g., Cu, Cd, Co, and Mo, and (3) mostly in roots with very little in shoots, e.g., Pb, Sn, Ti, Ag, Cr, and V (Wallace and Romney, 1977). This grouping however, can change according to plant species and soil and environmental conditions.

Plants can accumulate Pb from soil or foliar application. However, the results vary according to plant species, source term, and experimental conditions. There has been a general consensus that Pb primarily accumulates in the roots and is poorly translocated to other plant parts. For example, only < 3 ppm Pb accumulated in the foliage of barley plants but up to 800 ppm accumulated in the roots of some plants when grown in soil containing up to 800 ppm Pb (Keaton, 1937). Lead accumulation by bromegrass from sandy loam soils with a range in Pb contents of 12 to 680 ppm occurred only with the soil having the 680 ppm level (Marten and Hammond, 1966). The maximum accumulation was only 34 ppm, enhanced by the addition of a chelate. Testing several crop species in contaminated soil and in acid-washed sand to which soluble Pb was added at low concentrations, data showed that Pb can be absorbed through the root system with only limited translocation to other plant parts (Motto et al., 1970); most Pb taken up by the plants remained in the roots.

In a field experiment where Pb acetate had been applied to soil at rates ranging from 0 to 3200 kg Pb ha^{-1}, Pb concentration in corn foliage was significantly increased by Pb application (highest level was 27.6 ppm in leaves at tasseling at the 3200 kg Pb ha^{-1} level vs. 3.6 ppm in the control) (Baumhardt and Welch, 1972). The Pb content of corn grain was not affected by added Pb. The results indicate that Pb was absorbed by the roots and translocated to the plant tops, but it was not translocated from stover to the grain. Uptake studies of forage crops indicate that only a very small fraction (0.3 to 0.50%) of added Pb was utilized by the plants (Karamanos et al., 1976). However, corn grown in sand culture absorbed considerable Pb and growth was retarded at high levels of added Pb (Miller and Koeppe, 1970). Lead uptake by eight tree species grown on soil treated with five Pb levels (0 to 600 ppm) was significantly affected by soil Pb concentration, with higher uptake associated with higher soil Pb levels (Rolfe, 1973). Most of the Pb accumulated in the roots, with significantly lower levels in stems and leaves.

Apparently, Pb uptake by plants from soil is a passive process that proceeds rapidly until exchange sites in the root-free space are equilibrated with solution concentration (Zimdahl and Arvik, 1973). Solute transport from the external root environment to the central root xylem and to the shoot takes place through two major pathways: apoplastic (cell wall space between cell membranes) and symplastic (crossing many cell membranes along the path) (see also Section 3.2 of the chapter on bioavailability). The presence of the lipophilic Casparian strip at the root endodermis in the apoplastic pathway disrupts the apoplastic water flow and directs it to cross the plasma membranes at least twice, where selective transport as well as passive permeation of solutes occur (Wu et al., 1999). Under normal plant physiological conditions, the Casparian strip–guarded pathway accounts for over 99% of water flow through the roots. However, under chelate-induced uptake, this is somewhat uncertain. Hydrophilic compounds favor the apoplastic pathway, whereas lipophilic compounds favor the symplastic pathway. Materials that are very hydrophilic, as in the case of EDTA, are not favored for maximal plant transport; neither are compounds that are too hydrophobic (e.g., the gasoline additive tetraethyl Pb).

In addition to the relative lipophilicity and molecular size of the compound, accumulation of Pb in plants is dependent on the total flux of solute up to the shoot, i.e., the plant transpiration rate. The effect of soil chelate addition on plant transpiration rate, which in turn affects plant Pb accumulation, is also critical. Additionally, a reduced xylem flow might also help explain some of the increases in xylem Pb concentration observed after soil EDTA treatment (Wu et al., 1999).

Corn plants exposed to Pb in hydroponic solution showed that roots generally accumulated Pb as surface Pb precipitate and slowly accumulated Pb crystals in the cell walls (Malone et al., 1974). The root-surface precipitate formed without the apparent influence of cell organelles, whereas Pb taken up by the roots was concentrated in dictyosome vesicles, fused with one another to encase the Pb deposit. Eventually, the Pb deposits concentrate in the cell wall outside the plasmalemma, representing the final stage of entry of Pb. Lead was likely transported to the stems and leaves in a similar manner. Studies of the mode of entry and localization of Cd and Pb in rice roots indicate that initially, a small amount of Pb was found

TABLE 10.7. Translocation of lead from roots, as indicated by concentrations (in ppm) in various plant parts, in selected crop species.

Crop	Plant part	Plant Pb concentration		
		Control	200 ppm Pb in soil	1000 ppm Pb in soil
Leaf lettuce	Leaf	2.5	3.0	54.2
	Root	5.8	84.5	867.7
Spinach	Leaf	0.7	7.9	39.2
	Root	4.7	73.3	—
Broccoli	Leaf	7.2	8.4	18.4
	Root	6.5	83.0	745.6
Cauliflower	Leaf	5.3	6.3	11.8
	Root	2.5	55.1	532.2
Oats	Grain	3.2	4.4	4.9
	Leaf	6.0	6.8	20.1
	Stalk	1.6	2.5	9.2
	Root	4.5	82.0	396.6
Radish	Tops	3.7	9.9	14.3
	Tuber	6.3	7.0	44.6
Carrot	Tops	2.3	8.0	17.6
	Tuber	1.9	5.3	41.0

Source: Modified from John and Van Laerhoven, 1972a.

scattered broadly within the root tissues (Biddappa and Chino, 1981). After 180 min, Pb was concentrated more at the root surface than in the cortex. After 240 min, concentrations were sharply increased both at the surface and in the cortex, but the concentrations were still higher at the surface. It was concluded that Pb was more slowly absorbed into the roots than Cd and subsequently translocated less to the cytoplasm.

Lead also may form insoluble complexes with cell wall constituents. Only little translocation occurred in wild oat plants grown in nutrient solution in spite of Pb addition in chelated form (Rains, 1975). Lead exclusion from the shoots (3500 ppm in roots vs. 65 ppm in shoots) was due to precipitation or insoluble complexation of Pb with xylem elements and cell walls, thereby immobilizing it and greatly reducing its transfer from roots to shoots. The immobile nature of Pb in plant roots is illustrated in Table 10.7, where very little translocation occurred from roots to the vegetative parts of crops.

Lead values reported for crops from several countries are displayed in Appendix Table A.30. These values may be considered as background levels.

5.2 Foliar Deposition and Uptake of Lead

To a large extent, the following factors can affect Pb retention by vegetation: (1) aerosol properties (particle size, chemical composition), (2) leaf surface characteristics (roughness, pubescence, moisture, and stickiness), and (3) meteorological factors (relative humidity, cloud density, and wind speed). These factors influence the degree in which Pb particulates are deposited and retained on the surface of vegetation.

It is a general consensus that atmospheric Pb increases the Pb content of vegetation primarily by particulate deposition. Particulate Pb is quite insoluble and therefore is not expected to enter the leaf surface in substantial amounts. There are two primary sources of environmental Pb particulates: stack emissions from mining and smelting operations, and exhaust emission from automobiles. The chemical form, level, and particulate size of atmospheric Pb, as well as the nature of the soil Pb that can be resuspended into the atmosphere, play an important role in the uptake of Pb by plant tissue. Particle size is variable, with maximum dimensions ranging from 1 to 25 μm (Koslow et al., 1977). While Pb sulfate, phosphate, and oxide have been detected in motor vehicle exhaust, Pb chlorobromide is generally assumed to be the primary Pb salt introduced into the atmosphere from the tailpipe (Ter Haar and Bayard, 1971). Lead particles leaving the exhaust system are generally in the range of < 1 to 5 μm in size.

Generally, high percentages of the Pb deposited on vegetation surface can be removed by water wash indicating that Pb is externally located. For example, as much as 80% or more of Pb was washed off from grasses by water (Kloke and Riebartsch, 1964). Between 45 and 80% of the total Pb concentration in grasses originated from airborne deposition (Crump and Barlow, 1980). The question is how much of the deposited particulates can be assimilated internally by plants? Carlson et al. (1976) obtained an unusually high (95%) removal of topically applied Pb from soybean leaves using simulated rainfall. The leaves were experimentally fumigated with $PbCl_2$ aerosol particulate.

Exposing lettuce plants to automobile exhaust almost doubled the Pb content (17.2 ppm DW vs. 9.3 ppm in control) of foliar tissue (Bassuk, 1986). Washing with water alone recovered only little of the deposited Pb, while washing with vinegar (acetic acid) or liquid detergent reduced the Pb content to the same level as those in unexposed plants. These results infer that washing leafy vegetables as part of kitchen preparation should substantially decrease the dietary intake of Pb. Ample evidence has already been produced to this end (Höll and Hampp, 1975; Page et al., 1971; Rains, 1975).

Large differences in aerosol deposition may occur between plant species because of differences in their surface structure (Page et al., 1971; Wedding et al., 1975). Plant tissues with rough pubescent surface morphology can accumulate considerably more—up to at least seven times more—Pb than tissues with smooth surfaces. Thus, it can be reasonably expected that Pb accruing on rough pubescent tissues is not as easily washable.

Since it has been shown that Pb applied or deposited on the foliage remained as a topical coating on the surface, the barrier mechanism of foliar uptake of Pb should be elaborated. The barriers are composed primarily of epicuticular waxes and cuticular membrane (Arvik and Zimdahl, 1974a). The epicuticular waxes are extruded through the developing cuticle and form structures that may be typical of a given species or set of environmental conditions. The cuticular membrane, which must be penetrated for entry into the foliage, is a noncellular structure covering the epidermal layers of leaves, stems, and fruits of higher plants, and is composed primarily of cutin, within the matrix of which can be found varying amounts and kinds of waxes, fatty acids, alcohols, and sugars. The waxes are external to the cuticle as well as being interspersed within the cutin structure and serve to make wetting of the surface more difficult. The chemical and physical nature of the waxes may act as a barrier to the entry of Pb into the leaf, i.e., particulates may be trapped within the wax matrix. Consequently, only small amounts of Pb can penetrate the cuticle even after extended exposure and exceptional conditions of Pb solubility and pH (Arvik and Zimdahl, 1974a).

The Pb particulates that accumulate on leaf surfaces may cause adverse physiological effects, either by blocking the stomates and thus interfering with the normal exchange of gases between the leaf and its surrounding air, or by disrupting metabolic pathways after entry into the leaf. However, Carlson et al. (1976) did not find any evidence that Pb particulates applied on soybean leaves interfered with photosynthesis.

5.3 Interactions of Lead with Other Compounds

As discussed earlier, the complexation of Pb with phosphate is an attractive mechanism of immobilizing Pb in the environment. Consequently, less uptake of Pb by plants from soils having high levels of phosphate has been demonstrated (Adriano, 1986). Likewise, in sand culture, corn plants virtually did not take up Pb as long as there was enough phosphate in the solution to precipitate Pb (Miller and Koeppe, 1970). A reduction of about 50% in uptake of Pb by several tree species occurred when high levels of phosphate were added (Rolfe, 1973).

A potential consequence of Pb-phosphate complexation is the likelihood of P deficiency in plant nutrition when phosphates are precipitated in the formation of pyromorphites. For example, under high levels of Pb but marginal levels of soil phosphate, phosphate deficiency may be exhibited by plants because of the formation of insoluble Pb phosphate precipitates in the root zone (Koeppe and Miller, 1970; Lee et al., 1976).

The interaction of Pb with phosphate occurs outside of the roots but is expressed as lower Pb content in plant tissue, i.e., it is a geochemical process and not a physiological one. For example, in soil containing 635 ppm Pb, addition of ≥ 100 ppm P [as $Ca(H_2PO_4)_2$] significantly decreased the amount of Pb taken up by lettuce plants (Fig. 10.5). In the 3520 ppm Pb soil about 215 ppm Pb was necessary to alleviate Pb uptake. Similar results were obtained by Chlopecka and Adriano (1996) where substantial reduction in

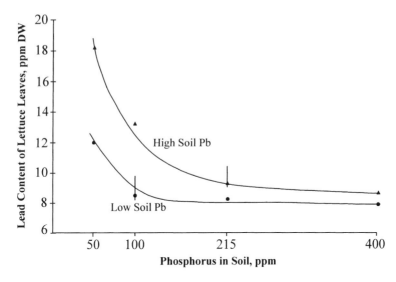

FIGURE 10.5. Influence of soil phosphorus on the lead uptake (n = 6) by lettuce plants. The soils had 635 and 3520 ppm Pb for the low and high Pb levels, respectively. (Modified from Bassuk, 1986.)

the uptake of Pb by test species occurred upon amending the contaminated soil with a phosphate compound (i.e., hydroxyapatite).

5.4 Phytotoxicity of Lead

Phytotoxicity of Pb is relatively low compared with other trace elements. In rice plants, the order of phytotoxicity (Chino, 1981) in decreasing order is:

$$Cd > Cu > Co \sim Ni > As \sim Cr > Zn > Mn \sim Fe \geq Pb$$

In sugar beets grown in sand culture, Cu, Co, and Cd were highly and about equally active in causing chlorosis due to toxicity, followed by $Cr(VI) > Zn(II) \geq V(II) \geq Cr(III) > Mn(II) \geq Pb(II)$ (Hewitt, 1953). This order of toxicity is just the reverse of the affinity sequence for the metals (see earlier section) where Pb generally has the highest affinity, as evaluated by sorption to various soil components. Simply stated, the strong sorption of Pb onto soils renders it less bioavailable for plant uptake.

Crop susceptibility to Pb toxicity varies greatly. For oats and tomato plants, a concentration of 25 mg L^{-1} of Pb as $Pb(NO_3)_2$, was required to cause toxicity (Berry, 1924). At a concentration of 50 mg L^{-1}, plant death occurred. Similarly, 25 mg L^{-1} of Pb used in sand culture solution was the critical level for barley growth, producing an average concentration of 35 ppm DW in the shoot of these plants (Davis et al., 1978). In rice plants, the following toxic levels for Pb were found (Chino, 1981): 50 to 2000 ppm in shoots and 300 to 3000 ppm in roots. Total Pb amounting to 400 to 500 ppm in soils in a polluted area in Japan was found to be toxic to the plants. In radish plants grown on soil, Pb toxicity was manifested as stunted growth—more pronounced in roots than in shoots (Khan and Frankland, 1983). In general, Cd proved to be about 20 times more toxic than Pb.

Excess Pb in plants may alter several physiological and biochemical processes within the plants, including mitochondrial respiration in corn (Koeppe and Miller, 1970), photosynthesis in soybean (Bazzaz et al., 1974; Huang et al., 1974), and photosynthetic electron transport in isolated spinach chloroplasts (Miles et al., 1972). Furthermore, corn seedlings grown in media treated with Pb (0 to 250 mg Pb as $PbCl_2$ per plant, or 0 to 4000 ppm Pb in solution) exhibited decreased net photosynthesis and transpiration with increasing Pb treatment level (Bazzaz et al., 1974) (Fig. 10.6). Similar reductions in photosynthesis and transpiration of 7-wk-old seedlings of loblolly pine (*Pinus taeda*) and autumn olive (*Elaegnus umbellata*) were observed after a 4-wk exposure to 320 ppm Pb as $PbCl_2$ in potting medium (Rolfe and Bazzaz, 1975). The general simultaneous rise and decline in photosynthesis and transpiration of the plants in response to increased Pb level suggests that rate changes of the two processes are related to changes in leaf stomatal resistance to CO_2 and water vapor diffusion. Indirect rather than direct Pb toxicity response is also possible. This includes interference with ion uptake and translocation, growth retardation due to inhibition of mitochondrial respiration, and inhibition of chloroplast activity. Soybean seedlings grown in culture solution spiked with $Pb(NO_3)_2$ (0 to 100 mg L^{-1} of Pb), had increased respiration rate, increased activities of the enzymes acid phosphatase, peroxidase, and alpha-amylase, and increased levels of soluble protein and ammonia with Pb treatment (Lee

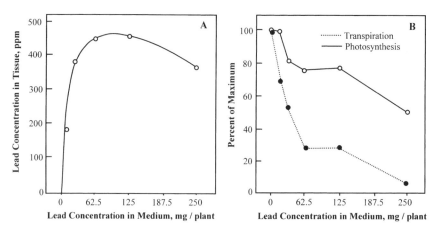

FIGURE 10.6. Lead accumulation in leaves of corn and its effect on photosynthesis and transpiration. (Modified from Bazzaz et al., 1974.)

et al., 1976). No changes in malic dehydrogenase and total free amino acids were observed. However, decreases were observed for glutamine synthetase activity and nitrate. Increased activities of the hydrolytic enzymes and peroxidase indicate that the Pb treatment enhanced senescence.

Although Pb is toxic to plants under experimental conditions, it is seldom phytotoxic under field conditions unless the soil levels are excessively high as in areas near point sources or when soil physical and chemical conditions (e.g., sandy, low CEC acidic soils) favor accumulation.

6 Ecological and Health Effects of Lead

6.1 Effects of Lead in Plants and Microbiota

Because of the high affinity of Pb for soil OM, Pb may not be a serious problem in agricultural soils. The exception however, is in old orchard soils where lead arsenate was used for a number of decades. Even in soils heavily sludged, Pb should not be a problem because of the OM and other constituents of the sludge. If any toxicity arises, it would be due to the additive effect of the metals from sewage sludge. Phytotoxicity in this case may be expected to arise primarily from Zn, Cd, and Ni.

Aside from agricultural settings, plants may be subjected to unusually high levels of Pb in old mining and smelting sites. In this situation, phytotoxicity from the metals may occur in areas acidified by SO_x emissions causing greater solubility and bioavailability of Pb (Dudka and Adriano, 1997). Oxidation of pyrite embedded in mining waste may also elevate soil acidity, causing similar effects as the SO_x. In areas by two zinc smelters that operated for over 80 yr, regeneration of plant communities has been primarily hampered by high levels of Zn in the litter and mineral soil (Storm et al., 1994).

Except in rare occasions discussed above, Pb phytotoxicity should not be encountered in ordinary environmental settings. It may well be that microorganisms are more sensitive to soil Pb pollution than plants. For example, the abundance of arthropods in soil litter was substantially reduced near the smelters compared with reference sites (Strojan, 1978). Altered composition of the microbial community, i.e., lower diversity, may be an early indication of Pb effects. For example, altered composition may occur at 750 ppm Pb in soil and nitrification inhibited at 1000 ppm (U.S. EPA, 1986). The effects are generally less in clayey and high-OM soils than in infertile sandy soils.

6.2 Effects of Lead in Livestock, Fish, and Wildlife

Lead is a nonspecific toxin that acts at the molecular level and inhibits many enzymatic activities that regulate normal biological functioning. As in humans, the most studied effects in animals include the hematological, central nervous system (CNS), learning and behavior, and reproduction and survival effects (Pain, 1995).

Lead has been singled out as the most frequent source of accidental poisoning in domestic animals (NAS, 1972). The most important sources of Pb relevant to animal intoxication are dust fallout on pastures near Pb smelters or Pb mines, Pb-based paints, and improperly disposed wastes such as oil wastes and storage battery casings. Fallout from automotive exhausts has been shown to increase Pb intake, but has not yet been demonstrated to account for reported cases of lead poisoning. Risks from Pb poisoning may exist in grazing cattle near Pb smelters or mines, where Pb fallout or soil Pb are extremely high. Numerous fatal intoxications have been reported from such areas (Merian, 1991). Acute signs characteristic for Pb poisoning include CNS disorders, excitement, depression, motor abnormalities, and blindness. Some animals may die without showing any of these signs. In beef cattle intoxicated with Pb-contaminated silage, clinical signs included ruminal atony, depression, head pressing, blindness, ataxia, circling, and convulsions (McEvoy and McCoy, 1993). The Pb intake by these animals was estimated at 96 to 120 g of Pb over ~10 days. Humphreys (1991) estimated that 50 to 100 g of Pb as Pb acetate, or 300 to 400 g of Pb from other Pb salts are required to poison adult cattle.

In East Helena, Montana where a primary Pb smelter has been in continuous operation for over 100 years, concentrations of Pb in blood from cattle ($n = 222$) representing nine herds were significantly ($p < 0.05$) elevated in herds near the smelter compared with a background herd (Fig. 10.7). The herd blood Pb (or PbB) levels were significantly correlated with distance from the smelter ($r = -0.86$) and with soil Pb concentration ($r = 0.96$). They were also moderately correlated with vegetation Pb levels ($r = 0.61$). Soil may be more important than forage as a source of Pb in these cattle. Young cattle, i.e., < 1 yr old, exhibited higher PbB levels than mature cattle, where in some young animals the PbB levels were > 35 µg Pb dL^{-1} (Neuman and Dollhopf, 1992). However, the majority (~80%) of the cattle tested near the smelter and all cattle from the background herd had PbB levels in the normal diagnostic range (1 to 21 µg dL^{-1}).

Buck et al. (1976) suggested that cattle blood Pb levels from 10 to 35 µg dL^{-1} were significant as a primary etiological agent or as a predisposing or contributing factor in Pb toxicity. In another study of 142 cattle (Buck, 1975), 52 animals exhibited clinical symptoms of Pb toxicosis and had PbB levels from 10 to 380 µg dL^{-1} (mean of 81 µg dL^{-1}). It was concluded that concentrations above 35 µg dL^{-1} should be considered a result of unusual exposure. In acute Pb poisoning in cattle, PbB levels were never < 35 µg dL^{-1} (Hammond and Aronson, 1964). The highest concentration of Pb in cattle blood at which toxicosis was not noted was 29 µg dL^{-1}.

Other animals can be expected to accumulate more Pb in areas adjacent to smelters because of their greater exposure to the contaminant. For example,

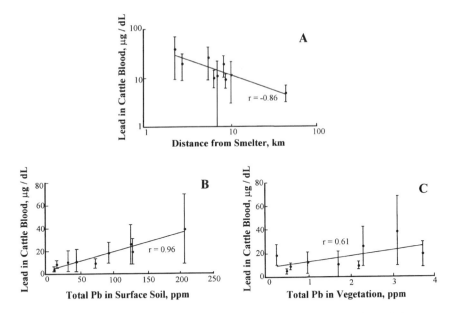

FIGURE 10.7. Lead concentrations in the cattle blood in East Helena, Montana smelter area as a function of (A) distance, (B) surface soil Pb level, and (C) vegetation Pb levels. (Modified from Neuman and Dollhopf, 1992.)

Beyer et al. (1985) reported that mammals and birds contained higher concentrations of Pb and Cd at the Palmerton, Pennsylvania smelter area than in reference sites. Metal concentrations were higher in shrews (*Sorex araneus*) than in birds or mice. Levels of Pb, Zn, and Cu from tissues of rabbits (*Sylvilagus* spp.) from Palmerton were 2 to 3 times higher than those reported in muskrats (*Ondatra zibethicus*) affected by heavy industrial activity in southeastern Pennsylvania (Storm et al., 1994). These metals occurred at more excessive levels in amphibians in Palmerton than in other contaminated and noncontaminated sites. These recent findings indicate that even after 6 yr of cessation of smelting, the absence or low abundance of wildlife at Palmerton due to the toxic effects of the smelting operation continues to linger.

Substantially higher levels of Pb in the kidney, liver, and bone tissues of small mammals captured in an abandoned shooting range having acidic sandy soils were observed (Ma, 1989). Shrews and voles (*Clethrionomys glareolus*) had higher kidney-to-body-weight ratios than the control animals, an indication of likely Pb poisoning. In his synthesis of the literature with an attempt to predict Pb residues in small mammals from soil Pb concentrations, Shore (1995) found significant relationships between Pb residues in soils and body organs of wood mice (*Apodemus sylvaticus*) and field voles (*Microtus agrestis*). Thus, small mammals can serve as bioindicators of metal pollution.

The U.S. Fish and Wildlife Service in the Coeur d'Alene River basin, Idaho has observed that passerines have elevated Pb levels in comparison with controls (Johnson et al., 1999). Mean liver Pb in song sparrows (*Melospiza melodia*) captured in the basin was 4.76 ppm WW, compared with 0.18 ppm in song sparrows captured in the noncontaminated St. Joe River basin. Liver Pb values in American robins (*Turdus migratorius*) from the Coeur d'Alene River basin averaged 4.25 ppm, compared with 0.80 ppm in robins from the control site. Robin ingesta Pb levels in the Coeur d'Alene River basin also were significantly higher (22.13 ppm) than in the control site (0.90 ppm). Recent studies indicate that 43% (95% confidence interval CI = 12.9–77.5%) of the song sparrows and 83% (95% CI = 41.8 – 99.2%) of the American robins inhabiting the floodplain along the Coeur d'Alene River basin are being exposed to Pb at levels sufficient to inhibit δ-aminolevulinic acid dehydratase (ALAD) by >50% (Johnson et al., 1999). Hematocrit values did not indicate highly significant Pb exposure; however, ALAD inhibition is considered the most sensitive indicator of Pb exposure (Scheuhammer, 1987).

Endangered species of wildlife are also exposed to Pb intoxication. Steller's sea eagles (*Haliaeetus pelagicus*) and white-tailed sea eagles (*H. albicilla*), the third and fourth largest eagles in the world, are endangered species. Populations of these species are declining in most parts of their habitats. Steller's sea eagles breed in Kamchatka, northeastern Siberia and along its neighboring coast. White-tailed sea eagles in the Far East are distributed in Kamchatka, Okhotsk, Sakhalin, and Hokkaido, Japan and winter in Japan and Korea. A survey of poisoning in these eagles in Hokkaido indicates that they are suffering from secondary poisoning through ingestion of Pb shot embedded in the tissue of their prey (Kim et al., 1999).

Waterfowl can also be exposed to Pb intoxication from ingesting Pb shot. Indeed the main cause of Pb poisoning in waterfowl in high hunting pressure areas is from the ingestion of Pb shot (Tirelli et al., 1996). Mallards (*Amas platyrhynchos*) that have ingested Pb shot can be expected to have high PbB levels of 40 μg dL^{-1} or higher (Pain, 1989). A threshold level of 100 μg dL^{-1} of Pb in blood in waterfowl may be used for nonbackground exposure (Daury et al., 1993).

Pintail ducks (*Anas acuta*) intoxicated from ingestion of Pb shot deposited on a tidal meadow as a result of trap and skeet shooting manifested muscular weakness (i.e., inability to fly) and were anemic. The PbB levels were ≥1.2 ppm in these poisoned ducks (Roscoe et al., 1989). It has been estimated that since the 1986 hunting season in the United States when the use of nontoxic shot became mandatory, over 6×10^{6} more ducks have been saved by converting to steel shot (Pain, 1995).

The various biological effects and symptoms of Pb poisoning in fish, avian species, and small and large mammals have been summarized by Pain (1995). Consumption of Pb-contaminated salmonids (salmon and trout) from Lake Ontario should not pose any exposure risk to consumers in the region (Cappon, 1987). The main metal of concern in aquatic systems is Hg, especially in the organic form, because it biomagnifies in the food chain (see later chapter on mercury).

6.3 Effects of Lead on Human Health

Lead poisoning in humans, especially in young children, is an environmental and public hazard of global proportions. It is local in nature as the causes and intensity of Pb poisoning vary from region to region, country to country, and community to community. It is also greatly affected by socioeconomic factors and age. Some generalities about Pb exposure and poisoning include the following:

Lead poisoning is the most prevalent environmental disease among children, which is totally eradicable through public awareness and exposure prevention.

In the absence of an industrial point source, both blood and environmental Pb levels are likely to be higher in urban than in suburban and rural areas.

Childhood Pb poisoning is typically more severe in developing countries due to lower or inadequate pollution control technologies, unregulated housing industry, and the use of folk medicines containing Pb.

Populations near a Pb-related industry, i.e., as a point source, are often exposed to unsafe levels of Pb in the air and soil with corresponding elevated levels of Pb in their blood.

Segments of populations in some developed countries, including the United States, have levels of Pb in their blood higher than what may be currently considered acceptable.

While the use of leaded gasoline, considered to be the most important source of environmental Pb especially in urban areas, has been strictly regulated in most developed countries, its use in developing countries goes unregulated.

The U.S. EPA has classified Pb as a probable human carcinogen. This agency has not established a RfD (reference dose) or RfC (reference concentratrion) for Pb (U.S. EPA, 1993).

6.3.1 Exposure Pathways of Lead to Humans

Lead is a naturally occurring element that has been used since the beginning of civilization. Because of the many industrial activities that have brought about its wide distribution, today Pb is ubiquitous in the environment. All humans have Pb in their bodies primarily as a result of exposure to man-made sources (Table 10.8).

Today, the major environmental sources of metallic Pb and its salts are paint, automobile exhaust, food, and water (Fig. 10.8). For the general

TABLE 10.8. Sources of lead exposure to humans.

Occupational	Environmental	Hobbies and related activities	Substance use
Plumbers, pipe fitters	Lead-containing	Glazed pottery	Folk remedies
Lead miners	paint	making	Health foods
Auto repairers	Soil/dust near	Target shooting at	Cosmetics
Glass manufacturers	lead	firing ranges	Moonshine whiskey
Shipbuilders	Industries,	Lead soldering	Gasoline "huffing"
Printers	roadways	(e.g., electronics)	
Plastic manufacturers	Lead-painted homes	Painting	
Lead smelters and	Plumbing leachate	Preparing lead shot,	
refiners	Ceramicware	fishing sinkers	
Police officers	Leaded gasoline	Stained-glass making	
Steel welders or cutters		Car or boat repair	
Construction workers		Home remodeling	
Rubber product			
manufacturers			
Gas station attendants			
Battery manufacturers			
Bridge reconstruction			
workers			
Firing range instructors			

Source: ATSDR (1992).

public and especially children, the most important pathways are ingestion of paint chips from Pb-painted surfaces, inhalation of Pb from automobile emissions, food from Pb-soldered cans, drinking water from Pb-soldered plumbing, and medications in the form of folk remedies (ATSDR, 1992). The problem of children's exposure to Pb-based paint hazards is especially acute in poor urban areas with older housing in poor condition, exposed soil in yards and vacant lots, and ongoing redevelopment that includes demolition of aging and abandoned housing.

Automobile emissions have been an important source of Pb exposure for urban residents, particularly in areas with congested traffic. Although inhalation of Pb from gasoline is no longer considered a public health problem, the Pb from dust in automobile emissions has been deposited in the soil. Children playing near roads and freeways may come in contact with contaminated soil.

The Pb content of paint was not regulated until 1977 (ATSDR, 1992). Many older structures, residential and commercial, have leaded paint that is peeling, flaking, and chipping. Children can ingest loose paint as a result of pica (compulsive eating of nonfood items) and through mouthing of items contaminated with Pb from paint, dust, and soil. High levels of Pb in soil and house dust have been associated with increased blood levels in children.

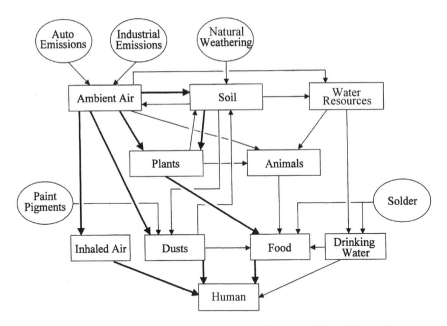

FIGURE 10.8. Sources and pathways of lead to humans from the environment. (Modified from NAS, 1993.)

Food may contain Pb from the environment or from containers. Since agricultural vehicles and equipment are not required to use unleaded gasoline in the United States, eventually Pb can be deposited on and retained by crops, particularly leafy vegetables. Acidic foods have been found to leach Pb from Pb solder in cans and Pb glazes used in making pottery and ceramicware. Water from leaded pipes, Pb-soldered plumbing, or water coolers is another potential source of Pb exposure. Stationary or point sources of Pb include mine and smelters.

Several folk remedies used in some countries have been shown to contain large amounts of Pb. Two Mexican folk remedies are azarcon and greta, which are used to treat empacho, a colic-like illness (ATSDR, 1992). Azarcon and greta are also known as liga, Maria Luisa, alarcon, coral, and rueda. Lead-containing remedies and cosmetics used by some Asian communities are chuifong tokuwan, pay-looah, ghasard, bali goli, and kandu. Middle Eastern remedies and cosmetics include alkohl, kohl, suma, saoott, and cebagin.

In addition to these environmental sources, many occupations, hobbies, and other activities result in potential exposures to high levels of Pb and can put the entire family at risk of Pb poisoning. Lead-glazed pottery, particularly if it is imported, is a potential source of exposure that is often overlooked. Even "safe" ceramicware can become harmful since dishwashing may chip or wear off the protective glaze and expose lead-containing pigments.

Inorganic Pb enters the body primarily through inhalation and ingestion, and basically does not undergo biological transformation (ATSDR, 1992). In contrast, organic Pb found primarily in gasoline as tetraethyl Pb enters the body through inhalation and skin contact, and is metabolized in the liver. In 1976 and 1984, government regulations drastically reduced the amount of Pb in gasoline, and today organic Pb in gasoline is not as great an environmental concern in the United States as it is in other countries, where it remains a serious hazard.

6.3.2 Global Variations in Lead Exposure of Humans

During the last three decades, the majority of PbB studies have been conducted in North America and western Europe. Among adult populations, average PbB levels were least likely to exceed the level of concern in western Europe and most likely to exceed 25 µg dL^{-1} in Africa (Alliance/EDF, 1994). Adult populations in the Middle East, and Central and South America are also likely to exceed the level of concern. In a nonoccupational survey of Pb exposure of adult women in five southeast Asian cities (Bangkok, Thailand; Kuala Lumpur, Malaysia; Manila, Philippines; Nanning and Tainan, China) the PbB levels ranged from 32.3 µg L^{-1} to 65.4 µg L^{-1} respectively for Bangkok and Kuala Lumpur (Zhang et al., 1999).

Among populations of children, the potential for Pb poisoning globally remains alarmingly high. The only region with more than half of the population averaging below 10 µg Pb dL^{-1} was North America (i.e., primarily Canada and the United States and excluding Mexico, which is included with Central America).

6.4 Bioavailability and Physiological Effects of Lead in Humans

The rate at which Pb is absorbed depends on its chemical and physical form and on the physiological characteristics of the exposed person (e.g., nutritional status and age). Inhaled Pb deposited in the lower respiratory tract is completely absorbed. The amount of Pb absorbed from the GI tract of adults is typically 10 to 15% of the total amount ingested; for pregnant women and children, the amount absorbed can increase to as much as 50%. The quantity absorbed increases significantly under fasting conditions and with Fe and Ca deficiencies. Once in the blood, Pb is distributed primarily among three compartments—blood, soft tissue (kidney, bone marrow, liver, and brain), and mineralizing tissue (bones and teeth) (ATSDR, 1992, 1993). Mineralizing tissue contains about 95% of the total body burden of Pb in adults. In bone, there is both a labile component, which readily exchanges Pb with the blood, and an inert pool. The Pb in the inert pool poses a special risk because it is a potential endogenous source of Pb. When the body is under physiological stress, such as pregnancy, lactation, or chronic disease, this normally inert Pb can be mobilized, increasing the Pb level in blood. Because of these mobile Pb pools, significant drops in a person's PbB level can take several months or sometimes years, even after complete removal from the source of Pb exposure.

Of PbB, 99% is associated with erythrocytes; the remaining 1% is in the plasma where it is available for transport to the tissues. The PbB not retained is either excreted by the kidneys or through biliary clearance into the GI tract. In single-exposure studies with adults, Pb has a half-life in blood of approximately 25 days; in soft tissue, about 40 days; and in the nonlabile portion of bone, more than 25 yr. Consequently, after a single exposure, a person's PbB level may begin to return to normal; the total body burden however, may still be elevated.

Major acute exposures to Pb need not occur for Pb poisoning to develop. The body accumulates this metal over a lifetime and releases it slowly, so even small doses over time can cause Pb poisoning. It is the total body burden of Pb that is related to the risk of adverse effects. The most sensitive target of Pb poisoning in humans is the nervous system (ATSDR, 1992, 1993). In children, neurological deficits have been documented at exposure levels once thought to cause no harmful effects. In addition to the lack of a precise threshold, childhood Pb toxicity may have permanent effects. One study showed that damage to the CNS that occurred as a result of Pb exposure at age 2 resulted in continued deficits in neurological development, such as lower IQ scores and cognitive deficits at age 5. In another study that measured total body burden, primary school children with high tooth Pb levels but with no known history of Pb poisoning had larger deficits in psychometric intelligence scores, speech and language processing, attention, and classroom performance than children with lower levels of Pb.

Adults may also experience CNS effects at relatively low PbB levels, manifested by subtle behavioral changes, fatigue, and impaired concentration (ATSDR, 1992, 1993). Peripheral nervous system damage, primarily motor, is seen mainly in adults. Peripheral neuropathy with mild slowing of nerve conduction velocity has been reported in asymptomatic Pb workers.

The body's ability to make hemoglobin is inhibited by Pb by interfering with several enzymatic steps in the heme pathway. Ferrochelatase, which catalyzes the insertion of Fe into protoporphyrin IX, is quite sensitive to Pb. A decrease in the activity of this enzyme results in an increase of substrate, erythyrocyte protoporphyrin (EP), in the red blood cells. Recent data indicate that the EP level, which has been used to screen for Pb toxicity in the past, is not sufficiently sensitive at lower levels of PbB and is therefore not as useful a screening test for Pb poisoning as previously thought.

Two types of anemia can be induced by Pb. Acute high-level Pb poisoning has been associated with hemolytic anemia. In chronic Pb poisoning, Pb induces anemia by both interfering with erythropoiesis and by diminishing red blood cell survival. It should be emphasized however, that anemia is not an early manifestation of Pb poisoning and is evident only when the PbB level is significantly elevated for prolonged periods.

Nephropathy is a direct effect on the kidney of long-term Pb exposure. Impairment of proximal tubular function manifests in aminoaciduria, glycosuria, and hyperphosphaturia. There is also evidence of an association between Pb exposure and hypertension, an effect that may be mediated through renal mechanisms. Gout may develop as a result of Pb-induced hyperuricemia, with selective decreases in the fractional excretion of uric acid before a decline in creatinine clearance. Renal failure accounts for 10% of deaths in patients with gout.

Increasing evidence indicates that Pb not only affects the viability of the fetus, but its development as well. Developmental consequences of prenatal exposure to low levels of Pb include premature birth, reduced birth weight, and slowed postnatal neurobehavioral development. Lead is an animal teratogen; however, most studies in humans have failed to show a relationship between Pb levels and congenital malformations (ATSDR, 1992).

6.5 Lead Toxicity in Humans

Because of differences in susceptibility among people, symptoms of Pb poisoning and their onset may vary. With increasing exposure, the severity of symptoms can be expected to increase. Those symptoms most often associated with varying degrees of Pb toxicity are listed in Table 10.9. In symptomatic Pb intoxication, PbB levels generally range from 35 to 50 μg dL^{-1} in children and 40 to 60 μg dL^{-1} in adults (Fig. 10.9). Severe toxicity is frequently found in association with PbB levels of 70 μg dL^{-1} or more in children and 100 μg dL^{-1} or more in adults. For biological monitoring in humans, reference values for blood and urine are reported as 7 to 22 μg/100 ml and 10 to 40 μg L^{-1}, respectively (Chang, 1996).

A PbB level is the most useful screening and diagnostic test for Pb exposure. It reflects Pb's equilibrium among absorption, excretion, and deposition in soft- and hard-tissue compartments. For chronic exposures, PbB levels may underrepresent the total body burden; nevertheless, it is the most widely accepted and commonly used measure of Pb exposure. Blood Pb levels respond rather rapidly to abrupt or intermittent changes in Pb intake (e.g., ingesting of Pb paint chips by children) and, within a limited range, bear a linear relationship with intake levels.

In the 1960s the defined toxic dose for Pb was 60 μg dL^{-1} (Needleman, 1996). When screening studies in eastern U.S. cities indicated that substantial numbers of children had PbB exceeding 40 μg dL^{-1}, the question of unrecognized toxicity came under scrutiny. As indicated earlier, PbB level can be affected by the type of exposure and age. Remote rural areas, such as those in the Himalayas and Papua New Guinea, should serve as good reference values for unexposed humans.

TABLE 10.9. Typical signs and symptoms associated with lead toxicity in humans.

Mild toxicity	Moderate toxicity	Severe toxicity
Myalgia or paresthesia	Arthralgia	Paresis or paralysis
Mild fatigue	General fatigue	Encephalopathy–may abruptly lead to seizures, changes in consciousness, coma, and death
Irritability	Difficulty concentrating	
Lethargy	Muscular exhaustibility	
Occasional abdominal discomfort	Tremor	Lead line (blue-black) on gingival tissue
	Headache	Colic (intermittent, severe abdominal cramps)
	Diffuse abdominal pain	
	Vomiting	
	Weight loss	
	Constipation	

Source: ATSDR (1992).

Lead is most harmful to children under age 6. Children are more sensitive because they absorb and retain more Pb ($\sim$50%) in proportion to their body weight than do adults (Mushak et al., 1989). Every child who has a developmental delay, behavioral disorder, or speech impairment, or who may have been Pb-exposed, should be considered for a PbB test. Equally important, siblings, housemates, and playmates of children with suspected Pb toxicity probably have similar exposures to Pb and should be promptly screened.

7 Factors Affecting Mobility and Bioavailability of Lead

7.1 pH and Redox Potential

Soil reaction is the single most important factor affecting the solubility, speciation, and bioavailability of Pb. It was recognized early on that Pb toxicity was inversely proportional to soil pH (Griffeth, 1919). Increasing soil pH decreased Pb in plant roots, although no consistent trends occurred in the foliage. Lime reduced the uptake of Pb by oat and alfalfa plants via repression of solubility of Pb because of the greater capacity of OM to complex Pb with increasing pH (MacLean et al., 1969). Application of lime to Pb-contaminated soils reduced the foliar Pb content of treated crops but had little effect on Pb content in roots (Cox and Rains, 1972; John and Van Laerhoven, 1972b). Although liming the soil did not have a consistent effect on uptake by corn plants, Pb translocation in the shoot appeared to have been restricted by liming (Zimdahl and Foster, 1976).

In rice, Reddy and Patrick (1977) found that Pb uptake by both roots and tops decreased with increasing pH and suspension redox potential (Fig. 10.10). Redox potential and pH had a pronounced effect on the uptake of Pb by the roots. Accumulation of Pb in the shoot was also influenced by pH, increasing with a decrease in pH from 8 to 5. Highly significant correlations of water-soluble Pb with both total Pb uptake

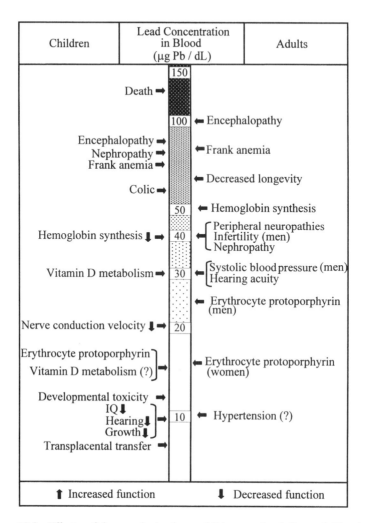

FIGURE 10.9. Effects of inorganic lead on children and adults and blood levels showing lowest observable adverse effects (LOAEL). (From ATSDR, 1988.)

($r = 0.83$, $p \leq 0.01$) and Pb uptake by the roots ($r = 0.86$, $p \leq 0.01$) were observed. At a given pH, Pb translocation from roots to the shoot increased with increasing redox potential. At low redox potential and at high pH levels more Pb accumulated in the roots. This was attributed to precipitation of Pb presumably as sulfides on the root surface. Based on the above observations, it appears that the primary effect of liming acidic soils may be on the translocation of Pb rather than on its uptake by roots. The effect on uptake may be partly due to competition between Pb and other cations for available exchange sites on the soil matrix.

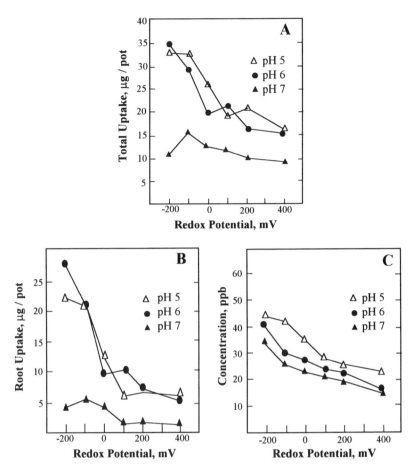

FIGURE 10.10. Effects of pH and redox potential on uptake of lead by (A) rice shoots and (B) roots and on the (C) amount of water-soluble lead in soils. (Modified from Reddy and Patrick, 1977.)

In flooded soils, especially in slightly acidic to acidic soils, interactions between Pb and P, and Pb and Fe can occur. Increased uptake by the rice plants of P and Fe, as well as Mn, under reducing soil conditions may result in the formation of Pb complexes with compounds of these elements in or on roots, retarding Pb translocation and resulting in more Pb accumulation in roots (Reddy and Patrick, 1977).

In groundwater, anoxic conditions should favor the formation of PbS. But collection and analysis by advanced spectroscopic techniques of water samples from monitoring wells (screened intervals were between 2 and 19 m deep) from a Superfund site did not reveal the occurrence of this mineral phase in the aquifer (Hesterberg et al., 1997).

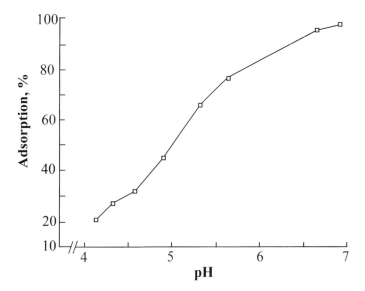

FIGURE 10.11. Adsorption of lead on kaolinite (KGa-2) as a function of pH (0.01 M NaClO₄). (Modified from Puls et al., 1991.)

Also influenced by pH are the stability of Pb species in aqueous solutions, the forms of Pb in soil components, and the intensity of sorption onto soil components. For example, the pH-dependent edge charge on kaolinite becomes increasingly negative as pH increases (Fig. 10.11). Furthermore, Basta and Tabatabai (1992b) observed significant positive correlations between Pb sorption and soil pH as well as CEC.

The degree of hydrolysis of Pb in soil solutions is largely influenced by the pH. For example, assuming the absence of large quantities of ligands other than OH^-, HCO_3^-, and CO_3^{2-}, Pb exists primarily as Pb^{2+} from pH 4 to 7; at pH~8, $PbOH^+$ predominates; and above pH 8, $Pb(OH)_2^0$ assumes importance (see also Fig. 10.4) (Harter, 1983). In conditions simulating alkaline suspension from scrubber fly ash (pH 12), the solubility of Pb increased above pH 11, where the predicted predominant species were $Pb(OH)_4^{2-}$, $Pb(OH)_3^-$, and to a lesser extent, $Pb(OH)_2^0$ (Roy et al., 1993).

There is an evident delineation of exchangeable Pb in contaminated soils from Upper Silesia, Poland by soil pH. In soils having pH <5.6, more exchangeable Pb (4.7% of total Pb) occurred than in soils with pH >5.6 (negligible exchangeable Pb) (Chlopecka et al., 1996). This again indicates greater bioavailability of Pb at lower pH values.

7.2 Organic Matter

It was indicated in Section 4.3 that in soil profile, Pb occurs considerably higher in the surface than in subsurface horizons. Lead accumulation in the surface horizon is associated with high OM content, partly due to reactions involving insoluble OM content. It may well be that OM is universally more important than precipitation by carbonates or sorption by hydrous oxides since the majority of Pb immobilized by soils is associated with OM.

A common mechanism of complexation by OM for metal cations is by ion exchange, i.e., the adsorption of metal ions onto humic materials occurs through an ion exchange process between the ion adsorbed and the H^+ ions from the OM (see Chapter 2). Because the CEC of soils is often correlated with the OM content, an increase in CEC is often associated with reduced plant uptake but increased soil sorption of Pb (Bunzl, 1974; Bunzl et al., 1976). Since they coexist in soils, OM may exhibit additive effects with the hydrous oxides of Fe, Mn, and Al, which are important soil constituents for sorption of metals.

The sorption capacity of OM for Pb is confirmed by several plant uptake studies. Additions of organic amendments such as animal wastes or peat (Fig. 10.12) to soils reduced plant uptake of Pb (Scialdone et al., 1980; Zimdahl and Foster, 1976). As a result, it was claimed that adsorption maxima for Pb by soils were primarily due to the soils' OM and clay contents (Riffaldi et al., 1976; Soldatini et al., 1976). In contrast, no significance could be attributed to the role of OM on the adsorption of Pb by surface soils from several northeastern states (Harter, 1979).

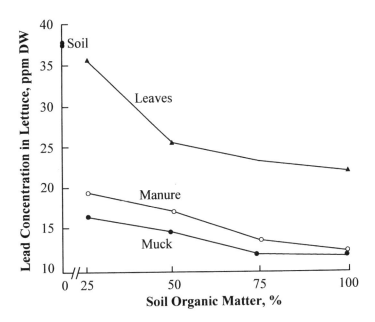

FIGURE 10.12. Influence of soil amendment with organic matter on lead uptake by lettuce grown in soil containing 3520 ppm lead ($n = 3$). (Modified from Bassuk, 1986.)

While OM may serve as an immobile complexer for Pb, it can also facilitate the transport of this element in soils via the formation of soluble complexes with OM (Stevenson and Welch, 1979). Indeed, the downward movement of Pb in the soil profile was attributed to the formation of soluble organic complexes. Employing batch-type reaction experiments, the time-dependency of metal lability was demonstrated suggesting that decomposition of OM in soils could increase metal solubility and bioavailability (Martinez and McBride, 1999). In general, the ability of OM to mobilize or immobilize metals depends on the nature of the OM, pH, redox, and competing ions and ligands. For example, 30 to 50% of dissolved Pb in a contaminated orchard soil treated with leaf compost was present as soluble OM complexes at low pH and up to 80 to 99% at pH~7 (Sauve et al., 1998). The solubility of Pb linearly decreased from pH 3 to 6.5 and is independent of soil OM in that pH range. From pH 6.5 to 8, higher pH promoted the formation and dissolution of organo-Pb complexes thereby increasing Pb solubility.

7.3 Soil Type and Clay Mineral

Soil type, in particular the soil's clay content, has been shown to influence plant uptake of Pb as well as sorption of Pb by soil materials. Soils with higher clay contents, assuming other soil constituents are constant, generally have higher CEC and, therefore, higher binding capacity for cations.

In a growth chamber study, Pb concentrations were higher in soil solutions obtained from soils with the lowest clay and OM contents, causing generally higher Pb concentrations in the tops of alfalfa and bromegrass (Karamanos et al., 1976). In soils treated with sewage sludge, metal extractability by DTPA solution and plant uptake also correlated with soil texture (Gaynor and Halstead, 1976). Extractable Pb and Pb uptake by lettuce and tomato plants were higher in sandy loam soils than in clay soils. Among various soil properties influencing plant (oats and lettuce) uptake of Pb, soil texture and soil pH were significantly related to the amounts of Pb in plant parts (John, 1972). Organic matter in this case had only minor influence on Pb phytoavailability.

Studies of Pb sorption by soils indicate that clay, OM, CEC, and pH are important soil parameters determining the sorption capacity of soils (Hassett, 1974; Riffaldi et al., 1976; Zimdahl and Skogerboe, 1977). The affinity of Pb and other metals for mineral soil components is typically as follows: clay > silt > sand. Clay mineralogy can also exert an influence on soil sorption of Pb. Adsorption of Pb^{2+} was favored over Ca^{2+} on model clay minerals, i.e., montmorillonite, illite, and kaolinite, with the most preference being onto kaolinite (Bittell and Miller, 1974). Similarly, in the presence of Al^{3+}, Pb^{2+} was adsorbed more strongly onto kaolinitic and montmorillonitic soils than onto illitic soil (Lagerwerff and Brower, 1973). Harter (1979) found that sorption of Pb by soils sampled from the B horizon in eight northeastern states was significantly correlated with vermiculite content.

7.4 Plant Factors

Since atmospheric deposition and root uptake are the two major exposure pathways for plants to Pb, plant species and cultivars as well as leaf morphology and characteristics are important factors underlying Pb retention and uptake by plants. Accordingly, foliar retention of Pb by plants partly depends on leaf surface characteristics, such as roughness and pubescence (see also previous Section 5.2). Since atmospheric Pb is primarily in a particulate form, plant tissue with a rough pubescent surface can intercept and accumulate considerably more Pb than tissue with smooth surface.

Interspecific variations in uptake of Pb by crops should also be expected, where Pb uptake from Pb-contaminated soil by crops was in the order soybeans $\gg$ clover > wheat $\geq$ corn $\geq$ oats (Adriano, 1986). Other interspecific differences in Pb uptake were also shown in crops (Czuba and Hutchinson, 1980; Sheaffer et al., 1979) and in tree species (Rolfe, 1973).

Differences in Pb concentration in various plant parts are also not uncommon, partly because of the unique translocation characteristics of Pb once it is taken up by the roots. Lead primarily concentrates in the roots and is poorly translocated to vegetative parts and particularly to reproductive organs (see Table 10.7).

Plant age also influences Pb uptake. Lead in plant tops decreased as a function of plant age in four of five crops (Cox and Rains, 1972). Similar trends were observed in young spring vegetables (higher) than in mature autumn tissues (Czuba and Hutchinson, 1980). Lead tends to decline during the rapid stage of growth, partly because of dilution of Pb by increased biomass. However, plant tissues may accumulate Pb again when they reach the senescent or dormant stage of growth which can be partly due to the loss of plant matter (10 to 20% loss), thereby concentrating Pb in remaining tissue (Rains, 1975).

7.5 Source Term

The forms and species of Pb vary according to environmental setting (see also Section 7.1 of the chapter on bioavailability). For example, the forms of Pb are expected to be more soluble in roadside soils than in soils near smelters or in mining sites. Roadside soils contain a higher proportion of Pb in more readily soluble form ($PbClBr$ and $PbSO_4$) from automobile emissions, compared with Pb oxide species (e.g., PbO_x, $PbO \cdot PbSO_4$) in soils contaminated by smelter operations (Davis et al., 1992). Galena, anglesite, and cerussite are the dominant forms of Pb associated with ore processing and hence they are the usual forms found in mining sites. In paints, the basic Pb carbonate $[2PbCO_3 \cdot Pb(OH)_2]$ and chromate $(PbCrO_4)$ dominate (Davies and Wixson, 1988; Ruby et al., 1994).

The lower bioavailability of Pb from mining sites (e.g., Butte, Montana mining district) compared with urban environments is due partly to the relative solubility of Pb-bearing phases and associated mineral assemblages (Ruby et al., 1992). *In vivo* swine studies indicate that Pb bioavailability ranges from <5% for mine tailing materials where the majority of total Pb occurs as galena (PbS) to 45% for surface soils where total Pb is dominated by Fe–Mn–Pb oxides (Casteel, 1997). In residential soils containing mine waste, two dominant Pb mineral phases were identified: sulfide/sulfate and oxide/phosphate (Davis et al., 1993). The former phase consisted primarily of galena, anglesite, and jarosite, while the latter phase was primarily MnPb oxide, Pb phosphates, and Pb oxides.

In soils collected near a Pb–Zn smelter, Pb was found to be divalent and coordinated to O, OH ligands (Manceau et al., 1996). In the vicinity of a battery recycling complex, Pb was found as Pb sulfate and was silica-bound; in soil contaminated by tetraethyl Pb, Pb was divalent and complexed to salicylate and catechol-type functional groups of humic substances.

7.6 Other Factors

In Section 4.5, it was pointed out that hydrous oxides of Fe and Mn, can play an important role in the sorption of Pb in soils. The clay fraction of soils and sediments is known to contain appreciable amounts of these materials (Deshpande et al., 1968). These hydrous oxides have very high sorption capacity for trace elements compared with that of clays themselves (Jenne, 1968; Stumm and Morgan, 1970). Metal retention by hydrous Fe oxide follows the general order: $Pb \geq Cu \gg Zn = Cd$ (see Chapter 2).

As indicated in Section 5.3, elevating phosphate levels through fertilization can immobilize Pb in soils. Because phosphate is a major nutrient applied to land before the growing season, low bioavailability of Pb to plants is almost always ensured. In heavily contaminated soils, other sources of phosphate that are plentiful and relatively inexpensive, such as rock phosphate or hydroxyapatite, should be considered (Chlopecka and Adriano, 1996). This method inactivates Pb *in situ* rendering it immobile and nonbioavailable. However, other remediation technologies require removal of the contaminant from the soil. In this case, a soil additive that solubilizes Pb to enhance phytoextraction or soil washing is more desirable. For example, addition of synthetic chelates can facilitate the uptake of Pb by plants from contaminated soils (Wu et al., 1999).

8 Sources of Lead in the Environment

Worldwide, the general public is exposed to Pb through contact with air and dust, and ingestion of food and water. Children are normally exposed more via ingestion of soil and dust. The risk of Pb poisoning in U.S. children is highest among urban minority children from low-income households and among children living in older housing units built before 1946 (CDC, 1997). Residential paint, dust, and soils containing Pb have been identified as major sources of exposure in U.S. children primarily via the hand-to-mouth route of ingestion (ATSDR, 1988; CDC, 1991). About 64 million privately-owned and occupied U.S. housing units are estimated to contain some Pb-based paint (U.S. HUD, 1995). Urban soils can be contaminated with Pb from vehicle emissions in past years, lead-based paint, and industrial sources.

Humans may be exposed to Pb via air, drinking water, food, paint, and industrial releases. For example, residents in Shanghai, China are chronically exposed to Pb from automobile emission, cement industry, coal and oil combustion, and the metallurgical industry (Wang et al., 2000).

8.1 Air

The U.S. EPA requires that the concentration of Pb in air the general public may breathe shall not exceed 1.5 μg m^{-3} averaged over a calendar quarter (ATSDR, 1992). This standard will probably be lowered. To reduce the amount of Pb released into the environment, the EPA regulations now limit the level of Pb in unleaded gasoline to 0.013 g L^{-1}. The Occupational Safety and Health Administration (OSHA) has set a PEL (permissible exposure limit) of Pb in workroom air at 50 μg m^{-3} averaged over an 8-hr workday for workers in general industry. For those exposed to air concentrations at or above the action level of 30 μg m^{-3} for more than 30 days yr^{-1}, OSHA mandates periodic determination of PbB levels. If a PbB level is found to be greater than 40 μg dL^{-1}, the worker must be notified in writing and provided with a medical examination. If a worker's PbB level reaches 60 μg dL^{-1} (or averages 50 μg dL^{-1} or more), the employer is obligated to remove the employee from excessive exposure until the employee's PbB level falls below 40 μg dL^{-1}.

Vehicular emissions can also enrich the air and areas by roadways. Only about 10% of Pb from automobile emissions settles out in the immediate vicinity (within 100 m) of the roadway. Another 45% settles out within 20 km, 10% between 20 and 200 km, and the remaining 35% is carried on long-range atmospheric transport systems (OECD, 1993). Evidence of this long-range transport is the enhanced level of Pb from gasoline and other sources found in snow in Greenland (Boutron et al., 1991). Most of the Pb on the surfaces of vegetation is suspendible particulates and the Pb in the soil is primarily concentrated in the surface (0 to 5 cm) (Page and Ganje, 1970).

A large-scale British study of Pb concentrations in garden soils and in street and road dust has shown that most urban gardens and streets are heavily contaminated from industrial and vehicular emissions (Thornton and Culbard, 1986). Generally, house dust appear to be more heavily contaminated with Pb than the soil outside, especially in old houses that were painted with leaded pigments. Urban environments can also be enriched with Pb from vehicular emissions. Soils, edible plants, and woody plants have been reported to be contaminated by deposited Pb-laden dusts and particulates (Adriano, 1986).

Historical atmospheric Pb depositions in Europe and North America have been assessed using a number of archives, such as peat bogs (MacKenzie et al., 1997), lake sediments (Renberg et al., 1994), wood from tree rings, and herbarium plants (Bacon et al., 1996). These studies greatly helped to assess anthropogenic perturbations of the natural biogeochemical cycle of Pb. To identify the sources and to quantify their relative contributions, Pb isotope ratios are widely used. Anthropogenic Pb in the air, derived from high-temperature industrial processes (e.g., steel and nonferrous metal production), fuel combustion (e.g., leaded gasoline, oil, and coal), and incineration (e.g., municipal solid waste) normally has an isotopic signature distinct from naturally occurring Pb supplied by rock weathering. Gasoline has been the dominant source of recent atmospheric Pb deposition (Farmer et al., 1996), with Pb isotope ratios far from the preanthropogenic values seen during the middle of the Holocene (Shotyk et al., 1998). For example, the ^{206}Pb/^{207}Pb ratios reported for atmospheric aerosols (Monna et al., 1997), snow (Rosman et al., 1993), sediments (Moor et al., 1996), and surface soils (Bacon et al., 1996) typically vary between 1.1 and 1.2. Recent lake sediment and peat deposits and surface soil horizons in Sweden have a similarly low isotope ratio, typically 1.15 to 1.18 (Bindler et al., 1999). In contrast, the natural ^{206}Pb/^{207}Pb ratios in unpolluted sediment, peat, and mineral soil horizons in Sweden are clearly higher (mean = 1.53 ± 0.08; range = 1.28 to 3.11; $n = 50$ sites). The low ratios in recent sediments and surface soils are caused by depression and accumulation of atmospheric Pb pollution.

Brannvall et al. (1999) used Pb concentrations and stable Pb isotopes (^{206}Pb/^{207}Pb ratios) of annually laminated sediments from four lakes in northern Sweden (lat ∼ 65°N) to provide a decadal record of atmospheric Pb pollution for the last 3000 years. They found a clear signal in the sediments of airborne pollution from Greek and Roman cultures 2000 years ago, followed by a period of "clean" conditions for 400 to 900 A.D. From 900 A.D. there was a conspicuous, permanent increase in atmospheric Pb pollution fallout. The sediments reveal peaks in atmospheric Pb pollution at 1200 and 1530 A.D. comparable to present-day levels. These peaks match the history of metal production in Europe. This study indicates that the contemporary atmospheric pollution climate in northern Europe was established in Medieval times, rather than in the Industrial era.

Additionally, atmospheric Pb deposition since the Industrial Revolution was studied in Switzerland using rural peat bogs (Weiss et al., 1999). Similar temporal patterns were found in western and central Switzerland, with two distinct periods of Pb enrichment relative to the natural background: between 1880 and 1920 with enrichments ranging from 40 to 80 times, and between 1960 and 1980 with enrichments ranging from 80 to 100 times. The fluxes also were generally elevated in those time periods: in western Switzerland between 1.16 and 1.55 μg cm^{-2} yr^{-1} during the first period. Between the Industrial Revolution and 1985, nonradiogenic Pb became increasingly important in all cores because of the replacement of coal by oil after about 1920, the use of Australian Pb in industry, and the extensive combustion of leaded gasoline after 1950. The introduction of unleaded gasoline in 1985 had a pronounced effect on the Pb deposition in all cores.

Similarly, rates of Pb deposition were determined over the past 150 to 200 years for eight sites throughout the Czech Republic using ^{210}Pb-dated, sphagnum-derived peat cores (Vile et al., 2000). Maximum historical Pb deposition was greater at sites in the northern and western parts of the country than at sites in the southern part (peak values averaging 57, 21, and 16 mg m^{-2} yr^{-1}, respectively). Lead deposition patterns generally reflect increasing industrialization over the past 100 to 200 years, especially in the post–World War II era. Maximum Pb deposition occurred between 1965 and 1992, corresponding to a period of peak production and burning of lignite coal. A decrease in Pb deposition rates since 1975 to 1980 was evident. The most recent Pb deposition rates (1992), estimated from the uppermost peat core sections, averaged 32, 11, and 7 mg m^{-2} yr^{-1} for the northern, western, and southern sites, respectively, and are higher than current Pb deposition in the eastern United States of 4 mg m^{-2} yr^{-1}.

8.2 Drinking Water

The U.S. EPA estimates that about 20% of the U.S. population (including 3.8 million children) consumes drinking water with Pb levels above 20 μg dL^{-1} (ATSDR, 1992). The U.S. EPA is required to set drinking water standards with two levels of protection. The primary standards define contaminant levels in drinking water as levels above which the water source requires treatment. These maximum contaminant levels (MCLs) are limits enforceable by law and are set as close as possible to the maximum contaminant level goals (MCLGs), the levels determined to be safe by toxicological and biomedical considerations, independent of feasibility. The U.S. EPA has promulgated a final rule for Pb in drinking water: the rule does not establish an MCL; the MCLG is zero and the action level is set at 15 μg L^{-1}. If more than 10% of targeted tap water samples exceed the action level, certain actions are required of water system administrators.

Although Pb is typically a minor constituent of surface waters and aquifers, drinking water is often contaminated by water distribution systems

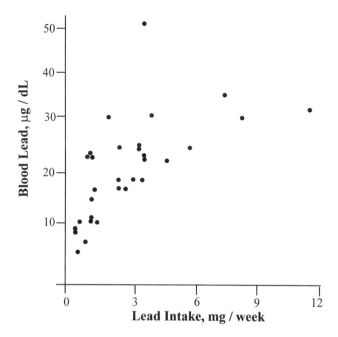

FIGURE 10.13. Relationship of blood lead to lead intake via drinking plumbosolvent water delivered through lead pipes. (Modified from Thornton and Culbard, 1987.)

that contain Pb in their pipes, solder, or leaded-brass plumbing fixtures. Water is most likely to absorb Pb from plumbing if it is hot, has low mineral content (i.e., soft water), or is acidic. Infants fed formula made from hot tap water are particularly at risk of Pb poisoning from this source. The contribution of water delivery systems in dwellings to PbB is demonstrated in Figure 10.13. The subjects were adult women drinking plumbosolvent water delivered through lead pipes.

Many OECD nations now limit the use of Pb-containing materials in drinking water systems (OECD, 1993). However, Pb water mains, service lines, and other plumbing components installed long ago in Europe and the United States can continue to present hazards.

While much is known about plumbing Pb in developed countries, there is little information available on Pb in plumbing in developing countries (Alliance/EDF, 1994). Much of the developing world lacks plumbing infra-structure, Pb-containing or otherwise, with over 1 billion people lacking access to clean and plentiful water (World Bank, 1992).

8.3 Food

In the United States, regulating Pb contamination in foods is the respon-sibility of the Food and Drug Administration (FDA). The U.S. FDA has set a goal of less than 100 μg day^{-1} as the total Pb intake by children ages 1 to 5 (ATSDR, 1992). Lead in food and beverages is encountered by virtually this entire age group in the United States. The U.S. FDA has estimated that about 20% of all dietary Pb comes from canned food; about two-thirds of that amount results from lead solder in cans. High levels of Pb may leach from lead-soldered cans into food, particularly if the food is acidic (e.g., tomatoes, citrus fruits). The number of food cans that are lead-soldered continues to decline. In 1979, over 90% of all food cans were Pb-soldered; in 1986, this figure was 20%, or fewer than about 2 million cans. Imported glazed ceramics and Pb-containing pottery are also potential sources of dangerously high levels of Pb. In 1991, the U.S. FDA reduced the guideline levels for ceramicware that had been issued in 1980 as follows (U.S. FDA, 1992): from 7 to 3 ppm for plates, saucers, and other flatware; from 5 to 2 ppm for small hollowware such as cereal bowls (but not cups and mugs); from 5.0 to 0.5 ppm for cups and mugs; from 2.5 to 1.0 ppm for large (> 1.1 L) hollowware such as bowls (but not pitchers); and from 2.5 to 0.5 ppm for pitchers.

It can be expected that the concentrations of Pb in most foods in North America and western Europe are low. In the United Kingdom, it is generally < 50 μg kg^{-1} DW (Thornton and Culbard, 1987). The average dietary intakes of Pb in the United Kingdom, have ranged from 24 to 65 μg day^{-1}. Hard evidence from western Europe and the United States indicates that in areas with high Pb concentrations in soils and dusts, certain vegetables, notably the leaf crops (e.g., *Brassicas* and lettuce) and leeks have higher surface contamination and that some root crops (e.g., parsnips and radishes) can also have elevated Pb contents (Davies and Wixson, 1988). Thus, for people who have large gardens or allotments and who grow a substantial portion of their own diet, Pb intake could increase substantially.

Anthropogenic Pb aerosols account for approximately 40% of the Pb in food (U.S. EPA, 1986). This estimate is based on calculations of the contributions of different sources to the total Pb content of livestock and food crops (Wolnik et al., 1983, 1985), relative to natural concentrations of Pb in the environment. The atmospheric Pb content of crops correlates with their proximity to anthropogenic sources of Pb. Numerous studies have shown that Pb levels on the surface of vegetation are proportional to air Pb concentrations, especially in the particle size range below 1 μm.

8.4 Paint

Today, dust contaminated by lead-based paint is considered the most significant source of moderate and severe Pb poisoning in children in the United States (Alliance/EDF, 1994). While the magnitude of the problem may not be as great as in most other countries, children are potentially at risk wherever lead-based paint has been used. Whenever leaded paint deteriorates, is disturbed during renovation, or is abraded on surfaces such as floors or window tracks, Pb is released into interior dust or exterior soil, where it remains indefinitely unless properly abated.

While leaded-paint has been recognized as a high-risk source for children in the United States and a few European countries (e.g., United Kingdom, France, and Belgium), it is largely ignored by the rest of the world. Many countries allow up to 2% Pb (20000 ppm Pb) in paints even in interior pigments (Alliance/EDF, 1994). In the United States, the limit for paint is 0.06% (or 600 ppm) Pb.

8.5 Industrial Releases

Mining, smelting, refining, manufacturing, and recycling and disposal of Pb-containing products (e.g., batteries) can release Pb into air, soil, and water at, oftentimes, extraordinarily high levels. Consequently, Pb ranks first on the priority list of hazardous substances found at Superfund sites in the United States (i.e., based on its frequent presence at cleanup sites, its toxicity, and its potential for human exposure).

Today, the average PbB level in the U.S. population is below 10 μg dL^{-1}, down from an average of 16 μg dL^{-1} (in the 1970s), the level before the legislated removal of Pb from gasoline. A PbB level of 10 μg dL^{-1} is about 3 times higher than the average level found in some remote populations.

The level defining Pb poisoning has been progressively declining. It was 30 μg dL^{-1} in 1970 to 1985, 25 μg dL^{-1} in 1986 to 1991, and has stayed at 10 μg dL^{-1} since 1991 (ATSDR, 1992). *Maybe there is no safe blood Pb level (above the background)!* Currently, the consensus level of concern for children is 10 to 14 μg dL^{-1} (Table 10.10). Effects on stature have been reported to begin at levels as low as 4 μg dL^{-1}, the present limit for accurate PbB measurement. Taken together, effects occur over a wide range of PbB concentrations, with no indication of a threshold. No safe level has yet been found for children! Even in adults, effects are being discovered at lower and lower levels as more sensitive analyses and measures are developed.

Most studies in the 1970s and 1980s used PbB as an exposure marker. But because Pb has a relative short residence time in blood, which may become normal in exposed individuals after losing contact with the contaminant, Needleman et al. (1979) measured Pb in shed deciduous teeth. High-Pb children had significantly lower IQ scores, poor attention, and diminished speech and language function.

There appears to be convincing evidence that the effects of early exposure to Pb are likely to be irreversible and may affect general life success (Needleman, 1996). For example, Bellinger et al. (1984) found that high-Pb 5th-grade subjects in the Boston area had lower intelligence scores on a school-based test, more need for remedial services, and more grade retentions. Findings on their 18th year further indicated that high-Pb subjects were 7 times more likely to fail out of high school than the lowest Pb group, and 6 times more likely to

TABLE 10.10. Classification of children in the United States based on blood lead concentration and the interpretation of blood lead test results and follow-up activities.

Class	Blood lead concentration ($\mu g\ dL^{-1}$)	Comment
I	≤9	A child in Class I is not considered to be lead-poisoned.
IIA	10–14	The presence of many children (or a large proportion of children) with blood Pb levels in this range should trigger communitywide childhood Pb poisoning prevention activities. Children in this range may need to be rescreened frequently.
IIB	15–19	A child in Class IIB should receive nutritional and educational interventions and more frequent screening. If the blood Pb level persists in this range, environmental investigation and intervention should be done.
III	20–44	A child in Class III should receive environmental evaluation and remediation and a medical evaluation. Such a child needs pharmacological treatment of lead poisoning.
IV	45–69	A child in Class IV will need both medical and environmental interventions, including chelation therapy.
V	≥70	A child with Class V lead poisoning is a medical emergency. Medical and environmental management must begin immediately.

Source: ATSDR (1992).

have a reading disability. Additionally, they ranked lower in class placement in their final year of high school, had more absenteeism, and poorer vocabulary, reasoning, and fine motor function (Needleman et al., 1990). Indeed, the U.S. National Academy of Sciences states, "The weight of the evidence from the 1980s clearly supports the conclusion...that PbB concentrations of around 10 $\mu g\ dL^{-1}$ in children are associated with disturbances in early physical and mental growth and in later intellectual functioning and academic achievement" (NAS, 1993). A PbB level at around 10 $\mu g\ dL^{-1}$ has also been considered by Australia, New Zealand, and the EU.

References

Adriano, D.C. 1986. *Trace Elements in the Terrestrial Environment.* Springer-Verlag, New York.
Ainsworth, C.C., and D. Rai. 1987. *Chemical Characterization of Fossil Fuel Combustion Wastes.* EPRI Rep EA-5321. Electric Power Res Inst, Palo Alto, CA.
Al-Chalabi, A.B., and D. Hawker. 2000. *Water Air Soil Pollut* 118:299–310.
[Alliance/EDF] Alliance to End Childhood Lead Poisoning and Environmental Defense Fund. 1994. *The Global Dimensions of Lead Poisoning: An Initial Analysis.* Washington, DC.

Alloway, B.J., ed. 1990. *Heavy Metals in Soils.* Blackie, London.

Alloway, B.J., I. Thornton, G.A. Smart, and M.J. Quinn. 1988. *Sci Total Environ* 75:41–69.

Andersson, A. 1977. *Swedish J Agric Res* 7:7–20.

Archer, F.C. 1980. *Minist Agric Fish Food* (Great Britain) 326:184–190.

Arvik, J.H., and R.L. Zimdahl. 1974a. *J Environ Qual* 3:369–373.

Arvik, J.H., and R.L. Zimdahl. 1974b. *J Environ Qual* 3:374–376.

[ATSDR] Agency for Toxic Substances and Disease Registry. 1988. *The Nature and Extent of Lead Poisoning in the United States: A Report to Congress.* ATSDR, Washington, DC.

ATSDR (Agency for Toxic Substances and Disease Registry). 1992. *Lead Toxicity.* U.S. Dept Health and Human Services, Atlanta, GA.

ATSDR (Agency for Toxic Substances and Disease Registry). 1993. *Toxicological Profile of Lead.* U.S. Dept Health and Human Services, Atlanta, GA.

Aubert, H., and M. Pinta. 1977. *Trace Elements in Soils.* Elsevier, New York.

Bacon, J.R., K.C. Jones, S.P. McGrath, and A.E. Johnson. 1996. *Environ Sci Technol* 30:2511–2518.

Bargar, J.R., S.N. Towle, G.E. Brown, and G.A. Parks. 1996. *Geochim Cosmochim Acta* 60:3541–3547.

Barzi, F., R. Naidu, and M.J. McLaughlin. 1996. In R. Naidu et al., eds. *Contaminants in the Soil Environment in the Australasia – Pacific Region.* Kluwer, Dordrecht, Netherlands.

Bassuk, N.L. 1986. *Hort Science* 21:993–995.

Basta, N.T., and M.A. Tabatabai. 1992a. *Soil Sci* 153:331–337.

Basta, N.T., and M.A. Tabatabai. 1992b. *Soil Sci* 153:108–114.

Baumhardt, G.R., and L.F. Welch. 1972. *J Environ Qual* 1:92–94.

Bazzaz, F.A., G.L. Rolfe, and P. Windle. 1974. *J Environ Qual* 3:156–158.

Bellinger, D., H.L. Needleman, R. Bromfield, and M. Mintz. 1984. *Biol Trace Elem Res* 6:207–223.

Berrow, M.L., and R.L. Mitchell. 1980. *Trans Roy Soc Edinburgh Earth Sci* 71:103–121.

Berrow, M.L., and G.A. Reaves. 1984. In *Proc Intl Conf Environmental Contamination.* CEP Consultants, Edinburgh.

Berry, R.A. 1924. *J Agric Sci* 14:58–65.

Berthelsen, B.O., E. Steinnes, W. Solberg, and L. Jingsen. 1995. *J Environ Qual* 24:1018–1102.

Beyer, W.N., O.H. Pattee, L. Sileo, D.J. Hoffman, and B.M. Mulhern. 1985. *Environ Pollut* 38:63–86.

Biddappa, C.C., and M. Chino. 1981. *Soil Sci Plant Nutr* 27:93–103.

Bindler, R., M.J. Brannvall, I. Renberg, O. Emteryd, and H. Gripp. 1999. *Environ Sci Technol* 33:3362–3367.

Bittel, J.E., and R.J. Miller. 1974. *J Environ Qual* 3:250–253.

Boon, D.Y., and P.N. Soltanpour. 1992. *J Environ Qual* 21:82–86.

Boutron, C.F., U. Gorlach, J.P. Candelone, M.A. Bolshov, and R.J. Delmas. 1991. *Nature* 353:153–156.

Bowen, H.J.M. 1979. *Environmental Chemistry of the Elements.* Academic Pr, New York.

Brannvall, M.J., R. Bindler, I. Renberg, J. Emteryd, J. Bartnicki, and K. Billstrom. 1999. *Environ Sci Technol* 33:4391–4395.

Brookins, D.G. 1988. *Eh-pH Diagrams for Geochemistry*. Springer-Verlag, New York.

Buck, W.B. 1975. *J Am Vet Med Assoc* 166:222–226.

Buck, W.B., G.D. Osweiler, and G.A. Van Gelder. 1976. In *Clinical and Diagnostic Veterinary Toxicology*. Kendall-Hunt Publ, Dubuque, IA.

Bunzl, K. 1974. *J Soil Sci* 25:517–534.

Bunzl, K., W. Schmidt, and B. Sansoni. 1976. *J Soil Sci* 27:32–41.

Cannon, H.L. 1974. In P.L. White and D. Robbins, eds. *Environmental Quality and Food Supply*. Futura Publ, Mt. Kisko, NY.

Cappon, C.J. 1987. *Bull Environ Contam Toxicol* 38:695–699.

Casteel, S.W. 1997. *Bioavailability of Lead in Soil and Mine Waste from the California Gulch NPL Site, Leadville, Colorado*. U.S. EPA draft report. U.S. Environmental Protection Agency, Washington, DC.

Carlson, R.W., F.A. Bazzaz, J.J. Stukel, and J.B. Wedding. 1976. *Environ Sci Technol* 10:1139–1142.

CDC (Centers for Disease Control). 1991. *Preventing Lead Poisoning in Young Children*, U.S. Dept Health and Human Services, Atlanta, GA.

CDC (Centers for Disease Control). 1997. *Update: Blood Levels—United States, 1991–1994*. U.S. Dept Health and Human Services, Atlanta, GA.

Chang, A.C., A.L. Page, J.E. Warneke, and E. Grgurevic. 1984a. *J Environ Qual* 13:33–38.

Chang, A.C., J.E. Warneke, A.L. Page, and L.I. Lund. 1984b. *J Environ Qual* 13: 87–91.

Chang, L.W. 1996. *Toxicology of Metals*. CRC Pr, Boca Raton, FL.

Chen, J., F. Wei, Y. Wu, and D.C. Adriano. 1991. *Water Air Soil Pollut* 57–58: 699–712.

Chino, M. 1981. In K. Kitagishi and I. Yamane, eds. *Heavy Metal Pollution in Soils of Japan*. Japan Sci Soc Pr, Tokyo.

Chisholm, D., and R.F. Bishop. 1967. *Phytoprotection* 48:78–81.

Chlopecka, A., and D.C. Adriano. 1996. *Environ Sci Technol* 30:3294–3303.

Chlopecka, A., J.R. Bacon, M.J. Wilson, and J. Kay. 1996. *J Environ Qual* 25: 69–79.

Cox, W.J., and D.W. Rains. 1972. *J Environ Qual* 1:167–169.

Creger, T.L., and F.J. Peryea. 1997. *Hort Sci* 27:1277–1278.

Crump, D.R., and P.J. Barlow. 1980. *Sci Total Environ* 15:269–274.

Czarnowska, K., and B. Gworek. 1991. *Environ Geochem Health* 12:289–290.

Czuba, M., and T.C. Hutchinson. 1980. *J Environ Qual* 9:566–575.

Daury, R.W., F.E. Schwab, and M.C. Bateman. 1993. *J Wildlife Dis* 29:577–581.

Davies, B., and B.G. Wixson. 1988. *Lead in Soil: Issues and Guidelines*. Science Reviews, Northwood, United Kingdom.

Davies, B.E. 1983. *Geoderma* 29:67–75.

Davis, A., M.V. Ruby, and P.D. Bergstrom. 1992. *Environ Sci Technol* 26:461–468.

Davis, A., J.W. Drexler, M.V. Ruby, and A. Nicholson. 1993. *Environ Sci Technol* 27:1415–1425.

Davis, R.D., P.H.T. Beckett, and E. Wollan. 1978. *Plant Soil* 49:395–408.

Deshpande, T.L., D.J. Greenland, and J.P. Quirk. 1968. *J Soil Sci* 19:108–122.

Dudka, S., and D.C. Adriano. 1997. *J Environ Qual* 26:590–602.

Elkhatib, E.A., G.M. Elshebiny, and A.M. Balba. 1991. *Environ Pollut* 69:269–276.

Elsokkary, I.H., M.A. Amer, and E.A. Shalaby. 1995. *Environ Pollut* 87:225–233.

Erel, Y., J.J. Morgan, and C.C. Patterson. 1991. *Geochim Cosmochim Acta* 55:7067–7719.

Farmer, J.G., L.J. Eades, A.B. MacKenzie, A. Kirika, and T.E. Bailey Watts. 1996. *Environ Sci Technol* 30:3080–3083.

Fisenne, I.M., G.A. Welford, P. Perry, R. Baird, and H.W. Keller. 1978. *Environ Inter* 1:245–246.

Frank, R., K. Ishida, and P. Suda. 1976. *Can J Soil Sci* 56:181–196.

Friedland, A., and A.H. Johnson. 1985. *J Environ Qual* 14:332–336.

Friedland, A., B.W. Craig, E.K. Miller, G.T. Herrick, T.G. Siccama, and A.H. Johnson. 1992. *Ambio* 21:400–403.

Friedland, A.J., A.H. Johnson, and T.G. Siccama. 1984. *Water Air Soil Pollut* 21:161–170.

Gaynor, J.D., and R.L. Halstead. 1976. *Can J Soil Sci* 56:1–8.

Griffeth, I.I. 1919. *J Agric Sci* 9:366–395.

Griffin, R.A., and A.K. Au. 1977. *Soil Sci Soc Am J* 41:880–882.

Gzyl, J. 1990. *Sci Total Environ* 96:199–209.

Hahne, H.C.H., and W. Kroontje. 1973. *J Environ Qual* 2:444–450.

Hammond, P.B., and A.C. Aronton. 1964. *Ann NY Acad Sci* 111:595–611.

Harter, R.D. 1979. *Soil Sci Soc Am J* 43:679–683.

Harter, R.D. 1983. *Soil Sci Soc Am J* 47:47–51.

Hassett, I.I. 1974. *Commun Soil Sci Plant Anal* 5:499–505.

Heinrichs, H., and R. Mayer. 1977. *J Environ Qual* 6:402–407.

Heinrichs, H., and R. Mayer. 1980. *J Environ Qual* 9:111–118.

Heinrichs, H., B. Schulz-Dobrick, and K.H. Wedepohl. 1980. *Ceochim Cosmochim Acta* 44:1519–1532.

Hesterberg, D., D.E. Sayers, W. Zhou, G.M. Plummer, and W.P. Robarge. 1997. *Environ Sci Technol* 31:2840–2846.

Hewitt, E.I. 1953. *J Exp Bot* 4:59–64.

Höll, W., and R. Hampp. 1975. *Residue Rev* 54:79–111.

Huang, C.Y., F.A. Bazzaz, and L.N. Vanderkoef. 1974. *Plant Physiol* 54:122–124.

Humphreys, D.J. 1991. *British Vet J* 147:18.

[ILZSG] International Lead and Zinc Study Group. 1992. *Principal Uses of Lead and Zinc,* 1960–90. Metro House, London.

Jenne, E.A. 1968. *Adv Chem Series* 73:337–387.

John, M.K. 1971. *Environ Sci Technol* 12:119–1203.

John, M.K. 1972. *J Environ Qual* 1:295–298.

John, M.K., and C.I. Van Laerhoven. 1972a. *Environ Lett* 3:111–116.

John, M.K., and C.I. Van Laerhoven. 1972b. *J Environ Qual* 1:169–171.

Johnson, A.H., T.G. Siccama, and A.I. Friedland. 1982. *J Environ Qual* 11:577–580.

Johnson, C.E., T.G. Siccama, C.T. Driscoll, G.E. Likens, and R.E. Moeller. 1995. *Ecol Applic* 5:813–822.

Johnson, G.D., D.J. Audet, et al. 1999. *Environ Toxicol Chem* 18:1190–1194.

Jones, L.H.P., C.R. Clement, and M.J. Hopper. 1973a. *Plant Soil* 38:403–414.

Jones, L.H.P., S.C. Jarvis, and D.W. Cowling. 1973b. *Plant Soil* 38:605–619.

Jurinak, J.J., and J. Santillan-Medrano. 1974. *The Chemistry and Transport of Lead and Cadmium in Soils.* Res Rep 18. Utah State Univ Agric Exp Sta, Logan, UT.

Kabata-Pendias, A. 2000. Polish Academy of Sciences, Warsaw, personal communication.

Karamanos, R.E., J.R. Bettany, and J.W.B. Stewart. 1976. *Can J Soil Sci* 56:485–494.

Keaton, C.M. 1937. *Soil Sci* 43:401–411.

Khan, D.H., and B. Frankland. 1983. *Plant Soil* 70:335–345.

Kim, E.Y., R. Goto, H. Iwata, Y. Masuda, S. Tanabo, and S. Fujita. 1999. *Environ Toxicol Chem* 18:448–457.

Kloke, A., and K. Riebartsch. 1964. *Naturwissenschaften* 51:367–368.

Knox, A., and D.C. Adriano. 2000. University of Georgia. Unpublished data.

Koeppe, D.E., and R.J. Miller. 1970. *Science* 167:1376–1378.

Koslow, E.E., W.H. Smith, and B.J. Staskawicz. 1977. *Environ Sci Technol* 11:1019–1021.

Kucharski, R., E. Marchwinska, and J. Gzyl. 1994. *Ecol Engr* 3:299–312.

Lagerwerff, J.V., and D.L. Brower. 1973. *Soil Sci Soc Am Proc* 37:11–13.

Laperche, V., S.J. Traina, P. Gaddam, and T.J. Logan. 1996. *Environ Sci Technol* 30:3321–3325.

Laperche, V., T.J. Logan, P. Gaddam, and S.J. Traina. 1997. *Environ Sci Technol* 31:2745–2753.

Lee, D.Y., and H.C. Zheng. 1994. *Plant Soil* 164:19–23.

Lee, K.C., B.A. Cunningham, K.H. Chung, G.M. Paulsen, and G.H. Liang. 1976. *J Environ Qual* 5:357–359.

Ma, Q.Y., S.J. Traina, T.J. Logan, and J.A. Ryan. 1993. *Environ Sci Technol* 27:1803–1810.

Ma, Q.Y., S.J. Traina, T.J. Logan, and J.A. Ryan. 1994a. *Environ Sci Technol* 28:1219–1228.

Ma, Q.Y., S.J. Traina, T.J. Logan, and J.A. Ryan. 1994b. *Environ Sci Technol* 28:408–418.

Ma, Q.Y., T.J. Logan, and S.J. Traina. 1995. *Environ Sci Technol* 29:1118–1126.

MacKenzie, A.B., et al. 1997. *Sci Total Environ* 203:115–127.

MacLean, A.L., R.L. Halstead, and B.J. Finn. 1969. *Can J Soil Sci* 49:327–334.

Malone, C., D.E. Koeppe, and R.J. Miller. 1974. *Plant Physiol* 53:388–394.

Manceau, A., M.C. Boisset, G. Sarret, J.L. Hazemann, M. Mench, P. Cambier, and R. Prost. 1996. *Environ Sci Technol* 30:1540–1552.

Marten, G.C., and P.B. Hammond. 1966. *Agron J* 58:553–554.

Martinez, C.E., and M.B. McBride. 1991. *Environ Sci Technol* 33:745–750.

McEvoy, J.D., and M. McCoy. 1993. *Vet Record* (Jan.).

McKeague, J.A., and M.S. Wolynetz. 1980. *Geoderma* 24:299–307.

McKenzie, R.M. 1970. *Aust J Soil Res* 8:97–106.

McKenzie, R.M. 1975. *Aust J Soil Res* 13:177–188.

McKenzie, R.M. 1980. *Aust J Soil Res* 18:61–73.

Merian, E. 1991. *Metals and their Compounds in the Environment: Occurrence, Analysis, and Biological Relevance.* VCH Publ, Weinheim, Germany.

Milberg, R.P., I.V. Lagerwerff, D.L. Grower, and G.T. Biersdorf. 1980. *J Environ Qual* 9:6–8.

Miles, C.D., I.R. Brandle, D.I. Daniel, O. Chu-Der, P.D. Schnare, and D.I. Uhlik. 1972. *Plant Physiol* 49:820–825.

Miller, E.K., and A.J. Friedland. 1994. *Environ Sci Technol* 28:662–669.

Miller, R.J., and D.E. Koeppe. 1970. *Trace Subs Environ Health* 4:186–193.

Misra, S.G., and G. Pandey. 1976. *Plant Soil* 45:693–696.

Monna, F., et al. 1997. *Environ Sci Technol* 31:2227–2286.

Moor, H., T. Schaller, and M. Sturm. 1996. *Environ Sci Technol* 30:2928–2933.

Motto, H.S., R.H. Daines, D.M. Chilko, and C.K. Motto. 1970. *Environ Sci Technol* 4:231–237.

Mushak, P., J.M. Davis, A.F. Crocetti, and L.D. Grant. 1989. *Environ Res* 50:11–36.

[NAS] National Academy of Sciences. 1993. *Measuring Lead Exposure in Infants, Children, and Other Sensitive Populations.* National Academy Pr, Washington, DC.

Needleman, H.L. 1996. In L.W. Chang, ed. *Toxicology of Metals.* CRC Pr, Boca Raton, FL.

Needleman, H.L., C. Gunnoe, and A. Leviton. 1979. *N Engl J Med* 300:689–695.

Needleman, H.L., A. Schell, D. Bellinger, A. Leviton, and E.N. Allred. 1990. *N Engl J Med* 322:83–89.

Neuman, D.R., and D.J. Dollhopf. 1992. *J Environ Qual* 21:181–184.

Nilsson, I. 1972. *Oikos* 23:132–136.

Norrish, K. 1975. In D.J.D. Nicholas and A.R. Egan, eds. *Trace Elements in Soil-Plant-Animal Systems.* Academic Pr, New York.

Nriagu, J.O., ed. 1978. *The Biogeochemistry of Lead in the Environment.* Elsevier, Amsterdam.

[OECD] Organization for Economic Cooperation and Development. 1993. *Risk Reduction.* Monograph 1: Lead. OECD, Paris.

Page, A.L., and T.J. Ganje. 1970. *Environ Sci Technol* 4:140–142.

Page, A.L., T.J. Ganje, and M.S. Joshi. 1971. *Hilgardia* 41:1–31.

Pain, D.J. 1989. *Environ Pollut* 60:67–81.

Pain, D.J. 1995. D.J. Hoffmann et al., eds. In *Handbook of Ecotoxicology.* Lewis Publ, Boca Raton, FL.

Pinheiro, J.P., A.M. Mota, and M.F. Benedetti. 1999. *Environ Sci Technol* 33:3398–3404.

Puls, R.W., R.M. Powell, D. Clark, and C.J. Eldred. 1991. *Water Air Soil Pollut* 57–58:423–440.

Rabinowitz, M.B. 1993. *Bull Environ Contam Toxicol* 51:438–444.

Rains, D.W. 1975. *J Environ Qual* 4:532–536.

Ramos, L., L.M. Hernandez, and M.J. Gonzales. 1994. *J Environ Qual* 23:50–57.

Reaves, G.A., and M.L. Berrow. 1984a. *Geoderma* 32:1–8.

Reaves, G.A., and M.L. Berrow. 1984b. *Geoderma* 32:117–129.

Reddy, C.N., and W.H. Patrick Jr. 1977. *J Environ Qual* 6:259–262.

Reiners, W.A., R.H. Marks, and P.M. Vitousek. 1975. *Oikos* 26:264–275.

Renberg, I., M. Persson, and O. Emteryd. 1994. *Nature* 368:323–326.

Riffaldi, R., R. Levi-Minzi, and G.F. Soldatini. 1976. *Water Air Soil Pollut* 6:119–128.

Robinson, I.M. 1978. In J.O. Nriagu, ed. *The Biogeochemistry of Lead in the Environment.* Elsevier, Amsterdam.

Rolfe, G.L., 1973. *J Environ Qual* 2:153–157.

Rolfe, G.L., and F.A. Bazzaz. 1975. *Forest Sci* 21:33–38.

Roscoe, D.E., L. Widjeskog, and W. Stansley. 1989. *Bull Environ Contam Toxicol* 42:226–233.

Rosman, K.J.R., et al. 1993. *Nature* 362:333–335.

Roy, W.R., I.G. Krapac, and J.D. Steele. 1993. *J Environ Qual* 22:537–543.

Ruby, M.V., A. Davis, J.H. Kempton, J.W. Drexler, and P.D. Bergstrom. 1992. *Environ Sci Technol* 26:1242–1248.

Ruby, M.V., A. Davis, and A. Nicholson. 1994. *Environ Sci Technol* 28:646–654.

Saar, R.A., and J.H. Weber. 1980. *Environ Sci Technol* 14:877–880.

Santillan-Medrano, J., and J.J. Jurinak. 1975. *Soil Sci Soc Am Proc* 39:851–856.

Sauve, S., M. McBride, and W. Hendershot. 1998. *Soil Sci Soc Am J* 62:618–621.

Sauve, S., C.E. Martinez, M. McBride, and W. Hendershot. 2000. *Soil Sci Soc Am J* 64:595–599.

Scheuhammer, A.M. 1987. *Toxicology* 45:155–163.

Schulthess, C.P., and C.P. Huang. 1990. *Soil Sci Soc Am J* 54:679–688.

Schwartz, S.E. 1989. *Science* 243:753–763.

Scialdone, R., D. Scognamiglio, and A.U. Ramunni. 1980. *Water Air Soil Pollut* 13:267–274.

Sheaffer, C.C., A.M. Decker, R.L. Chaney, and L.W. Douglas. 1979. *J Environ Qual* 8:455–459.

Shore, R.F. 1995. *Environ Pollut* 88:333–340.

Shotyk, W., A.K.D. Weiss, P.G. Appleby, A.K. Cheburkin, et al. 1998. *Science* 281:1635–1640.

Smith, W.H., and T.G. Siccama. 1981. *J Environ Qual* 10:323–333.

Soldatini, G.F., R. Riffaldi, and R. Levi-Minzi. 1976. *Water Air Soil Pollut* 6:111–118.

Sposito, G., L.I. Lund, and A.C. Chang. 1982. *Soil Sci Soc Am J* 46:260–264.

Stevenson, F.J., and L.F. Welch. 1979. *Environ Sci Technol* 13:1255–1259.

Storm, G.L., G.J. Fosmire, and E.D. Bellis. 1994. *J Environ Qual* 23:508–514.

Strawn, D.G., A.M. Scheidegger, and D.L. Sparks. 1998. *Environ Sci Technol* 32:2596–2601.

Strojan, C.L. 1978. *Oikos* 31:41–46.

Stumm, W., and J.J. Morgan. 1970. *Aquatic Chemistry*. Wiley, New York.

Swaine, D.J. 1955. Tech Commun 48. Commonwealth Bureau of Soil Science, Harpenden, England.

Swaine, D.J. 1978. *J Royal Soc New South Wales* 111:41–47.

Swaine, D.J., and R.L. Mitchell. 1960. *J Soil Sci* 11:347–368.

Taylor, R.M., and R.M. McKenzie. 1966. *Aust J Soil Res* 4:29–39.

Ter Haar, G., and M.A. Bayard. 1971. *Nature* 232:533–534.

Thornton, I. and E. Culbard. 1986. *Lead in the Home Environment*. Science Reviews, Northwood, United Kingdom.

Tirelli, E., N. Maestrini, S. Govoni, E. Catelli, and R. Serra. 1996. *Bull Environ Contam Toxicol* 57:729–733.

Tyler, G. 1972. *Ambio* 1:52–59.

U.S. Bureau of Mines. 1993. *Lead: Annual Report 1991*. U.S. Dept Interior, Washington, DC.

[U.S. EPA] United States Environmental Protection Agency. 1986.

U.S. EPA. 1993. *Integrated Risk Information System (IRIS) on lead*. Office of Health and Environ Assess, Cincinnati, OH.

[U.S. FDA] United States Food and Drug Administration. 1992. *Fed. Register* 57(129):29734–29736.

[U.S. HUD] United States Department of Housing and Urban Development. 1995. *Guidelines for the Evaluation and Control of Lead-Based Paint Hazards in Housing.* HUD, Washington, DC.

Van Hook, R.I., W.F. Harris, and G.S. Henderson. 1977. *Ambio* 6:281–286.

Vile, M.A., R.K. Wieder, and M. Novak. 2000. *Environ Sci Technol* 34:12–21.

Wallace, A., and E.M. Romney. 1977. In H. Drucker and R.E. Wildung, eds. *Biological Implications of Metals in the Environment.* CONF-750929. NTIS, Springfield, VA.

Wang, E.X., F.H. Bormann, and G. Benoit. 1995. *Environ Sci Technol* 29:735–739.

Wang, J., P. Guo, et al. 2000. *Environ Sci Technol* 34:1900–1905.

Warren, H.V., R.E. Delavault, and C.H. Cross. 1969. *Trace Subs Environ Health* 3:9–19.

Wedding, J.B., R.W. Carlson, J.I. Stukel, and F.A. Bazzaz. 1975. *Environ Sci Technol* 9:151–153.

Weiss, D., W. Shotyk, P.G. Appleby, J.D. Kramers, and A.K. Cheburkin. 1999. *Environ Sci Technol* 33:1340–1352.

Wiersma, D., B.J. Van Goor, and N.G. Van der Veen. 1986. *J Agric Food Chem* 34:1067–1074.

Williams, D.E., I. Vlamis, A.H. Pukite, and I.E. Corey. 1984. *Soil Sci* 137:351–359.

Wolnik, K.A., F.L. Fricke, et al. 1983. *J Agric Food Chem* 31:1240–1244.

Wolnik, K.A., F.L. Fricke, et al. 1985. *J Agric Food Chem* 33:807–811.

World Bank. 1992. *World Development Report 1992. Development and the Environment.* Oxford Univ Pr, New York.

Wu, J., F.C. Hsu, and S.D. Cunningham. 1999. *Environ Sci Technol* 33:1898–1904.

Zhang, P., J.A. Ryan, and L.T. Bryndzia. 1997. *Environ Sci Technol* 31:2673–2678.

Zhang, Z.W., S. Shimbo, et al. 1999. *Sci Total Environ* 226:65–74.

Zimdahl, R.L., and I.H. Arvik. 1973. *CRC Critical Rev Environ Control* 3(2):213.

Zimdahl, R.L., and I.M. Foster. 1976. *J Environ Qual* 5:31–34.

Zimdahl, R.L., and R.K. Skogerboe. 1977. *Environ Sci Technol* 11:1202–1207.

11
Mercury

1 General Properties of Mercury

Mercury, also called liquid silver, has atom. no. 80, atom. wt. of 200.6, melting pt. of -38.8 °C, specific gravity of 13.55, vapor pressure of 1.22×10^{-3} mm at 20 °C (2.8×10^{-3} mm at 30 °C), and solubility in water of 6×10^{-6} g/100 ml (25 °C). It is a heavy, glistening, silvery-white metal that is a liquid at room temperature, a rather poor conductor of heat but a fair conductor of electricity. It has seven stable isotopes with the following percentages of abundance: ^{195}Hg (0. 15%), ^{198}Hg (10.1%), ^{199}Hg (17.0%), ^{200}Hg (23.3%), ^{201}Hg (13.2%), ^{202}Hg (29.6%), and ^{204}Hg (6.7%). There are many minerals of Hg; the most common are the sulfides cinnabar and metacinnabar. Mercury is recovered almost entirely from cinnabar (α-HgS, 86.2% Hg); less important sources are livingstonite (HgS · 2Sb$_2$S$_3$), meta-cinnabar (β-HgS), and about 25 other Hg-containing minerals. Its unusual high volatility, which increases with increasing temperature, accounts for its presence in the atmosphere in appreciable amounts.

There are three stable oxidation states of Hg: 0 (elemental), I (mercurous), and II (mercuric). The properties and behavior of Hg depend on the oxidation state. Most of the Hg encountered in the atmosphere is elemental Hg vapor. Most of Hg in water, soils, sediments, or biota is in the form of inorganic salts and organic complexes.

2 Production and Uses of Mercury

World Hg production was fairly constant at about 3.6×10^3 tonnes yr^{-1} from 1900 to 1939 (Gavis and Ferguson, 1972). Since the 1960s, however, production has more than doubled but declined to 1.76×10^3 tonnes yr^{-1} in 1994 (U.S. Bureau of Mines, 1994). World production in 1979 was about 6.7×10^3 tonnes, considerably lower than rates in the early 1970s. The leading producers of Hg in 1994 in descending order were: China, Algeria, Spain, and Kyrgyztan. The principal deposits are found in Almaden, Spain; Idria, Yugoslavia; and Monte Amiita, Italy. On the American continent, significant Hg deposits are found in the Pacific Coast area, particularly in Peru, Mexico, and California.

The U.S. consumption of Hg is about double its production capacity resulting in a net import of the element roughly equivalent to that produced domestically (Drake, 1980). Consumption in the United States and abroad has been declining during the last few years. For example, Hg consumption in chlor-alkali plants in Brazil dropped from 80 tonnes yr^{-1} in 1979 to only 18 tonnes yr^{-1} in 1985 as a direct result of substitution of Hg cells, responsible for 90% of Hg utilization in chlor-alkali plants prior to 1980 (Lacerda and Salomons, 1998). In the United States the overall decline was led by reduced demand for Hg use in catalysts, paints, dental fillings, electrical equipment, and for several laboratory purposes. However, there was increased Hg use for electrolysis. Methyl Hg has no industrial uses.

Because of its diverse properties, Hg has wide applications in science, industry, and agriculture. In minute quantities, Hg conducts electricity, responds to temperature and pressure changes, and forms alloys with almost all other metals. Mercury serves an important role as a process or product ingredient in several industrial sectors. Its biggest consumers are chlor-alkali industry (electrolysis), electrical and control instruments industry, laboratory products, dentistry (dental amalgams), and agriculture. It is also widely used in paints, catalysis, pharmaceutical products, wood processing (as an antifungal agent) and in the paper and pulp industry (as slimicides).

The chlor-alkali industry, which manufactures chlorine and caustic soda, has been one of the biggest Hg users and emitters. Its use in agriculture also poses a threat to quality of the food chain because of its use as a seed dressing in grain, potatoes, flower bulbs, sugar cane, cotton, etc., and as a foliar spray against plant diseases. Thus, contamination of foodstuff and wildlife is often inevitable.

3 Mercury in Nature

The earth's crust contains approximately 50 ppb Hg, mainly as sulfide. Mercury occurs in all types of rocks; Hg content of most igneous rocks is generally <200 ppb (Fleischer, 1970). Except for shales, most sedimentary rocks have also Hg content of <200 ppb. Shales high in OM are particularly enriched in Hg. Sedimentary rocks in general tend to contain more Hg than igneous rocks (Table 11.1). Mercury minerals consist essentially of cinnabar, metacinnabar, or both (polymorphs of HgS), and one or more of native Hg, stibnite, native S, quartz, fluorite, and carbonates (NRCC, 1979).

Mercury is present naturally in soils at concentrations ranging from a few ppb to a few hundred ppb. Soils may be considered normal if their Hg contents fall within the <100 ppb level. In the vicinity of gold, molybdenum, and base-metal deposits, soils may contain from 50 to 250 ppb Hg, and in some instances as much as 2000 ppb Hg (Warren et al., 1966). In the vicinity of Hg deposits, soil levels of Hg can be expected to be considerably higher (range of 1000 to 10,000 ppb) and in some cases at the 100,000 ppb level.

Mercury concentrations in environmental samples collected near abandoned Hg mines in southwestern Alaska are elevated over those in background samples (Bailey and Gray, 1997). Vegetation samples collected from the mines contain as much as 970 ppb Hg whereas background vegetation samples contain no more than 190 ppb Hg. Soil samples collected from the mines contain as much as 1500 ppm Hg, but background samples contain no more than 1.2 ppm Hg. All stream-water Hg concentrations are well below the 2.0 ppb drinking water standard recommended by the state of Alaska.

TABLE 11.1. Commonly observed mercury concentrations (ppb) in various environmental media.

Material	Average concentration	Range
Igneous rocks[a]	—	5–250
Limestone[a]	40	40–220
Sandstone[a]	55	<10–300
Shale[a]	—	5–3250
Petroleum[a]	—	20–2000
Coal[a]	—	10–8530
Fly ash[b]		
Bituminous	100	—
Sub-bituminous	40	—
Lignite	100	—
Rock phosphate[a]	120	—
Peat[a]	—	60–300
Soil (normal)[a]	70	20–250
Sediment (normal)[c]	—	5–100
Soils near Hg deposits[a]	—	Up to 250 ppm
Soil horizon (normal)[a]		
A	161	60–200
B	89	30–140
C	96	25–150
Terrestrial plants (DW)[a]	—	30–700
Marine plants (WW)[a]	—	0.01–37
Groundwaters (normal)[a]	0.05	0.01–0.10
Rain water, snow[a]	0.2	0.01–0.48
Lake[c]	—	1–5
River[c]	—	0.2–0.7
Seawater[c]	—	0.3–1.0
Food (normal)[d]		
Vegetables and cereals	—	1–50
Meat	—	1–50
Fish	—	30–1500
Drinking water[d]		
Permitted limit	—	1

Sources: Extracted from [a]NRCC (1979); [b]Adriano et al. (1980); [c]Draback (1992); [d]Cox (1995).

Gavis and Ferguson (1972) reported that most river and lake waters from many parts of the world have Hg concentrations in the range of <0.1 to 6.0 ppb. In central Illinois public sewer systems, total Hg concentrations range from 1.3 to 1.8 ppb (Evans et al., 1973). Nearly all public water supplies in the United States contain <2 ppb of Hg (NAS, 1977). Groundwaters in Sweden showed Hg levels in the range of 0.020 to 0.070 ppb ($n = 34$) with a mean of 0.050 ppb (Wicklander, 1969). The WHO has an upper guideline limit of 1 ppb for total Hg content of water for human consumption (WHO, 1984). For drinking water in the U.S., the proposed limit is 2 ppb (U.S. EPA, 1976).

Some environments, like that of Hawaii, are naturally high in Hg. Measurements of Hg contents of air, water, rainfall, soils, rocks, and biological materials from the major islands indicate that the source of environmental Hg in Hawaii is geothermal (Siegel et al., 1973).

Soils and sediments retain Hg (up to 90% of Hg currently deposited on terrestrial landscapes) as OM complexes, on sorption sites of clays, or as accumulating particulates (Nater and Grigal, 1992). Because the mode of deposition of Hg particulate fraction differs strongly from that of Hg^0, being dominated by dry deposition and precipitation scavenging, it has been hypothesized that Hg in soils would show clear patterns of distribution paralleling trends of increasing industrialization and density of anthropogenic sources. Indeed, a significant gradient in concentrations and total amounts of Hg in organic litter and surface mineral soil occurred along a transect of forested sites [balsam fir (*Abies balsamea*), sugar maple (*Acer saccharum*), jack pine (*Pinus banksiana*), red pine (*P. resinosa*), and aspen (*Populus tremuloides*)] across the north central U.S. from northwestern Minnesota to eastern Michigan (Fig. 11.1). Mercury concentrations in forest floor material generally increased from west (zone 1 in western Minnesota) to east (zone 5 in Michigan). This pattern of increase has also been shown for Pb and Cd in the same samples.

4 Mercury in Soils/Sediments

4.1 Total Mercury in Soils

Normal soils are expected to have Hg levels not exceeding 100 ppb. Although Warren and Delavault (1969) reported very high concentrations of Hg in certain U.K. soils ranging as high as 15,000 ppb, they estimated that levels of Hg in most British arable soils would probably range between 10 and 60 ppb. Based on 912 surface soil samples collected in the continental United States, Shacklette et al., (1971) reported a mean of 112 ppb (geometric mean = 71 ppb). The mean for western states was 83 ppb ($n = 492$) and for eastern states, 147 ppb ($n = 420$). Soils of China were found to contain a mean of 65 ppb ($n = 4041$) and a geometric mean of 40 ppb (Chen et al., 1991). In relatively remote Tibet, the geometric mean for soil Hg was 21 ppb ($n = 199$). The Hg contents of soils from several countries are shown in Table 11.2. Swedish soils have Hg contents ranging from 20 to 920 ppb, with an average of 70 ppb (Andersson, 1967). For Canadian soils, a mean of 59 ppb ($n = 173$), with values ranging from 50 to 100 ppb was reported (McKeague and Wolynetz, 1980). A more typical concentration of Hg for world soils is reported at about 60 ppb (Berrow and Reaves, 1984). Other values for soil Hg are indicated in Appendix Table A.24.

In the North Ossetian Plains of the former Soviet Union, background levels of Hg in surface soils averaged 60 ppb (Zyrin et al., 1981). More than

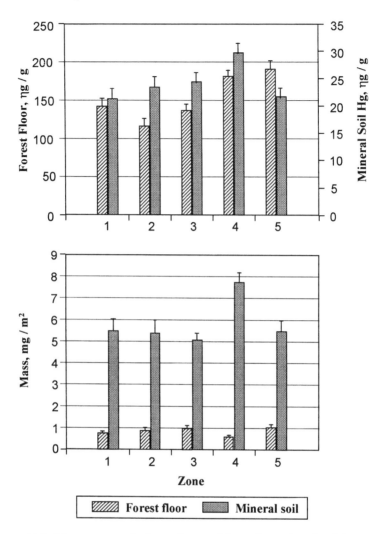

FIGURE 11.1. Mean mercury in forest floor and in surface mineral soil (0 to 25 cm) in five zones along a transect from northwestern Minnesota to eastern Michigan. Standard errors indicated. Differences in concentrations among zones were significant for both forest floor and surface soil. Bayes least significant difference for forest floor concentration was 30 ng g^{-1}, and for surface soil concentration was 7.5 ng g^{-1}. (Reprinted with permission from Nater and Grigal, *Nature*, 358:140. Copyright © 1992, Macmillan Magazines Limited.)

50% of total Hg content of these soils was found in soil particles <0.005 mm in size. Some uncontaminated virgin soils in Kyushu, Japan had an average Hg concentration of 40 ppb (range of 3 to 245 ppb) (Gotoh et al., 1979). Paddy soils however, showed Hg surface enrichment of up to 245 ppb due to past mercurial applications. Soils near the Nifu Hg mine in Japan had Hg

TABLE 11.2. Mercury contents of soils from several countries.

| | Hg content, mg kg^{-1} soil DW | | | |
n	Minimum	Maximum	Mean	Country
707	0.00	31	0.16 (median = 0.07)	Netherlands, soils used for agriculture and horticulture
671	0.00	0.32	0.08	Netherlands, same soils, without the highest 5% (field)
155	0.02	7.2	0.36 (median = 0.16)	Netherlands, greenhouse soils
147	0.02	1.1	0.24	Netherlands, same soils, without the highest 5%
248			0.2	Netherlands, clay
63			0.2	Netherlands, sand
273	0.004	0.922	0.06	Sweden
221			0.11	Germany (Hessen)
1041			0.12	Germany (Baden–Württemberg)
472			0.06	Germany (Nordrhein–Westfalen)
40	0.005	0.340	0.095	Austria
296	0.01	1.14	0.11	Canada (Ontario)

Source: Modified from Wiersma et al. (1986).

concentrations as high as 100 ppm (Morishita et al., 1982). The area in the vicinity of this mine is reportedly one of the most heavily polluted areas in the world by a Hg mining activity.

For arable soils, Archer (1980) reported an average value of 40 ppb ($n = 53$) for the United Kingdom, with a range of 8 to 190 ppb. Except in rare instances, agricultural practices apparently do not substantially enrich soils and edible portions of crops with Hg. For example, a survey of farm soils and grain in 16 major wheat-growing states in the United States indicated that Hg levels in samples collected from sites where Hg compounds had been used were not significantly different from those where no Hg was applied (Gowen et al., 1976). Mercury compounds used in amounts ranging from 0.005 to 0.01 kg ha^{-1} were used for seed treatment. Similar results were obtained for surface arable soils collected from 29 eastern U.S. states where there were no statistical differences between Hg levels in cropland (mean = 80 ppb, $n = 275$) and noncropland (mean = 70 ppb, $n = 104$) soils (Wiersma and Tai, 1974). For soils collected from north central states, the results of analysis for Hg residues indicate that for all practical purposes, Hg levels in soils were comparable for sites where Hg compounds were used or not used (Sand et al., 1971).

Geogenic processes of parent material may influence the Hg content of soils. In China, the highest Hg contents were found in Lithosols and Ultisols with Inceptisols and Aridisols having the lowest (Chen et al., 1991).

In Manitoba, Canada similar levels of Hg in horizon A soils of cultivated (mean = 36 ppb) and noncultivated areas (mean = 35 ppb) were found (Mills and Zwarich, 1975). However, in certain arable areas in Ontario, Canada soils from apple orchards where phenylmercuric acetate was used for foliar treatment had Hg levels of 290 ppb (Frank et al., 1976). This value is much higher than the overall mean value of 110 ppb Hg (n = 296) for arable soils, with extreme values of 10 to 1140 ppb. Total Hg contents of surface soils from experimental plots maintained for about 63 years at the University of Illinois were rather unusually high and variable, ranging from 100 to 3920 ppb (Jones and Hinesly, 1972). The high variation among samples was attributed to changes in management strategies affecting drainage differences between plots and possible sample contamination.

4.2 Bioavailable Mercury in Soils

There are only scanty data on extractable forms of Hg in soil and their relationships to plant uptake. Among various extractants ($CaCl_2$, NH_4OAc, EDTA, and several HCl solutions) used, the amount of Hg bioavailable for root uptake can be estimated from the amount of Hg in 0.01 M $CaCl_2$ extracts of soil samples taken after a lapse of about 5 mo of Hg application (Gracey and Stewart, 1974a). The soils were treated with 5% mercuric fungicide in a field experiment and samples taken at 4, 34, 49, and 111 days after treatment. An increase in amount of Hg detectable in $CaCl_2$ extract was observed between days 4 and 34, and a decrease by next sampling date. The $CaCl_2$ extractant does not include organically bound Hg. Using similar extractants, Hogg et al. (1978a) found that extraction of 0- to 10-cm soil layer by either $CaCl_2$, NH_4OAc, EDTA, or DTPA-TEA removed <1.0% of applied amounts of Hg. However, larger portions of Hg were removed by acid extraction (0.5 M HCl solution), which amounted to as much as 11% of total Hg applied. This indicates strong binding of Hg to soil components. This phenomenon was largely associated with the organic fraction of soil, as indicated by data for surface soils collected near a Hg mining/refining operation in Almaden, Spain (Lindberg et al., 1979). This conclusion was based on two observations: (1) the 0.5 M $NaHCO_3$-extractable Hg, which is the organic associated fraction, accounted for ~200 times more Hg than 1 M NH_4OAc-extractable Hg, which is the exchangeable fraction; and (2) soil density gradient fractionation experiments indicated ~2 times more total Hg associated with the light-density organoclay fraction in Almaden soil than in control soil.

4.3 Mercury in Soil/Sediment Profile

Two generalizations can be made regarding the profile distribution of Hg: (1) below the surface layer, Hg is fairly mobile in the soil profile; and (2) Hg tends to accumulate in surface horizons. The Hg content of a soil horizon has been reported to be related to OM and/or clay content of that horizon. Early on, Goldschmidt (1954) attributed higher levels of Hg in A horizons to accumulation of decayed plant material. Subsequently, Warren et al. (1966) reported that soil samples taken from horizons with either a high OM or clay content generally contain substantially higher amounts of Hg than the average for whole profile. Although Hg levels are expected to accumulate primarily in surface horizons, there are exceptions. The contents of Hg in A and C horizons from soils in Manitoba were not systematically different (Mills and Zwarich, 1975). However, 10 of the 16 soils reported in the study contained noticeably lower levels of Hg in the A horizon in comparison with amounts in the respective C horizon. In contrast, the Hg contents of some Saskatchewan soils had no significant differences in levels between A and C horizons (Gracey and Stewart, 1974b).

Gotoh et al. (1978) found that in northern Kyushu, Japan Hg in soil profiles was generally higher than in underlying parent material (igneous rock, mean ≤10 ppb Hg). A generally steep concentration gradient of Hg down the profiles, especially in paddy soils, occurred with surface soils having a mean Hg level of 197 ppb (range of 64 to 459 ppb).

In Chernozemic soils in Alberta, Canada content of Hg in surface horizons was considerably lower than content in the respective C horizon (Dudas and Pawluk, 1976). For eluviated soils, highest contents of Hg were found in the B horizon, with lowest levels in the A horizon. Two possibilities may have caused the decreased content of Hg in surface horizons: (1) Hg may have been translocated from the surface horizon to the C horizon due to leaching, and (2) portions of Hg in the surface horizon may have been mobilized and lost to the atmosphere. An extensive soil survey involving virgin soils from 234 horizons from Canada indicates that in more than half of the samples, particularly Podzolic and Gleysolic soils, the highest Hg concentrations were in zones of OM accumulation (McKeague and Kloosterman, 1974). However, for a few profiles the highest Hg levels occurred in the C horizon, and in some, Hg concentrations were similar in all horizons.

The behavior of Hg in soil profiles is typified in Figure 11.2 in diverse European soils and demonstrates that numerous factors influence Hg level and mobility in the profile. Soils 1, 4, 5, 6, and 7 represent the reference soils where no Hg had been introduced and no disturbances in the profile had occurred during the last 20 yr. The Hg values in profiles of these soils virtually represent natural background levels. Soil 11 represents soils from a bulb-growing area in the Netherlands, where application of Hg as a fungicide has been a common practice for about 60 yr, with its use being intensified about 30 yr ago. Thus, the behavior of Hg in soil profiles is immensely influenced by OM, clay content and mineralogy, and oxides of Fe and Mn. In acidic soils, the role of OM may predominate but in alkaline and calcareous soils, clay mineralogy and oxides of Fe may be more dominant (Andersson, 1979).

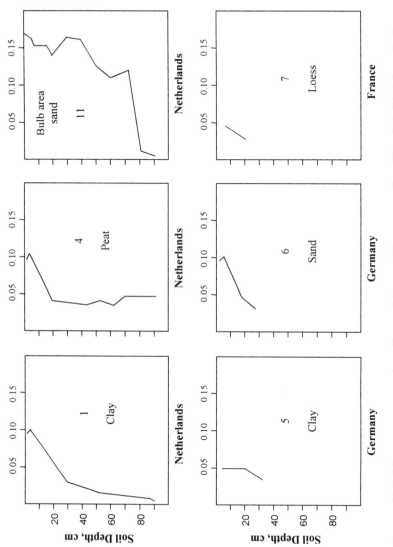

FIGURE 11.2. Mercury concentrations (in ppm) in some European soil profiles. (Modified from Poelstra et al., 1974.)

Profile distribution of Hg in sediments is similar to that for soil, i.e., decreasing concentration with depth. For example, in remote lakes in central Brazil where atmospheric transport is the only pathway of Hg contamination (90 to 120 μg Hg m^{-2} yr^{-1}), Hg concentrations were highest at the surface of the sediment (62 to 80 μg kg^{-1} DW) decreasing to values of 20 to 30 μg kg^{-1} in deeper layers (Fig. 11.3). The Hg content in these Brazilian lakes attenuated with the OM content. Similarly, Hg in sediment cores from high-latitude lakes in northern Canada accumulated in the surface (see Fig. 11.3). From extrapolation of dates downward to deeper cores assuming a constant sedimentation rate in northern Canadian lakes, the inputs increased slowly in the 1500s, more rapidly after 1750, and even more rapidly during the 1900s (Lockhart et al., 1995).

4.4 Sorption of Mercury in Soils/Sediments

When a Hg compound is added to soil or sediment, the adsorption process is initially dominant in determining the fate of this element. The remaining Hg in soil solution eventually will volatilize, precipitate, leach, or be taken up by organisms. Adsorption of Hg is dependent on several factors, among them the chemical form of Hg applied, reactivity of inorganic and organic soil colloids, soil pH, type of cations and anions on the exchange complex, OM, and redox potential. Among factors controlling adsorption of Hg, pH is one of the most important. Both surface charge characteristics of soil particles and metal speciation in solution are affected by pH. Because of its strong affinity for Hg, Cl$^-$ also could be an important factor affecting Hg adsorption. Barrow and Cox (1992) reported that addition of Cl$^-$ decreased Hg(II) adsorption on a loamy sand soil at low pH, but had little effect at high pH. An abrupt increase in Hg release from sediments when Cl$^-$ ($> 2 \times 10^{-2}$ M strength) was added (Wang et al., 1991) confirms the strong Hg–Cl complexation.

The effects of Hg(II) complexation with Cl$^-$ on Hg mobility are reflected in several adsorption studies. A drastic reduction in Hg(II) adsorption by inorganic colloids occurred at Cl$^-$ concentrations of 10^{-3} M or higher (Andersson, 1970). Additionally, HgCl$_2$ salts were adsorbed only slightly by soils and clays (Aomine and Inoue, 1967). Chlorides sharply reduced Hg(II) sorption by bentonite clay, especially at low pH (Newton et al., 1976). At pH 6 or lower, increasing the CaCl$_2$ concentration from 10^{-5} to 10^{-4} M depressed adsorption; higher CaCl$_2$ levels were required to decrease adsorption at neutral pH. Similar inhibitory effects of Cl$^-$ on Hg(II) adsorption were found for hydrous iron oxide gel (Kinniburgh and Jackson, 1978) and precipitated iron oxide (Lockwood and Chen, 1974).

The presence of other ligands can also affect Hg sorption. The presence of SO$_4^{2-}$ decreased Hg(II) retention by gibbsite [Al(OH)$_3$], which was attributed to the formation of the Hg(OH)$_2$SO$_4^{2-}$ ion pair; however, the presence of

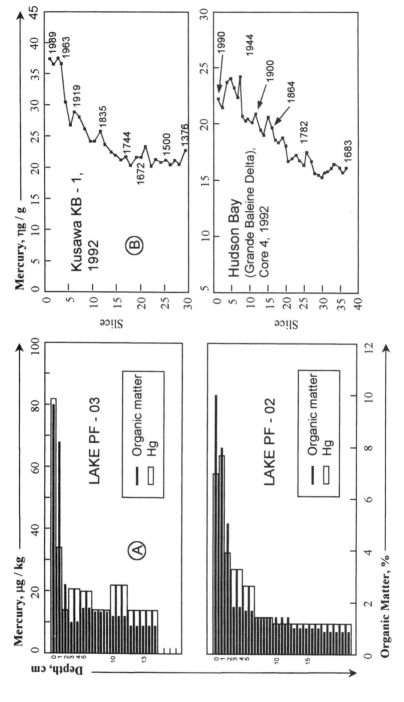

FIGURE 11.3. Mercury concentrations in sediment cores: (A) Mato Grosso state, central Brazil and (B) Kusawa Lake and Hudson Bay, Canada. (Modified from Lacerda et al., 1991; Lockhart et al., 1995.)

PO_4^{2-} increased retention by gibbsite due to the formation of a phosphate bridge [$\equiv AlOPO_3Hg(OH)_2^{2-}$] (Sarkar et al. 1999).

Mercury(II) behaves differently from most other divalent metals in that the extent of adsorption is greatest in acidic media (Sarkar et al., 1999; Yin et al., 1996). For instance, adsorption of Cd in 15 diverse New Jersey soils was <10% at pH <5 but increased with increasing pH reaching a maximum of >90% (Allen et al., 1994). With Hg(II), maximum adsorption in the same soils ranged from 86 to 98% of added Hg in the pH range of 3 to 5 (Yin et al., 1996). Further increasing the pH significantly decreased adsorption of Hg(II), e.g., from 89% at pH 4 to 39% at pH 8.5. With quartz (SiO_2) and gibbsite, the adsorption maximum for Hg(II) occurs at pH ~4.5, which is comparable to the pKa (3.2) for the hydrolysis of Hg^{2+} to form $Hg(OH)_2^0$ (Sarkar et al., 1999).

Another important factor in the behavior of Hg is its complexation by DOC whose concentration increases with increasing soil pH. Among metals, Hg and Fe generally exhibit the strongest affinity for humic acid (Kendorff and Schnitzer, 1990). Accordingly, when soil was treated with H_2O_2 to oxidize OM, adsorption of Hg(II) was significantly reduced at low pH (Yin et al., 1996). Two hypotheses are plausible: (1) OM generally has stronger affinity for Hg(II) than do soil inorganic components; and (2) when soil OM is removed, some inorganic surface areas are exposed. Relative to OM, soil inorganic components generally have smaller surface areas. Hence, total surface area is expected to have decreased upon OM removal, contributing to decreased adsorption. Similarly, adsorption of monomethyl Hg by soils is subject to the influence of the same factors that control inorganic Hg(II). Thus, adsorption of monomethyl Hg by soils is also highly pH dependent. Both soil OM and DOC as well as Cl^- exert influence on monomethyl Hg adsorption (Yin et al., 1997).

Sorption of Hg(II) has been demonstrated to occur on aluminosilicate clays (Farrah and Pickering, 1978; Reimers and Krenkel, 1974), oxides of Fe and Mn (Kinniburg and Jackson, 1978; Lockwood and Chen, 1973), and OM (Reimers and Krenkel, 1974). Adsorption of Hg by soils follows either the Langmuir (Hogg et al., 1978a) or Freundlich adsorption isotherm (Fang, 1978). Adsorption of Hg by synthetic $Fe(OH)_3$ follows the Freundlich isotherm (Lockwood and Chen, 1974). Adsorption rate coefficients by soils follow a one-site second order kinetic model, which is characterized by a biphasic pattern, a fast step followed by a slow step (Yin et al., 1997). Adsorption of Hg by soils depends not only on soil properties but also on the chemical form of Hg (Hogg et al., 1978a). The highest adsorption maxima for all Hg compounds tested were found for soils that had the most OM and clay contents. Adsorption maxima increased in the order, methylmercuric chloride < phenylmercuric acetate < mercuric chloride. Fang (1978) observed that clay mineralogy also plays an important role in the sorption of Hg by soils. Among clay minerals, illite had the highest sorption capacity and kaolinite the lowest.

4.5 Transformation of Mercury in Soils/Sediments

Mercury in nature and from industrial discharges is largely in inorganic form although organic(aryl) mercurials are used in agriculture. In general, Hg is very unstable in the environment because it is subject to chemical, biological, and photochemical reactions (Kaiser and Tölg, 1980). Many mercurial compounds, both organic and inorganic, decompose to yield elemental Hg, which may volatilize or be converted to HgS or complex with inorganic ligands. A simplified scheme for the environmental/biological interconversion of Hg is shown in Figure 11.4.

Wood (1974) has proposed a scheme whereby various forms of Hg entering the aquatic environment are transformed into mercuric ions and then methylated, resulting in the formation of toxic and volatile monomethyl or dimethyl Hg (Adriano, 1986). These transformations depend largely on biological processes (Jay et al., 2000). Consequently, a generalized scheme of the reactions involved in the formation of methyl Hg compounds has been proposed: (1) the precursor Hg^{2+} can be methylated by either aerobic and anaerobic bacteria, although methylation is usually enhanced under anaerobic (but not completely anaerobic) conditions; (2) monomethyl Hg, $(CH_3)Hg^+$, can be formed from Hg^{2+} by methane-producing bacteria; (3) in sediments and soils like paddy soils, reducing conditions produce sulfide ions forming HgS, which is quite insoluble and hence resistant to methylation; (4) under aerobic conditions, HgS may be oxidized to sulfate form, which can then undergo methylation; (5) dimethyl Hg, $(CH_3)_2Hg$, may be transformed into monomethyl Hg at low pH and in the presence of Hg^{2+}, or in the presence of ultraviolet light; (6) monomethyl Hg may attach to -SH groups of OM and may then decompose photolytically to Hg(0) and CH_4; (7) analogous reactions may occur with other alkyl, alkoxyalkyl, and aryl mercury compounds; (8) methyl Hg salts may be volatilized into the air by the action of H_2S; and (9) bacterial action can also cause deme-

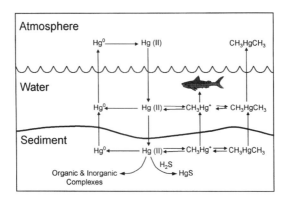

FIGURE 11.4. Cycling and interconversion of various mercury species in freshwater lakes. (Modified from Winfrey and Rudd, 1990.)

thylation of methyl Hg compounds. These reactions are discussed in more detail elsewhere (Kaiser and Tölg, 1980; Löfroth, 1970; Summers and Silver, 1978).

In aquatic environments, microorganisms in lake sediments could methylate inorganic Hg; likewise, cell-free extracts from methanogenic bacteria could also methylate Hg (Wood et al., 1968). Thus, methyl Hg could be formed by both enzymatic and nonenzymatic reactions (Bisogni, 1979). The processes by which Hg can be methylated chemically or biologically are discussed by Beijer and Jernelöv (1979).

While methylation of inorganic Hg in aquatic systems is the norm, it can also occur in the terrestrial environment. Indeed, methyl Hg was found in desert soils that had been amended with mercuric nitrate (Beckert et al., 1974). This form of Hg has also been detected in the atmosphere above a soil amended with mercuric chloride (Braman and Johnson, 1974). Lexmond et al. (1976) indicated that methyl Hg may be formed from mercuric Hg by both anaerobic and aerobic microorganisms and that dimethyl Hg may be formed by further methylation of methyl Hg or by methylation of Hg^0. It was found that inorganic Hg was methylated in alkaline agricultural soils and that it is possible that the methylation process could in part be abiotic in nature (Rogers, 1977). However, Van Faasen (1975) found little methylation of inorganic Hg in soils. Others have also shown that mercuric ions can be abiotically methylated by a variety of substances (Bertilsson and Neujahr, 1971; DeSimone, 1972; Imura et al., 1971).

In the terrestrial environment, Hg can be mobilized by volatilization, leaching, and/or plant uptake. It has been shown that volatilization is a result of chemical reaction (Frear and Dills, 1967; Toribara et al., 1970) and/or microbial activity (Landa, 1978; Rogers and McFarlane, 1979). Both appear to be operative in natural systems. Toribara et al. (1970) proposed that the mechanism of Hg loss follows the chemical reduction of Hg(II) to Hg(I), which disproportionates spontaneously to Hg(II) and Hg^0.

Volatilization of Hg from turf grass and soil has been considered as the dominant pathway in the loss of Hg in this type of landscape, more important than leaching or plant uptake. It accounted for as much as 56% loss of total Hg added over a period of 57 days (Gilmour and Miller, 1973) and in another case for as much as 45% of total Hg added to a soil in 1 wk (Rogers and McFarlane, 1979). Hg(0) is likely to be the predominant Hg species evolved.

Several environmental factors may influence methylation rates of Hg, most probably by altering its bioavailability. Complexation of Hg by DOC in Canadian lakes decreased Hg methylation rates (Miskimmin et al., 1992). The pH has been observed to influence methylation rate by controlling organic C compounds with functional groups that would otherwise bind Hg (Gilmour and Henry, 1991). Also, mercuric sulfide (HgS) does not facilitate methylation, which generally decreases with increasing sediment sulfide content (Gilmour and Capone, 1987). However, precipitation of S^{2-} with the addition of Fe^{2+} has dramatically increased the amount of Hg available methylation in anoxic sediments (Choi and Bartha, 1994).

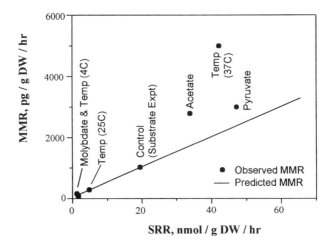

FIGURE 11.5. Predicted and observed Hg methylation rate (MMR) as affected by sulfur reduction rate (SRR) in anoxic slurry incubations using salt marsh sediments from the southeastern United States. (Modified from King et al., 1999.)

The high correlation between Hg methylation rate and sulfate reduction rate emphasizes the sensitivity of these processes to redox potential, which in turn influences sulfate reduction (Fig. 11.5). Both methylation and S reduction rates are similarly affected by temperature, presence of organic compounds, and microbial inhibitors (King et al., 1999).

4.6 Chemical Speciation of Mercury in Soils/Sediments

Mercury undergoes complexation with organic and inorganic ligands in natural systems. In the presence of OM, organo-complexes may occur in soluble or colloidal form that differ in mobility. There is evidence that Hg could be strongly chelated by soil OM. Humic substances containing S (e.g., in cysteine) are believed to keep Hg in soluble form (Andersson and Wiklander, 1965). Added Hg(II) in soils may be immobilized by its incorporation with –SH groups of OM, or depending on redox potential of soil, it may precipitate as HgS (Landa, 1978). Gambrell et al. (1980) noted that most dissolved Hg in sediments is probably bound as inorganic, nonionic, or negatively charged complexes, or complexed with organics. Lindberg and Harriss (1974) indicated that relatively soluble polysulfide complexes may form under some conditions, but that under reducing conditions, very insoluble mercuric sulfide (cinnabar) may form.

Such prediction of Hg speciation is displayed in Figure 11.6, which indicates the strong dependency of speciation on pH and Cl^- concentration. Figure 11.6A indicates that in soil solution having Cl^- concentrations of about 1×10^{-6} to 1×10^{-5} M, the major species are positively charged at low pH, i.e., Hg^{2+}, $HgCl^+$, and neutral $HgCl_2$. The larger the Cl^- concentration, the larger the fraction of $HgCl_2$ (see Fig. 11.6B), and the

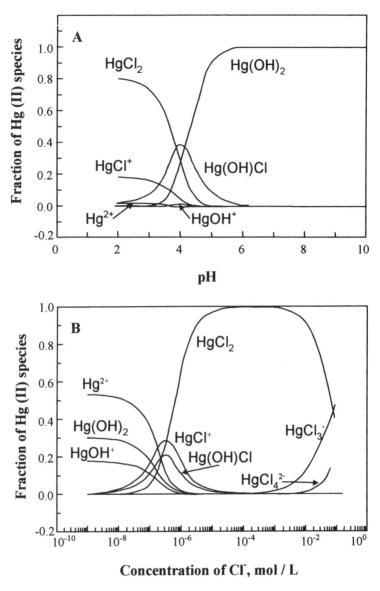

FIGURE 11.6. Speciation of Hg(II) as a function of pH and Cl^- concentration. The fraction of Hg(II) species was calculated using $MINEQL^+$. (A) $Cl^- = 10^{-6}M$; (B) pH = 3. (Modified from Yin et al., 1996.)

higher the pH the more the hydroxo-containing Hg species become predominant over the Hg–Cl complexes. The species $HgCl_2$ are poorly adsorbed by most of the inorganic particles (Yin et al., 1996).

Recently the apparent formation of complexes between Hg and polysulfides (S_x^{2-}, where x = 3 to 6) in sulfidic aquatic systems has been discussed (Jay et al., 2000; Paquette and Helz, 1997). It was proposed that the complex, $Hg(S_x)_2^{2-}$, dominates the speciation of Hg(II) in such waters.

The chemical speciation of Hg(II) largely determines the bioavailability and toxicity of this element to biota. In turn, Hg speciation is a function of the type and concentrations of all potential coordinating ligands in the solution. For example, in testing the acute toxicity of Hg(II) to *Pseudomonas fluorescens* cultured in synthetic growth media, the complexes of Hg with Cl^- (principally $HgCl^+$, $HgCl_2^0$, $Hg(OH)Cl^0$) were the species primarily responsible for Hg acute toxicity (Farrell et al., 1993). Additionally, it was found that the toxicity of Hg(II) was inversely related to Ca^{2+} and Mg^{2+} concentration in media, inferring that these ions have a moderating effect on the lethal effects of Hg(II). Thus, it is very important to recognize that biological effects of Hg and other metals are more closely related to chemical form than to total concentration of metals.

Chlorides occur in all natural soil and water systems and may be regarded as one of the most mobile and effective complexing agents for heavy metals. High Cl^- concentrations in soil solution can be expected in saline and saline sodic soils. Chlorides may complex with Hg(II) at Cl^- concentrations above 10^{-9} M (35×10^{-6} ppm) (see also Fig. 11.6B); $HgCl_2^0$ may form above $10^{-7.5}$ M Cl^- (1.1×10^{-3} ppm), peaking at about 10^{-4} M (~3.5 ppm Cl^-), and $HgCl_3^-$ and $HgCl_4^{2-}$ may form above 10^{-2} M Cl^- (350 ppm); $HgCl^+$ peaks at Cl^- concentration of 10^{-7} M (~0.0035 ppm). At Cl^- concentrations above 3550 ppm, $HgCl_4^{2-}$ may predominate (Hahne and Kroontje, 1973a).

5 Mercury in Plants

5.1 Uptake and Translocation of Mercury in Plants

Mercury and its compounds are absorbed by roots and translocated to a limited extent to other plant parts. Uptake has been demonstrated for both economic crops and trees. Additionally, evidence exists to show that Hg compounds applied to some parts of plants and trees can be readily translocated to other parts (Smart, 1968). Movement within foliage, stems, fruits, and tubers is greater than that from the roots upward. In general, the bioavailability of soil Hg to plants is low and there is a general tendency for Hg to accumulate in roots, i.e., the roots serve as a barrier to Hg uptake. For example, very high concentrations of Hg were found in alfalfa roots (up to 133 ppm DW) from several forms of Hg compared with <1 ppm in alfalfa foliage (Gracey and Stewart, 1974a).

Using solution culture, the uptake of inorganic Hg (as $HgCl_2$) by higher plants, such as peas, was shown to be a function of external concentration (Beauford et al., 1977). Over the concentration range (up to 10 ppm Hg) tested, the accumulation of Hg in roots was linear on a log–log scale. Translocation of Hg to the shoot appeared to be two-phased, i.e., up to 0.01 ppm of external Hg, a low threshold level of the element seemed to be maintained, and between 0.1 and 10 ppm, the transport of Hg to the shoot was like the roots, showing a linear trend. In addition, the proportion of Hg retained in the roots remained fairly constant (about 95%). Lindberg et al. (1979) found a dual mechanism of Hg uptake by alfalfa; roots accumulated

Hg in proportion to the soil level, while aerial plant material absorbed Hg vapor directly from the atmosphere. Volatilization of Hg from soil contributed to foliar uptake of Hg, which may occur in plants inhabiting environments near smelters.

The form of Hg could influence the absorption of Hg and its subsequent transport in plants. Huckabee and Blaylock (1973) reported that ^{203}Hg from methyl Hg accumulated more in snap beans than ^{203}Hg from an inorganic form. In the bean, the actual Hg concentration was 890 ppb from the organic vs. only 80 ppb from the inorganic form. Furthermore, they reported that inorganic Hg was not readily translocated in cedar trees.

Mercury compounds such as phenylmercury acetate can be translocated and redistributed in plants when applied as fungicides to the foliage of apple trees (Ross and Stewart, 1960, 1962). Following single cover sprays of phenylmercury acetate applied from June to July, Hg residues in apple fruits declined until early August and then progressively increased until harvest time, when the residues became greater than the initial deposits (Stewart and Ross, 1960). Thus, when apple trees are sprayed, some of the fungicide Hg is absorbed by leaves and may translocate later to growing fruit or new foliage. The fruit of apple trees seems to accumulate more Hg than other parts of the plant. Similarly, when foliage of potatoes is sprayed, the Hg can move downward and accumulate in the tuber (Smart, 1964).

5.2 Mercury in Economic Crops

It is reasonable to expect that Hg concentrations in agronomic and horticultural crops are at the ppb level. In general, total Hg concentrations in common agronomic plants and products derived from these plants range from <1 to 300 ppb, with higher levels caused by natural Hg deposits in soil (NAS, 1978). The actual level depends on location, plant species, and other factors. The results for crops grown in Saskatchewan field plots with normal soils containing <40 ppb Hg indicate that plant tissues generally contain Hg in <50 ppb levels (Gracey and Stewart, 1974b). Higher Hg levels were recorded for straw than for grain or seed of cereal and oil seed crops. Mercury was not detected in seeds of flax and rape. The levels in straw were approximately equal to those in soil with the exception of barley in which levels in straw were higher than levels found in soil. Later analysis for wheat grain sampled at harvest in 16 major wheat-producing states, however, indicates much higher levels of Hg (mean = 290 ppb with range of 70 to 1060 ppb; $n = 49$) (Gowen et al., 1976). The rather high concentrations in wheat grain are not surprising since Hg compounds can be absorbed by wheat plants and translocated into grain (Saha et al., 1970) producing levels of Hg similar to those found by Gowen et al. (1976). In addition, Selikoff (1971) reported that grain seeds treated with 20,000 ppb methyl Hg produced a crop containing 100 ppb in grain and 70 to 80 ppb in leaves.

Although Hg compounds have been demonstrated to be taken up by the roots of vegetable crops (e.g., broccoli, beans, lettuce, and carrot) from nutrient solution or soil with only little translocation to aerial parts, there

TABLE 11.3. Mercury contents of crops from various countries.

Crop	n	Minimum	Maximum	Mean	Median	Country
		Hg content, mg kg^{-1} WW				
Lettuce	75	0.0005	0.011	0.002	0.002	Netherlands
	15		0.0014			Germany
	6	0.001	0.002	0.001		Finland
	7	0.0005	0.0028	0.0014		Austria
Tomato	40	0.0001	0.008	0.0013	0.0004	Netherlands
	13		0.01			United Kingdom
	13		0.002			Germany
	6	0.0003	0.001	0.0006		Austria
Cucumber	45	0.0001	0.0015	0.0003	0.0002	Netherlands
	5	0.0002	0.002	<0.001		Finland
Spinach	82	<0.001	0.029	0.005	0.002	Netherlands
	13		0.005			Germany
	23	0.005	0.035	0.020		Switzerland
	8	0.002	0.023	0.008		Austria
Carrot	100	0.0006	0.005	0.002	0.002	Netherlands
	8	0.0008	0.0028	0.0014		Austria
Potato	94	<0.0001	0.017	0.003	0.002	Netherlands
	20	0.0004	0.010	0.002		Finland
	44	0.0008	0.015	0.004		Germany
Wheat	84	<0.0001	0.029	0.005	0.003	Netherlands
	75	0.002	0.009	0.004		Finland
	403	0.0005	0.016	0.002		Germany
	223	0.002	0.012	0.008		Canada
	12	0.01	0.03	0.02		United States
Barley	45	0.001	0.030	0.006	0.003	Netherlands
	46	0.002	0.004	<0.004		Finland
Oats	39	<0.0001	0.020	0.008	0.006	Netherlands
	36	0.002	0.004	<0.004		Finland
Apple Golden Delicious	50	0.0006	0.002	0.001	0.001	Netherlands
Cox's O. Pippin	50	0.0004	0.003	0.001	0.001	Netherlands
	31		0.002			Germany

Source: Modified from Wiersma et al. (1986).

are exceptions. Loam soil treated with up to 10 ppm Hg as methyl Hg dicyandiamide, caused very high concentrations in plant stems and leaves of potatoes (1045 ppb) and tomatoes (341 ppb) (Bache et al., 1973). Onion bulbs absorbed up to about 1087 ppb Hg from this soil. Other edible parts usually contained <100 ppb. The Hg contents of crops from several countries are shown in Table 11.3.

5.3 Bioaccumulation and Phytotoxicity of Mercury

Since Hg compounds have been used in agriculture and other industries for control of plant diseases (e.g., the use of mercuric oxide, mercuric chloride, or mercurous chloride for control of common scab of potatoes, club root in

Brassica species, white rot of onion, and brown patch, dollar spot, and snow mold in turf and others), there is some concern not only about the transfer of Hg to the food chain but also about the direct effect of Hg on plant growth.

Early on, Booer (1951) reported that metallic Hg and Hg compounds in soil can retard growth of plants. Using emergence tests, he found that growth of monocotyledons was only slightly retarded. Oats and barley behaved similarly to wheat, and lawn grasses were only slightly retarded. The percentage emergence of onions was unaffected by amounts of Hg of up to 100 ppm in soil. In the preemergence stage, most of dicotyledons appear to be more sensitive but more variable in their reaction to Hg. Severe preemergence losses might have been caused by as little as 50 ppm Hg in soil, particularly to sensitive species, such as lettuce and carrot. In contrast, percentage germination of numerous vegetable seeds was not severely affected by presence of Hg vapor, with subsequent growth only slightly retarded (Gray and Fuller, 1942).

In nutrient culture containing 0.06 to 028 ppm phenylmercuric acetate (PMA), roots of dwarf bean seedlings accumulated about 10 ppm Hg and terminal tissues from 0.09 to 0.90 ppm in 21 days (Pickard and Martin, 1959). The seedling biomass was reduced 50% by 0.11 ppm PMA in the solution. Beauford et al. (1977) observed that 5 ppm of Hg as $HgCl_2$ in solution culture inhibited growth of higher plants (*Pisum sativum* and *Mentha spicata*) and affected both physiological and biochemical processes in the plant. Siegel et al. (1984) observed that seed germination and early growth of several crop species and cultivars were not significantly affected by their exposure to Hg vapor at air saturation levels (14 ppb). The general toxic effect of Hg vapor on growing plants was premature senescence, which may not be associated with leaf necrosis or chlorosis. Young plants grown on nutrient culture spiked with $HgCl_2$ exhibited toxicity symptoms at levels as low as 10 ppb Hg in nutrient culture (Mhatre and Chaphekar, 1984). Toxic effects included reduced yield and reduced chlorophyll content.

In sand culture experiments, Davis et al. (1978) found that the critical level of Hg in dry matter tissue of barley was about 3 ppm. The critical concentration of Hg in solution was 4 ppm Hg(II), as $HgCl_2$. Visual symptoms of Hg toxicity in barley include yellowing of leaves and presence of reddish stems. Chino (1981) reported that in rice, the critical level of Hg was 0.5 ppm in stem and leaf, and 1000 ppm in roots. In soils near the Nifu mine in Japan, roots of rice plants contained 0.1 to 7.23 ppm Hg (Morishita et al., 1982). The leaves and stems of these plants contained 0.031 to 0.113 ppm, while rice grain contained 0.010 to 0.060 ppm. Relationships between Hg in highly polluted soils and Hg in roots of rice plants growing on them yield a high correlation coefficient of $r = 0.93$.

Other species are more tolerant to Hg. For example, Shacklette (1970) reported that as much as 3.5 ppm Hg in dry matter of Labrador tea was not injurious to the plant, whose roots were above cinnabar mineral deposits.

Velvet bentgrass showed no effect when grown on soils with Hg levels of 450 ppm, which accumulated from organomercurial fungicides applied annually over a period of 15 yr (Estes et al., 1973). These turf plants had an average Hg concentration of 1.7 ppm.

Growth of common Bermuda grass was reduced by 50 ppm of Hg in soil applied as $HgCl_2$ (Weaver et al., 1984). Apparently, excess levels of Hg in plants inactivated both biochemical and physiological processes. Significant portions of Hg in plants were associated with the cell wall. Beauford et al. (1977) have shown that Hg inhibits synthesis of proteins in leaves of plants. They also suggested that Hg could affect the water-absorbing and transporting mechanisms in plants. Bolli (1947) found that photosynthetic activity in several plant species was reduced by treatment with $HgCl_2$. Similarly, Harriss et al. (1970) noted that Hg compounds, such as organomercurial fungicides, reduced photosynthesis in plankton. Because Hg has a strong affinity for sulfhydryl or thiol groups involved in enzyme reactions, this element presumably can affect metabolic activity by producing structural changes in enzymes following mercaptide formation with –SH groups.

A review of Hg concentrations in plant species reveals the following general sequence: grassland herbs (~20 ppb median value) < shrubs and trees (~29 ppb) < aquatic macrophytes (40 ppb) < *Sphagnum* mosses (69 ppb) < lichens (170 ppb) < fungi (280 ppb) (Moore et al., 1995). In terms of ecological grouping, total and methyl Hg levels increased the closer the plant's habitat to the water table. In general, plants in nature can be expected to contain Hg in the range of 10 to 300 ppb DW, excluding samples from areas of point-source Hg pollution.

6 Ecological and Health Effects of Mercury

6.1 Effects of Mercury in Aquatic Ecosystems

The methylation of Hg is a key step in the transfer of Hg into the food chain. The biotransformation of inorganic Hg species to methylated organic species in bodies of water can occur in sediments and water columns. All Hg compounds entering an aquatic ecosystem however, are not methylated; demethylation reactions as well as volatilization of dimethyl Hg decrease the amount of methyl Hg available in aquatic environment. There is a large degree of uncertainty regarding the rates at which these processes take place. There is general scientific agreement, however, that there is significant variability among bodies of water concerning the environmental factors that influence the methylation of Hg (U.S. EPA, 1996).

The lipophilic nature of methyl Hg enhances its ability to be bioaccumulated relative to inorganic Hg. This results in the biomagnification of methyl Hg in the food chain (U.S. EPA, 1996). Predators of fish, such as the American bald eagle, tend to bioaccumulate more Hg than other organisms in the food web (Fig. 11.7).

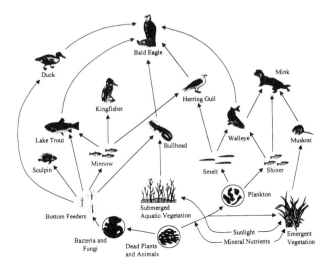

FIGURE 11.7. Simplified aquatic food web in North America where methyl Hg bioaccumulation follows the ascent in the food web. (From U.S. EPA, 1996.)

Methyl Hg accumulates in aquatic biota to concentrations several orders of magnitude greater than found in the water column. The ratio of methyl Hg to total Hg found in fish is much greater compared with the ratios found in sediment, the water column, or precipitation (Downs, 1998). The main discrimination between methyl Hg and inorganic Hg occurs during trophic transfer between phytoplankton and zooplankton; however, the major enrichment of absolute levels of Hg occurs between the water and phytoplankton, approximately $10^{5.5}$ between water and phytoplankton (Mason et al., 1994). The greater solubilization of methyl Hg than inorganic Hg from sediment by digestive fluid confirms previous observations that methyl Hg is generally more bioavailable from sediments than inorganic Hg. There is a strong dependence of methyl Hg desorption from particles on the sediment OM content which subsequently affects its bioavailability.

Basically all of the Hg that bioaccumulates in fish tissue is methylated (Bloom, 1992; U.S. EPA, 1996). A relationship exists between the methyl Hg content in fish and lake pH, with higher methyl Hg content in fish tissue typically found in more acidic lakes. Numerous factors can influence the bioaccumulation of Hg in aquatic biota in addition to low pH. These include the length of the aquatic food chain, temperature, and DOC. Physical and chemical characteristics of a watershed affect the amount of Hg that is translocated from soils to bodies of water. Interrelationships among these factors are poorly understood, however, and there is no single factor (including pH) that has been consistently correlated with Hg bioaccumulation in all cases examined (U.S. EPA, 1996).

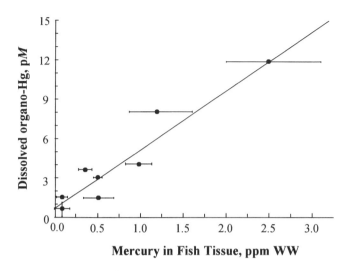

FIGURE 11.8. Correlation ($r^2 = 0.89$, $p < 0.005$) between dissolved organo-Hg contents of several lakes and reservoirs and historical reports of Hg contents of piscivorous fish tissues from the same water bodies. (Modified from Gill and Bruland, 1990.)

The effect of lower pH on Hg accumulation in aquatic invertebrates is poorly understood. Mercury levels in fish tissues are higher in lakes with low pH (Wren et al., 1995). Both water hardness and pH influence the rate of uptake and bioavailability of Hg to aquatic biota within a lake. As a reflector of methylated Hg species, lakes in California, Nevada, and other areas having high dissolved organo-Hg levels also had high levels of Hg in fish tissues (Gill and Bruland, 1990). In this case, a high correlation coefficient ($r^2 = 0.89$, $p < 0.005$) suggests the predictability of Hg in fish tissues using dissolved organic Hg from the water column (Fig. 11.8).

Mercury accumulates in organisms when the rate of uptake exceeds that of elimination. Although all forms of Hg can accumulate to some extent, methyl Hg generally accumulates the most. Inorganic Hg can also be absorbed but is generally taken up at a slower rate and with lower efficiency than is methyl Hg. Elimination of methyl Hg takes place very slowly, resulting in tissue half-lives (i.e., the time in which half of the mercury in the tissue is eliminated) ranging from months to years. Elimination of methyl Hg from fish is so slow that long-term reductions of Hg concentrations in fish are often due mainly to growth of the fish. In comparison, other Hg compounds are eliminated relatively quickly, resulting in reduced levels of accumulation.

The production and accumulation of methyl Hg in freshwater systems is an efficient process for accumulating Hg through ingestion by piscivores including birds, mammals (nonhuman), and humans. In addition, methyl

Hg generally comprises a relatively greater percentage of the total Hg content at higher trophic levels. Accordingly, Hg exposure and accumulation is of particular concern for animals at the highest trophic levels in aquatic food webs and for animals and humans that feed on these organisms.

In most cases, dietary Hg is the dominant exposure pathway for fish. Methyl Hg is taken up 10 to 100 times more rapidly than inorganic Hg and can be absorbed directly through the gills (Glass et al., 1990). Methyl Hg is also absorbed from food in preference to inorganic Hg: 70 to 90% and 5 to 15%, respectively. The ingested Hg may affect Na and Ca regulation in some organisms (Jagoe et al., 1996). Additionally, Hg can cause gill abnormalities in fish at sublethal levels. For example, Hg caused changes in gill morphology of mosquito fish (*Gambusia holbrooki*) exposed to 75 to 300 nM Hg(II) in natural stream water of low ionic strength (Jagoe et al., 1996). This exposure resulted in altered ion fluxes due to gill damage.

As mentioned earlier, methyl Hg biomagnifies through the aquatic food chain, hence predatory aquatic wildlife species are generally at greater risk of Hg exposure and toxicity (Meyer et al., 1995; Scheuhammer, 1991). Exposure of these biota to Hg is enhanced in reservoirs and in low-alkalinity, low-pH ecosystems. For example, the concentration of Hg in blood of great egrets (*Ardea albus*) from southern Florida reported by Sepulveda et al. (1999) is higher than that reported in great blue herons (*Ardea herodias*) from California, common terns (*Sterna hirundo*) from New York, common loons (*Gavia immer*) from Ontario, and bald eagles (*Haliaeetus leucocephalus*) from Oregon, Washington, and Florida. Common loons and common mergansers (*Mergus merganser*) from eastern Canada had Hg generally highest in liver, followed by kidney, then breast muscle (Scheuhammer et al., 1998). As total Hg concentrations increased in liver and kidney tissues, methyl Hg decreased. Liver and kidneys with the highest total Hg concentrations (>100 μg g^{-1} DW) had only 5 to 7% of the total as methyl Hg. In contrast, the proportion of methyl Hg in breast muscle remained high (80 to 100%) in both species, regardless of total Hg concentration. It is believed that elevated Hg concentrations in tissues of loons may have contributed to their emaciated condition.

Methyl Hg is an increasingly abundant environmental contaminant in wetlands around the world (Bouton et al., 1999). Freshwater fish in the Florida Everglades have been estimated to contain a mean Hg concentration of 0.41 mg kg^{-1} (Frederick et al., 1999). Piscivorous birds feeding on these contaminated fish represent the top of the food chain in wetland systems and, as such, potentially are good indicators of the effects of such chronic exposure. Consequently, the decline in the population of wading birds [e.g., great white herons (*Ardea herodias occidentalis*)] in southern Florida is partly attributed to the effect of Hg on the reproductive capacity of this wildlife (Spalding et al., 1994). It has been estimated that at the 0.41 mg Hg kg^{-1} level in the diet, great egret nestlings in the

Everglades would ingest 4.32 mg total Hg during an 80-day nestling period (Frederick et al., 1999). Captive egret feeding studies inferred that this level of exposure in the wild could be associated with reduced fledging mass, increased lethargy, decreased appetite, and possibly poor health and juvenile survival.

Massive poisoning of birds and wildlife from methyl Hg–treated seed grains were identified during the decades preceding the 1970s (U.S. EPA, 1996). These findings resulted in substantial limitation on use of methyl Hg to treat seed grains. However, methyl Hg contamination of the aquatic food chain from many sources continues to adversely affect wildlife and domestic mammals and wild birds. In Minamata, Japan from about 1950 to 1952 (prior to recognition of human poisoning), severe difficulties with flying and other grossly abnormal behavior were observed among birds. In birds, symptoms of acute methyl Hg poisoning include reduced food intake leading to weight loss; progressive weakness in wings and legs; difficulty flying, walking, and standing; and an inability to coordinate muscle movements. Signs of neurological disease, including convulsions, fits, and highly erratic movements (mad running, sudden jumping, bumping into objects), were observed among domestic animals, especially cats that consumed contaminated seafood. Whole-body residues of Hg in acutely poisoned birds usually exceed 20 μg g^{-1} FW. Although sublethal effects include a number of different organ systems, reproductive effects are the primary concern. These occur at concentrations far lower than those that cause overt toxicity.

Smaller animals (e.g., minks, monkeys) are generally more susceptible to Hg poisoning than are larger animals (e.g., mule deer, harp seals). Smaller mammals eat more per unit body weight than larger mammals; thus smaller mammals may be exposed to larger amounts of methyl Hg on a body weight basis.

Fish-eating (piscivorous) animals and those that prey on other fish-eaters accumulate more Hg than if they consumed food from the terrestrial food chain. In a study of fur-bearing animals in Wisconsin, the species with the highest tissue levels of Hg were otter and mink, which are top mammalian predators in the aquatic food chain (U.S. EPA, 1996). Top avian predators of the aquatic food chain include raptors such as the osprey and bald eagle. Smaller birds feeding at lower levels in the aquatic food chain also may be exposed to substantial amounts of Hg because of their high food consumption rate (consumption/day/g of body weight) relative to larger birds. Wildlife criteria have been estimated for three piscivorous birds and two mammals (Table 11.4). These values refer to the level of Hg in water that is expected to be harmless for the species and considers the bioaccumulation of Hg in large and small fish eaten by the mammals or birds. A bioaccumulation factor was used in the calculation; the bioaccumulation factor in turn was based on Hg data in fish and water from which they were taken.

TABLE 11.4. Estimated wildlife criteria based on Hg level in water that is expected to be harmless for selected species.

Organism wildlife criterion	(pg Hg L^{-1})
Mammals	
Mink	415
River otter	278
Piscivorous birds	
Kingfisher	193
Osprey	483
Bald eagle	538

Source: Modified from U.S. EPA (1996).

Methyl Hg toxicity in mammals is primarily manifested as central nervous system (CNS) impairment, including sensory and motor deficits and behavioral disorders (Wolfe et al., 1998). Animals initially become anorexic and lethargic. Muscle ataxia, motor control deficits, and visual impairment develop as toxicity progresses, with convulsions preceding death. Smaller carnivores are more sensitive to methyl Hg toxicity than are larger species, as reflected in shorter time of onset of toxic signs and time to death. Dietary concentrations of 4.0 to 5.0 µg g^{-1} methyl Hg were lethal to mink and ferrets within 26 to 58 days, whereas otters receiving the same concentration survived an average of 117 days.

6.2 Effects of Mercury in Terrestrial Ecosystems

Compared with aquatic systems, bioaccumulation and toxicity of Hg in terrestrial ecosystems are low. Once Hg has entered the soil, it is strongly sequestered by soil components, and uptake and transfer to plants does not constitute an important exposure pathway to animals. In general, the transfer of Hg from soil media to roots, and from roots to shoot is very low, yielding transfer coefficients from 0.01 to 1.0 (Wren et al., 1995). The extent of transfer to foliage depends on a multitude of factors, including Hg species, soil properties (primarily pH and OM), and plant species.

It should be recognized that Hg does not biomagnify in the terrestrial food chain, unlike in the aquatic food chain. This holds true even in the worst possible scenario. For example, vegetation collected in the vicinity of the Almaden, Spain mercury mine had only one- to threefold Hg levels in their foliage (compared with samples from unimpacted areas) in spite of >100 times elevation of Hg in the corresponding soil (Lindberg et al., 1979). This, then, would pose little threat to the well-being of herbivores.

The concentration of Hg in tissues of terrestrial herbivores is usually low (Wren et al., 1995). With deer, moose, and caribou, Hg concentrations in liver tissue are typically <0.15 µg g^{-1}, while muscle tissue concentrations are <0.10 µg g^{-1}. Likewise, the muscle tissue concentrations of Hg in terrestrial carnivores (coyotes, foxes, and wolves) are usually <0.20 µg g^{-1}.

Effects of Hg on fish include death, reduced reproductive success, impaired growth, and developmental and behavioral abnormalities. Exposure to Hg can also cause adverse effects in plants, birds, and mammals. Reproductive effects are the primarily concern for avian Hg poisoning and can occur at dietary concentrations well below those which cause overt toxicity. Sublethal effects of Hg on birds include liver and kidney damage, and neurobehavioral effects. Effects of Hg on plants include death and sublethal effects. Sublethal effects on aquatic plants can include plant senescence, growth inhibition, and decreased chlorophyll content. Sublethal effects on terrestrial plants include decreased growth, leaf injury, root damage, and inhibited root growth and function (U.S. EPA, 1996).

Although clear causal links between Hg contamination and population declines in various wildlife species have not been established, Hg may be a contributing factor to population declines of the endangered Florida panther and the common loon. Some researchers have concluded, however, that Hg levels in most areas are not high enough to adversely affect bird populations (U.S. EPA, 1996).

6.3 Mercury in the Amazon (Brazil): A Case Study

Widespread contamination of aquatic organisms, particularly fish, by Hg released in the environment by gold-mining activity has been reported in various Amazonian rivers (Lacerda and Salomons, 1998). Recent estimates show that from 70 to 130 tonnes of Hg are released annually into the Amazonian environment. About 30 to 60 tonnes are released into the soil and rivers annually. These high levels of Hg can be consequential as a result of major environmental conditions of most Amazonian aquatic systems, i.e., slight acidity, tropical climate, and moderate to high OM content, that favor high methylation rates, and therefore greater bioaccumulation potential, especially at high tropic levels (Lacerda et al., 1989, 1991a,b).

The mining activity in the Amazon is accentuated by as many as 350,000 garimpeiros (so-called surface miners) who use Hg for the amalgamation of gold. Because as much as 55% of the Hg is released to the atmosphere as metallic vapor (Hg^0) during the frying of the Au–Hg alloy, it has become both an occupational and environment hazard (Magalhaes and Tubino, 1995). The released Hg can contaminate not only the garimpeiros themselves but also the inhabitants on the borders of the Amazonian rivers and lakes, and the aborigines for whom fish is a very important part of their diet. Consequently in the Madeira River, the largest tributary of the Amazon River in terms of water flow and sediment load, Hg concentrations of up to 2.7 µg g^{-1} WW in carnivorous fish have been reported (Pfeiffer et al., 1989). In the Tapajos River in the southeastern Amazon, Hg concentrations in five fish species ranged from 0.15 to 0.73 µg g^{-1} WW, with the highest concentrations also found in carnivorous fish. Methyl Hg comprised about 95% of the total Hg content in the fish muscle tissue.

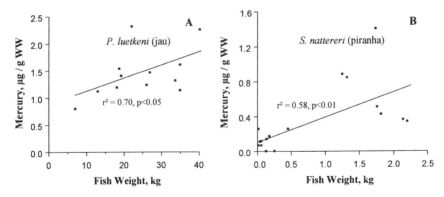

FIGURE 11.9. Relationship between body weight and size and total Hg concentrations in muscle tissues of carnivorous fish from the Carajás region in the Amazon. (Modified from Lacerda et al., 1994.)

Extensive studies in the river network of the Amazon indicate that carnivorous fish accumulated the most Hg, frequently surpassing the maximum permissible concentration of 0.50 µg g^{-1} WW of Hg in fish for human consumption (Lacerda and Salomons, 1998; WHO, 1976). There is strong evidence of methylation of Hg in these aquatic ecosystems. One example is the strong correlation between Hg concentration in fish muscle tissue and fish size (Fig. 11.9). Methylated Hg has very slow excretion rates, tending to accumulate in large, older fish. For example, Lacerda et al. (1994) showed Hg distribution in two top predators: the jau (*Paulicea luetkeni*) and the piranha (*Serrasaumus nattereri*) in the Carajás mining region. In both species, Hg concentrations were significantly higher in larger individuals following a logarithmic pattern.

Concentrations of Hg in muscle tissue of large carnivorous fish from gold-mining areas in the Amazon are compared with those from other contaminated ecosystems (Table 11.5). Results indicate that Hg concentra-

TABLE 11.5. Mercury concentrations in muscle tissue (WW) of large carnivorous fish from contaminated aquatic environments compared with those from the Amazonian rivers.

Location	Hg (µg g^{-1})
Lake Erie, Canada	0.20–0.79
Niigata, Japan	2.60–6.60
Tapajós, Amazon	0.15–0.73
Madeira, Amazon	0.21–2.70
Finnish lakes	0.21–1.80
Swedish lakes	0.68–0.86
Tyrrhenian Sea, Italy	0.40–2.20
Carajás, Amazon	0.30–2.30

Source: Modified from Lacerda et al. (1994).

tions in fish from the Amazonian mining areas are in the same order of magnitude as those found in areas heavily contaminated with Hg from industrial activities, such as Niigata, Japan and the Tyrrhenian Sea, Italy.

Despite the high Hg levels in fish tissues in the Amazonian mining areas, a survey on Hg contents in hair samples from the local human population showed relatively low Hg concentrations (0.9 to 4.8 μg g^{-1} hair DW) among fish eaters (Lacerda et al., 1994). These low values can be partly attributed to the low fish intake of the local population and their food habits. Also, noncarnivorous fish species have typically lower Hg contents in their muscle tissue. Thus, exposure of the regional population through fish consumption offers low risk. However, special groups, such as the native Indians, may have been exposed differently. Not all Hg in fish and other aquatic biota in the Amazon is due to gold mining, according to Veiga et al. (1994), but is also due to deforestation activities, which result in decomposition of vegetative biomass, releasing Hg.

Based on the recent assessment of the entire Amazonian Basin, the contribution of gold mining to atmospheric Hg dominates largely over that of forest fires, with respective emissions of 65 to 170 and 6 to 9 tonnes Hg yr^{-1} (Roulet et al., 1999). The latter estimate supports the conclusion of Lacerda (1995) that the contribution of fires to anthropogenic Hg emissions to the atmosphere is relatively small. However, data of Roulet et al. (1999) show that an important effect of human habitation may be the erosion of deforested soils, releasing an estimated 500 to 3000 μg Hg m^{-2} cm^{-1} of soil eroded. This source presumably causes significant Hg contamination of Amazonian rivers. Eroded soils may lose up to 30,000 μg Hg m^{-2}, most of which presumably reaches the aquatic environment in particulate form. In newly colonized Amazonian areas, soil erosion may be the major process increasing the loading of natural Hg to adjacent aquatic ecosystems.

6.4 Exposure of Humans to Mercury

The Hg poisoning in Minamata Bay, Japan demonstrates the potential hazards associated with chronic exposure to Hg, particularly methyl Hg (Hamada and Osame, 1996). A 20-yr survey that addressed methyl Hg exposure in 514 native communities across Canada indicates that many people in these areas are at risk of Hg poisoning (Wheatley and Paradis, 1995).

Mercury contamination of the fishery affects numerous lakes and rivers around the Great Lakes (Glass et al., 1990). Current levels of Hg in freshwater fish in the United States are such that advisories have been issued in 35 states that warn against the consumption of certain amounts and species of fish that are contaminated with Hg (U.S. EPA, 1996). Mercury has been listed as a pollutant of concern to U.S. EPA's Great Lakes Waters Program due to its persistence in the environment, potential to bioaccumulate, and toxicity to humans and the environment (U.S. EPA, 1994b). In

Minnesota alone, advisories restricting consumption of Hg-contaminated fish were issued in the 1980s for 285 bodies of water. Similar scenarios were observed in the 1980s in Wisconsin and Michigan. More recently, six states (based on 1994 data) have statewide advisories that are posted on every freshwater system in that state and are intended for people who catch or eat fish from those bodies of water. These advisories are based on the results of sampling surveys that measure Hg levels in representative fish species collected from bodies of water. The states have the discretion of establishing action levels that are different from those of the U.S. Food and Drug Administration (FDA).

The U.S. FDA has jurisdiction over commercial fish and issues action levels for concentrations of Hg in fish and shellfish. The current action level is 1 ppm Hg WW based on a consideration of health impacts (U.S. EPA, 1996). The typical U.S. consumer of commercial seafood is not in danger of ingesting harmful levels of methyl Hg in seafood. The existing U.S. FDA management and consumer advice protects commercial food. In some areas, freshwater fish can have Hg levels that exceed the U.S. FDA action limit of 1 ppm. The concentration of methyl Hg in commercially important marine species is lower on the average than the U.S. FDA action level.

In terms of human exposure, the most important Hg compound is methyl Hg. Fish and fish-product consumption is the most important exposure pathway for humans. Exposure may occur through other routes as well (e.g., the ingestion of methyl Hg–contaminated drinking water and food sources other than fish, and dermal uptake through soil and water); however, the fish consumption pathway dominates these other pathways for people who eat fish.

A general level of total Hg in fish typically ranges from 0.20 to 1.0 mg kg^{-1} WW in cinnabar (HgS)-enriched habitats, with a minimum for freshwater fish of about 0.035 mg kg^{-1} (Downs et al., 1998). More recent studies have quoted background concentrations of 0.15 mg kg^{-1} for the preindustrial period and between 0.05 to 0.30 mg kg^{-1} depending on the humic content of the water. Of equal importance in determining fish Hg concentrations is the species of fish. In various rivers of the East Anglian region of the United Kingdom, a survey of eel (*Anguila anguila*) and roach (*Rutilus rutilus*) clearly demonstrates that species behavior can affect the level to which Hg is bioaccumulated. Permissible safe consumption levels for Hg in fish are set by various international, national, and regional health authorities. Highest international permissible limits for Hg in fish are set between 0.30 and 1.0 mg kg^{-1} (CEC, 1993).

Wide variability exists among individuals in fish-eating populations with respect to food sources and fish consumption rates (U.S. EPA, 1996). The populations most highly exposed are those located in areas where the concentration of methyl Hg in freshwater fish is elevated, in part as a result of anthropogenic releases from industrial and combustion sources. Methyl Hg exposure rates among children who consume fish are expected to be higher than for adults who consume fish because of their lower body weight. Humans could also be exposed to inorganic Hg through inhalation, and consumption of contaminated water and food. A major source of exposure for elemental Hg is through inhalation in occupational settings. Another source of exposure to low levels of elemental Hg in the general population is elemental Hg released in the mouth from dental amalgam fillings (U.S. EPA, 1994).

Mercury is a human toxicant that has been associated with occupational exposure and with exposure through consumption of contaminated food (U.S. EPA, 1996). Generally, the effect seen at the lowest exposure level for elemental and methyl Hg is neurotoxicity. The range of neurotoxic effects can vary from subtle decrements in motor skills and sensory ability to tremors, inability to walk, convulsions, and death.

6.5 Toxicokinetics and Effects of Mercury in Humans

The toxicokinetics (i.e., absorption, distribution, metabolism, and excretion) of Hg is highly dependent on the form of Hg to which a receptor has been exposed (U.S. EPA, 1996). The absorption of elemental Hg vapor occurs rapidly through the lungs, but it is poorly absorbed from the GI tract. Once absorbed, elemental Hg is readily distributed throughout the body; it crosses both placental and blood–brain barriers. The distribution of absorbed elemental Hg is limited primarily by the oxidation of elemental Hg to the mercuric ion, as the mercuric ion has a limited ability to cross the placental and blood–brain barriers. Once elemental Hg crosses these barriers and is oxidized to the mercuric ion, its return to the general circulation is impeded, and Hg can be retained in brain tissue. Elemental Hg is eliminated from the body via urine, feces, exhaled air, sweat, and saliva. The pattern of excretion changes depending upon the extent to which the elemental Hg has been oxidized to mercuric Hg.

Absorption of inorganic Hg through the GI tract varies with the particular mercuric salt involved; absorption decreases with decreasing solubility (U.S. EPA, 1996). Estimates of the percentage of inorganic Hg that is absorbed vary—as much as 20% may be absorbed. Inorganic Hg has a reduced capacity for penetrating the blood–brain and placental barriers. There is some evidence indicating that mercuric Hg in the body following oral exposures can be reduced to elemental Hg and excreted via exhaled air. Because of the relatively poor absorption of orally administered inorganic Hg, the majority of the ingested dose in humans is excreted through the feces.

Among Hg forms, methyl Hg is rapidly and efficiently absorbed through the GI tract (U.S. EPA, 1996). Methyl Hg is readily absorbed from the GI tract (90 to 95%), whereas inorganic salts of Hg are less readily absorbed (7 to 15%) (Wolfe et al., 1998). In the liver, Hg binds to glutathione, cysteine, and other sulfhydryl-containing ligands. These complexes are secreted in the bile, releasing the Hg for reabsorption in the gut. In blood, methyl Hg distributes 90% to red blood cells and 10% to plasma. Inorganic Hg distributes approximately evenly or with a cell/plasma ratio of ≥ 2. Methyl Hg readily crosses the blood–brain barrier, whereas inorganic Hg does so poorly. The transport of methyl Hg into the brain is mediated by its affinity for the anionic form of sulfhydryl groups. This form of Hg is distributed throughout the body and easily penetrates the blood–brain and placental barriers in humans and animals. Methyl Hg in the body is considered to be relatively stable and is only slowly demethylated to form mercuric Hg in rats. Methyl Hg has a relatively long biological half-life in humans; estimates range from 44 to 80 days. Excretion occurs via the feces, breast milk, and urine.

Blood, urine, and scalp hair are the most common biological samples analyzed for Hg. Exposure to Hg has been expressed in these units: µg/kg body weight/day; concentrations of Hg in tissues such as blood, hair, feathers, liver, kidney, brain, etc.; g of fish per day; number of fish meals per time interval (e.g., per week). Reference values for Hg concentrations (expressed as total Hg) in biological materials commonly used to indicate human exposures to Hg have been published by WHO (1990). The mean concentration values of Hg is approximately 8 µg L^{-1} in whole blood, about 2 µg g^{-1} in hair, and approximately 4 µg L^{-1} in urine. The WHO report found that Hg concentrations in hair increased with increasing frequency of fish consumption (Table 11.6).

When exposure occurs to the developing embryo/fetus during pregnancy as well as when adults and children are exposed to methyl Hg, neurotoxicity becomes a great concern. Two major epidemics of methyl Hg poisoning through fish consumption have occurred. The best known epidemic occurred near Minamata City on the shores of Minamata Bay, Kyushu, Japan. The source of methyl Hg was a chemical factory that used Hg as a catalyst; methyl Hg in the factory waste sludge was drained into Minamata Bay. Once discharged into the bay, the methyl Hg accumulated in the tissue of shellfish and fish that were subsequently consumed by wildlife and humans. Fish was a routine part of the diet in the bay's populations; average

TABLE 11.6. Data synthesis on mercury content of human (head) hair as influenced by the frequency of fish consumption.

Fish consumption frequency	Average Hg concentration in hair (µg Hg g^{-1} of hair)
No unusual Hg exposure	2
Less than one fish meal per month	1.4 (range 0.1–6.2)
Fish meals twice a month	1.9 (range 0.2–9.2)
One fish meal a week	2.5 (range 0.2–16.2)
One fish meal each day	11.6 (range 3.6–24.0)

Source: WHO (1990).

fish consumption was in excess of 300 g day^{-1}, which is a much greater level of fish consumption than is typical for the general U.S. population (U.S. EPA, 1996).

The first poisoning case occurred in 1956 in a 6-yr-old girl diagnosed with symptoms characteristic of impairment of the nervous system. It took several years before people in the area became aware that they were developing the signs and symptoms of methyl Hg poisoning. Symptoms of *Minamata* disease in humans include the following: impairment of peripheral vision; disturbances in sensations ("pins and needles" feelings, numbness), usually in the hands and feet and sometimes around the mouth; incoordination of movements, as in writing; impairment of speech, hearing, and walking; and mental disturbances.

Over the next 20 yr the number of people known to be affected with what became internationally known as *Minamata* disease increased to thousands. Initially, deaths occurred among both adults and children. The nervous system damage of severe methyl Hg poisoning among infants was very similar to congenital cerebral palsy. Efforts were made to reduce the release of Hg into the bay, and after 1969, the average Hg concentrations in fish had fallen below 0.50 ppm WW.

Another methyl Hg poisoning outbreak occurred in Niigata, Japan in 1965. As in Minamata, the source was identified again to be a chemical factory releasing methyl Hg into the Agano River. The signs and symptoms of disease in Niigata were those of methyl Hg poisoning and were similar to the *Minamata* disease.

In general, infants born to women who ingest high levels of methyl Hg can be expected to exhibit mental retardation, ataxia, constriction of the visual field, blindness, and cerebral palsy. Human studies are inconclusive regarding elemental Hg and cancer, and no human studies are available on the carcinogenic effects of methyl Hg. The U.S. EPA has classified inorganic Hg and methyl Hg as possible human carcinogens, and elemental Hg as not classifiable as to human carcinogenicity (U.S. EPA, 1994a).

In summary, chronic (long-term) exposure to elemental Hg and methyl Hg in humans affects the central nervous system. Effects such as erythrism (increased excitability), irritability, excessive shyness, and tremor have been noted from elemental Hg exposure, and symptoms such as paresthesia (a sensation of pricking on the skin), blurred vision, malaise, speech difficulties, and constriction of the visual field from methyl Hg exposure. The major effect from chronic exposure to inorganic Hg is kidney damage (ATSDR, 1992).

6.6 Exposure of the General U.S. Population to Mercury

Three groups are potentially at increased risk from methyl Hg: pregnant women, women of childbearing age (i.e., between the ages of 15 and 44), and children aged 14 and younger (U.S. EPA, 1996). Pregnant women are of concern because of the adverse effects of methyl Hg on the fetal nervous system. Women of childbearing age are of concern for two reasons. The first is that methyl Hg persists in tissues; the half-life of methyl Hg in tissue is typically over 2 mo. The second reason is that women may not know they are pregnant until the pregnancy is past many of the critical stages of fetal development. Children may be at a higher risk of methyl Hg exposure than are adults because they have higher exposures on a per kilogram of body weight basis, and they may be inherently more sensitive than adults given the developmental state of the nervous system. In the methyl Hg poisoning epidemics in Japan and Iraq, the effects were seen not only in children exposed to methyl Hg *in utero,* but included children exposed through ingesting methyl Hg from food.

The analysis of at-risk population eating above average amounts of fish focuses on that part of the population that consumes on the average at least 100 g of fish or shellfish per day. The Hg contents of the top 10 types of fish consumed by the general U.S. population are shown in Appendix Table A.31. The basis for this focus on persons eating 100 g or more is a recommendation made by WHO's International Programme for Chemical Safety. In a subpopulation that consumes large amounts of fish (e.g., 100 g day^{-1}), WHO's recommendation is that as a preventive measure, hair levels for women of childbearing age should be monitored for methyl Hg. The 100 g day^{-1} recommendation by WHO can be used as a screening analysis to identify populations potentially at increased risk. The significance of the risk is, as mentioned above, also a function of the methyl Hg concentrations of the fish consumed. Typical concentrations of Hg in fish muscle are listed in Appendix Table A.32.

Populations that consume fish at 100 g day^{-1} or higher include several segments of the U.S. population (U.S. EPA, 1996). Subsistence fishers, who rely heavily on fish as a major source of protein, are a special group with high fish intake. Another group is people who consume high levels of fish due to health-based considerations pertaining to the reduction of certain diseases, particularly of the cardiovascular system. There are also large numbers of people who simply prefer fish and shellfish as a source of protein. These consumers are represented by these special groups: recreational anglers, members of some Native American tribes, members of ethnic groups who consume higher than typical intakes of fish, persons who preferentially select fish for health-promotion purposes, individuals who prefer the taste of fish, and persons who rely on self-caught fish from local sources because of limited money to buy food. These groups may be more reliant on local sources of self-caught fish than is the general population.

In summary, the U.S. EPA analysis indicates that the commercial U.S. fish supply is safe for the U.S. population who consume <100 g day^{-1} of fish and shellfish, and a wide variety of fish types (U.S. EPA, 1996). Those consumers who eat large quantities of predatory marine species may be at some level of risk from exposure to methyl Hg. Consumers of freshwater fish are also, in general, not expected to be at an elevated risk level unless their sources of fish are contaminated with more than average levels of Hg.

6.7 Permissible Intake and Level of Mercury in Humans

The current U.S. EPA reference dose (RfD) for methyl Hg was based on data on neurological changes in 81 Iraqi children who were exposed *in utero* —their mothers had eaten bread contaminated with methyl Hg during pregnancy. The incidence of several end points (including late walking, late talking, seizures or delayed mental development, and scores on clinical tests of nervous systems function) were mathematically modeled to determine a Hg level in hair (measured in all the mothers in the study) which was associated with no adverse effects. These effects include delays in motor and language development (U.S. EPA, 1996).

In calculating the Hg level in hair that was associated with no observable adverse effects (NOAEL)—the U.S. EPA chose a benchmark dose (in this instance, the lower bound for 10% risk of neurological changes, which is 11 ppm methyl Hg in hair) based on modeling of all effects in children. A dose-conversion equation was used to estimate a daily intake of 1.1 μg methyl Hg/kg body weight/day such that when ingested by a 60-kg person will maintain a blood concentration of approximately 44 μg L^{-1} of blood or a hair concentration of 11 μg g^{-1} of hair.

The reference concentration (RfC) for elemental Hg is 0.0003 mg m^{-3}. The U.S. EPA estimates that inhalation of this concentration or less, over a lifetime, would not likely result in the occurrence of chronic noncancer effects (U.S. EPA, 1994a). The RfD for methyl Hg is 0.0001 mg/kg/day. The U.S. EPA estimates that consumption of this dose or less, over a lifetime, would not likely result in the occurrence of chronic noncancer effects.

7 Factors Affecting Mobility and Bioavailability of Mercury

Variations in fish tissue Hg concentrations are influenced by, to varying degrees, biological variability associated with the species themselves (age, size, physiology, diet), geologic influences (bedrock and sediments), chemical variability (water quality and Hg biogeochemistry), physical variability (e.g., water temperature, lake and watershed size), and of other influences, such as climate and atmospheric deposition. Freshwater systems most at risk of Hg bioaccumulation in fish are those with low acid-neutralizing capacity and low pH, high DOC, high water temperatures, large watersheds, and high shoreline wetland lake and high exposed shoreline lake ratios (Evers et al., 1998; Vaidya et al., 2000). Newly created reservoirs and bog lakes commonly fit these criteria.

7.1 pH

The pH affects the speciation and stability of Hg in soil/sediment environments as has been shown in Figure 11.6. In addition, the nature of species that may predominate in a solution depends upon ambient pH and redox potential and upon the nature of anions and other groups, such as ligands, present. These can be easily described by means of an Eh–pH diagram for both undissolved and soluble species (Fig. 11.10).

It is predicted that as pH (or Cl^-) concentration increased, $Hg(OH)_2$ or $HgCl_2$ becomes dominant, while $Hg(OH)Cl$ serves as a transitional species (Gilmour, 1971) (see also Fig. 11.6). The Hg(II) species present in four Wisconsin rivers were $Hg(OH)_2$ (63 to 92%), $Hg(OH)Cl$ (8 to 35%), and $HgCl_2$ (<1 to 2%). Likewise, the stability of Hg in soil is also influenced by pH. In a limed loamy soil treated with $HgCl_2$, reduction of Hg salts to Hg^0 increased, i.e., an increase in volatile losses as soil pH increased from 5.3 to 6.4 (Frear and Dills, 1967). Additionally, Landa (1978) found increasing losses of Hg with increasing pH in Montana soils (pH 6.6 to 8.3) treated with $Hg(NO_3)_2$. However, Alberts et al. (1974) found that the loss of Hg via chemical reduction of mercuric ion from $Hg(NO_3)_2$-humic acid suspension

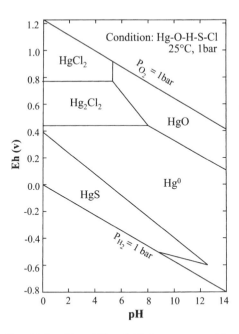

FIGURE 11.10. Predicted Eh–pH stability field for mercury. The assumed activities for dissolved species are $Hg = 10^{-8}$, $Cl = 10^{-3.5}$, $S = 10^{-3}$. (Modified from Brookins, 1988; with input from Dr. J. King, Savannah River Technology Center, Savannah River Site, Aiken, SC.)

was reduced by about 9% over a 290-hr period when the pH was increased from 6.5 to 8.2.

The pH can also affect the sorption of Hg by clay minerals and soil materials. In the absence of ligands, sorption of Hg by illite and kaolinite changed little with pH (Farrah and Pickering, 1978); with montmorillonite however, Hg sorption decreased with increasing pH (see also Section 4.4 on sorption). The influence of pH on the sorption of Hg from solutions by soils and soil constituents is shown Figure 11.11. The samples were adjusted to desired pH by adding $Ca(OH)_2$; then 10^{-5} M $HgCl_2$ was added and suspensions equilibrated for 2 hr. Andersson's data (see Fig. 11.11) indicate that the more effective sorbent for Hg in acid soils (pH <4) is OM, whereas at higher pH (pH >5.5) certain clay minerals may assume importance.

7.2 Forms and Chemical Species of Mercury

Carriers of Hg can also influence the stability of this element in soils. Rogers (1979) found that over a 6-day period, the Hg compounds most soluble in water [e.g., $Hg(NO_3)_2$ $HgCl_2$ and $Hg(C_2H_3O_2)_2$] had greater losses of Hg from tested soils (12 to 38%), followed by less-soluble HgO (6 to 19%), with insoluble HgS losing only insignificant amounts of applied Hg (0.2 to 0.3%). These results indicate that the form of Hg added to soil has an effect on potential volatilization of applied Hg.

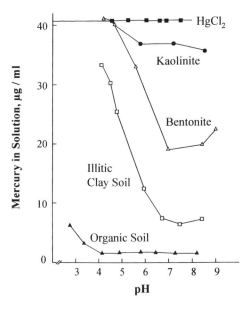

FIGURE 11.11. Retention of mercury by soils and soil components as influenced by pH. (Modified from Andersson, 1979.)

Mobility of Hg in soil columns was generally greater for methyl Hg chloride (MMC) than either the $HgCl_2$ or phenylmercuric acetate (PMA) (Hogg et al., 1978b). In addition, Hg levels in bromegrass foliage were significantly higher from MMC-treated soil than from either PMA- or $HgCl_2$-treated soils. This is expected since the extent of adsorption by soils increased in the order MMC < PMA < $HgCl_2$ (Hogg et al., 1978a).

Ionic strength and electrolyte cations have been shown to affect adsorption of metals to soils. This can be attributed to competitive binding of electrolyte cations or the indirect effect of these cations on dissolution of soil OM. Their effects on sorption however, is largely pH-dependent. For example, adsorption of Hg(II) in New Jersey soils is similar in both $NaNO_3$ and $Ca(NO_3)_2$ electrolytes at pH <4 (Fig. 11.12A). Increasing the pH above 4 resulted in significant decrease in adsorption in $NaNO_3$ due to complexation of Hg(II) by increasing dissolved OM. With $Ca(NO_3)_2$ background, adsorption of Hg(II) was much higher than in $NaNO_3$. This was not caused by the competitive binding of Ca^{2+} for sorption sites; rather, it is the decrease in concentration of soluble OM, which reduced the effect of soluble OM on adsorption (Yin and Allen, 1997). Calcium can reduce dissolution of OM by either coagulation or complexation in which Ca^{2+} serves as a bridge between sorption sites and OM. With methyl Hg (see Fig. 11.12B), the decrease in its adsorption at pH >6 is due primarily to its complexation by soluble OM. The increased adsorption of methyl Hg at low pH (<6) probably resulted from pH-selective dissolution of S-containing organic groups, which was reduced by Ca^{2+} at low pH (<6) but not affected at higher pH.

It is evident from earlier sections that Hg forms strong complexes with Cl^- ions and soluble organic C, the extent of which is pH-dependent. In turn, speciation largely influences the mobility and bioavailability of Hg.

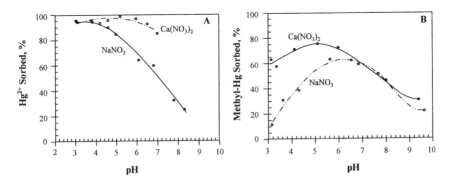

FIGURE 11.12. Adsorption of (A) Hg^{2+} by a loamy soil and (B) methyl mercury (CH_3Hg^+) by a sandy loam soil as affected by pH and electrolyte cations. (Modified from Yin and Allen, 1995.)

7.3 Organic Matter and Soil Type

The importance of OM in the retention of Hg against leaching, volatiliza-tion, or plant uptake is well documented. In an incubation experiment where PMA was added to various soils, volatilization of Hg ranged from 59% of that applied in sandy (mineral) soil to only 3% of that applied in a peat. Landa (1978) found the same preventive effect of OM from volatile losses of Hg from soils. This effect can be attributed to higher adsorption of Hg by soils onto both OM and clays. Gotoh et al. (1978) found that total Hg and OM content of forest soils were very highly correlated ($r^2 = 0.99$). Thus, in a given soil profile, the highest Hg accumulations are usually in horizons containing high OM contents.

Soil type or clay content also influences the stability of Hg in soils. In general, volatile losses of Hg can be expected to occur more in coarse-textured than in fine-textured soils. For example, in soils amended with 1 ppm Hg as $Hg(NO_3)_2$, 20% of applied Hg was lost from a silty clay loam soil during the first week compared with 45% loss from a loamy sand soil (Rogers, 1979; Rogers and McFarlane, 1979). In addition, Dudas and Pawluk (1976) observed that for any given soil sample, the highest quan-tity of Hg was associated with the clay fraction compared with other size particles.

In aquatic systems, DOC significantly decreases methyl [203]Hg uptake and accumulation in blackfish (Choi et al., 1998). Methyl [203]Hg-DOC complex-ation occurs even at low concentrations of both DOC and methyl [203]Hg. This study demonstrates that interactions between DOC and methyl Hg may be of primary importance for the mobility, bioavailability, and toxicity of methyl Hg in aquatic environments.

7.4 Redox Potential

The degree of aeration also influences the stability of Hg in soils/sediments. In sediments, a common hypothesis has been that reducing conditions are necessary for optimal formation of methyl Hg. However, other evidence is available indicating that methyl Hg formation occurs at a higher rate in oxic aqueous system than in reducing ones. Losses of methyl Hg could be greater under aerobic than anaerobic conditions in soils (Rogers, 1976) and in ocean sediments (Olson and Cooper, 1976). The results appear contradictory to the trend of increased methyl Hg loss with increasing soil moisture. However, a greater rate of methyl Hg production than the rate of loss under reducing conditions cannot be ruled out.

Soil moisture can also influence the ability of soil to sorb Hg vapor. In general, sorption of Hg vapor increases as the initial soil moisture content increases. Fang (1981) found that sorption of Hg vapor increased almost linearly as soil moisture content increased from about 2 to 20%. The moisture contents that gave maximal Hg vapor sorption coincided closely with the soil water-holding capacity values at 1/3 bar. Further increases in moisture content could decrease the Hg vapor sorption due to decreased soil pore volume, which is decreased by the filling of soil pore space with water.

Redox potential also influences the stability of solid phases of Hg. For example, HgS is stable in flooded soils (e.g., paddy soils), but can become unstable when the soil become aerobic (Aomine et al., 1967; Engler and Patrick, 1975).

8 Sources of Mercury in the Environment

Mercury can cycle in the environment as part of both natural and anthropogenic activities. Results to date indicate that the amount of Hg mobilized and released into the biosphere has increased since the beginning of the industrial age.

Several types of emission sources contribute to the total atmospheric loading of Hg (U.S. EPA, 1996). Once in the air, Hg can be widely dispersed and transported thousands of miles from emission sources. Studies indicate that the residence time of Hg in the atmosphere may be on the order of a year, allowing its distribution over long distances, both regionally and globally, before being deposited back to earth. Even after it deposits, Hg commonly is emitted back to the atmosphere either as a gas or in association with particulates to be redeposited elsewhere. Mercury undergoes a series of complex chemical and physical transformations as it cycles among the atmosphere, land, and water. Humans, plants, and animals are routinely exposed to Hg and accumulate it during this cycle, potentially resulting in a variety of ecological and human health impacts.

Three sources of methyl Hg to aquatic systems appear important: (1) precipitation, (2) in-lake methylation; and (3) runoff from wetlands (Downs et al., 1998). The contribution of methyl Hg in runoff after atmospheric deposition becomes more significant in locations with large catchment/lake or river area ratios and where atmospheric inputs are high. Concentrations of methyl Hg have been shown to increase in streams with increasing wetland areas and were further enhanced during intense precipitation events in the summer months.

Sources of Hg emissions in the United States are ubiquitous. The types of emission are (1) Natural Hg emissions—the mobilization or release of geologically bound Hg by natural processes, with mass transfer of Hg to the atmosphere; (2) Anthropogenic Hg emissions—the mobilization or release of geologically bound Hg by human activities, with mass transfer of Hg to the atmosphere; and (3) Re-emitted Hg—the mass transfer of Hg to the atmosphere by biological and geological processes drawing on a pool of Hg that was deposited on the earth's surface after initial mobilization by either anthropogenic or natural activities.

Releases from human activities today are adding to the Hg reservoirs that already exist in land, water, and air both naturally and as a result of previous human activities. Given the present understanding of the global Hg cycle, the flux of Hg from the atmosphere to land or water at any one location is

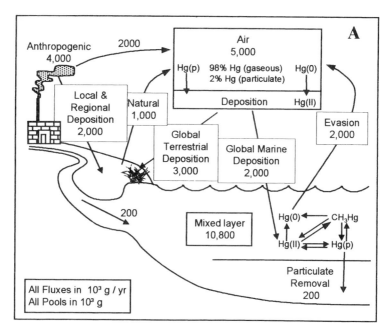

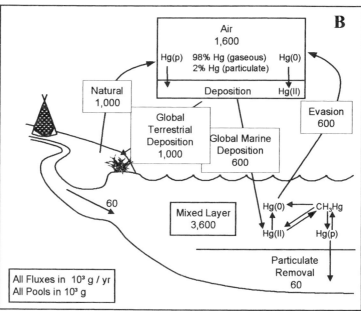

FIGURE 11.13. Comparison of (A) current and (B) preindustrial mercury budgets and fluxes. (Reprinted from *Geochimica et Cosmochimica Acta*, Mason et al., copyright © 1994, with permission from Elsevier Science.)

composed of contributions from the following: natural global cycle, global cycle perturbed by human activities, regional sources, and local sources.

A good example of a regional source of Hg is the widespread use of municipal sewage sludge on land in the United States. Routine application of municipal sludge to croplands has significantly increased both total and methyl Hg in surface soil from 80 to 6100 μg kg^{-1} and 0.30 to 8.3 μg kg^{-1}, respectively (Carpi et al., 1997). Using a dispersion model, sludge-amended soils in the United States may increase regional atmospheric methyl Hg concentrations by about 5%.

Knowledge of the global Hg cycle has improved significantly with sustained study of source emissions, Hg fluxes to the earth's surface, and the magnitude of Hg reservoirs that have accumulated in soils, watersheds, and ocean waters. It has become increasingly evident that anthropogenic emissions of Hg to the air may rival or exceed natural inputs. Recent estimates indicate that annual amounts of Hg released into the air by human activities are between 50 and 75% of the total yearly input to the atmosphere from all sources. Estimates indicate that of the approximately 200,000 tonnes of Hg emitted to the atmosphere since 1890, ~95% resides in terrestrial soils, ~3% in the ocean surface waters, and 2% in the atmosphere (U.S. EPA, 1996). It is believed that < 50% of the oceanic emission is from Hg originally mobilized by natural sources. Similarly, a potentially large fraction of terrestrial and vegetative emissions consists of recycled Hg from previously deposited anthropogenic and natural emissions.

Comparison of contemporary and historical records indicates that the total global atmospheric Hg burden has increased since the beginning of the industrialized period by a factor of between 2 and 5 (Fig. 11.13), where anthropogenic emissions now well exceed natural emissions, with 70 to 80% of total Hg in the atmosphere originating from man-made sources (Mason et al., 1994). Therefore, atmospheric deposition will constitute a significant source of Hg to aquatic systems in the absence of point-source discharges. For example, analyses of sediments from Swedish lakes show Hg concentrations in upper layers that are 2 to 5 times higher than those associated with preindustrialized times. Indeed, the deposition of atmospheric Hg may have doubled since the beginning of 19th century based on analyses of dated soil, peat bog, and lake sediments (Slemr and Langer, 1992). Thus, ample evidence exists to indicate that, at present, anthropogenic rather than natural sources are more important in the global Hg cycle.

References

Adriano, D.C. 1986. *Trace Elements in the Terrestrial Environment*. Springer-Verlag, New York.

Adriano, D.C., A.L. Page, A.A. Elseewi, A.C. Chang, and I. Straughan. 1980. *J Environ Qual* 9:331–344.

Alberts, J.J., J.E. Schindler, R.W. Miller, and D.E. Nutter, Jr. 1974. *Science* 184:895–897.

Allen, H.E., S.Z. Lee, C.P. Huang, and D.L. Sparks. 1994. *Final report to NJ Dept Environmental Protection*, Trenton, NJ.

Andersson, A. 1967. *Oikos Suppl* 9:13–15.

Andersson, A. 1970. *Grundförbättring* 23:31–40.

Andersson, A. 1979. In J.O. Nriagu, ed. *The Biogeochemistry of Mercury in the Environment*. Elsevier, Amsterdam.

Andersson, A., and L. Wiklander. 1965. *Grundförbättring* 18:171–177.

Aomine, S., and K. Inoue. 1967. *Soil Sci Plant Nutr* 13:195–200.

Aomine, S., H. Kawasaki, and K. Inoue. 1967. *Soil Sci Plant Nutr* 13:l86–188.

Archer, F.C. *1980. Minist Agric Fish Food* (Great Britain) 326:184–190.

[ATSDR] Agency for Toxic Substances and Disease Response. 1992. *Toxicological Profile for Mercury*. U.S. Dept Health and Human Services, Atlanta, GA.

Bache, C.A., W.H. Gutenmann, L.E. St. John, Jr., et al. 1973. *J Agric Food Chem* 21:607–613.

Bailey, E.A., and J.E. Gray. 1997. In J.A. Dumoulin and J.E. Gray, eds. *Geologic Studies in Alaska by the U.S. Geological Survey, 1995*. U.S. Geological Survey (USGS) Prof Paper 1574. USGS, Washington, DC.

Barrow, N.J., and V.C. Cox. 1992. *J Soil Sci* 43:305–312.

Beauford, W., J. Barber, and A.R. Barringer. 1977. *Physiol Plant* 39:261–265.

Beckert, W.F., A.-A. Moghissi, F.H.F. Au, E.W. Bretthauer, and J.C. McFarlane. 1974. *Nature* 249:674–675.

Beijer, K., and A. Jernelöv. 1979. In J.O. Nriagu, ed. *The Biogeochemistry of Mercury in the Environment*. Elsevier, Amsterdam.

Berrow, M.L., and G.A. Reaves. 1984. In *Proc Intl Conf Environmental Contamination*. CEP Consultants, Edinburgh.

Bertilsson, L., and H.Y. Neujahr. 1971. *Biochemistry* 10:2805–2808.

Bisogni, J.J., Jr. 1979. In J.O. Nriagu, ed. *The Biogeohemistry of Mercury in the Environment*. Elsevier, Amsterdam.

Bloom, N. 1992. *Can J Fish Aquat Sci* 49:1010–1017.

Bolli, M. 1947. *Ann Fac Agrar Univ Stud Perugia* 4:l80–186, cited by Beauford et al. (1977).

Booer, J.R. 1944. *Ann Appl Biol* 31:340–359.

Booer, J.R. 1951. *Ann Appl Biol* 38:334–347.

Bouton, S.N., P.C. Frederick, M.G. Spalding, and H. McGill. 1999. *Environ Toxicol Chem* 18:1934–1939.

Braman, R.S., and D.L. Johnson. 1974. In E.D. Copenhauer, ed. *Proc NSF RANN Trace Contaminants in the Environment*. Oak Ridge National Laboratory, Oak Ridge, TN.

Brookins, D.G. 1988. *Eh–pH Diagrams for Geochemistry*. Springer-Verlag, New York.

Carpi, A., S.E. Lindberg, E.M. Prestbo, and N.S. Bloom. 1997. *J Environ Qual* 26:1650–1655.

Chen, J., F. Wei, Y. Wu, and D.C. Adriano. 1991. *Water Air Soil Pollut* 57–58:699–712.

Chino, M. 1981. In K. Kitagishi and I. Yamane, eds. *Heavy Metal Pollution in Soils of Japan*. Japan Sci Soc Pr, Tokyo.

Choi, M.H., J.H. Cech, and M. Laguras-Solar. 1998. *Environ Toxicol Chem*. 17:695–701.

Choi, S.C., and R. Bartha. 1994. *Bull Environ Contam Toxicol* 53:805–812.

[CEC] Comm. European Communities. 1993. *Off J Eur Comm*. L144:23 (93/351/EEC).

Cox, P.A. 1995. *The Elements on Earth*. Oxford Univ Pr, New York.

Davis, R.D., P.H.T. Beckett, and E. Wollan. 1978. *Plant Soil* 49:395–408.

DeSimone, R.E. 1972. *J Chem Commun* (Chem Soc London) 13:780–781.

Downs, S.G., C.L. Macleod, and J.N. Lester. 1998. *Water Air Soil Pollut* 108: 149–187.

Drake, H.J. 1980. In *Minerals Yearbook, 1978–79*. U.S. Bureau Mines, U.S. Dept Interior, Washington, DC.

Dudas, M.J., and S. Pawluk. 1976. *Can J Soil Sci* 56:413–423.

Ebinghaus, R., R. Hinstelmann, and R.D. Wilken. 1994. *F J Anal Chem* 350:21–29.

Engler, R.M., and W.H. Patrick, Jr. 1975. *Soil Sci* 119:217–221.

Estes, G.O., W.E. Knoop, and F.D. Houghton. 1973. *J Environ Qual* 2:451–452.

Evans, R.L., W.T. Sullivan, and S. Lin. 1973. *Water Sewage Works* 120:74–76.

Evers, D.C., J.D. Kaplan, M.W. Meyer, and A.M. Scheuhammer. 1998. *Environ Toxicol Chem* 17:173–183.

Fang, S.C. 1978. *Environ Sci Technol* 12:285–288.

Fang, S.C. 1981. *Arch Environ Contam Toxicol* 10:193–201.

FAO/WHO. See WHO.

Farrah, H., and W.F. Pickering. 1978. *Water Air Soil Pollut* 9:23–31.

Farrell, R.E., J.J. Germida, and P.M. Huang. 1993. *Appl Environ Microbiol* 59:1507–1574.

Fleischer, M. 1970. In *Mercury in the Environment*. U.S. Geological Survey (USGS) Prof Paper 713. USGS, Washington, DC.

Frank, R., K. Ishida, and P. Suda. 1976. *Can J Soil Sci* 56:181–196.

Frear, D.E.H., and L.E. Dills. 1967. *J Econ Entom* 60:970–974.

Frederick, P.C., M.J. Spalding, M.S. Sepulveda, and R. Robins. 1999. *Environ Toxicol Chem* 18:1940–1947.

Gambrell, R.P., R.A. Khalid, and W.H. Patrick, Jr. 1980. *Environ Sci Technol* 14:431–436.

Gavis, J., and J.F. Ferguson. 1972. *Water Res* 6:989–1008.

Gill, G.A., and K.W. Bruland. 1990. *Environ Sci Technol* 24:1392–1400.

Gilmour, C.G., and D.G. Capone. 1987. *EOS Trans Geophys Union* 68:1718–1725.

Gilmour, C.G., and E.A. Henry. 1991. *Environ Pollut* 71:131–169.

Gilmour, J.T. 1971. *Environ Letters* 2:143–152.

Gilmour, J.T., and M.S. Miller. 1973. *J Environ Qual* 2:145–148.

Glass, G.E., J.A. Sorensen, K.W. Schmidt, and G.R. Rapp. 1990. *Environ Sci Technol* 24:1059–1069.

Goldschmidt, V.M. 1954. *Geochemistry*. Oxford Univ Pr, London.

Gotoh, S., S. Tokudome, and H. Koga. 1978. *Soil Sci Plant Nutr* 24:391–406.

Gotoh, S., H. Otsuka, and H. Koga. 1979. *Soil Sci Plant Nutr* 25:523–537.

Gowen, J.A., G.B. Wiersma, and H. Tai. 1976. *Pesticides Monit J* 10:111–113.

Gracey, H.I., and J.W.B. Stewart. 1974a. In *Proc Conf Land Waste Management*. Ottawa, Ontario, Canada.

Gracey, H.I., and J.W.B. Stewart. 1974b. *Can J Soil Sci* 54:105–108.

Gray, N.E., and H.J. Fuller. 1942. *Am J Bot* 29:456–459.

Hahne, H.C.H., and W. Kroontje. 1973a. *J Environ Qual* 2:444–450.

Hahne, H.C.H., and W. Kroontje. 1973b. *Soil Sci Soc Am Proc* 37:838–843.

Hamada, R., and M. Osame. 1996. In L.W. Chang, ed. *Toxicology of Metals*. CRC Pr, Boca Raton, FL.

Harriss, R.C., D.B. White, and R.B. MacFarlane. 1970. *Science* 170:736–737.

Hogg, T.J., J.W.B. Steward, and J.R. Bettany. 1978a. *J Environ Qual* 7:440–444.

Hogg, T.J., J.W.B. Steward, and J.R. Bettany. 1978b. *J Environ Qual* 7:445–450.

Huckabee, J.W., and B.G. Blaylock. 1973. *Adv Exp Med Biol* 40:125–160.

Imura, N., E. Sukegawa, S. Pan, K. Nagae, J. Kim, T. Kwan, et al. 1971. *Science* 172:1248–1249.

Jagoe, C.H., A. Faivre, and M.C. Newman. 1996. *Aquat Toxicol* 34:163–183.

Jay, J.A., F.M.M. Morel, and H.F. Hemond. 2000. *Environ Sci Technol* 34:2196–2200.

Jones, R.L., and F.D. Hinesly. 1972. *Soil Sci Soc Am Proc* 36:921–923.

Kaiser, G., and G. Tölg. 1980. In O. Hutzinger, ed. *Handbook of Environmental Chemistry*, vol. 3, part A. Springer-Verlag, New York.

Kendorff, H., and M. Schnitzer. 1990. *Geochim Cosmochim Acta* 44:1701–1708.

King, J.K., F.M. Saunders, R.F. Lee, and R.A. Jahnke. 1999. *Environ Toxicol Chem* 18:1362–1369.

Kinniburgh, D.G., and M.L. Jackson. 1978. *Soil Sci Soc Am J* 42:45–47.

Lacerda, L.D. 1995. *Nature* 374:20.

Lacerda, L.D., and W. Salomons. 1998. *Mercury from Gold and Silver Mining: A Chemical Time Bomb?* Springer-Verlag, Berlin.

Lacerda, L.D., W.C. Pfeiffer, A.T. Ott, and B.G. Silveira. 1989. *Biotropica* 21:91–93.

Lacerda, L.D., R.V. Marins, C.M. Souza, S. Rodrigues, W.C. Pfeiffer, and W.R. Bastos. 1991a. *Water Air Soil Pollut* 55:283–294.

Lacerda, L.D., W. Salomons, W.C. Pfeiffer, and W.R. Bastos. 1991b. *Biogeochemistry* 14:91–97.

Lacerda, L.D., E.D. Bidone, A.F. Guimaraes, and W.C. Pfeiffer. 1994. *An Acad Bras Ci* 66:373–379.

Landa, E.R. 1978. *J Environ Qual* 84–86.

Lexmond, T.M., F.A.M. de Haan, and M.J. Frissel. 1976. *Netherlands J Agric Sci* 24:79–97.

Lindberg, S.E., and R.C. Harriss. 1974. *Environ Sci Technol* 8:459–462.

Lindberg, S.E., D.R. Jackson, J.W. Huckabee, S.A. Janzen, M.J. Levin, and J.R. Lund. 1979. *J Environ Qual* 8:572–578.

Lockhart, W.L., P. Wilkinson, B.N. Billeck, R.V. Hunt, R. Wagemann, and G.J. Brunskill. 1995. *Water Air Soil Pollut* 80:603–610.

Lockwood, R.A., and K.Y. Chen. 1973. *Environ Sci Technol* 7:1028–1034.

Lockwood, R.A., and K.Y. Chen. 1974. *Environ Lett* 6:151–166.

Löfroth, G. 1970. In *Methylmercury: A Review of Health Hazards and Side Effects Associated with the Emission of Mercury Compounds into Natural Systems*. Ecol Res Comm Bull 4. Swed Natl Sci Res Council, Stockholm, Sweden.

Magalhaes, M.E.A., and M. Tubino. 1995. *Sci Total Environ* 170:229–239.

Martinez-Cortizas, A., X. Panteredra-Pombal, E. Garcia-Rodeja, J.C. Novoa-Munoz, and W. Shotyk. 1999. *Science* 284:939–942.

Mason, R.P., W.F. Fitzgerald, and F.M.M. Morel. 1994. *Geochim Cosmochim Acta* 58:3191–3198.

McKeague, J.A., and B. Kloosterman. 1974. *Can J Soil Sci* 54:503–507.

McKeague, J.A., and M.S. Wolynetz. 1980. *Geoderma* 24:299–307.

Meyer, M.W., D.C. Evers, T. Daulton, and W.E. Braselton. 1995. *Water Air Soil Pollut* 80:871–880.

Mhatre, G.N., and S.B. Chaphekar. 1984. *Air Soil Pollut* 21:1–8.

Mills, J.G., and M.A. Zwarich. 1975. *Can J Soil Sci* 55:295–300.

Miskimmin, B.M., J.W.M. Rudd, and C.A. Kelly. 1992. *Can J Fish Aquat Sci* 49:17–22.

Moore, T.R., J.L. Bubier, A. Heges, and R.J. Flett. 1995. *J Environ Qual* 24:845–850.

Morishita, T., K. Kishino, and S. Idaka. 1982. *Soil Sci Plant Nutr* 28:523–534.

Nater, E.A., and D.F. Grigal. 1992. *Nature* 358:139–141.

[NAS] National Academy of Sciences. 1977. *Drinking Water and Health*. NAS, Washington, DC.

[NAS] National Academy of Sciences. 1978. *An Assessment of Mercury in the Environment*. NAS, Washington, DC.

[NRCC] National Research Council Canada. 1979. *Effects of Mercury in the Canadian Environment* NRCC 16739. NRCC, Ottawa.

Newton, D.W., R. Ellis, Jr., and G.M. Paulsen. 1976. *J Environ Qual* 5:251–254.

Olson, B.H., and R.C. Cooper. 1976. *Water Res* 10:113–116.

Paquette, K.E., and G.R. Helz. 1997. *Environ Sci Technol* 31:2148–2153.

Pfeiffer, W.C., L.D. Lacerda, O. Malma, C.M. Souza, E.G. Silveira, and W.R. Bastos. 1989. *Sci Total Environ* 7/88:233–240.

Pickard, J.A., and J.T. Martin. 1959. In *Annual Rept Long Ashton Res Sta, Bristol, UK,* as cited by Smart (1964).

Poelstra, P., M.J. Frissel, N. Van der Klugt, and W. Tap. 1974. In *Comparative Studies of Food and Environmental Contamination*. Proc Series IAEASM-175/46. IAEA, Vienna.

Reimers, R.S., and P.A. Krenkel. 1974. *J Water Pollut Control Fed* 46:352–365.

Rogers, R.D. 1976. *J Environ Qual* 5:454–458.

Rogers, R.D. 1977. *J Environ Qual* 6:463–467.

Rogers, R.D. 1979. *Soil Sci Soc Am J* 43:289–291.

Rogers, R.D., and J.C. McFarlane. 1979. *J Environ Qual* 8:255–260.

Rose, J., M.S. Hutchenson, C.R. West, and A. Screpetis. 1999. *Environ Toxicol Chem* 18:1370–1379.

Ross, R.G., and D.K.R. Stewart. 1960. *Can J Plant Sci* 40:117–122.

Ross, R.G., and D.K.R. Stewart. 1962. *Can J Plant Sci* 42:280–285.

Roulet, M., M. Lucotte, N. Farella, and M. Amorim. 1999. *Water Air Soil Pollut* 112:297–313.

Russell, L.H. Jr. 1979. In F.W. Oehme, ed. *Toxicity of Heavy Metals in the Environment,* Marcel Dekker, New York.

Saha, J.G., Y.W. Lee, R.D. Tinline, S.H.F. Chinn, and H.M. Austenson. 1970. *Can J Plant Sci* 50:597–599.

Sand, P.F., G.B. Wiersma, H. Tai, and L.J. Stevens. 1971. *Pesticides Monit J* 5:32–33.

Sarkar, D., M.E. Essington, and K.C. Misra. 1999. *Soil Sci Soc Am J* 63:1626–1636.

Scheuhammer, A.M. 1991. *Environ Pollut* 71:329–375.

Scheuhammer, A.M., A.H.K. Wong, and D. Bond. 1998. *Environ Toxicol Chem* 17:197–201.

Selikoff, I.J. 1971. *Environ Res* 4:1–69.

Sepulveda, M.S., P.C. Frederick, M.G. Spalding, and G.E. Williams. 1999. *Environ Toxicol Chem* 18:985–992.

Shacklette, H.T. 1970. In *Mercury in the Environment*. U.S. Geological Survey (USGS) Prof Paper 713. USGS, Washington, DC.

Shacklette, H.T., J.G. Boerngen, and R.L. Turner. 1971. U.S. Geological Survey (USGS) Circ 644. USGS, Washington, DC.

Siegel, B.Z., M. Lasconia, E. Yaeger, and S.M. Siegel. 1984. *Water Air Soil Pollut* 23:15–24.

Siegel, S.M., B.Z. Siegel, A.M. Eshleman, and K. Bachmann. 1973. *Environ Biol Med* 2:81–89.

Slemr, F., and E. Langer. 1992. *Nature* 355:434–437.

Smart, N.A. 1964. *J Sci Food Agric* 15:102–108.

Smart, N.A. 1968. *Residue Rev* 23:1–36.

Somers, E. 1974. *J Food Sci* 39:215–217.

Spalding, M.G., R.D. Bjork, G.V. Powell, and S.F. Sundlof. 1994. *J Wildl Manage* 58:735–739.

Stewart, D.K.R., and R.G. Ross. 1960. *Can J Plant Sci* 40:659–665.

Summers, A.O., and S. Silver. 1978. *Ann Rev Microbiol* 32:617–672.

Toribara, F.Y., C.P. Shields, and L. Koval. 1970. *Talanta* 17:1025–1028.

U.S. Bureau of Mines. 1994. *Minerals Yearbook*. U.S. Dept Interior, Washington, DC.

[U.S. EPA] United States Environmental Protection Agency. 1976. *Quality Criteria for Water*. U.S. EPA, Washington, DC.

U.S. EPA. 1994a. *Integrated Risk Information System (IRIS) on Mercury*. Office of Health and Environ Assess, Cincinnati, OH.

U.S. EPA. 1994b. *Deposition of Air Pollutants to the Great Waters*. Office of Health and Environ Assess, Cincinnati, OH.

U.S. EPA. 1996. In *Mercury: Study Report to Congress, vol 1: Executive Summary*. US-EPA/R-96-001a, NTIS, Springfield, VA.

Vaidya, O.C., G.D. Howell, and D.A. Leger. 2000. *Water Air Soil Pollut* 117:353–369.

Van Faasen, H.G. 1975. *Plant Soil* 44:505–509.

Veiga, M.M., J.A. Meech, and N. Ovate. 1994. *Nature* 368:816–817.

Wang, J.S., P.M. Huang, W.K. Liaw, and U.T. Hammer. 1991. *Water Air Soil Pollut* 56:533–542.

Warren, H.V., and R.E. Delavault. 1969. *Oikos* 20:537–539.

Warren, H.V., R.E. Delavault, and J. Barakso. 1966. *Econ Geol* 61:1010–1028.

Weaver, R.W., J.R. Melton, D. Wang, and R.L. Duble. 1984. *Environ Pollut* (A) 33:133–142.

Wheatley, B., and S. Paradis. 1995. *Water Air Soil Pollut* 80:3–11.

[WHO] World Health Organization. 1976. *Environmental Health Criteria 1: Mercury*. WHO, Geneva.

WHO. 1984. *Guidelines for Drinking-Water Quality, vol 1:* Recommendations. WHO, Geneva.

WHO. 1990. *Environmental Health Criteria 101: Methylmercury*. WHO, Geneva.

Wiersma, D., B.J. Van Goor, and N.G. Vander Veen. 1986. *J Agric Food Chem* 34:1067–1074.

Wiersma, G.B., and H. Tai. 1974. *Pesticides Monit J* 7:214–216.

Wiklander, L. 1969. *Geoderma* 3:75–79.

Winfrey, M.R., and J.W.M. Rudd. 1990. *Environ Toxicol Chem* 9:853–861.

Wolfe, M.F., S. Schwarzbach, and R.A. Sulaiman. 1998. *Environ Toxicol Chem* 17:146–160.

Wood, J.M. 1974. *Science* 183:1049–1052.

Wood, J.M., F.S. Kennedy, and C.G. Rosen. 1968. *Nature* 220:173–174.

Wren, C.D. 1995. In D.J. Hoffman et al., eds. *Handbook of Ecotoxicology*. Lewis Publ, Boca Raton, FL.

Yin, Y., and H.E. Allen. 1997. In *Proceedings 4th Intl Conf on Biogeochemistry of Trace Elements,* June 1997, Berkeley, CA. Univ California, Berkeley, CA.

Yin, Y., H.E. Allen, Y. Li, C.P. Huang, and P.F. Sanders. 1996. *J Environ Qual* 25:837–844.

Yin, Y., H.E. Allen, C.P. Huang, and P.F. Sanders. 1997. *Environ Toxicol Chem* 16:2457–2462.

Yin, Y., H.E. Allen, C.P. Huang, D.L. Sparks, and P.F. Sanders. 1997. *Environ Sci Technol* 31:496–503.

Zyrin, N.G., B.A. Zvonarev, L.K. Sadovnikova, and N.I. Voronova. 1981. *Sov Soil Sci* 13:44–52.

12
Boron

1 General Properties of Boron

Boron belongs to Group III-A of the periodic table and is the only nonmetal among plant micronutrients. It has an atom. wt. of 10.81, a melting pt. of 2080 °C, with specific gravity of 2.34 for crystals. Boron has two stable isotopes in nature, ^{10}B (19.78%) and ^{11}B (80.22%). At room temperature, B is inert except to strong oxidizing agents such as HNO_3. When fused with oxidizing alkaline mixtures such as NaOH and $NaNO_3$, it forms borates. The only important oxide is boric oxide (B_2O_3), which is acidic, soluble in water, and forms boric acid $B(OH)_3$, a very weak acid. In nature, B is fairly rare and occurs primarily as the borates of Ca and Na. Borax ($Na_2B_4O_7 \cdot 10H_2O$) is the most common compound along with boric or boracic acid. By far the most important source of B is the mineral kernite ($Na_2B_4O_7 \cdot 4H_2O$) (also known as rasorite), an evaporite deposit found in the Mojave desert of California. Boron does not form B^{3+} cations because of its high ionization potential; thus, B complexes mainly involve covalent rather than ionic bonds. Boron chemistry more closely resembles Si than Al chemistry. Boron may be entrapped in the clay lattice by substituting for Al^{3+} and/or Si^{4+} ions. There are numerous borosilicate minerals in soils, but only tourmaline and axinite may be of significance. Boric acid has a trigonal structure while borate anion has a tetrahedral structure.

2 Production and Uses of Boron

World production of B_2O_3 in 1988 was about 1.3×10^6 tonnes, with about 58% (0.75×10^6 tonnes) of this total produced by the United States (Mortvedt and Woodruff, 1993). Other major B-producing countries include Turkey (37% of total), the former Soviet Union, and Argentina. Known world reserves exceed 91×10^6 tonnes of B (292×10^6 tonnes of B_2O_3) and reserves in the United States were estimated to be 210×10^6 tonnes of ore.

Domestically, California (Mojave desert) is the chief source of B minerals, which are mostly in the form of sodium borate, but also sometimes as calcium borate (colemanite) and sodium–calcium borates (ulexite). Recently, insulation products, fire retardants, and glass fiber–reinforced plastics are the most important consuming sectors of B (mostly borax pentahydrate and ulexite-probertite). Borates (colemanite, orthoboric acid, and anhydrous boric acid) are also extensively used in the manufacture of textile-grade glass fibers for use in aircraft, autos, and sports equipment.

Borates are also used in the manufacture of special borosilicate glasses (glasses, pottery, and enamels) and in cleaning and bleaching compounds, especially in the production of sodium perborate detergents. Boron compounds are used in the manufacture of biological growth control agents for use in water treatment, algicides, herbicides, insecticides, and also in fertilizers since B is a plant micronutrient. Other uses of B compounds include metallurgical work and electroplating, and the nuclear industry. Boric acid is useful medicinally as a mild antiseptic, especially as an eyewash. It was used as a food preservative in the past.

Boron is very important technologically because of the applications of B compounds to synthetic organic chemistry. Organoborane compounds are used as chemical intermediates in the manufacture of drugs as well as in new classes of pesticides.

3 Boron in Nature

Readily available sodium and calcium borates occur naturally in most soils. These come from the very slow dissolution of tourmaline, a mineral that contains about 3 to 4% B. Tourmaline $[H_2MgNa_9Al_3(BO)_2Si_4O_{20}]$ is present in soils formed from acid rocks and metamorphosed sediments. Because of its slow dissolution, it is not a very good source of B for plants (Graham, 1957). Boron can substitute for tetrahedrally coordinated Si in some minerals and it is likely that much of the B in rocks and soils is dispersed in the silicate minerals in this manner (Norrish, 1975).

The average concentration of total B in the earth's crust is about 10 ppm; it ranks 37th in abundance among the elements (Krauskopf, 1979). Concentrations in igneous rocks are in the same range but increase several times for sedimentary rocks (Table 12.1). Soils derived from marine clays and shales generally have sufficient B for plant production, since B and its compounds are soluble and accumulative in seawater. Some sedimentary rocks of marine origin can contain 500 ppm B or more (Aubert and Pinta, 1977). In Belorussia and Amur region of the former Soviet Union, B levels in the range of 35 to 70 ppm were found in glacial clays, lacustrine alluvium, and stratified plain deposits. However, low levels of around 2 ppm B were reported for ancient alluvial sands in these regions. As a general rule, marine clays contain more B than clays accumulating in lakes or floodplains.

In general, the B contents of soils are higher than the B contents of the rocks. The B content of phyllosilicate clay minerals is distinctly higher than that of other minerals (Goldberg, 1993); illite can have 100 to 2000+ ppm B, followed by muscovite and montmorillonite, then by kaolinite, chlorite, and biotite (1 to 6 ppm) in decreasing order. Weathering of B-containing rocks yields borate in solution, chiefly as the nonionized boric acid $[B(OH)_3]$. Boric acid is a weak acid (pK $= 9.14$) noted for its volatility even at low temperature. In natural environments, B solutions contain primarily $B(OH)_3$ and $B(OH)_4^-$, and polymeric forms, such as tetraborate $B_4O_7^{2-}$, are likely to occur only in concentrated solutions.

In nature, B is found as a constituent of borax ($Na_2B_4O_7 \cdot 10H_2O$), kernite ($Na_2B_4O_7 \cdot 4H_2O$), colemanite ($Ca_2B_6O_{11} \cdot 5H_2O$), ulexite ($NaCaB_5O_9 \cdot 8H_2O$), tourmaline, and axinite.

TABLE 12.1. Commonly observed boron concentrations (ppm) in various environmental media.

Material	Mean (or Median)	Range
Basalt[b]	5	—
Granite[b]	15	—
Limestone[b]	20	—
Sandstone[b]	—	2–35
Shale[b]	100	—
Phosphorites[d]	<50	—
Soils[c]	—	2–100
Coal[c]	—	4–200
Petroleum	0.002	—
Lignite[d]	70	1–400
Fly ash[f]	—	10–1900
Animal manure	24	—
Hog manure	61	6–180[e]
Sewage sludge	—	20–2000
Fertilizers	—	1.2–382[e]
Common crops	—	20–100
Herbaceous vegetables[d]	—	8–200
Dolomitic lime	4	<1–21
Lichens[d]	4.1	—
Fungi[d]	16	—
Ferns[d]	77	—
Groundwaters	—	0.01–0.22[e]
Irrigation waters	—	<0.1–0.3
Freshwater[d]	15[a]	7–500[a]
Seawater[d]	4440	—

Sources:
[a]µg L^{-1}.
[b]Krauskopf (1972).
[c]Trudinger et al. (1979).
[d]Bowen (1979).
[e]Komor (1997).
[f]Adriano and Weber (1998).

4 Boron in Soils

The biogeochemical pathways of B in the soil–plant system is graphically represented on Figure 12.1. It shows the various solid constituents of the soil that can retain B against leaching or plant uptake.

4.1 Total Boron in Soils

The total B contents of normal soils range from 2 to 100 ppm, with an average of about 30 ppm (Swaine, 1955). The total B content of soils depends largely on the parent material. The relationships between soil content and parent rock have been reported for U.S. soils (Whetstone et al.,

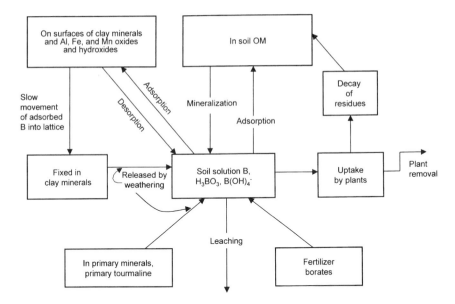

FIGURE 12.1. Schematic representation of the biogeochemical cycling of boron in the soil–plant system. (Modified from Mengel, 1980.)

1942). An average content of 30 ppm ($n = 200$) and a range of 4 to 98 ppm were found. Average values of 14 and 40 ppm were found for soils formed from igneous and sedimentary rocks, respectively. The highest B values occurred in arid saline soils; fine-textured humid soils had 30 to 60 ppm B, whereas sandy soils had as low as 2 to 6 ppm. Soils of marine shale origin were particularly high in B. Similar relationships were noted in soils of California. Soils weathered from alluvium eroded from marine shales in Kern County had B contents ranging from 25 to 68 ppm, whereas soils from granitic material contained 10 ppm B or less (Bingham et al., 1970a). More recently, a national survey indicated that surface soils of the conterminous United States have an average B content of 33 ppm (geometric mean = 26; range of <20 to 300 ppm).

Low B content can be expected of soils derived from acid igneous rocks, freshwater sedimentary deposits, and in coarse-textured soils low in OM. Data for total B content of soils are available from several countries. The average B content of Russian soils including all horizons but excluding salinized southern soils is 6 ppm (Vinogradov, 1959). However, upon inclusion of the salinized southern soils, the average becomes 50 ppm. In the former Soviet Union, total B contents are highest in soils of the dry steppe, semi-desert, and desert zones compared with soils in the central region, where levels of 200 to 400 ppm are not uncommon (Il'in and Anikina, 1974). Soils of China have higher B contents ($x = 68$ ppm, geometric mean = 39, $n = 834$) than those of other countries (Chen et al., 1991).

The highest content of total B was found in the Himalayan region where soils were of marine sediment origin. However, tropical Asian paddy soils have a mean B content of 96 ppm, with a high of 197 ppm for Burmese soils and a low of 52 ppm for Indonesian soils (Domingo and Kyuma, 1983). Soils of central and northern Europe reportedly have lower B contents than soils of the United States, especially the western and central states. Soils of England and Wales have B contents in the range of 7 to 71 ppm, with a median of 33 ppm (Archer, 1980). A number of soils from eastern Canada have total B in the range of 45 to 124 ppm (Gupta, 1968). Soils from Nigerian cacoa-growing areas contain 8 to 54 ppm total B with some areas showing deficiency levels (Chude, 1988).

4.2 Bioavailable Boron in Soils

Total B is an unreliable index of the bioavailability of B in soils because of the insolubility of indigenous B. Consequently, the extractable B is the universal form of this element used for diagnostic purposes. Hot water-soluble B is generally regarded as the best index of phytoavailability of B (Sims and Johnson, 1991). Only a small fraction of total B, usually <5%, is found in extractable form (Berger and Truog, 1940). The procedure for determining bioavailable soil B, extracted by boiling a 1:2 soil/water suspension for 5 min and then filtering it, has basically remained the same since Berger and Truog (1939) adopted it.

Extraction of some Australian soils by shaking for 1 hr at 20 °C with 0.01 M $CaCl_2$ + 0.05 M mannitol was a more effective soil test for plant-available B than the standard hot water-soluble method and proved to be as reliable in predicting the response of B uptake by plants (Cartwright et al., 1983). However, this method can be expected to extract less B from soils having low B contents but more B from soils having potentially toxic concentrations than the hot water-soluble method. In the U.S., the average water-soluble B contents of Coastal Plain and Piedmont soils were 0.24 and 0.27 ppm, respectively (Piland et al., 1944). Water-extractable B in several cropped Michigan soils ranged from 0.16 to 0.95 ppm, with the highest levels due to recent B fertilization (Robertson et al., 1975). Soils in Georgia had 0.01 to 0.65 ppm water-soluble B and Illinois soils had 0.20 to 1.22 ppm (DeTurk and Olson, 1941).

In Colorado, soils having hot water-soluble B levels between 0.10 and 6.47 ppm did not produce any deficient or toxic symptoms in greenhouse-grown alfalfa (Gestring and Soltanpour, 1984). Water-soluble B in soils of six geographically diverse areas of Egypt had 0.30 to 3.4 ppm B, with an average of 1.3 ppm (Elseewi and Elmalky, 1979). In eastern Canada, hot water-soluble B ranged from 0.38 to 4.67 ppm (Gupta, 1968). In England and Wales, hot water-soluble B ranged from 0.1 to 4.7 ppm with a median of 1.0 ppm (Archer, 1980).

Another water-soluble form of B is determined from soil saturation extract. The values obtained by this procedure are usually lower than the hot water-soluble B, are more variable, and generally give narrower ranges, thus making the test more difficult to interpret. For example, an average of 0.50 ppm with saturation extracts vs. 0.74 ppm B with hot water extract was obtained by Robertson et al. (1975). Similar results were observed for Egyptian soils where B from saturated extracts averaged only 0.6 ppm vs. 1.3 ppm from the other method (Elseewi and Elmalky, 1979).

The importance of levels of hot water-soluble soil B in plant nutrition and potential disorders are demonstrated in Figure 12.2. Plant uptake of B can be predicted from water-soluble B in soil (Fig. 12.2B) as can the incidence of internal damage due to B deficiency (Fig. 12.2A). In general, highly significant positive correlations between plant B concentration and hot water-soluble B concentration in soils can be expected.

4.3 Boron in Soil Profile

In general, concentrations of total B are usually higher in the surface horizon of soil profiles than in the underlying layers, partly because occurrence of B is often associated with OM. Higher OM content usually occurs in the surface layer. Examination of 26 soil profiles in the Egyptian Nile Delta, divided into 30 cm increments up to a depth of 90 cm, revealed that total B (ranging from 25 to 336 ppm) usually was higher in the top 30 cm, but only occasionally high in the next 30 to 60 cm layer (Elseewi and Elmalky, 1979). On the average, total B was twice as high in clay than in sandy samples, and decreased by about 32% beyond the 60 cm depth of the profile. However, hot water-soluble B in saturation extracts was more or less evenly distributed in the soil profile. Additional studies of Egyptian soils near the Nile Delta but more toward the western desert showed that water-soluble B, even at much higher concentrations (1.8 to 5.7 ppm), was practically evenly distributed in the soil profile (0 to 90 cm) (Awad and Mikhael, 1980).

Others found different profile distributions of B. Soils supporting sugar beets in Michigan had water-soluble B ranging from nondetectable in the subsoil (120 cm and shallower) to 0.54 ppm in surface soil (0 to 15 cm) (Robertson et al., 1975). Similarly, in arable Polders region of the former Soviet Union, water-soluble B concentrations were usually higher in surface soil, decreasing to trace levels in the subsoil (Shirokov and Panasin, 1972); water-soluble B ranged from 0.13 to 6.80 ppm in surface soil and from trace to 9.84 ppm in subsoil. This region is below or close to sea level with mostly peat bog and alluvial bog soils influenced by marine processes. In the vineyard region of Georgia Republic, water-soluble B concentrations were higher (0.34 to 2.4 ppm) in surface soil (0 to 20 cm) than in the subsoil (Bagdasarashvili et al., 1974).

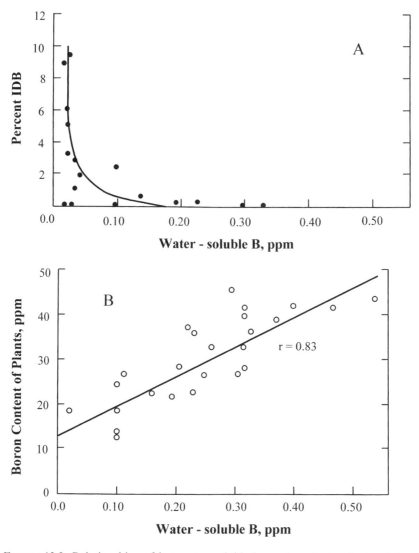

FIGURE 12.2. Relationships of hot water–soluble boron in soils to plant uptake of boron and to incidence of internal damage (IDB) from boron deficiency in peanuts. (From (A) Hill and Morrill, 1974; (B) Baird and Dawson, 1955.)

In saline–alkaline soils of arid and semiarid regions, B distribution in the soil profile is variable. In Coachella Valley of southern California, B concentration in the saturation extract decreased from 50 ppm in surface (0 to 15 cm) to 5 ppm B at the 120 to 150 cm depth (Reeve et al., 1955). In contrast, in the San Joaquin Valley of California, total B and saturation-extract B accumulated in the subsoil at much higher levels than in surface soil (Bingham et al., 1970a). In arid regions of India, the pattern is similarly inconsistent as in California. In the Haryana and Punjab regions, total B increased while water-soluble B decreased with depth (Singh, 1970). However, in the Rajasthan area the inverse pattern occurred in the profile distribution of B (Talati and Agarwal, 1974).

Most of saline–alkaline soils are unfit for agricultural production because of the presence of excessive salts, undesirable soil structural properties, and occasionally excessive B levels. Reclaiming these soils with leaching can render them productive since most of these salts move with the leaching water. In salt- and B-affected soils, about 30 cm of water or more maybe required per 30 cm of soil profile for mitigating salt content to tolerable levels; B removal may require three or more times water (Reeve et al., 1955), because B does not leach out of the profile as readily as chloride, nitrate, and sulfate salts.

Natural leaching from precipitation can also mobilize B down the soil profile. Leaching of applied B from the soil surface has been found by several investigators (Adriano, 1986) to be fairly rapid, especially in well-drained sandy soils. Winsor (1952) reported that more than 85% of the added B as borax was leached below the 105 cm depth in Florida's Lakeland sand after 40 cm of rainfall in 4 mo. Thus, it can be expected that B is higher in surface soils during dry periods and low during rainfall season.

4.4 Forms and Chemical Speciation of Boron in Soils

There are four main forms of B in soils: water-soluble, adsorbed, organically bound, and fixed in the clay and mineral lattice (see Fig. 12.1). The water-soluble fraction is regarded as a plant-response indicator and is determined either in a soil saturation or hot water extract. The latter is usually designated as bioavailable B. Adsorbed B represents the fraction that is bound onto surface of soil constituents and is in equilibrium with soluble B. Under normal conditions, field soils contain only small amounts of soluble and adsorbed B. Soils in arid and semiarid regions could be expected to contain from 5 to 16% of total B in water-soluble form (Aubert and Pinta, 1977). Soils from several countries were reported to contain about 10% of the total B in the water-soluble form (Vinogradov, 1959). In areas of high salinity, this form could rise to as much as 80% of total B. Soils in the Nile Delta of Egypt had water-soluble B ranging up to only 6.4% of total (Elseewi and Elmalky, 1979). Soils in the eastern United States had only minimal amounts of water-soluble B (from 0.02 to 0.34% of the total B) as well as $CaCl_2$-extractable B (from 0.05 to 0.23% of total B) (Table 12.2). Similarly, in the Russian central region, water-soluble B comprises only about 2 to 5% of the total B (Il'in and Anikina, 1974) rising to as much as 10% of the total in saline soils.

In soils having unusually high B such as the alluvial soils in Kern County, California, as much as 30% of the total B is in the adsorbed form (Bingham et al., 1970a). This fraction is considered to be easily leachable from the soil

TABLE 12.2. Fractionation data for native boron (ppm DW) in soils from several U.S. regions.

				B fraction				
					NH$_4$-oxalate extraction			
Soil series	Water soluable	CaCl$_2$ extraction	Mannitol extraction	NH$_2$OH · HCl extraction	Dark	UV light	Residue	Total
Coastal Plain region								
Craven	0.07	0.02	0.02	0.17	2.9	15.3	25.0	43.5
Dragston	0.03	<0.01	0.03	0.12	2.7	12.3	14.0	29.2
Emporia	0.18	0.10	0.06	0.27	2.4	19.0	31.4	53.3
Kemps-ville	<0.01	0.03	0.07	0.14	2.6	6.1	25.7	34.6
Myatt	0.04	<0.01	0.03	0.14	2.2	6.8	16.8	26.1
Rains	<0.01	0.09	0.06	0.17	1.8	9.2	26.9	38.3
Rumford	<0.01	0.04	0.03	0.10	2.8	10.1	8.5	21.5
Slagle	0.12	0.02	0.09	0.48	3.8	11.4	28.7	44.5
Tarboro	0.05	<0.01	0.02	0.06	3.5	8.0	13.0	24.7
Piedmont region								
Dyke	0.13	0.12	0.17	1.31	18.6	39.1	26.8	86.3
Fauquier	0.22	0.12	0.20	0.74	23.8	42.5	1.7	69.3
Starr	0.13	0.17	0.14	1.13	11.9	61.5	8.2	83.1
Ridge and Valley region								
Christian	0.02	0.07	0.05	0.37	2.6	16.9	76.3	96.3
Lodi	0.02	0.15	0.19	0.72	3.4	15.3	44.3	64.3

Source: Modified from Jin et al. (1987).

profile. Boron is also contained in the organic fraction of soils, although most of the B is associated with minerals such as tourmaline that are resistant to weathering. In general, soils high in OM are usually high in B (Fleming, 1980). The complexation and availability of B in OM remain rather unclear. However, B is known to exhibit an affinity for alpha-hydroxy aliphatic acids and ortho-dihydroxy derivatives of aromatic compounds. Boron is also known to react with sugars such as the type produced from microbial breakdown of soil polysaccharides.

The acid-soluble fraction represents precipitated B and that incorporated in OM (Jackson, 1965). It is otherwise known as maximum available B since it represents a capacity factor potentially capable of supplying soluble B to plants through equilibrium among the various forms of B in soils (Eaton and Wilcox, 1939). In Egyptian soils, this fraction ranged from trace to as much as 68% of total B (Elseewi and Elmalky, 1979).

Only two soluble B species in ordinary soils can be expected (Lindsay, 1972). The nonionized species, $B(OH)_3$, is the predominant species expected in soil solution. At pH greater than 9.2, $B(OH)_4^-$ becomes predominant. Polymeric forms of B are unstable in soils unless the concentrations exceed 10^{-4} M, which are seldom present in soils. Most of the B fertilizers are of the

TABLE 12.3. Distribution (percentage) of boron species in simulated irrigation water.[a]

pH	H_3BO_3	$B(OH)_4^-$	$NaB(OH)_4^0$	$CaB(OH)_4^+$	$MgB(OH)_4^+$
7.0	99.0	0.9	<0.1	<0.1	<0.1
8.4	79.9	17.9	0.8	0.7	0.7
9.2	39.2	55.3	2.4	0.9	2.2

[a]Irrigation water composition: B = 0.3 mM, Na$^+$ = 45 mM, Ca^{2+} = 2.5 mM, Mg^{2+} = 2.5 mM, Cl$^-$ = 50 mM, and CO_3^{2-} = 2.5 mM. The computer program GEOCHEM was used to calculate the chemical speciation.
Source: Keren and Bingham (1985).

form $B_4O_7^{2-}$ that hydrolyze to $B(OH)_3$. In saline soils, $B(OH)_3$ is the expected predominant species at pH 7 to 8.4, but at higher pH, $B(OH)_4^-$ becomes predominant (Table 12.3). Even under these conditions, the concentrations of the cation B complexes are low.

4.5 Sorption of Boron in Soils

The following plausible mechanisms are responsible for the chemical interaction of B with soils: anion exchange, precipitation of insoluble borates with sesquioxides, sorption of borate ions or molecular boric acid, formation of organic complexes, and fixation of B in clay lattice. The major inorganic adsorption sites for B in soils are: (1) Fe- and Al-hydroxy compounds present as coating on or associated with clay minerals, (2) Fe- and Al-oxides in soils, (3) clay minerals, especially the micaceous type, and (4) Mg-hydroxy coating on the surface of ferromagnesian minerals. Adsorption sites are associated with broken Si–O and Al–O bonds exposed at edges of aluminosilicate minerals, and also with surface of amorphous hydroxide materials present in weathered soils such as allophane (Bingham, 1973). In layer silicate clays, most of the adsorption can be attributed to sesquioxide coating on the surface of clay rather than to exposed Si–O and Al–O bonds.

The oxides of Fe and Al are likely the major soil constituents that sorb B. Freshly precipitated oxides of Fe and Al have been shown to adsorb large amounts of B (Sims and Bingham, 1968a). Adsorption was pH-dependent with the maximum occurring at pH 7 and 8.5 for oxides of Al and Fe, respectively. Retention of B by hydroxy Al material was an order of magnitude greater than retention of B by hydroxy Fe material. The pH for maximum adsorption suggests that adsorption is of the borate ion. The decrease in B retention at pH values above 8 (or 9) could be expected partly from OH$^-$ ion competition for sorption sites but also from the increase in negative charge of the hydrous oxides resulting in the repulsion of the borate ions and possible dissolution of the adsorbent, forming aluminate ions (Fig. 12.3).

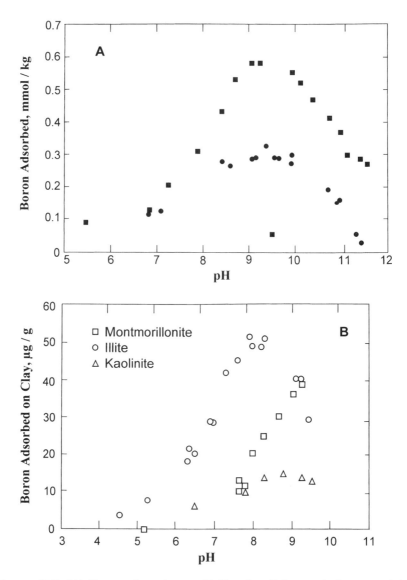

FIGURE 12.3. (A) Boron adsorption on Fallbrook soil from solutions containing 5 mg B L^{-1}. Squares denote surface samples (0 to 25 cm), circles denote 25- to 51-cm samples. (B) Boron adsorption on clay minerals from solutions containing 2 mg B L^{-1}. (From [A] Goldberg and Glaubig, 1986; [B] Hingston, 1964.)

The retention of B by hydroxy Fe and Al compounds is facilitated by two possible mechanisms (Sims and Bingham, 1967; 1968a,b): anion exchange of borate ions with hydroxyl ions, and the formation of diol–borate complexes. The consensus has favored the inner-sphere ligand exchange between the

[B(OH)$_3$ and borate] ions and the hydrous (Fe and Al) oxides, forming metal–O–B bonds (Ellis and Knezek, 1972; Harter, 1991). More recently, Goldberg et al. (1993) suggested an inner-sphere adsorption mechanism for goethite, gibbsite, and kaolinite and an outer-sphere adsorption mechanism for montmorrilonite and soils.

The borate ion may also be bonded through exchange of a single OH$^-$ ion. The reaction appears to occur on sesquioxide surface coating or to reactive sites at mineral edges (Evans and Sparks, 1983; Goldberg and Glaubig, 1986). With pyrophyllite, Keren and Sparks (1994) suggested that B is specifically adsorbed via ligand exchange to the structural Al located on the edge surface. Other findings also indicate the specificity of B adsorption (Goldberg et al., 1996).

The pH dependence of B adsorption can be explained by the hydrolysis reaction:

$$B(OH)_3 + 2H_2O \rightleftharpoons B(OH)_4^- + H_3O^+$$

and the assumption that B(OH)$_3$, B(OH)$_4^-$, and OH$^-$ are all competing for the same adsorption sites (Keren and Gast, 1981). The pH at which maximum adsorption occurs is then a function of the relative affinity of B(OH)$_3$, B(OH)$_4^-$, and OH$^-$ for the reactive surface. It has been shown that B(OH)$_4^-$ and OH$^-$ have a much greater affinity for the reactive surface than B(OH)$_3$ (Keren et al., 1981). The latter species predominates at lower pH however, and consequently there is relatively less total B adsorbed. As pH increases, the B(OH)$_4^-$ concentration increases rather rapidly and since the OH$^-$ concentration is still relatively low, the amount of B adsorbed dramatically increases. A further increase in pH results in increased OH$^-$ concentration relative to B(OH)$_4^-$, and B adsorption decreases rapidly due to the OH$^-$ competition for the sorption sites (see Fig. 12.3). For example, from allophanic soils from Mexico, Chile, and several soils from California, Goldberg (1993) indicated that adsorption increased over the pH range of 5 to 8.5, exhibited a peak at around pH 8.5 to 10, and decreased in the pH range of 10 to 11.5 (see Fig. 12.3A). Similarly, adsorption by layer silicate clays peaked at pH of 8.5 to 9.5 after which adsorption decreased (see Fig. 12.3B).

Adsorption of B has been described by the Langmuir isotherm for a wide range of soils, including amorphous soils, resulting in the usual adsorption maxima in the range of 10 to 100 μg B g^{-1} of soil. However, the Langmuir isotherm was found to be applicable over only a limited B concentration range for New Mexico soils, whereas the Freundlich isotherm was applicable over a much wider B concentration range (0 to 100 μg ml^{-1}) for all the 10 soils used (Elrashidi and O'Connor, 1982). In general, adsorption of B can be modeled using the Langmuir or the Freundlich adsorption isotherm equation (Evans, 1987; Goldberg, 1993). Adsorption maxima by amorphous soils were found to be considerably greater than those observed for other soils (Bingham et al., 1971; Schalscha et al., 1973). These soils, associated with volcanic ash deposits, contain amorphous materials such as allophane and hydrous oxides of Fe and Al and can be found in the western portion of North and South America and on many islands in the Pacific Ocean.

Although allophanic, amorphous soils are characterized by their strong affinity for other anions such as Cl^-, NO_3^-, SO_4^{2-}, and HPO_4^{2-}, B adsorption by these soils is typically specific, being independent of the presence of other anions (Bingham and Page, 1971). Among clay minerals, in general, the micaceous-type clays including vermiculite will adsorb the most B, followed by montmorillonite and kaolinite (see Fig. 12.3B). In this case, B may be fixed in the clay lattice likely through the substitution of B for Al.

5 Boron in Plants

5.1 Essentiality of Boron in Plants

Boron has been recognized as an essential element in higher plant nutrition since early in the 20th century. Its requirement varies markedly within the plant kingdom. It is essential for the normal growth of monocots, dicots, conifers, ferns, and several diatom species but is not essential for fungi and some algae (Shelp, 1993). There is no hard evidence that B is a nutritional requirement for animals and bacteria. Despite the essentiality of B in plant nutrition, its precise biochemical role remains rather vague. More recent evidence favors the involvement of B at the membrane level where it interacts with a number of processes involving enzyme function and the transport of ions, metabolites, and hormones (Shelp, 1993). Its role has been deduced indirectly from physiological experiments, especially with B-deficient plants. For example, bean plants suffering from B deficiency show decrease in root elongation, followed by degeneration of meristematic tissues possibly due to limited cell division.

Since B is relatively immobile in plants, once utilized in actively growing tissues, it is hardly retranslocated to other parts (Fig. 12.4). However, evidence suggests that B is present in the phloem and that it is generally retranslocated in the phloem to satisfy the demands of sink organs that do not readily transpire (Shelp et al., 1995). Indeed, Oertli (1993) concluded that only little B is remobilized and transported to plant tops, whereas a small but adequate amount was remobilized and transported to the roots. Therefore, it is necessary to have a rather continuous source of B available to the plant throughout its growth cycle. For example, sunflower plants grown in nutrient solution containing a total of 50 µg B per plant developed deficiency symptoms at about the 18th or 19th day (Skok, 1957). Boron is required for the formation of new tissues but apparently is not required for the maintenance of older tissues. Thus, it is well known that actively

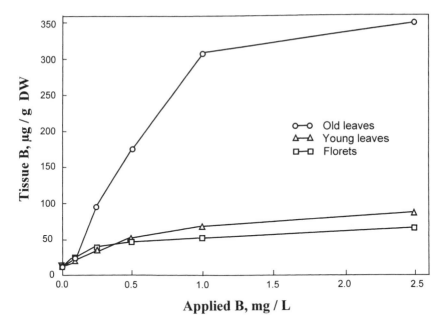

FIGURE 12.4. Boron concentrations in various parts of "Commander" broccoli plants grown to commercial maturity in a greenhouse under a continuous supply of boron. (Reprinted from Shelp, in Gupta, *Boron and Its Role in Crop Production*. Copyright © 1993, CRC Press, Boca Raton, FL.)

growing plants require larger amounts of B than slowly growing or mature plants. Similarly, deficiency symptoms appear more readily in younger than in older tissues.

Boron is also thought to facilitate the transport of sugars through membranes, to be involved in auxin metabolism and in the synthesis of nucleic acids, and in protein and possibly phosphate utilization. It is also implicated in carbohydrate metabolism and in the synthesis of cell wall components such as polyphenolic compounds (Jackson and Chapman, 1975; Price et al., 1972). It is also believed to play a part in photosynthetic and transport processes in plants (Bonilla et al., 1980; Gauch, 1972). Other functions include root growth (Wildes and Neales, 1969), regulation of seed dormancy (Cresswell and Nelson, 1972), pectin synthesis, water balance relationships, resynthesis of ATP, and pollen, anther, and grain development (Cheng and Berkasem, 1993; Romheld and Marschner, 1991).

5.2 Uptake and Translocation of Boron in Plants

Boron uptake by plants exhibits similar two-phase patterns typical for micronutrients. Early absorption is rapid, followed by a steady linear phase. Boron may be absorbed in one of its ionic forms, but since $B(OH)_3$ is more likely to exist at pH below 9.2 (at pH 9.2 there are equal concentrations of undissociated acid and dissociated anion), molecular uptake is most likely. Being neutral, the $B(OH)_3$ species seems to encounter the least resistance upon passage through charged root membranes. Evidence indicates that the cell membranes are more readily permeable to $B(OH)_3$ and much less to $B(OH)_4^-$ (Shelp, 1993). Thus, B is more favorably absorbed as $B(OH)_3$.

In alkaline conditions, the reduced B absorption is possibly due in part to excessive OH^- concentration through a competitive action on the root surface, indicating absorption in an anionic form (Bingham et al., 1970b). Boron absorption by excised barley roots is noncumulative, rapid, physical, and a nonmetabolic process acting in response to B concentration gradient and unaffected by pH variations of the substrate in the acid range (Bingham et al., 1970b). However, increases in pH above 7.0 may result in sharp reduction. At pH 6.0, changes in substrate temperature, salt composition, and level or addition of the respiratory inhibitors (KCN and 2,4-DNP) failed to exert any influence on B uptake. Recently, it was reported that a passive equilibrium exists between B concentrations in tomato cells and the medium (up to 0.42 mM) (Seresinhe and Oertli, 1991), which indicates that B enters through passive uptake.

Because of the ability of $B(OH)_3$ to form complexes with sugars and phenols containing *cis*-diol groups in *in vitro* studies, B has been implicated in facilitating sugar transport across cell membranes (Follett et al., 1981). Phenylboric acid forms complexes with sugars on a 1:1 mole basis. Because of their higher polarity, these complexes may be better able to move through cell membranes.

Sesame plants grown in solution culture accumulated B in decreasing order: leaf blades > petioles > stems (Yousif et al., 1972). In sugar beet leaves, B accumulated in decreasing order: leaf margin > central part > petiole (Oertli and Roth, 1969). This pattern of B accumulation once again supports the view that B uptake is passive and that B is transported in the transpiration stream to accumulate in the leaves. As water is lost to the atmosphere, residual B accumulates and eventually a level is reached that is toxic to the plant. The mechanism of B injury is not well known, although an osmotic mechanism has been hypothesized (Oertli and Roth, 1969).

In short, B absorption by plants can be easily explained by the unassisted permeation of $B(OH)_3$ (i.e., passive transport) and the subsequent formation of *cis*-diol complexes. Long-distance transport of B from the roots to the shoot is confined to the xylem via response to transpiration.

5.3 Deficiency of Boron in Plants

Boron deficiency is considered one of the most widespread deficiencies of the seven micronutrients. Boron deficiency commonly occurs in sandy soils, which have low CEC and OM content, where leaching and heavy cropping have diminished the soil B reserves and in limed acidic soils. In the tropics, where soil is predominantly acidic and prone to leaching by rain, B deficiency commonly occurs in infertile soils (low CEC and OM content). Crops remove B from the soil and it must be replenished for top yields and quality.

Boron deficiency has been reported for one or more crops in 43 states in the United States (Sparr, 1970); in alfalfa, deficiency has been reported in 38 states where it is also of widespread occurrence in tree and food crops. However, there is no apparent geographic pattern in the occurrence of B deficiency in crops in the United States (Kubota and Allaway, 1972) and the United Kingdom (Thornton and Webb, 1980) although in the United States B deficiency tends to occur more often on low-OM, acid, sandy and silt loam soils in the South Atlantic and Pacific states. Field and vegetable crops are affected as are fruit and nut trees. Deficiency is also reported in certain soils of Nigeria (Chude and Obigbesan, 1983; Chude, 1988), and B is one of the most common micronutrient deficiencies in *Pinus radiata* in New Zealand and eastern Australia (Hopmans and Flinn, 1984).

The amount of B required by plants for normal nutrition varies with plant species and cultivars, ensuring variable sensitivity to available B (Tables 12.4 and 12.5). In general, monocotyledons require only about a fourth as much B for normal growth as do dicotyledons (Berger, 1949). Grasses do not seem to require as much B as do broadleaf plants, but there are variations within these categories (Bowen, 1977). Legumes seem to have an exceptionally large requirement for this nutrient. In contrast, most of the fruit and nut trees do not need as much (see Table 12.4). Amounts needed for responsive crops, such as alfalfa, beets, and celery, may cause damage to other crops, such as small grains, beans, peas, and cucumbers.

The range between beneficial and toxic B concentrations in the growth substrate is narrow. For example, solution culture concentrations of 0.05 to 0.10 μg B ml^{-1} are ordinarily safe and adequate for many plants, whereas concentrations of 0.50 to 1.0 μg B ml^{-1} are often excessively high for B-sensitive plants (Bingham, 1973). Concentrations <0.05 μg B ml^{-1} may produce deficiency. Similarly, a B level of 0.50 ppm in sand culture solution was adequate for sunflower growth, but 1.0 ppm was toxic (Eaton, 1940).

As a diagnostic tool, either soil tests or plant tests for B can be used. The hot water-soluble B in soils proposed by Berger and Truog (1939) is the most commonly used. However, this value of B cannot be used universally because of the complicating effects of plant and soil factors, and other nutrients on B availability. In general, there is a direct positive correlation between B content of plants and water-soluble B in soils. The critical range of 0.10 to 0.70 ppm of hot water-soluble B in soils has been offered for most crops (Cox and Kamprath, 1972). At least two factors, soil texture and pH, must also be considered in the interpretation of soil tests based on this form of B. Consequently, critical levels of hot water-soluble B vary among locations and crops. For example, heavier soils in Illinois containing <0.50 ppm water-soluble B appear to be deficient for alfalfa and other

TABLE 12.4. Relative sensitivity of plants to boron.

Sensitive	Semitolerant	Tolerant
Alfalfa	Birdsfoot trefoil	Artichoke
Apple	Broccoli	Asparagus
Apricot	Cabbage	Athel
Avocado	Calendula	Barley
Blackberry	California poppy	Bean
Cauliflower	Carrot	Blueberry
Celery	Clover	Broadbean
Cherry	Field pea	Chard
Cotton	Hops	Corn
Cowpea	Kentucky bluegrass	Cucumber
Elm	Lettuce	Gladiolus
Fig	Lima bean	Grass
Grape	Millet	Mangel
Grapefruit	Milo	Muskmelon
Jerusalem artichoke	Olive	Mustard
Kidney bean	Parsley	Oat
Kola	Parsnip	Onion
Larkspur	Peanut	Oxalis
Lupine	Pepper	Palm
Navy bean	Radish	Pasture grass
Orange	Rice	Pea
Pansy	Rose	Peppermint
Peach	Rutabaga	Potato
Pear	Spinach	Rye
Pecan	Sweet corn	Sesame
Persimmon, Japanese	Sweet pea	Sorghum
Plum	Sweet potato	Soybean
Strawberry	Timothy	Spearmint
Sugar beet	Tobacco	Sudan grass
Sunflower	Tomato	Sweet clover
Table beet	Vetch	Table beet
Turnip	Zinnia	Wheat
Violet		
Walnut		

Sources: Bradford (1965); Bingham (1973); Martens and Westermann (1991); Robertson et al. (1976); USDA (1954).

legumes (Stinson, 1953). For coarse sandy soils, the critical level was about 0.30 ppm. In sandy Podzolic soils in Alabama, about 0.15 ppm hot water-soluble B was reported sufficient to support good growth of alfalfa and other legumes (Rogers, 1947). In Michigan, alfalfa needs B fertilization when the soil tests indicate <0.90 to 1.0 ppm hot water-soluble B (Baker and Cook, 1956).

In Oklahoma, no internal damage in peanuts due to B deficiency (see Fig. 12.2) appeared when the hot water-soluble B was above 0.15 ppm (Hill and Morrill, 1974). With rutabaga, brown-heart symptoms due to B deficiency occurred when hot water-soluble B in soil ranged between 0.40

and 1.3 ppm (Gupta and Munro, 1969). For other crops, such as broccoli, Brussels sprouts, and cauliflower, levels of this form of B in the range of 0.34 to 0.49 ppm in soil were sufficient for optimum growth (Gupta and Cutcliffe, 1973). For apple, 0.50 ppm water-soluble B in soil would be sufficient for normal production (Woodbridge, 1937).

For plant diagnostic criteria encompassing a wide spectrum of crops, B deficiency may occur when the B levels are <15 ppm in the dry matter (Jones, 1972). For field-grown wheat, plants may be B deficient when the flag leaf B content at the boot stage is <5 ppm and sufficient when the B content of the flag leg is >7 ppm (Rerkasem and Loneragan, 1994). Adequate but not excessive B may occur between 20 to 100 ppm and toxicity can be expected when the plant levels of B exceed 100 ppm, although much lower levels may cause toxicity to B-sensitive plants. For example, the accepted range for B concentrations for pecan leaflet is around 50 to 100 mg B kg^{-1} DW (Picchioni et al., 2000).

In pine species (*Pinus radiata, P. elliottii*, and *P. pinaster*) levels below 10 ppm in the foliage proved deficient (see Table 12.5) (Raupach, 1975; Will, 1971). Specific examples of crops and trees showing deficient, sufficient, and toxic levels of B in plant tissues have been presented in Table 12.5. The data in this table indicate the importance of plant species, and plant parts and age in interpreting plant test results.

Boron deficiency may be expected to be more prevalent on leached acid soils. Soil types deficient in B are usually the sandy loams, fine-textured lakebed soils, and acid organic soils (Vitosh et al., 1973). Sandy and silty loam soils in humid regions such as the acid sandy soils of the Mississippi Valley and Atlantic Coastal Plain in the United States, are more prone to B deficiency. On the other hand, liming to above pH 6.5 increases the incidence of B deficiency.

Since B is hardly translocated from old to young plant parts, the first symptoms of B deficiency appear in growing points such as the stem tips, flower buds, and axillary buds. Boron-deficient plants exhibit a breakdown of the growing tip tissue or a shortening of the terminal growth. This may appear as a rosetting of the plant, or twisting and distortion of the growing points and the youngest leaves of the plant. The growing points of stems generally show varying degrees of necrosis, sometimes accompanied by splitting along the length of the stem. More specifically, B deficiency in apical meristematic tissue of tomato can cause enlarged cell size, nuclear enlargement, and tissue disintegration (Brown and Ambler, 1973). In leaf tissue, B deficiency caused necrosis and collapse of the upper epidermal and palisade cells, and accumulation of phenolic materials. Internal tissues of beets, turnips, and rutabagas show breakdown and corky dark coloration.

In pines, needles are fused and necrotic shoots suffer from dieback in late summer or autumn, with foliage immediately behind dieback abnormally short and yellowish (Will, 1971). Trees that are B deficient can have branches that may crook or resin flowing from the stems (Raupach, 1975).

TABLE 12.5. Deficient, sufficient, and toxic levels of boron in plants.

Plant	Plant part	B in tissue (ppm DW)		
		Deficient	Sufficient	Toxic
Crops				
Rutabaga	Leaf tissue at harvest	20–38; <12 severely deficient	38–40	>250
	Leaf tissue when roots begin to swell	32–40, moderately deficient; <12 severely deficient	40	—
Sugar beets	Blades of recently matured leaves	12–40	35–200	—
	Middle fully developed leaf without stem taken at end of June/early July	<20	31–200	>800
Cauliflower	Whole tops before the appearance of curd	3	12–23	—
	Leaves	23	36	—
	Leaf tissue when 5% heads formed	4–9	11–97	—
Broccoli	Leaves	—	70	—
	Leaf tissue when 5% heads formed	2–9	10–71	—
Brussels sprouts	Leaf tissue when sprouts begin to form	6–10	13–101	161
Carrots	Mature leaf lamina	<16	32–103	175–307
	Leaves	18	—	—
Tomato	Mature young leaves from top of plant	<10	30–75	>200
	Whole plant; 15 cm tall	<12	51–88	>172
Celery	Petioles	16	28–75	—
	Leaflets	20	68–432	720
Potato	32-day old plants	—	12	>180
	Fully developed first leaf at 75th day after planting	<15	21–50	>50
Ryegrass	Whole plants at rapid growth	—	6–12	>39–42
Radish	Whole plants when roots began to swell	<9	96–127	—
Soybean	Mature trifoliate leaves at bloom	9–10	—	63
Strawberry	Old/young leaves at active growth	—	—	123
Tobacco	Leaves from first priming	—	—	169
Beans	43-day old plants	—	12	>160
Dwarf kidney beans	Leaves and stems (plants cut 50 mm above the soils)	—	44	132
	Pods	—	28	43
White pea beans	Aerial portion of plants 1 mo after planting	—	36–94	144

TABLE 12.5 (*continued*)

Plant	Plant part	B in tissue (ppm DW)		
		Deficient	Sufficient	Toxic
Cucumber	Mature leaves from center of stem 2 wk after first picking	<20	40–120	>300
Spanish peanuts	Young leaf tissue from 30-day old plants		54–65	>250
			18–20b	
Alfalfa	Whole tops at early bloom	<15	20–40	200
	Top one-third of plant shortly before flowering	<20	31–80	<100
	Whole tops at 10% bloom	8–12	39–52	99
Red clover	Whole tops at bud stage	12–20	21–45	>59
	Top one-third of plant at bloom	—	20–60	>60
Birdsfoot trefoil	Whole tops at bud stage	14	30–45	>68
Timothy	Whole plants at heading stage	—	3–93	>102
Pasture grass	Aboveground part at first bloom at first cut	—	10–50	>800
Corn	Whole plants, 25 cm tall	—	8–38	>98
	Whole shoot at vegetative stage until ear formation	<9	15–90	>100
Wheat	Boot stage tissue	2.1–5.0	8	>16
	Straw	4.6–6.0	17	>34
Winter wheat	Whole shoot when plants 40 cm high	<0.3	2.1–10.1	>10
Oats	47-day old plants	—	—	<105
	Boot stage tissue	<1–3.5	6–50	>30–400
	Straw	3.5–5.6	14–24	>50
Barley	Boot stage tissue	1.9–3.5	10	>20
				50–70
	Straw	7.1–8.6	21	>46
Conifers				
Arucaria angustifolia (Bert.) O. Ktze			6–19	
Picea abies, many locations		<4	5–18	
Pinus caribaea Morelet var. *hondurensis*		6	9–21	
Pinus elliottii Engelm. *elliottii*, nursery		1.8–4.2	8–21	
Pinus khasya Royle		7	13–36	
Pinus palustris Mill.		6	6–10	
Pinus patula Schl. & Cham.		6	7–22	

480   12 Boron

TABLE 12.5 (*continued*)

Plant	Plant part	B in tissue (ppm DW)		
		Deficient	Sufficient	Toxic
Pinus ponderosa Laws.			20–31	
Pinus radiata, nursery				
6 yr		8	10–12	
2 yr		4–5		101
Pinus radiata, 18–20 yr			10–36	
Pinus resinosa Ait.			22–27	75
Pinus sabiniana Dougl.			32	400
Pinus strobus				355
Pinus sylvestris L.			32	480
20-yr planation;				>53
many locations		<4–9.9		
Pinus taeda L. nursery		1.8–4.2	11–25	
Pseudotsuga menziesii (Mirb.) Franco		<4	28–82	
Angiosperms				
Acer platanoides L.			134	400–1200
Acer saccharum Marsh.				1560
Betula spp.		8.3 ± 1.5	21.3 ± 2.7	
Castanopsis chrysophylla (Dougl.)				
Eucalyptus deglupta Blume			~30–70 (toxic?)	
Eucalyptus tereticornis Sm.		14–16		
Quercus stellata Wang.			35	

[a]Considered critical; [b]Considered high; [c]B in foliage (DW).
Sources: From various sources, as summarized by Gupta (1979, 1993); Stone (1990).

More specifically, B deficiency in *Pinus radiata* causes the terminal buds to die. Leader growth is arrested and permanent malformation of the stem occurs (Hunter et al., 1990).

In summary, B deficiency causes many anatomical, physiological, and biochemical changes but the majority of these represent secondary effects, complicated by physiological age between normal and deficient tissues.

5.4 Phytotoxicity of Boron

The following soils are likely to cause excess B uptake in plants (Bradford, 1965): soils of marine sediment and, in general, those derived from parent materials rich in B; arid and semi-arid soils; and soils derived from geologically young deposits. Boron usually does not exist in toxic quantities in most arable soils unless it has been added in excessive amount.

As discussed earlier, plants have varying degree of tolerance to B excess in the growth substrate (see Table 12.4). Members of the grass family are usually tolerant to high B levels in soils. In general, when the level of B in plant dry matter exceeds 100 ppm, B toxicity can be expected to occur. Tissue B concentrations of 108, 92, and 95 ppm in field-grown alfalfa, red clover, and timothy, respectively, produced B toxicity symptoms on the foliage but without any substantial yield declines (Gupta, 1984). There are instances, however, when toxicities can occur even at much lower B concentrations in semitolerant species. Some tree species, such as red pine, white pine, and Scotch pine, are highly sensitive to excess B. Levels of about 55 to 100 ppm B in the needles of red pines produced toxicity symptoms (see Table 12.5) (Neary et al., 1975; Stone and Baird, 1956).

Boron is toxic to many plants when present in soil solution at levels not much greater than the trace amounts needed for normal growth. Boron concentrations <0.50 µg ml^{-1} in soil solution are probably safe for most plants (Fig. 12.5), but many plants may be adversely affected when B levels are in the range of 0.50 to 5.0 µg ml^{-1} (Eaton and Wilcox, 1939; Wilcox, 1960). Such toxic levels of B have been found in soils and irrigation waters in many arid regions of the world, including irrigated areas in the western United States (Rhoades et al., 1970). For example, toxic levels of B occur in several alkaline areas of eastern Oregon (Powers and Jordan, 1950) as well as in many parts of California (Bingham, 1973). Being a soluble salt, B accumulates where other soluble salts, such as chloride, accumulate. Thus, in semiarid areas, the subsoil B concentration often exceeds that in surface soil (Leyshun and Jame, 1993); high B levels can also be prevalent in soils with little or no drainage or in areas with a shallow water table. In general, irrigation waters with B levels above 4 µg ml^{-1} are unfit for plants.

Irrigation waters can be classified as to their effect on crop nutrition (Wilcox, 1960): sensitive crops, 0.30 to 1.0 µg ml^{-1}; semi-tolerant crops, 1.0 to 2.0 µg ml^{-1}; and tolerant crops, 2.0 to 4.0 µg ml^{-1} of B. Toxicity symptoms are manifested similarly in most plants, first appearing as a browning of leaf tips and margins. The yellowing (chlorosis) rapidly increases in severity, quickly followed by necrosis and tissue death. In severely affected pine trees, necrosis is most pronounced on needles near the end of shoot and in the upper half of the tree. The browning can be noticed also on the distal 1 to 2 cm of 1 yr-old needles (Stone and Baird, 1956).

In summary, B toxicity is not as widespread as B deficiency. Toxicity may be induced by the following practices under field conditions: in soils being irrigated with water high in B content and salinity; in soils receiving high rates of coal fly ash, especially unweathered ones; and in soils being used as disposal sites for B-enriched waste by-products.

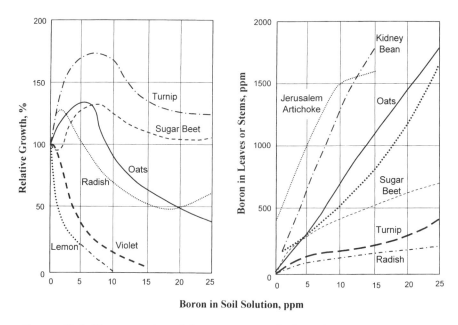

FIGURE 12.5. Plant growth and boron content of plant tissues as influenced by boron level in soil solution. (From Sprague, 1972.)

6 Factors Affecting Mobility and Bioavailability of Boron

6.1 pH

Soil reaction influences the mobility and phytoavailability of B. In general, B becomes more available to plants with decreasing pH. Thus, liming acid soils will likely reduce the B concentration in plant tissues of crops; in rare cases, liming may even induce B deficiency or mitigate B toxicity in crops. Consequently, most of the reported studies relating to soil pH effects involve the interaction of B and the pH itself, or lime constituent. For example, in the absence of added B, the B levels in alfalfa shoots decreased with increasing pH (Su et al., 1994); pH elevation in this soil (from pH 4.2) was induced by lime addition. Growth stimulation by lime addition may result in B deficiency when there is significant dilution of B in plant shoot tissue. There are exceptions, however. Uptake of either indigenous or added B by tall fescue was not affected by pH in the range 4.7 to 6.3, but was substantially decreased at about pH 7.4 (Peterson and Newman, 1976). Similarly, no discernible trend on the relationships between soil pH and plant B below pH 6.5 was observed by other investigators (Gupta and MacLeod, 1977). Furthermore, there was no effect of pH over the range 5.4 to 7.5 on B mobility in soils, but instead, OM, clay, and Fe and Al oxide contents proved to be more influential (Parker and Gardner, 1982).

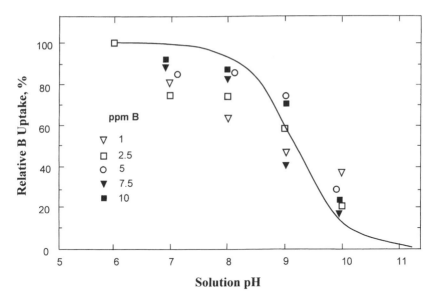

FIGURE 12.6. Relative uptake of boron as a function of the external solution pH. Uptake at pH 6 = 100. Solid line = % of undissociated $B(OH)_3$. (From Oertli and Grgurevic, 1975.)

Interactions between soil pH and hot water-soluble B on the severity of brown-heart disorder in rutabaga have been observed (Gupta and Cutcliffe, 1972). The degree of this disorder was more severe at high soil pH than at low pH. In contrast, no significant relationships were found between water-soluble B and pH in eastern Canada soils in the pH range 4.5 to 6.5 (Gupta, 1968) and in Michigan soils (Robertson et al., 1975).

Boron uptake by excised barley roots was unaffected by pH of the substrate in the acid range. However, increases in pH above 7.0 resulted in sharp reduction in absorption (Fig. 12.6) (Bingham et al., 1970b; Oertli and Grgurevic, 1975). The mechanism of B uptake by plants as related to pH has been discussed previously in Section 5.2. Boron adsorption by soils and soil constituents was also shown to be largely pH-dependent, with adsorption maxima occuring in the alkaline range.

6.2 Parent Material

As discussed previously in section 3, the natural level of B in soils largely depends upon the soil parent material. In general, soils derived from igneous rocks and those of tropical and semitropical regions of the world are considerably lower in B content compared with soils derived from sedimentary rocks and those of arid and semiarid regions. The content of total B in the latter group may range up to 200 ppm, particularly in alkaline, calcareous soils, while that for the former group is usually <10 ppm.

6.3 Clay Mineral and Soil Texture

In the pH range commonly found in soils, illite tends to adsorb more B than kaolinite or montmorillonite, with kaolinite adsorbing the least (see Fig. 12.3) (Fleet, 1965; Hingston, 1964). It has been proposed that most of the B in clay fraction of sedimentary rocks is contained in the illite fraction (Frederickson and Reynolds, 1959). Adsorption by these clays is quite pH-dependent, with the maxima in the alkaline range (Hingston, 1964); both physical (molecular adsorption) and anion adsorption mechanisms are involved.

In general, B is more rapidly leached to lower depths in sandy than in clay soils; the possibility of toxic B levels accumulating in soils is far less in coarse-textured than in fine-textured soils. In addition, the former soils, especially those of the humid regions, often are low in OM and therefore, are inherently low in B. Thus, B deficiency can be expected to be more prevalent in sandy than in clay soils. For example, of the 266 alfalfa fields surveyed in Quebec, Canada B deficiency was found in 119. Of that number, 61% were located on light-textured and only 14% on heavy-textured soils (Ouellette and Lachance, 1954). Similarly, in Michigan the occurrence of B deficiency in alfalfa was more prevalent in coarse-textured soils where water-soluble B was lower than in fine-textured soils (Baker and Cook, 1956). In eastern Canada, greater amounts of hot water-soluble B were present in fine- than in coarse-textured soils (Gupta, 1968). This can be attributed to the fact that more of the B was adsorbed on the clay and, therefore, less subjected to leaching loss than B found in sandy soils. The relationships between the B content of plants and hot water-soluble B in soils, showing the importance of soil texture, is illustrated on Figure 12.7.

6.4 Oxides of Iron and Aluminum

Boron adsorption has been demonstrated to be greater for Fe- and Al-coated kaolinite or montmorillonite than for uncoated clays (Sims and Bingham, 1967, 1968b). The presence of hydroxy Fe and Al compounds enhance the affinity of B for clay minerals. Boron adsorption by certain soils is exacerbated by the presence of free Al oxide. For example, positive significant correlations were obtained between adsorption intensity of B by Mexican and Hawaiian soils and their amorphous Al_2O_3 content (Fig. 12.8). The role of these compounds in B retention by soils has been discussed previously in Section 4.5.

6.5 Organic Matter

Organic matter is a major source of B in acid soils. Several workers found positive relationships between soil OM content and available B. Thus,

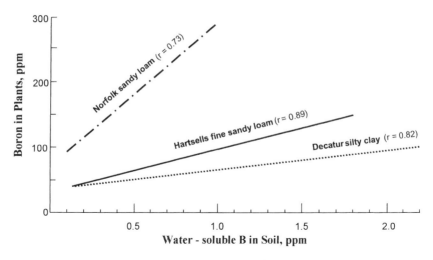

FIGURE 12.7. Importance of soil texture on the relationship between boron in plants and hot water-soluble boron in soil. (From Wear and Patterson, 1962.)

the significance of OM as a reservoir for plant B cannot be ignored (see Fig. 12.1).

6.6 Interaction with Other Elements

The higher incidence of B deficiency on overlimed or alkaline soils than on acid soils suggests the possible effects of Ca^{2+} and Mg^{2+} ions on B nutrition of plants. These cations are primary constituents of agricultural limestone. Thus, studies have reported an apparent relationship between the level of Ca in soils and the phytoavailability of B (Adriano, 1986). This often results in antagonistic interactions between Ca and B in plant nutrition, producing a high Ca/B ratio in plant tissues. To this end, the use of the Ca/B ratio in plant tissue as a diagnostic criterion in B fertilization has been proposed (Drake et al., 1941; Gupta, 1972a). It can be expected that liming would increase the Ca/B ratio in plant tissues. Conversely, high water–soluble B content in soils would decrease the Ca/B ratio. Thus, the ideal Ca/B ratio was determined to be 1200 for tobacco; 500 for soybeans, and 100 for sugar beets (Jones and Scarseth, 1944). In rutabagas, Ca/B ratios over 3300 were associated with very severe brown-heart conditions; boot stage tissue Ca/B ratio of 1370 was related to severe B deficiency in barley (Gupta and Cutcliffe, 1972; Gupta, 1972a). A Ca/B ratio of about 180 appeared optimal for barley, while ratios of 10 to 45 were in the severe B toxicity range. Oats showed B toxicity when the ratio was 200 (Jones and Scarseth, 1944).

There appear to be two schools of thought regarding liming effect on B bioavailability: direct pH effect vs. Ca^{2+} and B interaction. First, the effect

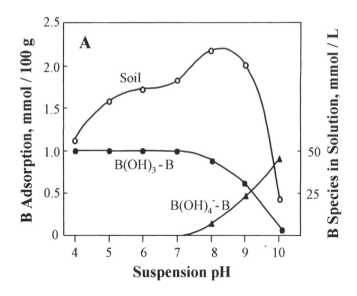

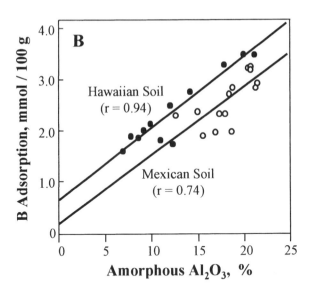

FIGURE 12.8. Boron retention by soil related to suspension pH and Al_2O_3 content of soil. (From Bingham et al., 1971.)

of lime in reducing B uptake by plants is due to soil pH effect rather than the Ca^{2+} level in the soil (Gupta and MacLeod, 1981). Growing barley and peas on three podzol soils limed with three different sources of Ca^{2+} ($CaSO_4$, $CaCO_3$, and $CaCl_2$), it was noted that at equivalent amounts of Ca^{2+}, tissue B concentrations were much higher using $CaSO_4$ rather than $CaCO_3$. The lower uptake with $CaCO_3$ source corresponded to much higher increases in soil pH. Similarly, no consistent effects of variable levels of Ca and K were noted on B absorption by excised barley roots (Bingham et al., 1970b). Second, Fox (1968) found that at high pH, high Ca^{2+} concentrations restricted B uptake of cotton and alfalfa by about 50%.

Interactions of B have also been observed with the nutrition of Mg, K, N, and P (Adriano, 1986). However, the interactions of these elements with B are inconsistent, i.e., sometimes synergistic or antagonistic. The presence of anions can have a competitive, although small, effect on the adsorption of B. These competitive effects on B adsorption decreased in the order phosphate > molybdate > sulfate (Goldberg et al., 1996).

6.7 Plant Factors

As discussed in section 5.3, plant species vary in their B requirement for normal nutrition (see Tables 12.4 and 12.5). Grasses require less B than do broadleaf plants; legumes seem to need larger amounts of B. The *Pinus* spp. and pecan (Picchioni et al., 2000) are generally sensitive to B; pistachio, on the other hand, displayed no B foliar injury even when leaf B concentrations were 220 to 235 ppm, but did so at about 1000 ppm (Ashworth et al., 1985). Similarly, B-tolerant ornamentals could withstand up to 8.3 mg L^{-1} of soil-extractable B before foliar toxicity symptoms appeared (Handreck, 1990). In general, tolerant species accumulate B at a much slower rate than the sensitive species.

As with other micronutrients, B nutrition varies among cultivars. An adequate supply of B for one cultivar may be toxic to another. For example, differential uptake of B by two tomato genotypes in both soil and solution culture have been noted (Brown and Ambler, 1973). The sensitivity between species and between genotypes appears to depend on the ability of tolerant species to reduce B uptake, or the inability of susceptible (inefficient) species in translocating B from roots to shoot, or a combination of both processes (Bellaloui and Brown, 1998; Pant et al., 1998; Xue et al., 1998). In general, the most susceptible genotypes contained considerably more B in both roots and shoots than tolerant genotypes (Nable et al., 1990; Nable, 1991; Shann and Adriano, 1988).

Boron also accumulates in varying degrees in plant parts as was noted in section 4.5 and Figure 12.4, i.e., B accumulates more in leaf blades than in other plant parts and more in leaf margin than in other leaf parts. In general, B tends to accumulate in the margin of leaves.

6.8 Other Factors

Certain environmental factors appear to influence the B nutrition in plants. Boron deficiency seems to be more prevalent under droughty conditions (Scott et al., 1975; Stewart and Axley, 1956). By the same token, drought can also alleviate toxicity symptoms in plants (Gupta et al., 1976). This can be explained by the reduction of B uptake by plants since mass flow in the rhizosphere is more restricted. The loss of water from the plants via transpiration induces the mass flow or diffusion of ions from the bulk soil to the root surface.

Seasonal variation in B level in plants has also been observed. In citrus, B levels tend to increase with the age of the leaves (Bradford and Harding, 1957; Labanauskas et al., 1959). Increase of B with tissue age was also observed in corn leaves (Clark, 1975) and cucumber leaves (Alt and Schwarz, 1973).

7 Boron in Animal and Human Nutrition

To date, there is no known essentiality role of B in animal and human nutrition. The only regulatory restriction on B use pertains to irrigation water for crops and loading rate of coal fly ash on arable land. The U.S. EPA (1976) has set a limit of 0.75 $\mu g \, ml^{-1}$ B in water intended for long-term irrigation of sensitive crops. The loading rate of B from the application of fly ash is due to usually high B content of unweathered ash that can affect initial establishment and growth of B-sensitive plant species (Adriano and Weber, 1998).

The average daily human intake for B has been estimated at 3.0 mg (WHO, 1973). In the United Kingdom, the daily intake from diet was estimated at 2.83 ± 1.55 mg compared with ICRP (International Commission on Radiological Protection) II's 6.0 mg intake value (Hamilton and Minski, 1972). Intake by sixth-grade children from five geographic regions in the United States from Type A school lunches averaged 0.50 mg/lunch (Murphy et al., 1970).

Toxicologically, B content in foodstuff is not a great concern compared with other trace elements. In fact in the past, $B(OH)_3$ had been used as a food preservative but is now declared unacceptable by the FAO/WHO Expert Committee on Food Additives (WHO, 1973). Only at excessively high intake of 4000 mg day^{-1} and higher can B prove toxic (Bowen, 1979).

8 Sources of Boron in the Environment

Construction of B fluxes and budget in the global troposphere suggests that the most important anthropogenic sources of B are coal combustion and agricultural burning (Fogg and Duce, 1985). On a regional scale, the anthropogenic inputs may be quite significant. The data of Fogg (1983) support the belief that the sea is the net sink for atmospheric gaseous B.

8.1 Fertilizers

Solubor (20% B) and sodium tetraborate ($Na_2B_4O_7 \cdot 5H_2O$) (14% B) are the most commonly used B fertilizers (Table 12.6) (Martens and Westermann, 1991). Other sodium borates such as sodium pentaborate ($Na_2B_{10}O_{16} \cdot 10H_2O$) (18% B) are also used. Boric acid $B(OH)_3$ (17% B) is used only sparingly.

Boron deficiency is usually corrected by application of B fertilizers to the soil although some carriers such as $B(OH)_3$ and Solubor are used as foliar spray. The optimal rate of soil-applied B depends on plant species, soil type, soil pH, as well as other factors. In general, recent results indicate that a rate of 1.5 to 4.0 kg B ha^{-1} is needed for soil treatments on legumes and certain root crops, while lower rates are needed for other crops to avoid toxicity problems (see Table 12.6). Applying 2 kg B ha^{-1} could maintain sufficient B levels in soils under eastern Canada conditions for alfalfa and red clover for 2 yr or longer (Gupta, 1984).

In Michigan, rates of up to 3.4 kg B ha^{-1} are recommended for certain soil types and crops (Robertson et al., 1975). For soil application, rates of 1.7 to 3.4 kg ha^{-1} are recommended for highly responsive crops, while 0.6 to 1.1 kg ha^{-1} are recommended for low to medium responsive crops (Lucas, 1967). However, band application of 5.0 kg B ha^{-1} on pea beans caused decreased yields and toxicity symptoms (Rieke and Davis, 1964). Because of these small amounts, B fertilizers are usually mixed in the desired proportion with other fertilizers before application (Mortvedt, 1968; Mortvedt and Osborn, 1965).

TABLE 12.6. Rates of boron application for correction of boron deficiency in various crops.

Crop	B source	Range (kg ha^{-1})	Optimum (kg ha^{-1})	Application method
Alfalfa	$Na_2B_4O_7$	1.12–4.48	2.24	Broadcast
Alfalfa	Fly ash	1.7–3.4	1.7	Broadcast
Cole crops	$Na_2B_4O_7$	—	0.45	Banded
Grape	Solubor	0.38–3.05	0.76	Broadcast
Pine tree[a]	$Na_2B_4O_7 \cdot 10H_2O$	5.7–17.0	5.7	Broadcast
Rapeseed	H_3BO_3	0–2.8	1.4	Broadcast
Soybean	$Na_2B_4O_7 \cdot 5H_2O$	0.28–2.24	1.12	Broadcast
Strawberry	$Na_2B_4O_7$	0.56–2.24	1.12	Broadcast
Sugar beet	Solubor	2.2–6.6	2.2	Banded
Table beet	Solubor	3.4–6.7	3.4	Broadcast

[a]*Pinus radiata* D. Don.
Source: As summarized by Martens and Westermann (1991).

Foliar application of B has gained popularity recently for fruit crops such as apple because it can be mixed with pesticide spray. Rates as low as 0.2 kg B ha^{-1} in 360 L of water was as effective in providing enough B to apple as was application of 0.02 to 0.04 kg B/tree every 3 yr (Burrell, 1958). In general, foliar B applications to annual crops are superior to broadcast applications, and equivalent or slightly more effective than banded applications.

8.2 Coal Fly Ash

Coal contains B that can range up to 400 ppm and higher, with an average of about 50 to 60 ppm (Adriano, 1986). Upon combustion, B in coal becomes concentrated in combustion residues, primarily in fly ash, in which levels up to 1900 ppm have been reported (Table 12.7). Losses in the neighborhood of 30% of total B in coal through emissions have been quoted (Gladney et al., 1978). Boron is considered one of the most highly enriched elements in power plant emissions partly because of the volatility of $B(OH)_3$ and B-halide species. These levels are of environmental significance in the United States considering the fact that electric power utilities produce over 100×10^6 tonnes of fly ash annually. Because of the B-enriched status of fly ash, one of the main constraints of land utilization and/or disposal of fly ash is the B content, along with Mo, Se, and As (Adriano et al., 1980; Carlson and Adriano, 1993; Wong et al., 1998).

Boron in precipitator fly ash collected throughout the midwestern states was leachable by water in the range of 17 to 64% of the total content (James et al., 1982). Since many plant species can only tolerate a narrow range of B concentration in soils, the toxic effects of B to plants from fly ash may be expected to be severe during the initial period after land application, after which the B concentration would have substantially diminished to tolerable levels.

8.3 Irrigation Waters

Irrigation waters with B levels above 0.3 μg B ml^{-1} can damage certain crops, particularly sensitive species (see Section 5.4). Although B levels in some waters may exceed this level, the majority of surface water supplies in the western United States have B concentrations in the range of 0.1 to 0.3 μg ml^{-1} (USDA, 1954). In arid and semiarid regions, the B content of irrigation waters, especially groundwaters, are often elevated and in some cases may be as high as 5 μg ml^{-1} (Paliwal and Mehta, 1973). In other bodies of natural waters, B levels are in the neighborhood of 100 μg L^{-1}, as shown for natural waters in Florida (Carriker and Brezonik, 1978), streams in the southeastern United States (Boyd and Walley, 1972), and rivers of Norway and Sweden (Ahl and Jonsson, 1972).

Agricultural drainage effluents are often high in salinity and B content. For example, in the Central Valley of California where irrigation water quality simulated field drainage effluents (electrical conductivity of 2 to 28

TABLE 12.7. Boron contents (total and hot-water extractable, ppm DW) in coal fly ash and bottom ash.

Ash source	Total	Hot water extractable
Fly ash:		
United States ($n = 21$)	10–600	
Big Sandy (Kentucky)	319	—
Clinch River (Virginia)	342	—
Glen Lyn (Virginia)	307	—
Kanawha River (West Virginia)	391	—
Muskingum River (Ohio)	301	—
Muskingum River (West Virginia)	340	—
Mt. Storm (West Virginia)	278	—
Albright (West Virginia)	264	—
Riversville (West Virginia)	236	—
Ft Marten (West Virginia)	415	—
Crawford Ed. (Illinois)	613	—
John Sevier (Tennessee)	323	—
Ernest C. Gaston (Alabama)	428	—
Ernest C. Gaston (Alabama)	367	—
Hoot Lake (Minnesota)	618	—
Lewis and Clark (Montana)	414	—
Illinois Power Co.	1900	855
Illinois Power Co.	1320	449
Columbia I (Wisconsin)	608	—
Columbia II (Wisconsin)	272	—
Alma (Wisconsin)	678	—
Colstrip (Montana)	645	—
Lansing (Iowa)	327	—
Ashland (Wisconsin)	1900	—
White Bluff (Arkansas)	860 (820–970)	—
Beech Island (South Carolina)	—	21–25[a]
Bottom ash:		
United States ($n = 34$)	3–1356	<1–187
Illinois Power Co.	960	0.96
Edgewater-5 (Wisconsin)	—	290

[a]Adriano and Weber (1998).
Source: From various sources, as summarized by Horn (1995).

dS m^{-1} and B concentrations from 1 to 30 mg L^{-1}), B injury was correlated with increasing irrigation water B in only older leaves of *Eucalyptus camaldulensis* (Poss et al., 1999). Once again, because of the immobility of B in this species, the interactive effects of B and salinity on foliar injury depends on the physiological age of the leaf.

8.4 Sewage Sludge and Effluent

Because of the use of borates and perborates in detergents as buffering, softening, and bleaching agents, B is usually present in sewage effluent and sludge. Irrigating forested areas with sewage effluent over an extended period was found injurious to pine species (Neary et al., 1975; Sopper and Kardos, 1973). The B levels in sewage effluent from several sewage treatment plants in southern California ranged from 0.3 to 2.5 µg ml^{-1}, with a mean of 1.0 µg B ml^{-1} (Bradford et al., 1975).

Boron is usually more concentrated in sewage sludge, with levels ranging from a few to several hundred ppm (Bradford et al., 1975; Berrow and Webber, 1972). When applied to soils, however, B did not accumulate in plant tissues even at rates above 112 tonnes ha^{-1} of sludge containing 13 ppm B (Dowdy and Larson, 1975).

References

Adriano, D.C. 1986. *Trace Elements in the Terrestrial Environment*. Springer-Verlag, New York.

Adriano, D.C., and J. Weber. 1998. *Coal Ash Utilization for Soil Amendment to Enhance Water Relations and Turf Growth*. EPRI Rep TR-111318. Electric Power Res Inst, Palo Alto, CA.

Adriano, D.C., A.L. Page, A.A. Elseewi, A.C. Chang, and I. Straughan. 1980. *J Environ Qual* 9:333–344.

Ahl, T., and E. Jonsson. 1972. *Ambio* 1:66–70.

Alt, D., and W. Schwarz. 1973. *Plant Soil* 39:277–283.

Archer, F.C. 1980. *Minist Agric Fish Food* (Great Britain) 326:184–190.

Ashworth, L.J. 1985. *Phytopath* 75:1084–1091.

Aubert, H., and M. Pinta. 1977. *Trace Elements in Soils*. Elsevier, New York.

Awad, F., and M.I. Mikhael. 1980. *Egypt J Soil Sci* 20:89–98.

Bagdasarashvili, Z.G., K.A. Abashidze, N.V. Koberidze, D.V. Abashidze, and N.M. Korkotadze. 1974. *Sov Soil Sci* 7:576–580.

Baird, G.B., and J.E. Dawson. 1955. *Soil Sci Soc Am Proc* 19:219–222.

Baker, A.S., and R.L. Cook. 1956. *Agron J* 48:564–568.

Bellaloui, N., and P.H. Brown. 1998. *Plant Soil* 198:153–158.

Berger, K.C. 1949. *Adv Agron* 1:321–351.

Berger, K.C., and E. Truog. 1939. *Ind Eng Chem Anal Ed* 11:540–545.

Berger, K.C., and E. Truog. 1940. *J Am Soc Agron* 32:297–301.

Berger, K.C., and E. Truog. 1946. *Soil Sci Soc Am Proc* 10:113–116.

Berrow, M.L., and J. Webber. 1972. *J Sci Food Agric* 23:93–100.

Bingham, F.T. 1973. *Adv Chem Ser* 123:130–138.

Bingham, F.T., and A.L. Page. 1971. *Soil Sci Soc Am Proc* 35:892–893.

Bingham, F.T., R.J. Arkley, N.T. Coleman, and G.R. Bradford. 1970a. *Hilgardia* 40:193–204.

Bingham, F.T., A.A. Elseewi, and J.J. Oertli. 1970b. *Soil Sci Soc Am Proc* 34:613–617.

Bingham, F.T., A.L. Page, N.T. Coleman, and K. Flach. 1971. *Soil Sci Soc Am Proc* 35:546–550.

Bingham, F.T., A.W. Marsh, R. Branson, R. Mahler, and R. Ferry. 1972. *Hilgardia* 41:195–211.

Bonilla, I., C. Cadahia, and O. Carpena. 1980. *Plant Soil* 57:3–9.

Bowen, H.J.M. 1979. *Environmental Chemistry of the Elements*. Academic Pr, New York.

Bowen, J.E. 1977. *Crops Soils* (Aug–Sept):12–14.

Boyd, C.E., and W.W. Walley. 1972. *Am Midl Nat* 88:1–14.

Bradford, G.R. 1965. In H.D. Chapman, ed. *Diagnostic Criteria for Plants and Soils.* Quality Printing, Abilene, TX.

Bradford, G.R., and R.B. Harding. 1957. *Proc Am Soc Hart Sci* 70:252–256.

Bradford, G.R., A.L. Page, L.J. Lund, and W. Olmstead. 1975. *J Environ Qual* 4:123–127.

Brown, J.C., and J.E. Ambler. 1973. *Soil Sci Soc Am Proc* 37:63–66.

Burrell, A.B. 1958. *Proc Am Soc Hort Sci* 71:20–25.

Carlson, C.L., and D.C. Adriano. 1993. *J Environ Qual.* 22:227–247.

Carriker, N.E., and P.L. Brezonik. 1978. *J Environ Qual* 7:516–522.

Cartwright, B., K.G. Tiller, B.A. Zarcinas, and R.L. Spouncer. 1983. *Aust J Soil Res* 21:321–332.

Chen, J., F. Wei, Y. Wu, and D.C. Adriano. 1991. *Water Air Soil Pollut* 51–58:699–712.

Cheng, C., and B. Rerkasem. 1993. *Plant Soil* 155/156:313–315.

Chude, V. 1988. *Plant Soil* 107:293–295.

Chude, V., and G.O. Obigbesan. 1983. *Plant Soil* 74:145–147.

Clark, M.C., and D.J. Swaine. 1962. CSIRO Tech Commun 45. CSIRO, Chatswood, Australia.

Clark, R.B. 1975. *Commun Soil Sci Plant Anal* 6:451–464.

Cook, R.L., and C.E. Millar. 1939. *Soil Sci Soc Am Proc* 4:297–301.

Cox, F.R., and E.J. Kamprath. 1972. In J.J. Mortvedt, P.M. Giordano, and W.L. Lindsay, eds. *Micronutrients in Agriculture.* Soil Sci Soc Am, Madison, WI.

Cresswell, C.F., and H. Nelson. 1972. *Proc Grass Soc South Afr* 7:133–137.

DeTurk, E.E., and L.C. Olson. 1941. *Soil Sci* 52:351–357.

Domingo, L.E., and K. Kyuma. 1983. *Soil Sci Plant Nutr* 29:439–452.

Dowdy, R.H., and W.E. Larson. 1975. *J Environ Qual* 4:278–282.

Drake, M., D.H. Sieling, and G.D. Scarseth. 1941. *J Am Soc Agron* 32:454–462.

Eaton, F.M., and L.V. Wilcox. 1939. In *The Behavior of Boron in Soils.* USDA Tech Bull 696. U.S. Dept Agriculture, Washington, DC.

Eaton, S.V. 1940. *Plant Physiol* 15:95–107.

Ellis, B.G., and B.D. Knezek. 1972. In J.J. Mortvedt, P.M. Giordano, and W.L. Lindsay, eds. *Micronutrients in Agriculture.* Soil Sci Soc Am, Madison, WI.

Elrashidi, M.A., and G.A. O'Connor. 1982. *Soil Sci Soc Am J* 46:27–31.

Elseewi, A.A., and A.E. Elmalky. 1979. *Soil Sci Soc Am J* 43:297–300.

Evans, C.M., and D.L. Sparks. 1983. *Commun Soil Sci Plant Anal* 14:827–846.

Evans, L.J. 1987. *Can J Soil Sci* 67:33–42.

Fairhall, L.T. 1941. *N Engl Water Works Assoc J* 55:400–410.

Fleet, M.E.L. 1965. *Clay Miner Bull* 6:3–16.

Fleming, G.A. 1980. In B.E. Davis, ed. *Applied Soil Trace Elements.* Wiley, New York.

Fogg, T.R. 1983. *Sources and Sinks of Atmospheric Boron.* PhD dissertation, Univ Rhode Island, Kingston, RI.

Fogg, T.R., and R.A. Duce. 1985. Am Geophys Union (AGU) Paper 4DI385. AGU, Washington, DC.

Follett, R.H., L.S. Murphy, and R.L. Donahue. 1981. *Fertilizers and Soil Amendments.* Prentice-Hall, Englewood Cliffs, NJ.

Fox, R.H. 1968. *Soil Sci* 106:435–439.

Frederickson, A.F., and R.C. Reynolds, Jr. 1959. *Clays Clay Miner* 8:203–213.

Furr, A.K., T.F. Parkinson, R.A. Hinrichs, D.R. Van Campen, et al. 1977. *Environ Sci Technol* 11:1104–1112.

Gauch, H.G. 1972. In *Organic Plant Nutrition.* Dowden, Hutchinson and Ross, Strousburg, PA.

Gestring, W.D., and P.N. Soltanpour. 1984. *Soil Sci Soc Am J* 48:96–100.

Gladney, E.S., L.E. Wangen, D.B. Curtis, and E.T. Jurney. 1978. *Environ Sci Technol* 12:1084–1085.

Goldberg, S. 1993. In U.C. Gupta, ed. *Boron and Its Role in Crop Production.* CRC Pr, Boca Raton, FL.

Goldberg, S., and R.A. Glaubig. 1986. *Soil Sci Soc Am J* 50:1442–1448.

Goldberg, S., H.S. Forster, and E.L. Herck. 1993. *Soil Sci Soc Am J* 57:704–708.

Goldberg, S., H.S. Forster, S.M. Lesch, and E.L. Heich. 1996. *Soil Sci* 161:99–103.

Graham, E.R. 1957. *Soil Sci Soc Am Proc* 21:505–508.

Gupta, U.C. 1968. *Soil Sci Soc Am Proc* 32:45–48.

Gupta, U.C. 1972a. *Soil Sci Soc Am Proc* 36:332–334.

Gupta, U.C. 1972b. *Commun Soil Sci Plant Anal* 3:355–365.

Gupta, U.C. 1979. *Adv Agron* 31:273–307.

Gupta, U.C. 1984. *Soil Sci* 137:16–22.

Gupta, U.C., ed. 1993. In *Boron and Its Role in Crop Production.* CRC Pr, Boca Raton, FL.

Gupta, U.C., and J.A. Cutcliffe. 1972. *Soil Sci Soc Am Proc* 36:936–939.

Gupta, U.C., and J.A. Cutcliffe. 1973. *Can J Soil S*ci 53:275–279.

Gupta, U.C., and J.A. MacLeod. 1973. *Commun Soil Sci Plant Anal* 4:389–395.

Gupta, U.C., and J.A. MacLeod. 1977. *Soil Sci* 124:279–284.

Gupta, U.C., and J.A. MacLeod. 1981. *Soil Sci* 131:20–25.

Gupta, U.C., and D.C. Munro. 1969. *Soil Sci Soc Am Proc* 33:424–426.

Gupta, U.C., J.A. MacLeod, and J.D.E. Sterling. 1976. *Soil Sci Soc Am J* 40:723–726.

Hamilton, E.I., and M.J. Minski. 1972. *Sci Total Environ* 1:375–394.

Handrek, R.A. 1990. *Commun Soil Sci Plant Anal* 21:2260–2280.

Harter, R.D. 1991. In J.J. Mortvedt et al., eds. *Micronutrients in Agriculture,* SSSA 4, Soil Sci Soc Am, Madison, WI.

Hatcher, J.T., and C.A. Bower. 1958. *Soil Sci* 85:319–323.

Hatcher, J.T., C.A. Bower, and M. Clark. 1967. *Soil Sci* 104:422–426.

Hill, W.E., and L.G. Morrill. 1974. *Soil Sci Soc Am Proc* 38:791–794.

Hill, W.E., and L.G. Morrill. 1975. *Soil Sci Soc Am Proc* 39:80–83.

Hingston, F.J. 1964. *Aust J Soil Res* 2:83–95.

Hopmans, P., and D.W. Flinn. 1984. *Plant Soil* 79:295–298.

Horn, M.E. 1995. *Land Application of Coal Combustion By-Products: Use in Agriculture and Land Reclamation.* EPRI Rep TR-103298, Electric Power Res Inst, Palo Alto, CA.

Hunter, I.R., G.M. Will, and M.F. Skinner. 1990. *Forest Ecol Manage* 37:77–82.

Il'in, V.B., and A.P. Anikina. 1974. *Sov Soil Sci* 6:68–75.

Jackson, J.F., and K.S.R. Chapman. 1975. In D.J.D. Nicholas and A.R. Egan, eds. *Trace Elements in Soil-Plant-Animal Systems.* Academic Pr, New York.

Jackson, M.L. 1965. *Soil Chemical Analysis.* Prentice-Hall, Englewood Cliffs, NJ.

James, W.D., C.C. Graham, M.D. Glascock, and A.G. Hanna. 1982. *Environ Sci Technol* 16:195–197.

Jones, H.E., and G.D. Scarseth. 1944. *Soil Sci* 57:15–24.

Jones, J.B., Jr. 1970. *Commun Soil Sci Plant Anal* 1:27–34.

Jones, J.B., Jr. 1972. In J.J. Mortvedt, P.M. Giordano, and W.L. Lindsay, eds. *Micronutrients in Agriculture*. Soil Sci Soc Am, Madison, WI.

Keren, R., and F.T. Bingham. 1985. *Adv Soil Sc*i 1:229–276.

Keren, R., and R.G. Gast. 1981. *Soil Sci Soc Am J* 45:478–482.

Keren, R., and D.L. Sparks. 1994. *Soil Sci Soc Am J* 58:1095–1100.

Keren, R., R.G. Gast, and B. Bar-Yosef. 1981. *Soil Sci Soc Am J* 45:45–48.

Komor, S.C. 1997. *J Environ Qual* 26:1212–1222.

Krauskopf, K.B. 1972. In J.J. Mortvedt, P.M. Giordano, and W.L. Lindsay, eds. *Micronutrients in Agriculture*. Soil Sci Soc Am, Madison, WI.

Krauskopf, K.B. 1979. *Introduction to Geochemistry,* 2nd ed. McGraw-Hill, New York.

Kubota, J., and W.H. Allaway. 1972. In J.J. Mortvedt, P.M. Giordano, and W.L. Lindsay, eds. *Micronutrients in Agriculture*. Soil Sci Soc Am, Madison, WI.

Kubota, J., K.C. Berger, and E. Truog. 1948. *Soil Sci Soc Am Proc* 13:130–134.

Labanauskas, C.K., W.W. Jones, and T.W. Embleton. 1959. *Proc Am Soc Hort Sci* 74:300–307.

Leyshun, A.J., and Y.W. Jame. 1993. In U.C. Gupta, ed. *Boron and Its Role in Crop Production*. CRC Pr, Boca Raton, FL.

Lindsay, W.L. 1972. In J.J. Mortvedt, P.M. Giordano, and W.L. Lindsay, eds. *Micronutrients in Agriculture*. Soil Sci Soc Am, Madison, WI.

Lucas, R.E. 1967. Ext Bull E-486. Michigan State Univ, East Lansing, MI.

Martens, D.C., and D.T. Westermann. 1991. In J.J. Mortvedt et al., eds. *Micronutrients in Agriculture,* SSSA 4, Soil Sci Soc Am, Madison, WI.

Mengel, D.V. 1980. *Boron in Soils and Plant Nutrition*. US Borax Plant Food Conf, Lafayette, IN.

Mortvedt, J.J. 1968. Soil *Sci Soc Am Proc* 32:433–437.

Mortvedt, J.J., and G. Osborn. 1965. *Soil Sci Soc Am Proc* 29:187–191.

Mortvedt, J.J., and J.R. Woodruff. 1993. In U.C. Gupta, ed. *Boron and Its Role in Crop Production*. CRC Pr, Boca Raton, FL.

Murphy, E.W., B.K. Watt, and L. Page. 1970. *Trace Subs Environ Health* 4:194–205.

Murphy, L.S., and L.M. Walsh. 1972. In J.J. Mortvedt, P.M. Giordano, and W.L. Lindsay, eds. *Micronutrients in Agriculture*. Soil Sci Soc Am, Madison, WI.

Murphy, L.S., R. Ellis, Jr., and D.C. Adriano. 1981. *J Plant Nutri* 3:593–613.

Neary, D.G., G. Schneider, and D.P. White. 1975. *Soil Sci Soc Am Proc* 39:981–982.

Nable, R.O. 1991. *J Plant Nutr* 14:453–461.

Nable, R.O., R.C. Lance, and B. Cartwright. 1990. *Ann Bot* 66:83–90.

Norrish, K. 1975. In D.J.D. Nicholas and A.R. Egan, eds. *Trace Elements in Soil-Plant-Animal Systems*. Academic Pr, New York.

Oertli, J.J., 1993. *Plant Soil* 155/156:301–304.

Oertli, J.J., and E. Grgurevic. 1975. *Agron J* 67:278–280.

Oertli, J.J., and H.C. Kohl. 1961. *Soil Sci* 92:243–247.

Oertli, J.J., and J.A. Roth. 1969. *Agron J* 61:191–195.

Offiah, O.O., and J.H. Axley. 1993. In U.C. Gupta, ed. *Boron and Its Role in Crop Production*. CRC Pr, Boca Raton, FL.

Okazaki, E., and T.T. Chao. 1968. *Soil Sci* 105:255–259.

Ouellette, G.J., and R.O. Lachance. 1954. *Can J Agric Sci* 34:494–503.

Paliwal, K.V., and K.K. Mehta. 1973. *Indian J Agric Sci* 43:766–772.

Pant, J., B. Rerkasem, and R. Noppakoonwong. 1998. *Plant Soil* 202:193–200.

Parker, D.R., and E.H. Gardner. 1982. *Soil Sci Soc Am J* 46:573–578.

Peterson, L.A., and R.C. Newman. 1976. *Soil Sci Soc Am J* 40:280–282.

Picchioni, G.A., H. Karaca, L.G. Boyse, B.D. McCaslin, and E.A. Herrera. 2000. *J Environ Qual* 29:955–963.

Piland, J.R., C.F. Ireland, and H.M. Reisenauer. 1944. *Soil Sci* 57:75–84.

Poss, J.A., S.R. Grattan, C.M. Grieve, and M.C. Shannon. 1999. *Plant Soil* 206:237–245.

Powers, W.L. 1941. *Better Crops Plant Food* 15(6):17–19;36–37.

Powers, W.L., and J.V. Jordan. 1950. *Soil Sci* 70:99–107.

Price, C.A., H.E. Clark, and E.A. Funkhouser. 1972. In J.J. Mortvedt, P.M. Giordano, and W.L. Lindsay, eds. *Micronutrients in Agriculture*. Soil Sci Soc Am, Madison, WI.

Raupach, M. 1975. In D.J.D. Nicholas and A.R. Egan, eds. *Trace Elements in Soil-Plant-Animal Systems*. Academic Pr, New York.

Reeve, R.C., A.F. Pillsbury, and L.V. Wilcox. 1955. *Hilgardia* 24:69–91.

Rerkasem, B., and J.F. Loneragan. 1994. *Agron J* 86:887–890.

Rhoades, J.D., R.D. Ingvalson, and J.T. Hatcher. 1970. *Soil Sci Soc Am Proc* 34:871–875.

Rieke, P.E., and J.F. Davis. 1964. *Mich Agric Exp Sta Quart Bull* 46:401–406. East Lansing, MI.

Robertson, L.S., B.D. Knezek, and J.O. Belo. 1975. *Commun Soil Sci Plant Anal* 6:359–373.

Robertson, L.S., R.E. Lucas, and D.R. Christenson. 1976. In Mich Coop Ext Bull E-1037. East Lansing, MI.

Rogers, H.T. 1947. *J Am Soc Agron* 39:914–928.

Romheld, V., and H. Marschner. 1991. In J.J. Mortvedt et al., eds. *Micronutrients in Agriculture*, 2nd ed. SSSA 4, Soil Sci Soc Am, Madison, WI.

Schalscha, E.B., F.T. Bingham, G.G. Galindo, and H.P. Galvan. 1973. *Soil Sci* 116:70–76.

Scott, H.D., S.D. Beasley, and L.F. Thompson. 1975. *Soil Sci Soc Am Proc* 39:1116–1121.

Seresinhe, P.S.J., and J.J. Oertli. 1991. *Physiol Plant* 81:31–36.

Shann, J.R., and D.C. Adriano. 1988. *Environ Exp Bot* 28:289–299.

Shelp, B.J. 1993. In U.C. Gupta, ed. *Boron and Its Role in Crop Production*. CRC Pr, Boca Raton, FL.

Shelp, B.J., E. Marentes, A.M. Kitheca, and P. Yivekanandan. 1995. *Physiol Plant* 94:356–361.

Shirokov, V.V., and V.I. Panasin. 1972. *Sov Soil Sci* 4:341–344.

Sims, J.R., and F.T. Bingham. 1967. *Soil Sci Soc Am Proc* 31:728–732.

Sims, J.R., and F.T. Bingham. 1968a. *Soil Sci Soc Am Proc* 32:364–369.

Sims, J.R., and F.T. Bingham. 1968b. *Soil Sci Soc Am Proc* 32:369–373.

Sims, J.T., and G.V. Johnson. 1991. In J.J. Mortvedt et al., eds. *Micronutrients in Agriculture*, SSSA 4, Soil Sci Soc Am, Madison, WI.

Singh, M. 1970. *J Indian Soc Soil Sci* 18:141–146.

Skok, J. 1957. *Plant Physiol* 32:648–658.

Sopper, W.E., and L.T. Kardos. 1973. In *Recycling Treated Wastewater and Sludge Through Forest and Cropland*. Penn State Univ, University Park, PA.

Sparr, M.C. 1970. *Commun Soil Sci Plant Anal* 1:241–262.

Sprague, R.W. 1972. *The Ecological Significance of Boron*. US Borax Corp, Anaheim, CA.

Stewart, F.B., and J.H. Axley. 1956. *Agron J* 48:259–262.

Stinson, C.H. 1953. *Soil Sci* 75:31–36.

Stone, E.L. 1990. *Forest Ecol Manage* 37:49–75.

Stone, E.L., and G. Baird. 1956. *J Forest* 54:11–12.

Su, C., L.J. Evans, T.E. Bates, and G.A. Spiers. 1994. *Soil Sci Soc Am J* 58:1445–1450.

Swaine, D.J. 1955. The Trace Elements Content of Soils. Commonwealth Bureau Soil Sci (GB), Tech Commun 48. Herald Printing Works, York, United Kingdom.

Talati, N.R., and S.K. Agarwal. 1974. *J Indian Soc Soil Sci* 22:262–268.

Thornton, I., and J.S. Webb. 1980. In B. Davies, ed. *Applied Soil Trace Elements*. Wiley, New York.

Trudinger, P.A., D.J. Swaine, and G.W. Skyring. 1979. In P.A. Trudinger and D.J. Swaine, eds. *Biogeochemical Cycling of Mineral-Forming Elements*. Elsevier, Amsterdam.

[U.S. EPA] United States Environmental Protection Agency. 1976. *Quality Criteria for Water*. U.S. EPA, Washington, DC.

[U.S. PHS] United States Public Health Service. 1970. *Community Water Supply Study, Analysis of National Survey Findings*. U.S. Dept Health, Education, and Welfare, Washington, DC.

Vinogradov, A.P. 1959. *The Geochemistry of Rare and Dispersed Chemical Elements in Soils*. Consultants Bureau, New York.

Vitosh, M.L., D.D. Warncke, and R.E. Lucas. 1973. Ext Bull E-486. Michigan State Univ, East Lansing, MI.

Wear, J.I. 1965. In C.A. Black, ed. *Methods of Soil Analysis Agron* 9:1059–1063.

Wear, J.I., and R.M. Patterson. 1962. *Soil Sci Soc Am Proc* 26:344–346.

Whetstone, R.R., W.O. Robinson, and H.G. Byers. 1942. USDA Tech Bull 797. U.S. Dept Agriculture, Washington, DC.

[WHO] World Health Organization. 1973. WHO Tech Rep Ser 532. WHO, Geneva.

Wilcox, L.V. 1960. USDA Info Bull 211. U.S. Dept Agriculture, Washington, DC.

Wildes, R.A., and T.E. Neales. 1969. *J Exp Bot* 20:591–603.

Will, G.M. 1971. *NZ Forest Serv Leaflet* 32:1–4.

Williams, C.H. 1977. *J Aust Inst Agric Sci* (Sept–Dec):99–109.

Winsor, H.W. 1952. *Soil Sci* 74:459–466.

Wong, J.W.C., R.F. Jiang, and D.C. Su. 1998. *Water Air Soil Pollut* 106:137–147.

Woodbridge, C.G. 1937. *Sci Agric* 18:41–48.

Xue, J., R.W. Bell, R.D. Graham, and Y. Yang. 1998. *Plant Soil* 204:155–163.

Yousif, Y.H., F.T. Bingham, and D.M. Yermanos. 1972. *Soil Sci Soc Am Proc* 36:923–926.

13
Copper

1 General Properties of Copper

Copper (atom. no. 29), one of the most important metals to society, is reddish, takes on a bright metallic luster, is malleable, ductile, and a good conductor of heat and electricity (second only to silver in electrical conductivity). It belongs to Group I-B of the periodic table, has an atom. wt. of 63.55, a melting pt. of 1083 °C, and a specific gravity of 8.96. It consists of two natural isotopes, ^{63}Cu and ^{65}Cu, with relative abundances of 69.1 and 30.9%, respectively. The radioactive isotope ^{64}Cu with a half-life of 12.8 hr, is the most suitable for tracer work.

In nature, Cu occurs in the I and II oxidation states with ionic radii of 0.96 and 0.72 Å, respectively. In the II state, it is isomorphous with Zn^{2+}, Mg^{2+}, and Fe^{2+} ions. It is found in minerals such as cuprite, malachite, azurite, chalcopyrite, and bornite. The most important Cu ores are the sulfides, oxides, and carbonates.

2 Production and Uses of Copper

The principal uses of Cu are in the production of wire and its alloys brass (with zinc) and bronze (with tin). Copper is also alloyed with Au, Pb, Cd, Cr, Be, Ni, Al, and Mn. The electrical industry is one of the major users of Cu in the production of electrical wires and other electrical apparatus. Because of its high thermal conductance and relative inertness, Cu is extensively used in containers such as boilers, steam pipes, automobile radiators, and kitchenware. The metal is widely used in water delivery systems. It is also extensively used in agriculture in the form of fertilizers, bactericides and fungicides, and as algicides in water purification. It is used as a feed additive, such as in antibiotics, drugs, and selected chemical compounds, as a growth promoter, and as an agent for disease control in livestock and poultry production.

Chile has been the world's leader in mine output of Cu with about 23% of the total in 1994, followed by the United States, Canada, Russia, and Australia. World consumption of refined Cu rose to 9.50×10^6 tonnes in 1978, to 9.86×10^6 tonnes in 1979, and 11.1×10^6 tonnes in 1994 (Jolly, 1980; U.S. Bureau of Mines, 1995).

3 Copper in Nature

In nature, Cu forms sulfides, sulfates, sulfosalts, carbonates, and other compounds and also occurs as the native metal. Chalcopyrite ($CuFeS_2$, 34%

Cu) is by far the most abundant Cu mineral, being found widely dispersed in rocks and concentrated in the largest Cu ore deposits. Estimates for the average crustal abundance of Cu range from 24 to 55 ppm (Cox, 1979); Bowen (1979) quoted 50 ppm. It is ranked 26th among elements in crustal abundance behind Zn (Krauskopf, 1979).

The following mean values, in ppm, are given for world soils (McKeague and Wolynetz, 1980): China, 23 (geometric mean = 20); Canada, 22 (range of 5 to 50); United States, 25; world soils, 20. These are considered background levels of Cu in soils, i.e., Cu derived from the soil parent material and/or redistributed in the profile via pedogenesis. Levels of Cu in various geologic materials, plants, and other environmental media are given in Table 13.1.

Long-distance transport of air pollutants can affect remote areas, including forest ecosystems. There has been substantial enrichment of Cu in atmospheric particulates collected at several remote locations (Nriagu, 1979). For example, high depositions of metals in the New England mountain region in the United States (Reiners et al., 1975) and in the Solling mountains in central Germany have been reported (Heinrichs and Mayer, 1980). Similarly, in Sweden (Ruhling and Tyler, 1971) and recently in Norway (Berthelsen et al., 1995) distinct regional gradients for heavy metals indicate that the airborne metals originate from outside these countries.

Substantial decreases were observed in atmospheric deposition of Cu, Zn, Pb, and Cd in southern Norway and of Pb in central Norway from 1982 to 1992 (Berthelsen et al., 1995). However, except for Pb, no significant decreases in plant Zn, Cu, or Cd concentrations were observed either in southern or central Norway from 1982 to 1992, despite the large reductions (i.e., 26 to 40%) in atmospheric Zn, Cu, and Cd deposition during this period (Table 13.2). Evidently, higher surface soil and peat levels and with that higher plant-available fractions of Zn, Cu, and Cd in soils and peat in southern Norway than in central Norway resulted in significantly higher plant Zn, Cu, and Cd uptake in southern Norway. Thus, atmospheric deposition of Zn, Cu, and Cd has influenced plant levels of these elements indirectly, through long-term surface soil contamination and subsequent increased rates of root uptake compared with regions with a low degree of surface soil Zn, Cu, and Cd contamination. The response of changes in atmospheric Pb deposition on Pb levels in vegetation is much greater due to a high degree of direct atmospheric Pb deposition on plants.

The total aerial burden of metals is presumed to originate from both present day urban-industrial sources (particulate, aerosols, smokes, automobile exhaust, etc.), power generation (combustion of fossil fuel), and former industry (via resuspension of contaminated soil and industrial wastes). Lead, Cd, Zn, and Cu are usually the principal metals of concern in forest ecosystems because they are often the main metal constituents of air pollutants and because of their known toxicities to biota. Combustion of fossil fuel is apparently a major source of aerial Cu (Kneip et al., 1970). In the Solling mountains in Germany, the annual precipitation input of Cu is

TABLE 13.1. Commonly observed copper concentrations (ppm) in various environmental media.

Material	Average concentration	Range
Igneous rocks[b]	—	10–100
Limestone and dolomite[c]	6	0.6–13
Sandstone[c]	30	6–46
Shale and clay[c]	35	23–67
Petroleum[d]	1.3	—
Coal[c]	17	1–49
Coal ash[e]		
Fly ash	185	45–1452
Bottom ash	78	27–146
FGD sludge	32	8–121
Oil ash[e]	692	344–995
Sewage sludges[f]	690	100–1000
Soils[d]	30	2–250
Common crops[g]	—	6–40
Common fruit trees[h]	—	4–20
Ferns[d]	—	15
Fungi[d]	—	7–160
Lichens[d]	—	9–24
Freshwater[a,d]	3	0.2–30
Seawater[a,d]	0.25	0.05–12

Sources: Extracted from
[a]μg L^{-1}.
[b]Krauskopf (1972).
[c]Cox (1979).
[d]Bowen (1979).
[e]Ainsworth and Rai (1987).
[f]Page (1974).
[g]Follett et al. (1981).
[h]Reuther and Labanauskas (1965).

about 2 times greater than at an urban Indiana site in the United States (Parker et al., 1978). In virtually nonpolluted forest ecosystems in western Switzerland, vegetation and soil profile samples were collected from an acid brown soil (Typic Dystrochrept) and a podzol (Typic Haplorthod) (Keller and Vedy, 1994). The former site is primarily vegetated by dense Norway spruce while the latter has mixed coniferous forest consisting of Norway spruce, larch, silver fir, and Swiss stone pine. The distribution of Cu and Cd in the profile largely depended on the soil type rather than the metal itself, even though the concentrations of Cu were at least 50 times more than those of Cd. At the Dystrochrept site, the litter layer had the largest concentration of Cu (13.5 mg kg^{-1}). The Cu concentrations declined rapidly with depth and only 5.7 mg kg^{-1} Cu occurred in the A2 layer. The distribution of Cu in the mineral soil of the Haplorthod is consistent with the leaching from the E horizon (eluvial) and accumulation in the Bh and Bs horizons, observed also for Fe and Al. On a mass basis, the A horizon of Haplorthod and the A1 and BC horizons of the Dystrochrept, respectively, have the largest Cu

TABLE 13.2. Copper concentrations (ppm DW) in different plant species and plant compartments collected from forests in southern and central Norway in 1982 and 1992, given as arithmetic mean ± SD.

Species	Plant compartment	Southern Norway		Central Norway	
		1982	1992	1982	1992
Spruce	twig	5.5 ± 0.5	5.0 ± 0.9	4.8 ± 0.2	4.9 ± 0.7
	needle	2.1 ± 0.2	2.4 ± 0.4	1.8 ± 0.1	2.0 ± 0.2
Birch	twig	4.9 ± 0.1	5.5 ± 0.5	4.7 ± 0.2	4.9 ± 0.0
	leaf	4.7 ± 0.2	5.1 ± 0.3	4.8 ± 0.1	4.1 ± 0.1
Scotch pine	twig	5.8 ± 0.2	5.3 ± 0.6	4.0 ± 0.4	3.7 ± 0.2
	needle	3.6 ± 0.3	4.2 ± 1.2	2.0 ± 0.3	2.6 ± 0.1
Juniper	twig	4.2 ± 0.5	4.6 ± 0.3	3.7 ± 0.0	3.6 ± 0.3
	needle	3.9 ± 0.6	3.5 ± 0.5	3.1 ± 0.4	2.8 ± 0.4
Bilberry	stem	7.0 ± 0.8	6.3 ± 0.8	8.0 ± 0.4	5.3 ± 0.1
	leaf	6.0 ± 0.5	5.7 ± 0.5	7.1 ± 0.2	4.5 ± 0.2
Mountain cranberry	stem	8.2 ± 0.6	6.1 ± 0.8	7.8 ± 0.8	5.9 ± 0.3
	leaf	4.6 ± 0.1	3.5 ± 0.1	3.3 ± 0.1	3.1 ± 0.2
Heather		6.9 ± 0.2	6.8 ± 0.4	5.5 ± 0.7	5.1 ± 0.5
Crowberry		7.2 ± 0.5	6.0 ± 0.6	6.7 ± 0.4	6.1 ± 0.2
Feather moss		11.1 ± 2.0	8.7 ± 0.3	3.8 ± 0.3	4.3 ± 0.6
Bog whortleberry	stem	9.1	9.9 ± 0.8	—	7.6 ± 0.9
	leaf	5.1	4.1 ± 0.5	—	4.7 ± 0.5

Source: Extracted from Berthelsen et al. (1995).

concentrations. These larger concentrations within the humified organic horizons are typical for forest soils. The values in the upper horizons at both sites are less than the Swiss guide value for the amount of pollutants contained in soil (Cu = 50 mg kg^{-1}) and lie within the typical values found in Switzerland for coniferous soils. Estimates indicate more than half of the metal load found in the Oi and A1 horizons of the Dystrochrept and in the Oi and Oa horizons of the Haplorthod could be of anthropogenic origin even though these soils are considered to be nonpolluted.

Total metal pools (Fig. 13.1) show that in the Dystrochrept, total Cu pools were larger than those in the Haplorthod because the concentrations were greater and the soil profile deeper. Compared with the whole ecosystem, the mineral soil contained the most metal: the amounts stored in the Dystrochrept were 98% of total Cu; in the Haplorthod they were 90% of total Cu. This confirms the consensus that the vegetation is not an important compartment of the ecosystem for storing Cu. Water- and NaNO$_3$-extractable fractions of the above soils contained very little Cu. The amounts of water- and NaNO$_3$-extractable Cu were strongly correlated ($r = 0.95$) and were greatest in the Oi and A1 horizons, which are acidic and contain much organic C. Indeed, a negative correlation was observed between NaNO$_3$- extractable Cu and pH, and a positive correlation with the organic C and total Cu, indicating that the amount of metal solubilized by NaNO$_3$ depends strongly on soil properties.

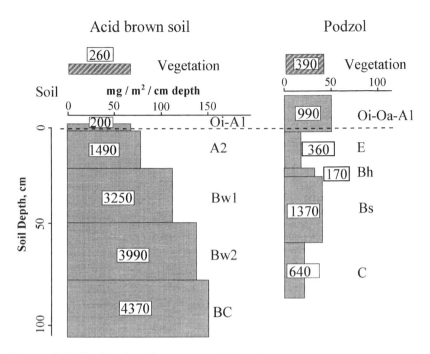

FIGURE 13.1. Partitioning of copper pools in various horizons of acid brown soil and podzol in forest ecosystems in western Switzerland. (Extracted from Keller and Vedy, 1994.)

Studies in forest ecosystems indicate that metals accumulate primarily in the forest floor humus and that atmospheric input exceeds output (Siccama and Smith, 1978; Van Hook et al., 1977). In the Solling forest ecosystems (Heinrichs and Mayer, 1980), Cu accumulated primarily in the OM, with only about 8% in the total ecosystem stored in the biomass, and 2% in the organic topsoil. Similarly, in the Bärhalde watershed in central Europe, increased atmospheric input of metals has resulted in an enrichment and rise of turnover for Cu as well as Cd and Pb (Stahr et al., 1980). These three metals are adsorbed primarily by the humus and oxides of Fe in the soil.

4 Copper in Soils

4.1 Total Copper in Soils

An average value of 30 ppm (range of 2 to 250 ppm) for total Cu content of world soils has been reported (Adriano, 1986). For normal arable soils, total Cu is expected to range from 1 to about 50 ppm (Gilbert, 1952). In the United States the severely weathered, leached, and acidic soils of the southeastern states average much lower in total Cu than soils from western and midwestern states. Total Cu averaged 25 ppm (range of <1 to 700 ppm) in surface soils of the conterminous United States. A mean of 33 ppm total Cu was reported for tropical Asian paddy soils, ranging from 11 ppm for Malaysian soils to 50 ppm for Sri Lanka soils (Domingo and Kyuma, 1983). In the European part of the former Soviet Union, total Cu content of surface soils ranged from 5 ppm in sod-podzolic sandy soils to 55 ppm in meadow-chernozemic soils (Zborishchuk and Zyrin, 1978). Most of the values for other soil groups fall within 10 to 30 ppm. In China, a total Cu content (mean = 20 ppm, n = 3938) of soils having a geometric mean of 23 ppm was reported (Chen et al., 1991). The Mollisols and Oxisols have the lowest Cu contents (10 and 11 ppm, respectively) while the Lithosols have the highest (i.e., 26.5 ppm). For Japanese soils, Shiha (1951) reported an average of 93 ppm total Cu, the range being 26 to 151 ppm. In Canada, the mean value for total Cu in soils was 22 ppm, ranging from 5 to 50 ppm (McKeague and Wolynetz, 1980). Similar mean values were reported for soils in Manitoba (Mills and Zwarich, 1975) and in Ontario (Whitby et al., 1978). The average total Cu content of British surface soils has been reported as 20 ppm (Swaine and Mitchell, 1960). Similarly, an average value of 20 ppm total Cu has been determined for soils of France (Aubert and Pinta, 1977).

Copper contents of world soils have been characterized according to climatic zones as follows (Aubert and Pinta, 1977): (1) temperate and boreal regions—total Cu ranges from trace to several hundred ppm, with the lowest contents (trace to 8 ppm) reported on the fluvio-glacial and alluvial sands of the lower Volga Valley in the former Soviet Union; some of the highest values (50 to 200 ppm) in the former Soviet Union are found in the Kola Peninsula; (2) arid and semiarid regions—soils in these regions generally contain from average to high total Cu; and (3) tropical humid regions—soils have fluctuating total Cu contents, with values from trace to upwards of 250 ppm.

4.2 Bioavailable Copper in Soils

It is widely accepted that micronutrient uptake by crops can be correlated with some extractable fraction of the element in soils. However, variability in extractability with time, chemical extractant employed, and other extraction conditions must be taken into account. Various extraction techniques for Cu can be summarized into three groups: (1) water-soluble and exchangeable, e.g., hot water and salts of NH_4^+, Mg^{2+}, and Ca^{2+} ions such as NH_4NO_3, $MgCl_2$, etc.; (2) dilute acid, e.g., dilute HCl and HNO_3; and (3) chelating agents, e.g., EDTA, DTPA, etc. While total Cu in normal arable soils varies from 1 to 50 ppm, bioavailable Cu may range from only 0.1 to 10 ppm depending on the kind of extractant used (Baker, 1974). For example, Follett and Lindsay (1970) found that Cu determined by the DTPA method ranged from 0.14 to 3.18 ppm, while total Cu ranged from 2 to 92 ppm. In this case, total Cu and DTPA-extractable Cu were correlated ($r = 0.67$ to 0.76). Mitchell (1964) found that Scottish soils extractable by 0.5 M EDTA contained between 0.30 and 10.0 ppm Cu.

Soluble Cu (extracted with 0.01 M $CaCl_2$) from a suite of soils is proportional to the total Cu content alone, irrespective of pH (Sauve et al., 1997). However, the free Cu^{2+} can be predicted using total Cu and soil pH.

Extractable Cu is often used as a diagnostic tool to identify Cu-deficient conditions in plants. However, unlike Zn, documented plant responses to Cu are rare and occur in specific situations (e.g., high-OM soils) (Sims and Johnson, 1991). Consequently, soil tests for Cu have not received much individual effort and extractable Cu is generally determined by the same procedures as for the other micronutrients. The critical levels for some extractants have been given as: 0.005 M DTPA, 0.20 ppm (Lindsay and Norvell, 1978); NH_4HCO_3-DTPA solution (AB-DTPA), 0.50 ppm (Soltan-pour and Schwab, 1977); 1 N HCl, 1.9 ppm for citrus culture (Fiskell and Leonard, 1967); and 0.05 M EDTA, 0.60 ppm (Mitchell, 1964).

For coastal plain soils of North Carolina, the following critical levels of soil Cu were identified: double acid (Nelson et al., 1953), 0.26 ppm; Mehlich-Bowling reagent (Mehlich and Bowling, 1975), 0.62 ppm; NH_4HCO_3-DTPA (Soltanpour and Schwab, 1977), 0.53 ppm; and Meh-lich-3 (Makarim and Cox, 1983), 0.37 ppm. A summary (Table 13.3) of various soil test methods, factors influencing their interpretation and typical ranges in critical levels for Cu has been provided (Sims and Johnson, 1991). As indicated, soil properties, such as pH, OM, $CaCO_3$ content, etc., are often used with extractable Cu to predict the likelihood of a Cu response.

Because of the overriding importance of organically bound Cu in regulating Cu supply to plant roots, chelating agents are becoming more widely accepted as extractants to measure bioavailable Cu (Baker, 1974; Robson and Reuter, 1981). Initial efforts in selecting a chelate for micronutrient tests centered around EDTA. Trierweiler and Lindsay (1969) developed an EDTA-$(NH_4)_2CO_3$ extractant that was highly correlated with plant-available Zn in calcareous soils ($r > 0.81$, critical value = 1.4 ppm). The DTPA soil test, developed for near neutral and calcareous soils by Lindsay and Norvell (1978) is an evolution of the EDTA method that simultaneously extracts four micronutrient cations (Fe, Mn, Cu, and Zn).

TABLE 13.3. Soil test methods, soil factors influencing their interpretation, and typical ranges in the critical levels for copper.

Method	Critical level	Interacting factors
Mehlich-1, Mehlich-3	0.1–10.0	Crop species, OM, pH,
DTPA, AB-DTPA	0.1–2.5	$CaCO_3$ content
EDTA	0.4–1.0	
0.1 M HCl	1.0–2.0	
Modified Olsen's	0.3–1.0	

Source: Extracted from Sims and Johnson (1991).

Norvell (1984) modified the DTPA soil test to increase its applicability to acidic and metal-rich soils. Modifications included buffering to pH 5.3 and increasing the solution/soil ratio from 2:1 to 5:1 to provide greater chelating capacity and pH control.

Walsh et al. (1972) found highly significant negative correlations between 0.1 N HCl, EDTA, or DTPA and the yield of snapbeans. In other words, they found highly significant positive correlations between these extractants and Cu levels in plant tissues up to phytotoxic levels. Gupta and MacKay (1966) recommended the use of 0.2 M ammonium oxalate (pH 3) over 0.1 N HCl. Fiskell (1965) used citrate-EDTA for a quick test to determine toxic Cu levels exceeding 50 ppm, even if the extractant NH_4OAc (pH 4.8) was the one generally used. For Wisconsin soils, the extractant 1 N NH_4OAc (pH 7)/0.01 M EDTA had been applied in soil tests for the simultaneous diagnosis of Cu, Zn, and Mn (Dolar and Keeney, 1971; Dolar et al., 1971). In general, acidic extractants such as HCl may extract Cu from pools that are not bioavailable to plants, are less reproducible, and are unsuitable for extracting calcareous soils (Robson and Reuter, 1981).

In the Netherlands, the extractant 0.05 M $Mg(NO_3)_2$ has been effective, especially in soils having high Cu contents (Lexmond and de Haan, 1977). When the level of Cu extracted by this solution exceeds 2 ppm, yield depressions of some crops can be expected to occur (Fig. 13.2). In comparing soil and plant analysis as predictors of crop response, Karamanos et al. (1986) studied the responses of wheat, flax, barley, and canola to Cu applications for 2 yr at six locations in the sandy transitional grassland soils of Saskatchewan, Canada. The DTPA-extractable Cu was found to be more effective than plant analysis in diagnosing Cu deficiency. The critical DTPA-extractable level of soil Cu was 0.40 ppm, identical to the value proposed by Kruger et al. (1985) in an earlier study in Saskatchewan that involved 21 field trials with wheat, barley, rapeseed and alfalfa, and growth chamber studies with wheat grown in 11 soils (Fig. 13.3).

In paddy soils, the EDTA extractant was recommended to be the most suitable for Cu, followed by NH_4OAc and HCl (Selvarajah et al., 1982). There were a few instances however, when plant uptake of Cu could not be predicted by extractable Cu in soil even if chelating agents were used (Gough et al., 1980; Jarvis, 1981b).

Robson and Reuter (1981) proposed that, in order to more or less standardize soil testing procedures, the following tests may be adopted to identify Cu-responsive conditions:

1. For acid and near-neutral soils (containing no free $CaCO_3$)—0.02 M EDTA in 1 M NH_4OAc (pH 7) as recommended by Borggaard (1976).
2. For neutral and calcareous soils—0.005 M DTPA containing 0.01 M $CaCl_2$ and 0.1 M triethanolamine (pH 7.3) as recommended by Lindsay and Norvell (1978).
3. For polluted or potentially toxic soils—0.1 M EDTA adjusted to pH 6 with NH_4OH as recommended by Clayton and Tiller (1979).

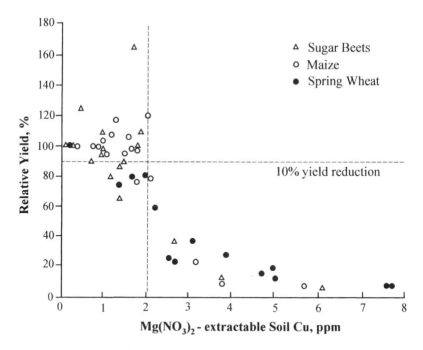

FIGURE 13.2. Relationships between Mg(NO$_3$)$_2$-extractable copper in soil and crop yield. (From Lexmond and de Haan, 1977.)

In summary, the most consistent predictor of plant response to Cu has been with EDTA- and DTPA-based extractants, with critical values ranging from 0.30 to 0.60 ppm.

4.3 Copper in Soil Profile

Applied or deposited Cu will persist in soil because it is strongly complexed or sorbed by OM, oxides of Fe and Mn, and clay minerals (Adriano, 1986). Thus, it is one of the least mobile trace metals, thereby rendering it more uniformly distributed in many soil profiles. For instance, Jones and Belling (1967) observed virtually no downward movement of Cu in silty and clay soils, and only slight movement (1 to 3 cm) in sandy soils with low CEC. Copper accumulates in litter overlying the soil profile or in the top few centimeters of the surface soil, as observed in soils near a Cu smelter (Kuo et al., 1983), or in soils heavily impacted by industrial emissions (Miller and McFee, 1983). Similarly, Cu accumulates in the surface layer of arable soils from fungicide applications, soil amendments, and by accumulation from crop residues (Thornton, 1979; Walsh et al., 1972). However, considerable leaching of Cu in the soil profile can occur in exceptional cases such as in humus-poor, acidic peat (Histosol) upon application of high levels of Cu (1500 ppm or more) (Mathur et al., 1984) or in very acidic mineral soils, such

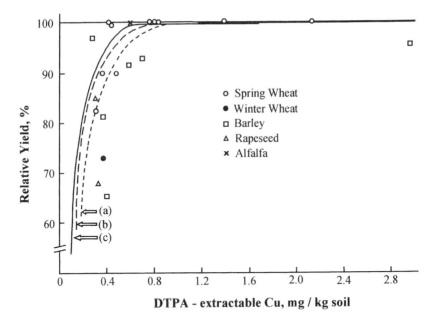

FIGURE 13.3. Results of DTPA soil test for critical values for copper for a number of crops: (a) barley, (b) all crops, and (c) spring wheat. (From Kruger et al., 1985.)

as those around Ni and Cu smelters in Coniston, Ontario, Canada (Hazlett et al., 1984). Soils around these smelters have been acidified to as low as pH 2.4 by the aerial emission of sulfur-containing compounds during smelting.

The level and distribution of total and extractable Cu in the soil profile can be expected to vary with soil type and parent material from which the soils are derived. The vegetation may also exert some influence on the profile distribution of Cu. For example, in forest ecosystems in western Switzerland, Cu accumulates in the forest litter, tapers off in the surface soil, and becomes uniformly distributed in the soil profile (Keller and Vedy, 1994). Distribution of total Cu can be expected to be fairly uniform in profiles developed *in situ* on igneous rock or on unsorted sediment such as sandstone or loess (Mitchell, 1964). Little variation can also be expected in freely drained soils (Thornton and Webb, 1980), but Cu may be mobile in profiles with restricted drainage, with extractable Cu possibly accumulating in gleyed horizons of these soils (Swaine and Mitchell, 1960). However, profile distribution of Cu can be altered by various pedological processes. Thornton (1979) reported that there is strong evidence that Cu is redistributed in the profile as a result of podzolization, with depletion in the A2 and accumulation in the B horizon. Examples have been cited from podzols in the United States, the United Kingdom, and Canada. Other factors, such as OM accumulation and pH, can also influence the distribution of Cu in the soil profile.

4.4 Forms and Chemical Speciation of Copper in Soils

Copper in soils can occur in several forms that are partitioned between the solution and solid phases. From the standpoint of plant nutrition, the dissolved and exchangeable forms are of special importance. Copper in the soil solution (i.e., free and complexed) and exchangeable form is considered labile fractions; those specifically adsorbed on OM and oxide surfaces as semilabile, and the residual fraction as nonlabile. There have been indications that organically bound Cu (and other metals) is more phytoavailable than other fractions retained in primary minerals (residual) and inorganic precipitates (e.g., sulfides and carbonates) (McLaren and Crawford, 1973a; Stevenson, 1991).

Distribution of Cu between soil constituents can be influenced considerably by the presence of OM, and Mn and Fe oxides. The bulk of the bioavailable soil Cu resides in the organically bound fraction (McLaren and Crawford, 1973a). This makes Cu somewhat unique among metals because of its strong affinity for OM. The organic-fraction Cu is high compared with that for other metals even though the absolute amounts are low (Shuman, 1991). The importance of OM in influencing metal behavior is in the order $Cu > Zn > Mn$ (McGrath et al., 1988). Accordingly, organic-bound Cu is the dominant form in Histosols (Mathur and Levesque, 1983) and in sludged soils (Sposito et al., 1982). The organic-bound form of Cu also can be expected to dominate in ordinary arable soils. For example, Shuman (1979) obtained 1.0 to 68.6% Cu in organically bound form in 10 representative southeastern U.S. soils. Enrichment of soil with anthropogenic Cu can be reflected by increases in most fractions with most of the increases occurring in the organic fraction. For example, in the heavily metal-contaminated soils in Doñana National Park, Spain the organic Cu fraction dominated the total Cu (Ramos et al., 1994). Similarly, in heavily manured soils in the Netherlands and the United States, Cu in soil solutions increased with increasing DOC (del Castillo et al., 1993); Cu increased in the OM fraction (Payne et al., 1988). In some cases, however, Cu associated with Fe and Mn oxides may constitute the major fractions of total Cu in soils (Kuo et al., 1983; Miller and McFee, 1983).

In soil solutions from sludged soils, it has been predicted that only small amounts (2 to 19%) of the total Cu occur as free ionic form (Behel et al., 1983; Emmerich et al., 1982); even smaller amounts were complexed with sulfate and phosphate, while up to about 80% of the total amount of Cu occurs in the fulvic-metal complexes (Behel et al., 1983). McBride and Bouldin (1984) estimated that at least 99.5% of the Cu in soil solution of a surface loamy soil (5.3% OM content) was in an organically complexed form. Similarly, Kerven et al. (1984) found that most of the Cu in soil solutions of acid peaty soils was present as organic complexes.

In solutions containing relatively high concentrations of chlorides, up to 80% of Cu^{2+} can be complexed with this ligand (Doner et al., 1982).

Mattigod and Sposito (1977) predicted the major inorganic form of complexed Cu^{2+} in neutral and alkaline soil solutions to be $CuCO_3^0$. In highly alkaline soils, $Cu(OH)_4^{2-}$ and $Cu(CO_3)_2^{2-}$ become the predominant soluble species.

4.5 Sorption of Copper in Soils

The most important sink in soils and sediments for metals is the oxides of Fe and Mn, OM, sulfides, and carbonates (Jenne, 1968, 1977). Of lesser importance are the phosphates and clay-size aluminosilicate minerals. Since $CaCO_3$ is abundant in soils of arid and semiarid regions, it may be an important sink, particularly for Zn and Cd. Sulfides, which are prevalent in paddy and wetland soils, are an important sink because they effectively retain metals by coprecipitation.

Relative to specific adsorption, it can be generalized that the oxides of Fe and Mn, and OM are the most important soil constituents (Jenne, 1977; McLaren et al., 1981). The adsorption maxima among soil constituents decrease in the order: Mn oxide > OM > Fe oxide > clay minerals. However, in soils where clay minerals and Fe oxides are dominant, their contribution can override those of OM and Mn oxides (McLaren and Crawford, 1973b). Stevenson and Fitch (1981) indicated however, that organic colloids and clays play a major role in Cu retention by soils. In most mineral soils, Cu may be bound as clay-metal-organic complex, since in these soils OM is intimately bound to the clay. Their pesticide retention results infer that in soils containing up to 8% OM, both organic and mineral surfaces are involved in adsorption; at higher OM contents, adsorption occurs mostly on organic surfaces. Thus, it appears that for the relatively small amount of Cu normally present in soils, specific adsorption, particularly to OM, is a more important process than nonspecific adsorption (i.e., cation exchange process) in controlling the concentration of Cu in soil solution. Indeed, sorption isotherms indicate preferential adsorption of Cu onto OM associated with the clay fraction of the soil (Wu et al., 1999).

The relative adsorptive capacities of various soil constituents are depicted in Figure 13.4, which demonstrates that Mn oxide and OM are the most likely to bind Cu^{2+} in a nonexchangeable form in soils. With the possible exception of Pb^{2+}, Cu^{2+} is the most strongly adsorbed of all transition metals on Fe and Al oxides, and oxyhydroxides (McBride, 1981). In general, metal affinity for soils follows the order: Pb > Cu > Zn = Cd ≥ Ni (Basta and Tabatabai, 1992; Harter, 1983). The Cu sorption capacity of soils may follow closely the extent of CEC and the amounts of the oxides of Fe and Mn, and OM in the soil. However, OM alone may not be a good indicator of the sorption capacity of soil (Harter, 1979; Kuo and Baker, 1980).

The sorption of Cu onto soils and soil constituents has been shown to follow either the Langmuir and the Freundlich isotherms (Adriano, 1986; Harter, 1991). From Langmuir isotherm plots, montmorillonite surface

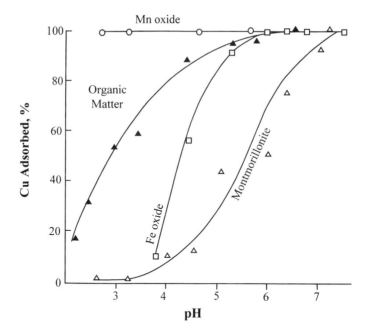

FIGURE 13.4. Adsorption of copper by soil constituents as a function of pH (1 g sample equilibrated with 200 ml of 0.05 M CaCl$_2$ containing 5 ppm Cu). (Modified from McLaren and Crawford, 1973b.)

appears to provide three types of adsorption sites for Cu (Takahashi and Imai, 1983). Adsorption onto site 1 corresponds to an ion-exchange reaction accounting for 70 to 85% of the adsorption maximum; precipitation of the metal hydroxides in the bulk solution corresponds to an site 3. Adsorption onto site 2 is probably due to formation of metal-hydroxy species although it occurred even at pH 4.

On the relative adsorption of Cu and Cd by mineral soils, Cavallaro and McBride (1978) summarized their findings as follows:

1. Cu^{2+} is much more strongly adsorbed than Cd^{2+} in soils so that Cd^{2+} is likely to be more mobile than Cu^{2+} in the soil profile.
2. The soil solutions are generally undersaturated with respect to the least-soluble mineral phases of these heavy metals, indicating that precipitation does not control metal solubility in the surface soils.
3. Competing ions such as Ca^{2+} shift the adsorption equilibria for Cd^{2+} drastically, with less effect upon Cu^{2+}, suggesting that ion exchange is responsible for Cd^{2+} adsorption, but Cu^{2+} may be bonded more specifically.
4. Low-pH soils are much less effective in removing Cu^{2+} and Cd^{2+} from solution than are neutral soils or soils containing calcium carbonate.

5. Soluble metal-organic complexes of Cu^{2+} are more prevalent than those of Cd^{2+} in soil solutions, but acidic soils demonstrate much less evidence of the occurrence of these soluble complexes.

In acidic soils, a greater fraction of functional groups in the OM is associated with protons or Al^{3+} thereby reducing the soil's ability to adsorb Cu^{2+}.

Hodgson et al. (1966) showed that more than 98% of the Cu in soil solution was in organic complexed form, suggesting that only small quantities of free Cu^{2+} ions are available for adsorption in neutral soils. The hydrolysis constant for Cu^{2+} (for the reaction $Cu^{2+} + H_2O \rightleftharpoons CuOH^+ + H^+$) is $10^{-7.6}$, indicating that at pH above 7 the hydrolyzed species $CuOH^+$ would be more important than Cu^{2+} for adsorption (Ellis and Knezek, 1972). Although $CuOH^+$ has been suggested to be the adsorbed hydroxy ion (Menzel and Jackson, 1950), solution hydrolysis data suggest that $Cu_2(OH)_2^{2+}$ and $Cu(OH)_2^0$ are probably more important species (Baes and Mesmer, 1976). Indeed, Harter (1991) indicates that at neutral to alkaline pH range, $Cu(OH)_2^0$ predominates.

4.6 Complexation of Copper in Soils

Early on, Coleman et al. (1956) predicted that because of the stability of Cu-organic complexes, especially at low salt concentrations, only extremely small quantities of free Cu^{2+} ions should exist in soil solution. A number of studies have shown that Cu^{2+} in soil solution, especially at high pH, exists primarily in a form complexed with soluble organics (Hodgson et al., 1965, 1966; Stevenson, 1991). In soils amended with organic residues, such as sewage sludge, organically complexed Cu may be expected to be the predominant form (Behel et al., 1983). Stevenson and Fitch (1981) reported that, based on a commonly used soil Cu fractionation scheme, organically bound Cu accounts for about 20 to 50% of the total soil Cu. The amount of organically complexed Cu in solution generally increases above pH 7 because of the greater solubility of soil OM at higher pH while the concentration of free Cu^{2+} at higher pH is much lower, usually in the range of 10^{-9} to 10^{-8} M (Hodgson et al., 1965; McBride and Blasiak, 1979). Total native Cu in soil solution is often in the range 10^{-7} to 10^{-6} M (Bradford et al., 1971) so that the ratio of total dissolved Cu to free Cu^{2+} can be greater than 100. However, this ratio may be only as low as 4 in soil solutions of organically enriched soils (Behel et al., 1983).

In Section 3, it was indicated that organic-enriched surface horizons usually contain higher concentrations of Cu than the lower horizons, which usually contain less OM. Thus, complexation by OM in the form of humic and fulvic acids has long been recognized as an effective mechanism of Cu retention in soils. Ample evidence exists for the complexation of Cu^{2+} by humic and fulvic acids (Stevenson and Fitch, 1981): (1) inability of monovalent cations (e.g., K^+, Na^+, etc.) to replace adsorbed Cu^{2+} from mineral and organic soils, (2) strong correlation between Cu^{2+} retention and OM content, (3) ability of known chelating agents to extract Cu while solubilizing part of the soil humus, and (4) selective retention of Cu^{2+} by humic and fulvic acids in the presence of a cation exchange resin.

In addition, it has been shown that among the metals, Cu is most extensively complexed by humic materials (Baker, 1974; Bloomfield and Sanders, 1977). Bloom (1981) indicates the following preference series for divalent ions for humic acids and peat: Cu > Pb ≫ Fe > Ni = Co = Zn > Mn = Ca. The complexing ability of humic and fulvic acids is due to their high oxygen-containing functional groups such as carboxyl, phenolic hydroxyl, and carbonyls of various types (Schnitzer, 1969; Stevenson, 1972). Because of high acidities and relatively lower molecular weights, metal complexes with fulvic acids are more mobile than those with humic acids. Fulvic acids are also more efficient in complexing metals than humic acids (Kawaguchi and Kyuma, 1959; Stevenson, 1972) and may also be more available to plants than humic acid–complexed metals (Stevenson and Fitch, 1981).

Stevenson and Fitch (1981) summarized the following effects of the formation of Cu-OM complexes on soils: (1) soil solution Cu concentration can be decreased through complexation to clay-humus or via the formation of insoluble complexes with humic acids; soluble ligands may be of considerable importance in transforming solid-phase forms of Cu into dissolved forms; (2) in high pH soils, complexation may promote maintenance of Cu in dissolved forms; (3) in the presence of excess Cu, complexation may reduce the concentration of Cu^{2+} to nontoxic levels; and (4) natural complexing agents (i.e., humic and fulvic acids) may be involved in the transport and mobility of Cu in soils. From the above discussions, it is clear that OM provides a dual contrasting role in the chemistry of Cu: as a specific adsorption site (as particulate OM) decreasing the mobility of Cu, and as a natural complexing agent (as soluble OM) increasing the solubility of Cu, especially in alkaline pH.

Like the natural complexing agents that enhance the migration of metals, synthetic chelating agents (e.g., EDTA, DTPA, etc.) combine with metals to increase the levels of these metals in soil solution, thereby enhancing their bioavailability. However, the stability of metal-synthetic chelating agents depends largely on soil pH. Copper-DTPA like ZnDTPA, is expected to be unstable in acid soils, moderately stable in slightly acid soils, and stable in alkaline and calcareous soils (Norvell, 1991). In contrast, CuEDTA is most stable in slightly acidic to near neutral soils (pH 6.1 to 7.3). In acid soils (pH 5.7 and below) CuEDTA becomes unstable since Fe displaces Cu, while in alkaline soils (pH 7.85 and above) Ca displaces Cu.

5 Copper in Plants

5.1 Essentiality of Copper in Plants

Copper is one of seven traditional micronutrients (Zn, Cu, Mn, Fe, B, Mo, and Cl) essential for normal plant nutrition. The essentiality of Cu was firmly established in the 1930s primarily through the work of Sommer (1931) and Lipman and MacKinney (1931) using nutrient culture. During the ensuing 20 yr, a great mass of evidence was gathered confirming the essentiality of Cu for the growth of plants (Reuther and Labanauskas, 1965).

Copper is required only in very small amounts; 5 to 20 ppm in crop tissue is adequate for normal growth, while <4 ppm is considered deficient, and 20+ ppm is considered toxic (Jones, 1972). Copper has been established as a constituent of a number of plant enzymes. It occurs as part of the prosthetic groups of enzymes, as an activator of the enzyme systems, and as a facultative activator in enzyme systems (Gupta, 1979). Insufficient Cu affects a multitude of physiological processes in plants: carbohydrate metabolism (photosynthesis, respiration, and carbohydrate distribution), N metabolism (N_2 fixation, and protein synthesis and degradation), cell wall metabolism (especially lignin synthesis), water relations, seed production (especially pollen viability), and disease resistance (Bussler, 1981; Romheld and Marschner, 1991). It may also affect ion uptake and plant differentiation in early developmental stages. Copper deficiency in legumes depresses nodulation and N_2 fixation, leading to N deficiency symptoms that can be corrected by application of N fertilizers. In general, Cu deficiency depresses reproductive growth (i.e., formation of seeds and fruits) more than vegetative growth. One possible cause of sterility is the occurrence of starch deficiency in the pollen (Graham, 1975; Misuno et al., 1983).

5.2 Deficiency of Copper in Plants

A micronutrient survey conducted between 1961 and 1967 revealed that 16 states recognized and made recommendations for Cu application for major field, forage, vegetable, fruit, and nut crops (Sparr, 1970). The world distribution of Cu deficiency closely follows that of Zn deficiency occurring in the Americas, Africa, Asia, Oceania, Europe, and other parts of the world (Adriano, 1986; Bould et al., 1953). Copper deficiency in grain crops is also known as reclamation disease, wither-tip, yellow-tip, or blind ear. In woody species, particularly in citrus, the malady is known as dieback or exanthema. Prevalent Cu deficiency can be expected to occur in: peat and muck soils; alkaline and calcareous soils, especially sandy types or those with high levels of free $CaCO_3$; highly leached sandy soils, such as acidic sandy soils in Florida; highly leached acid soils, in general; calcareous sands, as in the south coast of Australia; soils heavily fertilized with N, P, and Zn; and old corrals.

Copper deficiency in plants has frequently appeared in many parts of the world whenever acid Histosols are brought into agricultural production (Kubota and Allaway, 1972). Sorption reactions between Cu and OM cause Cu deficiency in peat and mucks especially under acidic conditions. Low Cu concentrations in parent materials such as quartz led to low available Cu in some sandy-textured soils (Alloway and Tills, 1984). Soils that developed from limestone often contain inadequate Cu due to its inherently low total Cu content coupled with an alkaline pH (Gartrell, 1981). The term *reclamation disease* refers to severe Cu deficiency that appears in newly drained and developed acid Histosols. Peats and mucks have commonly produced Cu-deficient crops throughout northern and western Europe, some midwestern and eastern states in the United States, and in New Zealand and Australia (Gartell, 1981).

In general, soils most commonly found to be deficient in Cu are poorly drained mineral soils, mineral soils high in OM, and Histosols (Barnes and Cox, 1973). Thus, application of Cu fertilizers on organic soils is usually recommended for normal plant nutrition. Total Cu in soil should exceed 4 to 6 ppm in mineral soils and 20 to 30 ppm in organic soils to sustain maximum yields of responsive crops (Lucas and Knezek, 1972). The responsiveness of crops to Cu fertilization is indicated in Table 13.4, with some major cereal crops noticeably highly responsive. Crops that are highly sensitive to Cu deficiency can serve as indicator species for detecting Cu-deficient soils. Some of the sensitive species, with their respective Cu concentrations are shown in Table 13.5.

Copper deficiency in conifers can occur in trees growing in sandy soils low in Cu content and in soils likely to immobilize Cu, such as peat soils, sandy Podzols, and calcareous soils. Copper deficiency in *Pseudotsuga menziesii* occurs on acid humus Podzols in the Netherlands, in *Pinus radiata* on alkaline soils in Greece, and on alkaline soils in South Australia (Turvey, 1984). Copper deficiency in *Pinus radiata* in Victoria, Australia has been associated with acid, organic-rich sandy Podzols. Very acid, infertile, humus Podzols in forests in northeastern Netherlands are also prone to cause Cu deficiency in conifers (Van den Burg, 1983). In eucalypts (*Eucalyptus globulus*), Cu deficiency is associated with tree deformation and reduced wood production from plantations (Gherardi et al., 1999). Copper is required in the biosynthesis of lignin, which is mediated by catechol oxidase (or polyphenol oxidase, a Cu-dependent enzyme). In general, conifers are more susceptible to Cu deficiency than broadleaf woody species.

In most plant species, Cu deficiency is characterized by chlorosis, necrosis, leaf distortion, and terminal dieback, with symptoms occurring first in young shoot tissues (Robson and Reuter, 1981). Symptoms of Cu deficiency in spring wheat includes rolling and wilting of young leaves and twisting and terminal dieback (Owuoche et al., 1994a,b). Once absorbed, Cu is poorly translocated. Hence, the terminal growth of most plants is the first to be affected. Specific symptoms often depend on plant genotypes and the

TABLE 13.4. Relative sensitivity of crops to copper deficiency.

Low	Moderate	High
Beans	Barley	Wheat
Peas	Broccoli	Lucerne
Potato	Cabbage	Carrots
Asparagus	Cauliflower	Lettuce
Rye	Celery	Spinach
Pasture grasses	Clover	Table beets
Lotus spp.	Parsnips	Sudan grass
Soybean	Radish	Citrus
Lupine	Sugar beets	Onion
Rape	Turnips	Alfalfa
Pines	Pome and stone fruits	Oats
Peppermint	Vines	Barley
Spearmint	Pineapple	Pangolagrass
Rice	Cucumber	Millet
Rutabagas	Sugar beets	Sunflower
	Corn	Dill
	Cotton	Watermelon
	Sorghum	
	Sweet corn	
	Tomato	
	Apple	
	Peaches	
	Pears	
	Blueberries	
	Strawberries	
	Tung-oil	
	Mangels	
	Swiss chard	

Sources: Follett et al. (1981); Gartrell (1981); Jones (1991); Lucas and Knezek (1972).

stage of deficiency. In general, deficiencies in crops produce abnormal coloring and development, curling of leaves, general absence of flower formation in severe cases, wilted appearance, lowered quality in fruit and grain, and reduced grain yield (Haque et al., 1993; Murphy and Walsh, 1972).

5.2.1 Correction of Copper Deficiency

Occurrence of Cu deficiency frequently reflects the management practices during crop production. Overliming acid soils may induce Cu deficiency by increasing the amounts of Cu complexed by soil OM, sorption capacity due to increase in surface charge of inorganic soil constituents, and immobilization by hydrolysis reactions. Application of macronutrients may cause rapid plant growth, thereby accentuating Cu deficiency by depleting Cu in the soil solution (Martens and Westermann, 1991). Plants differ widely in susceptibility to Cu deficiency (see Tables 13.4 and 13.5). For example, Cu

TABLE 13.5. Deficient, sufficient, and toxic levels of copper in crops.

Plant species	Part or plant tissue sampled	ppm Cu DW		
		Deficient	Sufficient	Toxic
Spring wheat	Boot stage tissue	1.9	3.2	
	Kernels	0.8	2.3	
	Straw	2.4	3.9	
	Grain	2.5[a]	—	
Winter wheat	Aboveground vegetative plant tissue	—	5–10	>10[b]
Barley	Boot stage tissue	2.3	4.8	
	Kernels	0.5	2.0	
	Straw	1.5	3.0	
	Top 4 leaves at bloom	—	6–12	>12[b]
Oats	Boot stage tissue	2.3	3.3	
	Kernels	0.7	1.8	
	Straw	1.2	2.3	
	Leaf lamina	<4.0	6.8–16.5	
Soybeans	Upper fully mature trifoliate leaves before pod set	<4.0	10–30	>50
Corn	Leaf at or opposite and below ear level at tassel stage	5[a]	—	
	Ear leaf sampled when in initial silk	<2	6–20	>50
	Middle of first ear leaf at tasseling	<2	6–50	>70
Alfalfa	Upper stem cuttings in early flower stage	7[a]	—	
	Top 15 cm of plant sampled before bloom	<5	11–30	>50
	Top 1/3 of plant shortly before flowering	<2	8–30	>60
Red clover	Tops	—	7–16.4	
	Top of plants at bloom	<3	8–17	>17[b]
	Top of plants at bloom		9.8–11.5	
Timothy	Tops	—	5.7–11.7	
Pasture grasses	Aboveground part at first cut	<5	5–12	>12[b]
Potato	Aboveground part 75 days from planting	<8	11–20	>20[b]
Cucumber	Mature leaves from center of stem 2 weeks after picking	<2	7–10	>10[b]
Tomato	Mature leaves from top of plant	<5	8–15	>15[b]
	Mature leaves	—	3.1–12.3	

[a]Considered critical level.
[b]Considered high but not toxic level.
Source: As summarized by Gupta (1979).

deficiency may occur in wheat (a highly sensitive species) but not in soybeans (a highly insensitive plant) when these plants are used in a crop rotation on the same soil.

Copper deficiency is most frequently corrected by soil application of Cu rather than by foliar application (Murphy and Walsh, 1972). Commercial Cu fertilizers are available as inorganic carriers or organic carriers (Martens and Westermann, 1991). Cupric sulfate is the most common Cu compound used for correction of Cu deficiency in crops. The general use of $CuSO_4$ reflects its high water solubility, relatively low cost, and wide availability.

Copper deficiency is generally corrected by applying 3.3 to 14.5 kg Cu ha^{-1} as broadcast $CuSO_4$ (Table 13.6). Differing broadcast rates of Cu required for correcting Cu deficiency reflect variations in soil properties, plant requirements, and concentrations of extractable soil Cu.

TABLE 13.6. Recommended application rates of $CuSO_4$ for correction of copper deficiency in crops.

Crop	Application rate (kg Cu ha^{-1})	Application method	Remarks
Citrus, tung	7	Broadcast	Every 5 yr
Citrus	0.09	Foliar	Annual application using 100 L H_2O ha^{-1}
Vegetables	1.1–4.5	Band	Maximum accumulative
	4.5–14.5	Broadcast	application ≤68 kg Cu ha^{-1}
Carrot, cauliflower, celery, clover, alfalfa, corn, oats, potato, radish, sudan grass, wheat	13.4	Broadcast	Recommendation for mucks; use 4 kg Cu ha^{-1} for sand
	3.4	Band	Recommendation for mucks; use 1 kg Cu ha^{-1} for sand
Dryland wheat	4.0	Foliar	Two sprays of 2 kg Cu ha^{-1} at Feekes growth stages 1 and 10
Corn, soybean, wheat	2.2–3.3	Band	Band with row fertilizer
	3.3–6.6	Broadcast	
	2.2	Foliar	Use basic $CuSO_4$
Highly sensitive crops	6.6	Band	Soils with 1.0 M HCl-extractable Cu, <9 mg kg^{-1}
	3.3	Band	Soils with 1.0 M HCl-extractable Cu, 10 to 20 mg kg^{-1}
Barley, flax, oats, wheat	0.3	Foliar	Organic soils with DTPA-extractable Cu, <2.5 mg kg^{-1}
Alfalfa, barley, flax, lettuce, oats, onion, spinach, table beet, wheat	6.7–13.4	Broadcast	Organic soils with DTPA-extractable Cu, <2.6 mg kg^{-1}
	6.7	Broadcast	Organic soils with DTPA-extractable Cu, 2.6 to 5.0 mg kg^{-1}

Source: As summarized by Martens and Westermann (1991).

Copper application is required at lower rates for correcting Cu deficiency with banded than with broadcast $CuSO_4$ (see Table 13.6). Rates of band-placed $CuSO_4$ required for correcting Cu deficiency are as low as 1.1 kg Cu ha^{-1} for vegetables and as high as 6.6 kg Cu ha^{-1} for highly sensitive crops. The 6.6 kg Cu ha^{-1} is needed for highly sensitive crops grown on organic soils when the 1.0 M HCl-extractable Cu concentrations are <9 mg kg^{-1}. Organic soils that contain 1.0 M HCl-extractable Cu concentrations between 10 and 20 mg kg^{-1} should be provided with about 3.3 kg Cu ha^{-1}.

The relatively high residual availability of applied Cu limits the usual recommended application rates. Fiskel and Younts (1963) suggested reapplication of Cu every 5 yr for correcting Cu deficiency in citrus. An interval of 5 to 8 yr for Cu application depending on the sensitivity of the crop to Cu deficiency and on its severity is generally recommended (Powell, 1975). Studies in Michigan indicate that Cu addition to organic soils is not necessary after a total of 22 and 45 kg Cu ha^{-1} is applied to responsive and highly responsive crops, respectively, or when 1.0 M HCl-extractable soil Cu exceeds 20 mg kg^{-1} (Robertson et al., 1981). The total cumulative Cu application recommended for correcting Cu deficiencies ranges from 33.6 to 45 kg Cu ha^{-1}.

5.3 Accumulation and Phytotoxicity of Copper

Absorption rates of Cu by plant roots are among the lowest for the micronutrients (Graham, 1981). The metabolic requirement for Cu^{2+} absorption is still unclear (Kochian, 1991). There is considerable evidence that free Cu^{2+} ion is the absorbed species. A number of researchers have demonstrated that Cu is absorbed more rapidly from Cu^{2+} solutions than from solutions of Cu complexed with synthetic chelators such as DTPA or EDTA (Kochian, 1991). Using electron paramagnetic resonance spectroscopy to study root Cu absorption, Goodman and Linehan (1979) presented evidence consistent with the dissociation of Cu from EDTA during absorption from a Cu(II)-EDTA solution. Once absorbed, Cu accumulates in roots even in cases where roots have been damaged by toxicity (Adriano, 1986).

In a national survey, Kubota (1983) found a median concentration of 8.4 ppm Cu ($n = 2399$) in dry matter for U.S. legumes (range of 1 to 28 ppm); the median value for grasses was much lower—4 ppm (range of 1 to 16 ppm). Most of the forage samples were from soils with about 25 ppm total Cu (range of 1 to 200 ppm total Cu). About 20 ppm Cu in the shoot of spring barley grown in sand culture (4 ppm Cu in solution) was critical for its growth (Davis et al., 1978). Chino (1981) reported that for rice, toxic levels in plant tops were in the range of 20 to 30 ppm Cu; in roots, toxic range was from 100 to 300 ppm. Mathur and Levesque (1983) consider 20 to 30 ppm Cu as toxic for potatoes. In general, 20 to 100 ppm in tissues of most crops is considered the toxic range (Jones, 1991).

Ranking of metals in inducing visual toxicity symptoms in barley has been reported by Agarwala et al. (1977) as: $Ni^{2+} > Co^{2+} > Cu^{2+} > Mn^{2+} > Zn^{2+}$. In rice, Chino and Kitagishi (1966) found that the toxicity of metals follows the order: $Cu > Ni > Co > Zn > Mn$, which correspondingly follows the order of the metal electronegativity. Hewitt (1953) observed that Cu consistently induced Fe chlorosis in crops that were susceptible to Cu toxicity.

Phytotoxicity of Cu could be predicted by Cu concentrations in soils, either on total or extractable basis. Copper can accumulate in soils from continued applications of Cu in excess of that needed for normal plant growth. In Florida, problems of excessive Cu in citrus soils are a much greater problem than Cu deficiency (Leonard, 1967). Surface soil in some citrus groves in Florida contained Cu in excess of 300 ppm, which had accumulated from repeated use of Bordeaux sprays. Generally, Cu toxicities have been associated with soil Cu levels of 150 to 400 ppm (Baker, 1974). The most phytotoxic form of Cu to citrus rootstock seedlings grown on soils collected from citrus groves is the readily soluble Cu (i.e., exchangeable and sorbed) and is largely dependent on soil pH (Alva et al., 2000). In pot experiments using tomato plants, the near maximum Cu concentration in plant tissue without yield depression occurred at a soil level of 104 ppm Cu (Rhoads et al., 1989); at pH below 6.5 and at a soil level of 150 ppm, reduced plant growth was obtained.

In a field study, significant reductions in yield of snapbeans were noted when >20 ppm of Cu was extracted from soil with HCl or DTPA solution and when >15 ppm Cu was extracted with EDTA (Walsh et al., 1972). In Japan, the limit of Cu in the soils has been set at 125 ppm extractable by 0.1 N HCl for rice cultivation (Chino, 1981). In soils with pH of 4.5 to 6.5, soil Cu may become phytotoxic when 20 to 30 ppm of HNO_3-extractable Cu is present for each 1% of soil organic C (Lexmond and de Haan, 1980). The threshold of soil Cu phytotoxicity is about 25 ppm of NH_4-exchangeable Cu for sandy soils and 100 ppm in clay soils (Delas, 1980). Similarly, Drouineau and Mazoyer (1962) reported a limit of 50 ppm NH_4-exchangeable Cu in soils with pH 5 and 100 ppm in soils with pH 6 to 7. Mathur and Levesque (1983) indicate that soil Cu could become phytotoxic when its total level in certain organic soils is equivalent to more than 5% of the soil's CEC—in their case, 16 ppm Cu for each meq of CEC per 100 g soil.

Reuther and Smith (1954) emphasized the importance of OM and clay in Cu retention and found that, in general, phytotoxicity in orchard trees began when total soil Cu level reached about 1.6 mg for each meq of CEC per 100 g of soil having pH 5.0. In soils with pH of about 5, the loading limit for total soil Cu can be estimated to equal 5% of soil CEC (Leeper, 1978).

In Australian apple and pear orchards where sprays containing Pb arsenate and Cu have been used, soils may have 25 to 35 times the normal concentrations of Cu, Pb, and As (Merry et al., 1986). The Cu concentrations in pasture species grown from such sites were usually high (20 to 60 ppm) and potentially pose some toxicity risk to grazing animals.

The most common Cu toxicity symptoms include reduced growth vigor, poorly developed and discolored root system, reduced tillering, and leaf chlorosis (Jones, 1991; Robson and Reuter, 1981). In addition, Cu toxicity causes stunting, reduced branching, thickening, and unusually dark coloration in the rootlets of many plants (Reuther and Labanauskas, 1965). Similarly, in woody plants, Cu toxicity symptoms include senescent leaves, necrotic stems, and bulbous and stunted roots (Arnold et al., 1994). The chlorotic symptoms in shoots often resemble those of Fe deficiency. At the physiological level, excess Cu^{2+} in plant nutrition can cause breakdown of chlorophyll and carotenoid as well as increases in membrane permeability, acid RNAse activity, and rates of lipid peroxidation (Lidon and Henriques, 1993a; Lura et al., 1994).

6 Ecological and Health Effects of Copper

6.1 Effects of Copper in Terrestrial Ecosystems

A series of field investigations was conducted in northwest England to elaborate on the pathways of Cu and Cd to the food webs of contaminated grasslands subject to aerial deposition from an active Cu refinery complex (Hunter et al., 1987a,b,c). Land within 1 km of the facility perimeter is dominated by housing, grassland, recreational open spaces, and a park and woodland estate. The refinery complex has been established for over 60 yr. The industrial plant includes a copper fire refinery, a copper electrolytic refinery, a brass foundry, and a plant for the production of high-grade continuously cast copper rod and copper–cadmium alloys. The findings can be summarized as follows:

For soil and vegetation contamination (Hunter et al., 1987a): (1) Emissions of Cd and Cu from the refinery caused widespread contamination of soil and vegetation. Dispersion followed an exponential decay away from the source area with elevated levels persisting within 3 km of the refinery; (2) Soil profiles were characterized by retention of metals on the surface soils, which was more prominent for Cu than for Cd; (3) Indigenous flora in the complex was of low diversity and dominated by metal-tolerant populations of *Agrostis stolonifera* and *Festuca rubra*; (4) Vegetation levels of Cu and Cd showed marked seasonal variation in contaminated sites, with peak values occurring in the winter months; the increased levels were from a combination of root absorption and accumulation of particulates adhering on leaf surfaces; and (5) Soil extractions and vegetation analysis indicate a much greater mobility of Cd compared with Cu.

For invertebrate contamination (Hunter et al., 1987b): (1) Invertebrates from contaminated and semicontaminated grasslands in the vicinity of the refinery showed significant elevation of total body Cu and Cd concentrations relative to control values; (2) Copper concentrations in the major taxa were ranked in decreasing order as follows: Isopoda, Collembola, Oligochaeta, Linyphiidae, Opiliones, Lycosidae, Chilopoda, Curculionidae, Carabidae, Staphylinidae, and Orthoptera; (3) Detritivorous soil macrofauna showed accumulation of Cu (2 to 4 times) and Cd (10 to 20 times) relative to concentrations in organic surface soil and plant litter in the refinery site; Oligochaeta and Isopoda both showed significant reductions in population size at the refinery site; (4) Herbivorous invertebrates showed body/diet concentration factors of 2 to 4 times for Cu and 3 to 5 times for Cd; results showed a degree of homeostatic control over accumulation of Cu in contrast to the pattern for Cd; (5) Biotransfer of metals to carnivorous invertebrates reveals marked differences in metal accumulation by predatory beetles and spiders; the Cu/Cd concentration ratio in predatory beetles was 30 to 35:1 compared with 9 to 11:1 in spiders; and (6) Seasonal changes in the abundance, species composition, and age structure of invertebrate populations caused marked variation in metal contamination levels throughout the year.

For contamination of small mammals (Hunter et al., 1987c, 1989): (1) Metal contamination levels in small mammal diets followed the order *S. araneus* > *M. agrestis* > *A. sylvaticus* and varied significantly throughout the year; (2) At relatively low levels of Cd exposure, *M. agrestis* and *A. sylvaticus* appeared to regulate Cd accumulation; in contrast, *S. araneus* showed considerable accumulation of Cd; total body/diet concentration factors were greater than 1 at control and refinery sites; (3) The highly contaminated nature of the invertebrate diet of *S. araneus* resulted in ingestion of 3 times more Cu and 12 times more Cd than for *M. agrestis* occupying the same contaminated grasslands; (4) Cd was highly mobile in the soil–plant–invertebrate–small mammal system; in contrast, Cu was mobile through the soil–plant–invertebrate pathway but accumulation was effectively regulated in small mammals at all levels of dietary intake; (5) Despite high ingestion rates for Cu at the refinery sites, accumulation appeared to be effectively regulated by homeostatic mechanisms; (6) At low levels of Cd intake, accumulation was centered in the kidney and liver; however, at the high ingestion rates of 25 μg g^{-1} per day recorded for *S. araneus*, storage of Cd in the liver increased compared with storage in the kidney; (7) In the refinery population of *S. araneus,* liver Cd concentrations were 2 to 3 times greater than kidney concentrations; this represents storage of 87 and 6% of the total body burden of Cd in the liver and kidney, respectively; (8) Cadmium concentrations of 800 to 1200 μg g^{-1} in the liver and 400 to 550 μg g^{-1} in kidneys are among the highest recorded in terrestrial wildlife; and (9) Cadmium accumulation was significantly positively correlated with the age in *S. araneus*.

Thus, it appears that because Cd is more mobile in the soil–plant system than Cu and since the latter is an essential element in animal nutrition whose accumulation is homeostatically controlled, the potential risk in this type of environmental setting is greater from Cd than from Cu.

While the above scenario depicts environmental enrichment with Cu from industrial emission, soil acidification from long-term input of acid precipitation can cause leaching of nutrients from the soil, inducing deficiency levels in forage tissue. For example, in southwestern Sweden (Frank et al., 1994) the incidence of moose mortality from Cu deficiency (and imbalance of Mn, Mg, and Ca) was attributed largely to low Cu intake from herbage by the animals, manifesting decreased hepatic Cu concentrations. Apparently Cu has leached from surface soils under the acidic condition, producing deficient soil levels of Cu and consequently deficient Cu in the herbage.

6.1.1 Accumulation of Copper in Coffee Plantations: A Case Study

Copper-based fungicides are relatively inexpensive, readily available in developing countries, and are of proven efficacy in controlling a broad range of tropical crop diseases. Unlike the situation in temperate regions, they have not been replaced by synthetic products (Dickinson et al., 1988). As a consequence, widespread and long-term usage of Cu has led to excessively high crop and soil residues of the metal and occurrence of phytotoxicity symptoms in a number of major tropical crops including coffee, tea, cocoa, and bananas (Ayanlaja, 1983; Lepp and Dickinson, 1987).

Heavy usage of Cu fungicides in tropical and tropical-like plantations has been well documented. In central Florida, excess Cu in soils originated from application of fertilizers containing high levels of Cu and continuous use of Cu fungicides on citrus (Zhu and Alva, 1993). In Brazil (Bahia region), Cu from fungicidal use on cocoa has resulted in excess levels in some plantation soils to the point where the soil's retention capacity for the metal has been exceeded, resulting in high Cu concentrations in soil solutions and leachates (Lima, 1994). Similarly, in Costa Rica, some banana plantation soils have Cu levels in surface soils exceeding 1000 ppm, caused by intensive use of Cu fungicides (as high as 100 kg ha^{-1} yr^{-1}) (Cordero and Ramirez, 1979).

This case study dealt with coffee plantations in Kenya (Lepp and Dickinson, 1994). Introduced from Mauritius in the beginning of the 20th century, coffee has been in continuous cultivation since 1915 as a plantation crop in Kenya. Coffee bushes normally have a productive life span of 30 to 40 years, punctuated by routine pruning to maintain the population of young, productive shoots on a 4- to 7-yr rotation.

The distribution of Cu in the soil–plant system of a 68 yr-old coffee (*Coffea arabica*) plantation is displayed in Figure 13.5. It is obvious that

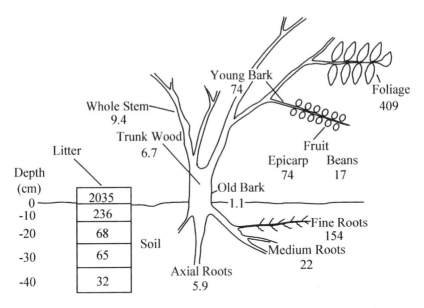

FIGURE 13.5. Distribution of copper (ppm DW) in plant components and soils in a 68 yr-old coffee plantation in Kenya. (Modified from Dickinson and Lepp, 1983.)

particular tissues accumulate significant amounts of Cu, whereas concentrations in other tissues did not appear to be significantly above normal background levels; the root system serves as a good example. Coffee possesses three main types of root: the primary tap root, which may penetrate down to 3 to 6 m; secondary roots, which act as anchors; and an extensive network of branched feeding roots, which form a dense mat in surface soil. These fine roots show significant Cu accumulation compared with the larger roots from lower down the soil profile. Copper also accumulates in the bark, in unwashed foliage, and in young shoots. In associated soils, Cu levels are very high in the litter (2035 ppm) and significantly elevated in soils to a depth of 40 cm (236 ppm at 0 to 10 cm; 32 ppm at 30 to 40 cm depth). Results of a comparative study of Cu accumulation in three different-aged coffee stands (24, 14, and 4 yr old), located on the same plantation and subject to a common management regime are summarized in Table 13.7 (Lepp et al., 1984). Copper accumulated in litter, soil horizon down to 80 cm, fine roots, and bark. The lack of mobility of Cu from soil to the plant, and from bark to the remainder of the plant via lateral transport is indicated by the lack of increase in Cu content with time in trunk wood, and the lack of differences between foliar Cu contents in the bushes from different age stands. The results clearly indicate that much of the Cu was bound in surface tissues, suggesting that contamination is surficial in nature.

TABLE 13.7. HNO₃-extractable copper concentrations (ppm DW) in plant tissues, soil profile, and surface litter in different age stands of coffee in Kenya (values are means ± SD).

Tissue	Stand age (yr) 24	Stand age (yr) 14	Stand age (yr) 4	Unsprayed control site
Fine root	333 ± 15	119 ± 5.5	46 ± 1.0	
Foliage				
Unopened leaves	36 ± 0.34	38 ± 8.6	54 ± 1.7	
Young leaves	72 ± 3.0	41 ± 2.6	51 ± 3.4	
Mature leaves	104 ± 3.2	121 ± 2.5	129 ± 3.3	
Lateral branches	301 ± 16	151 ± 13	33 ± 1.1	
Main trunk[a]				
Bark	415 ± 52	283 ± 14	96 ± 3.4	
(unwashed)	691 ± 64	517 ± 51	249 ± 12	
Wood	9.3 ± 0.80	24 ± 1.1	8.2 ± 0.36	
	12 ± 2.0	10 ± 2.8	8.2 ± 0.95	
Soil depth, cm				
0–20	136 ± 4.7	47 ± 4.8	24 ± 0.67	11 ± 0.53
20–40	56 ± 1.5	25 ± 1.3	16 ± 0.31	10 ± 0.95
40–60	23 ± 1.1	25 ± 0.57	19 ± 0.34	9.7 ± 0.46
60–80	30 ± 0.33	27 ± 0.89	13 ± 0.39	9.6 ± 0.08
Litter fraction				
Leaf	424 ± 38	211 ± 2.5	170 ± 5.6	
Twig	220 ± 14	100 ± 5.3	158 ± 14	
Total litter[a]	884 ± 397	320 ± 150	—[b]	

[a]Total litter including amorphous fraction.
[b]Insufficient litter had accumulated to obtain samples for analysis.
Source: Modified from Lepp et al. (1984).

Budgets for Cu have been established for the different age coffee stands discussed below (Fig. 13.6). It is clear that patterns of retention and loss of Cu differ with the age of the stand to which Cu is applied (Dickinson and Lepp, 1985). In the oldest stand (24 yr), having a long history of treatment, the bulk of the applied Cu is retained in the soil and losses from the system are negligible. The younger stands show significant losses of Cu from the system, with lower proportional retention by soil. In each case, <1% of the applied Cu is removed by harvesting and pruning. This can be mainly attributed to two reasons: (1) the rate of fungicide input became more intense in the late 1960s and 1970s where the younger stands received greater doses in shorter time intervals, and (2) the litter layer in the older stand is better established and contains significantly more Cu than in the two younger stands. This lack of an organic layer coupled with the higher rate of application may account for more rapid movement into the soil horizons and subsequent leaching or losses in runoff.

In summary, using coffee plantations as a case in point, long-term use of Cu fungicides to control crop diseases has a great potential for excessive accumulation of the metal in soils that can lead to phytotoxic levels to the

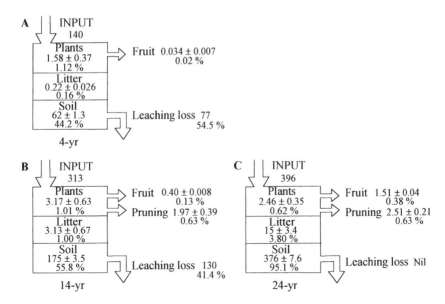

A
INPUT
140

Plants
1.58 ± 0.37
1.12 %   → Fruit 0.034 ± 0.007
                    0.02 %

Litter
0.22 ± 0.026
0.16 %

Soil
62 ± 1.3
44.2 %   → Leaching loss 77
                    54.5 %

4-yr

B
INPUT
313

Plants
3.17 ± 0.63
1.01 %   → Fruit 0.40 ± 0.008
                    0.13 %
            → Pruning 1.97 ± 0.39
                    0.63 %

Litter
3.13 ± 0.67
1.00 %

Soil
175 ± 3.5
55.8 %   → Leaching loss 130
                    41.4 %

14-yr

C
INPUT
396

Plants
2.46 ± 0.35
0.62 %   → Fruit 1.51 ± 0.04
                    0.38 %
            → Pruning 2.51 ± 0.21
                    0.63 %

Litter
15 ± 3.4
3.80 %

Soil
376 ± 7.6
95.1 %   → Leaching loss Nil

24-yr

FIGURE 13.6. Copper budgets for three different-aged stands of coffee (values are kg Cu ha^{-1} ± SE). (Modified from Dickinson and Lepp, 1985.)

coffee plants themselves. Indeed, Dickinson et al. (1988) predicted that Cu loading in the surface soils (0 to 40 cm) of coffee plantations would double every 9.5 yr based on the current usage of fungicides.

6.2 Effects of Copper in Aquatic Ecosystems

The Cu^{2+} ion is the most environmentally relevant species to aquatic systems and is generally considered the most toxic form of Cu to aquatic life. A large body of literature indicates that bioavailability or toxicity of trace metals in aquatic systems is directly correlated to concentrations of free metal ions rather than to total or complexed metal concentrations (Allen, 1996; Campbell, 1995; Ma et al., 1999). The dominant role of free Cu^{2+} ions on the bioavailability of Cu to aquatic organisms is reconfirmed by recent studies where the toxicity of this element to *Ceriodaphnia dubia* decreased with increasing Cu-DOC reaction time; the lethal concentration (LC_{50}) of Cu increased (i.e., toxicity decreased) linearly with increasing total available binding sites (Kim et al., 1999; Ma et al., 1999). In many cases, the metal dissolved in water but complexed with DOC is not immediately available. Inorganic ion pairs sometimes manifest some degree of toxicity, but most of the responses can usually be explained by using free metal activity. Indeed, studies indicate that addition of humic acids to test waters elevated the incipient lethal concentrations of Cu for the Atlantic salmon

(*Salmo salar*) and the survival time for rainbow trout (*Oncorhynchus mykiss*) to toxic levels of Cu (Winner, 1984).

Copper is present in both soluble and particulate forms in aquatic systems. For example, Cu oxide is very insoluble, whereas Cu hydroxide is reasonably soluble and potentially bioavailable (Hall et al., 1998). Bioavailability of Cu in aerobic environments is largely controlled by the presence of Fe and Mn oxides as well as DOC. In anaerobic systems, the role of sulfides predominates. In freshwater environments, an increase in hardness is known to reduce toxicity. Particulate forms of Cu may be deposited in bedloads near the source or distributed into adjacent environments. Particle size, currents, and density determine the final deposition of Cu in aquatic systems. Aquatic biota have a moderate to high potential to bioconcentrate Cu (Hall et al., 1998). Copper and its salts are soluble in water, are persistent, and may bind to particulates. Aquatic biota bioconcentrate Cu in their tissue. Bioconcentration factors (BCFs) as high as 2000 for freshwater algae and 28,200 for saltwater bivalves have been reported (U.S. EPA, 1985).

Copper is also an essential micronutrient for animals but is known to be toxic to aquatic life at concentrations approximately 10 to 50 times higher than normal. As indicated in the preceding section, Cu toxicity in animals is avoided through homeostatic regulation of the metal once taken up.

In fish, Cu effectively interferes with branchial ion transport, plasma ion concentration, hematologic parameters, and enzyme activities (Stouthart et al., 1996). In addition, Cu may cause immunosuppression, vertebral deformities, and neurological disorders. Reproductive effects from Cu exposure, such as reduced egg production in females, abnormalities in newly hatched fry, and reduced survival of the young, have also been reported (Sorensen, 1991). A review of acute toxicity data (Hall et al., 1998) reveals that adverse effects of Cu on aquatic biota have been reported at concentrations as low as 1.3 μg L^{-1} for *Daphnia* in freshwater and 1.2 μg L^{-1} for bivalves in saltwater.

6.2.1 Potential Risk from Copper in Surface Waters: A Case Study

An ecological risk assessment designed to characterize Cu and Cd exposure of aquatic biota in the Chesapeake Bay watershed was undertaken by Hall et al. (1998). This assessment is of relavance in view of the 1987 Chesapeake Bay Agreement where the enhancement of water quality is a most critical factor in the restoration and protection of the bay (U.S. EPA, 1988). The focus has been on Cu and Cd because they have been identified as major toxins of concern in the watershed. Anthropogenic activities that contribute to Cu loading in Chesapeake Bay include municipal and industrial effluents (particularly from smelting, refining, or metal plating industries), nonpoint source runoff (e.g., poultry manure–based fertilizer and pesticides), atmospheric deposition, commercial and recreational boating, and water treatment for algae control. In 1985, the estimated annual urban loading of Cu in

Chesapeake Bay was 104,550 kg. Total annual atmospheric deposition loads of Cu to tidal waters of Chesapeake Bay were estimated to be 10,910 kg.

The primary toxicity benchmark used for this risk assessment was the 10th percentile of species sensitivity (i.e., protection of 90% of the species) from acute exposures (Hall et al., 1998). The inferred hypothesis when using this benchmark is that protecting a large percentage of the species assemblage will preserve ecosystem structure and function. This rather high level of species protection may not be universally accepted especially if the unprotected 10% are keystone species that have commercial or recreational values. However, protection of the species at 90% of the time (10th percentile) has been recommended by the Society of Environmental Toxicology and Chemistry, and others (Hall et al., 1998). A summary of acute and chronic Cu toxicity results by water type (freshwater vs. saltwater) is presented in Table 13.8.

Acute toxicity of copper: Acute freshwater Cu toxicity data were available for 121 species, primarily benthos and fish. Hardness data were available for 73 species. The range of acute toxicity values was from 1.3 μg L^{-1} for *Daphnia* to 13,000 μg L^{-1} for an aquatic sowbug. Within the fish trophic group, the Cyprinidae and Salmonidae families contained species that were more sensitive to acute exposures than the other eight families of freshwater fish. The benthic species most sensitive to acute exposure were gastropods, followed by amphipods. Despite the variability in sensitivities of the various

TABLE 13.8. The 10th percentile intercepts for freshwater and saltwater copper toxicity data by test duration and trophic group (values represent protection of 90% of the test species).

Water type	Data type	Trophic group	n	10th Percentile (μg Cu L^{-1})
Freshwater[a]	Acute	All species	73	8.3
		Zooplankton	4	7.0[b]
		Benthos	31	6.9
		Fish	36	10.8
	Chronic	All species	21	3.8
		Zooplankton	3	0.8[b]
		Benthos	7	3.8[b]
		Fish	10	3.9
Saltwater	Acute	All species	57	6.3
		Phytoplankton	3	2.1[b]
		Zooplankton	7	9.3[b]
		Benthos	30	4.1
		Fish	15	16.1
	Chronic	All species	4	6.4[b]

[a]Hardness-adjusted values (50 mg L^{-1}) were used.
[b]Because of the small data sets ($n < 8$), these values have a high degree of uncertainty and were therefore not used for risk estimates.
Source: Hall et al. (1998).

species and trophic groups, the acute freshwater 10th percentile values for all species together (8.3 μg L^{-1}) and by trophic group (6.9 to 10.8 μg L^{-1}) are fairly similar (see Table 13.8).

Saltwater toxicity data were available for 57 species; most of the data were available for benthos and fish. Acute Cu toxicity values ranged from 1.2 μg L^{-1} for a bivalve to 346,700 μg L^{-1} for a crab species. The fish families with the species most sensitive to saltwater Cu exposure were Pleuronectidae, Antherinidae, and Moronidae. The acute saltwater 10th percentile values for all species, phytoplankton, zooplankton, benthos, and fish were 6.3, 2.1, 9.3, 4.1, and 16.1 μg L^{-1}, respectively (see Table 13.8).

Chronic toxicity of copper: Chronic Cu toxicity data were available for 35 freshwater species, and hardness data were available for 21 of these species. Chronic values ranged from 3.9 μg L^{-1} for the brook trout to 60.4 μg L^{-1} for the northern pike. The lowest freshwater 10th percentile value (0.8 μg L^{-1}) was for zooplankton (see Table 13.8). The 10th percentile values for all species, benthos, and fish were similar (~3.8 μg L^{-1}). The 10th percentile value (3.8 μg L^{-1}) for all species from chronic tests was approximately half the 10th percentile value (8.3 μg L^{-1}) reported for all freshwater species from acute tests.

Saltwater chronic toxicity data were limited to 12 species, and actual chronic values were reported only for the mysid (54 μg L^{-1}) and a copepod (64 μg L^{-1}). The 10th percentile value for the saltwater chronic toxicity data was 6.4 μg L^{-1} (see Table 13.8).

In summary (Hall et al., 1998), potential ecological risk from Cu (and Cd) exposure was estimated by using freshwater acute effects data for freshwater areas and saltwater effects data for saltwater areas. The highest potential ecological risk area for Cu exposure in the Chesapeake Bay watershed was reported in the Chesapeake Bay and the Delaware Canal; the probability of exceeding the acute freshwater 10th percentile value for all species was 86%. For the most sensitive trophic group (based on acute freshwater exposures), the probability of exceeding the 10th percentile value was even higher (90%) for benthos.

6.3 Effects of Copper on Human Health

Worldwide, Cu deficiency in humans is rare; it is usually associated with long-term intake of cow's milk or with severe malnutrition in infants and young children (Van Campen, 1991). Copper-catalyzed enzymes include (1) ferroxidase I and II to catalyze the oxidation of Fe^{2+} to Fe^{3+}, (2) cytochrome C oxidase—the terminal oxidase in the respiratory chain, (3) lysyl oxidase—for the synthesis of collagen and elastic proteins, (4) monoamine oxidase—catalyzing the oxidation of epinephrine and serotonin, and (5) tyrosinase—catalyzing the conversion of tyrosine to melanin pigment.

Clinical manifestations of Cu deficiency in humans include (1) anemia, (2) bone and cardiovascular disorders, and (3) mental and/or nervous systems deterioration and defective keratinization of hair (Davis and Mertz, 1987). Copper deficiency also results in reduction in levels of neurotransmitters, dopamine and norephinephrine, and in defective myelination in the brain stem and spinal cord.

Recent studies suggest that a daily dietary intake of ≤2 mg Cu per day for adults may suffice (Van Campen, 1991). Approximately 2% of body Cu is lost daily; thus a 70-kg male needs to ingest about 1.0 to 1.5 mg Cu per day to maintain balance. Much less Cu is needed for younger people, especially infants. Because of uncertainty about human requirements of Cu, the Food and Nutrition Board (1989) provides estimated safe and adequate daily dietary intakes instead of recommended daily dietary allowances (Table 13.9).

The Cu content of leafy plants consumed as food rarely exceeds 25 ppm and usually ranges between 10 and 15 ppm of dry matter (Underwood, 1973). Whole cereal grains normally contain 4 to 6 ppm Cu, and white flour and bread, 1 to 2 ppm. Root vegetables, nuts, and fruits have usually <10 ppm Cu (Quarterman, 1973). Dairy products usually contain <1 ppm Cu. Crustacea and shellfish, especially oysters, and organ meats (kidney and liver) contain as high as 200 to 400 ppm Cu, but these food items comprise a very small proportion of the total diet (Underwood, 1973). Most other foods of animal origin contain between 2 and 4 ppm (Russell, 1979).

Copper toxicity in humans is very rare and is primarily confined in individuals consuming acidic food or drink that has had prolonged contact with Cu containers (Van Campen, 1991). Zinc/Cu ratio may have some clinical significance in that it has been associated with a variety of epidemiological features of coronary heart disease and with the metabolism of cholesterol (Klevay, 1975). Thus, it is possible that diets deficient in Cu may contribute to the increase in risk of coronary heart disease associated with high Zn/Cu ratio.

Absorption of Cu in humans ranges from 30 to 50% but is affected by several dietary and physiological factors. Zinc, Fe, and Ca as well as vitamin C, fiber, organic acids, and amino acids can interact with Cu utilization.

TABLE 13.9. Estimated safe and adequate daily dietary intake for copper.

Category	Age (yr)	Cu (mg day^{-1})
Infants	0.0–0.5	0.4–0.6
	0.5–1	0.6–0.7
Children and adolescents	1–3	0.7–1.0
	4–6	1.0–1.5
	7–10	1.0–2.0
Adults	11+	1.5–2.5
		1.5–3.0

Source: Food and Nutrition Board (1989).

7 Factors Affecting Mobility and Bioavailability of Copper

7.1 pH

Solubility and bioavailability of Cu to plants are largely dependent on soil pH. Copper bioavailability is drastically reduced over pH 7 and is most readily available below pH 6 (Locascio, 1978), especially at pHs below 5 (Lucas and Knezek, 1972). The more acidic organic soils are, the greater would be the relative response to Cu and the greater would be the number of crop species that are likely to respond to Cu application (Harmer, 1945). Increasing the pH of a soil solution, which can be effected by liming, can lead to the formation of hydrolysis products that have different affinities for exchange sites (Cavallaro and McBride, 1980; Jarvis, 1981b). Adsorption of Cu increased appreciably as the soil pH was increased from 4 to 7 (Cavallaro and McBride, 1984; Msaky and Calvet, 1990). Increasing the pH can also hydrolyze Al^{3+} ions that occupy exchange sites, thus decreasing the fraction of sites occupied by Al^{3+} ions and creating more available exchange sites. In short, Cu solubility in soils depends largely on pH due to hydrolysis of Cu^{2+} at pH values above 6 and to the removal of Al^{3+} and H^{+} ions from exchange sites as the pH is increased in acidic soils.

Soil pH also influences the degree of complexation of Cu in soil solution. Increasing the pH of the soil solution (or peat extract) increases the degree of complexation of soluble Cu (Fig. 13.7). Increasing the pH of the peat extracts to about 6 resulted in almost complete complexation of Cu (see

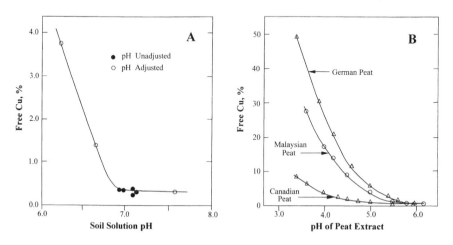

FIGURE 13.7. Influence of pH on the amount of total soluble copper present in (A) soil solution and (B) peat extracts as uncomplexed Cu^{2+} ions. The peat extracts were adjusted to have a total solution copper concentration of 2 μ*M*. (From [A] McBride and Bouldin, 1984; [B] Kerven et al., 1984.)

Fig. 13.7). In contrast, only 60 to 90% of total Cu in solution of these acidic peats was complexed at its natural pH value of about 3.4. The total soluble Cu concentration in the soil solution however, should remain relatively insensitive to pH change (Jeffery and Uren, 1983; McBride and Bouldin, 1984).

Sludge-treated soils should have their pH maintained at a certain pH as a safeguard against excessive uptake of toxic metals. For example in England and Wales, soil pH should be above 6.0 for grassland and 6.5 for arable land (DOE, 1981). Similarly, soil pH management is a mandate for the rest of western Europe, Canada, and the United States. Where sludge is spread on soils with a pH value less than 6.0, member states of the European Union (EU) are required to consider the increased bioavailability of metals to crops and to reduce the maximum limit values of metals in soils (CEC, 1986). The benefits of growing crops on sludged soils with proper pH was demonstrated when ryegrass yields were increased with increasing pH due to controlled phytotoxicity primarily from Zn and Ni (Sanders et al., 1986; Smith, 1994); Cu was less sensitive to changing pH compared with Zn and Ni, with greater uptake occurring at low pH values. Similarly, no adverse effects on corn grown on three field soils occurred from the application of 265 kg Cu ha^{-1} from Cu-enriched swine manure when soil pH was maintained above 6.1 (Payne et al., 1998).

In general, liming lowers Cu concentration in plants. There is usually a negative correlation between Cu uptake and soil pH or a negative correlation between bioavailable Cu and soil pH. Thus, the predictability of Cu uptake by plants can best be achieved by inclusion of soil pH.

7.2 Organic Matter

Discussions on the role of OM in the incidence of deficiency, and magnitude of complexation and sorption of Cu explain why peats and mucks (Histosols) commonly produce Cu-deficient crops (see Chapter 2). The high sorption capacity of OM for Cu (see Fig. 13.7) indicates that among soil constituents it is probably as important as the oxides of Mn and Fe in sorbing Cu. Among divalent cations, Cu forms the most stable complexes with humic and fulvic acids, more so than Pb, Fe, and Zn. Additionally, it has been observed that Cu and Pb complexed more strongly with OM than Zn and Cd (Martinez and McBride, 1999; Stevenson, 1976). Such complexation, especially insoluble Cu-humic complexes may decrease the solubility of Cu in soil solution. However, organic ligands can play a major role in the leaching of Cu in waste sites. For example, >95% of the dissolved Cu was organically bound in leachates from waste incinerator bottom ash (Meima et al., 1999).

Fractionation studies show that appreciable amounts of soil Cu are associated with OM (Jarvis, 1981b; Kishk et al., 1973; Kuo et al., 1983) and it has been calculated that on a global basis, 36% of total Cu in soils is in this form (Nriagu, 1979). High OM content in soils may substitute for high pH in immobilizing metals and thus sewage sludge can be applied to soils that have pH values lower than the suggested value of pH 6.5 for metal application from municipal sewage sludge (King and Dunlop, 1982). This indicates the buffering effect of OM from sludge relative to bioaccumulation of metals by crops.

The ecological importance of OM on the cycling of Cu was demonstrated in the Tantramar copper swamp located on the isthmus between New Brunswick and Nova Scotia, Canada (Dykeman and de Sousa, 1966). Soils in the swamp, which support a luxurious growth of larch, black spruce, and ground cover species contain up to 7% Cu, which is believed to have come from the bedrock beneath a 60-m layer of glacial drift through deep circulating water and upward migration to the ground surface. The lush growth of vegetation in the swamp despite the large amounts of total Cu in soils is attributed largely to Cu binding by OM.

7.3 Oxides of Iron and Manganese

The role of Fe and Mn oxides in the environmental chemistry of trace metals has been discussed earlier. Briefly, it has been concluded that the hydrous oxides of Mn and Fe, and OM are the principal soil constituents controlling the sorption of metals (see Fig. 13.4), and are probably more important than clay minerals. The microcrystalline and noncrystalline oxides in clay fraction of soils provide chemisorption sites for Cu and other metals. Leaching studies of diverse soil types also indicate the relevant role the oxides of Fe and Mn provide in estimating metal release from soils (Korte et al., 1975). Sorption studies of estuarine particulate matter indicate that Mn-Fe oxide coatings may play a relatively greater role in the binding of Pb, while adsorption of Cu and Cd may be controlled to a greater extent by organic coatings (Lion et al., 1982). Since these hydrous oxides are virtually ubiquitous in soils, clays, and sediments, their significance in metal behavior cannot be overlooked.

7.4 Soil and Clay Type

Copper deficiencies in plants occur most commonly on Histosols and soils high in OM (i.e., peats and mucks) but also occur on coarse-textured mineral soils. In general, soils derived from coarse-grained materials (i.e., sands and sandstone) or from acid igneous rocks contain lower concentrations of Cu than those developed from fine-grained sedimentary rocks (i.e., shales and clays) or from basic igneous rocks (Jarvis, 1981b; Moraghan and Mascagni, 1991).

In England and Wales, Cu deficiencies are primarily restricted to crops grown on organic or peaty soils, on highly leached and podzolic sands, or on shallow rendzina soils developed on chalk (Caldwell, 1971). In Australia, Cu deficiencies occur on calcareous sands and on lateritic podzolic soils (Donald and Prescott, 1975).

Soil texture, particularly the clay and silt fractions, has been positively correlated with the sorption capacity of soils for Cu (Dhillon et al., 1981), with Cu bioavailability (Dragun and Baker, 1982; Rai et al., 1972), and with total soil Cu content (Neelakantan and Mehta, 1961; Zborishchuk and Zyrin, 1978). Among clay minerals, their sorption capacity generally relates to their CEC values (i.e., montmorillonite > illite > kaolinite) (McLaren et al., 1981; Riemer and Toth, 1970).

7.5 Other Factors

7.5.1 Interactions with Other Nutrients

Adverse effects of N or P interactions with micronutrients on plant growth and nutrition have been frequently reported during the past 30 years. In general, interactions of metals with a macronutrient are expressed as an intensification of micronutrient deficiency when supplemental macronutrients are applied to soils where micronutrients are deficient or marginal. Good examples of these macronutrient–micronutrient interactions are the N–Cu and P–Cu couples.

Nitrogen–Cu interactions have been shown for several crops. Increased protein concentrations in roots as a result of increased N application, could lead to increased retention of Cu by increased formation of protein-Cu complexes (Gilbert, 1951). Cereal genotypes with higher grain protein contents are potentially more susceptible to Cu deficiency than those with lower grain protein (Nambiar, 1976b). Nitrogen-induced Cu deficiency may be explained by the two effects of N on growth — a large increase in total growth, and a marked increase in shoot relative to root growth, both contributing to the dilution of Cu contents in plant tissues (Chaudry and Loneragan, 1970).

Phosphorus–Cu interactions have also been shown in several crops and tree species (Adriano, 1986). Smilde (1973) reported that in the Netherlands, P can accentuate Cu deficiency in trees, especially in the presence of N. Copper has also been shown to antagonistically interact with Zn (Adriano et al., 1971; Kausar et al., 1976; Levesque and Mathur, 1983), with Mo (Gilbert 1952; MacKay et al., 1966), and with Mn (Gilbert, 1952; Gupta, 1972). Using *Silene vulgaris* as a test plant, Cu–Zn interaction on the root–solution interface was demonstrated as a function of root internal and external concentrations of the metals (Sharma et al., 1999).

7.5.2 Crop Species

Genotypic differences in Cu nutrition among cereal crops have been reported. Concentration of Cu is often greater in pasture legumes than in grasses growing in the field under similar conditions (Jarvis and Whitehead,

1983). In testing spring wheat cultivars in the field (Owuoche et al., 1994a), there exists a cultivar $\times$ Cu interaction ($p \leq 0.05$) indicating differential response of the cultivars to Cu for yield. Cultivar differences in growth among cabbage seedling exposed to Cu^{2+} stress have also been observed (Rousos et al., 1989; Rousos and Harrison, 1987). The cultivar having lower tolerance to Cu might be due to significantly high shoot Cu levels than in the more tolerant plants.

7.5.3 Placement and Carrier of Fertilizer

Copper nutrition by crops can be also affected by the method of placement of Cu fertilizers (band, broadcast, or foliar) and carrier of fertilizer (organic vs. inorganic) although results may be affected by soil type, crop species, application rates, and other factors (Martens and Westermann, 1991; Murphy and Walsh, 1972).

8 Sources of Copper in the Environment

8.1 Copper Fertilizers

The most widely used source of Cu in agriculture is $CuSO_4 \cdot 5H_2O$ (25.5% Cu), which is water soluble (Table 13.10). Other sources, such as copper oxides (CuO, 75% Cu; Cu_2O, 89% Cu) and basic copper sulfate [$CuSO_4 \cdot 3Cu(OH)_2$, (13 to 53% Cu)] have been used widely also. A single application of 2.25 to 9.0 kg ha^{-1} of Cu on mineral soils or 22.5 to 45.0 kg ha^{-1} on a peat or muck soil is generally adequate for several years (Locascio, 1978).

Copper chelates are another source of Cu in agriculture and are more effective in alkaline than in acid soils. Chelates are fairly stable in soils, keeping Cu in soluble and bioavailable forms to plants. Examples of

TABLE 13.10. Copper sources used for correcting copper deficiency in plants.

Source	Formula	g Cu kg^{-1}
Inorganic carriers		
Basic copper(ic) sulfate	$CuSO_4 \cdot 3Cu(OH)_2$	130–530
Copper(ic) chloride	$CuCl_2$	470
Copper(ic) sulfate monohydrate	$CuSO_4 \cdot H_2O$	350
Copper(ic) sulfate pentahydrate	$CuSO_4 \cdot 5H_2O$	250
Cuprous oxide	Cu_2O	890
Cupric oxide	CuO	750
Organic carriers		
Copper chelate	$Na_2CuEDTA$	130
Copper chelate	NaCuHEDTA	90
Copper lignosulfonate	—	50–80
Copper polyflavonoid	—	50–70

Source: As summarized by Martens and Westermann (1991).

synthetic Cu chelates are $Na_2CuEDTA$ (13% Cu) and NaCuHEDTA (9% Cu), and natural organic complexes (lignosulfonates and polyflavonoid), which are highly efficient Cu sources.

Foliar spray of chelated Cu or low rates of $CuSO_4$ (0.30 kg ha^{-1}) can also be used to prevent or alleviate Cu deficiency. However, multiple sprays may be required to obtain maximum production, since Cu is immobile in plants and hardly moves from older to younger tissues. Under alkaline Egyptian conditions, foliar application of CuEDTA is recommended for cotton production (Sawan et al., 1993). Common sources of Cu currently used for soil and foliar applications and the various application rates are available (Martens and Westermann, 1991). Like Zn, Cu is most commonly applied to the soil by broadcast or banded application.

In a soil survey in Ontario, Canada the highest total Cu levels were found in organic soils (65 ± 27 ppm) due to the production practices of adding $CuSO_4$ once every 3 to 4 yr to prevent Cu deficiency and using Cu fungicides on vegetables (Frank et al., 1976).

8.2 Fungicides and Bactericides

Copper compounds have been used for many years as fungicides and bactericides. In the past, $CuSO_4$ was mixed with lime (Bordeaux mixture) to control many plant pathogens, while $Cu(OH)_2$ has been used extensively in sprays as a blanket preventive treatment (Walsh et al., 1972). These fungicides are used on pome, stone, and citrus fruits as well as on grapevines, hops, and vegetables to counter downy mildew (Tiller and Merry, 1981).

Copper fungicides are the mainstay for control of foliar fungal disease of citrus in Florida (Timmer and Zitko, 1996). The compounds are highly effective in controlling melanose, greasy spot, and scabs in citrus. It is not uncommon to find total Cu levels of 370 kg Cu ha^{-1} in soils of old citrus groves. For all uses, annual rates of application often exceeded 30 to 40 kg Cu ha^{-1}. The most common impact of the excess soil Cu is phytotoxicity, especially to young trees, and severe Fe deficiency (Alva, 1993). For example, Cu sprays have been recommended for disease control on vegetable crops (snapbeans, potatoes, and cucumbers) grown on irrigated sandy soils in central Wisconsin. These sprays can enrich Cu in the soil by as much as 11.2 to 16.8 kg ha^{-1}. Copper toxicity could develop in these sandy soils because they have a low CEC (<5 meq/100 g) and often contain <1% OM.

Copper fungicides have been used extensively in Kenya since the 1930s to control leaf rust and since the 1960s to control the berry disease in coffee (Lepp and Dickinson, 1994). In a large-scale coffee plantation near Nairobi, Cu concentrations as high as 236 ppm were found in the surface 10 cm of the soil. Established in 1915, the plantation is being sprayed 10 to 12 times per year at the rate of 5 kg ha^{-1} Cu as copper oxychloride. Copper fungicides are also intensively used in the production of cocoa (Lima, 1994), banana (Cordero and Ramirez, 1979), tea and other beverage crops (Lepp and Dickinson, 1994). Similar increases of Cu in surface soils were observed in vineyards in Portugal (Dias and Soveral-Diaz, 1997). The soil level increased with the vineyard age, indicating an accumulative effect from Cu fungicides. Copper-containing fungicides and bactericides are now used extensively for disease control on staked tomatoes in North Florida (Rhoads et al., 1989). Fungicides containing $Cu(OH)_2$ are routinely applied to commercial tomato crops at rates supplying as much as 15 mg Cu kg^{-1} soil annually in the rooting zone.

In summary, continued overuse of Cu-containing fungicides and bactericides can result in accumulation of Cu in soil at phytotoxic levels.

8.3 Livestock Manure

Copper as $CuSO_4$ is a typical feed additive for swine and is used to increase feed efficiency, weight gains, and to control dysentery (Cromwell et al., 1978; Lucas et al., 1962; Stahly et al., 1980). Dietary Cu levels of 125 to 250 ppm are usually sufficient to stimulate growth of swine and poultry, but higher levels can be toxic (Dalgarno and Mills, 1975; Hedges and Kornegay, 1973). Zinc and Fe are usually added simultaneously with Cu to prevent possible Cu intoxication in animals and to ensure prevention of Fe- and Zn-deficiencies in animals. Zinc as $ZnSO_4$ is added at a level of 80 ppm. Most (80 to 95%) of this dietary Cu is excreted in the manure. Purves (1977) reported that concentrations up to 800 ppm Cu (dry basis) are not unusual in swine wastes. Therefore, the potential accumulation and toxic effects of Cu from high rates of manure application to croplands are of environmental concern just like in sewage sludge.

Here and abroad, the continual use of swine manure on land has been shown to have the potential to elevate soil Cu levels and eventually cause phytotoxicity (Batey et al., 1972; Kornegay et al., 1976; Sutton et al., 1983). For example, repeated applications of swine manure on grassland in the Netherlands caused Cu accumulation in the soil layer (Lexmond and de Haan, 1977). There is evidence of Cu leaching in the profiles of these Dutch soils, especially in Podzols, due to their low OM contents.

Surface application on pastures of manure slurry from pigs fed 250 ppm of $CuSO_4$ in the diet resulted in Cu toxicity to grazing sheep (Blaxter, 1973). Underwood (1973) reported that sheep are very susceptible to Cu poisoning (a level of 25 ppm in the diet can be toxic to sheep), while monogastric animals such as swine and poultry can tolerate much higher levels of Cu.

8.4 Municipal Sewage Sludge

Metal concentrations in sewage sludge vary widely. Among the heavy metals, Cu, Ni, Zn, and Cd are the ones usually singled out as critical to agricultural use of sewage sludge (Page, 1974; Sommers, 1977). Although sludge often contains appreciable levels of Cu, application of sludge to soils generally results in only slight to moderate increases in Cu contents of plants (CAST, 1976). In general, sound management practices could render Cu in sludge harmless to plants and pose no hazard to the food chain.

8.5 Industrial Emissions

Among worldwide anthropogenic sources of Cu, metallurgical processing (smelting) for Cu, iron and steel production, and coal combustion are among the major sources of Cu (Nriagu, 1979). Environmental contamination due to heavy metals from industrial emissions including mining and smelting is already well documented. An estimate suggests that about 40% of total Cu emitted from the Sudbury (Ontario, Canada) smelter was deposited within a 60-km radius (Dudka and Adriano, 1997; Freedman and Hutchinson, 1980). In general, the deposition rate of metals from smelters is a function of distance. In the Alcacer do Sal region of Portugal, Cu contamination in the lower reaches of Sado River has imperiled rice production as well as fishery and oyster farming in the estuary (Lidon and Henriques, 1993b).

References

Adriano, D.C. 1986. *Trace Elements in the Terrestrial Environment*. Springer-Verlag, New York.
Adriano, D.C., G.M. Paulsen, and L.S. Murphy. 1971. *Agron J* 63:36–39.
Agarwala, S.C., S.S. Bisht, and C.P. Sharma. 1977. *Can J Bot* 55:1299–1307.
Ainsworth, C.C., and D. Rai. 1987. In *Chemical Characterization of Fossil Fuel Combustion Wastes*. EPRI Rep EA-5321, Electric Power Res Inst, Palo Alto, CA.
Allen, H.E., and D.J. Hansen. 1996. *Water Environ Res* 68:42–54.
Alloway, B.T., and A.R. Tills. 1984. *Outlook Agric* 13:32–42.
Alva, A.K. 1993. *Bull Environ Contam Toxicol* 51:857–864.
Alva, A.K., B. Huang, and S. Paramasivam. 2000. *Soil Sci Soc Am J* 64:955–962.
Arnold, M.A., R.D. Lineberger, and D.K. Struve. 1994. *J Am Soc Hort Sci* 119:74–79.
Aubert, H., and M. Pinta, eds. 1977. *Trace Elements in Soils*. Elsevier, New York.
Ayanlaja, S.A. 1983. *Plant Soil* 73:403–409.
Baes, C.F., and R.E. Mesmer. 1976. *The Hydrolysis of Cations*. Wiley, New York.
Baker, D.E. 1974. *Proc Fed Am Soc Exp Biol* 33:1188–1193.
Barnes, J.S., and F.R. Cox. 1973. *Agron J* 65:705–708.
Basta, N.T., and M.A. Tabatabai. 1992. *Soil Sci* 153:331–337.
Batey, T., C. Berryman, and C. Line. 1972. *J Br Grassl Soc* 27:139–143.
Behel, D., Jr., D.W. Nelson, and L.E. Sommers. 1983. *J Environ Qual* 12:181–186.
Berthelsen, B.O., E. Steinnes, W. Solberg, and L. Jingsen. 1995. *J Environ Qual* 24:1018–1026.

Blaxter, K.L. 1973. *Vet Rec* 92:383–386.

Bloom, P.R. 1981. In R.H. Dowdy, J.A. Ryan, V.V. Volk, and D.E. Baker, eds. *Chemistry in the Soil Environment*, ASA 40, Am Soc Agron, Madison, WI.

Bloomfield, C., and J.R. Sanders. 1977. *J Soil Sci* 28:435–444.

Borggaard, K. 1976. *Acta Agric Scand* 26:144–149.

Bould, C., D.J.D. Nicholas, J.A.H. Tolhurst, and J.M.S. Potter. 1953. *J Hort Sci* 28:268–277.

Bowen, H.J.M. 1979. *Environmental Chemistry of the Elements*. Academic Pr, New York.

Bradford, G.R., F.L. Bair, and V. Hunsaker. 1971. *Soil Sci* 112:225–230.

Bussler, W. 1981. In J.F. Loneragan, A.D. Robson, and R.D. Graham, eds. *Copper in Soils and Plants*. Academic Pr, New York.

Caldwell, T.H. 1971. In *Trace Elements in Soils and Crops*. Tech Bull 21. Her Majesty's Sta Office, London.

Campbell, P.G.C. 1995. In A. Tessier and D.R. Turner, eds. *Metal Speciation and Bioavailability in Aquatic Systems*. Wiley, New York.

[CAST] Council for Agricultural Science and Technology. 1976. *Application of Sewage Sludge to Cropland: Appraisal of Potential Hazards of the Heavy Metals to Plants and Animals*. EPA-430/9-76-013. General Serv Admin, Denver, CO.

Cavallaro, N., and M.B. McBride. 1978. *Soil Sci Soc Am J* 42:550–556.

Cavallaro, N., and M.B. McBride. 1980. *Soil Sci Soc Am J* 44:729–732.

Cavallaro, N., and M.B. McBride. 1984. *Soil Sci Soc Am J* 48:1050–1054.

[CEC] Comm. European Community. 1986. *Off J. European Community* No. L 181/ 6–12.

Chaudry, F.M., and J.F. Loneragan. 1970. *Aust J Agric Res* 21:865–879.

Chen, J., F. Wei, Y.Wu, and D.C. Adriano. 1991. *Water Air Soil Pollut* 57–58:699–712.

Chino, M. 1981. In K. Kitagishi and I. Yamane, eds. *Heavy Metal Pollution in Soils of Japan*. Japan Sci Soc Pr, Tokyo.

Chino, M., and K. Kitagishi. 1966. *J Sci Soil Manure* 37:342–347.

Clayton, P.M., and K.G. Tiller. 1979. Div Soil Tech Paper 41, CSIRO, Australia.

Coleman, N.T., A.C. McClung, and D.P. Moore. 1956. *Science* 123:330–331.

Cordero, A., and G.F. Ramirez. 1979. *Agron Costa* 3:63–78.

Cox, D.P. 1979. In J.O. Nriagu, ed. *Copper in the Environment*. Wiley, New York.

Cromwell, G.L., V.W. Hays, and T.L. Clark. 1978. *J Anim Sci* 46:692–698.

Dalgarno, A.C., and C.F. Mills. 1975. *J Agric Sci* 85:11–18.

Davis, G., and W. Mertz. 1987. In W. Mertz, ed. *Trace Elements in Human and Animal Nutrition*, 5th ed. Academic Pr, San Diego, CA.

Davis, R.D., P.H.T. Beckett, and E. Wollan. 1978. *Plant Soil* 49:395–408.

Delas, J. 1980. In *Problems Encountered with Copper*. Europ Econ Commun Workshop, Bordeaux, France, Oct 8–10.

del Castillo, P., W.J. Chardon, and W. Salomons. 1993. *J Environ Qual* 22:679–689.

Department of the Environment. 1981. U.K. DOE/Natl. Water Council Comm Rep 20. London.

Dhillon, S.K., P.S. Sidhu, and M.K. Sinha. 1981. *J Soil Sci* 32:571–578.

Dias, R.S., and J.C. Soveral-Dias. 1997. In *Modern Agriculture and Environment*. Kluwer, Dordrecht, Netherlands.

Dickinson, N.M., and N.W. Lepp. 1983. In *Proc Intl Conf Heavy Metals in Environ*. CEP Consultants, Edinburgh.

Dickinson, N.M., and N.W. Lepp. 1985. *Agric Ecosys Environ* 14:15–23.

Dickinson, N.M., N.W. Lepp, and G.T.K. Surtan. 1988. *Agric Ecosys Environ* 21:181–190.

Dolar, S.G., and D.R. Keeney. 1971. *J Sci Food Agric* 22:273–278.

Dolar, S.C., D.R. Keeney, and L.M. Walsh. 1971. *J Sci Food Agric* 22:282–286.

Domingo, L.E., and K. Kyuma. 1983. *Soil Sci Plant Nutr* 24:439–452.

Donald, C.M., and J.A. Prescott. 1971. In D.J.D. Nicholas and A.R. Egan, eds. *Trace Elements in Soil-Plant-Animal Systems*. Academic Pr, New York.

Doner, H.E., A. Pukite, and E. Yang. 1982. *J Environ Qual* 11:389–394.

Dragun, J., and D.E. Baker. l982. *Soil Sci Soc Am J* 46:921–925.

Drouineau, G., and R. Mazoyer. 1962. *Ann Agron Paris* 13:31–53.

Dudka, S., and D.C. Adriano. 1997. *J Environ Qual* 26:590–602.

Dykeman, W.R., and A.S. de Sousa. 1966. *Can J Bot* 44:871–878.

Ellis, B.G., and B.D. Knezek. 1972. In J.J. Mortvedt, P.M. Giordano, and W.L. Lindsay, eds. *Micronutrients in Agriculture*. Soil Sci Soc Am, Madison, WI.

Emmerich, W.E., L.J. Lund, A.L. Page, and A.C. Chang. 1982. *J Environ Qual* 11:182–186.

Fiskel, J.G., and S.E. Younts. 1963. *Plant Food Rev* 9:6–10.

Fiskell, J.G.A. 1965. In C.A. Black, ed. *Methods of Soil Analysis,* part 2. Am Soc Agron, Madison, WI.

Fiskell, J.G.A., and C.D. Leonard. 1967. *J Agric Food Chem* 15:350–353.

Follett, R.H., and W.L. Lindsay. 1970. In *Profile Distribution of Zinc, Iron, Manganese, and Copper in Colorado Soils*. Colorado Agric Exp Sta Tech Bull 110. Fort Collins, CO.

Follett, R.H., L.S. Murphy, and R.L. Donahue. 1981. *Fertilizers and Soil Amendments*. Prentice-Hall, Englewood Cliffs, NJ.

Food and Nutrition Board. 1989. *Recommended Dietary Allowances,* 10th ed. National Research Council, Washington, DC.

Frank, A., V. Galgan, and L. Petersson. 1994. *Ambio* 23:315–317.

Frank, R., K. Ishida, and P. Suda. 1976. *Can J Soil Sci* 56:181–196.

Freedman, B., and T.C. Hutchinson. 1980. *Can J Bot* 58:1722–1736.

Gartrell, J.W. 1981. In J.F. Loneragan, A.D. Robson, and R.D. Graham, eds. *Copper in Soils and Plants*. Academic Pr, New York.

Gherardi, M.J., B. Dell, and L. Huang. 1999. *Plant Soil* 210:75–81.

Gilbert, F.A. 1952. *Adv Agron* 4:147–177.

Gilbert, G.S. 1951. *Plant Physiol* 26:398–405.

Goodman, B.A., and D.J. Linehan. 1979. In *The Soil-Root Interface*. Academic Pr, London.

Gough, L.P., J.M. McNeal, and R.C. Severson. 1980. *Soil Sci Soc Am J* 44:1030–1036.

Graham, R.D. 1975. *Nature* 254:514–515.

Graham, R.D. 1981. In J.F. Loneragan, A.D. Robson, and R.D. Graham, eds. *Copper in Soils and Plants*. Academic Pr, New York.

Gupta, U.C. 1972. *Soil Sci* 114:131–136.

Gupta, U.C. 1979. In J.O. Nriagu, ed. *Copper in the Environment*. Wiley, New York.

Gupta, U.C., and D.C. MacKay. 1966. *Soil Sci* 101:93–97.

Hall, L.W., M.C. Scott, and W.D. Killen. 1998. *Environ Toxicol Chem* 17:1172–1189.

Haque, I., E.A. Aduayi, and S. Sibanda. 1993. *J Plant Nutr* 16:2149–2212.

Harmer, P.M. 1945. *Soil Sci Soc Am Proc* 10:284–294.

Harter, R.D. 1979. *Soil Sci Soc Am J* 43:679–683.

Harter, R.D. 1983. *Soil Sci Soc Am J* 47:47–51.

Harter, R.D. 1991. In J.J. Mortvedt et al., eds. *Micronutrients in Agriculture,* SSSA 4, Soil Sci Soc Am, Madison, WI.

Hazlett, P.W., G.K. Rutherford, and G.W. Van Loon. 1984. *Geoderma* 32:273–285.

Heinrichs, H., and R. Mayer. 1980. *J Environ Qual* 9:111–118.

Hedges, J.D., and E.T. Kornegay. 1973. *J Anim Sci* 37:1147–1154.

Hewitt, E.J. 1953. *J Exp Bot* 4:59–64.

Hodgson, J.F., H.R. Geering, and W.A. Norvell. 1965. *Soil Sci Soc Am Proc* 29:665–669.

Hodgson, J.F., W.L. Lindsay, and J.F. Trierweiler. 1966. *Soil Sci Soc Am Proc* 30:723–726.

Hunter, B.A., M.S. Johnson, and D.J. Thompson. 1987a. *J Appl Ecol* 24:573–586.

Hunter, B.A., M.S. Johnson, and D.J. Thompson. 1987b. *J Appl Ecol* 24:587–599.

Hunter, B.A., M.S. Johnson, and D.J. Thompson. 1987c. *J Appl Ecol* 24:601–614.

Hunter, B.A., M.S. Johnson, and D.J. Thompson. 1989. *J Appl Ecol* 26:89–99.

Jarvis, S.C. 1978. *J Sci Food Agric* 29:12–18.

Jarvis, S.C. 1981a. *J Soil Sci* 32:257–269.

Jarvis, S.C. 1981b. In J.F. Loneragan, A.D. Robson, and R.D. Graham, eds. *Copper in Soils and Plants.* Academic Pr, New York.

Jarvis, S.C., and D.C. Whitehead. 1983. *Plant Soil* 75:427–434.

Jeffery, J.J., and N.C. Uren. 1983. *Aust J Soil Res* 21:479–488.

Jenne, E.A. 1968. *Adv Chem* 73:337–387.

Jenne, E.A. 1977. In W.R. Chappel and K.L. Petersen, eds. *Molybdenum in the Environment,* vol 2. Marcel Dekker, New York.

Jolly, J.H. 1980. In U.S. Bureau of Mines. *Minerals Yearbook,* 1978–79. U.S. Dept of Interior, Washington, DC.

Jones, J.B. Jr. 1972. In J.J. Mortvedt, P.M. Giordano, and W.L. Lindsay, eds. *Micronutrients in Agriculture.* Soil Sci Soc Am, Madison, WI.

Jones, J.B. 1991. In J.J. Mortvedt et al., eds. *Micronutrients in Agriculture,* SSSA 4, Soil Sci Soc Am, Madison, WI.

Jones, J.B., and G.B. Belling. 1967. *Aust J Agr Res* 18:733–740.

Karamanos, R.E., G.A. Kruger, and J.W.B. Stewart. 1986. *Agron J* 78:317–323.

Kausar, M.A., F.M. Chaudry, A. Rashid, A. Latif, and S.M. Alam. 1976. *Plant Soil* 45:397–410.

Kawaguchi, K., and K. Kyuma. 1959. *Soil Plant Food* 5:54–63.

Keller, C., and J.C. Vedy. 1994. *J Environ Qual* 23:987–999.

Kerven, G.L., D.G. Edwards, and C.J. Asher. 1984. *Soil Sci* 137:91–99.

Kim, S.D., H. Ma, H.E. Allen, and D.K. Cha. 1999. *Environ Toxicol Chem* 18:2433–2437.

King, L.D., and W.R. Dunlop. 1982. *J Environ Qual* 11:608–616.

Kishk, F.M., M.N. Hassan. 1973. *Plant Soil* 39:497–505.

Kishk, F.M., M.N. Hassan, I. Ghanem, and L. El-Sissy. 1973. *Plant Soil* 39:487–496.

Klevay, L.M. 1975. *Nutr Reports Intl* 11:237–242.

Kneip, T.J., M. Eisenbud, C.D. Strehlow, and P.C. Freudenthal. 1970. *J Air Pollut Cont Assoc* 20:144–149.

Kochian, L.V. 1991. In J.J. Mortvedt et al., eds. *Micronutrients in Agriculture,* SSSA 4, Soil Sci Soc Am, Madison, WI.

Kornegay, E.T., J.D. Hedges, D.C. Martens, and C.Y. Kramer. 1976. *Plant Soil* 45:151–162.

Korte, N.E., J. Skopp, E.E. Niebla, and W.H. Fuller. 1975. *Water Air Soil Pollut* 5:149–156.

Krauskopf, K.B. 1972. In J.J. Mortvedt, P.M. Giordano, and W.L. Lindsay, eds. *Micronutrients in Agriculture*. Soil Sci Soc Am, Madison, WI.

Krauskopf, K.B. 1979. *Introduction to Geochemistry,* 2nd ed. McGraw-Hill, New York.

Kruger, G.A., R.E. Karamanos, and J.P. Singh. 1985. *Can J Soil Sci* 65:89–99.

Kubota, J. 1983. *Agron J* 75:913–918.

Kubota, J., and W.H. Allaway. 1972. In J.J. Mortvedt, P.M. Giordano, and W.L. Lindsay, eds. *Micronutrients in Agriculture*. Soil Sci Soc Am, Madison, WI.

Kuo, S., and A.S. Baker. 1980. *Soil Sci Soc Am J* 44:969–974.

Kuo, S., P.E. Heilman, and A.S. Baker. 1983. *Soil Sci* 135:101–109.

Leeper, G.W. 1978. *Managing the Heavy Metals on the Land.* Marcel Dekker, New York.

Leonard, C.D. 1967. *Farm Technol* 23(6).

Lepp, N.W., and N.M. Dickinson. 1987. In *Pollution Transport and Fate in Ecosystems*. Blackwell, Oxford.

Lepp, N.W., and N.M. Dickinson. 1994. In S.M. Ross, ed. *Toxic Metals in Soil-Plant Systems*. Wiley, New York.

Lepp, N.W., N.M. Dickinson, and K.C. Ormand. 1984. *Plant Soil* 77:263–270.

Levesque, M.P., and S.P. Mathur. 1983. *Soil Sci* 135:88–100.

Lexmond, T.M., and F.A.M. de Haan. 1977. In *Proc. Intl Sem Soil Environment and Fertility Management in Intensive Agriculture*. Soc Sci Soil Manure, Tokyo, Japan.

Lexmond, T.M., and F.A.M. de Haan. 1980. In J.K.R. Gasser, ed. *Effluents from Livestock*. Applied Science Publ, London.

Lidon, F.C., and F.S. Henriques. 1993a. *J Plant Nutr* 16:1449–1464.

Lidon, F.C., and F.S. Henriques. 1993b. *J Plant Nutr* 16:1619–1630.

Lima, J.S. 1994. *Agric Ecosys Environ* 48:19–25.

Lindsay, W.L., and W.A. Norvell. 1978. *Soil Sci Soc Am J* 42:421–428.

Lion, L.W., R.S. Altmann, and J.O. Leckie. 1982. *Environ Sci Technol* 16:660–666.

Lipman, C.B., and G. MacKinney. 1931. *Plant Physiol* 6:593–599.

Locascio, S.J. 1978. *Solutions* 30–42.

Locascio, S.J., P.H. Everett, and J.G.A. Fiskell. 1968. *Proc Am Soc Hort Sci* 92:583–589.

Lucas, I.H.M., R.M. Livingston, A.W. Boyne, and I. McDonald. 1962. *J Agric Sci* 58:201–208.

Lucas, R.E., and B.D. Knezek. 1972. In J.J. Mortvedt, P.M. Giordano, and W.L. Lindsay, eds. *Micronutrients in Agriculture*. Soil Sci Soc Am, Madison, WI.

Lura, C.L., C.A. Gonzales, and V.S. Trippi. 1994. *Plant Cell Physiol* 35:11–15.

Ma, H., S.D. Kim, D.K. Cha, and H.E. Allen. 1999. *Environ Toxicol Chem* 18:828–837.

Makarim, A.K., and F.R. Cox. 1983. *Agron J* 75:493–496.

Martens, D.C., and D.T. Westermann. 1991. In J.J. Mortvedt et al., eds. *Micronutrients in Agriculture,* SSSA 4, Soil Sci Soc Am, Madison, WI.

Martinez, C.E., and M.B. McBride. 1999. *Environ Sci Technol* 33:745–750.

Mathur, S.P., and M.P. Levesque. 1983. *Soil Sci* 135:166–176.

Mathur, S.P., R.B. Sanderson, A. Belanger, M. Valk, E.N. Knibbe, and C.M. Preston. 1984. *Water Air Soil Pollut* 22:277–288.

Mattigod, S.V., and G. Sposito. 1977. *Soil Sci Soc Am J* 41:1092–1097.

McBride, M.B. 1981. In J.F. Loneragan, A.D. Robson, and R.D. Graham, eds. *Copper in Soils and Plants*. Academic Pr, New York.

McBride, M.B., and J.J. Blasiak. 1979. *Soil Sci Soc Am J* 43:866–870.

McBride, M.B., and D.R. Bouldin. 1984. *Soil Sci Soc Am J* 48:56–59.

McGrath, S.P., J.R. Sanders, and M.H. Shalaby. 1998. *Geoderma* 42:177–188.

McKeague, J.A., and M.S.Wolynetz. 1980. *Geoderma* 24:299–307.

McLaren, R.G., and D.V. Crawford. 1973a. *J Soil Sci* 24:172–181.

McLaren, R.G., and D.V. Crawford. 1973b. *J Soil Sci* 24:443–452.

McLaren, R.G., and D.V. Crawford. 1974. *J Soil Sci* 25:111–119.

McLaren, R.G., R.S. Swift, and J.G. Williams. 1981. *J Soil Sci* 32:247–256.

Mehlich, A., and S.S. Bowling. 1975. *Commun Soil Sci Plant Anal* 6:113–128.

Meima, J.A., A. Van Zomeren, and R.N.J. Comans. 1999. *Environ Sci Technol* 33:1424–1429.

Menzel, R.G., and M.L. Jackson. 1950. *Soil Sci Soc Am Proc* 15:122–124.

Merry, R.H., K.G. Tiller, and A.M. Alston. 1986. *Plant Soil* 91:115–128.

Miller, W.P., and W.W. McFee. 1983. *J Environ Qual* 12:29–33.

Mills, J.G., and M.A. Zwarich. 1975. *Can J Soil Sci* 55:295–300.

Mitchell, R.L. 1964. In F.E. Bear, ed. *Chemistry of the Soil*. Am Chem Soc 160. Reinhold, New York.

Mizuno, N., O. Ivazu, and K. Dobashi. 1983. *Soil Sci Plant Nutr* 29:1–6.

Moraghan, J.T., and H.J. Mascagni. 1991. In J.J. Mortvedt et al., eds. *Micronutrients in Agriculture*, SSSA 4, Soil Sci Soc Am, Madison, WI.

Msaky, J.J., and R. Calvet. 1990. *Soil Sci* 150:513–522.

Murphy, L.S., and L.M. Walsh. 1972. In J.J. Mortvedt, P.M. Giordano, and W.L. Lindsay, eds. *Micronutrients in Agriculture*. Soil Sci Soc Am, Madison, WI.

Nambiar, E.K.S. 1976a. *Aust J Agric Res* 27:453–463.

Nambiar, E.K.S. 1976b. *Aust J Agric Res* 27:465–477.

Neelakantan, V., and B.V. Mehta. 1961. *Soil Sci* 91:251–256.

Nelson, W.L., A. Mehlich, and E. Winters. 1953. In W.H. Pierre and A.G. Norman, eds. *Soil and Fertilizer phosphorus in Crop Nutrition*. Academic Pr, New York.

Norvell, W.A. 1984. *Soil Sci Soc Am J* 48:1285–1292.

Norvell, W.A. 1991. In J.J. Mortvedt et al., eds. *Micronutrients in Agriculture*, SSSA 4, Soil Sci Soc Am, Madison, WI.

Nriagu, J.O., ed. 1979. *Copper in the Environment*. Wiley, New York.

Owuoche, J.O., K.G. Briggs, G.J. Taylor, and D.C. Penney. 1994a. *Can J Plant Sci* 74:25–30.

Owuoche, J.O., K.G. Briggs, G.J. Taylor, and D.C. Penney. 1994b. *Can J Plant Sci* 75:405–411.

Page, A.L. 1974. *Fate and Effects of Trace Elements in Sewage Sludge When Applied to Agricultural Lands*. EPA-670/2-74-005. U.S. Environmental Protection Agency. Cincinnati, OH.

Parker, G.R., W.W. McFee, and J.M. Kelly. 1978. *J Environ Qual* 7:337–342.

Payne, G.G., D.C. Martens, E.T. Kornegay, and M.D. Lindemann. 1988. *J Environ Qual* 17:740–746.

Powell, R.D. 1975. Univ Wisconsin Coop Ext Pub A2527. Univ Wisconsin, Madison, WI.

Purves, D. 1977. *Trace Element Contamination of the Environment.* Elsevier, Amsterdam.

Quarterman, J. 1973. *Qual Plant Pl Fds Hum Nutr* 23:171–190.

Rai, M.M., J.M. Dighe, and A.R. Pal. 1972. *J Indian Soc Soil Sci* 20:135–142.

Ramos, L., L.M. Hernandez, and M.J. Gonzales. 1994. *J Environ Qual* 23:50–57.

Reiners, W.A., Marks, R.H., and P.M. Vitousek. 1975. *Oikos* 26:264–275.

Reuther, W., and C.K. Labanauskas. 1965. In H.D. Chapman, ed. *Diagnostic Criteria for Plants and Soils.* Quality Printing, Abilene, TX.

Reuther, W., and P.F. Smith. 1952. *Proc Fla State Hort Soc* 65:62–69.

Reuther, W., and P.F. Smith. 1954. *Proc Soil Sci Soc Fla* 14:17–23.

Rhoads, F.M., S.M. Olson, and A. Manning. 1989. *J Environ Qual* 18:195–197.

Riemer, O.N., and S.J. Toth. 1970. *Am Water Works Assoc J* 62:195–197.

Robertson, L.S., D.D. Warncke, and B.D. Knezek. 1981. Michigan Coop Ext Service Bull E-1519. East Lansing, MI.

Robson, A.D., and D.J. Reuter. 1981. In J.F. Loneragan, A.D. Robson, and R.D. Graham, eds. *Copper in Soils and Plants.* Academic Pr, New York.

Romheld, V., and H. Marschner. 1991. In J.J. Mortvedt et al., eds. *Micronutrients in Agriculture,* SSSA 4, Soil Sci Soc Am, Madison, WI.

Rousos, P.A., and H. Harrison. 1987. *J Am Hort Sci* 112:928–931.

Rousos, P.A., H. Harrison, and K.L. Steffen. 1989. *J Am Hort Sci* 114:149–152.

Rühling, A., and G. Tyler. 1971. *J Appl Ecol* 8:497–507.

Russell, L.H., Jr. 1979. In F.W. Oehme, ed. *Toxicity of Heavy Metals in the Environment.* Marcel Dekker, New York.

Sanders, J.R., S.P. McGrath, and T.M. Adams. 1986. *J Sci Food Agric* 37:961–968.

Sauve, S., M.B. McBride, W.A. Norvell, and W.H. Hendershot. 1997. *Water Air Soil Pollut* 100:133–149.

Sawan, Z.M., M.H. Mahmoud, and B.R. Gregg. 1993. *J Agric Sci* 121:199–204.

Schnitzer, M. 1969. *Soil Sci Soc Am Proc* 33:75–81.

Selvarajah, N., V. Pavanasasivam, and K.A. Nandasena. 1982. *Plant Soil* 68:309–320.

Sharma, S.S., H. Schat, R. Vooijs, L.M. Van Herrwaarden. 1999. *Environ Toxicol Chem* 18:348–355.

Shiha, K. 1951. *J Sci Soil Manure* 22:26–28.

Shuman, L.M. 1979. *Soil Sci* 127:10–17.

Shuman, L.M. 1991. In J.J. Mortvedt et al., eds. *Micronutrients in Agriculture,* SSSA 4, Soil Sci Soc Am, Madison, WI.

Siccama, T.G., and W.H. Smith. 1978. *Environ Sci Technol* 12:593–594.

Sims, J.T., and G.V. Johnson. 1991. In J.J. Mortvedt et al., eds. *Micronutrients in Agriculture,* SSSA 4, Soil Sci Soc Am, Madison, WI.

Smilde, K.W. 1973. *Plant Soil* 39:131–148.

Smith, S.R. 1994. *Environ Pollut* 85:321–327.

Soltanpour, P.N., and A.P. Schwab. 1977. *Commun Soil Sci Plant Anal* 8:195–307.

Sommer, A.L. 1931. *Plant Physiol* 6:339–345.

Sommers, L.E. 1977. *J Environ Qual* 6:225–232.

Sorensen, E.M.B. 1991. *Metal Poisoning in Fish.* CRC Pr, Boca Raton, FL.

Sparr, M.C. 1970. *Commun Soil Sci Plant Anal* 1:241–262.

Sposito, G., L.J. Lund, and A.C. Chang. 1982. *Soil Sci Soc Am J* 46:260–264.

Spratt, E.E., and A.E. Smid. 1978. *Agron J* 70:633–638.

Stahly, T.S., G.L. Cromwell, and H.J. Monegue. 1980. *J Anim Sci* 51:1347–1351.

Stahr, K., H.W. Zöttl, and Fr. Hädrich. 1980. *Soil Sci* 130:217–224.

Stevenson, F.J. 1972. *BioScience* 22:643–650.

Stevenson, F.J. 1991. In J.J. Mortvedt et al., eds. *Micronutrients in Agriculture,* SSSA 4, Soil Sci Soc Am, Madison, WI.

Stevenson, F.J., and A. Fitch. 1981. In J.F. Loneragan, A.D. Robson, and R.D. Graham, eds. *Copper in Soils and Plants.* Academic Pr, New York.

Stouthart, X.J.H.X., J.L.M. Haans, R.A.C. Lock, and S.E.W. Bonga. 1996. *Environ Toxicol Chem* 15:376–383.

Sutton, A.L., D.W. Nelson, V.B. Mayrose, and D.T. Kelly. 1983. *J Environ Qual* 12:198–203.

Swaine, D.J., and R.L. Mitchell. 1960. *J Soil Sci* 11:347–368.

Takahashi, Y., and H. Imai. 1983. *Soil Sci Plant Nutr* 29:111–122.

Thornton, I. 1979. In J.O. Nriagu, ed. *Copper in the Environment.* Wiley, New York.

Thornton, I., and J.S. Webb. 1980. In B.E. Davies, ed. *Applied Soil Trace Elements.* Wiley, New York.

Tiller, K.G., and R.H. Merry. 1981. In J.F. Loneragan, A.D. Robson, and R.D. Graham, eds. *Copper in Soils and Plants.* Academic Pr, New York.

Timmer, L.W., and S.E. Zitko. 1996. *Plant Dis* 80:166–169.

Trierweiler, J.F., and W. Lindsay. 1969. *Soil Sci Soc Am Proc* 33:49–54.

Turvey, N.D. 1984. *Plant Soil* 77:73–86.

Underwood, E.J. 1973. In *Toxicants Occurring Naturally in Foods.* National Academy of Sciences, Washington, DC.

U.S. Bureau of Mines. 1995. In *Minerals Yearbook.* U.S. Dept Interior, Washington, DC.

[U.S. EPA] United States Environmental Protection Agency. 1985. *Ambient Aquatic Life Criteria for Copper.* EPA 440/5-84-031. U.S. EPA, Washington, DC.

US-EPA. 1988. In *Chesapeake Bay Agreement Paper.* Annapolis, MD.

Van Campen, D.R. 1991. In J.J. Mortvedt et al., eds. *Micronutrients in Agriculture,* SSSA 4, Soil Sci Soc Am, Madison, WI.

Van den Burg, J. 1983. *Plant Soil* 75:213–219.

Van Hook, R.I., W.F. Harris, and G.S. Henderson. 1977. *Ambio* 6:281–286.

Walsh, L.M., W.H. Erhardt, and H.D. Seibel. 1972. *J Environ Qual* 1:197–200.

Winner, R.W. 1984. *Aquat Toxicol* 5:267–274.

Whitby, L.M., J. Gaynor, and A.J. MacLean. 1978. *Can J Soil Sci* 58:325–330.

Wu, J., A. Liard, and M.L. Thompson. 1999. *J Environ Qual* 28:334–338.

Zborishchuk, Y.N., and N.G. Zyrin. 1978. *Sov Soil Sci* 10:27–33.

Zhu, B., and A.K. Alva. 1993. *J Plant Nutr* 16:1837–1845.

14
Manganese

1 General Properties of Manganese

Manganese (atom. no. 25; atom. wt. 54.94; melting pt, 1244 ± 3 °C, and specific gravity, 7.2), is a member of Group VII-A of the periodic table. It is next to Fe in the atomic series, is similar to it in chemical behavior, and is often closely associated with it in its natural occurrence (i.e., Mn ores and Fe ores often coexist). Its compounds can exist in the oxidation states of I, II, III, IV, VI, and VII. Its most stable salts are those of oxidation states II, IV, VI, and VII. The lower oxides (MnO and Mn_2O_3) are basic; the higher oxides are acidic. Manganese is a whitish-gray metal, harder than Fe but quite brittle. Manganese metal oxidizes superficially in air and rusts in moist air.

Manganese is an ubiquitous metal in the earth's crust, the 12th most abundant element and makes up about 0.10% of the earth's crust (Krauskopf, 1979). It is the principal metallic component of nodules deposited on the ocean floor. Manganese minerals are widely distributed, the most common being the oxides, carbonates, and silicates. The main ores are pyrolusite (MnO_2), rhodochrosite ($MnCO_3$), manganite ($Mn_2O_3 \cdot H_2O$), hausmannite (Mn_3O_4), braunite ($3Mn_2O_3 \cdot MnSiO_3$), and rhodonite ($MnSiO_3$).

Divalent Mn is very stable in acid solutions and MnO_2 is very stable in alkaline solutions in the presence of O_2; Mn^{2+} does not form complexes with ligands as strongly as other micronutrients.

2 Production and Uses of Manganese

Total worldwide production of Mn ores in 1993 was $21{,}757 \times 10^3$ tonnes, with the Ukraine, South Africa, Brazil, Gabon, Australia, and China as the leading producers. China, South Africa, and Ukraine together produced about 60% of the world's mined Mn in 1993 (U.S. Bureau of Mines, 1993). Although not a producer, the United States is a big consumer of imported Mn. For example, 347×10^3 tonnes of Mn ferroalloys and metal were consumed in 1994.

Manganese ore deposits are widespread throughout the tropical, subtropical, and warmer temperate zones of the earth (NAS, 1973). The largest deposits are found in the Ukraine, China, and South Africa. The biggest suppliers of Mn in international trade are Australia, Brazil, South Africa, and Gabon.

Since 1939, Mn has been used primarily in metallurgical industry and its use increased markedly after the introduction of the Bessemer process (Mena, 1980). As an essential ingredient of steel, it neutralizes the harmful effects of sulfur, serves as an antioxidant, and provides strength, toughness, and hardness. For these reasons, Mn is used also in the production of alloys of steel, aluminum, and copper. These alloys are used in the electrical industry and for ship propellers. Manganese or its compounds are used quite extensively in alkaline batteries, electrical coils, ceramics, matches, welding rods, glass, dyes, paints, and drying industries. Black manganese oxide (pyrolusite, MnO_2) is used as a depolarizer in dry cell batteries. For various chemical reactions, Mn(IV) oxide, manganese chloride, and manganese stearate are used as catalysts. Other compounds of Mn are used as driers for paints, varnishes and oils, fertilizers, disinfectants, and animal food additives. Some organic manganese compounds of importance include: the fungicide maneb [manganese ethylene-*bis*(dithiocarbamate)] and the antiknock agent methylcyclopentadienyl manganese tricarbonyl (MMT).

3 Manganese in Nature

The average content of Mn in the lithosphere is about 1000 ppm; the lithosphere contains about 50 times as much Fe, one-fifth as much Ni, and one-tenth as much Cu. Elemental Mn does not exist in nature. Manganese is widely distributed in metamorphic, sedimentary, and igneous rocks. One reason for its wide distribution in different types of rocks is its similar ionic size to Mg and Ca, enabling it to replace the two elements in silicate structures. Manganese also replaces ferrous Fe in magnetite. There are more than 100 minerals (e.g., sulfides, oxides, carbonates, silicates, phosphates, arsenates, tungstates, and borates) that contain Mn as an essential component, and it is an accessory element in perhaps more than 200 others (NAS, 1973). The most important Mn mineral is the native black manganese oxide, MnO_2 (pyrolusite). In igneous rocks, Mn is usually present in silicates in the II oxidation state. During weathering, Mn oxidizes and separates out as Mn oxides in soils, appearing as dark nodules and coatings (Norrish, 1975).

Normal soils contain an average Mn concentration of 850 ppm (range of 100 to 4000 ppm); recently, world soils were reported to contain about 1000 ppm (range of 20 to 10,000 ppm) (Table 14.1).

In nature, Mn chemically behaves similarly to Fe. In its naturally occurring compounds, Mn displays three oxidation states (II, III, and IV) in contrast to only two exhibited by Fe. Oxidation of Mn(II) compounds requires higher redox potential than the oxidation of Fe(II) compounds; MnS is more soluble than FeS, and MnS_2 is far less stable than FeS_2 (pyrite) (Krauskopf, 1979). In reducing environments, the Mn(II) species are most stable, while in oxidizing environments the most stable compound is the dioxide, MnO_2.

4 Manganese in Soils

4.1 Total Manganese in Soils

Surficial soils in the conterminous United States have an arithmetic mean of 560 ppm Mn, with a range of <1 to 7000 ppm ($n = 863$) (Shacklette et al., 1971). A mean of 520 ppm (range of 100 to 1200 ppm) has been reported for Canadian soils ($n = 173$ from 53 soils) (McKeague and Wolynetz, 1980).

Similarly, a mean of 530 ppm Mn with a range of 90 to 3000 ppm has been reported for Ontario, Canada soils (Frank et al., 1976). For soils collected from the plow layer in the central Russian region, an average Mn content of 740 ppm (range of 472 to 1250 ppm) was obtained (Krupskiy et al., 1978). The total Mn content of soils of China average 582 ppm (geometric mean $= 432$; $n = 3990$) (Chen et al., 1991). The Alfisols have the highest Mn content ($\bar{x} = 1127$ ppm; $n = 186$); the Ultisols and Oxisols have the lowest contents; 323 ppm ($n = 796$) and 197 ppm ($n = 262$), respectively.

Swaine (1955) reported a range of 200 to 3000 ppm for total Mn content in most soils; Bowen (1979) reported a mean value of 1000 ppm for world soils with a range of 20 to 10,000 ppm (see Table 14.1); Berrow and Reaves (1984) reported a lower mean of 450 ppm Mn for world soils. Normal soils can be expected to have a maximum Mn content of ~4000 ppm.

TABLE 14.1. Commonly observed manganese concentrations (ppm) in various environmental media.

Material	Average concentration	Range
Igneous rocks[b]	—	390–1620
Limestone	620	—
Sandstone[c]	460	—
Shale	850	—
Manganese nodules[b]	160,000	—
Coal[d]	100	—
Coal ash[d]		
Fly ash	357	44–1332
Bottom ash	426	56–1940
FGD sludge	138	27–312
Oil ash[d]	707	113–1170
Sewage sludge[e]	—	60–3900
Normal soils[e]	850	100–4000
Soils[c]	1000	20–10,000
Soil solution[f]	10^{-7}–10^{-5} M	10^{-9}–10^{-3} M
Common crops[e]	—	15–100
Herbaceous vegetables[c]	—	0.3–1000
Ferns[c]	700	—
Fungi[c]	—	10–400
Lichens[c]	—	3–190
Freshwater[a,c]	8	0.02–130
Seawater[a,c]	0.2	0.03–21

Sources: Extracted from
[a] μg L^{-1}.
[b] NAS (1973).
[c] Bowen (1979).
[d] Ainsworth and Rai (1987).
[e] Freeman and Hutchinson (1981).
[f] Robson (1988).

4.2 Bioavailable Manganese in Soils

As with the other micronutrients, there is no universal method for determining bioavailable Mn in soils that can accurately predict plant response to this element. It is universally agreed that total soil Mn is an unreliable predictor of Mn bioavailability. Methods used in various countries largely depend on plant species and properties of the soil.

Various extractants have been used to estimate the forms of Mn: water-soluble, exchangeable, acid-soluble, easily reducible, and complexed forms of Mn. Water-soluble and exchangeable Mn could be used to predict uptake and potential toxicity; exchangeable Mn has been shown to relate closely with Mn uptake and in general, has correlated closely with the occurrence of Mn deficiency in plants (Adriano, 1986). The level of easily reducible Mn in soil [can be determined with 0.2% hydroquinone in 1 N NH$_4$OAc, pH 7] has been used to evaluate available Mn for a number of years (Jones and Leeper, 1951; Leeper, 1947; Sherman et al., 1942). Several chelating agents have recently become more popular especially DTPA (Lindsay and Norvell, 1978).

The sensitivity of soil tests for extractable Mn to soil properties and plant factors is reflected by divergent results obtained by various investigations. Relationships between soil Mn extracted by two commonly used methods (water-soluble and NH$_4$OAc-extractable) and plant tissue Mn indicate good correlations between the extractable level and tissue level of Mn in red kidney beans, French beans, and peas (White, 1970). However, relationships between extractable fraction and plant response have been extremely variable. For example, of the eight methods (two exchangeable extractants—1 N KCl and 1 N NH$_4$OAc at pH 3; reducible extractant 0.2% hydroquinone in 1 N NH$_4$OAc at pH 7; four soluble extractants—0.01 M CaCl$_2$, 0.1 N HOAc, 0.1 N H$_3$PO$_4$, and 0.002 N HCl; total Mn extraction), a 16-hr extraction with 0.01 M CaCl$_2$ gave the best estimate of plant-available Mn with barley ($r = 0.76$), rape ($r = 0.73$), and alfalfa ($r = 0.86$) as plant indicators (Hoyt and Nyborg, 1971).

In addition, 0.01 M CaCl$_2$-soluble Mn produced better fit with the crop data than 0.1 N H$_3$PO$_4$-soluble, water-soluble, reducible, or exchangeable Mn (Hoyt and Webber, 1974). In another instance, Hoff and Mederski (1958) found that, of the nine extraction methods employed, hydroquinone, 3 N NH$_4$H$_2$PO$_4$, and 0.1 N H$_3$PO$_4$ gave the highest correlation coefficients with Mn in soybean plants. Salcedo et al. (1979b) noted that the correlation between extractable Mn and uptake of Mn by soybean plants decreased in the order: 0.1 N H$_3$PO$_4$ > steam/1 N NH$_4$OAc > 1.5 M NH$_4$H$_2$PO$_4$ > 0.1 N HCl > 1 N NH$_4$OAc > 0.005 M DTPA. In oats, Mn extracted by 0.1 N H$_3$PO$_4$ correlated better with Mn levels in plant tissue than did Mn extracted by 1.5 M NH$_4$H$_2$PO$_4$ or 1 N H$_3$PO$_4$ (Hammes and Berger, 1960).

In general, extractable Mn has been used successfully in many studies, especially for soybeans and small grains to predict crop response to Mn fertilization (Sims and Johnson, 1991). Most of the soil testing laboratories in the United States use acidic extractants (e.g., Mehlich-1, Mehlich-3, 0.1 M HCl, or 0.03 M H$_3$PO$_4$) to estimate bioavailable Mn. Critical values (in ppm) for the most commonly used extractants are reported as follows (Sims and Johnson, 1991): Mehlich-1, 5 at pH 6, or 10 at pH 7; Mehlich-3, 4 at pH 6, or 8 at pH 7; 0.1 M HCl, 1 to 4; 0.03 M H$_3$PO$_4$, 10 to 20; and DTPA, 1 to 5. In tropical regions, the following extractants are being employed in addition to the above extractants: 0.05 M H$_2$SO$_4$, 1.0 M NH$_4$OAc, modified Olsen's, and water. For example, in Oxisol soil solution, Mn produced a better fit with cowpea tissue Mn than did DTPA-extractable Mn (Fig. 14.1).

Among soil factors, soil pH plays an important role in the selection of a suitable extractant. For example, for acidic southeastern U.S. soils, DTPA may produce the best measure of plant-available Mn at pH 5.8 to 6.8, whereas for soils at pH 4.8, water-soluble Mn value may be the best measure (Shuman and Anderson, 1974). Because the correlations between amount of Mn extracted from soils and the amount found in plants are dependent on

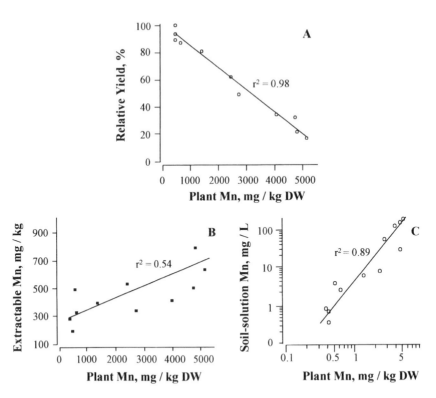

FIGURE 14.1. Relationships of manganese content of cowpea plants grown on an Oxisol with (A) yield, (B) DTPA-extractable manganese, and (C) soil-solution manganese. (Extracted from Vega et al., 1992.)

soil pH, prediction equations should include soil pH (Randall et al., 1976; Salcedo et al., 1979b).

Cox (1968) found that the degree of response by soybeans to Mn fertilization could be predicted from the soil pH and the level of double acid–extractable (0.05 M HCl + 0.0125 M H_2SO_4 mixture) Mn in the soil. A predictive model for yield response of soybeans has been developed based on the above soil test over a soil pH range of 5.2 to 7.1 and a range of 0 to 10 ppm extractable Mn. As an example, Mn fertilization can produce a soybean yield response of at least 400 kg ha^{-1} in soils having low levels of extractable Mn (≤2 ppm) and pH values >6.0. Cox's (1968) results were then extended to develop a probability scale for yield response. Thus, it can be predicted that yield response to Mn fertilization would occur when soybeans contain ≤20 ppm in the uppermost developed leaf, or in the mature seed.

4.3 Manganese in Soil Profile

As a general rule, Mn can be expected to be highest in the surface soil, reaching a minimum in the B horizon, and then increasing in the C horizon. To a large extent, the accumulation of Mn in the surface horizon could be attributed to plant root uptake of Mn and deposition on the surface upon decay of foliage. For example, acid Ultisols and Alfisols in North Carolina displayed surface accumulation of Mn and a tendency to diminish with depth in the pedon of a very strongly acid soil (Fig. 14.2); whereas in moderately acid soil, the Mn content declined and then increased again with depth. The difference in the subsoil content of the two pedons is apparently due to greater leaching loss in the very strongly acid soil. In contrast to Mn behavior, Fe content was lowest at the surface and tended to accumulate with depth (McDaniel and Buol, 1991).

In major soil groups of Rajasthan, India Mn (also Fe and Cu) was uniformly distributed in the profile of relatively less weathered desert and old alluvial soils (Lal and Biswas, 1974). Accumulation of $CaCO_3$ in the lowest horizons of desert, gray, brown, and black soils substantially reduced the content of total Mn. Extractable micronutrients were concentrated in surface horizons of well-drained soils but were concentrated in the lower horizons in poorly drained soils. In other Indian soils, both total and extractable Mn generally decreased with depth (Agrawal and Reddy, 1972; Patel et al., 1972). Similarly, the concentration of DTPA-extractable Mn (and Zn, Fe, and Cu) decreased with depth in the profile of Colorado soils (Follett and Lindsay, 1970). A similar pattern was observed for extractable Mn in Scottish soil profiles (Berrow and Mitchell, 1980). In contrast, water-soluble and exchangeable Mn showed little variation within profile or between soils of Fraser Valley, British Columbia, Canada (Safo and Lowe, 1973). These soils were derived from alluvial and marine-deposited parent materials.

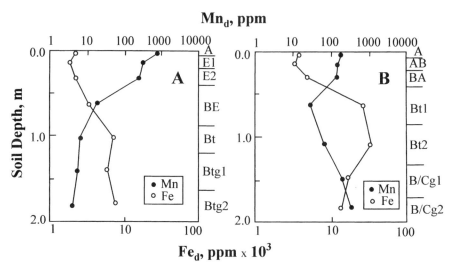

FIGURE 14.2. Distribution of secondary manganese and iron in the soil profile of moderately well-drained soils as influenced by subsoil pH; (A) = very strong acid, fine-loamy, mixed, thermic Typic Hapludalf; (B) = moderately acid, coarse loamy, mixed, thermic Typic Paleudalf. Secondary Fe or Mn content is approximately equal to the total minus the residual fraction. (From McDaniel and Buol, 1991.)

4.4 Forms, Chemical Speciation, and Transformation of Manganese

The biogeochemistry of Mn is very complex due to the following reasons: Mn in soils can exist in several oxidation states; the oxides of Mn can exist in several crystalline or pseudocrystalline states; the oxides can form copre-cipitates with Fe oxides; Fe and Mn hydroxides exhibit amphoteric behavior and show a tendency to interact specifically with anions as well as with cations; and oxidation-reduction reactions involving Mn are influenced by physical, chemical, and microbiological factors. In most acid and alkaline soils, Mn^{2+} is the predominant solution species (Fig. 14.3). In neutral and alkaline soils, soluble Mn silicates may become an important source of Mn supply to the roots (Boxma and de Groot, 1985).

The forms of soil Mn can be delineated and operationally defined by chemical extractants and may be classified as: water-soluble, exchangeable, organic, easily reducible (Mn oxide), Fe oxide, and residual. The organic and Mn oxide forms are more soluble and therefore easier to redistribute to plant-available forms than the Fe oxide and residual forms. Such forms can be determined by sequential fractionation as indicated in Chapter 2. Results of such fractionation are presented in Table 14.2 for two acidic well-drained soils in the North Carolina Piedmont. The Ultisol has much higher Mn content in the exchangeable, OM, Mn oxide, amorphous, and Fe oxide

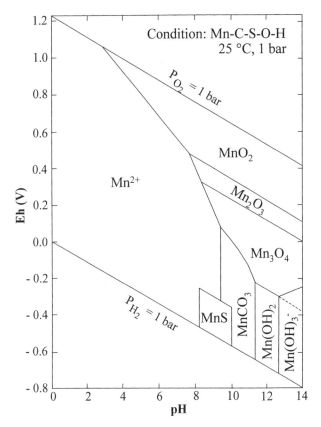

FIGURE 14.3. Predicted Eh–pH stability field for manganese; Mn $= 10^{-6}$, C $= 10^{-3}$, S $= 10^{-3}$. (From Brookins, 1988.)

fractions in the A layer than did the Alfisol. It was the reverse in the deeper B layer. The difference in the Mn distribution pattern between the two pedons can be attributed to high OM and clay contents of the Ultisol compared with the Alfisol. Indeed, Shuman (1985) found that most of the total Mn in 16 surface soils was in the organic and Mn oxide fractions.

In eight Kentucky soils, the reducible fraction accounted for an average of nearly 45% of the total Mn (Sims et al., 1979). The exchangeable, organic, and Fe-bound fractions each contained intermediate amounts (13 to 19%), and the water-soluble had <2% of the total. However, the distribution varied among the soils and was generally related to soil pH. Soils of low pH generally had the highest amounts of water-soluble, exchangeable, and organic Mn whereas soils with high pH contained large amounts of reducible Mn. Using a similar extraction scheme, about 75% of the total

TABLE 14.2. Manganese distribution among soil fractions as determined by a sequential extraction procedure.

Soil	Horizon	Exch[†1]	OM[‡2]	Mn oxide[3]	Amor- phous[4]	Fe oxides[§5]	Total secondary	Residual[6]	Total
					mg kg^{-1}				
Alfisol	A	33	71	43	33	3	183	49	232
(1-10)	BA	8	nd	78	29	2	117	131	248
	Bt2	8	nd	27	16	4	55	245	300
	B/Cg2	24	nd	466	149	29	668	936	1604
Ultisol	A	90	380	160	56	7	693	nd	693
(11A-6)	E1	25	23	183	32	4	267	nd	267
	BE	6	nd	6	6	3	21	nd	21
	Bt	1	nd	3	5	2	11	50	61
	Btg2	1	nd	4	7	3	15	74	89

[†]Exch=exchangeable Mn; [‡]OM=Mn associated with OM; [§]Mn associated with Fe oxide; [*]nd= none detected. Fractions 1,....,6 were respectively extracted by 1 M Mg(NO$_3$)$_2$ (pH 7), 0.7 M NaOCl (pH 8.5), 0.1 M NH$_2$OH · HCl (pH 2), 0.2 M (NH$_4$)$_2$C$_2$O$_4$ · H$_2$O–0.2 M H$_2$C$_2$O$_4$ (pH 3), CBD: 0.3 M Na$_3$C$_6$H$_5$O$_7$ · · · 2H$_2$O/0.1 M NaHCO$_3$/NaS$_2$O$_4$ (1 g), and HF; Pedons 1-10 and 11A-6 were respectively, fine-loamy, mixed, thermic Typic Hapludalf (Alfisol) and coarse loamy, mixed, thermic Paleudult (Ultisol).
Source: Extracted from McDaniel and Buol (1991).

indigenous soil Mn found in selected U.K. soils is in the oxide-bound (easily reducible, Mn oxide) and resistant fractions (Goldberg and Smith, 1984). Less than 3% of the total is found in the water-soluble plus exchangeable and EDTA (organic-bound) fractions. The fate of soil-applied Mn fertilizers can be deduced from ^{54}Mn tracer studies (Goldberg and Smith, 1984). Most of the applied isotope, well over 70% in some cases, was recovered by the first three extractants (i.e., water-soluble plus exchangeable, EDTA-extractable, and easily reducible) in most of the soils used, but the distribution is variable among soil samples. In general, the easily reducible (Mn oxide) fraction contained the most ^{54}Mn.

Using a similar fractionation scheme for flooded soil, the water-soluble Mn was observed to be inversely related to soil pH (4.5 to 7.5) and Eh (-150 to $+500$ mV) (Sims and Patrick, 1978). In general, greater amounts of Mn (also Fe, Cu, and Zn) were found in either the exchangeable or organic fraction at low pH and Eh than at high pH or Eh, indicating that under acidic and reducing conditions, micronutrients that were precipitated and occluded as oxides and hydroxides were eventually solubilized. Later results by Patrick and Jugsujinda (1992) confirm the greater solubilization of both Mn and Fe at lower redox potentials (Fig. 14.4). As was shown on Figure 14.3, Mn is immobilized as oxide phases (mainly MnO$_2$) under oxidizing and near neutral conditions. Like Fe, Mn is soluble under acid and reducing conditions.

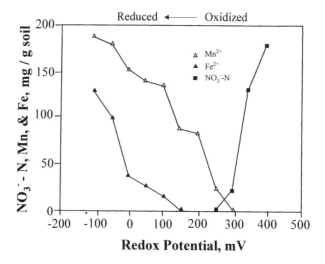

FIGURE 14.4. Soluble NO_3^-, Mn(II), and Fe(II) in suspensions of Crowley silt loam soil (pH 6.5) during an oxidized-to-reduced incubation half cycle under controlled redox potential conditions. (Modified from Patrick and Jugsujinda, 1992.)

Conditions around tile drains may become more favorable for solubilization of Mn at Eh $\leq$400 mV (Grass et al., 1973a,b). In displaced soil solutions obtained from A horizons from several areas of the United States, about 84 to 99% of the total Mn was in the complexed form (Geering et al., 1969), with the Mn in soil solution occurring in the II oxidation state. In contrast, Sanders (1983) found that free Mn^{2+} ions comprised the major Mn species in soil solution at pH 5, and the proportion of Mn in this form decreased as the pH increased. In general, Mn occurs primarily as the divalent ion in waters and soil solutions [i.e., Mn(II) $\gg$ Mn(IV) > Mn(III)] (NAS, 1977). The levels of Mn(IV) are very low because of the low solubility of manganese dioxide minerals, whereas the concentrations of Mn(III) are low because they readily reduce to Mn(II).

In flooded soils, the main transformation of Mn involves the reduction of Mn(IV) to Mn(II), resulting in an increase in the concentration of Mn^{2+} ions, precipitation of manganous carbonate (see also Fig. 14.3), and reoxidation of Mn^{2+} ions diffusing or moving by mass flow to aerobic interfaces in the soil (Ponnamperuma, 1972). This transformation is represented below:

$$MnO_{2(s)} + 4H^+ + 2e^- \rightleftharpoons Mn^{2+} + 2H_2O$$

It can be expected that within 1 to 3 wk of flooding, almost all organic and oxide-bound Mn present in most soils is reduced. The reduction is biologically mediated. The reduction of oxidized inorganic redox components in soils that serve as electron acceptors is generally sequential, with O_2 being first, followed by NO_3^-, Mn(IV) compounds, Fe(III) compounds, sulfate, and CO_2 (Patrick and Jugsujinda, 1992). Acid soils high in Mn and OM may build up Mn^{2+} levels as high as 90 ppm within about 2 wk of flooding, then exhibit a rapid decline to a fairly stable level of about 10 ppm. On the other hand, alkaline soils rarely contain over 10 ppm water-soluble Mn^{2+} at any period of flooding. Manganese is present in reducing soil solutions as Mn^{2+}, $MnHCO_3^+$, and as organic complexes. Most flooded soils contain sufficient water-soluble Mn for the growth of rice plants, and Mn toxicity is generally not known to occur in paddy soils.

Under flooded conditions, a sharp increase in the level of water-soluble plus exchangeable Mn accompanied by significant decrease in the level of reducible Mn has been observed (Mandal and Mitra, 1982). Manganous ions have been oxidized and precipitated (along with Fe) in soil tile drains (Grass et al., 1973a,b). The black precipitates in the drains are primarily Mn oxide compounds. This oxidation appears to be microbiological in nature and that it was increased with increasing level of bicarbonate and O_2 (Meek et al., 1973). In the Imperial Valley, California oxidation of Mn in tile lines occurs at pH as low as 6. Tyler and Marshall (1967) concluded that the oxidation was primarily microbial in nature after demonstrating the absence of oxidation in autoclaved or azide-treated systems. Ehrlich (1968) reported that the increase in oxidation of Mn by bacteria may be the result of enzymatic catalysis, biodegradation of an Mn chelate followed by autooxidation of Mn^{2+} ions.

Secondary Mn oxides (usually along with Fe oxides) occur in soils in several forms, including concretion, pans, coating, and mottle. Their occurrence is fairly common in soils throughout the world and, in some cases, they constitute a considerable part of the soil. Some soils contain 29 to 40% (by weight) of Fe–Mn concretion (Childs, 1975). Iron–Mn concretions are of interest in geochemical cycling of trace elements because of their ability to influence the distribution of metal ions in soils and waters (Jenne, 1968; Taylor and McKenzie, 1966).

The concentration of Mn^{2+} in soil solution (varying from 10^{-9} to 10^{-3} M, with most soils having 10^{-7} to 10^{-5} M) is largely controlled by redox reactions. The oxidation of Mn in acidic and neutral soils is almost entirely a microbial process (Robson, 1988). However, Mn is reduced in soils both abiotically and biotically—the relative importance of the process depending upon several factors: pH, organic reductants, Eh, and type of Mn oxide. Because the redox status of soils can change rather rapidly, Mn^{2+} concentration can fluctuate rapidly leading to transient deficiencies and toxicities in plants. It is the sensitivity of Mn^{2+} concentration to environmental conditions that renders the predictability of available Mn using chemical extractants fairly difficult.

The soil is a heterogeneous environment for transformation of Mn. Within the same soil, oxidation and reduction could be occurring simultaneously; oxidation within pores and reduction at anaerobic microsites within aggregates. Transformations of Mn can occur in the rhizosphere (Robson, 1988): (1) sterile roots can reduce Mn and the extent of reduction varies with plant species; (2) some genotypes can support greater populations of Mn-oxidizing bacteria than others; (3) rhizosphere acidification associated with cation uptake exceeding anion uptake can ameliorate Mn deficiency and presumably accentuate Mn toxicity; (4) plant roots can produce reductant organic acids that can influence the phytoavailability of Mn; and (5) some species (e.g., rice) can transport oxygen into their rhizosphere, decreasing the availability of Mn to plants.

The role of microorganisms in transforming Mn has not been fully elucidated. Bacteria and fungi are capable of oxidizing Mn(II) in soils. Organisms capable of oxidizing Mn(II) include strains of the bacterial genera *Arthrobacter, Bacillus,* etc., and of the fungal genera *Cladosporium, Curvularia,* etc. (Alexander, 1977). In tile lines having water containing Fe(II) and Mn(II), oxidizing organisms colonizing on the inner surface and joints of the tiles may cause the gradual accumulation of Fe and Mn deposits (Peterson, 1966; Spencer et al., 1963), resulting in the sealing of tile drainage joints. Some of these microorganisms are heterotrophs, which grow on organic compounds and accumulate Fe(III) and Mn(III) salts (Alexander, 1977). Others derive all or part of their energy from the oxidation of soluble inorganic substances such as Mn and Fe. In some cases, Mn and Fe may be precipitated in tile lines as a result of chemical oxidation due to increases in pH or Eh (Meek et al., 1968). Similarly, microorganisms mediate the reduction of Mn. For example, aerobic microbes may deplete the soil of O_2, causing a reduction in redox potential and the solubilization of Mn. Certain microbes may be directly involved in the reduction of MnO_2 through an enzymatic reaction (Ghiorse and Ehrlich, 1976).

Chemical analysis and speciation of Mn and Fe in a contaminated groundwater plume below an industrial landfill in Denmark (Jensen et al., 1998) indicated that the free metal ions were the dominant species in the dissolved fraction [dissolved fraction averaged 90% for Fe(II) and 91% for Mn(II)] as calculated by MINTEQ A2, followed by the bicarbonate complexes (colloids) (Table 14.3). The groundwater was 0 to 3 m deep, had a pH of 6.5, and apparently was supersaturated with respect to the carbonate. Organic complexes were not significant.

In general, the solution chemistry of Mn is dominated by Mn^{2+}, with neither complexed nor hydrolysis species altering it. The dissolution of Mn compounds is generally favored by low pH and redox potential.

4.5 Sorption of Manganese in Soils

Among the micronutrients, Mn adsorption is relatively more complicated since it forms insoluble oxides in response to pH–Eh condition. This results in rather meager information on adsorption reactions; however, data have

TABLE 14.3. Distribution of Fe(II) and Mn(II) (in groundwater samples obtained from anaerobic polluted plume) among the four species of the dissolved fractions determined by the ion exchange method and by speciation with MINTEQ A2.[a]

	Concentration (mg L^{-1})	Distribution (%)
Experimental speciation:[b]		
Fe(II)		
free	79	79
labile	21	21
slowly labile	0.1	0
stable	0.1	0
Mn(II)		
free	2.0	77
labile	0.6	23
slowly labile	0	0
stable	0	0
MINTEQ A2 speciation:		
Fe(II)		
Fe^{2+}	55	55
$FeCO_3$	1.4	1
$FeHCO_3^+$	43	43
Mn(II)		
Mn^{2+}	1.4	54
$MnCO_3$	0.1	4
$MnHCO_3^+$	0.9	36
Mn-DOM	0.1	5

[a]Groundwater samples were obtained 8.5 m below the initial groundwater depth, pH 6.5, specific conductivity of 2.24 mS cm^{-1}, total alkalinity of 21 meq/L.
[b]Speciation by ion exchange method resulted into separations of free divalent metal ions and three different complexed fractions (labile, slowly labile, and stable).
Source: Extracted from Jensen et al. (1998).

been generated on the influence of pH–Eh on Mn bioavailability and plant uptake.

Equilibration studies have shown that the adsorption of Mn by soils conforms to the Langmuir or Freundlich isotherms (Curtin et al., 1980; Willett and Bond, 1995). For example, adsorption of Mn by a highly weathered sandy soil conforms to the Freundlich model (Fig. 14.5). The isotherms indicate that the magnitude of adsorption was influenced by the soil pH as well as other soil properties. The enhanced adsorption with increasing pH is consistent with the increased hydrolysis of Mn^{2+}, increased likelihood of Mn precipitation, and increased negative charge on the exchange complex. The higher adsorption capacity of the surface Ao (0 to 4 cm) soil over the A2 (29 to 45 cm) is due to the higher CEC, and higher OM and amorphous Fe-oxide contents of that soil.

Adsorption–desorption data by Curtin et al. (1980) indicate that the amount of exchangeable Mn ranged from 6 to 76% of adsorbed Mn and that the exchangeable fraction increased as the amount of adsorbed Mn

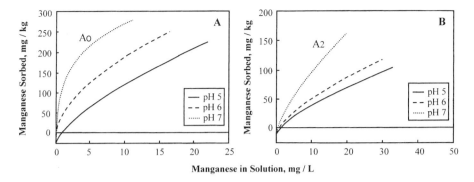

FIGURE 14.5. Adsorption of manganese by soils from the Ao (A) and A2 (B) horizons of a Grossameric Kandiustalf. Ao is 0–4 cm and has pH 6.0, organic C = 15.8%, crystalline Fe = 8350 ppm and amorphous Fe = 414; A2 is 29–45 cm, pH 5.8, organic C = 2.5%, crystalline Fe = 19,170 ppm and amorphous Fe = 240 ppm. (Extracted from Willett and Bond, 1995.)

increased. The DTPA fraction also indicates that part of the adsorbed Mn was due to complexation with OM. The Mn portion not extracted by $CaCl_2$ and DTPA varied from 12 to 93% of total adsorbed Mn, indicating that the quantity of this tightly bound form can be significant.

In high-OM soils, most of the retention of Mn in an unavailable form was attributed to OM complexation (Pavanasasivam, 1973); the sorption of heavy metals by organic soil is less pH-dependent than sorption by mineral soil (Andersson, 1977). However, organic soils are more effective sorbents than mineral soils in acid environments; above neutrality, some of the OM could dissolve together with the metals, increasing the concentration of the metal in the equilibrium solutions (see also Section 7 of the chapter on copper). Changes in the bioavailability of Mn with pH were probably not due to the formation of higher oxides of Mn or to biological transformation of Mn, but to complexation of Mn by OM (Page, 1962). Indeed, Geering et al. (1969) found that the extent of complexation of Mn in soil solutions was intermediate between Zn and Cu. Bloom (1981) reported the following affinity series of divalent metal ions for humic acids and peat: Cu > Pb ≫ Fe > Ni = Co = Zn > Mn = Ca. In flooded soils, the extent of complexation with OM was: Mn < Fe < Cu < Zn (Sims and Patrick, 1978). The complexing ability of OM was discussed in Chapter 2.

Manganese can be strongly adsorbed by clay minerals and the amount held in nonexchangeable form by kaolinite, illite, and bentonite increases with increasing pH (Reddy and Perkins, 1976). Illite and bentonite sorbed significant amounts of Mn under wetting and drying conditions. Soils kept at moisture saturation resulted in considerably less Mn sorbed than when subjected to repeated wetting and drying. Sorption of Mn onto soils can be facilitated by the following mechanisms: oxidation of Mn to higher-valence oxides and/or precipitation of insoluble compounds in soils subjected to drying and wetting, physical entrapment in clay lattice, and adsorption on exchange sites. In calcareous soils, chemisorption onto $CaCO_3$ surfaces may be important in retaining Mn (McBride, 1979); this can be followed by precipitation of $MnCO_3$.

Norvell and Lindsay (1972) indicated that soluble Mn could be lost from solutions following addition of MnDTPA to soils. At low pH, Fe dissolves from the soil and displaces Mn from MnDTPA. In calcareous soils, Mn can easily be displaced by Ca. At neutral pH, both Fe and Ca can displace Mn. In the case of MnEDTA, the loss of Mn can be rapid in soils and essentially complete in less than 1 day from soil suspensions at pH 6.1 to 7.8 (Norvell and Lindsay, 1969). This instability could seriously limit the usefulness of this chelate in micronutrient fertilizers. None of the commonly used chelating agents—EDTA, DTPA, EDDHA, etc.—could form stable Mn chelates in soils because either Fe or Ca, or both, can substitute for Mn (Norvell, 1972) The Mn^{2+} ions can then be complexed by OM or precipitated as MnO_2.

5 Manganese in Plants

5.1 Essentiality of Manganese in Plants

Manganese is an essential micronutrient in plant nutrition and has several functions. It activates many enzyme reactions involved in the metabolism of organic acids, P, and N. In higher plants, Mn activates the reduction of nitrite by hydroxylamine to ammonia. The most well-known function of Mn is its involvement in photosynthetic O_2 evolution (Hill reaction) in chloroplasts (Romheld and Marschner, 1991). The electrons are liberated by the water-splitting enzyme S, which contains four Mn atoms that are transferred to photosystem II. It is also a constituent of some respiratory enzymes and of other enzymes responsible for protein synthesis. As an activator in enzymes involved in carboxylic acid cycle and carbohydrate metabolism, it may be replaced by Mg. Manganese functions along with Fe in the formation of chlorophyll.

5.2 Deficiency of Manganese in Plants

Manganese deficiencies have been observed in various crops throughout the world. Manganese deficiency is otherwise known as gray speck in oats, yellow disease in spinach, speckled yellows in sugar beets, marsh spot in peas, crinkle leaf in cotton, stem streak necrosis in potato, streak disease in sugar cane, mouse ear in pecan, and internal bark necrosis in apple. Deficiencies are common in cereal grains, beans, corn, potatoes, sugar beets, soybeans, and many vegetable crops (Rumpel et al., 1967).

The following soil conditions may promote Mn deficiency: thin, peaty soils overlying calcareous subsoils; alluvial soils and marsh soils derived from calcareous parent materials such as calcareous silts and clays; poorly drained calcareous soils with a high OM content; calcareous black sands and reclaimed acid heath soils; calcareous soils newly tilled from old grassland; black garden soils where manure and lime have been applied regularly for many years; and acid sandy mineral soils that are low in total native Mn content. Manganese deficiency in cereals has been found in many parts of the world, especially in calcareous soils (Reuter et al., 1988). Deficiency usually results from low Mn bioavailability rather than a low content in the soil, as calcareous soils often contain large reserves of total Mn (Graham, 1988). Manganese deficiency in crops is widespread in Canada and the United States, especially on reclaimed calcareous soils, peats, and mucks found in the Great Lakes region and sandy and poorly drained soils in the Atlantic Coastal Plain states (Welch et al., 1991). In Mexico and in Central and South America, Mn deficiencies have been reported for certain crops, especially those growing in calcareous soils.

In England, Mn deficiency has been observed in crops grown on peaty soils and mineral soils that contain high OM. However, in some European countries, Mn deficiency is more strongly correlated with soil texture than with OM content (Welch et al., 1991). For example, in Sweden, Mn deficiency is associated with coarse-textured soils, and in the Netherlands, inadequate levels of available Mn occur in both coarse- and fine-textured soils. In Denmark, Mn deficiency is likely to occur in excessively limed, coarse-textured soils and in some calcareous soils high in OM.

Manganese is reportedly the most common micronutrient deficiency in soybeans (Scott and Aldrich, 1970), which has been widely observed in the southeastern United States, predominantly in poorly drained soils with pH above 6.0 (Anderson and Mortvedt, 1982). In Michigan, Mn deficiency is closely related to soil reaction in mineral soils whose pH is >6.5. Response to Mn may occur in organic soils with pH >5.8 (Rumpel et al., 1967). In general, Mn deficiency is most often associated with naturally wet areas that have been drained and brought into cultivation (NAS, 1977). This is apparently associated with the losses of bioavailable soil Mn upon leveling the soil surface and the exposure of calcareous subsoils.

The relative sensitivity of crops to Mn deficiency is indicated on Table 14.4. In general, members of the bean family are very sensitive to Mn deficiency. Among grain crops, oats is one of the most sensitive to Mn deficiency. Other crops that respond well to Mn application include sugar beets, navy beans, and soybeans, as well as tomatoes, spinach, peas, onions, lettuce, and potatoes.

Table 14.4. Relative sensitivity of various crops to manganese deficiency.

Low	Medium	High
Asparagus	Alfalfa	Apple
Blueberry	Barley	Beans
Cotton	Broccoli	Cherry
Rye	Cabbage	Citrus
	Carrot	Lettuce
	Cauliflower	Oats
	Celery	Onion
	Clover	Peach
	Corn	Peas
	Cucumber	Pecan
	Grass	Potato
	Parsnips	Radish
	Peppermint	Sorghum
	Spearmint	Soybean
	Sugar beet	Spinach
	Sweet corn	Sudan grass
	Tomato	Table beet
	Turnip	Wheat

Source: From various sources and as summarized by Martens and Westermann (1991).

Data in Table 14.5 indicate the deficiency, sufficiency, and toxicity ranges of Mn in crops grown in various growth media. Jones (1972) generalized the following levels of Mn (in ppm DW of mature leaf tissue) for most crops: deficient, <20; sufficient, 20 to 500; and toxic, >500. For selected crops, the following critical levels of Mn (in ppm) were reported: corn, 15; soybeans, 20; wheat, 30; and alfalfa, 25 (Melsted et al., 1969). The above plant tissues analyzed were: corn—leaf at or opposite and below ear level at tassel stage, soybeans—the youngest mature leaves and petioles on the plant after pod formation, wheat—the whole plant at the boot stage, and alfalfa—upper stem cuttings in early flower stage.

Manganese deficiency has been reported to delay crop maturity, observed in barley, lupin, wheat, and cotton (Longnecker et al., 1991). In Mn-deficient barley, tillering is delayed and tillers are fewer than normal plants. A common consequence of Mn deficiency in grain crops such as soybeans is reduced yield associated with lower pod number and seed weight. But it may also advance flower and pod initiation in certain crops such as soybeans. It may also cause an increase in soybean seed protein and a decrease in seed oil (Graham et al., 1994).

Manganese deficiency symptoms typically include interveinal chlorosis (yellowish to olive green) with dark green veins. In most plants, deficiency symptoms appear first in the young leaves. The deficient plant appears to grow normally for a short period, then the first symptom to appear is the loss of green color in the web of the leaf that is farthest from the vein. The veins tend to stay green, giving a mottled appearance to the leaf. Eventually, the foliage may become completely yellow. Under severe deficiency, leaves develop brown speckling and bronzing in addition to interveinal chlorosis, with abscission of developing leaves (Anderson and Ohki, 1977). The pattern of chlorosis may be confused with those for Fe, Mg, or N deficiency. Wheat and barley often exhibit colorless spots. In corn, deficient leaves are manifested by lighter green leaves with parallel, yellowish stripes.

The following soil conditions and soil management practices may induce or alleviate Mn deficiency:

1. Soils with pH above 6.5 favor the oxidation of Mn(II) to Mn(IV), which limits the solubility of Mn and its bioavailability to plants.
2. Manganese is prone to leach from strongly acidic soils, resulting in deficiency.
3. Overliming of acidic soils is a common cause of Mn deficiency.
4. Burning of organic soils, especially those that are limed, produces an alkaline condition and may cause Mn deficiency.
5. Some soils become more alkaline with irrigation, favoring development of Mn deficiency.
6. Manganese deficiency is often produced in areas with fluctuating redox conditions: waterlogging solubilizes Mn and subsequent percolation removes the Mn^{2+} ions; after the soils dry, conditions for rapid oxidation of Mn may prevail and Mn deficiency may follow.
7. Liming of poorly buffered sandy soils often causes Mn deficiency.
8. Manganese deficiency in apple can be more prevalent in areas with dry weather than those with ample rainfall.
9. Manganese deficiency in peach trees (first noted in mid-June in the U.S.) can be aggravated by dry weather in late spring and early summer.

5.3 Phytotoxicity of Manganese

Worldwide, Mn toxicity is considered more important than Mn deficiency in crops. Under field conditions, Mn toxicity in plants could occur in poorly drained, acidic soils. Manganese toxicity occurs frequently with Al toxicity. Aluminum and Mn toxicities are the most important growth-limiting factors in many acid soils (Foy and Campbell, 1984). In well-drained soils, Mn toxicity is generally found in soils having pH below 5.5; in flooded soils, reducing conditions can produce levels of Mn^{2+} ions approaching toxic levels even at much higher pH. Manganese toxicities in soybeans have been reported in Georgia and Mississippi, mainly on well-drained soils, with pH levels <5.5 (Anderson and Mortvedt, 1982). On acidic mineral soils, the risk of Mn toxicity in plants may be exacerbated by the application of livestock manure without affecting the bulk soil pH (Lee and MacDonald, 1977). Elevated Mn concentrations in the soil solution and a drop in rhizosphere pH caused by N mineralization were likely factors inducing this effect.

TABLE 14.5. Deficient, sufficient, and toxic concentrations of manganese in plants.

Plant	Type of culture	Tissue	Remark	Deficient	Sufficient	Toxic[a]
			Horticulture crops			
Apple	Field	Leaves	—	15	30	—
Apple	Soil	Leaves	Interveinal bark necrosis	—	—	>400
Apricot	Field	Leaves	—	10	86–94	4300–6000
Avocado	Solution	Leaves	—	—	1300	—
Avocado	Field	Leaves	September	—	366–655	—
Banana	Field	Leaves	—	<10	—	—
Bean	Field soil	Tops	pH 4.7	—	40–940	1104–4201
Bean	Soil	Leaves	pH 4.7	—	—	600–800
Brussels sprouts	Field	Leaves, petioles	Youngest fully expanded	—	78–148	760–2035
Carrot	Solution	Tops	Reduced yields	—	—	7100–9600
Cassava	Soil	Leaves	Vegatative	<50	—	—
Lima beans	Field	Tops	—	32–68	207–1340	—
Lettuce	Solution	Leaves	27 days	—	—	1000
Onion	Limed organic soil	Tops	Maturity	—	34	—
Orange	Field	Leaves	7-mo bloom cycle; leaves from non-fruiting terminals	<19	20–90	>100
Peas	Podzol soil	Leaves	pH 4.7	—	—	500
Potato	Field	Leaves	—	7	40	—
Spinach	Field	Plant	—	23	34–60	—
Spinach	Solution	Leaves	Maturity	—	—	100
Tomato	Solution	Leaves	—	5–6	70–398	—
Turnip	Field	Leaves	—	—	75	—
			Cereals/others			
Barley	Solution	Old leaves	Moderate to severe necrosis	—	—	305–410
Barley	Acid soil	Tops				80–100
Barley	Soil	Tops	Symptoms	—	14–76	—
Barley	Soil	Tops	Maturity	—	—	120–300
Oats	Soil	Tops	—	8–12	30–43	—
Oats	Acid soil	Tops	—	—	301–370	—
Rice	Solution	Leaves	—	—	—	4000–5000
Rye	Soil	Mature tops	—	—	10–50	—
Rye	Solution	Leaves	—	—	—	1400
Wheat	Field	Plants	—	4–10	75	—
Wheat	Field	Leaves	Plant grown at pH 4.9 and 6.9		108–113	356–432

TABLE 14.5. (*continued*)

Plant	Type of culture	Tissue	Remark	ppm Mn DW		
				Deficient	Sufficient	Toxic[a]
Wheat	Solution	Tops	Grown with different Mn contents	—	181–621	396–2561
Corn	Field	Ear leaf	Single cross hybrids	—	116–214	—
Corn	Soil	Whole leaf	Moderately fertile soil	—	76–213	—
Cotton	Solution	Tops	var. Pima S-2	—	196–924	1740–8570
Cotton	Field	Leaves, petioles	16 varieties on 3 soils	—	58–238	—
Cowpeas	Solution	Tops	Toxicity symptoms	—	—	1224
Lupins	Field	Leaves	Vegetative	17–30	—	—
Peanut	Sand	Leaves	—	—	110–440	890–10,900
Soybean	Soil	Leaves	—	<20	—	—
Soybean	Solution	Leaves	30 days	2–3	14–102	173–199
Sugar beet	Field	Leaves	—	5–30	7–1700	1250–3020
Safflower	Field	Leaves	Vegetative	9–13	—	—
Sugarcane		Leaves	Vegetative	<10	—	—
Tobacco	Soil	Tops	var. Burley 21	10	45	—
Tobacco	Pots	Tops		—	—	933–1130
Forage crops—temperate group						
Alfalfa	Field	Tops	—	—	—	477–1083
Alfalfa	Soil (pots)	Tops	Several soils of southeastern USA	—	65–240	651–1970
Alfalfa	Sand	Tops	—	—	—	175–400
Lespedeza	Soil	Tops	—	—	—	>570
Ryegrass	Solution	Leaves	—	—	—	800
Sweet clover	Solution	Tops	Toxic symptoms	—	—	321–754
Vetch	Field	Tops	—	—	—	500–1117
White clover	Soil	Tops	var. New Zealand	—	—	650
Forage crops—tropical group						
Trifolium sub-terraneum	Soil (pots)	Tops	—	4–25	30–300	—
Trifolium sub-terraneum	Solution	Tops	—	—	200	—
Phaseolus lathyroides	Solution	Tops	Cultivar Murray	—	—	840
Glycine javanica	Solution	Tops	Cultivar Jineroo	—	—	560

[a]Toxicity threshold levels defined as Mn concentrations found when yields were 5% below the maximum.

Sources: From various sources; summarized by NAS (1973); Hannam and Ohki (1988).

Manganese toxicity can be associated with Mn concentrations in plant tissues exceeding 500 ppm (see Table 14.5). However, most crop species appear to tolerate as much as 200 ppm Mn in their tissues without showing toxicity effects. Levels of Mn in potato foliage in excess of 250 ppm Mn may cause toxicity (White et al., 1970). In rice plants, the toxicity levels were reported at 300 to 1000 ppm in shoots and 200 to 600 ppm in roots (Chino, 1981). In soybeans, the critical Mn deficiency (causing a 10% reduction in growth from maximum due to deficiency) and toxicity (causing a 10% reduction in growth from maximum due to toxicity) levels for leaf Mn are 20 and 250 ppm, respectively, at the 90% relative yield level (Anderson and Mortvedt, 1982). A considerable number of yield depressions were obtained when leaf Mn exceeded 150 ppm, but near-maximum yields have also occurred with leaf Mn up to 600 ppm. Hence, the critical toxicity level for soybean leaf Mn varies considerably among cultivars, locations, and years.

On red basaltic soils in New South Wales, Australia Mn toxicity is common in French beans and lettuce on soils with pH below 4.5 (Siman et al., 1974). In southern New South Wales, the establishment of pasture legumes, mainly lucerne, is restricted on some soils by the existence of toxic levels of bioavailable Mn after heavy rain on slightly acid (pH 4.7 to 5.5) soils or dry, hot conditions. In the Goulburn Valley in Victoria, waterlogging–induced Mn toxicity has caused substantial damage to young apple trees.

In potato-growing areas, the soil is routinely maintained at a pH of 5.4 or below to minimize the incidence of common scab. However, under soil conditions of low pH concomitant with high fertility, concentrations of soluble Mn and Al may be sufficiently high to depress potato yields (Langille and Batteese, 1974). In some cases, soil pH has been allowed to drop as low as 4.5 to effectively control the scab (Berger and Gerloff, 1948). These low pH values often produce stem streak necrosis in potato plants, which causes premature death of the vines, thereby reducing yield. However, this condition could easily be corrected by lime application.

Manganese toxicity symptoms in plants are characterized by marginal chlorosis and necrosis of leaves (alfalfa, kale, lettuce, and rape), leaf puckering (cotton, snapbeans, soybeans), and necrotic spots on leaves (barley, lettuce, and soybeans). More specifically, physiological disorders associated with excess Mn are manifested as stem streak necrosis in potato, internal bark necrosis in apple trees, leaf burning or growth retardation in carnation, and fruit cracking at the blossom end of musk melon (Foy et al., 1995). In general, affected plants have deformed leaves, chlorotic areas, dead spots, stunted growth, and depressed yield.

Manganese toxicity has been associated with the following biochemical effects: destruction of indoleacetic acid auxin (IAA) by increased activity of IAA oxidase; possible amino acid imbalance; increased activity of peroxidases; decreased activities of catalase, ascorbic acid oxidase, glutathione oxidase, and cytochrome C oxidase; and lower ATP content and respiration rates (Foy et al., 1995).

Several extraction procedures (see Section 4.2) have been tested to diagnose soil toxic conditions, with inconsistent results. As pointed out earlier, no single extractant can be universally adopted. The most dependable approach is to perform both plant and soil tests to diagnose Mn status.

In summary, the following soil conditions and soil management practices can result in excess Mn in soils: (1) Mn is frequently present in toxic concentrations in strongly acid soils; since soil pH is the controlling factor in the bioavailability of Mn, the more acidic the soil, the greater the solubility of Mn; (2) Waterlogging of soil can increase soluble Mn^{2+} concentrations caused by the reduction of Mn(III) and Mn(IV) to Mn(II); and (3) Fertilizers that lower the pH of soils may increase the severity of Mn toxicity. Thus liming of acid soils to raise the pH to 5.5 and above may alleviate Mn toxicity conditions. In general, Mn toxicity can be alleviated by the application of lime and by drainage to alleviate reducing conditions.

5.4 Role of Manganese in Plant Disease Resistance

Among the micronutrients, perhaps Mn is the most important in the development of resistance of plants to root and foliar diseases of fungal origin (Graham and Webb, 1991). The question often asked about the role of Mn in suppressing these diseases centers on whether Mn is effective only in the deficiency range. Although the effects of nutrition on disease are normally limited to the deficiency range, there are indications that the combating effect of Mn operates well into the sufficiency range for the host plant (Graham and Webb, 1991).

An inverse relationship exists between the incidence of diseases and the Mn concentration in host tissues, i.e., affected plants tend to have low Mn content. Two main reasons have been offered: the fungi are capable of immobilizing the available Mn by oxidizing the Mn in the rhizosphere into insoluble oxides; and environmental conditions favoring the occurrence of the diseases also favor the immobilization of Mn in the soil (Graham and Webb, 1991). Perhaps the importance of Mn in disease control in crops can be exemplified by the take-all in wheat and common scab in potato.

Some evidence indicates that wheat plants growing on Mn-deficient soils are more susceptible to infection by take-all (*Gaeumannomyces graminis* var. tritici or Ggt), a worldwide fungal disease in wheat that can be devastating (Graham and Rovira, 1984). A hypothesis has been proposed that the susceptibility of wheat to the disease depends in part on the bioavailability of Mn in the soil and is inversely related to the concentration of Mn in host tissue. The following mechanisms by which Mn may exert its influence on the severity of take-all have been offered: (1) Mn(II) may be directly toxic to the free inoculum of the fungus in the soil as shown for *Streptomyces scabies* in potato; (2) Through Mn metabolism in the plants, i.e., Mn nutrition affects photosynthesis, which in turn controls the rate of exudation of soluble organic compounds by roots; these exudates affect the rhizosphere microflora, including the growth of the take-all fungus; and (3) Through lignin production, which is controlled by Mn-activated enzymes. Lignin can provide partial defense against take-all and may be more poorly developed in Mn-deficient plants. More detailed and additional mechanisms for the role of Mn in disease resistance have been discussed by Graham and Webb (1991).

In recent years, several studies have demonstrated that in Mn-deficient conditions, addition of Mn fertilizer significantly decreased the incidence of take-all (Wilhelm et al., 1988, 1990).

6 Manganese in Animal and Human Nutrition

Manganese was recognized as essential for animals in 1931 and since then, deficiency effects of Mn in animals have been well characterized (Van Campen, 1991). Information from human studies is rather limited.

In humans, Mn behavior is different from other essential trace elements in that its concentration differences among organs, and between infants and adults are small (Casey and Robinson, 1978). There is little evidence for Mn deficiency occurring in humans consuming normal diets; thus the estimated current intakes of 2 to 5 mg Mn day^{-1} appear adequate (Table 14.6).

The ordinary foods consumed by humans differ considerably in Mn content (Table 14.7). On a dry weight basis, leafy vegetables contain higher concentrations of Mn than other foodstuffs. Total dietary Mn intakes by adults vary from 2 to 8 mg day^{-1}, depending upon the amounts and proportions of cereals, nuts, green leafy vegetables, and tea consumed (Underwood, 1973). Tea and cloves are exceptionally rich in Mn; one cup of tea has been reported to contain 0.30 to 1.3 mg Mn, compared with a very much lower amount in a cup of coffee. However, diets high in milk, sugar, and refined cereals might provide inadequate Mn particularly in growing

TABLE 14.6. Estimated safe and adequate daily dietary intakes for manganese.

Category	Age (yr)	Mn (mg day^{-1})
Infants	0.0–0.5	0.3–0.6
	0.5–1.0	0.6–1.0
Children	1–3	1.0–1.5
	4–6	1.5–2.0
	7–10	2.0–3.0
Others	11+	2.0–5.0

Source: Food and Nutrition Board (1989).

TABLE 14.7. Manganese contents of groups of principal foodstuff.

Class of food	n	Mn concentration (ppm WW)		
		Mininum	Maximum	Average
Nuts	10	6.3	41.7	22.7
Cereal products	23	0.5	91.1	20.2
Dried legume seeds	4	10.7	27.7	20.0
Green leafy vegetables	18	0.8	12.6	4.5
Dried fruits	7	1.5	6.7	3.3
Roots, tubers, and stalks	12	0.4	9.2	2.1
Fresh fruits (including blueberries)	26	0.2	44.4	3.7
Fresh fruits (excluding blueberries)	25	0.2	10.8	2.0
Nonleafy vegetables	5	0.8	2.4	1.5
Animal tissues	13	0.08	3.8	1.0
Poultry and by-products	6	0.30	1.1	0.5
Dairy products	7	0.03	1.6	0.5
Fish and seafoods (excluding oysters)	7	0.12	2.2	0.5
Fish and seafoods (including oysters)	6	0.12	0.4	0.25

Source: NAS (1973).

children, pregnant women, and persons with diabetes and rheumatoid arthritis. These diets may possess a slower turnover rate of Mn (NAS, 1973). In general, whole grains and nuts are the richest source of Mn, followed by fruits and vegetables. Dairy products, meats, and seafoods contain only small amounts.

In Japan, the daily intake by adults is about 2.8 mg Mn day^{-1}, with more than 70% of Mn supplied by plant products (Murakami et al., 1965). In Canada, daily intake for Mn was at 2.9 to 3.6 mg, with most of the Mn being supplied by cereals, potatoes, and fruits (Kirkpatrick and Coffin, 1974). Other intake values reported include the United Kingdom, 2.7 ± 0.8 mg day^{-1} (Hamilton, 1979) and India, 8.3 mg day^{-1} (Soman et al., 1969).

Manganese serves as a cofactor for a number of enzymes: hydrolases, kinases, decarboxylaces, transferases, and several metalloenzymes (e.g., arginase, pyruvate carboxylase, superoxide dismutase, etc.). In the absence of widespread deficiencies in normal populations, its practical significance in human nutrition is difficult to assess (Van Campen, 1991).

Cases of Mn toxicity in humans are generally the result of chronic inhalation of airborne Mn in occupational settings (e.g., factories or mines). Manganese toxicity to humans is related to the route of exposure. Ingested Mn is relatively low in toxicity (Davis and Elias, 1996). Inhaled Mn however, has been known to be neurotoxic as well as toxic to the respiratory and reproductive systems. Manganism is characterized by various psychiatric and movement disorders, with some resemblance to Parkinson's disease in terms of impairment in movement control, facial expression, and certain neurochemical functions. The greater toxicity of inhaled Mn is probably due to differences in intake route—inhaled Mn first passes the brain whereas ingested Mn first passes the liver, which has a high capacity to metabolize and eliminate Mn.

The use of Mn as a gasoline additive (primarily as MMT) to boost the octane rating of gasoline is potentially an important source for environmental exposure of the general population when its use becomes more widespread. It can replace organic compounds of Pb (e.g., tetraethyl and tetramethyl Pb) as gasoline additives (Davis and Elias, 1996).

7 Factors Affecting Mobility and Bioavailability of Manganese

7.1 pH

Manganese is generally considered as one of the two most important toxic metals in acid soils, Al being the most important. Several investigators found that pH has the greatest effect on the bioavailability of Mn, followed in order by OM and moisture (Christensen et al., 1951; Sanchez and Kamprath, 1959). Manganese uptake by plants is more closely related to the soil pH than is the uptake rate of any other micronutrient (Marschner, 1988). Marked increases in the bioavailability of Mn should be expected when soil pH decreases below 5.5 (Sims and Johnson, 1991). Morris (1949) found that water-soluble Mn in 25 naturally acid soils averaged 2.1 ppm in soils with pH below 5.2, 1.0 ppm in soils with pH of 5.2 to 5.4, and 0.5 ppm in soils having pH of 5.4. The highest content of Mn in carrot shoot (grown in acid sphagnum peat soil under greenhouse conditions) occurred at pH 4.4 to 5.0, and the lowest at pH 6.2 to 6.4 (Gupta et al., 1970). However, the Mn content was lower at pH 4.0 to 4.1 than at pH 4.4 to 5.0. In general, the amount of extractable Mn can be expected to be inversely related to soil pH.

In soils, both plant uptake and level of extractable Mn increase with acidity. However, in nutrient culture, Mn uptake tends to be maximal at pH 6.5 and decreases as the pH changes in either direction (Godo and Reisenauer, 1980). The reduction in Mn uptake at low pH may be caused by the competition for absorption sites among Mn^{2+}, H^+, and Fe^{2+} ions, the last two being more available at lower solution pH. Soil Mn bioavailability is not controlled solely by soil or plant characteristics per se, but by the combined effects of plant characteristics, soil properties, and the biogeochemical processes in the rhizosphere. Root exudates may influence plant uptake of soil Mn (this also applies to Fe) in that exudate compounds, such as hydroxycarboxylates, increase soil Mn solubility by reducing MnO_2 and by complexing the Mn^{2+} ion (Warden and Reisenauer, 1991).

A series of 38 soybean experiments were conducted by Mascagni and Cox (1985) to evaluate a single Mn availability index incorporating bioavailable Mn (Mehlich-1 and Mehlich-3) and soil pH (Fig. 14.6). Manga-

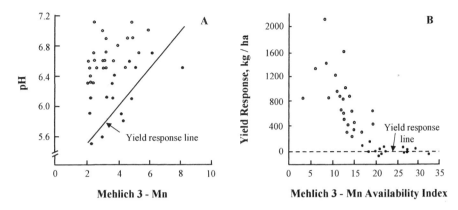

FIGURE 14.6. Prediction of yield response in soybeans using (A) soil pH and Mehlich-3 extractable manganese (open circles indicate significant yield response to Mn fertilization); and (B) the Mehlich-3 manganese availability index. (Extracted from Mascagni and Cox, 1985.)

nese fertilization resulted in significant yield responses in 25 studies. Integrating both the pH and bioavailable Mn information produced significant r^2 values of 0.56 and 0.59 for the Mehlich-1 and Mehlich-3 extractants, respectively. Critical values obtained from Mehlich-1 were 4.7 and 9.7 ppm at pH 6 and 7, respectively; critical values for Mehlich-3 were 3.9 and 8.0 ppm, respectively, at these pH values. A Mn availability index (MnAI) was developed using the empirical relationship among yield responses, soil pH, and extractable Mn. The critical index was set at 25, with probable responses for MnAI values of 20 to 25, and likely responses at MnAI below 20.

Results from field experiments conducted in the southeastern United States indicate that th e relative yield of soybeans consistently decreased with increasing soil pH above pH 6.0 (Fig. 14.7A) (Anderson and Mortvedt, 1982). The results further indicate that very little response to Mn fertilizer application can be expected at lower soil pH levels and underline the increasing need for Mn at higher soil pH levels, especially those soils with pH above 6.0.

Manganese toxicity in acidic soils can easily be alleviated or completely eliminated by the application of lime at rates sufficient to raise the soil pH to about 6.5 (Snider, 1943; White et al., 1970). Calcium carbonate, however, can have a strong negative effect on the uptake of Mn through adsorption and precipitation reactions, or on the formation of manganocalcite (Jauregui and Reisenauer, 1982). Therefore, other sources for lime should be considered.

7.2 Manganese–Iron Interaction

Antagonism between Mn and Fe is a well-documented interaction in higher plants. Ohki (1975) noted that as Mn concentration in cotton tissue increased

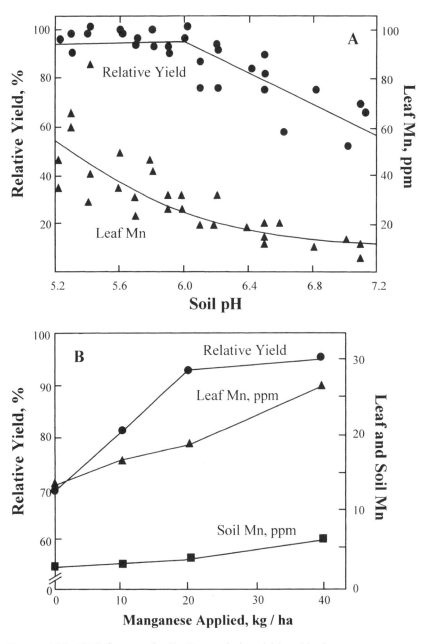

FIGURE 14.7. (A) Influence of soil pH on relative yield and leaf manganese content of soybeans in the southeastern United States not receiving applied manganese, (B) Influence of manganese application rate on relative yield and leaf manganese content of soybeans and double acid–extractable soil manganese on manganese-deficient soils in the southeastern United States. (Extracted from Anderson and Mortvedt, 1982.)

from the critical Mn level to a high but nontoxic level, Fe concentration in plant tissue decreased from high to moderate level. In severe cases, high levels of Mn supply can reduce the Fe concentration and induce Fe deficiency, as observed in pineapple culture (Sideris and Young, 1949). Others have also reported reciprocal relationships of Fe and Mn in rice and barley (Vlamis and Williams, 1964), soybeans (Somers and Shive, 1942), and other plants (Dokiya et al., 1968; Gerald et al., 1959). In nutrition of most plant species, the ratio of Fe to Mn in the nutrient medium must be maintained between 1.5 and 2.5 in order to obtain healthy plants (Mulder and Gerretsen, 1952). If the ratio is above 2.5, symptoms of Fe toxicity may occur; if it is below 1.5, the plant may suffer from Mn toxicity.

There are other aspects of the interaction. In some plants, Fe deficiency leads to rhizosphere acidification and increased reductant (organic acids) production (Robson, 1988). In these plants, it can be expected that these responses to Fe deficiency could lead to enhanced Mn uptake from soil. In other plants, Fe deficiency leads to the production of phytosiderophores by regions of the roots close to the tips. Because of the much greater specificity of binding of the siderophores with Fe than with Mn, it is likely that Fe deficiency will not lead to greater Mn absorption from solution in these species. The Fe:Mn interaction may also be species-dependent. For example, Fe–Mn antagonism in flax occurred because flax plants are more susceptible than other plants to interference of Mn with Fe function in the plant and the roots of flax plants are especially efficient at mobilizing Mn from calcareous soils (Warden and Reisenauer, 1991). Where mobilization of Mn and Fe in soils is effected by similar root processes, Mn and Fe uptake by plants may be positively interrelated.

Addition of Fe to the growing medium may reduce Mn toxicity to plants (Heenan and Campbell, 1983; Hiatt and Ragland, 1963). Addition of Fe chelates to soils has induced Mn deficiency in bush beans (Holmes and Brown, 1955), reduced the Mn content in plants (Wallace, 1958; Wallace et al., 1957), and alleviated Mn toxicity (Shannon and Mohl, 1956; Wallace, 1962). The addition of MnEDTA or FeEDTA to the soil intensified Mn deficiency symptoms, depressed growth, and reduced Mn uptake, while the Fe concentration remained relatively constant resulting in a wide Fe/Mn ratio in plants (Knezek and Greinert, 1971). The ineffectiveness of MnEDTA in overcoming plant Mn deficiency was due to a rapid substitution of Fe for Mn in the EDTA molecule, with the released Mn probably complexing with OM or precipitating as MnO_2. The Mn in the MnEDTA added to limed or calcareous soils may be replaced by Ca^{2+} ions. In soybeans grown on 23 calcareous soils (pH 7.7 to 8.4, 18 to 46% $CaCO_3$ equivalent) supplied with FeEDDHA, the interference of Fe with Mn nutrition was attributed to restricted Mn translocation from soil to roots and/or from roots to the plant shoots (Roomizadeh and Karimian, 1996).

Other reported Mn interactions include P–Mn (Neilsen et al., 1992), Mn–B (Ohki, 1973), Mn–Ca (LeBot et al., 1990), and Mn–Mg (Genon et al., 1994; Goss and Carvalho, 1992) couples.

7.3 Redox Potential

The redox status of soils influences the solubility and bioavailability of Mn. Manganese in the III and IV oxidation states occurs as precipitate in oxidized alkaline environments, whereas Mn in the II oxidation state is dominant in solution and solid phases under reducing conditions (see also Fig. 14.3). In general, flooding a soil enhances Mn solubility and its bioavailability to plants. However, in some instances, flooding may decrease Mn uptake by plants due to increased solubility of Fe and its competition with Mn for plant absorption. Saturation of air-dried surface soils for 5 days resulted in complete reduction of Mn to Mn^{2+} in the soil matrix, confirmed by X-ray absorption near-edge structure (XANES) spectroscopy (Schulze et al., 1995).

In flooded soils, both pH and Eh largely control Mn behavior (see Figs. 14.3 and 14.4). Soluble and exchangeable Mn should increase with decrease in both pH and Eh. Between pH 6 and 8, the conversion of insoluble soil Mn to the water-soluble and exchangeable forms was dependent on both pH and Eh, while at pH 5, Eh had little effect due to the overriding effect of acidity (Gotoh and Patrick, 1972). In general, reducing conditions resulting from waterlogging, soil compaction, or OM accumulation can produce toxic levels of Mn^{2+} in soils with pH $\geq$ 5.5.

7.4 Organic Matter

The role of OM in complexing Mn has already been discussed (Section 4.5). Addition of OM to soils can alleviate Mn toxicity symptoms (Cheng and Ouellette, 1971). In addition to its complexing ability, OM can affect the redox status of soils. In flooded conditions, microbial decomposition of OM present in soil as plant debris, humus, animal wastes, etc., leads to reducing condition by utilizing free oxygen in the soil atmosphere, producing CO_2 and organic acids. Subsequently, microorganisms utilize oxygen associated with oxidized forms of Mn and Fe to transform insoluble to soluble forms of Mn and Fe. This effect of OM in lowering the redox potential of soil and enhancing the solubility of Mn has been demonstrated in the field (Mandal and Mitra, 1982; Meek et al., 1968).

Application of farmyard manure may induce solubilization of Mn in soils due to soil acidification caused by mineralization of organic N and ammonium N in the manure. In general, decomposition of OM, especially that enriched in ammonium (e.g., sewage sludge), may induce Mn phytotoxicity.

7.5 Management and Fertilization Practices

Type of Mn carriers and method of fertilizer placement can affect Mn bioavailability to plants. For example, soil application of MnEDTA may even cause a higher incidence of Mn deficiency (Knezek and Greinert, 1971; Mortvedt, 1980). Besides, it is an impractical Mn fertilizer since it is highly unstable in soils having pH above 6 (Norvell and Lindsay, 1969).

The effect of soil-applied Mn on yield and Mn concentration of soybean leaves in instances where yields were increased by Mn fertilization was shown in Figure 14.7B. Leaf and soil Mn increased with each increment of applied Mn, and near maximum yields were obtained with the 20 kg Mn ha^{-1} rate. Leaf Mn and soil Mn associated with near maximum yields were 19 ppm and 3.6 ppm, respectively. Of the various application methods, banding and foliar applications of Mn may provide higher efficiency of Mn utilization (Table 14.8) (Randall et al., 1975). Foliar application is a common practice for nut and fruit trees.

The use of certain fertilizers that produce acidic soil reactions, such as $(NH_4)_2SO_4$, can cause reduced pH and subsequently Mn toxicity to plants. Heavy and prolonged applications of NH_4NO_3 and $(NH_4)_2SO_4$ fertilizers can substantially lower soil pH and enhance Mn uptake by plants (Sillanpaa, 1972; Siman et al., 1971). In addition, the application of an acid-forming N fertilizer, such as $(NH_4)_2SO_4$ and NH_4Cl, was more effective than other N sources (e.g., urea and various urea-phosphate formulations) in increasing the Mn concentration and alleviating Mn deficiency in barley and oats (Petrie and Jackson, 1984b). In laboratory experiments, Petrie and Jackson (1984a) demonstrated that application of either $(NH_4)_2SO_4$ or NH_4Cl alone or with $Ca(H_2PO_4)_2$, decreased the soil solution pH. The pH depression was attributed to nitrification of NH_4^+ ions and solubilization products of $Ca(H_2PO_4)_2$. Higher increases in soil solution Mn concentration were obtained from application of $(NH_4)_2SO_4$ or NH_4Cl in combination with $Ca(H_2PO_4)_2$ than from only single application of $(NH_4)_2SO_4$, NH_4Cl, or $Ca(H_2PO_4)_2$.

7.6 Plant Factors

Plant species and genotypes differ widely in their susceptibility to Mn deficiency and likewise in their tolerance to excess available Mn (Foy et al., 1978, 1981; Foy and Campbell, 1984; Ohki et al., 1981; Parker et al., 1981). Barley is less susceptible to Mn deficiency than oats and wheat and hence is the preferred crop on Mn deficient soils (e.g., usually in calcareous soils). Differences in sensitivity among genotypes to Mn deficiency appear to be largely related to differences in uptake from soil rather than differences in internal Mn requirement (Robson, 1988). Moreover, it is likely that these

TABLE 14.8. Rates of foliar-applied and soil-applied manganese for correcting manganese deficiency in crops.

Crop	Source or method	Total applied, kg Mn ha^{-1}	Comments
Foliar-applied			
Barley	MnSO$_4$	3.9–5.4	Three sprays of 1.3 to 1.8 kg Mn ha^{-1} each at tillering, stem elongation, and after ear emergence
Soybean	MnSO$_4$	0.34	Two sprays of 0.17 kg Mn ha^{-1} at early blossom and early pod set growth stages
Soybean	MnSO$_4$	4.4	Two sprays of 2.2 kg Mn ha^{-1} at the 2–3 and 5–6 trifoliolate leaf growth stages
Soybean	MnSO$_4$	2.2	Two sprays of 1.1 kg Mn ha^{-1}, the first at pre-bloom
Peanut	MnSO$_4$	4.5	2.3 kg Mn ha^{-1} in mid-June, and 1.1 kg Mn ha^{-1} in mid-July and late August
Corn	MnSO$_4$	1.2–2.2	Two sprays of 0.6 to 1.1 kg Mn ha^{-1} at the 4- and 8-leaf growth stage
Soybean	MnSO$_4$ MnCl$_2$ MnEDTA MnDTPA Mn-lignosulfonate	0.3	Three sprays of 0.1 kg Mn ha^{-1}, the first at the V4 growth stage and then when symptoms reappear
Safflower	MnSO$_4$	0.5	One spray at the rosette to end of stem elongation growth period
Soybean	MnSO$_4$	1.1	First application at the V5 growth stage
	MnEDTA	0.4	3 additional applications at 2-wk intervals
Soil-applied			
Barley	Broadcast	6	Foliar Mn was also needed for optimum yield
Soybean	Band	5–22	Applied with starter fertilizer
Soybean	Broadcast	45	—
Peanut	Broadcast	45	—
Soybean	Broadcast	20–40	—
Corn	Band	5	Applied with 170 kg diammonium phosphate ha^{-1}
Soybean	Broadcast	14	—
	Band	3	Applied without starter fertilizer

Source: Martens and Westermann (1991).

differences arise through plant effects on the availability of Mn for absorption (e.g., contact reduction, suitability of rhizosphere for the activity of Mn oxidizers) rather than through differences in Mn absorption from solution. Tolerance of Mn deficiency is heritable, but the genetics appear to be complex.

8 Sources of Manganese in the Environment

Anthropogenic sources of Mn contamination in terrestrial environments are primarily associated with certain industrial activities, such as metal smelting and refining. Other sources come from agricultural practices (fertilizer use, sewage sludge, and animal waste disposal), and atmospheric deposition from fossil fuel combustion and municipal incinerators.

8.1 Manganese Fertilizers

Manganese deficiency can be corrected by the addition of Mn fertilizers to the soil, either singly or mixed with macronutrient fertilizers, or by foliar application. Manganese is available as inorganic and organic fertilizers for agricultural use. The most commonly used Mn fertilizer is $MnSO_4 \cdot 4H_2O$ (24% Mn), which has high water solubility in both acid and alkaline conditions. Manganous oxide (MnO, 78% Mn) is also widely used. Organic carriers of Mn are seldom used because of their high costs and instability of certain products such as MnEDTA. The rates of application depend on the soil type, soil pH, application method, and type of crops. The residual value of Mn in alkaline soils is very low. Application rates may range from about 7 kg Mn ha^{-1} for low- to medium-responsive crops growing in mineral soils with pH of 6.5 to 7.2 to as high as 45 kg Mn ha^{-1} for high-responsive crops growing in peats and mucks with pH of 7.3 to 8.5 (Lucas, 1967). Reuter et al. (1988) reported an optimal range of 2 (side-banded for cotton on acidic soils) to 72 kg Mn ha^{-1} (broadcast for soybeans on neutral sand). In foliar applications, only 2 to 4.5 kg ha^{-1} is required to correct Mn deficiency.

On certain alkaline, calcareous, and highly organic soils, applied Mn is rapidly immobilized and therefore Mn deficiency cannot be entirely corrected by soil application alone—in this case, foliar application is also required.

8.2 Sewage Sludge and Coal Combustion Residues

A report identified Cd, Cu, Ni, Zn, and Mo in sludge as the elements with the greatest potential to accumulate in plants and pose a threat to animals and humans (CAST, 1976). Cadmium and Mo may pose greater risks to the plant–animal–human pathway. The concern with Mo from sewage sludge arises from its increased bioavailability to crops when the amended soil is limed (primarily to limit the uptake of Cd). Thus, in spite of the high levels of Mn in certain sludges (Appendix Table A.9) and reported enrichment of some soils and crops from sludge application (Andersson and Nilsson, 1972; Cunningham et al., 1975), Mn in sludge is generally not a concern in the food chain.

Although fly ash is enriched in Mn compared with coal, it is not considered a threat to the soil–plant–animal pathway, while Se and Mo are identified as potential risks to the food chain pathway (Adriano et al., 1980). Burning of coal and fuel oil contributes to the Mn burden of the atmosphere (Smith, 1972). Other sources of Mn emission into urban air include municipal incinerators and diverse industrial manufacturing processes.

8.3 Other Sources

Of localized importance is Mn emission from mining and smelting. Usually a majority of the Mn particulates resuspended from the ground or emitted to the atmosphere settle within a few kilometers of the originating source (Hutchinson and Whitby, 1974).

From an evaluation of the global cycle of Mn, Garrels et al. (1973) noted that the major exchange of Mn between the atmosphere and the prehuman earth surface was due to continental dust being swept into the atmosphere by winds and then falling back onto the earth's surface. Today, this dust flux is complemented by Mn emitted to the atmosphere in particulate form by industrial activities (e.g., iron and steel industry). The total river flux of Mn to the ocean today is nearly 3 times the prehuman flux. This increase represents principally an increase in the denudation rate of the land's surface from about 100×10^{14} g yr^{-1} prehuman to today's rate of about 225×10^{14} g yr^{-1}. Because this increase in denudation reflects an increase in the suspended load of rivers from deforestation and agricultural activities, and because Mn is concentrated in the ferric oxide coatings on suspended material and in the suspended particles, the land-to-ocean flux is higher today than in the past. Manganese in particulate emissions from industrial activities rivals the natural input of continental dust to the atmosphere (30×10^{10} vs. 43×10^{10} g). Most particulate Mn falls out of the atmosphere near industrial sources. The mining of Mn ore has resulted in a net gain for the land reservoir and a net loss from the sediment reservoir. There is no evidence for change with time of dissolved Mn in the oceanic reservoir.

References

Adriano, D.C. 1986. *Trace Elements in the Terrestrial Environment.* Springer-Verlag, New York.

Adriano, D.C., A.L. Page, A.A. Elseewi, A.C. Chang, and I. Straughan. 1980. *J Environ Qual* 9:333–344.

Agrawal, H.P., and C.J. Reddy. 1972. *J Indian Soc Soil Sci* 20:241–247.

Ainsworth, C., and D. Rai. 1987. *Chemical Characterization of Fossil Fuel Combustion Wastes.* EPRI Rep EA-5321, Electric Power Res Inst, Palo Alto, CA.

Alexander, M. 1977. *Introduction to Soil Microbiology.* Wiley, New York.

Anderson, O.E., and J.J. Mortvedt, eds. 1982. *Soybeans: Diagnosis and Correction of Manganese and Molybdenum Problems.* Southern Coop Series Bull 281. Univ. Georgia, Athens, CA.

Anderson, O.E., and K. Ohki. 1977. *J Fert Solutions* 21(6):30.

Andersson, A., and K.O. Nilsson. 1972. *Ambio* 1:176–179.

Berger, K.C., and G.C. Gerloff. 1948. *Soil Sci Soc Am Proc* 12:310–314.

Berrow, M.L., and R.L. Mitchell. 1980. *Trans Royal Soc (Edinburgh) Earth Sci* 71:103–121.

Berrow, M.L., and G.A. Reaves. 1984. In *Proc Intl Conf Environ Contamination.* CEP Consultants, Edinburgh.

Bloom, P.R. 1981. In R.H. Dowdy, J.A. Ryan, V.V. Volk, and D.E. Baker, eds. *Chemistry in the Soil Environment.* ASA Spec Publ 40. Am Soc Agro, Madison, WI.

Bowen, H.J.M. 1979. *Environmental Chemistry of the Elements.* Academic Pr, New York.

Boxma, R., and A.J. de Groot. 1985. *Plant Soil* 83:411–417.

Brookins, D.C. 1988. *Eh–pH Diagrams for Geochemistry.* Springer-Verlag, New York.

Casey, C.E., and M.F. Robinson. 1978. *Br J Nutr* 39:639–646.

Chen, J., F. Wei, Y. Wu, and D.C. Adriano. 1991. *Water Air Soil Pollut* 57–58:699–712.

Cheng, B.T., and G.J. Ouellette. 1971. *Plant Soil* 34:165–181.

Childs, C.W. 1975. *Geoderma* 13:141–152.

Chino, M. 1981. In K. Kitagishi and I. Yamane, eds. *Heavy Metal Pollution in Soils of Japan.* Japan Sci Soc Pr, Tokyo.

Christensen, P.D., S.J. Toth, and F.E. Bear. 1951. *Soil Sci Soc Am Proc* 15:279–282.

[CAST] Council for Agricultural Science and Technology. 1976. *Application of Sewage Sludge to Cropland: Appraisal of Potential Hazards of the Heavy Metals to Plants and Animals.* EPA-430-9-76-013. General Serv Admin, Denver, CO.

Cox, F.R. 1968. *Agron J* 60:521–524.

Cunningham, J.D., D.R. Keeney, and J.A. Ryan. 1975. *J Environ Qual* 4:448–454.

Curtin, D., J. Ryan, and R.A. Chaudhary. 1980. *Soil Sci Soc Am J* 44:947–950.

Davis, J.M., and R.W. Elias. 1996. In L.W. Chang, ed. *Toxicology of Metals.* CRC Pr, Boca Raton, FL.

Dokiya, Y., N. Owa, and S. Mitsui. 1968. *Soil Sci Plant Nutr* 14:169–174.

Ehrlich, H.L. 1968. *Appl Microbiol* 16:197–202.

Follett, R.H., and W.L. Lindsay. 1970. In *Profile Distribution of Zinc, Iron, Manganese and Copper in Colorado Soils.* Colorado Agric Exp Sta Tech Bull 110. Fort Collins, CO.

Food and Nutrition Board. 1989. *Recommended Dietary Allowances,* 10th ed. National Research Council, Washington, DC.

Foy, C.D., and T.A. Campbell. 1984. *J Plant Nutr* 7:1365–1388.

Foy, C.D., R.L. Chaney, and M.C. White. 1978. *Ann Rev Plant Physiol* 29:511–566.

Foy, C.D., H.W. Webb, and J.E. Jones. 1981. *Agron J* 73:107–111.

Foy, C.D., R.R. Weil, and C.A. Coradetti. 1995. *J Plant Nutr* 18:685–706.

Frank, R., K. Ishida, and P. Suda. 1976. *Can J Soil Sci* 56:181–196.

Freedman, B., and T.C. Hutchinson. 1981. In N.W. Lepp, ed. *Effect of Heavy Metal Pollution in Plants, vol. 2: Metals in the Environment.* Applied Science Publ, London.

Garrels, R.M., F.T. MacKenzie, and C. Hunt, eds. 1973. *Chemical Cycles and the Global Environment: Assessing Human Influences.* W. Kaufmann, Los Angeles, CA.

Geering, H.R., J.F. Hodgson, and C. Sdano. 1969. *Soil Sci Soc Am Proc* 33:81–85.

Genon, J.G., N. de Hepcee, J.E. Duffy, B. Delvaux, and P.A. Hennebert. 1994. *Plant Soil* 166:109–115.

Gerald, H.R., J.F. Hodgson, and C. Sdano. 1959. *Plant Physiol* 34:608–613.

Ghiorse, W.C., and H.L. Ehrlich. 1976. *Appl Environ Microbiol* 31:977–985.

Godo, G.H., and H.M. Reisenauer. 1980. *Soil Sci Soc Am J* 44:993–995.

Goldberg, S.P., and K.A. Smith. 1984. *Soil Sci Soc Am J* 48:559–564.

Goss, M.J., and M.J. Carvalho. 1992. *Plant Soil* 139:91–98.

Gotoh, S., and W.H. Patrick, Jr. 1972. *Soil Sci Soc Am Proc* 36:738–742.

Graham, M.J., C.D. Nickell, and R.G. Hoeft. 1994. *J Plant Nutr* 17:1333–1340.

Graham, R.D. 1988. In R.D. Graham, R.J. Hannam, and N.C. Zeren, eds. *Manganese in Soil and Plants.* Kluwer, Dordrecht, Netherlands.

Graham, R.D., and A.D. Rovira. 1984. *Plant Soil* 78:441–444.

Graham, R.D., and M.J. Webb. 1991. In J.J. Mortvedt et al., eds. *Micronutrients in Agriculture,* SSSA 4, Soil Sci Soc Am, Madison, WI.

Grass L.B., A.J. MacKenzie, B.D. Meek, and W.F. Spencer. 1973a. *Soil Sci Soc Am Proc* 37:14–17.

Grass, L.B., A.J. Mackenzie, B.D. Meek, and W.F. Spencer. 1973b. *Soil Sci Soc Am Proc* 37:17–21.

Gupta, U.C., E.W. Chipman, and D.C. Mackay. 1970. *Soil Sci Soc Am Proc* 34:762–764.

Hamilton, E.I. 1979. *Trace Subs Environ Health* 13:3–15.

Hammes, J.K., and K.C. Berger. 1960. *Soil Sci Soc Am Proc* 24:361–364.

Hannam, R.J., and K. Ohki. 1988. In R.D. Graham, R.J. Hannam, and N.C. Zeren, eds. *Manganese in Soils and Plants.* Kluwer, Dordrecht, Netherlands.

Heenan, D.P., and L.C. Campbell. 1983. *Plant Soil* 70:317–326.

Hiatt, A.J., and J.L. Ragland. 1963. *Agron J* 55:47–49.

Hoff, D.J., and H.J. Mederski. 1958. *Soil Sci Soc Am Proc* 22:129–132.

Holmes, R.S., and J.C. Brown. 1955. *Soil Sci* 80:167–179.

Hoyt, P.B., and M. Nyborg. 1971. *Soil Sci Soc Am Proc* 35:241–244.

Hoyt, P.B., and M.D. Webber. 1974. *Can J Soil Sci* 54:53–61.

Hutchinson, T.C., and L.M. Whitby. 1974. *Environ Conserv* 1:123–132.

Jauregui, M.A., and H.M. Reisenauer. 1982. *Soil Sci* 134:105–110.

Jenne, E.A. 1968. *Adv Chem* 73:337–387.

Jensen, D.L., J.K. Boddum, S. Redemann, and T.H. Christensen. 1998. *Environ Sci Technol* 32:2657–2664.

Jones, J.B., Jr. 1972. In J.J. Mortvedt, P.M. Giordano, and W.L. Lindsay eds. *Micronutrients in Agriculture.* Soil Sci Soc Am, Madison, WI.

Jones, L.H.P., and G.W. Leeper. 1951. *Plant Soil* 3:141–153.

Jugsujinda, A., and W.H. Patrick, Jr. 1992. *Agron J* 69:705–710.

Kirkpatrick, D.C., and D.E. Coffin. 1974. *J Inst Can Sci Technol Aliment* 7:56–58.

Knezek, B.D., and H. Greinert. 1971. *Agron J* 63:617–619.

Krauskopf, K.B. 1972. In J.J. Mortvedt, P.M. Giordano, and W.L. Lindsay, eds. *Micronutrients in Agriculture.* Soil Sci Soc Am, Madison, WI.

Krauskopf, K.B. 1979. *Introduction to Geochemistry,* 2nd ed. McGraw-Hill, New York.

Krupskiy, N.K., L.P. Golovina, A.M. Aleksandrova, and T.I. Kisel. 1978. *Sov Soil Sci* 10:670–675.

Kubota, J., and W.H. Allaway. 1972. In J.J. Mortvedt, P.M. Giordano, and W.L. Lindsay, eds. *Micronutrients in Agriculture.* Soil Sci Soc Am, Madison, WI.

Lal, F., and T.D. Biswas. 1974. *J Indian Soc Soil Sci* 22:333–346.

Langille, A.R., and R.I. Batteese. 1974. *Can J Plant Sci* 54:375–381.

LeBot, J., E.A. Kirkby, and M.L. van Beusichem. 1990. *J Plant Nutr* 13:513–525.

Lee, C.R., and M.L. MacDonald. 1977. *Soil Sci Soc Am J* 41:573–577.

Leeper, G.W. 1947. *Soil Sci* 63:79–94.

Lindsay, W.L., and W.A. Norvell. 1978. *Soil Sci Soc Am J* 42:421–428.

Longnecker, N.E., R.D. Graham, and G. Card. 1991. *Field Crops Res* 28:85–102.

Lucas, R.E. 1967. In *Micronutrients for Vegetables and Field Crops.* Ext Bull E-486. Michigan State Univ, East Lansing, MI.

Mandal, L.N., and R.R. Mitra. 1982. *Plant Soil* 69:45–56.

Marschner, H. 1988. In R.D. Graham, R.J. Hannam, and N.C. Zeren, eds. *Manganese in Soils and Plants.* Kluwer, Dordrecht, Netherlands.

Martens, D.C., and D.T. Westermann. 1991. In J.J. Mortvedt et al., eds. *Micronutrients in Agriculture,* SSSA 4, Soil Sci Soc Am, Madison, WI.

Mascagni, H.J., and F.R. Cox. 1985. *Soil Sci Soc Am J* 49:382–386.

McBride, M.B. 1979. *Soil Sci Soc Am J* 43:693–698.

McDaniel, P.A., and S.W. Buol. 1991. *Soil Sci Soc Am J* 55:152–158.

McKeague, J.A., and M.S. Wolynetz. 1980. *Geoderma* 24:299–307.

Meek, B.D., A.J. MacKenzie, and L.B. Grass. 1968. *Soil Sci Soc Am Proc* 32:634–638.

Meek, B.D., A.L. Page, and J.P. Martin. 1973. *Soil Sci Soc Am Proc* 37:542–548.

Melsted, S.W., H.L. Motto, and T.R. Peck. 1969. *Agron J* 61:17–20.

Mena, I. 1980. In H.A. Waldron, ed. *Metals in the Environment.* Academic Pr, New York.

Morris, H.D. 1949. *Soil Sci Soc Am Proc* 13:362–371.

Mortvedt, J.J. 1980. *Soil Sci Soc Am J* 44:621–626.

Mulder, E.G., and F.C. Gerretsen. 1952. *Adv Agron* 4:221–277.

Murakami, Y., Y. Suzuki, T. Yamagata, and N. Yamagata. 1965. *J Rad Res* 6:105–110.

[NAS] National Academy of Sciences. 1973. *Manganese.* In NAS. 1977. *Geochemistry and the Environment.* NAS, Washington, DC.

Neilsen, D., G.H. Neilsen, A.H. Sinclair, and D.J. Linehan. 1992. *Plant Soil* 145:45–50.

Norrish, K. 1975. In D.J.D. Nicholas, and A.R. Egan, eds. *Trace Elements in Soil–Plant–Animal Systems.* Academic Pr, New York.

Norvell, W.A. 1972. In J.J. Mortvedt, P.M. Giordano, and W.L. Lindsay, eds. *Micronutrients in Agriculture.* Soil Sci Soc Am, Madison, WI.

Norvell, W.A., and W.L. Lindsay. 1969. *Soil Sci Soc Am Proc* 33:86–91.

Norvell, W.A., and W.L. Lindsay. 1972. *Soil Sci Soc Am Proc* 36:778–783.

Ohki, K. 1973. *Agron J* 65:482–485.

Ohki, K. 1975. *Agron J* 67:204–207.

Ohki, K., D.O. Wilson, and O.E. Anderson. 1980. *Agron J* 72:713–716.

Page, E.R. 1962. *Plant Soil* 16:247–257.

Parker, M.B., F.C. Boswell, K. Ohki, L.M. Shuman, and D.O. Wilson. 1981. *Agron J* 73:643–646.

Patel, M.S., P.M. Mehta, and H.G. Pandya. 1972. *J Indian Soc Soil Sci* 20:79–90.

Patrick, W.H., and A. Jugsujinda. 1992. *Soil Sci Soc Am J* 56:1071–1073.

Pavanasasivam, V. 1973. *Plant Soil* 38:245–255.

Peterson, L. 1966. *Acta Agric Scan* 16:120–128.

Petrie, S.E., and T.L. Jackson. 1994a. *Soil Sci Soc Am J* 48:315–318.

Petrie, S.E., and T.L. Jackson. 1984b. *Soil Sci Soc Am J* 48:319–322.

Ponnamperuma, F.N. 1972. *Adv Agron* 24:29–96.

Randall, G.W., E.E. Schulte, and R.B. Corey. 1975. *Agron J* 67:502–507.

Randall, G.W., E.E. Schulte, and R.B. Corey. 1976. Soil Sci Soc *Am J* 40:282–287.

Reddy, M.R., and H.F. Perkins. 1976. *Soil Sci* 121:21–24.

Reuter, D.J., A.M. Alston, and J.D. McFarlane. 1988. In R.D. Graham, R.J. Hannam, and N.C. Zeren, eds. *Manganese in Soils and Plants*. Kluwer, Dordrecht, Netherlands.

Robson, A.D. 1988. In R.D. Graham, R.J. Hannam, and N.C. Zeren, eds. *Manganese in Soils and Plants*. Kluwer, Dordrecht, Netherlands.

Romheld, V., and H. Marschner. 1991. In J.J. Mortvedt et al., eds. *Micronutrients in Agriculture*, SSSA 4, Soil Sci Soc Am, Madison, WI.

Roomizadeh, S., and N. Karimian. 1996. *J Plant Nutr* 19:397–406.

Rumpel, J., A. Kozakiewicz, B. Ellis, G. Lessman, and J. Davis. 1967. *Mich Agric Exp Sta Quart Bull* 50(l):4–11. East Lansing, MI.

Safo, E.Y., and L.E. Lowe. 1973. *Can J Soil Sci* 53:95–101.

Salcedo, I.H., B.G. Ellis, and R.E. Lucas. 1979b. *Soil Sci Soc Am J* 43:138–141.

Sanchez, C., and E.J. Kamprath. 1959. *Soil Sci Soc Am Proc* 23:302–304.

Sanders, J.R. 1983. *J Soil Sci* 34:315–323.

Schulze, D.G., S.R. Sutton, and S. Bajt. 1995. *Soil Sci Soc Am J* 59:1540–1548.

Scott, W.O., and S.R. Aldrich. 1970. *The Farm Quarterly*. Cincinnati, OH.

Shacklette, H.T., J.C. Hamilton, J.C. Boerngen, and J.M. Bowles. 1971. USGS Prof Paper 574-D. U.S. Geological Survey, Washington, DC.

Shannon, L.M., and J.S. Mohl. 1956. In A. Wallace, ed. *Symp Use of Metal Chelates in Plant Nutrition*. Los Angeles, CA.

Sherman, G.D., and P.M. Harmer. 1942. *Soil Sci Soc Am Proc* 7:398–405.

Sherman, G.D., J.S. McHargue, and W.S. Hodgkiss. 1942. *Soil Sci* 54:253–257.

Shuman, L.M. 1985. *Soil Sci* 140:11–22.

Shuman, L.M., and O.E. Anderson. 1974. *Soil Sci Soc Am Proc* 38:788–791.

Sideris, C.P., and H.Y. Young. 1949. *Plant Physiol* 24:416–440.

Sillanpaa, M. 1972. *Trace Elements in Soils and Agriculture*. Soils Bull 17. United Nations Food and Agriculture Organization (FAO), Rome.

Siman, A., F.W. Cradock, P.J. Nicholls, and H.C. Kirton. 1971. *Aust J Agric Res* 22:201–214.

Siman, A., F.W. Cradock, and A.W. Hudson. 1974. *Plant Soil* 41:129–140.

Sims, J.L., and W.H. Patrick, Jr. 1978. *Soil Sci Soc Am J* 42:258–262.

Sims, J.L., P. Duangpatra, J.H. Ellis, and R.E. Phillips. 1979. *Soil Sci* 127:270–274.

Sims, J.T., and G.V. Johnson. 1991. In J.J. Mortvedt et al., eds. *Micronutrients in Agriculture*, SSSA 4, Soil Sci Soc Am, Madison, WI.

Smith, R.G. 1972. In D.H.K. Lee, ed. *Metallic Contaminants and Human Health*. Academic Pr, New York.

Soman, S.D., V.K. Panday, K.T. Joseph, and S.J. Raut. 1969. *Health Phys* 17:35–40.

Somers, I.I., and J.W. Shive. *Plant Physiol* 17:582–602.

Snider, H.J. 1943. *Soil Sci* 56:187–195.

Spencer, W.F., R. Patrick, and H.W. Ford. 1963. *Soil Sci Soc Am Proc* 27:134–141.

Swaine, D.J. 1955. *The Trace Element Content of Soils.* Commonwealth Bureau Soil Sci(GB), Tech Commun 48. Herald Printing Works, York, United Kingdom.

Taylor, R.M., and R.M. McKenzie. 1966. *Aust J Soil Res* 4:29–39.

Tyler, P.A., and K.C. Marshall. 1967. *Antonie Van Leeuwenhoek* 33:171–183.

Underwood, E.J. 1973. In *Toxicants Occurring Naturally in Foods.* National Academy of Sciences, Washington, DC.

U.S. Bureau of Mines. 1993. In *Minerals Yearbook,* U.S. Dept Interior, Washington, DC.

van Campen, D.R. 1991. In J.J. Mortvedt et al., eds. *Micronutrients in Agriculture,* SSSA 4, Soil Sci Soc Am, Madison, WI.

Vega, S., M. Calisay, and N.V. Hue. 1992. *J Plant Nutr* 15:219–231.

Vlamis, J., and D.E. Williams. 1962. *Plant Physiol* 37:650–655.

Vlamis, J., and D.E. Williams. 1964. *Plant Soil* 20:221–231.

Wallace, A. 1958. *Agron J* 50:285–286.

Wallace, A., ed. 1962. *Decade of Synthetic Chelating Agents in Inorganic Plant Nutrition.* Los Angeles, CA.

Wallace, A., L.M. Shannon, O.R. Lunt, and R.L. Impley. 1957. *Soil Sci* 84:2741.

Warden, B.T., and H.M. Reisenauer. 1991. *J Plant Nutr* 14:7–30.

Welch, R.M., W.H. Allaway, W.A. House, and J. Kubota. 1991. In J.J. Mortvedt et al., eds. *Micronutrients in Agriculture,* SSSA 4, Soil Sci Soc Am, Madison, WI.

White, R.P. 1970. *Soil Sci Soc Am Proc* 34:625–629.

White, R.P., E.C. Doll, and J.R. Melton. 1970. *Soil Sci Soc Am Proc* 34:268–271.

Wilhelm, N.S., R.D. Graham, and A.D. Rovira. 1988. *Aust J Agric Res* 39:1–10.

Wilhelm, N.S., R.D. Graham, and A.D. Rovira. 1990. In N. El Bassam et al., eds. *Genetic Aspects of Plant Mineral Nutrition.* Kluwer, Dordrecht, Netherlands.

Willett, I.R., and W.J. Bond. 1995. *J Environ Qual* 24:834–845.

15
Molybdenum

1 General Properties of Molybdenum

Molybdenum (atom. no. 42, atom. wt. 95.94) is in the second row of the transition metal elements and occurs as five isotopes. It is in Group VI-B with Cr and W and shares some chemical properties with each of these elements. It is a silvery white metal that is very hard, although softer and more ductile than W. It has a density of 10.22 at 20 °C, and a melting pt. of 2617 °C. Molybdenum has five possible oxidation states (II, III, IV, V, and VI), but in nature the IV and VI oxidation states predominate, with the latter being the most stable. At high oxidation states Mo has an affinity for oxides, and for sulfur- and oxygen-containing groups. The affinity of Mo for oxygen-containing groups is the reason for its predominant presence as dissolved anionic species in aqueous systems. The most important compound is the trioxide (MoO_3) from which most of the known Mo compounds can be prepared. Molybdenum is resistant to HCl, H_2SO_4, H_3PO_4, and HF solutions under many conditions of concentration and temperature. However, the metal is attacked by oxidizing acids and fused alkalis. It is rapidly oxidized in air at >500 °C. At moderate to high concentrations in solution, molybdate readily polymerizes into polymolybdates with a wide variety of highly complex structures. However, in dilute solutions, such as those found in soils or in most natural waters, the predominant form of soluble Mo is the molybdate anion (MoO_4^{2-}). Only under unusual conditions of very high enrichment will Mo be found as soluble polymolybdate in waters. In nature, the only important ore is molybdenite (MoS_2); however, some powellite and deposits containing wulfenite also occur.

2 Production and Uses of Molybdenum

About half of the world's Mo resources are in the United States, with about 75% occurring in deposits where Mo is the principal metal mined. Most of the Mo deposits in the United States are found in Colorado, the Sangre de Cristo Range in New Mexico, and in Utah. Major deposits are also found in Chile, China, Canada, and Russia. Together they accounted for 87% of the world's production of Mo in 1993 (Blossom, 1993).

The world's largest Mo mine is located in Climax, Colorado which produces approximately 60% of the Mo mined in the United States and 40% of that mined in the Western world (Chappell, 1975). World mine production of Mo increased from 95.4×10^3 tonnes in 1977 to an average of 101.8×10^3 tonnes in 1978 and 1979. The United States, Canada, Chile, and Russia continue to supply nearly all of the world's output. The United States accounted for about 60% of the world's output in 1978 and 1979 and exported over half of its production primarily to Japan and western Europe.

One of the most important reasons for the increased demand for Mo is that it is considered nontoxic to humans and can be substituted for Cr or other toxic metals used in steel alloys, corrosion inhibitors, and in pigments. Its chief industrial uses are in manufacture of various alloy steels and stainless steel. Molybdenum uses can be categorized into chemical and metallurgical. There are three major types of Mo chemicals that are commercially utilized: molybdenum disulfide (MoS_2), molybdic oxide (MoO_3), and the various molybdates [$Na_2MoO_4 \cdot 2H_2O$, $(NH_4)_2Mo_2O_7$, and $CoMoO_4$] (Lander, 1977). These chemicals are used as catalysts, corrosion inhibitors, or as ingredients in pigments, dyes, plastic and rubber parts, industrial gear oils, and high-pressure grease. With the exception of $CoMoO_4$, these chemicals are commonly used as sources of Mo fertilizers in agriculture (Murphy and Walsh, 1972). The metallurgical uses of Mo include production of stainless steels (1 to 6% Mo), high-strength steels (0.25 to 0.75% Mo), chrome-moly elevated temperature steels (0.5 to 1.0% Mo), superalloys (1 to 28% Mo), and molybdenum metal (99 to 100% Mo).

The domestic use of Mo is approximately 70% for production of steel; 22% for production of cast irons, superalloys, other alloys, and as a refractory metal; and 8% for use in catalysts, lubricants, pigments, and other chemical uses (Kummer, 1980).

3 Molybdenum in Nature

The average concentration of Mo in the earth's crust varies from 1.0 to 2.3 ppm ranking it 53rd in crustal abundance, following As (Krauskopf, 1979). For different rock types, the following Mo concentrations (ppm) are commonly reported (Table 15.1): igneous rock, 0.9 to 7; phosphorite, 5 to 100; shale, 5 to 90; limestone and dolomite, <3 to 30; and sandstone, <3 to 30. Black shales are usually high in Mo, with values up to 300 ppm. In the eastern Yukon Peninsula, paleozoic bedrock (shale, limestone, siltstone) Mo concentrations have been found ranging from <2 to 100 ppm, producing stream sediments with 3 to 35 ppm Mo, and soils with 4 to 29 ppm Mo (Doyle et al., 1972). Proterozoic and cretaceous bedrocks have much lower concentrations (<2 ppm). In the United Kingdom, black shales have Mo concentrations of <2 to 40 ppm, much higher than for other rocks (Thornton, 1977). They cause correspondingly high values for stream sediments (3 to 60 ppm) and subsoils (<2 to 85 ppm). High concentrations of Mo were found in soils developed on black shales in Korea (Kim and Thornton, 1993); plant uptake of Mo was pH-dependant with more uptake at increasing pH. Others found Mo-rich soils that were derived from molybdeniferous shale bedrock (Doyle and Fletcher, 1977). Shale and granite are major rocks contributing Mo to soil parent materials. Average

TABLE 15.1. Commonly observed molybdenum concentrations (ppm) in various environmental media.

Material	Average (or median)	Range
Basaltic igneous[a]	1.5	0.9–7
Granitic igneous[a]	1.6	1–6
Phosphorite[a]	30	5–100
Shale[b]	16	5–90
Black shale[c]	72	<22–290
Black shale (high C)[a]	10	1–300
Limestone and dolomite[c]	0.79	<3–30
Siltstone[b]	4	<2–5
Sandstone[c]	~3	<3–30
Coal[d]	3	0.3–30
Coal ash[j]		
Fly ash	44	7–236
Bottom ash	14	3–443
FGD sludge	12	<4–53
Oil ash	375	22–779
Lime[e]	—	1.3–3.8
Fertilizers[e]	—	<0.05–1.7
Sewage sludge[f]	15.6	1–76
Soils[d]	—	0.2–5
Common crops[g]	—	0.03–5
Legumes[g]	—	0.3–5
Common plants[h] (mostly weeds)	0.74	0.17–2.0
Ferns[i]	—	0.8–2.5
Fungi[i]	1.4	—
Lichens[i]	—	0.03–3
Woody angiosperm[i]	—	0.06–3
Freshwater ($\mu g\ L^{-1}$)[i]	0.5	0.3–10
Seawater ($\mu g\ L^{-1}$)[i]	10	4–10

Sources: Extracted from
[a]Fleischer (1972).
[b]Doyle et al. (1972).
[c]Connor and Shacklette (1975).
[d]Adriano et al. (1980);
  Carlson and Adriano (1993).
[e]Evans et al. (1950a).
[f]Jarrell et al. (1980)
[g]Gupta and Lipsett (1981).
[h]Robinson and Edgington (1954).
[i]Bowen (1979).
[j]Ainsworth and Rai (1987).

Mo concentration in igneous rocks may be as high as 15 ppm, although North American granites average only 2 ppm (Goldschmidt, 1954). Since Mo can substitute for Fe, Ti, and Al in silicate lattice structure, it tends to be uniformly distributed in the environment. High Mo geochemical environments produce vegetation with correspondingly high concentrations of Mo (Doyle et al., 1972; Thornton, 1977).

Several important primary minerals contain Mo. Among them, in order of importance are: molybdenite (MoS_2), powellite [$(Ca(MoW)O_4)$], ferri-molybdite [$Fe_2(MoO_4)_3$], wulfenite ($PbMoO_4$), ilsemanite (molybdenum oxysulfate), and jordisite (amorphous molybdenum disulfide). Molybdenite, ferrimolybdite, and jordisite currently are the most economically important of these minerals because they are the most abundant and are found in deposits that can be mined solely for their Mo content.

Molybdenum concentrations in normal soils average about 1 to 2 ppm, with some unusual concentrations as high as 24 ppm (Allaway, 1968; Aubert and Pinta, 1977). Highly weathered soils are more likely to be deficient in Mo. In contrast, soils derived from granite rocks, slates, or argillaceous schists tend to be high in Mo. Alkaline and poorly drained soils with high water table tend to produce plants high in Mo.

The average Mo concentration in surface waters in the United States is about 1 μg L^{-1} (Hem, 1989); river and lake waters from areas not impacted by pollution generally have Mo levels <1 μg L^{-1} (Smith et al., 1997). Surface waters having almost 5 μg Mo L^{-1} or higher as in Colorado may be considered as anomalous.

The concentration of Mo in plants varies quite widely. Levels as low as 0.01 ppm in common crops to several hundred ppm in legumes have been reported (Gupta, 1997).

4 Molybdenum in Soils/Sediments

4.1 Total Molybdenum in Soils

The Mo content of soils is dependent on the nature of parent material, degree of weathering, and OM content. Soil levels may be greater than that of the parent material. Accumulation of Mo in organic-rich zones of soils can be attributed to OM binding. Another possibility is that the inorganic soil constituents in the surface are more weathered, adsorbing more Mo. Most reports have associated high Mo levels in soils with sedimentary parent materials, especially shales (Swaine and Mitchell, 1960).

The primary source of Mo in soils is molybdenite (MoS_2). In Fe-rich soils, the most common molybdate is $Fe_2(MoO_4)_3$, and in soils high in Al, it is $Al_2(MoO_4)_3$. In general, levels of 0.5 to 5 ppm are considered normal and are in agreement with the relative abundance of Mo in the earth's crust (1.0 to 2.3 ppm). Robinson and Alexander (1953) suggested a mean soil total Mo concentration of 2.5 ppm, based on nearly 500 U.S. soil samples and 237 soils from other parts of the world. However, recent assessments that include a broader range of soils suggest that a median of 1.2 to 1.3 ppm Mo may be a more representative value, with the range from 0.1 to 40 ppm (Kubota, 1977). Levels of 0.5 ppm or less could be considered low (Williams, 1971). Shacklette and Boerngen (1984) gave a mean concentration of 0.97 ppm Mo (range of <3 to 15 ppm) for U.S. surface soils (Appendix Table A.24). For tropical Asian paddy soils, a mean concentration of 2.9 ppm Mo (range of trace to 8 ppm) was reported (Domingo and Kyuma, 1983).

The Mo contents of Nova Scotia (Canada) Podzols range from 0.05 to 12.1 ppm (MacLean and Langille, 1973). Some soils in France had Mo concentrations varying from 4.3 to 6.9 ppm (Sauchelli, 1969). The mean content of Mo in British surface soils has been reported at 1 ppm (Thornton, 1977); that of surface soils in western Russia was about 1.7 ppm, whereas the Russian Plain was reported to contain 2.6 ppm (Zborishchuk and Zyrin, 1974). In India, alluvial soils more recently developed from granites and basalts contain Mo in the range of 1.5 to 3.0 ppm as compared with 1.50 to 1.84 ppm in black soils formed from basalt and limestone (Pasricha et al., 1997). The lowest Mo contents were found in some alluvial soils, Alfisols, Spodosols, and Inceptisols. Soils in China have an average Mo content of 2.0 ppm ($n = 850$; geometric mean = 1.2) (Chen et al., 1991). Soils derived from loessial deposits in the loess plateau and along the Yellow River are usually Mo deficient (Liu et al., 1983). Also, acid Chinese soils are often deficient in Mo.

Molybdenum levels in soils from the western United States had a median of 6 ppm, while soils from the eastern states had a median of 0.5 ppm (Kubota, 1977). United States soils have a median Mo concentration of slightly over 1 ppm, with a range of 0.08 to >30 ppm. The trend toward less Mo in the eastern states is attributed to the fact that soils in the eastern region are formed from sandy materials of glacial and marine origin (Table 15.2), while shales and granites are the main source of Mo in many western U.S. soils. Areas of high (toxic) accumulation are usually sparsely distributed, whereas Mo-deficient soils may extend over larger areas particularly in acid soils. Because Mo accumulates in soils, it has been used as an indicator element for geochemical prospecting both for Mo ores and for Cu minerals with which it is oftentimes associated (Jarrell et al., 1980).

4.2 Bioavailable Molybdenum in Soils

Total Mo content of soils, except possibly when occurring at very low levels, is generally not a good indicator of Mo bioavailability. Extractable Mo may not be necessarily correlated with the total Mo content of soils, although soils richest in extractable Mo are generally those having high total Mo contents. A number of extractants have been used to evaluate the phytoavailable Mo ranging from mildly acid, neutral, and alkaline extractants including water, to strong acid and alkali solutions. The method first proposed by Grigg (1953) using acid ammonium oxalate is perhaps still the most commonly used soil test extractant for Mo (Sims and Johnson, 1991). Based on this test, a level of >0.20 ppm acid ammonium oxalate–extractable Mo is considered adequate (Cox and Kamprath, 1972).

Using the ammonium oxalate extractant, Walsh et al. (1952) found 0.04 to 0.12 ppm Mo in soils (pH 5 to 5.5) where Mo deficiency in plants occurred. However, soils of similar total Mo contents (0.4 to 3.5 ppm) with higher pH (pH 7 to 8) gave much higher extractable values of 0.2 to 0.7 and

TABLE 15.2. Molybdenum concentrations (ppm) in U.S. soils derived from various parent materials.

Soil parent material	n	Mean	Range
Mancos shale (Colorado)	16	2.5	0.3–8.1
Colorado shale (Montana)	8	2.1	0.5–3.9
Marine and nonmarine shales (Oregon)	15	2.2	1.0–7.0
Shales, interbedded (Maryland)	10	2.3	1.5–3.0
Shales and limestone			
Virginia	12	2.9	1.4–31.5
W. Virginia	80	1.8	0.8–3.3
Gneiss and schist (Connecticut)	30	1.2	0.8–2.1
Sandstone (Connecticut)	17	0.7	0.6–0.9
Glacial outwash			
New York	8	0.49	0.22–0.86
Michigan	10	0.96	0.16–1.76
Wisconsin	7	0.30	0.08–0.80
Wyoming	3	0.51	0.25–0.65
Montana	1	0.38	—
Glacial till			
New York	13	0.49	0.14–0.92
Michigan	6	1.81	0.16–4.68
Minnesota	13	0.66	0.48–0.88
Loess (Illinois)	15	2.53	0.75–6.40
Lacustrine (Ohio)	12	4.14	1.20–7.15
Granitic alluvial fan (Nevada)	35	11.5	4.4–42.5
Volcanic ash			
Oregon	33	1.2	0.4–1.8
Alaska	18	2.3	1.3–2.8

Source: Modified from Kubota (1977).

produced plants with levels of Mo potentially harmful to livestock. In New Zealand, Grigg (1953) obtained ammonium oxalate–extractable Mo of 0.03 to 0.22 ppm. The water-soluble Mo contents of these soils were generally low. Water extraction usually yields very low amounts of Mo, making this extractant analytically unattractive. However, in alkaline soils with fairly high Mo contents, water extraction may be considered a suitable soil test (Vlek and Lindsay, 1977). In general, concentrations of extractable Mo in soils are rarely higher than 1 ppm.

An intensive study of six forage legumes grown at 11 sites in five states of the southern United States has been conducted (Mortvedt and Anderson, 1982). Ammonium oxalate–extractable Mo in soil ranged from 0.11 to 0.49 ppm. Molybdenum fertilization increased extractable Mo, but seldom increased forage yield. Yield increases were more consistently obtained by liming than by Mo fertilization. No significant correlations between soil Mo and plant Mo were obtained. However, relative forage yield and plant Mo concentrations were highly correlated to soil pH. In general, a Mo level in the range of 0.1 to 0.3 ppm can be expected to produce Mo deficiency, depending on the soil pH and crop species. Widespread Mo deficiency has been observed in soybeans in northern Alabama (Burmester et al., 1988). Acid ammonium oxalate–extractable Mo was not correlated with yield increase or other plant parameter responses associated with Mo fertilization or liming. A high correlation ($r^2 = 0.88$) however, was obtained between relative yield and the ratio of Fe/Mo extractable by acid ammonium oxalate. Ratios of Fe/Mo >1540 were identified with Mo deficiency (Fig. 15.1A).

Soil Mo extraction with 1 M $(NH_4)_2CO_3$ at pH 9 has been tested to represent the more Mo-labile fraction (Vlek and Lindsay, 1977). The amount of Mo obtained using this extractant on soils having pH >7 amended with 0 to 2 ppm Mo showed a correlation of $r = 0.98$ with Mo uptake by alfalfa (Fig. 15.1B). However, inclusion of soils having pH <7 caused poor correlation due to extraction of some soil Mo by $(NH_4)_2CO_3$ that may not be available to plants. This method is considered good for testing soil Mo in alkaline soils. Williams and Thornton (1972) found that, when considered with soil pH and OM, Mo extractable by NH_4OAc and EDTA can provide useful information on the potential uptake of Mo by pasture plants.

Soils supplying excessive amounts of Mo (>10 ppm) usually show better correlation between extractable Mo and plant uptake than do soils having low extractable Mo. Good correlations between water-extractable soil Mo and plant Mo concentrations of up to 40 ppm have been obtained (Barshad, 1951; Kubota et al., 1963). Correlation coefficients between plant Mo and Mo extracted by acid oxalate and 0.1 M NaOH were 0.80 and 0.85, respectively, for plants having up to 11 ppm Mo (Haley and Melsted, 1957). Significant relationships between the Mo in soils extracted by the Grigg's reagent (acid ammonium oxalate) and the Mo concentration of the rice plant have also been demonstrated (Ahmed et al., 1982).

A much stronger correlation ($r^2 = 0.88$) of Mo to relative soybean yield has been obtained by using the ratio between oxalate-extractable Fe and Mo (see Fig. 15.1A). The data indicate the necessity to exclude results for plants grown in strongly acidic soils. In general, the degree of accuracy in predicting Mo uptake and yield of plants can be improved by taking into account the type of extractant suited to a particular soil, soil Mo level, soil pH, and soil Fe oxide (and to some extent Al oxide) content.

Molybdenum extracted by anion-exchange resin has been shown to satisfactorily correlate with plant Mo uptake over a wide range of soil pH (Adams, 1997; Jackson et al., 1975; Jarrell and Dawson, 1978).

4.3 Molybdenum in Soil Profile

Meager information is available on the distribution of Mo in the soil profile. In soils prone to leaching, soluble Mo may be lost from the soil profile especially in alkaline soils. In alkaline soils, Mo is more mobile and if not

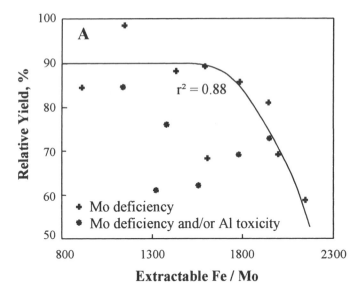

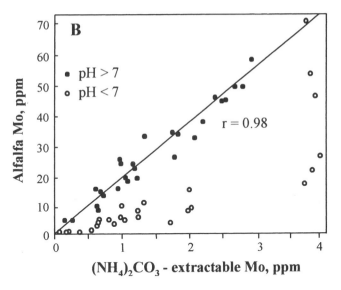

FIGURE 15.1. (A) Yields for untreated soybeans relative to yields with lime and Mo applied as a function of the extractable Fe/Mo ratio. The relationship shown is for only Mo-deficient conditions. (B) Relationship between ammonium carbonate–extractable molybdenum and molybdenum in alfalfa grown on Colorado soils in the greenhouse. ([A] Modified from Burmester et al., 1988; [B] reprinted from Vlek and Lindsay, 1977, p. 637, by courtesy of Marcel Dekker, Inc.)

leached from the profile by rain or irrigation water, the Mo may accumulate in plants (Johnson et al., 1952). As indicated by resin-extractable Mo, there was a net increase in Mo in subsoils due to irrigating a field planted in alfalfa (Jackson et al., 1975). Although the field was irrigated with water containing trace amounts of Mo, the increase of Mo in the subsoil (pH >8) was probably caused by leaching of Mo from the surface soil. Utilizing soil columns subjected to excessive precipitation, Mo has leached through most soils particularly soils with alkaline pH, but some accumulation of Mo occurred in the top 5 cm of the soil (Jones and Bellings, 1967). Using lysimeters in irrigated fields, Pratt and Bair (1964) found that in soils with leaching fractions of about 7 to 8%, Mo buildup in soils was about 75% of the amount added in the irrigation water. The soil Mo buildup was calculated as the difference between the Mo input from irrigation water and Mo output through crop removal and leaching.

In general, Mo tends to accumulate in the A horizon, plow layer, or the few top centimeters of the soil profile. Molybdenum tends to be high in layers of OM accumulation, usually corresponding to the A layer or plow layer.

4.4 Forms and Chemical Speciation of Molybdenum in Soils

Molybdenum exists in soils in the following forms: fixed within the crystal lattice of primary and secondary minerals (nonlabile), adsorbed by soil constituents as an anion, adsorbed by OM, exchangeable, and water-soluble. The last two forms are labile and should be easily bioavailable to plants. The distribution of these various forms of Mo will vary among soils of differing characteristics. These forms can be defined as follows: (1) water-soluble—includes readily soluble molybdates and water-soluble organic complexes, (2) exchangeable (extractable by NH_4OAc)—is part of the Mo adsorbed through ion exchange, (3) complexed (extractable by EDTA)—includes Mo loosely bound in organic complexes and associated with organic compounds soluble in EDTA, and (4) nonextractable—that fraction of Mo not extractable by the alkaline extractants (NaOH or NH_4OH) (Williams and Thornton, 1973). Data in Table 15.3 indicate that at least 45% of the total Mo in these soils is nonextractable and that soils with the lowest organic C contents (5.1 to 7.2%) have the most nonextractable Mo (74 to 79%). The nonextractable form may be considered fixed in the crystal lattice of silicate minerals. The data also indicate that except on the noncalcareous brown earth soil, between 5 and 7% of soil Mo was in the exchangeable form; EDTA extracted greater amounts of Mo from the organic (peaty) than from the inorganic soils; and the water-soluble fraction constitutes only a minor source of plant-available Mo, only around 1% of total Mo.

TABLE 15.3. Distribution of molybdenum fractions in selected soils.

Soil	Parent material	pH	Org. C (%)	Total Mo (ppm)	Residual	Exch.	Complex	H₂O
Peat	Lake alluvium (limestone/shale)	5.4	35.5	35	47	7	12	1
Peaty gley	Glacial drift (limestone/shale)	5.1	20.1	34	67	5	9	<1
Gley	Lower namurian shale	5.8	7.2	13	74	7	8	<1
Calcareous	Lower lias shale	7.1	6.9	39	79	6	1	3
Brown earth	Lower lias shale	6.4	5.1	31	76	1	4	1

The columns under "Form of Mo (% total Mo)[a]" are: Residual, Exch., Complex, H₂O.

[a]The exchangeable fraction was extracted by 1 M NH$_4$OAc, the complexable fraction by 0.05 M EDTA, and the residual fraction was represented by the amount of soil Mo not desorbed by NH$_4$OH and NaOH extractants.
Source: Modified from Williams and Thornton (1973).

Adsorption of the molybdate anions by hydrous oxides of Fe and Al partly accounts for some unavailable Mo in soils (Davies, 1956). Water-soluble Mo would normally be very low, except possibly under alkaline conditions (Barshad, 1951).

Molybdenum generally forms soluble anionic species in aqueous systems. The species of Mo comprising soluble Mo in soils are anionic, with MoO_4^{2-} as the predominant ion (Lindsay, 1972). The species of Mo in solution varies with pH (Fig. 15.2A) and Eh (Fig. 15.2B). Below pH 4.2, $HMoO_4^-$ and H_2MoO_4 are significant species; above pH 4.2, MoO_4^{2-} becomes the major solution species (Lindsay, 1979; Harter, 1991). With increasing pH, solution species generally decrease in the order: MoO_4^{2-} > $HMoO_4^-$ > H_2MoO_4 > $MoO_2(OH)^+$ > MoO_2^+. The first three species predominate the total Mo in solution in the pH range of 3 to 5. The latter two ions can be ignored in soils. In very dilute solutions (10^{-5} to 10^{-8} M) the following Mo species can be expected (Chojnacki, 1963): above pH 5, MoO_4^{2-} predominates; from pH 2.5 to 4.5, MoO_4^{2-}, $HMoO_4^-$, and H_2MoO_4 predominate; from pH 1.0 to 2.5, H_2MoO_4 predominates; and below pH 1.0, H_2MoO_4 disappears and cationic forms appear. Since most soils have a pH above 5.0, MoO_4^{2-} would be the predominant ion. Below this pH value, the MoO_4^{2-} ion becomes protonated. Dissolved Mo species in alkaline soil solutions are dominated by $MgMoO_4^0$, followed by $CaMoO_4^0$ and MoO_4^{2-} (Table 15.4).

In soil solution, these species are expected to exist in monomeric forms, since polymer formation of Mo does not occur in solutions having Mo(VI) concentrations less than 10^{-4} M (Baes and Mesmer, 1976; Jenkins and Wain, 1963). Saturation extracts of some Indiana soils have shown Mo concentrations in the neighborhood of 2 to 8×10^{-8} M (Lavy and Barber, 1964). Molybdate does not form strong complexes with major ions such as Na, K, Mg, or Ca (Turner et al., 1981).

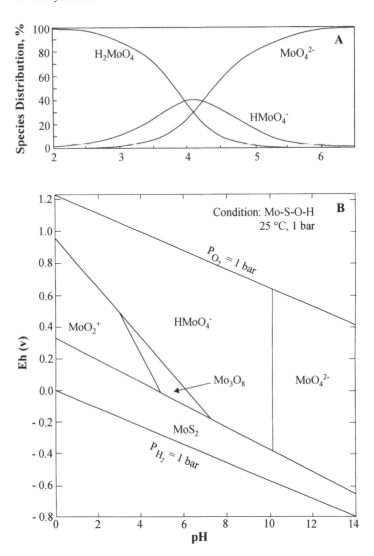

FIGURE 15.2. (A) Speciation of molybdenum as a function of solution pH and (B) predicted Eh–pH stability field for molybdenum. Assumed activities for dissolved species are: $Mo = 10^{-8}$, $S = 10^{-3}$. ([A] From Harter, 1991; [B] From Brookins, 1988.)

4.5 Sorption of Molybdenum in Soils

Molybdenum sorption onto soils has been used as an indicator of a soil's potential to lower the phytoavailability of Mo. Molybdate is sorbed strongly by Fe and Al oxides. Hydrous Fe oxide adsorbs Mo much more strongly than Al oxide, halloysite, nontronite, and kaolinite listed in decreasing order

TABLE 15.4. Speciation of dissolved molybdenum in alkaline soil solution with depth.

	Depth, cm		
	10	30	60
Total dissolved Mo (mg L^{-1})	0.06	0.18	0.14
pH	7.95	8.10	8.07
$MgMoO_4^0$	40%	66%	61%
$CaMoO_4^0$	37%	17%	21%
MoO_4^{2-}	20%	13%	14%
Other[a]	3%	4%	4%

[a]Na and K complexes.
Source: Extracted from Reddy and Gloss (1993).

of amounts adsorbed (Jones, 1956, 1957). Several studies have highlighted the major role of Fe oxides on the Mo sorption by soils (Jarrell and Dawson, 1978; Karimian and Cox, 1978). A portion of the sorbed Mo becomes unavailable to the plant, while the remainder is in equilibrium with the soil solution Mo. As plant roots deplete Mo from solution, more Mo is desorbed into the solution by mass action.

Strongly binding anions such as molybdate, sorb onto oxides by a ligand-exchange mechanism that involves the exchange of an oxide surface hydroxyl group (SOH) for an aqueous anion (A^{2-}) (Sposito, 1984; Stumm, 1992). This reaction yields the formation of an inner sphere complex and is illustrated as

$$SOH + A^{2-} + H^+ \rightleftharpoons SA^- + H_2O$$

This reaction indicates that sorption of the anion is a function of pH, density of surface sites, and concentration of the anion. A decrease in pH favors the removal of the anion from the solution.

Goldberg et al. (1996) were able to use the constant capacitance model to describe Mo adsorption onto Al and Fe oxides and clay minerals (Figs. 15.3 and 15.4). But the model was unable to describe Mo sorption onto soils (see Fig. 15.3A). This indicates an inner sphere adsorption mechanism for the Mo surface complex for Al and Fe oxides, and clay minerals. Molybdenum adsorption onto both Al and Fe oxides exhibited a maximum at low pH extending to about pH 4 to 5 (Goldberg et al., 1996). Above pH 5, adsorption decreased rapidly with only little adsorption occurring above pH 8 (see Fig. 15.3B and C). Molybdenum adsorption was higher for the oxide minerals having higher specific surface area and lower crystallinity (i.e., more amorphous). On the other hand, Mo adsorption onto clay minerals exhibited a peak near pH 3 and then decreased rapidly with increasing pH until adsorption was virtually nil near pH 7 (see Fig. 15.4). The intensity of Mo adsorption onto clays increased in the order: kaolinite < illite < montmorillonite. Molybdenum adsorption behavior on three arid-zone non-calcareous soils resembled that on clays, exhibiting a peak near pH 3 to 4 and decreasing with increasing pH up to pH 7. This behavior is expected

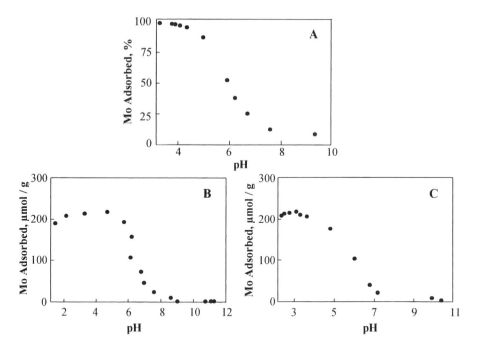

FIGURE 15.3. Adsorption of molybdenum on soil and soil constitutent as a function of pH: (A) Porterville soil, (B) gibbsite (Al oxide), and (C) goethite (Fe oxide). (Modified from Goldberg et al., 1996.)

since the oxide contents of these soils are low. However, Mo adsorption on calcite and two calcareous arid-zone soils was low, indicating that $CaCO_3$ is not a significant sink for Mo in soils.

In soils, adsorption of Mo was highly correlated with extractable Al (Barrow, 1970) and Fe (Jarrell and Dawson, 1978; Karimian and Cox, 1978). Removal of amorphous Fe oxides drastically reduced Mo adsorption by an Australian soil (Jones, 1957). The pH dependance of Mo adsorption on soil resembles that on oxides and clay minerals. Adsorption increased with increasing pH up to a peak near pH 4 and then decreased with increasing pH above 4.

Conclusively, the amount of Mo adsorbed by soil or hydrous Fe or Al oxide is strongly pH-dependent. For example, Mo sorption increased with decreasing pH from 7.7 to 4.4 (Reisenauer et al., 1962). Possible explanations for this effect on adsorption in this pH range are that the hydroxide and molybdate ions compete for adsorption sites, or Fe and Al oxides become more reactive as pH decreases. Recently, the maximum adsorption for Mo occurred below pH 4.5 for three Finnish mineral soils (Mikkonen and Tummavouri, 1993).

In addition to Fe oxide content, the amount of Mo adsorbed closely relates to soil OM content. In acid soils, Fe oxides carry positive charge

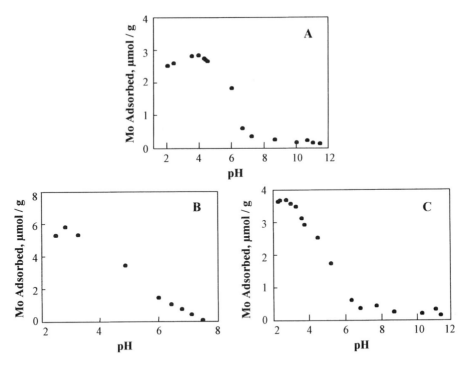

FIGURE 15.4. Adsorption of molybdenum on clay minerals as a function of pH: (A) kaolinite, (B) montmorillonite, and (C) illite. (Modified from Goldberg et al., 1996.)

(i.e., below the zero point of charge) and can react with molybdate. Adsorption isotherms for three Histosols (organic soils) indicate that the level of Mo adsorbed increases with OM content; isotherms for four mineral soils show that the amount of adsorbed Mo increases with Fe oxide content (Karimian and Cox, 1978). Since Mo in soils exists in anionic form, it is rather difficult to explain its adsorption by OM. However, based on the known reaction of OM and Fe in soil, it may well be that the Fe oxide serves as a bridge between the OM and the molybdate ions. In volcanic ash–derived soils, Mo adsorption is due mainly to the presence of allophane and amorphous Al, Si, and Fe compounds and may not be influenced by the OM content of the soil (Gonzales et al., 1974).

5 Molybdenum in Plants

5.1 Essentiality of Molybdenum in Plants

The role of Mo in biological processes was first established by Bortels (1930), who showed its requirement by free-living *Azotobacter* to fix atmospheric N_2. Subsequently, Mo was found to be required for symbiotic N_2 fixation by legumes (Anderson, 1942; Jensen, 1941). Arnon and Stout (1939) and Piper (1940) showed it to be essential for the growth of higher plants; that Mo was essential in pasture production in the process of N_2 fixation by the *Rhizobium*–legume complex (Anderson, 1942). Since that time, Mo has been demonstrated to be required by a variety of nonlegumes, including lettuce, spinach, beet, and species in the family Brassicaceae (Johnson, 1966).

In biological systems, Mo is a component of several enzymes that catalyze unrelated reactions (Table 15.5). Three to four of these enzymes (nitrate reductase, nitrogenase, sulfite oxidase, and xanthine dehydrogenase) are found in plants (Gupta and Lipsett, 1981; Srivastava, 1997). The most important functions of Mo in plants are associated with N metabolism. Its role is primarily involved in enzyme activation, mainly with nitrogenase and nitrate reductase enzymes. Nitrogenase catalyzes the reduction of atmospheric N_2 to NH_3 (see Table 15.5), the reaction by which *Rhizobium* bacteria in root nodules supply N to the host plant. Asymbiotic bacteria, such as *Azotobacter, Rhodospirulum, Klebsiella,* and blue-green algae also carry out the reaction. For this reason, Mo-deficient legumes often exhibit symptoms of N deficiency (Hagstrom, 1977). Nitrogenase contains Mo and Fe ions, both of which are required for activity of the enzyme (Allaway, 1977).

Molybdenum is needed by plants when N is absorbed in the nitrate (NO_3^-) form because it is a critical constituent of the nitrate reductase enzyme. This enzyme catalyzes the biological reduction of NO_3^- to NO_2^-, which is the first step toward the incorporation of N as NH_2 into proteins. The increased Mo requirement of most plants grown on NO_3^--N compared with NH_4^+-N attests to the catalytic role of Mo for nitrate reductase (Srivastava, 1997).

A molybdoenzyme, sulfite oxidase mediates the biochemical oxidation of sulfite to sulfate that has been reported in plants, animals, and certain bacteria (Srivastava, 1997). This element has a wide range of indirect effects

TABLE 15.5. Common molybdoenzymes in various organisms.

Enzyme	Reaction mediated	Organism
Nitrate reductase	$NO_3^- + NADH + H^+ \longrightarrow$ $NO_2^- + NAD + H_2O$	Bacteria, fungi, higher plants
Nitrogenase	$N_2 \xrightarrow[6H^+]{6e^-} 2NH_3$	Symbiotic and asymbiotic N_2-fixing bacteria
Xanthine oxidase	$XH \xrightarrow[H_2O]{O_2} XOH + H_2O_2$	Bacteria, insects, birds, mammals
Xanthine dehydrogenase	$XH \longrightarrow X \xrightarrow{H_2O} XOH$	Bacteria, insects, birds, mammals, higher plants(?)
Aldehyde oxidase	$RCHO \xrightarrow[H_2O]{O_2} RCOOH + H_2O_2$	Mammals
Sulfite oxidase	$SO_3^{2-} + H_2O \longrightarrow SO_4^{2-} + 2H^+ + 2e^-$	Birds, mammals, higher plants(?)
Formate dehydrogenase	$HCOO^- \longrightarrow CO_2$	Bacteria

in regulating other enzymes' activities that are not molybdoenzymes, such as in carbohydrate metabolism, reproductive physiology, anion balance, root exudation, plant water relations, and disease control in plants.

5.2 Deficiency of Molybdenum in Plants

Molybdenum deficiencies in field-grown plants were first recorded in Australia more than 60 years ago. In the United States, 27 states have reported Mo problems on 40 crops. Deficiencies have been reported on soybeans, alfalfa and other legumes, citrus, peanuts, tobacco, tomato, rice, onions, sweet potato, and several vegetable crops. Generally, Mo deficiencies are not widespread and are found in localized areas. Plant deficiencies often occur in acidic soils, partly due to the presence of hydrous oxides of Fe and Al. It can also occur in plants growing on coarse-textured soils and soils low in OM, such as the Oxisols in the tropics.

The uptake of Mo by plants is generally considered to be induced by mass flow, diffusion, and root interception. Plant roots absorb Mo in the form of the anion MoO_4^{2-} (Gupta, 1997). This is also the form that is mobile in the plant and the predominant species translocated between plant parts.

Among the micronutrients that are essential for plant growth, Mo is required in the smallest amount. The adequate concentrations of Mo in most crops range between 0.1 and 1.0 ppm, but the threshold for Mo toxicity in plants is much higher than that for any other micronutrient. Because of these low requirements, there is only a narrow range between the deficient and sufficient levels in most plants (Table 15.6). Crop species have widely varying Mo requirements and therefore have different sufficiency requirements (Johnson et al., 1952). For example, among crop species grown on a Mo-deficient soil, healthy barley, maize, oats, and wheat contained 0.03 to 0.07 ppm Mo; tomatoes, sugar beets, squash, and spinach contained as much as 0.10 to 0.20 ppm Mo while still being moderately to severely Mo-deficient.

Legumes appear to have a 2 to 3 times greater Mo requirement than nonlegumes (Jarrell et al., 1980). In general, plants that demand a high soil pH for satisfactory growth appear to suffer most from Mo deficiency (Williams, 1971). For example, cauliflower, broccoli, Brussels sprouts, spinach, lettuce, tomatoes, sugar beets, kale, and rape are relatively highly sensitive; celery, lucerne, carrot, flax, and the clovers are intermediate; while the cereals (including corn, rice, and small grains), grasses, cotton, and large-seeded legumes are the least sensitive (Lucas and Knezek, 1972; Williams, 1971). Among the orchard or fruit crops, citrus has shown a high requirement for Mo (Stewart and Leonard, 1952).

Overall, symptoms of Mo deficiency have been known to occur in more than 40 higher plant species (Hewitt, 1956). Thus, more sensitive plants, such as cauliflower, broccoli, cabbage, lettuce, and tomatoes, may be used as indicator crops for diagnostic purposes. Citrus may be useful in the tropics.

TABLE 15.6. Deficient and sufficient levels of molybdenum in plant tissues.

Plant	Plant parts	ppm Mo, DW	
		Deficient	Sufficient
Alfalfa	Leaves at 10% bloom	0.26–0.38	0.34
	Whole plants at harvest	0.55–1.15	
	Top 15 cm of plant prior to bloom	<0.4	1.5
	Upper stem cutting at early flowering stage		0.5[a]
	Shortly before flowering (top 1/3 of plants)	<0.2	0.5–5.0
	Whole tops at 10% bloom		0.12–1.29
Barley	Blades 8 wk old		0.03–0.07
	Whole tops at boot stage		0.09–0.18
	Grain		0.26–0.32
Beans	Tops 8 wk old		0.4
	Upper fully developed leaves	1–5	
Beets	Tops 8 wk old	0.05	0.62
Broccolsi	Tops 8 wk old	0.04	
Brussels sprouts	Whole plants when sprouts began to form	<0.08	0.16
	Leaves	0.09	0.61
Cabbage	Leaves	0.09	0.42
	Whole plants at first signs of curding	<0.26	0.68–1.49
Carrot	Whole tops at maturity		0.04–0.15
Cauliflower	Whole plants before curding	<0.11	0.56
	Young leaves showing whiptail	0.07	
	Leaves	0.02–0.07	0.19–0.25
Corn	Grain	0.03	0.05–0.08
	Stems	0.013–0.11	1.4–7.0
	At tassel middle of the first leaf; opposite and below the lower ear	<0.1	<0.2
	Ear leaf at silk	0.6–1.0	
	Fully developed leaves		0.2–0.5 (1.4–22)
Cotton	65-day-old leaves	0.5	
	Fully matured leaves at bloom		0.6–2.0
Cucumber	Mature whole leaves		0.5–1.0
Lettuce	Leaves	0.06	0.08–0.14
Oats	Leaves	0.07–0.40	0.10–0.60

Molybdenum deficiencies have been reported throughout the world, especially on acid soils. Sandy soils are more prone to Mo deficiency than are loamy or clay soils. Most Mo deficiencies are associated with legume crops because Mo is an essential constituent of enzymes necessary for the N_2 fixation by symbiotic bacteria dwelling in the nodules.

In the United Kingdom Mo deficiency has been more commonly reported in cauliflower and broccoli, but it only occurs sporadically (Williams, 1971). The deficiency can often be corrected by liming the soil, as indicated by response lines for soybeans and clover–grass pastures (Adriano, 1986). In soybeans, application of Mo resulted in an average yield increase of 4.2 hl ha^{-1} (hectoliter per hectare) when the soil pH was 5.0 (Sedberry et al.,

TABLE 15.6. (*continued*)

| Plant | Plant parts | ppm Mo, DW | |
		Deficient	Sufficient
Onions	Matured whole tops	<0.06	0.10
Pasture grass	First cut at first bloom		0.2–0.7
Peanuts	Upper fully developed leaves	<1.0	
	Upper fully developed leaves at blossom	—	—
Peas	Fully developed leaves at blossom		0.4–1.0
Potato	Leaf blades	0.15	
	Fully developed leaves at blossom		0.2–0.5
Red clover	Total aboveground plants at bloom	<0.15	0.3–1.59
	First cut at flowering		0.26
	Whole plants at the bud stage	0.1–0.2	0.45–1.08
Rice	Upper leaves before flowering		0.2–0.5
Soybeans	Plants 27 cm high	0.19	
	Fully developed leaves top at blossom		0.5–1.0
Spinach	Leaves 8 wk old	0.1	1.61
	Whole tops at normal maturity		0.15–1.09
Subt. clover	Leaflets and petioles		0.10
Sugar beets	Blades shortly after symptoms appear	0.01–0.15	0.2–20.0
	Fully developed stemless leaf; late June or early July	<0.1	0.2–2.0 >20[b]
Temperate legumes	Plant shoots		>0.1
Timothy	Whole tops at prebloom/bloom	0.11	0.15–0.50
Tobacco (also burley)	Leaves 8 wk old	0.38–0.42	1.08 (0.06–0.15)
Tomatoes	Leaves 8 wk old	0.13	0.68
Tropical legumes	Plant shoots		>0.02
Wheat	Whole tops; boot stage		0.09–0.18
	Grain		0.16–0.20
Winter wheat	Aboveground plants at ear emergence		>0.3

[a]Considered critical.
[b]Considered toxic.
Source: Extracted from various sources, mostly from summary of Gupta (1997).

1973). The response of soybeans to Mo application diminished as the pH increased up to 6.2, when no more response was obtained (Davies and Griggs, 1953).

A value of about 0.10 ppm Mo in dry matter of plant tops is regarded as the critical level below which Mo deficiency is likely to occur (Anderson, 1956; Jones, 1972). However, the value may vary widely (see Table 15.6). For example, good responses by legumes to Mo application have been found where the tissues contained 0.50 ppm Mo; on the other hand, clover may not respond even if the tissue content is only about 0.10 ppm (Anderson and

Oertel, 1946; Walker et al., 1955). In Illinois, the following critical levels of Mo (in ppm) were used for diagnostic interpretation (Melsted et al., 1969): corn, 0.20; soybeans, 0.50; wheat, 0.30; and alfalfa, 0.50. Growth reduction may be expected to occur if Mo levels in plants fall below these values. Responses to applications of Mo could be obtained when the ammonium oxalate–extractable Mo level in the soil is below 0.14 ppm provided the soil pH is lower than 6.3 (Grigg, 1953).

While most micronutrients display their deficiency symptoms on the young leaves at the top of the plant (i.e., young shoots), Mo deficiency symptoms are generally manifested on the whole plant. Reason: Mo is readily translocated to the entire plant whereas other micronutrients (e.g., Zn, Mn, Cu) are not as readily translocated. Since Mo is required in N metabolism, most symptoms resemble those of N deficiency.

Some of the most spectacular symptoms of Mo deficiency have been displayed in *Brassica* plants as whiptail disease (Sauchelli, 1969) and yellow spot disease in citrus (Stewart and Leonard, 1952, 1953). The most striking manifestation of Mo deficiency is leaf chlorosis that resembles N deficiency (Johnson et al., 1952). Affected plants display a yellow or orange-yellow chlorosis, with some brownish tints that start in the youngest leaves (Bergmann, 1992). In yellow spot in citrus, the symptoms appear on the leaves as large interveinal chlorotic spots. In whiptail, blade distortion is exhibited with some leaves showing perforations or marginal scorching. The chlorosis is followed by marginal curling, wilting, and finally necrosis and withering of the leaf. Symptoms usually appear first in the older leaves, showing up later in younger leaves until the growing point dies (Gupta, 1997). In wheat, Mo-deficient plants exhibit golden-yellow coloration of older leaves along the apex and apical leaf margins. Plants displayed reduced foliage and shorter internodes. Normal tillering may not occur and young plants have whitish, necrotic areas extending back along the leaves from the tips (Lipsett and Simpson, 1971). In onions, it shows up as a drying of the leaf tips (Lucas, 1967). Below the dead tip, the leaf shows about 2.5 to 5.0 cm of wilting and flabby formation. As the deficiency progresses, the wilting and dying advances down the leaves. In severe cases, the plant dies.

Soils commonly deficient in Mo include the following: (1) highly podzolized soils because of indigenously low total Mo or sequestering of Mo by the oxyhydroxides, (2) highly weathered soils as in the tropics, in which secondary minerals may sorb Mo, (3) soils with pH values below 6 in which the Mo is insoluble, and (4) well-drained sandy soils in which total Mo is low (Severson and Shacklette, 1988).

The following methods have all been shown effective in correcting Mo deficiency in horticultural and field crops (Adams, 1997; Murphy and Walsh, 1972): lime application, lime application plus Mo, foliar sprays of Mo, soil application of Mo, seed treatment with Mo, mixing Mo with potting compost, and treatment of seedbed with Mo.

5.3 Accumulation and Phytotoxicity of Molybdenum

In normal crop plants, Mo concentration usually ranges from 0.80 to 5.0 ppm in the tissue (Lucas, 1967). However, some plants have been found to contain >15 ppm. Deficient plants usually contain <0.50 ppm (see Table 15.6). Certain nonresponsive crops such as the grasses and corn, may contain as low as 0.10 ppm.

Molybdenum toxicity in plants has not been observed under field conditions. However, toxicity can be induced under extreme experimental conditions. In sand culture, the upper critical level of Mo for barley was about 135 ppm (Davis et al., 1978). Similarly, 200 ppm Mo was critical for oat plants (Hunter and Vergnano, 1953). Molybdenum was less toxic than Ni, Co, Cu, Cr, and Zn, in decreasing order. In solution culture, 710 ppm in bush bean leaves markedly reduced plant yield (Wallace et al., 1977). This tissue level was caused by 96 ppm Mo in the culture solution. On the other hand, turnips with up to 1800 ppm in the leaves showed normal growth pattern when grown in nutrient solution containing 20 ppm Mo (Lyon and Beeson, 1948). Soybeans grown on pot culture having 80 ppm Mo in leaves were normal (Golov and Kazakhkov, 1973) as were cotton plants containing 1585 ppm in leaf tissue (Joham, 1953). In general, the presence of large quantities (on the order of 100 to 200 ppm) of Mo in plants does not produce detrimental effects on crop yields or any toxicity appearance on the foliage. Thus, Mo phytotoxicity virtually does not occur under field condition.

Applied Mo is readily available to plants as indicated by 60% recovery by French beans grown in greenhouse (Widdowson, 1966). When an acid soil (pH 5.6) was limed to pH 7.5, Mo recovery rose to 80%. Plants suffering from excess levels of Mo exhibit chlorosis and yellowing, apparently due to interference with Fe metabolism in plants (Millikan, 1949; Warington, 1954).

Under field conditions, plant uptake of Mo is well correlated with the amount of Mo in the soil solution, which can range from 10^{-8} M in low Mo soils (Lavy and Barber, 1964) to 10^{-5} M in soils producing toxic herbage (Kubota et al., 1963). When livestock ingest forages containing relatively high Mo levels (10 to 20 ppm) they develop a condition known as *molybdenosis,* which can often be fatal. For this reason, high levels of Mo in plants are a concern to agriculturalists and animal nutritionists. Actually, *molybdenosis* (also known as teart or peat scour) is Mo-induced Cu deficiency in livestock and can be rapidly corrected by injections of Cu. An intake of high concentrations of Mo (e.g., herbage or fodder containing >5 ppm Mo DW) may interfere with the assimilation and metabolism of Cu in animals. The symptoms of the disease are severe diarrhea, followed by roughening of the coat, change of color, and loss in weight (Barshad, 1948). Young animals are more prone to this disease often showing stiffness in the hind legs. In severe cases, their hair sheds in spots, leaving reddened, dry, and cracked skin.

Most animals require <1.0 ppm Mo in their diet (Allaway, 1968). However, as much as 6 ppm dietary Mo is safe, with sheep tolerating slightly higher levels than cattle, and animals could tolerate substantially higher levels if the diet is amply augmented with Cu (Neathery and Miller, 1977). Where Cu is low (<4 ppm DW), 5 ppm Mo may adversely affect animals. Injury to cattle grazing herbage with 20 ppm Mo may occur (Barshad, 1948). In rare occasions, 10 ppm may be toxic (Marclise et al., 1970; Wynne and McClymont, 1955). In general, Mo toxicity is manifested much earlier in animals than in plants.

A U.S. regional pattern of Mo concentration in legumes has been established (Kubota, 1977). The principal legumes collected over a period of 10 yr were alfalfa, clovers, and biennial sweet clovers. The general pattern is one of decreasing Mo content from the west (~6 ppm) to the east (~0.50 ppm) across the United States. The pattern roughly coincides with changes in soil pH, from neutral to alkaline (often calcareous) soils in the western region to predominantly acid soils in the east.

In general, plant production in acidic soils concerns agronomists due to the likelihood of Mo deficiency; in alkaline, calcareous soils, bioaccumulation of Mo becomes more of a concern to animal nutritionists due to this element's enhanced bioavailability to plants. Thus, in the United States, there is greater concern over excess Mo in feedstuff and the associated trace element imbalance in animal nutrition than with Mo deficiency in plants.

6 Molybdenum in Animal and Human Nutrition

Molybdenum deficiency has not been observed in humans and is difficult to produce in experimental animals (Van Campen, 1991). Much of the interest centers on the role of Mo as a catalyst of various enzymes in the human and mammalian system and its effects on Cu nutrition of livestock.

High intake levels of Mo have been correlated with low incidence of dental caries in children (Anderson, 1966; Healy et al., 1961). However, high dietary intake levels may be harmful. A population in eastern Europe exposed to a high-Mo environment produced increased blood xanthine oxidase, increased concentrations of uric acid in blood and urine, and a high incidence of gout. In the absence of human deficiencies, the Food and Nutrition Board (1989) has suggested adequate daily dietary intakes of Mo based on reported average intake levels. These are summarized in Table 15.7 for various age groups.

Food is the major source of Mo for humans. Milk and milk products, legume seeds, organ meats, cereal grains, and baked goods are the primary contributors. Daily intake levels in the adult population of the United States

TABLE 15.7. Estimated safe and adequate daily dietary intakes of molybdenum.

Category	Age (yr)	Mo (µg day^{-1})
Infants	0.0–0.5	15–30
	0.5–1.0	20–40
Children	1–3	25–50
	4–6	50–75
	7–10	50–150
Others	11+	75–250

Source: Food and Nutrition Board (1989).

are estimated at 80 to 350 µg day^{-1} (Tsongas et al., 1980), which is well within the safe and adequate range.

Based on limited data, total dietary intake of Mo by U.S. adults is about 100 µg Mo day^{-1}, or about 0.13 ppm of the dry diet (Tipton et al., 1966). This value compares favorably with that for the United Kingdom at 128 ± 34 µg Mo day^{-1} and less than the International Commission on Radiological Protection average diet intake of 300 µg Mo day^{-1} (ICRP II = 450 µg Mo day^{-1}) (Hamilton and Minski, 1972). Underwood (1973) indicated that Mo toxicity in humans is impossible from the intake of Mo occurring naturally in ordinary foods. Molybdenum concentrations in vegetables commonly range from <5 to 70 ppm in ash (Connor and Shacklette, 1975). In North American wheat and flours, the Mo content of whole grain samples range from 0.30 to 0.66 ppm, with a mean of 0.48 ppm (Czerniejewski et al., 1964). Milk is usually low in Mo but is susceptible to levels of dietary intake of the element, with concentrations of 20 to 30 ppb Mo reported from cows fed ordinary rations (Hart et al., 1967).

The normal range of Mo concentrations in animal diet is trace to <5 ppm. Toxic effects of Mo have been found in cattle grazing on forage containing 15 to 300 ppm Mo in the United States, Canada, New Zealand, and Sweden (Dick, 1956). Kubota et al. (1967) indicated that the toxic range is 10 to 20 ppm for grazing animals. The National Research Council (1984) indicated that 6 ppm is the maximum tolerable level for Mo in the ration for cattle. Jarrell et al. (1980) indicate that Mo toxicity is dependant on Cu and SO_4^{2-} concentration in animal diet and that adding these two constituents to diet reduces Mo toxicity. Since molybdenosis is in reality a Mo-induced Cu deficiency (also referred to as hypocuprosis) the Cu/Mo ratio in the diet can better serve as an index to the disease than the Mo content alone. Miltimore and Mason (1971) claim that the critical Cu/Mo ratio in animal feeds is 2.0 and that feeds with lower ratios can be expected to result in Cu deficiency; however, studies of English pastures indicate that ratios lower than 4.0 are the ones likely to produce Cu deficiencies (Alloway, 1973).

A major concern in the land application of sewage sludge is the intoxication (i.e., *molybdenosis*) of cattle and other ruminants grazing on pastures that have accumulated Mo. Although long-term research by O'Connor (2000) indicates low risk from ingestion of sludge-derived Mo for ruminants when coupled with good management practices (e.g., augmenting cattle diets with Cu supplements), caution should be observed when planting legumes in limed soils or limed sludge (McBride et al., 2000).

7 Factors Affecting Mobility and Bioavailability of Molybdenum

7.1 pH

Bioavailability of Mo to plants is largely dependent on soil pH. Unlike most other micronutrients, bioavailability of Mo in soils is greatest under alkaline condition and less under acidic condition. Thus, soil reaction is viewed as the most important factor affecting Mo availability. Liming Mo-deficient acid soils often corrects the deficiency; liming may substitute for Mo fertilization by releasing Mo from soils into forms readily bioavailable for plant uptake. Conversely, plant response to application of Mo under field conditions is more effective on acid soils (Adriano, 1986). The amount of lime required for maximum production of Mo-responsive crops may be sharply reduced or eliminated by Mo fertilization (Ahlrichs et al., 1963; James et al., 1968). It is a common practice to control Mo deficiency by liming the soil to a pH of 6.0 to 6.5. However, the effect of lime on Mo deficiency varies with conditions (Anderson, 1956). It may partially or entirely correct the deficiency, it may have practically no effect, or it may increase the response to Mo.

The dependency of Mo solubility on soil pH is a concern when applying sewage sludge on land. Since pH of soils receiving sludge should be maintained at about pH 6.5 to limit the bioavailability of especially Cd, the availability of Mo then becomes a question especially if the sludge is highly contaminated with Mo. For one thing, Mo supplied from sludge is readily taken up particularly by legumes (McBride et al., 2000).

The importance of soil reaction on the availability of applied Mo to alfalfa and clover are displayed on Figure 15.5. The data indicate correlations between soil pH, available Mo, and plant uptake. In the southeastern United States, yields of alfalfa and clover were significantly correlated with soil pH and soil Mo, with soil pH having the greatest effect (Mortvedt and Anderson, 1982).

In tropical and subtropical regions, legumes are often grown in acidic soils and are dependent on N_2 fixation. Nodulation and N_2 fixation frequently are poor under these conditions. Beans generally do not respond to Mo when grown in acidic soils (pH <5.2) that are Mo-deficient (Franco

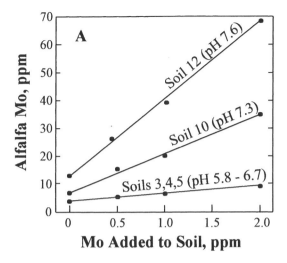

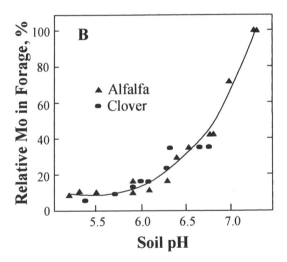

FIGURE 15.5. Influence of soil pH (A) on the molybdenum uptake by alfalfa and (B) on the relative molybdenum concentrations in alfalfa and clover grown in the southeastern U.S. region. ([A] Reprinted from Vlek and Lindsay, 1977, p. 635, by courtesy of Marcel Dekker, Inc.; [B] From Mortvedt and Anderson, 1982.)

and Munns, 1981). Applications of Mo did not increase crop yields from acid soils in Brazil limed to pH >6.0 (Franco and Day, 1980). Likewise, Boswell (1980) found that Mo application did not increase soybean yield from soils limed above pH 6.2. In general, Mo deficiency in tropical and subtropical soils is more widespread than elsewhere because soils in these areas are highly weathered, relatively rich in sesquioxides (i.e., oxides of Fe and Al), and acidic (Gupta, 1997). Acidic peat soils are widely distributed in southeast Asia; more than 20×10^6 ha of peat soils are located in Indonesia, Malaysia, and Thailand alone (Pasricha et al., 1997). Availability of Mo in these soils is known to increase with increase in pH up to 4.7, but may decrease with much further increase in pH.

7.2 Soil Type and Parent Material

Mitchell (1964) stated that the trace element content of soils is dependent almost entirely on that of the rocks from which the parent material was derived and on the weathering process to which the soil-forming materials have been subjected. The more mature and older the soil, the less may be the influence of the parent rock. Thus, pedogenic factors and the nature of parent material play important roles in determining the Mo levels in the solum.

Soils formed from shale generally have more Mo than those formed from other kinds of rocks (Kubota, 1977). In general, shale and granite are the major rocks contributing Mo to soil parent materials, like the alluvium in the western United States. The Appalachian region in the eastern states has extensive areas of soils formed on shale or interbedded shale and sandstone. A similar pattern of high-Mo soils might be expected in the Rocky Mountains, where shales are important rock sources of soil parent materials. Soils formed from Devonian black shales contained approximately 20 times as much Mo as any other soil (Massey et al., 1967). Among the other soils, the parent rock averages ranged from 0.44 ppm Mo for older Ordovician limestone (generally high phosphatic) to 0.08 ppm Mo for Pennsylvanian sandstones and shales.

Irrigation of alkaline soils developed from cretaceous-age shales in the western United States may increase the mobility and transport of Mo, resulting in higher amounts of Mo in arable soils, crops, and wetlands soils and biota (Smith et al., 1997). Data on streams, lake, and wetland sediments from more than 20 irrigation projects in western states show a range of total Mo from <2 to 120 ppm.

Soils in the north-central and northeastern United States reflect amounts of Mo in glacial drift and related deposits, such as loess and lacustrine deposits (see Table 15.2). Coarse-textured soils formed in glacial outwash of Wisconsin, Michigan, and New York have about 0.60 ppm Mo. The Mo contents of eight Canadian soils formed on glacial drift ranged from 0.20 to 0.82 ppm (Wright et al., 1955).

Properties of acid soils where responses to Mo applications are likely to occur have been summarized as (Williams, 1971): (1) high in oxides of Fe (e.g., red loams and clays), eroded laterites, and ironstone soils, (2) highly podzolized, derived from sedimentary material and often high in Fe, (3) high in Mo-retention capacity and with low pH, (4) highly leached or depleted in Mo supply through intensive cropping, (5) high in available Mn, and (6) heavily amended with sulfates, as with gypsum and $(NH_4)_2SO_4$.

Leaching removed approximately 10% of added Mo from two western Australian gray sands (pH 5.8 to 6.1) but, at lower pH (5.0 to 5.4) only negligible amounts of Mo were removed (Riley et al., 1987). Leaching does not appear to be a problem in the incidence of Mo deficiency in acidic yellow-brown sandy soils of western Australia's wheat belt. In general, Mo in acidic soil profiles should be resistant to leaching except in very sandy soils.

7.3 Iron and Aluminum Oxides

Adsorption and precipitation of molybdate ions by hydrous oxides of Fe and Al are major processes rendering Mo less labile. The solubilities of Fe and Al molybdates increase with increasing pH values (Davies, 1956). Oxides of Fe have been shown to be a major soil component responsible for retaining Mo (Reisenauer et al., 1962; Reyes and Jurinak, 1967). Freshly prepared Fe oxides and a soil high in Fe oxide removed large amounts of Mo from aqueous solution (Jones, 1956, 1957); removal of Fe oxide from the soil drastically reduced its effectiveness in adsorbing Mo. The oxides of Fe and Al are perhaps the next most important factor after soil pH regulating the solubility of Mo. For example, Mo adsorption in mineral soils increased with increasing content of Fe oxide (Karimian and Cox, 1978). Aluminum oxide can also remove Mo from aqueous solution but has lower affinity for Mo than Fe oxide under the same conditions (Jones, 1957).

The Fe oxide content of soil has been found to negatively correlate with Mo uptake by plants (Robinson and Edgington, 1954; Takahashi, 1972). More specifically, in eight western Oregon soils, the amorphous fraction of Fe oxide extracted with ammonium oxalate played a major role in Mo sorption by these soils (Jarrell and Dawson, 1978).

7.4 Redox Potential and Organic Matter

Soils that produce high-Mo plants are generally wet and poorly drained, frequently neutral to alkaline in reaction, and high in OM (Allaway, 1975). This phenomenon may be caused by the following: (1) the Fe in these soils is likely to be in the reduced state [i.e., Fe(II)], and ferrous molybdate is more soluble than ferric molybdate at pH values commonly found in soils, (2) loss of soluble Mo by leaching is restricted, causing accumulation of soluble Mo in the root zone, and (3) movement of Mo to the plant roots via mass flow is expected to be greater in wet than in dry soils. Molybdenum-enriched plants capable of intoxicating cattle and sheep are generally confined to poorly drained areas where soils have been derived from Mo-rich basalt parent materials, usually alluvium from shales and igneous rocks (Welch et al., 1991).

The teart areas in the United Kingdom are poorly drained (Lewis, 1943), as are the problem peat areas in New Zealand and Florida (Kretschmer et al., 1956). Molybdenum toxicity to animals is a problem in some isolated valleys with floodplains of Mo-rich alluvium (Kubota et al., 1961; Kubota, 1977). In Nevada, the highest Mo concentrations in forage samples (185 ppm) were from soils with poor drainage on granitic alluvial fans. Kubota et al. (1961) attributed Mo toxicity in animals to two distinctly different kinds of soil. One soil group includes poorly drained soils, such as those in California, Nevada, and the teart areas in the United Kingdom. The other group includes organic soils. Levels of Mo in forages for the latter group are in the range of 2 to 7 ppm. Peats and mucks are formed in wet environments and are associated with Mo toxicity in the soils of the California delta, the Klamath area in Oregon, and the Everglades in Florida (Kubota, 1972). High OM apparently increases soluble Mo due to its effect on redox potential and more rapid reduction of Fe rather than through formation of soluble organic forms of Mo.

Where redox potential is not a question, the amounts of Mo sorbed by soils is a function of not only the Fe oxide but also the OM content (Karimian and Cox, 1978). In this case, sorpton was primarily attributed to the Fe oxide content of the OM.

7.5 Phosphorus and Sulfur

Of the nutrient elements commonly used in agriculture, P and S can exert a major influence on the bioavailability of Mo by plants. Several workers reported synergistic effects of soil application of P on Mo uptake by plants; molybdates could be released from the anion-exchange complex by $H_2PO_4^-$ replacement and thus increase the concentration of Mo in the solution phase available for plant uptake (Stout et al., 1951). Indeed, the adsorption of arsenate and molybdate by soils was significantly reduced by the presence of phosphate, which is attributed to competitive interactions (Roy et al., 1986). The P stimulation of Mo uptake was suggested due to the formation of a complex phosphomolybdate anion more readily adsorbed by plants (Barshad, 1951). Molybdenum uptake was enhanced by P supplied as orthophosphoric acid, a P source that contains no S. Further evidence of the beneficial effects of P on Mo uptake have been summarized (Gupta, 1997). Bingham and Garber (1960) found that this relationship depends on soil pH; synergistic effects of P fertilization on Mo uptake by orange seedlings grown in acid soils occurred, whereas in alkaline soils excessive P reduced Mo availability.

In tropical Asian countries, large-scale use of cultures of N_2-fixing blue-green algae or *Azolla* is being implemented for paddy rice production to supplement N fertilizers. Application of both P and Mo are known to enhance the growth and N_2 fixation by these organisms. Applied P had a synergistic effect on the Mo uptake by the rice plant (Basak et al., 1982a,b).

While P application can synergize Mo uptake by plants, S application to soil may depress Mo uptake by plants. Since S is present in superphosphate fertilizers, application to soil of these fertilizers may decrease Mo uptake by plants. The inhibitory effects of SO_4^{2-} anions on Mo uptake have been suggested to occur primarily during the absorption process, with some antagonistic mechanism involved during translocation from roots to shoots. Since the MoO_4^{2-} and SO_4^{2-} are binegative anions and tetrahedral in structure, they can compete for sorption sites, and in biological systems they are absorbed, transported, and excreted along many of the same routes (Smith et al., 1997). Application of S has been found to antagonize the Mo content in crops, including Brussels sprouts, berseem, soybeans, tobacco, clover, and perennial grass (Gupta, 1997). Such antagonism has also been observed on sorption where sulfate reduced Mo sorption on soil (Smith et al., 1987). Similar competitive effects have been observed between arsenate and molybdate ions, especially below pH 6 (Manning and Goldberg, 1996).

Other antagonistic relationships of Mo with other elements in plant nutrition have been reported for Mn, Fe, Zn, and Cu (Adriano, 1986).

7.6 Plant Factors

Plant species are known to accumulate different amounts of Mo from the soil. Legumes normally contain higher Mo than do grasses growing in the same area, and this has been used to define areas of potential Mo toxicity to livestock (Allaway, 1977). Members of the Brassicaceae family (see Table 15.6), such as cauliflower and broccoli, are highly sensitive to Mo deficiency; however, they generally contain higher Mo than other species growing on the same soil if the Mo level is sufficient. The cereal crops and large-seeded legumes also contain less Mo, in general. Among seeds, soybeans and peas accumulate much higher amounts of Mo than wheat, barley, and oats (Gupta, 1997). More specifically, Mo uptake was in the order peas > barley > wheat. Others have shown significant plant interspecific variation in Mo contents (Barshad, 1951; Rinne et al., 1974).

Cultivars also vary in their uptake potential for Mo as demonstrated in alfalfa, cauliflower, maize, and soybeans (Gupta, 1997). Although plants may tend to decrease in trace element content with maturity, Mo tends to increase with age of several plant species, both legumes and nonlegumes, especially during periods of slow growth (Barshad, 1948, 1951). In contrast, the Mo content of leaves and stems of soybeans was found to decrease with maturity (Singh and Kumar, 1979).

Once absorbed by the roots, Mo is differentially translocated to various plant parts. Molybdenum tends to accumulate, in decreasing order: leaves > stems > seed or fruit. In general, the greatest concentrations of Mo are found in the blades of leaves and in actively growing plant parts (Barshad, 1948). More specifically, the interveinal areas of leaves are found to preferentially accumulate Mo (Stout and Meagher, 1948).

8 Sources of Molybdenum in the Environment

8.1 Molybdenum Fertilizers

Sodium molybdate ($Na_2MoO_4 \cdot 2H_2O$, 39% Mo) and $(NH_4)_6Mo_7O_{24} \cdot 4H_2O$ (54% Mo) are very soluble and are the most widely used sources of Mo in agriculture (Table 15.8). Molybdenum is present in the VI oxidation state as the MoO_4^{2-} ion. Less satisfactory carriers of Mo are MoO_3 (66% Mo), MoS_2 (60% Mo), and Mo frits (2 to 3% Mo).

The amount of Mo required to correct a deficiency varies with the soil, plant species, source, and application method. The amount of Mo needed to produce optimum yield is very small. Consequently, Mo is not commonly added to mixed fertilizers. Rates of field soil application producing optimum yields range from 28 g ha^{-1} in pasture to 444 g ha^{-1} of Mo for cauliflower (Murphy and Walsh, 1972).

TABLE 15.8. Application rates, methods, and sources of molybdenum for crops.

Plant[a]	Method of application	Source of Mo	Rate of application
Cauliflower	Seed	$Na_2MoO_4 \cdot 2H_2O$	65 g ha^{-1}
Cauliflower, cabbage, turnips, lettuce	Soil	$Na_2MoO_4 \cdot 2H_2O$	390–1560 g ha^{-1}
Corn	Seed	$Na_2MoO_4 \cdot 2H_2O$	25 g per 50 kg of seed
Cowpeas	Seed	"Nitramolybdenum" (4.8% Mo)	10–20 mg per 100 seed
	Foliar spray	$(NH_4)_6Mo_7O_{24} \cdot 4H_2O$	60 g ha^{-1}
Honeydew	Foliar spray	"Nitramolybdenum" (4.8% Mo)	20 g ha^{-1}
Lupin	Seed	"Moly-Gro" (38% Mo)	53 g ha^{-1}
Rice	Foliar spray	$(NH_4)_6Mo_7O_{24} \cdot 4H_2O$	110 g ha^{-1}
Peas	Seed	$Na_2MoO_4 \cdot 2H_2O$	55 g ha^{-1}
Peanuts	Foliar spray	$Na_2MoO_4 \cdot 2H_2O$	218 g ha^{-1}
	Seed	MoO_3	100 g ha^{-1}
Soybeans	Seed and foliar spray	$Na_2MoO_4 \cdot 2H_2O$	100–224 g ha^{-1}
	Foliar spray	$Na_2MoO_4 \cdot 2H_2O$	100 g ha^{-1}
	Soil	$Na_2MoO_4 \cdot 2H_2O$	220–2460 g ha^{-1}
Tobacco	Soil	$Na_2MoO_4 \cdot 2H_2O$	220–880 g ha^{-1}
Wheat	Seed	$Na_2MoO_4 \cdot 2H_2O$	117–137 g ha^{-1}
	Foliar spray	$Na_2MoO_4 \cdot 2H_2O$	137 g ha^{-1}
	Soil	$Na_2MoO_4 \cdot 2H_2O$	1310 g ha^{-1}

[a]Responses to Mo application were observed.
Source: Modified from Adams (1997).

In general, Mo deficiencies are corrected with a soil application or seed treatment. For example, over 452×10^3 ha of soybeans are planted with Mo-treated seeds each year on acid unlimed soils in Arkansas. However, a foliar spray can be employed when the deficiency occurs during the growing season. Because seed treatment can be more uniformly applied and proved to be more effective, it has become probably the most commonly used technique for correcting Mo deficiency in crops. However, seed treatment tends to increase the Mo content of the plant much more than soil treatment (Reisenauer, 1963). In acid soils, foliar, soil, or seed treatment proved to be equally effective in correcting Mo deficiency (Sedberry et al., 1973). In this case, seed treatment was more economical, because of the smaller amount of Mo required. However, seed treatments can vary from about 50 g ha^{-1} to as much as 900 g ha^{-1} of Mo (Murphy and Walsh, 1972). Most work indicates that soil treatments with 50 to 100 g ha^{-1} of Mo can satisfactorily correct most Mo deficiencies. Certain species, especially members of the Brassicaceae family such as cauliflower may need as much as 450 g ha^{-1} of Mo.

Except in acid soils with large amount of Fe and Al oxides, a single application of Mo at the rate of 143 g ha^{-1} can remain effective for several years for most crops (Anderson, 1956).

8.2 Sewage Sludge

Molybdenum is frequently found in municipal sewage sludge at levels above the natural soil concentration. The mean Mo concentration in sewage sludge from 16 U.S. cities was about 15 ppm, with the range of 1.0 to 40 ppm (Appendix Table A.9). Similar data were obtained for Canadian cities, with an average of 16 ppm (Van Loon and Lichwa, 1973). In a survey of nine states, Dowdy et al. (1976) reported a median of 30 ppm, with a range of 5 to 39 ppm. Recently, O'Connor (2000) indicated an average Mo content in U.S. sludge of 20 to 30 ppm DW. Since the average concentration of Mo in sludge is much higher than that in normal soils, it can be expected that application of high rates of sludge can increase soil Mo levels, and hence plant uptake of Mo. Molybdenum uptake by plants, especially those intended for livestock feedstuff, can be a major concern especially in limed sludged soils where the usual concern revolves around Cd and other heavy metals. Both Cd and Mo are highly pH sensitive in their behavior in soils but in contrasting order. In pot culture, adding sewage sludge up to 4% by weight of soil showed enrichment of plants in Mo (1.6 ppm in barley leaf with 4% addition vs. 0.5 ppm in control) (Jarrell et al., 1980). Because of the danger in ruminants grazing on forage crops grown on high-Mo-soils developing *molybdenosis,* recent research suggests that the cumulative Mo loadings on forages grown on nonacidic soils should not exceed 1.0 kg ha^{-1} for limed sludge, or 4 kg ha^{-1} for dewatered sludge (McBride et al., 2000).

8.3 Coal Combustion

As much as 1000 tonnes Mo yr^{-1} has been estimated to be potentially mobile from the combustion of coal for power generation in the United States (Straughan et al., 1978). As much as 15 tonnes Mo yr^{-1} can be emitted into the surrounding environment for every 1000 MW of power produced (Schwitzgebel et al., 1975). The amount of aerial release is largely determined by the type of emission control devices, with very little released when electrostatic precipitators are used.

The potential enrichment of the terrestrial environment with Mo arises primarily from coal combustion residues, now amounting to over 100×10^6 tonnes yr^{-1} (Adriano and Weber, 1998). Approximately 65×10^6 tonnes of this amount is fly ash. Most of these residues are destined for landfill disposal, with a fraction used for agricultural production. Its potential use in agriculture is primarily due to its nutrient content and liming potential. However, fly ash is known to provide readily available forms of B, Mo, Se, and As, which can result in toxic levels for plants or animals fed these plants (Carlson and Adriano, 1993). Substantial enrichment of plant tissues grown on fly ash–treated soils with Mo has been demonstrated many times over (Adriano and Weber, 1998). In addition, fly ash–treated soils may produce plants having low Cu/Mo ratios, suggesting problems for this type of plant when intended for animal use. A survey of 21 states revealed Mo average concentrations of 15.0 to 25.4 ppm in fly ash produced from various coal ranks (Furr et al., 1977). Fly ash derived from bituminous coals tends to contain higher levels of Mo. Molybdenum-deficient soils may benefit from low applications (<40 g kg^{-1} soil) of fly ash, but Mo levels in plants growing in such soils should be continuously monitored (Elseewi and Page, 1984).

8.4 Mining and Smelting

Research has demonstrated that cattle pastured near a Mo-processing complex in Chile suffered from severe *molybdenosis* (Parada et al., 1979). Parada and Covarrubias (1980) reported that the Mo contents of soils and plants surrounding this industrial area were excessive and inversely correlated with distance from the smokestack. Recent analysis of grass samples from the area indicates that cattle grazing in this area could suffer from *hypocuprosis* due to improper Cu/Mo ratio in the tissue (Schalscha et al., 1987).

Mining and processing of Mo is a potential source of Mo release into the environment. The source would be very likely from the drainage system adjoining the mine (Chappell, 1975). Mining activities in the U.S. Rocky Mountains have increased the Mo levels in some of the streams draining the mined areas. Concentrations of Mo in Colorado River waters higher than 5 ppb are generally associated with Mo mineralization, whereas levels above 60 ppb are associated with mining activities (Vlek and Lindsay, 1977). Use of high-Mo waters to irrigate pastures can adversely affect the quality of herbage produced as shown in Colorado (Jackson et al., 1975; Vlek and Lindsay, 1977).

References

Adams, J.F. 1997. *Molybdenum in Agriculture*. Cambridge Univ Pr, Cambridge.

Adriano, D.C. 1986. *Trace Elements in the Terrestrial Environment*. Springer-Verlag, New York.

Adriano, D.C., and J. Weber. 1998. *Coal Ash Utilization for Soil Amendment to Enhance Water Relations and Turf Growth*. EPRI Rep TR-111318, Electric Power Res Inst Palo Alto, CA.

Adriano, D.C., A.L. Page, A.A. Elseewi, A.C. Chang, and I. Straughan. 1980. *J Environ Qual* 9:333–344.

Ahlrichs, L.E., R.G. Hanson, and J.M. MacGregor. 1963. *Agron J* 55:484–486.

Ahmed, M.S., and L. Rahman. 1982. *Plant Soil* 69:287–291.

Allaway, W.H. 1968. *Adv Agron* 20:235–274.

Allaway, W.H. 1975. In *Proc Intl Conf Heavy Metals in the Environment*. Univ. of Toronto, Ontario, Canada.

Allaway, W.H. 1977. In W.R. Chappell and K.K. Petersen, eds. *Molybdenum in the Environment*. Marcel Dekker, New York.

Alloway, B.J. 1973. *J Agric Sci* 80:521–524.

Anderson, A.J. 1942. *J Aust Inst Agric Sci* 8:73–75.

Anderson, A.J. 1956. *Adv Agron* 8:163–202.

Anderson, A.J., and A.C. Oertel. 1946. *Council Sci Ind Res Bull* 198(Part 2):25–44.

Anderson, R.J. 1966. *Brit Dent J* 120:271–275.

Arnon, D.I., and P.R. Stout. 1939. *Plant Physiol* 14:599–602.

Aubert, H., and M. Pinta. 1977. *Trace Elements in Soils*. Elsevier, New York.

Baes, C.F., and R.E. Mesmer. 1976. *The Hydrolosis of Cations*. Wiley, New York.

Barrow, N.J. 1970. *Soil Sci* 109:282–288.

Barshad, I. 1948. *Soil Sci* 66:187–195.

Barshad, I. 1951. *Soil Sci* 71:297–313.

Basak, A., L.N. Mandal, and M. Haldar. 1982a. *Plant Soil* 68:261–269.

Basak, A., L.N. Mandal, and M. Haldar. 1982b. *Plant Soil* 68:271–278.

Bergmann, W. 1992. In *Nutritional Disorders of Plants: Visual and Analytical Diagnosis*. Gustav Fisher Verlag, Jena, Germany.

Bingham, F.T., and M.J. Garber. 1960. *Soil Sci Soc Am Proc* 24:209–213.

Blossom, J.W. 1993. In *U.S. Bureau of Mines Annual Report* 1993. U.S. Dept Interior, Washington, DC.

Bortels, H. 1930. *Arch Mikrobiol* 1:333–342.

Boswell, F. 1980. In *World Soybean Research Conference*. Westview Pr, Boulder, CO.

Bowen, H.J.M. 1979. *Environmental Chemistry of the Elements*. Academic Pr, New York.

Brookins, D.G. 1987. *Eh–pH Diagrams for Geochemistry*. Springer-Verlag, New York.

Burmester, C.H., J.F. Adams, and J.W. Odom. 1988. *Soil Sci Soc Am J* 52:1394.

Carlson, C., and D.C. Adriano. 1993. *J Environ Qual* 22:227–247.

Chappell, W.R. 1975. In P.A. Krenkel, ed. *Heavy Metals in the Aquatic Environment*. Pergamon Pr, New York.

Chen, J., F. Wei, Y. Wu, and D.C. Adriano. 1991. *Water Air Soil Pollut* 57–58:699–712.

Chojnacka, J. 1963. *Rocz Chem* 37:259–272.

Connor, J.J., and H.T. Shacklette. 1975. USGS Prof Paper 574-F. U.S. Geological Survey, Washington, DC.

Cox, F.R., and E.J. Kamprath. 1972. In J.J. Mortvedt, P.M. Geordano, and W.L. Lindsay, eds. *Micronutrients in Agriculture*. Soil Sci Soc Am, Madison, WI.

Czerniejewski, C.P., C.W. Shank, W.G. Bechtel, and W.B. Bradley. 1964. *Cereal Chem* 41:65–72.

Davies, E.B. 1956. *Soil Sci* 81:209–221.

Davies, E.B., and J.L. Grigg. 1953. *NZ J Exp Agric* 87:561–567.

Davis, R.D., P.H.T. Beckett, and E. Wollan. 1978. *Plant Soil* 49:395–408.

Dick, A.T. 1956. *Soil Sci* 81:229–236.

Domingo, L.E., and K. Kyuma. 1983. *Soil Sci Plant Nutr* 29:439–452.

Dowdy, R.H., W.E. Larson, and E. Epstein. 1976. In *Land Application of Waste Materials*. Soil Conserv Soc Am, Ankeny, IA.

Doyle, P., K. Fletcher, and V.C. Brink. 1972. *Trace Subs Environ Health* 6:369–375.

Doyle, P.J., and K. Fletcher. 1977. In W.R. Chappell and K.K. Petersen, eds. *Molybdenum in the Environment*. Marcel Dekker, New York.

Elseewi, A.A., and A.L. Page. 1984. *J Environ Qual* 13:394–398.

Evans, H.J., E.R. Purvis, and F.E. Bear. 1950a. *Soil Sci* 71:117–124.

Evans, H.J., E.R. Purvis, and F.E. Bear. 1950b. *Plant Physiol* 25:555–566.

Fleischer, M. 1972. *Ann NY Acad Sci* 199:6–16.

Food and Nutrition Board. 1989. *Recommended Daily Allowances*. National Research Council, Washington, DC.

Franco, A.A., and J.M. Day. 1980. *Turrialba* 30:99–105.

Franco, A.A., and D.N. Munns. 1981. *Soil Sci Soc Am J* 45:1144–1148.

Furr, A.K., T.F. Parkinson, R.A. Hinrichs, D.R. Van Campen et al. 1977. *Environ Sci Technol* 11:1104–1112.

Goldberg, S., H.S. Forster, and C.L. Godfrey. 1996. *Soil Sci Soc Am J* 60:425–432.

Goldschmidt, V.M. 1954. *Geochemistry*. Oxford Univ Pr, London.

Golov, V.I., and Y.N. Kazakhkov. 1973. *Sov Soil Sci* 5:551–558.

Gonzales, R., H. Appelt, E.B. Schalscha, and F.T. Bingham. 1974. *Soil Sci Soc Am Proc* 38:903–906.

Grigg, J.L. 1953. *NZ J Sci Technol* A 34:405–414.

Grigg, J.L. 1960. *NZ J Agric Res* 3:69–86.

Gupta, U.C. 1997. *Molybdenum in Agriculture*. Cambridge Univ Pr, Cambridge.

Gupta, U.C., and J. Lipsett. 1981. *Adv Agron* 34:73–115.

Hagstrom, G.R. 1977. *Fert Solutions* (July–Aug):18–28.

Haley, L.E., and S.W. Melsted. 1957. *Soil Sci Soc Am Proc* 21:316–319.

Hamilton, E.I., and M.J. Minski. 1972. *Sci Total Environ* 1:375–394.

Hart, L.I., E.C. Owen, and R. Proudfoot. 1967. *Br J Nutr* 21:617–630.

Harter, R.D. 1991. In J.J. Mortvedt et al., eds. *Micronutrients in Agriculture,* 2nd ed. SSSA 4, Soil Sci Soc Am, Madison, WI.

Healy, W.B., T.G. Ludwig, and F.L. Losee. 1961. *Soil Sci* 29:359–366.

Hern, J.D. 1989. *Study and Interpretation of Chemical Characteristics of Natural Waters,* 3rd ed. USGS Paper 2254. U.S. Geological Survey, Reston, VA.

Hewitt, E.J. 1956. *Soil Sci* 81:159–171.

Hunter, J.G., and O. Vergnano. 1953. *Ann Appl Biol* 40:761–777.

Jackson, D.R., W.L. Lindsay, and R.D. Heil. 1975. *J Environ Qual* 4:223–229.

James, D.W., T.L. Jackson, and M.E. Hayward. 1968. *Soil Sci* 105:397–402.

Jarrell, W.M., and M.D. Dawson. 1978. *Soil Sci Soc Am J* 42:412–415.

Jarrell, W.M., A.L. Page, and A.A. Elseewi. 1980. *Residue Rev* 74:1–43.

Jenkins, I.L., and A.G. Wain. 1963. *J Appl Chem (Lond)* 13:561–564.

Jensen, H.L. 1941. *Aust J Sci* 3:98–99.

Joham, H.E. 1953. *Plant Physiol* 28:275–280.

Johnson, C.M. 1966. In H.D. Chapman, ed. *Diagnostic Criteria for Plants and Soils.* Univ Calif Div Agric Sci, Berkeley, CA.

Johnson, C.M., G.A. Pearson, and P.R. Stout. 1952. *Plant Soil* 4:178–196.

Jones, G.B., and G.B. Belling. 1967. *Aust J Agric Res* 18:733–740.

Jones, J.B., Jr. 1972. In J.J. Mortvedt, P.M. Giordano, and W.L. Lindsay, eds. *Micronutrients in Agriculture.* Soil Sci Soc Am, Madison, WI.

Jones, L.H.P. 1956. *Science* 123:1116.

Jones, L.H.P. 1957. *J Soil Sci* 8:313–327.

Karimian, N., and F.R. Cox. 1978. *Soil Sci Soc Am J* 42:757–761.

Kim, K.W., and I. Thornton. 1993. *Environ Geochem Health* 15:119–133.

Krauskopf, K.B. 1979. *Introduction to Geochemistry,* 2nd ed. McGraw-Hill, New York.

Kretschmer, A.E., and R.J. Allen. 1956. *Soil Sci Soc Am Proc* 20:253–257.

Kubota, J. 1972. In H.C. Hopps and H.L. Cannon, eds. *Geochemical Environment in Relation to Health and Disease.* Geological Society of America, Boulder, CO.

Kubota, J. 1977. In W.R. Chappell and K.K. Petersen, eds. *Molybdenum in the Environment.* Marcel Dekker, New York.

Kubota, J., V.A. Lazar, L.N. Langan, and K.C. Beeson. 1961. *Soil Sci Soc Am Proc* 25:227–232.

Kubota, J., E.R. Lemon, and W.H. Allaway. 1963. *Soil Sci Soc Am Proc* 27:679–683.

Kubota, J., V.A. Lazar, G.H. Simonson, and W.H. Hill. 1967. *Soil Sci Soc Am Proc* 31:667–671.

Kummer, J.T. 1980. In *Minerals Yearbook, vol. 1: Metals and Minerals.* U.S. Bureau Mines, U.S. Dept Interior, Washington, DC.

Lander, H.N. 1977. In W.R. Chappell and K.K. Petersen, eds. *Molybdenum in the Environment.* Marcel Dekker, New York.

Lavy, T.L., and S.A. Barber. 1964. *Soil Sci Soc Am Proc* 28:93–97.

Lewis, A.H. 1943a. *J Agric Sci* 33:52–57.

Lewis, A.H. 1943b. *J Agric Sci* 33:58–63.

Lindsay, W.L. 1972. In J.J. Mortvedt, P.M. Giordano, and W.L. Lindsay, eds. *Micronutrients in Agriculture.* Soil Sci Soc Am, Madison, WI.

Lindsay, W.L. 1979. *Chemical Equilibria in Soils.* Wiley, New York.

Lipsett, J., and J.R. Simpson. 1971. *J Aust Inst Agric Sci* 37:348–351.

Liu, Z., Q.Q. Zhu, and L.H. Tang. 1983. *Soil Sci* 135:40–46.

Lucas, R.E. 1967. In *Micronutrients for Vegetables and Field Crops.* Ext Bull E-486. Michigan State Univ, East Lansing, MI.

Lucas, R.E., and B.D. Knezek. 1972. In J.J. Mortvedt, P.M. Giordano, and W.L. Lindsay, eds. *Micronutrients in Agriculture.* Soil Sci Soc Am, Madison, WI.

Lyon, C.B., and K.C. Beeson. 1948. *Botan Gaz* 109:506–520.

MacLean, K.S., and W.M. Langille. 1973. *Commun Soil Sci Plant Anal* 4:495–505.

Manning, B.A., and S. Goldberg. 1996. *Soil Sci Soc Am J* 60:121–131.

Marclise, N.A., C.B. Ammerman, R.M. Valsecchi, D.G. Dunavant, and G.K. Davis. 1970. *J Nutr* 100:1399–1406.

Massey, H.F., R.H. Lowe, and H.H. Bailey. 1967. *Soil Sci Soc Am Proc* 31:200–202.

McBride, M.B., B.K. Richards, T. Steenhuis, and G. Spiers. 2000. *J Environ Qual* 29:848–854.

Melsted, S.W., H.L. Motto, and T.R. Peck. 1969. *Agron J* 61:17–20.

Mikkonen, A., and J. Tummavouri. 1993. *Acta Agric Scand* 43:206–212.

Millikan, C.R. 1949. *Proc R Soc Victoria* 61:25–42.

Miltimore, J.E., and J.L. Mason. 1971. *Can J Anim Sci* 51:193–200.

Mitchell, R.L. 1964. In F.E. Bear, ed. *Chemistry of the Soil.* Am Chem Soc 160, Am Chem Soc, Reinhold, New York.

Mortvedt, J.J., and O.E. Anderson, eds. 1982. *Forage Legumes: Diagnosis and Correction of Molybdenum and Manganese Problems.* Southern Coop Ser Bull 278. Univ of Georgia, Athens, GA.

Murphy, L.S., and L.M. Walsh. 1972. In J.J. Mortvedt, P.M. Giordano, and W.L. Lindsay, eds. *Micronutrients in Agriculture.* Soil Sci Soc Am, Madison, WI.

National Research Council. 1984. In *Nutrient Requirements for Domestic Animals.* National Academy of Science, Washington, DC.

Neathery, M.W., and W.J. Miller. 1977. *Feedstuff* (Aug):18–20.

O'Connor, G. 2000. *Environ Sci Technol:* News p. 243 A.

Parada, R., and L. Covarrubias. 1980. *Arch Med Vet* 12:229–232.

Parada, R., J. Torres, L. Covarrubias, and A. Rivas. 1979. *Arch Med Vet* 11:76–79.

Pasricha, N.S., V.K. Nagger, and R. Singh. 1997. *Molybdenum in Agriculture.* Cambridge Univ Pr, Cambridge.

Piper, C.S. 1940. *J Aust Inst Agric Sci* 6:162–164.

Pratt, P.F., and F.L. Bair. 1964. *Agric Chem* 19:39.

Reddy, K.J., and S.P. Gloss. 1993. *Appl Geochem* 2:159–163.

Reisenauer, H.M. 1963. *Agron J* 55:459–460.

Reisenauer, H.M., A. Tabikh, and P.R. Stout. 1962. *Soil Sci Soc Am Proc* 26:23–27.

Reyes, E.D., and J.J. Jurinak. 1967. *Soil Sci Soc Am Proc* 31:637–641.

Riley, M.N., A.D. Robson, J.W. Gartrell, and R.C. Jeffery. 1987. *Aust J Soil Res* 25:179–184.

Rinne, S.L., M. Sillanpaa, E. Houkuna, and S.R. Hiivola. 1974. *Ann Agric Fenniae* 13:109–118.

Robinson, W.O., and L.T. Alexander. 1953. *Soil Sci* 75:287–291.

Robinson, W.O., and G. Edgington. 1954. *Soil Sci* 77:237–251.

Robinson, W.O., G. Edgington, W.H. Armiger, and A.V. Breen. 1951. *Soil Sci* 72:267–274.

Roy, W.R., J.J. Hassett, and R.A. Griffin. 1986. *Soil Sci Soc Am J* 50:1176–1182.

Sauchelli, V. 1969. *Trace Elements in Agriculture.* Reinhold, New York.

Schalscha, E.B., M. Morales, and P.F. Pratt. 1987. *J Environ Qual* 16:313–315.

Schwitzgebel, K., F.B. Meserole, R.G. Oldham, et al. 1975. In *Proc Intl Conf Heavy Metals in the Environment.* Univ of Toronto, Ontario, Canada.

Sedberry, J.E., T.S. Dharmaputra, R.H. Brupbacher, et al. 1973. *Louisiana Agric Exp Sta Bull* 670:3–39.

Severson, R.C., and H.T. Shacklette. 1988. In USG Circular 1017. U.S. Geological Survey Reston, VA.

Shacklette, H.T., and J.G. Boerngen. 1984. *Element Concentrations in Soils and Other Surficial Materials of the Conterminous United States.* USGS Prof Paper 1270. U.S. Geological Survey, Washington, DC.

Sims, J.T., and G.V. Johnson. 1991. In J.J. Mortvedt et al., eds. *Micronutrients in Agriculture,* 2nd ed. SSSA 4, Soil Sci Soc Am, Madison, WI.

Singh, M., and V. Kumar. 1979. *Soil Sci* 127:307–312.

Smith, C., K.W. Brown, and L.E. Deuel. 1987. *J Environ Qual* 16:377–382.

Smith, K.S., L.S. Balistrieri, S.M. Smith, and R.C. Severson. 1997. *Molybdenum in Agriculture.* Cambridge Univ Pr, Cambridge.

Sposito, G. 1984. *The Surface Chemistry of Soils.* Oxford Univ Pr, Oxford.

Srivastava, P.C. 1997. In U.C. Gupta, ed. *Molybdenum in Agriculture.* Cambridge Univ. Pr, Cambridge.

Stewart, I., and C.D. Leonard. 1952. *Nature* 170:714–715.

Stewart, I., and C.D. Leonard. 1953. *Am Soc Hort Sci Proc* 62:111–115.

Stout, P.R., and W.R. Meagher. 1948. *Science* 108:471–473.

Stout, P.R., W.R. Meagher, G.A. Pearson, and C.M. Johnson. 1951. *Plant Soil* 3:51–87.

Straughan, I., A.A. Elseewi, and A.L. Page. 1978. *Trace Subs Environ Health* 12:389–402.

Stumm, W. 1992. *Chemistry of the Solid-Water Interface.* Wiley, New York.

Swaine, D.J., and R.L. Mitchell. 1960. *J Soil Sci* 11:347–368.

Takahashi, T. 1972. *Plant Soil* 36:665–674.

Thornton, I. 1977. In W.R. Chappell and K.K. Petersen, eds. *Molybdenum in the Environment.* Marcel Dekker, New York.

Tipton, I.H., P.L. Stewart, and P.G. Martin. 1966. *Health Phys* 12:1683–1690.

Tsongas, T., R. Meglew, P. Walravens, and W. Chappell. 1980. *Am J Clin Nutr* 33:1103–1107.

Turner, D.R., M. Whitfield, and A.G. Dickson. 1981. *Geochim Cosmochim Acta* 45:855–881.

Underwood, E.J. 1973. In *Toxicants Occurring Naturally in Foods.* National Academy of Sciences, Washington, DC.

Van Campen, D.R. 1991. In J.J. Mortvedt et al., eds. *Micronutrient in Agriculture,* 2nd ed. SSSA 4, Soil Sci Soc Am, Madison, WI.

Van Loon, J.C., and J. Lichwa. 1973. *Environ Lett* 4:1–8.

Vlek, P.L.G., and W.L. Lindsay. 1977. In W.R. Chappell and K.K. Petersen, eds. *Molybdenum in the Environment.* Marcel Dekker, New York.

Walker, T.W., A.F.R. Adams, and H.D. Orchiston. 1955. *Plant Soil* 6:201–220.

Wallace, A., E.M. Romney, G.V. Alexander, and J. Kinnear. 1977. *Commun Soil Sci Plant Anal* 8:741–750.

Walsh, T., M. Neenan, and L.B. O'Moore. 1952. *J Dept Agric* (Ireland) 48:32–43.

Warington, K. 1954. *Ann Appl Biol* 41:1–22.

Welch, R.M., W.H. Allaway, W.A. House, and J. Kubota. 1991. In J.J. Mortvedt et al., eds. *Micronutrients in Agriculture,* 2nd ed. SSSA 4, Soil Sci Soc Am, Madison, WI.

Widdowson, J.P. 1966. *NZ J Agric* 9:59–67.

Williams, C., and I. Thornton. 1972. *Plant Soil* 36:395–406.

Williams, C., and I. Thornton. 1973. *Plant Soil* 39:149–159.

Williams, J.H. 1971. In *Trace Elements in Soils and Crops*. Minist Agric Fish Food (UK), Tech Bull 21. London.

Wright, J.R., R. Levick, and H.J. Atkinson. 1955. *Soil Sci Soc Am Proc* 19:340–344.

Wynne, K.N., and G.L. McClymont. 1955. *Nature* 175:471–472.

Zborishchuk, Y.N., and N.G. Zyrin. 1974. *Agrokhimiya* 3:88–93.

16
Zinc

1 General Properties of Zinc

Zinc (atom. no. 30) is a bluish white, relatively soft metal with a density of 7.13 g cm^{-3}. It belongs to Group II-B of the periodic table and has an atom. wt. of 65.38 and melting pt. of 420 °C. Zinc is divalent in all its compounds. It is a composite of five stable isotopes: ^{64}Zn, ^{66}Zn, ^{67}Zn, ^{68}Zn, and ^{70}Zn. Their relative abundances are: 48.89, 27.81, 4.11, 18.56, and 0.62%, respectively. Six radioactive isotopes have been identified: ^{62}Zn, ^{63}Zn, ^{65}Zn, ^{69}Zn, ^{72}Zn, and ^{73}Zn, with ^{65}Zn, ($t_{1/2} = 245$ days) and ^{69}Zn ($t_{1/2} = 55$ min) being the most commonly used.

The Zn^{2+} ion is colorless and exists in a hydrated form in acidic and neutral aqueous solutions; however, the hydroxide is precipitated in alkaline solution. With excess base, the hydroxide redissolves to form zincate ion, $Zn(OH)_4^{2-}$. Due to its amphoteric nature, Zn forms a variety of salts; the chlorates, chlorides, sulfates, and nitrates are readily soluble in water whereas the oxides, carbonates, phosphates, silicates, the sulfides are relatively insoluble in water. In dry air Zn oxidizes, and in moist air a basic carbonate ($2 \, ZnCO_3 \cdot 3H_2O$) is formed on the surface thereby protecting the metal from corrosion. The oxidation state of Zn in nature is II.

2 Production and Uses of Zinc

Most of the Zn produced globally comes from ores containing Zn sulfide minerals. Although more than 80 Zn minerals are known, there are only a few important commercial ores. The principal ores are the sulfides sphalerite and wurtzite and their weathering products, mainly smithsonite ($ZnCO_3$) and hemimorphite [$Zn_4Si_2O_7(OH)_2 \cdot H_2O$]. The more common ore minerals of Zn, their compositions, and their geographic locations are discussed in Adriano (1986).

In 1930, world mine production was about 1.6 million tonnes with Australia, Canada, Germany, Mexico, and the United States providing about 75% of the total (Adriano, 1986). By 1950, production was 2.15 million tonnes with Australia, Canada, Mexico, the United States, and the former

Soviet Union producing 65% of the total. By 1979 world mine production was almost 6.0 million tonnes with the countries above as major producers. The United States was the largest Zn metal producer in the world from 1901 to 1971 but has substantially depended on imports for its smelter operations since 1941. However, from 1950 to 1977, Canada's growth dramatically increased to become the world's largest producer.

Zinc ranks fourth among metals of the world in annual consumption behind Fe, Al, and Cu. Zinc is mainly used as a protective coating for iron and steel. The automobile industry accounts for almost one-third of United States slab Zn consumption. Zinc is also extensively used as a protective coating on a number of metals to prevent corrosion and in alloys such as brass and bronze. Galvanized metals have a variety of applications in the building, transportation, and appliance industries. Galvanized pipes are commonly used in domestic water delivery systems and Zn solubilized by corrosion is thought to make some contribution to Zn enrichment in waste waters. Zinc and its compounds are ingredients of many household items including utensils, cosmetics, powders, ointments, antiseptics and astringents, paints, varnishes, linoleum, and rubber. They are also used in the manufacture of parchment papers, glass, automobile tires, television screens, dry cell batteries, and electrical apparatus. Other uses include agricultural micronutrient fertilizers, insecticides, hardeners in cement and concrete, and in printing and drying of textiles. They are also used in the production of adhesives, as a flux in metallurgical operations, and as a wood preservative.

3 Zinc in Nature

Zinc is the 24th most abundant element in the earth's crust, with the average value at $\sim$70 ppm (Krauskopf, 1979). Zinc is infrequently present in metamorphic and igneous rocks as the sulfide (sphalerite) but most of it is distributed as a minor constituent of rock-forming minerals, especially those rich in iron such as magnetite (Fe_3O_4), the pyroxenes [$(Mg,Fe)_2Si_2O_6$ and $Ca(Mg,Fe)Si_2O_6$], the amphiboles such as $Ca_2(Mg,Fe)_5Si_8O_{22}(OH)_2$, biotite, spinel, garnet, and staurolite. Some of the rock-forming minerals have the following Zn concentrations, in ppm: magnetite (<25 to 2500), olivine (50 to 82), garnet (<30 to 5275), staurolite (2000 to 6000), pyroxene (<30 to 2250), amphibole (34 to 8900), and biotite (40 to 2540). Zinc abundance in different minerals is influenced by the Zn concentration of the magma and premetamorphic rock, and the ability of the crystal lattice to incorporate this element (Wedepohl, 1978). The most common impurities in Zn minerals are Fe, Cd, and Pb. Usually Cd is about 0.05% as abundant as

TABLE 16.1. Commonly observed zinc concentrations (ppm) in various environmental media.

Material	Average concentration	Range
Igneous rocks[b]	65	5–1070
Limestone[c]	20	<1–180
Sandstone[c]	30	5–170
Shale[c]	97	15–1500
Petroleum[d]	30	—
Coal[d]	50	3–300
Coal ash[e]		
Fly ash	449	27–2880
Bottom ash	127	4–575
FGD sludge	100	8–612
Oil ash[e]	1140	404–2310
Lime[f]	6	<5–8
Phosphate fertilizers[f]	305	40–600
Organic wastes[f]	390	8–1600
Sewage sludge[g]	2250	1000–10000
Soils[d]	90	1–900
Herbaceous vegetables[d]	—	1–160
Common crops[h]	—	6–200
Grasses and monocots[i]	—	10–60
Common fruit trees[h]	—	2–200
Conifers[i]	—	15–65
Ferns	—	30–40
Mosses[i]	—	20–45
Lichens[d]	—	20–60
Freshwater[d]	15[a]	<1–100[a]
Seawater[d]	5[a]	<1–48[a]

Sources: Extracted from
[a] $\mu g \, L^{-1}$.
[b] NRC (1979).
[c] Wedepohl (1978).
[d] Bowen (1979).
[e] Ainsworth and Rai (1987).
[f] Whitton and Wells (1974).
[g] Page (1974).
[h] Chapman (1966).
[i] Boardman and McGuire (1990).

Zn; processing 1 tonne of Zn ore produces about 3 kg of Cd. Examples of the extreme variability in Zn concentrations of soil-forming rocks are presented in Appendix Table A.5. Zinc is very variable in environmental media (Table 16.1).

Forest ecosystems can be viewed as nutrient-element conserving systems controlled by nutrient bioavailability and climatic constraints. Nutrient cycling processes in forests collectively constitute a closed cycle in which all major processes are critical for the system to be maintained. Disruption of one of these processes can adversely affect forest productivity, such as when trace elements in these ecosystems become enriched from anthropogenic input. These elements become airborne from industrial operations, are transported over distant areas, and are eventually deposited on the earth's surface, usually with precipitation. Cadmium, Zn, Hg, and Pb are some metals likely to be detected in elevated concentrations in natural ecosystems even in remote areas.

The distribution and cycling of Cd, Pb, and Zn in a mixed deciduous forest in eastern Tennessee have been determined (Van Hook et al., 1977). The major anthropogenic sources of atmospheric emissions in the vicinity of this watershed are three coal-fired electric power generating stations with a combined coal consumption of 7×10^6 tonnes yr^{-1}. Annual discharges of Zn from these plants were estimated at approximately 9.5 tonnes yr^{-1}. Despite this input, this system is still considered as relatively unpolluted. Concentrations of Zn in living vegetation generally followed the pattern: roots > foliage > branch > bole (Table 16.2). Zinc concentrations generally were higher in *Carya* spp. and *Quercus velutina* than the other species

TABLE 16.2. Zinc concentrations (ppm) of tree species from relatively nonpolluted areas.

Species	Needles or leaves	Twigs (wood & bark)	Branches (wood & bark)	Trunk (wood)	Trunk (bark)	Roots (bark & wood)
Hubbard Brook Experimental Forest, New Hampshire[a]						
Sugar maple	52	62	19	8	29	46
Yellow birch	334	226	154	32	259	118
American beech	25	40	21	7	10	47
Mt. maple	45	63	45	12	97	75
Red spruce	21	40	59	12	103	119
Pine cherry	24	22	27	9	30	28
University Forest, Maine[b]						
Red spruce	45	46	44	8	50	67
Balsam fir	50	67	48	11	45	40
Hemlock	10	29	28	2	15	13
White pine	52	68	55	11	65	22
White birch	77	91	100+	28	99	100+
Red maple	41	49	48	29	78	69
Aspen	100+	88	100+	17	97	47
Walker Branch Watershed, Tennessee[c]						
Shortleaf pine	10	—	6	4	—	29
Chestnut oak	14	—	1	<1	—	17
Red maple	19	—	13	8	—	—
Hickory	36	—	11	14	—	—
Yellow poplar	15	—	5	3	—	29
White oak	18	—	4	3	—	—
Northern red oak	17	—	3	3	—	—
Tupelo	15	—	10	6	—	—
Dogwood	16	—	12	5	—	—
Black oak	28	—	10	5	—	—

Sources: Extracted from
[a]Likens and Bormann (1970).
[b]Young and Guinn (1966).
[c]Van Hook et al. (1977).

examined. Generally, Zn concentrations in O2 litter were higher than those in the O1 litter (Adriano, 1986). The A1 soil horizon contained the highest concentration of Zn, which was associated with its high OM content. The standing pools of Zn for the soil profile (100 cm deep) averaged ~354 kg ha^{-1}. The metal inventory in the watershed indicates that the soil is the major sink for trace elements. Similar results were obtained in the Solling Mountains in central Germany where 262 to 315 kg ha^{-1} of Zn were found in the soil (0 to 50 cm), leaving only 1.5 to 4.5% of the total inventory contained in the vegetation (Heinrichs and Mayer, 1980). In tropical moist forest, soil (0 to 30 cm) inventory for Zn is much lower (average of 134 kg ha^{-1} for two sites), but the amount contained in the vegetation (~5% of total) is still comparable with the other ecosystems (Golley et al., 1975).

In another relatively unpolluted forested ecosystem in New Hampshire (Reiners et al., 1975), it was found that the concentrations of Zn in organic horizons of soils distributed over an elevational gradient fell within a broad range of levels to represent natural conditions. The Zn levels in these organic layers compare favorably with those from other ecosystems. Why Zn is sometimes lower in the mostly decomposed organic layer than the least decomposed layer in this particular ecosystem is not well understood. In forest stands along an elevational gradient on Camels Hump Mountain, northern Vermont, Friedland et al. (1984) found that Zn accumulations in the forest floor during the period 1966 to 1980 in the northern hardwood, transition, and boreal forests were 0.18, 0.41, and 0.09 kg Zn ha^{-1} yr^{-1}, respectively. The accumulated amounts of Zn in the forest floor prior to this period were estimated to be 6.08, 2.13, and 6.97 kg Zn ha^{-1} in the respective stands. Zinc concentration in the profile of the forest floor of the boreal forest decreased significantly with depth, with the highest concentration occuring in the F layer (103 ppm in 0 to 2 cm depth vs. 67 ppm in the 8 to 10 cm depth).

Zinc concentrations for various tree species in two forest ecosystems in New England are shown in Table 16.2. The data indicate that birch might be an accumulator of this element.

Forest ecosystems close to diffuse or point sources have markedly higher concentrations of trace elements than similar ecosystems in rural areas. For example, Zn concentrations in the surface litter (O1 + O2 horizons) for an urban site near Chicago (Parker et al., 1978) were about 10 times greater than those reported by Reiners et al. (1975) for the New Hampshire forests. The bulk of the metals in the urban site is retained in the soil layers that contain appreciable OM. Smith (1973) found that Zn in six woody species in New Haven, Connecticut was present in above-normal amounts. In the Belgian Kempen area, the affected forest surface soils had a mean Zn concentration of 38 ppm compared with the normal content of 5 to 15 ppm (Bosmans and Paenhuys, 1980). In white pine stands in central Massachusetts, Siccama et al. (1980) found that the total Zn content of the forest floor increased by 0.71 kg ha^{-1} after 16 yr. However, Zn concentrations in both the F and H layers decreased significantly, which was attributed a primarily to a significant increase in forest floor total dry weight.

Long-term enrichment of the soil surface from atmospheric deposition can contribute to higher uptake of Zn by plants. This is illustrated by data from central Norway and southern Norway, which is closer to the upwind source (Table 16.3). Rates of atmospheric heavy metal deposition in central Norway are several times lower than in southern Norway and both areas have experienced a strong decline in deposition during the last 15 yr (Berthelsen et al., 1995). This decline in input is not well reflected by the vegetation data for the 1982 and 1992 periods indicating that plant uptake from enriched surface soil predominates over direct deposition on the canopy.

In an Adirondack lake in New York, Zn did not appear to be significantly retained in this ecosystem (Fig. 16.1). Annual Zn influx closely approximates the efflux. Additionally, gross Zn deposition (D_g) and net Zn deposition (D_n) suggest very little in-lake retention of Zn (White and Driscoll, 1987). The lake water pH hovers around 5.3; soluble Zn concentrations averaged $0.30 \pm 0.15 \; \mu ML^{-1}$ ($n = 160$) which is higher than total metal concentrations measured in lakes in Norway and Sweden, which are lakes considered to represent background concentrations.

TABLE 16.3. Zinc concentrations (ppm DW, arithmetic mean ± SD) in different plant species and plant compartments collected from forests in southern and central Norway in 1982 and 1992.

Species		Southern Norway		Central Norway	
		1982	1992	1982	1992
Norway spruce	twig	82.3 ± 8.0	75.4 ± 1.2	60.7 ± 2.1	65.5 ± 13.1
	needle	52.9	58.7 ± 15.3	34.2 ± 7.4	43.2 ± 10.4
Birch	twig	279.6 ± 25.4	284.8 ± 44.5	156.6 ± 18.7	101.1 ± 11.9
	leaf	290.9 ± 7.4	243.0 ± 27.3	102.2 ± 40.1	67.1 ± 10.6
Scotch pine	twig	52.0 ± 3.1	40.7 ± 13.0	36.4 ± 3.4	31.5 ± 0.4
	needle	20.4 ± 1.0	60.9 ± 11.3	35.4 ± 2.9	37.5 ± 6.9
Juniper	twig	19.7 ± 1.7	16.4 ± 1.5	11.1 ± 0.2	10.0 ± 1.4
	needle	20.4 ± 1.0	22.0 ± 1.0	17.2 ± 2.2	16.9 ± 3.1
Bilberry	stem	59.2 ± 9.3	59.3 ± 4.4	45.1 ± 3.4	27.3 ± 4.2
	leaf	13.5 ± 0.9	14.5 ± 1.1	11.2 ± 0.6	1.3 ± 1.1
Mountain cranberry	stem	35.4 ± 5.1	41.3 ± 4.1	29.1 ± 1.3	22.2 ± 1.6
	leaf	39.8 ± 4.3	33.3 ± 2.8	25.6 ± 1.9	23.3 ± 0.9
Heather		26.4 ± 2.0	25.3 ± 2.0	15.3 ± 0.2	11.6 ± 0.4
Crowberry		23.2 ± 0.3	22.8 ± 2.7	18.7 ± 1.4	15.2 ± 2.7
Feather moss		54.5 ± 9.7	47.5 ± 3.2	17.8 ± 0.6	20.9 ± 2.8
Bog whortleberry	stem	76.8	119.6 ± 5.1	—	89.0 ± 12.9
	leaf	34.5	53.6 ± 11.0	—	45.7 ± 9.7

Source: Extracted from Berthelsen et al. (1995).

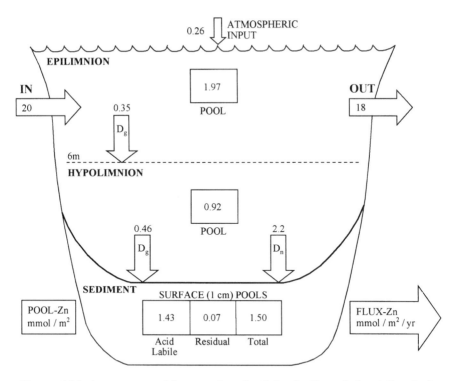

FIGURE 16.1. Average annual fluxes and pools of zinc for Darts Lake, Adirondack, New York. Atmospheric input was based on nearby Woods Lake. Gross deposition (D_g) was determined from sediment trap concentrations. Net deposition (D_n) was estimated from mass balance calculations. Surficial sediment pools (1-cm depth) of acid-labile and residual zinc were based on a sequence of chemical digestions, the sum yielding total zinc. (Adapted from White and Driscoll, 1987.)

4 Zinc in Soils/Sediments

4.1 Total Zinc in Soils

The Zn content of soils depends on the nature of the parent rocks, OM, texture, and pH. As soils develop from the parent material of the earth's surface, they acquire in varying degrees the elements present in the parent material. Soils formed from basic rocks are more enriched in Zn compared with soils from granites, gneisses, and the like (Vinogradov, 1959).

The most quoted range for total Zn in normal soils is 10 to 300 ppm (Swaine, 1955), although the wider range of 1 to 900 ppm with an average of 90 ppm has been given. For world soils, a mean of 50 ppm Zn was given (Vinogradov, 1959). More recently, a mean content of 40 ppm for world soils was reported (Berrow and Reaves, 1984) (see also Appendix Table A.24).

For Canadian soils, a range of 10 to 200 ppm Zn, with a mean of 74 ppm, has been given as contrasted with U.S. soils having a mean of 54 ppm (McKeague and Wolynetz, 1980). More specifically, in a survey of agricultural areas of Ontario, Canada that included 296 farms, the soils were found to have a mean of 54 ppm with a range of 5 to 162 ppm (Frank et al., 1976). Zinc content of the Ontario soils was related to soil texture, with sandy soils having much lower Zn (40 ppm) than loamy (64 ppm), clayey (62 ppm), and organic soils (66 ppm).

In England and Wales, Archer (1980) gave the median value of 77 ppm Zn ($n = 784$), with a range of 5 to 816 ppm. However, about 63% of the samples had Zn concentrations between 40 and 99 ppm. In the Ukraine, the total Zn content of soils averaged 40 ppm with a range of 14 to 95 ppm (Golovina et al., 1980). In China, total Zn in soils was found to average 74 ppm ($n = 3939$) with a geometric mean of 67 (Chen et al., 1991). The variation was primarily attributed to the influence of the underlying soil parent material. For example, Zn is highest in Lithosols (95 ppm) and lowest in Oxisols (34 ppm). Geographically, the Zn concentration in soils decreases from the southern to the northern region of China, with calcareous soils having lower Zn than acidic soils. The mean total Zn content of tropical Asian paddy soils ranges from 35 to 88 ppm (Domingo and Kyuma, 1983).

4.2 Bioavailable Zinc in Soils

Zinc in soil that is water soluble and adsorbed on exchange sites of colloidal materials is considered to be phytoavailable. The amount present in water-soluble form is usually negligible, while the amount removed by an exchanger, such as NH_4OAc is very small. However, in heavily contaminated soils, these Zn forms can be substantial. It is a consensus that phytoavailable Zn in soil can best be predicted by the use of extractants that remove only a fraction of the total amount.

Ample literature on extractants used to measure phytoavailable Zn (e.g., salt solutions, acids, bases, and chelates) exists. The most commonly used Zn extractants differ widely in their capacity to extract Zn, from almost zero to several parts per million (Bauer, 1971). The amounts extracted are only a small fraction of the total Zn content. For example, Stewart and Berger (1965) found 4.5 ppm Zn with 0.1 N HCl vs. 0.81 ppm with 2 N $MgCl_2$ in soils having an average total Zn content of 55.2 ppm. Similarly, Trierweiler and Lindsay (1969) found approximately 5, 1.6, and 1.0 ppm Zn using 0.1 N HCl, EDTA–$(NH_4)_2CO_3$, and NH_4OAc and dithizone, respectively, in soils averaging 59 ppm of total Zn. In routine soil tests, soils that have the highest total contents may not have the highest amounts of phytoavailable Zn since there are numerous factors, including inherent soil factors that affect its extractability.

A soil test using one of the extractants is done to (1) determine if the Zn level is sufficient and the soil requires no fertilization with Zn, and (2) determine if the Zn level is excessive and will cause phytotoxicity, as in the case of sludged soils or soils heavily contaminated. In the latter case, management strategies should be part of any operational scheme, including adjustment of soil pH, adding amendments, or selecting tolerant plant species. There is no universal Zn extractant for phytoavailability measurement. The needs vary by country, crop, and soil type, and even by state in the United States. The more popular extractable Zn tests are based on the amounts of Zn extracted by inorganic acids (Bidwell and Dowdy, 1987; Nelson et al., 1959; Wear and Evans, 1968), neutral salts (Bell et al., 1991; Sanders et al., 1987), and by chelating agents (Brown et al., 1971; Lindsay and Norvell, 1978; Trierweller and Lindsay, 1969).

Acid-extractable Zn is generally a poor indicator of bioavailable Zn in near-neutral and alkaline soils. Zinc extractable by EDTA, DTPA, and dithizone (diphenylthiocarbazone) has been delineated into Zn-deficient and Zn-sufficient categories in near-neutral and alkaline soils based on the growth of corn plants (Brown et al., 1971; Trierweiler and Lindsay, 1969). More recently, the DTPA-extractable Zn has gained more attention as a diagnostic tool in the Zn nutrition of plants. Whether the soil is sludged (Bell et al., 1991; Bidwell and Dowdy, 1987) or just ordinarily mineral arable soil (Dang et al., 1993; Singh et al., 1987), the DTPA solution is the extractant of choice. For example, the criteria for predicting Zn deficiency in Saskatchewan (Canada) soils is based on DTPA extraction, i.e., below 0.50 ppm (Singh et al., 1987).

The double-acid or Mehlich extractant has been widely utilized in the southeastern United States for Zn diagnosis in crops. Critical levels in soils have been established for a wide range of soils between pH 4.5 and 6.5 (Fig. 16.2). The results indicate the importance of soil pH and OM content in the bioavailability of Zn to crops (Junus and Cox, 1987). The lines shown in the figure represent the soil concentrations that predict a corn leaf concentration of 16.5 ppm, a deficient condition. Any point below or to the right of line would represent a sufficient condition.

In Cerrado vegetation in Brazil, Lopez and Cox (1977) found the median level for Zn extracted with double acids (0.05 N HCl + 0.025 N H$_2$SO$_4$) to be 0.60 ppm, with a range of 0.20 to 2.2 ppm. The critical level for these soils is set at about 0.80 ppm.

Plant needs for supplemental Zn can be determined by soil testing or by plant tissue analysis. An example is the DTPA-extractable Zn test (Table 16.4) adopted by several states in the United States. Through this test, critical levels of extractable Zn have been established for certain crops and soils. For example, 0.50 ppm DTPA-extractable Zn is the critical level for deficiency in soils for some midwestern states for growing corn, sorghum, soybeans, and pinto beans (Whitney et al., 1973; Whitney, 2000). In California, the critical level for DTPA-extractable Zn is also set at 0.50 ppm (Brown et al., 1971). In Virginia, 0.80 ppm EDTA-extractable Zn is considered critical for corn (Alley et al., 1972). Lindsay and Norvell (1978)

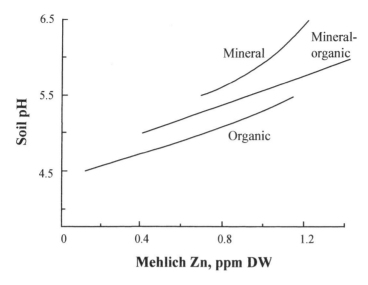

FIGURE 16.2. Effect of soil pH and Mehlich-3–extractable Zn concentration on dividing Zn-deficient (above a line) and nondeficient (below a line) conditions for growing corn on mineral, mineral–organic, and organic soils. (Modified from Junus and Cox, 1987.)

in Colorado found DTPA-extractable Zn to be critical for corn at below 0.80 ppm. In rice, the critical level using this extractant is at 1.65 ppm (Gangwar and Chandra, 1976). Sakai et al. (1982) found that the DTPA-extractable Zn in calcareous soils ranged from 0.34 to 3.42 ppm, which produced Zn levels in rice leaves of 15 to 50 ppm. The critical level of DTPA-extractable Zn in these soils is 0.78 ppm.

Havlin and Soltanpour (1981) have demonstrated that the NH_4HCO_3–DTPA extractant is as effective as the DTPA soil test of Lindsay and Norvell (1978) in predicting phytoavailable Zn in Colorado soils. The NH_4HCO_3–DTPA extractant simultaneously extracts both the macronutrients and

TABLE 16.4. Levels and interpretation of DTPA-extractable test for soil zinc.

DTPA-extractable zinc (ppm)	Test ranking	Comments
0–0.50	Low	Likelihood of a response to Zn is good on corn, sorghum, soybeans, and pinto beans with good management.
0.51–1.00	Medium	Questionable range where response to Zn may be obtained under adverse conditions.
>1.00	High	Response to Zn is not likely to occur.

Sources: Whitney, 2000 (personal communication, Dept of Agronomy, Kansas State Univ.). The DTPA extraction procedure was adapted from Lindsay and Norvell (1978).

micronutrients from soil (Soltanpour and Schwab, 1977). The critical Zn level for corn using this extractant is 0.90 ppm, slightly higher than the critical level set by the DTPA extractant for the same crops growing on Colorado soils.

Extractable Zn has been positively correlated with total Zn, OM, clay content, and CEC (Follett and Lindsay, 1970; John, 1974) and inversely correlated with free $CaCO_3$, soil pH, and base saturation (Adriano, 1986).

4.3 Zinc in Soil Profile

The major factors influencing the transport and redistribution of metals in soils are plant uptake and harvest removal, erosion, leaching, and tillage practices. In arable soils, all those factors are relevant, but in undisturbed areas, plant uptake and leaching would be more predominant. In the latter case, bioturbation by large mammals, such as wild boars, can accelerate distribution or natural attenuation of metals.

In soil profiles in Scotland, total Zn was distributed quite uniformly from horizon to horizon (Swaine and Mitchell, 1960). However, the distribution of extractable Zn showed greater variation with horizon than does total Zn content. In most well-drained arable soils and in typical Podzols, there was normally a decrease in extractable Zn with depth in the profile.

In Canada, total and extractable Zn contents in horizons of seven soil profiles generally declined with increasing soil depth (John, 1974). Similarly, Roberts (1980) noted that total Zn declined with profile depth in two Canadian serpentine soils. Inconsistencies in profile distribution of Zn occur also in forest soils (Nakos, 1983).

It can be generalized that extractable Zn decreases with depth while total Zn is more uniformly distributed throughout the profile. For example, in Colorado, soils had 62, 60, and 52 ppm total Zn for the surface soil, subsoil, and parent material (C horizon), respectively (Follett and Lindsay, 1970). However, values for the DTPA-extractable Zn for the respective horizons were 1.62, 0.59, and 0.29 ppm. Similarly, in Hawaii, the concentration of total Zn appeared to be less dependent on depth than was the acid-extractable Zn (Kanehiro and Sherman, 1967). The extractable Zn followed a general decreasing pattern with soil depth, while the total Zn remained essentially the same. In 50 sites in California, total Zn appear to be uniformly distributed in the profile (Bradford et al., 1967). And in Kansas, acid-extractable Zn from calcareous and noncalcareous soils was primarily concentrated in the A horizon (Travis and Ellis, 1965).

A decreasing trend of extractable Zn in the soil profile has important implications relative to plant nutrition. When the surface horizon is removed, as in the case of leveling the land for irrigation purposes, the exposed subsoils are typically deficient in Zn and supplemental Zn fertilizer will be required.

The content and distribution of Zn in the profile have been linked to various soil processes and factors, such as weathering, OM, clay content, and pH. Although soils within a given soil group may have similar properties, they can differ widely in total and extractable Zn contents because of variations in local pedological and biological soil-forming factors. It is a general belief that surface accumulation of Zn is a result of its acquisition by plant root. from deeper horizons, decay of plant dry matter, and subsequent immobilization at the surface. Cases of increased extractable Zn in deeper horizons are indicative of parent materials having high Zn content or of the overlying horizons being depleted because of Zn removal by plants.

In sludged soils, a question that often arises is the effect of the OM input on the vertical movement of the sludge-associated metals. Because of the apparent sequestration of the metals by the sludge OM and constituents, metals have invariably moved only slightly beyond the zone of application. Long-term sludge applications have lent credence to this premise (Baveye et al., 1999; Dowdy et al., 1991; Liang and Corey, 1993; Williams et al., 1987).

A more intriguing question pertains to the long-term migration of metals from heavily contaminated soils. Metals that have accumulated in surface soils from industrial emissions (e.g., smelting) may eventually leach to the groundwater, especially upon soil acidification. Such migration can be modeled as shown in Figure 16.3 and indicates the importance of the removal of the immediate source term, i.e., surface soil, on the transport of contaminants.

4.4 Forms and Chemical Speciation of Zinc in Soils

Trace elements can be partitioned into forms or pools (sinks) based primarily on ways that elements are bound onto soil/sediment or held against extraction by chemical agents. These forms (or fractions) have varying affinities for various soil constituents (i.e., solid phases) (see also Chapter 2). Factors controlling the distribution of trace elements between the solid phase and soil solution largely govern both their dissolution and fate in biological systems. Consequently, Zn in different soil fractions varies in plant bioavailability. Zinc present in water-soluble and exchangeable forms is readily bioavailable to plants; Zn in the other forms is either not bioavailable (nonlabile) or not as readily bioavailable to plants (semilabile). Exchangeable Zn is adsorbed as a cation by electrostatic attraction to negatively charged sites formed on mineral and organic colloids. Zinc that is fixed within the clay crystal lattice can become available only through weathering.

The amount of different Zn forms in soils varies considerably depending on the type and nature of soil constituents. Alkaline soils that are calcareous

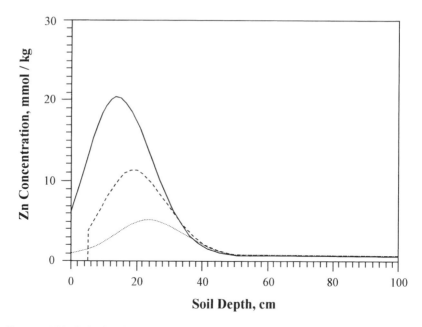

FIGURE 16.3. Calculated zinc depth profiles for a soil contaminated by a smelter 100 yr after application of different hypothetical remediation procedures in the year 2000. No action taken (solid line), removal of the top 5 cm of soil (dashed line), and replacement of the top 10 cm by noncontaminated soil (dotted line). (Modified from Cernik et al., 1994.)

can be expected to be high in the carbonate form; similarly, soils high in OM can be expected to be high in the organic form.

Shuman (1979) found that for southeastern U.S. soils, the exchangeable form was in the 1 to 7% range. Most of the Zn was in the colloidal fractions—40.4% in clay and 11.5% in silt; the remainder was in the following fractions: 12% in organic, 20% in noncrystalline Fe oxide, and about 12% in sand. The fine-textured soils had very high amounts of Zn in the clay compared with the other fractions, but the coarse-textured soils had proportionately more in the OM fraction. In another study of southeastern U.S. soils, Iyengar et al. (1981) found ∼0.4, 3.3, ∼2.5, 2.0, 25.4, and 69.6% of total Zn distributed among the following forms (Table 16.5): exchangeable (nonspecifically adsorbed), exchangeable (specifically adsorbed), organically bound, Mn oxide–bound, Al and Fe oxide–bound, and residual, respectively. These results demonstrate that most of the Zn in these soils is in the Al and Fe oxide–bound and residual forms; the other forms that are relatively bioavailable represent only a small fraction in the total soil content.

In agricultural settings, management practices can affect the distribution of metals among soil constituents. For example, in examining cultivated

TABLE 16.5. Concentrations (ppm) and distribution (%) of various forms of zinc for southeastern U.S. soils.

Soil type	Non specifically adsorbed ppm	$\%^b$	Specifically adsorbed ppm	$\%^b$	Organically bound ppm	$\%^b$	Mn-oxide bound ppm	$\%^b$	Al and Fe-oxide bound ppm	$\%^b$	Residual ppm	$\%^b$
					**Zn fraction**[a]							
Appalachian region												
Dunmore sil[c]	0.16	0.6	0.50	1.8	0.44	2.6	0.14	0.5	4.4	15.8	22.6	81.7
Emory cl	0.09	0.1	2.71	2.6	2.20	2.2	1.99	1.9	26.0	25.4	64.7	63.2
Frederick sil	0.39	1.0	0.83	2.1	0.76	1.9	0.71	1.8	8.5	21.5	19.8	61.7
Hayter sil	0.04	<0.1	0.77	1.2	4.24	6.4	1.36	2.1	13.9	21.1	38.7	59.0
Litz sil	0.04	<0.1	0.54	0.5	0.92	0.8	1.14	0.1	64.7	53.5	55.4	45.8
Westmoreland sicl	0.04	<0.1	2.41	1.9	3.35	2.6	2.78	2.2	55.6	43.7	72.7	57.1
Coastal Plain region												
Altavista sl(I)	0.21	0.4	6.67	14.0	0.97	2.0	2.15	4.5	4.7	9.8	33.2	69.5
Altavista sl(II)	0.14	0.4	3.84	12.0	2.24	7.0	1.10	3.4	7.3	22.9	19.8	61.7
Dragston sl	0.02	0.1	0.20	1.0	0.76	4.0	0.18	0.9	2.0	10.4	15.9	83.7
Ruston s	0.44	2.2	2.06	10.6	0.37	1.9	0.40	2.0	6.9	35.6	12.1	61.8
Piedmont region												
Cecil l	0.03	<0.1	1.41	2.0	0.46	0.6	1.09	1.5	16.1	22.7	54.9	77.3
Cullen cl	0.05	<0.1	0.75	1.3	0.71	1.2	0.65	1.1	14.5	24.6	43.3	69.0
Davidson c	0.02	<0.1	2.71	1.7	1.12	0.7	4.20	2.6	23.5	14.7	143.4	89.4
Penn sil	0.34	0.3	0.77	0.7	0.56	0.5	1.22	1.1	21.0	19.0	96.4	87.3
Starr sicl	0.10	<0.1	2.10	1.7	2.71	2.2	3.28	2.7	36.9	29.9	90.0	73.0
Tatum sil	0.06	<0.1	1.11	1.2	<0.01	<0.1	1.00	1.1	31.6	33.9	64.4	69.0
Average	0.16	~0.4	1.76	3.3	~1.51	~2.5	1.48	2.0	21.2	25.4	52.8	69.6

[a] The following extractants were used: nonspecifically adsorbed, 0.05 M CaCl$_2$; specifically adsorbed, 2.5% HOAc; organically bound, 0.1 M K$_4$P$_2$O$_7$; Mn oxide–bound, 0.1 M NH$_2$OH·HCl in 0.01 M HNO$_3$; Al and Fe oxide–bound, Tamm's solution; and residual, aqua regia-HF mixture.
[b] Percent of the total Zn.
[c] sil = silty loam; cl = clay loam; sicl = silty clay loam; sl = sandy loam; s = sand; l = loam; c = clay.
Source: Extracted from Iyengar et al. (1981).

Norwegian soils, Jeng and Singh (1993) found that most soil Zn was associated with the oxide and residual forms (~78% of total) in soils treated with farmyard manure (FYM), rock phosphate, or mixed fertilizers. The FYM soils exhibited more Zn in the organic fraction but had the least exchangeable form. Likewise, crop production in Alberta, Canada on land formerly under boreal forest resulted in the shift of soil Zn into hydrous oxide–associated forms (Soon, 1994). A greater portion of the decrease in labile Zn as a result of cropping was attributed to sorption by soil oxides. In prairie soils in Saskatchewan, the residual form of Zn constituted 67 to 91% of total Zn (Liang et al., 1990).

In polluted soils from Poland, most of the Zn (~65% of total) resided in the oxide and residual fractions as opposed to 15% associated with the organic form and 10% with the carbonate and exchangeable forms (Chlopecka et al., 1996). In the Polish case, the pH controls the residential preference for Zn: about 10 times more Zn was found in the exchangeable form where the soil pH <5.6 compared with soils with pH >5.6. The latter samples contain twice as much Zn in the carbonate form. Hickey and Kittrick (1984) found that a substantial amount of Zn in heavily polluted samples was associated with the Fe and Mn oxide fraction (39%), followed by the carbonate fraction containing approximately 28% of the Zn. Samples having high levels of inorganic C also had high levels of carbonate-bound Zn. Zinc in the exchangeable and organic fractions was also appreciable, but the residual fraction was only slightly enriched. The polluted samples included a soil that received metals from Cu smelting, a sludged soil, and a harbor sediment.

Chang et al. (1984) noted a shift in the chemical forms of Zn in soils after treatment with sewage sludge for 7 yr. In sludge-treated soils, the percentage of Zn in KNO_3–extracted (exchangeable), NaOH–extracted (organic-bound), and EDTA-extracted (carbonate-bound) fractions increased significantly. Prior to sludge application, the HNO_3–extractable (sulfide/residual) fraction accounted for >80% of the total soil Zn but was reduced to <50% after sludge treatment. Based on the same long-term study of Chang et al. in which soils were sequentially extracted, LeClaire et al. (1984) concluded that (1) Zn sequentially extracted by KNO_3 and H_2O is associated with a highly labile, soluble pool dominated by Zn^{2+}, which determines immediate bioavailability to barley, (2) Zn sequentially extracted by KNO_3, H_2O, and NaOH has a strong positive correlation with DTPA-extractable Zn and therefore is associated with the labile, phytoavailable Zn pool, (3) EDTA-extractable Zn represents a pool of potentially bioavailable Zn, and (4) HNO_3-extractable Zn is associated with nonlabile pool that is not bioavailable.

In lowland rice, Murthy (1982) suggested that the $Cu(OAc)_2$- and acidified ammonium oxalate-extractable Zn are the prevalent forms of Zn controlling the bioavailability of Zn to the rice plants. The $Cu(OAc)_2$ reagent extracted the exchangeable and complexed Zn fractions whereas the ammonium oxalate extracted the Zn bound by amorphous sesquioxides.

Sims and Patrick (1978) found that greater amounts of Zn were found in either the exchangeable or organic form at low pH and Eh values than at high pH or Eh. In contrast, amounts of Zn in the remaining forms (oxalate and residual) generally were greater at high than low pH or Eh, indicating that Zn precipitated/occluded as oxides and hydroxides was solubilized in low pH and reducing conditions.

The predicted dominant Zn species in soil solution below pH 7.7 is Zn^{2+} (Lindsay, 1979). Above this pH, $ZnOH^+$ is more prevalent. The neutral species $Zn(OH)_2^0$ is the primary species above pH 9.1, while $Zn(OH)_3^-$ and $Zn(OH)_4^{2-}$ are never significant solution species in the pH range of soils (Fig. 16.4). However, Zn in soil solution from the rhizosphere is likely to be

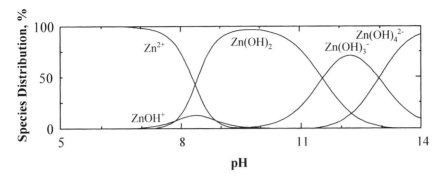

FIGURE 16.4. Predicted aqueous monomeric chemical speciation of zinc. These predictions assume that any metal-hydroxide precipitation did not occur. (Adapted from Schultess and Huang, 1990.)

dominated by the complexed form (Hamon et al., 1995), presumably from low-molecular-weight organic acids (i.e., root exudates).

4.5 Sorption of Zinc in Soils

Sorption of Zn in soils is an important factor governing Zn concentration in soil solution and Zn bioavailability to plants. The adsorption capacity of soils generally exceeds the few kg of Zn per hectare applied to soils to correct Zn deficiency. Apparently, the usual amounts of Zn applied are adsorbed almost completely within a short period of time and held against leaching and runoff.

Sorption of Zn in soils is influenced by several factors, such as pH, clay mineral, CEC, OM, and soil type. Clay minerals, hydrous oxides, carbonates, and OM are constituents known to influence soil sorption of added Zn. Clay minerals vary in their capacity to sorb Zn due to their differences in CEC, specific surface area, and basic structural makeup. Elgabaly (1950) suggested that some of the applied Zn is irreversibly fixed by clay. However, Bingham et al. (1964) found that Zn behaved as an exchangeable cation and amounts in excess of the CEC were retained as $Zn(OH)_2$. Reddy and Perkins (1974) inferred that the greater fixing capacity of bentonite and illite (2:1 clays) over kaolinite (1:1 clay) was caused by Zn^{2+} entrapment in the interlattice wedge zones of the clay structure when the zones expanded due to hydration (wetting) and contracted upon drying. Tiller and Hodgson (1962) characterized clay-bound Zn as dominantly reversible in association with clay surface groups, the rest existing in an irreversible nonexchangeable form associated with lattice entrapment.

Zinc sorption by carbonates, precipitation of Zn hydroxide or carbonates, or formation of insoluble calcium zincate are believed partly responsible for Zn unavailability in calcareous and alkaline soils (Adriano, 1986). Since soil reaction is the most important factor influencing Zn bioavailability in soils, calcareous soils, which inherently have high pH values, are more prone to Zn deficiencies. Thus, overliming acid soils, such as the highly leached and weathered Ultisols and Oxisols in the tropics, can induce Zn deficiency in crops.

The greater bioavailability of Zn to plants under acidic soil conditions can also be explained by the effect of pH on Zn sorption by soil constituents. Simply stated—more Zn can be adsorbed in alkaline than in acidic medium, which is partly attributed to decreased competition from protons for adsorption sites. However, solubilization of OM in soils with high pH may also reduce Zn adsorption due to the formation of metal–organic complexes. Similarly, the presence of EDTA in soil suspension can decrease Zn sorption by soils, indicating that Zn is strongly complexed with EDTA, decreasing its affinity for sorption sites (Elrashidi and O'Connor, 1982). In contrast, complex formation of Zn with Cl^-, NO_3^-, and SO_4^{2-} did not have significant effects on Zn sorption. Thus, the presence of synthetic chelates, such as EDTA and DTPA, could maintain most of the Zn in mobile form, thus increasing its potential of contaminating groundwater.

The surface charge on hydrous oxides is highly pH-dependent and increases with increasing pH. The retention of Zn in a nonexchangeable form in soils is partly due to the presence of oxide surfaces in soils whose clay fractions are dominated by layer silicates (McBride and Blasiak, 1979). For example, White (1957) found that large fractions of Zn (30 to 60%) in soils of the eastern United States are associated with the hydrous ferric oxides. Similarly, the increased Zn content in sand fraction in deeper soil profiles in South Africa was attributed to increased ferruginous concretions that had higher Zn than the surrounding soil from which they were taken (Stanton and Burger, 1967). Cavallaro and McBride (1984) concluded that the oxide constituent of clays could be more important than its organic constituent in sorbing Zn, based on their finding that pretreatment of these clays to remove OM did not decrease Zn sorption.

4.6 Complexation of Zinc in Soils

Soil OM influences the chemical behavior of trace elements in soils by way of increased buffering and CEC in the soil. More importantly, the humic and fulvic acids complex with metals, thereby affecting mobility and bioavailability. Three types of retention site for Zn by humic acid have been identified: the less stable complex is associated with phenolic hydroxyl groups and weakly acidic carboxyls, while the more stable complex is associated with strongly acidic carboxyls, (Randhawa and Broadbent, 1965). Tan et al. (1971) indicate that Zn complexes with fulvic acid extracted from sewage sludge involve the formation of coordinate covalent bonds between OH groups and Zn, and electrovalent linkages between COO^- and Zn. Zinc sorption in soils may be affected by fulvic acid due to competition for Zn between surface sorption sites on clay minerals and fulvic acid complexing sites. Additionally, coverage of clay surfaces by sorbed fulvic acid may mask Zn sorption sites. Indeed, the affinity of Zn-fulvic acid complex for montmorillonite is appreciably lower than that for $ZnOH^-$ (Bar-Tal et al., 1988), indicating the reducing effect of fulvic acid on Zn sorption. A similar reduction in Zn adsorption onto quartz in the presence of fulvic acid also occurred (Duker et al., 1995).

In comparing mineral and organic soils, Matsuda and Ikuta (1969) found that the percentage of added Zn in the exchangeable form was higher in mineral soil, while the percentage of organically complexed Zn was higher in organic soil. Zinc complexation increased with the humification of OM. Addition of organic amendment to soils in the form of solid OM (as opposed to soluble OM, such as humic and fulvic acids) in general, will tend to shift Zn more into nonbioavailable forms (Shuman, 1999).

Commercial chelating agents are available to supply micronutrients to plants. Chelated micronutrient cations in soils and nutrient solutions are soluble complexes. Experimental data on the chemical reaction and stability of Zn chelates in soils were generated in the late 1960s by Lindsay and colleagues; ZnEDTA and ZnDTPA are the most commonly used chelated Zn. Both are fairly stable at pH levels below 7, but above this pH, ZnDTPA is more stable (Lindsay and Norvell, 1969). At pH 7.5, about 12% of the EDTA and 82% of the DTPA are complexed with Zn. It can be generalized that ZnDTPA is unstable in acid soils, moderately stable in slightly acidic soils, and fairly stable in calcareous and alkaline soils (Lindsay and Norvell, 1969; Norvell, 1972). The low stability under acid conditions is apparently caused by the relatively high solubility of soil Fe at low pH, while at high pH the cation can be displaced by Ca^{2+} ions, followed by precipitation of Zn as sparingly soluble compounds.

Chelating agents, either natural or synthetic, are important in the mobility and plant nutrition of Zn. Because chelated Zn is generally soluble, it is subject to root surface contact or interception via diffusion and mass flow. Thus, chelating agents play an important role in transporting insoluble nutrients to roots.

Zinc also forms complexes with Cl^-, PO_4^-, NO_3^-, and SO_4^{2-} (Lindsay, 1979). A commonly asked question pertains to the dominant Zn species present in the rhizosphere. Since active plant roots release organic exudates, it can be expected that part of Zn in the rhizosphere soil solution is in the complexed form. Indeed, it can be generalized that metals in unplanted soils occur largely as uncomplexed (with regard to humic acid) while dissolved metals in planted soils occur largely as complexed. For example, Zn occurred as 97% free Zn^{2+} in nonrhizosphere soil solution vs. 78% in rhizosphere soil solution (Holm et al., 1995).

5 Zinc in Plants

5.1 Essentiality of Zinc in Plants

Zinc is an essential element for plant nutrition. Unlike the major nutrients ordinarily supplied in mixed fertilizers it is required in only minute amounts. For example, a hectare of healthy oat plants contains only about 70 g of Zn in the dry matter. Yet without this small amount, no crop will normally grow. Zinc is very important in plant nutrition because it is involved in a number of metallo-enzymes, is essential in the stability of cytoplasmic ribosomes and root cell plasma membrane, catalyzes the process of oxidation, is concerned with protein synthesis and the synthesis of auxin indole-acetic acid (IAA), and the transformation of carbohydrates (Kochian, 1993; Romheld and Marschner, 1991).

5.2 Deficiency of Zinc in Plants

Zinc deficiency is more common worldwide in both tropical and temperate climates than deficiency of any other micronutrient. Unlike Fe deficiency, Zn deficiency is quite common in acid soils (e.g., Cerrado soils in Brazil). Treated areas amount to several million hectares a year requiring thousands of tonnes of Zn compounds in the form of fertilizers and sprays. In the United States, Zn deficiency is known in 39 states (Sparr, 1970). Abroad, Zn deficiencies occur in Canada, Australia, New Zealand, Africa, Asia, South America, and Europe, and probably in all areas where commercial crops are grown. In the tropics, the most prominent documented areas for Zn deficiency occur in Brazil, Nigeria, India, and the Philippines (Sillanpaa and Vlek, 1985). In Africa, Zn deficiencies have been reported in Nigeria and other countries mostly occurring in places where lime is used to increase soil pH from about 4.5 to about 7.0 (Kang and Osiname, 1985).

Zinc deficiency is caused primarily by three factors: (1) low content of Zn in the soils; (2) unavailability of Zn present in the soil to the plant; and (3) management practices that depress Zn bioavailability or its uptake. Zinc is low in highly leached acid sandy soils such as those found in many coastal areas in the southeastern United States. The southeastern United States contains the largest contiguous area of Zn deficiency; the Florida citrus-producing Lakeland and associated soil types contain some of the largest single areas of Zn deficiency in the United States (Welch et al., 1991). In southern and western Australia, Zn deficiency has been found on large areas of sandy soils (Leeper, 1970). Widespread dieback of *Pinus radiata* occurs in the southeastern region of South Australia (Boardman and McGuire, 1990). In most cases, the total quantity of Zn present is high but most of the Zn occurs in an unavailable form and the extractable Zn present is insufficient to sustain normal plant growth. This can be expected in high-pH, high-OM soils where Zn would be mostly in unavailable forms.

Because plants vary in their optimum requirement for Zn even among various species and cultivars, it is difficult to establish a single critical value. However, plants with Zn contents below 20 ppm in dry tissue can be suspected of Zn deficiency, with normal values ranging from 25 to 150 ppm Zn (Jones, 1972). Lucas (1967) gave the following sufficiency ranges (ppm DW) for Zn: corn, ear leaf at tasseling, 20 to 70; alfalfa, top 15 cm, 20 to 70; soybeans, top fully open leaves, 20 to 50; and vegetables, top fully open leaves, 30 to 100. A survey of 182 citrus groves showed that the average Zn content of the leaves was 34 ppm (Stewart et al., 1955). However, about half of the groves had <30 ppm, which is considered low. Wheat grown on deficient black earths in Australia had critical Zn concentration of 20 ppm in the tops which corresponded to 0.60 ppm EDTA-extractable Zn in the soil (Radjagukguk et al., 1980).

In legumes, a critical concentration of 20 to 21 ppm in whole shoot of chickpea was observed (Khan et al., 1998). In soybeans, Ohki (1977) found that critical Zn concentration in blade 5 was 21 ppm, and Reuter et al. (1982) determined critical Zn concentration in youngest open leaf blade of subterranean clover as 18 ppm. Critical Zn concentrations reported by Leggett and Westermann (1986) for navy beans ranged from 17 to 20 ppm; in cowpea, the critical concentration was 20 ppm for upper leaves (Marsh and Waters, 1985).

In avocado the critical Zn level in leaf tissue is at 20 ppm, with 30 to 150 ppm considered as normal (Crowley et al., 1996). In pecan the critical leaf level is at 14 ppm (Hu and Sparks, 1991). The threshold value for pecan on an orchard basis has been set at $\sim$50 ppm that will not adversely affect nut yield and vegetative growth in an orchard (Sparks, 1993). Zinc deficiency is widespread in both avocado and pecan grown in alkaline soils.

In broadleaf plants, foliar symptoms of acute Zn deficiency are manifested in new leaf tissue by a reduction in leaf size and the development of malformed leaves or little leaf (also termed rosette in pecan), which is characterized by interveinal leaf yellowing caused by impaired chlorophyll synthesis (Brennan et al., 1993). Other symptoms include shortened internode length on the branches, reduced fruit size, and reduced leaf biomass. In calcareous soils, the visual leaf yellowing from Zn deficiency may be confused with Fe deficiency, which produces similar leaf symptoms (Crowley et al., 1996). Physiologically, Zn deficiency can adversely affect the leaf chlorophyll content, stomatal conductance, and net photosynthesis (Hu and Sparks, 1991).

Zinc deficiency is widespread in China, especially in calcareous soils and calcareous paddy soils (Liu et al., 1983). These soils are extensive on the loess plateau and North China Plain and have DTPA-extractable Zn averaging only 0.37 ppm (range of trace to 3 ppm). The average value of 0.37 ppm is below the considered critical value of 0.50 ppm for Zn deficiency (see Table 16.4). Zinc application on these soils has increased the yields of rice, corn, legumes, and fruit trees.

Zinc deficiency in lowland rice is widespread in Asia according to scientists at the International Rice Research Institute in the Philippines (Yoshida et al., 1973). Rice grown in the greenhouse on Zn-deficient soils collected from the Philippines, Korea, Thailand, Sri Lanka, Indonesia, Taiwan, and Pakistan indicates that with ~19 ppm or less of Zn in the shoot, rice responded to Zn application in the field. Furthermore, when Zn contents of the shoot were 9 to 12 ppm, growth was severely retarded; at 15 to 20 ppm Zn, visible symptoms showed up in the foliage. The following are criteria for diagnosing Zn deficiency in rice:

Zn in whole shoot (ppm)	Diagnosis
<10	Definite deficiency
10–15	Very likely deficiency
15–20	Likely deficiency
>20	Unlikely deficiency

5.3 Phosphorus–Zinc Interaction in Plants

Being a macronutrient, P is usually applied at or before planting in quantities determined by soil analysis or crop needs. When large amounts of P are supplied, particularly when placed close to the seed and soil Zn level is low, luxury uptake of P may occur. The resulting high concentrations of P in plant tissues (roots and shoots) are sometimes associated with reduced uptake of micronutrients, particularly Zn. Such an association of reduced Zn concentration with high P concentration is known as P-induced Zn deficiency. Phosphorus–Zn interaction has been the subject of intensive investigations in many countries. Results have been quite variable and conflicting, reflecting the complexity of the subject.

Many crops (e.g., beans, corn, cotton, potato, okra, soybeans, sorghum, subterranean clover, flax, citrus, rice, wheat, tomato, and hops) have been reported to show P–Zn interaction with consequential detrimental effects on plant growth. Several hypotheses have been offered as to the causes of P–Zn interaction: (1) P–Zn interaction in the soil, either on soil–root interface or on adsorption, (2) simple dilution effect on concentration of Zn in plant shoots due to growth response to P, (3) a slower rate of translocation of Zn from the roots to shoots compared with P, (4) a metabolic disorder within the plant cells related to an imbalance between P and Zn, and (5) the impact of P on mycorrhizal infection of roots.

Although most investigators reported significant P–Zn interaction in terms of induced Zn deficiency or lower plant Zn concentration (Ellis et al., 1964) other studies produced synergistic interactions in which P application enhanced Zn uptake (Chaudhry et al., 1977; Pauli et al., 1968; and Watanabe et al., 1965). For example, in a field experiment, Peck et al. (1980) found that fertilizer P when combined with Zn fertilizer increased Zn concentrations in peas, snap beans, cabbage, and table beet tissues as compared with no P fertilizer. However, applied fertilizer P without Zn decreased the Zn concentration in plant tissue. In addition, fertilizer Zn with P increased Zn in plant tissue more than fertilizer Zn without P.

Beans have exhibited P–Zn interaction under both controlled and field conditions. However, effects on yield were conflicting. The most adverse effects appear to occur when high levels of P were applied in the presence of low or deficient levels of available Zn in soil. Reductions in yield were associated with high P/Zn ratios in plant tissue (Lessman and Ellis, 1971; Khan and Soltanpour, 1978). Other field experiments however, failed to produce P-induced Zn deficiency even when conditions of low soil-available Zn were present, yielding poor relationships between yield and P/Zn ratio (Boawn et al., 1954).

Corn is another major crop prone to P–Zn interaction. The most severe antagonistic interactions have been observed in instances where high application rates of P were superimposed on marginal or deficient Zn levels in soil. Corn yield reductions or poor growth of plants have been attributed to impaired Zn nutrition as a result of P fertilization (Adriano and Murphy, 1970; Ganiron et al., 1969). Variations in results were noted in other studies where yields were unaffected despite depressed Zn concentrations associated with high tissue concentrations of P (Adriano et al., 1971; Edwards and Kamprath, 1974; Giordano and Mortvedt, 1978; Stukenholtz et al., 1966). Variations in grain and dry matter yields in field and greenhouse studies bear out the dependence of P–Zn interaction on a number of other factors, including climatic condition.

Under some conditions, high levels of available P in the growth medium have decreased the Zn concentrations in the aerial portion of plants compared with the roots (Stukenholtz et al., 1966). Lower concentrations of Zn in the aerial portions of plants could be attributed to difficulties in translocating Zn from the roots to the shoot. However, faster growth of plants receiving high rates of P could affect Zn concentrations by dilution.

Collectively, results suggest that P-induced Zn deficiency in crops cannot be attributed to precipitation of Zn as insoluble P–Zn compounds in soils. Olsen (1972) reviewed the various possible causes of such induced Zn deficiency and concluded that dilution of an already low Zn concentration by growth response to applied P was the chief cause of such induced Zn deficiencies. This is not the entire explanation, however, in instances where applied P caused a net reduction in Zn uptake by the shoot tissue. Evidence also exists for P inhibition of Zn absorption into the roots or interference with translocation of Zn from roots to metabolic sites in the leaves (Khan and Zende, 1977; Paulsen and Rotimi, 1968; Soltanpour, 1969; Youngdahl et al., 1977). Additional theories of physiological imbalance due to P–Zn interaction (Boawn and Brown, 1968; Watanabe et al., 1965) have not adequately described all aspects of the phenomenon and consequently, the explanation for the interaction remains rather vague.

More recently, results of studies on the mechanism of P-induced Zn deficiency in cotton together with studies using other plant species (Cakmak and Marschner 1987; Loneragan et al., 1982; Webb and Loneragan, 1990) may be summarized as shown in Figure 16.5. With increases in the supply, uptake, and transport of P, the uptake of Zn is not directly affected. The shift of the P nutritional status of plants from the deficient to the sufficient range is usually related to decreases in root growth. In soil-grown plants where Zn nutrition depends mainly on supply of Zn by diffusion, a decrease in root growth with improvement of the P nutritional status may well be an indirect factor in decreasing Zn concentration in plants. Recent evidence indicates that when plants are supplied high P with low Zn, they accumulate high P in their leaves, precipitating Zn and therefore increasing their internal need for Zn (Loneragan and Webb, 1993). This gives rise to a new phenomenon, P-enhanced Zn requirements.

A higher proportion of Zn may be retained in the roots at high P supply. Formation of sparingly soluble Zn phosphates in the apoplast of the root cortex might be involved in this shift in Zn distribution between roots and shoot. However, it is assumed that the main components in P–Zn interaction are the physiological inactivation of Zn (precipitation of sparingly soluble Zn phosphates) in the stem and leaf tissue and the enhancement of uptake and particularly, P translocation to the shoot induced by Zn deficiency. Although the latter mechanism is not yet fully understood, a specific inhibition of P retranslocation from shoot to roots (impairment of feedback control) by Zn deficiency is probably involved. As a result, Zn-

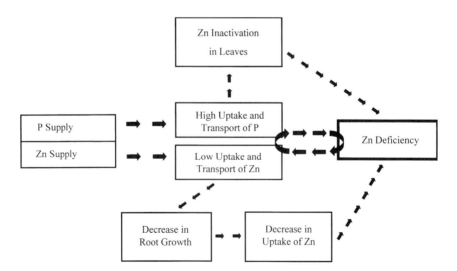

FIGURE 16.5. A model for P-induced Zn deficiency in plants. (Reprinted from Cakmak and Marschner. Copyright © 1987 Munksgaard International Publishers Ltd., Copenhagen Denmark.)

deficient plants continue to increase the P concentrations in the leaves up to toxic levels, a behavior that is otherwise observed only in P-deficient plants after a resupply of P. Increases in P concentrations in Zn-deficient leaves would simultaneously further decrease the physiological availability of Zn, accentuating Zn deficiency.

The enhancement of P uptake and translocation was specific for Zn deficiency and did not occur in plants deficient in Fe, Mn, or Cu. Thus, Zn seems to have specific functions in P uptake and transport of P from roots to shoot. Welch et al. (1982) showed that Zn-deficient wheat roots have increased membrane permeability for phosphate and chloride. This may indicate a specific increase in membrane permeability for anions in Zn-deficient plants. From the results of Parker et al. (1992), it appears that several physiological aspects including the link between uncontrolled tissue P accumulation and Zn-deficiency inducing increases in root membrane leakiness remain unresolved.

Phosphorus–Zinc interaction may affect fertilizer use and induce Zn deficiency with sensitive crops. Since increased negative surface charge with sorbed P has been reported (Gillman and Fox, 1980; Kuo and McNeal, 1984), P addition to soil may result in increased Zn sorption onto soils reducing its bioavailability to plants. Indeed, Xie and MacKenzie (1989) confirmed that P addition increased specific adsorption of Zn on soil Fe oxides.

5.4 Zinc–Cadmium Interaction in Plants

Because of the biogeochemical similarity of Zn and Cd, they can be expected to interact in the soil or within the plant. Although results have been variable, there appears to be a tendency for an antagonistic interaction between Zn and Cd to occur where Zn is applied in Zn-deficient or marginal soils (Oliver et al., 1994). The variable results are related to plant species, soil type and Zn status, source and level of applied Zn or Cd, and Cd/Zn ratio in growth media. The mitigative effect of Zn on Cd uptake by plants may serve as a means of improving the quality of foods, such as potato and cereal, with reference to Cd content. For example, in some areas of southern Australia, addition of Zn at planting (up to 100 kg Zn ha^{-1}) significantly reduced potato tuber Cd contents. While the antagonistic effect of Zn on Cd nutrition has often been observed, the converse is not so (Smilde et al., 1992). The antagonistic interaction between Zn and Cd in plant nutrition can be partly explained by their reduced adsorption when these two metals concurrently exist due to their competitive nature for sorption sites.

5.5 Phytotoxicity of Zinc

Potential Zn phytotoxicity from excess Zn input into soils exists because Zn is fairly immobile in soil and since reversion of Zn to unavailable forms occurs slowly in soils. Zinc phytotoxicity is undesirable because of decreased crop yield and quality, difficulty in correcting Zn phytotoxicity, and likelihood of Zn transfer into the food chain. Occurrence of Zn toxicity has been associated with Zn smelting (Singh and Lag, 1976), naturally high localized Zn concentrations (Staker and Cummings, 1941), or production practices that add extremely large quantities of Zn to soils.

For many plant species, Zn leaf concentrations >100 ppm may result in yield reduction or phytotoxicity symptoms similar to Fe chlorosis. For agronomic species, a normal range of 27 to 150 ppm Zn in mature leaf tissue is reported (Jones, 1991). Tissue Zn concentrations (ppm DW) for plants showing visual toxicity symptoms were: cotton and orange, 200; tung, 485; tomato, 526; and oats, 1700 (Chapman, 1966; Ohki, 1975). Once plant leaf levels of Zn exceed 400 ppm, toxicities can be expected (Jones, 1972).

Using a fine sandy loam, Benson (1966) found stunted apple seedlings with tissue concentration of 100 ppm, and at 200 ppm there was no top growth. In pot experiments with different soil types and rates of Zn of up to 250 ppm Zn in the soil, Zn concentrations of 792 ppm in corn, 523 ppm in lettuce, and 702 ppm in alfalfa were associated with yield depressions (MacLean, 1974). Some species however, are sensitive to even lower Zn levels in the tissue. For example, 50 ppm in pea tissue was toxic, reducing growth (Melton et al., 1970).

For rice, a 10% reduction in yield occurred when extractable (0.1 N HCl) Zn was 460 ppm in paddy soils in Japan (Chino, 1981b). Ichikura et al. (1970) reported that from 250 to 1000 ppm of total Zn in soil was harmful to rice plants. Chino (1981b) summarized the toxic Zn levels for rice tops as 100 to 300 ppm and for rice roots as 500 to 1000 ppm. Other crops can tolerate much higher application of Zn as shown in Table 16.6. The crops were grown on a noncalcareous fine sandy loam in which $ZnSO_4$ was broadcast at variable rates. Among the crops, only Swiss chard and spinach displayed stunted growth and accumulated much more Zn.

6 Ecological and Health Effects of Zinc

6.1 Effects of Zinc from Smelting Emissions

Research conducted near smelters and related sources of pollution have provide some of the most dramatic evidence of ecological damage caused by metals. Sudbury, Ontario (Canada) and Copperhill, Tennessee are known for the massive destruction of vegetation by smelting. The Palmerton, Pennsylvania Zn smelting has also produced well-documented severe ecological effects. For many years Palmerton was the site of two Zn smelters operated by New Jersey Zinc Company (Beyer and Storm, 1995). Soils, streams, microflora, and other biota (plants, salamanders, deer, horses, etc.) have been harmed by the emissions.

TABLE 16.6. Zinc concentrations (ppm) of leafy vegetables field-grown with normal and excessive rates of Zn (as $ZnSO_4$) fertilization.

Crop	Sample description	Zn treatment (kg ha⁻¹)				
		0	11	56	224	896
Head lettuce	Market size heads	38	45	64	144	248
Leaf lettuce	Market size plants	38	46	64	157	269
Romaine lettuce	Whole plant prior to heading	32	40	56	108	179
Romaine lettuce	Market size heads	48	50	62	100	122
Endive	Market size plants	32	38	73	247	343
Parsley	Market size plants	58	50	86	188	438
Swiss chard	Market size plants	80	72	153	615	862[a]
Spinach	Market size plants	139	119	148	240	340[a]
Chinese cabbage	Whole plant prior to heading	54	48	68	89	114
Chinese cabbage	Market size heads	46	42	60	112	389
Mustard	Market size plants	32	32	36	58	364
Collard	Rosette of young leaves	33	34	38	63	366
Cabbage	Market size heads	22	20	23	34	73
Brussels sprouts	Market size heads	50	47	56	62	79

[a]At the indicated rate of Zn fertilization plants showed normal color, but were stunted.
Source: Extracted from Boawn (1971).

Contaminants from the Palmerton smelters were emitted at phytotoxic levels. Zinc accompanied Cd pollution at about 50 to 100 kg Zn/1 kg Cd (Chaney et al., 1988). Total emissions during the years of operation were estimated at about 260,000 tonnes of Zn, 330 tonnes of Cd, and 6,800 tonnes of Pb.

Although destruction of the original forest in Palmerton can be largely attributed to sulfur dioxide emission, metal toxicity provided additive effects. Mature trees manifested chlorosis, marginal necrosis, and leaf curling and appeared prone to drought (Buchauer, 1973; Jordan, 1975). The canopies of stressed forests were more open than normal, and the densities of seedlings and saplings were unusually low. Soil Zn concentrations were high enough to inhibit root elongation of tree seedlings, preventing the forests from regenerating, which eventually led to soil erosion. The accumulation of excessive amounts of metals on the forest floor and surface soil has reduced microfaunal diversity and density; decomposition of plant material on forest floor has declined. Populations of soil microorganisms and litter invertebrates have been greatly reduced, resulting in undecomposed tree trunks, branches, and litter (Beyer and Storm, 1995). Densities of 23 taxonomic categories of soil and litter arthropods were lowest near the smelters compared with sites farther away (Strojan, 1978). High Zn concentrations were present in soils, and the lichen and moss communities were depauperated for at least 20 km along Blue Mountain by the smelters (Beyer and Storm, 1995).

Although Zn concentrations in Palmerton were shown to be toxic to plants and litter invertebrates, they apparently were not detrimental to most of the wildlife near the smelters (Beyer and Storm, 1995). Although incidences occurred where Zn was slightly elevated in some wildlife (e.g., songbirds, shrews, rabbits, etc.) by the smelters, high Zn concentrations probably mitigate Cd toxicity to wildlife. However, destruction of wildlife habitat on Blue Mountain has reduced wildlife populations.

Are there agronomic problems in farms and gardens at Palmerton? Initially there was, but since so much agronomic knowledge has been generated on how to wisely manage metal-stressed soils, food crops and feedstuff can be raised successfully and safely. For example, death of foals (i.e., young horses) on farms 2.5 and 10 km from smelters were reported in 1979 (Chaney et al., 1988). The no-longer persistent agronomic problems at Palmerton have been summarized by Chaney et al. (1988) this way: (1) Several agronomic problems in the Palmerton vicinity were caused by excessive Zn and Cd in soils; (2) Excessive Zn can cause phytotoxicity if soils are acidic. Zinc induced chlorosis of vegetable crops was observed in numerous acidic gardens. Zinc toxicity is a strong limiting factor in the yield of vegetable crops; (3) Zinc toxicity often occurs with common lawn grasses even if the yard soils are limed. "Merlin" red fescue has succeeded as a lawn grass in Palmerton, although it is not popular for lawn use; (4) Zinc toxicity reduced alfalfa nodulation on nearby farms. It is likely that some yield reductions could occur in alfalfa and some other forage crops raised by farmers in the Palmerton area; and (5) Zinc toxicity from locally produced forages has injured horses by Zn-induced Cu deficiency. This problem can be remedied by Cu supplementation in the diet that must be continued indefinitely.

6.2 Effects of Zinc from Sewage Sludge Application

A consensus exists that when too much metal-contaminated sludge is applied on farmland not only phytotoxicity but also alteration of soil processes can be expected to occur. For examples, negative effects have been reported by excessive soil metals on microbial activity (McGrath, 1993), leaf litter decomposition (Strojan, 1978), symbiotic N_2 fixation (Heckman et al., 1987), and mycorrhizal infection of clover roots.

Long-term sewage sludge studies at Rothamsted, England showed that only ineffective populations of *R. leguminosarum* bv. *trifolii* survived in soil amended with sludge 30 to 50 yr earlier (Giller et al., 1989; McGrath et al., 1988). A significant reduction in growth of white clover and a reduction in N_2 fixation compared with the unamended control were observed. It was concluded that sludge-borne metals, primarily Cd, Zn, and Cu were responsible for these observations. In contrast, Rother et al. (1993) reported only a minor decrease in nitrogenase activity, plant size, and nodulation of white clover grown in grassland soils contaminated with Cd, Pb, and Zn.

Only minimal effects were observed and plants and nodules were healthy and capable of fixing atmospheric N_2. Recent research indicates that when the soil pH is maintained at 6.0 or higher, effective nodulation and N_2 fixation by legumes can be expected even in sludged soils (Ibekwe et al., 1995). Support to this observation was lent by Kelly and Tate (1998) who inferred from using principal component analysis that although metal-stressed soils has altered the structure of microbial communities, a viable microbial consortium still remains.

Toxicologically, Zn in general is relatively inconsequential since there is a wide range between the usual environmental levels and toxic levels.

6.3 Dietary Intake of Zinc in Humans

Zinc is an essential element in human and animal nutrition. The dietary allowances (RDAs) for Zn recommended by the Food and Nutrition Board (1989) are indicated on Table 16.7. The RDAs are based on body weight or age. For adult males, the RDA is 15 mg day^{-1}, and assuming a 20% absorption efficiency by the gastrointestinal (GI) tract, this yields 3 mg Zn day^{-1} absorbed, higher than the required 2.5 mg day^{-1} absorbed Zn (Van Campen, 1991). Because of their lower body weight, the RDA for adult females is set at 12 mg day^{-1}. The RDAs for pregnant and lactating women include an allowance to meet the needs of the fetus and to compensate for Zn losses in milk. For infants and children, RDAs are adjusted to provide Zn required for growth. Marginal or deficient Zn intake rather than toxicity is the health concern for the general population.

Dietary intakes of Zn (mg day^{-1}) in several countries vary slightly: 14.3 in the United Kingdom; 16.1 in India; 14.0 in Japan, 18.0 in the United States, 4.7 to 11.3 in Italy, and 13.0 by the International Commission on Radiological Protection (Adriano, 1986). The variation in intake can be partly attributed to the different eating habits and preferences of the people of the world. For example, foods of Asians may be dominated by fish and poultry, grain and cereal, leafy and root vegetables, and fruits in contrast to the diet of an average American adult as shown in Table 16.8. Foods

TABLE 16.7. Recommended daily dietary allowances for zinc.

Category	Age (yr)	Zn (mg day^{-1})
Infants	0.0–1.0	5
Children	1.0–10	10
Males	11+	15
Females	11+	12
Pregnant		15
Lactating		
1st 6 mo		19
2nd 6 mo		16

Source: Food and Nutrition Board (1989).

TABLE 16.8. Estimated dietary intakes and concentrations of zinc in various food classes.

Class	Food intake[a]		Zn intake[b]		Zn concentration (ppm WW)[b]	
	g food/day	% of total diet	µg Zn/day	% of total diet	Average	Range
Dairy products	769	22.6	3837	21.4	5.1	3.4–7.3
Meat, fish, and poultry	273	8.0	6600	37.2	30	20–58
Grain and cereal	417	12.3	3370	18.8	7.4	5.1–8.2
Potatoes	200	5.9	1198	6.7	4	1.6–7.8
Leafy vegetables	63	1.9	136	0.8	3	1.9–7.5
Legume vegetables	72	2.1	542	3.0	8.1	4.7–16.8
Root vegetables	34	1.0	80	0.5	2.6	1.6–4.2
Garden fruits	89	2.6	267	1.5	3.4	1.9–7.5
Fruits	222	6.5	194	1.1	—	—
Oil and fats	51	1.5	314	1.8	—	—
Sugars and adjuncts	82	2.4	254	1.4	—	—
Beverages	1130	33.2	1066	6.0	—	—
Total	3402	100.0	17918 (~18 mg)	100.0		

[a]Based on 1972–1973 total diet survey by the U.S. Food and Drug Administration for an adult person (Tanner and Friedman, 1977).
[b]Based on 1973 total diet survey by the U.S. Food and Drug Administration (Russell, 1979).

vary not only by classes but also by individual items. Among the richest sources of Zn are red meat, wheat germ and bran (40 to 120 ppm), and seafoods (with oysters, up to 1000 to 1500 ppm); white sugar, pome, and citrus fruits are among the lowest sources, usually with <1 ppm of fresh edible portion (Underwood, 1973). In the United States, approximately 70% of the Zn consumed comes from meat and dairy products (Welch, 1993). Plant-based foods are not major contributors to the Zn requirements, except for vegetarians and infants being fed soy-based formula. The tolerance level given for Zn in fresh vegetables in Canada to comply with the 1970 Food and Drug Act and Regulations is 50 ppm on a wet weight basis (MacLean, 1974). Assuming a moisture content of 94% for lettuce, this tolerance level is equal to 833 ppm Zn on a dry matter basis. Only in very rare instances can this high concentration be encountered in agricultural practices.

Food processing can alter the Zn content in foods. Results from 22 countries indicate that unpolished rice contains, on the average, 16.4 ppm Zn vs. polished rice, which contains 13.7 ppm Zn (Suzuki et al., 1980). Zinc may be lost during polishing since it is contained mostly in the germ and bran, which are eliminated during milling. The average per capita consumption of polished rice in countries such as Bangladesh, Indonesia, Japan, Philippines, Singapore, and Taiwan is very high: 260, 300, 304, 235, 237, and 364 g day^{-1} respectively. In contrast, daily consumption in most European countries is very low, usually <10 g/person.

6.4 Zinc Deficiency in Humans

Zinc deficiency in humans is typically a result of inadequate dietary intake that occurs particularly at times of high mineral requirement, such as growth (as in children), pregnancy, and lactation (Ohnesorge and Wilhelm, 1991). Additionally vegetarianism, synthetic diet, and protein–calorie deficiency may induce Zn deficiency.

More than 200 Zn enzymes and proteins have been identified (Hambidge et al., 1986). Severe Zn deficiency depresses activities of several important enzymes, including alkaline phosphatase, alcohol dehydrogenase, thymidine kinase, and DNA and RNA polymerases (Sandstead and Evans, 1984).

Clinical signs of Zn deficiency in humans were first reported for adolescent males in the Middle East displaying growth impairment and sexual immaturity (Prasad et al., 1963). Poor children from the southeastern United States and certain institutionalized children as well as teenage and college-age women may be suspected to have marginal status relative to Zn nutrition (Van Campen, 1991).

Because severe Zn deficiency seldom occurs in humans, assessment of Zn deficiency, especially at the marginal scale, is difficult. Clinical signs include dermatitis, anemia, poor wound healing, impaired immunity, dwarfism, hypogonadism, impotence, and neuropsychological dysfunction (Prasad et al., 1961; Sandstead and Evans, 1984).

6.5 Zinc Toxicity in Humans

Zinc toxicity is generally limited to cases of accidental overdose or therapeutic use of high Zn doses (Van Campen, 1991). High intake of Zn coupled with low Cu intakes may adversely affect cholesterol metabolism. The lowest lethal dose of humans is estimated at $\sim$50 mg $ZnCl_2$ or 106 mg $ZnSO_4$ per kg body weight (Ohnesorge and Wilhelm, 1991). The provisional maximum tolerable daily intake for Zn for humans has been set at 0.30 to 1.0 mg Zn per kg body weight (WHO, 1983).

7 Factors Affecting Mobility and Bioavailability of Zinc

Since Zn^{2+} ions are intercepted by plant roots primarily through diffusion, any factor inducing changes in the solubility or affinity of Zn for soil constituents is important relative to its fate in the soil–plant system.

7.1 pH

Among factors affecting the solubility of metals in soils, pH is likely to be the most easily managed. In general, the solubility and phytoavailability of metals are inversely related to soil pH. At soil pH above 7.0 the bioavailability of Zn is substantially reduced. Severe Zn deficiency is often associated with alkaline soil pH. In calcareous soils, Zn deficiency can also be prevalent. It is difficult to correct Zn deficiency in calcareous soils because the added Zn precipitates in soil rapidly under this condition. There are numerous reports elucidating the depression of Zn uptake by crops in alkaline or calcareous soils (Adriano, 1986). Liming of an acid sandy loam soil (pH 4.9) to pH ~7 markedly reduced the amounts of extractable Zn (MacLean, 1974). Zinc activity in soil solution declined sharply when soils were limed from pH 4.3 to pH above 5.0 (Friesen et al., 1980). Liming a strongly acid sandy loam soil to pH 5.6 and above reduced the availability of Zn to forage crops (Gupta et al., 1971). Pepper et al. (1983) found that liming plots from 4.6 to 6.5 amended with anaerobically digested sewage sludge reduced Zn levels in corn tissue.

Zinc deficiency of rice is widespread throughout Asia on neutral to alkaline, calcareous soils especially those containing 1% or more OM. The incidence of Zn deficiency has been correlated positively with high pH (Forno et al., 1975; Yoshida et al., 1973). When soil pH is low, a low content of extractable Zn is not necessarily associated with Zn deficiency. In controlled experiments, Zn uptake by rice plants under either aerobic or anaerobic conditions progressively decreased as soil pH was raised from 5.0 to 8.0 (Jugsujinda and Patrick, 1977). There was a sharp decrease in uptake under aerobic conditions at pH 7.0 to 8.0. Lowland rice growing in calcareous or limed soils often exhibits Zn deficiency (Ponnamperuma, 1972).

In a greenhouse experiment, Chang et al. (1982) found that uptake by barley of sludge-borne Zn was much lower from a soil whose pH was 7.8 than from the other two soils (pH 7.1 and 6.0). In potato, lime application increased the foliage and tuber contents of Ca and concurrently decreased Zn uptake (Laughlin et al., 1974). In soybeans grown on strongly acid soils containing Zn from peach sprays, lime application decreased the Zn content of leaves from 229 ppm in the acid soil (pH 5.4) to 77 ppm in limed soil (pH 6.4) (Lee and Craddock, 1969). However, overliming soils, especially infertile sandy soils, can drastically reduce Zn availability and result in severe Zn deficiency among sensitive crops.

Under field conditions, positive yield response of corn to extractable Zn concentrations has occurred (Lins and Cox, 1988). The critical soil Zn levels were 0.90 and 1.1 ppm extractable Zn at pH 6.0 and 6.7, respectively (Fig. 16.6). The response is not linear but quadratic and indicates a trend for the critical level to increase with an increase in pH.

While elevated soil pH is expressed as decreased metal content in plant tissue, its effect is directly expressed in the soil capacity to adsorb metals. In general, there is direct correlation between adsorption and soil pH (Harter, 1991). High exchangeable Zn fraction is usually obtained in acidic conditions (Chlopecka et al., 1996; Liang et al., 1990). In contrast, high pH (i.e., >6.5) reverts most of the Zn into forms that are not phytoavailable (Payne et al., 1988). In oxide systems (i.e., Fe- and Mn-oxide), increasing

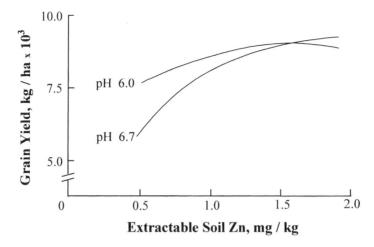

FIGURE 16.6. Relationships between corn grain yield obtained under field conditions as a function of Mehlich 1–extractable soil Zn and pH. (Modified from Lins and Cox, 1988.)

pH induced the retention of predominantly nonexchangeable Zn (Stahl and James, 1991a,b; Bruemmer et al., 1988). In soils, adsorption of Zn dramatically increased above pH 7.0 to 7.5 (Harter, 1983). At this pH range, Zn^{2+} and $ZnOH^+$ are expected to be the predominant species adsorbed.

The importance of pH on the chemical speciation of Zn and its adsorption by soil constituents has been discussed by Harter (1991). It is well known that raising the pH by liming apparently is the most efficient method for reducing plant uptake of potentially toxic metals (Albasel and Cottenie, 1985). The importance of pH in regulating the available form of Zn can best be demonstrated by Figure 16.7, showing the greater release of extractable Zn under acidic conditions. In calcareous soils where Zn deficiency of crops and fruit trees is common, acidification of the root zone may prove an efficient method to increase the bioavailability of Zn to plants (Fenn et al., 1990).

7.2 Organic Matter

Zinc bioavailability to plants is generally low in organic soils. Thus, Zn deficiency oftenly occurs in muck or peat soils or even in old barnyards and corral sites. This can be attributed to the formation of insoluble Zn-organic complexes or adsorption by insoluble soil OM that render Zn unavailable to plants. In comparing two Coastal Plain soils, Zn uptake by soybeans was lower from the higher OM soil than the lower OM soil (White and Chaney, 1980). The former soil had 3.8% OM and CEC of 16 meq/100 g, compared

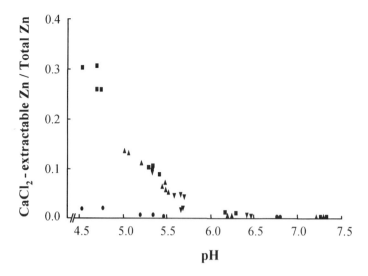

FIGURE 16.7. Influence of pH on the proportion of extractable zinc from a Rothamsted silty clay loam soil (500 g) spiked with Zn from $Zn(NO_3)_2$ or sewage sludge. ●, control; ■, 65 mg Zn as $Zn(NO_3)_2$; ▲, 65 mg Zn as $Zn(NO_3)_2$ + 2.5 g dry solid of uncontaminated sludge; ▼, 65 mg Zn from contaminated sludge. (Modified from Sanders and El Kherbawy, 1987.)

with 1.2% OM and 5.4 meq/100 g for the latter. Organic matter has been rated to be more important than hydrous oxides of Fe and Mn in moderating the effects of excessive soil Zn to soybeans. Other evidence linking the importance of OM to Zn retention by soils is the decreased adsorption capacity for soils where OM was removed (Shuman, 1991). Even paddy soils with high extractable Zn can be expected to produce Zn deficiency in rice if the soil contains over 1% OM (Yoshida et al., 1973). Addition of rice straw to a silt loam soil in Arkansas caused a significant decrease in soil solution Zn concentrations at 15 days after flooding, after which the differences between treatments dissipated (Yoon et al., 1975). Apparently Zn was temporarily immobilized by the bicarbonates, hydroxides, and organic ligands from the decomposition of straw. Another possibility is immobilization of Zn in microbial biomass during decomposition.

The role of OM on the mobility of Zn is influenced by pH. As discussed in Chapter 2, the solubility of soil OM increases with increasing pH, which in turn affects the stability of metal–humic complexes.

7.3 Redox Potential

Redox condition has also a pronounced effect on the solubility of Zn and hence its bioavailability to rice plants. The effect is not directly related with

Zn since it cannot be reduced under low redox conditions. In constantly flooded soils, insoluble zinc sulfide may be formed under strongly reducing conditions. This occurs at about -150 mV redox potential (Connell and Patrick, 1968), which may render Zn nonavailable to rice plants. However, the activity of rice plant roots may be involved in metal absorption and translocation from flooded soils averting potential nutritional problems (Chino, 1981a).

7.4 Management and Fertilization Practices

In addition to overliming, the following management practices can cause Zn deficiency:

7.4.1 Removal of Surface Soil

Removal of surface soil during the course of land leveling for irrigation purposes exposes the subsoils, which are often low in bioavailable Zn. Numerous instances are known where crops grown on newly leveled fields showed deficiency symptoms (Adriano, 1986). Zinc deficiency may be intense where sensitive crops are grown on areas where surface soil is removed by erosion, land leveling, or terracing (Martens and Westermann, 1991).

7.4.2 Zinc Carrier

Several inorganic compounds, natural organic complexes such as Zn lignosulfonate, and synthetic chelates are used to correct Zn deficiency. Zinc sulfate is used more than other inorganic carriers for correcting Zn deficiency (Table 16.9) (Martens and Westermann, 1991). In general, Zn applied as a chelate is more readily utilized by the plants than Zn applied as an inorganic salt. Chelated sources of Zn are believed to be about 5 times as effective as inorganic sources in overcoming Zn deficiency (Adriano, 1986). Boawn et al. (1957) noted that when Zn was applied to an alkaline silt loam soil, bean plants took up 3.5 times more Zn from ZnEDTA than from $ZnSO_4$. Based on Zn concentration in sweet corn, Wallace and Romney (1970) estimated that 0.90 kg Zn ha^{-1} as ZnEDTA was as effective as 9.0 kg ha^{-1} as $ZnSO_4$. Field tests in Michigan showed that 0.55 kg Zn ha^{-1} as chelate was as effective as 2.7 kg ha^{-1} applied as $ZnSO_4$ (Brinkerhoff et al., 1966; Judy et al., 1965). Both carriers were banded near the seeds. In field experiments with potato, Soltanpour et al. (1970) found ZnEDTA to be a more effective Zn source than $ZnSO_4$ when 1.8 kg ha^{-1} Zn as EDTA produced a comparable yield to that of 7.1 kg ha^{-1} Zn as $ZnSO_4$.

Zinc deficiency can be corrected by applying Zn to plant foliage or to soil by either band or broadcast procedures. Foliar application is often considered an emergency procedure for salvaging a crop when unexpected

TABLE 16.9. Zinc sources used for correcting zinc deficiency in crops.

Source	Formula	Zn (g kg^{-1})[a]
Inorganic carriers		
Zinc ammonia complex	Zn-NH$_3$	100
Zinc carbonate	ZnCO$_3$	520–560
Zinc chloride	ZnCl$_2$	480–500
Zinc frits	Fritted glass	100–300
Zinc nitrate	Zn(NO$_3$)$_2 \cdot$ 6H$_2$O	220
Zinc oxide	ZnO	500–800
Zinc oxysulfate	ZnO + ZnSO$_4$	400–550
Zinc sulfate monohydrate	ZnSO$_4 \cdot$ H$_2$O	360
Zinc sulfate heptahydrate	ZnSO$_4 \cdot$ 7H$_2$O	230
Basic zinc sulfate	ZnSO$_4 \cdot$ 4Zn(OH)$_2$	550
Organic carriers		
Zinc chelate	Na$_2$ZnEDTA	140
Zinc chelate	NaZnHEDTA	90
Zinc chelate	NaZnNTA	90
Zinc lignosulfonate	—	50–80
Zinc polyflavonoid	—	50–100

[a]Approximate concentration.
Source: As summarized by Martens and Westermann (1991).

Zn deficiency occurs during a growing season. The effectiveness of the Zn source may be influenced by the method of soil placement. Using pot culture, Brown and Krantz (1966) noted that Zn uptake by corn from ZnSO$_4$ and ZnEDTA was comparable when these materials were mixed into the soil, but when banded under the seeds ZnEDTA was more effective than ZnSO$_4$. Similarly, Chesnin (1963) reported that when the sources were banded under the corn seeds in pot culture, Zn chelates produced higher dry matter than when ZnSO$_4$ was used. Lower rates of band than broadcast application of ZnSO$_4$ can be used to correct Zn deficiency (Table 16.10). The greater efficiency of the banded Zn is due to the lower extent of soil-Zn contact and thus to slower reversion of the applied Zn to unavailable forms.

In rice plants grown in the greenhouse, application of equal amounts of ZnSO$_4$ and ZnEDTA showed leaf-burning symptoms with ZnEDTA but not with ZnSO$_4$, probably due to toxicity of excess EDTA or high accumulation of Zn in plant tissues (414 ppm) (Yoshida et al., 1973). In pot experiments, ZnEDTA gave the highest yield of paddy rice, followed by ZnSO$_4$ and ZnO when these materials were mixed with silty loam soil (Singh and Singh, 1980). The usual greater response to chelate sources may be attributed to greater diffusion of the Zn^{2+} ions into the root surface.

Sewage sludge has become a more popular source of metals in farmlands in Europe and the United States. Metals applied to arable soils as soluble salts are generally considered more phytoavailable on a short-term basis than equivalent amounts of metals applied from sewage sludge (Bell et al., 1991).

TABLE 16.10. Sources and methods for correcting zinc deficiency in crops.

Crop	Zn source	Zn rate, kg ha^{-1}	Application method	Comment
Red Mexican bean, sweet corn	Na$_2$ZnEDTA	0.9–1.8	Broadcast	Apply with macronutrient fertilizer and plow in.
Rice	ZnSO$_4$, ZnO	9.0	Broadcast	Surface apply before flooding.
Corn	ZnSO$_4$	34.0	Broadcast	Reapply every 4 to 5 yr.
		6.6	Band	
Corn	ZnSO$_4$, ZnO	11.0	Broadcast	Use finely divided ZnO.
	Zn chelate	2.2–3.3	Broadcast	
Corn	ZnSO$_4$, ZnO	3.3	Band	Band with starter fertilizer, use finely divided ZnO.
	Zn chelate	0.6–1.1	Band	
Corn, snapbean, vegetables	ZnSO$_4$	4.5–9.0	Broadcast	Apply with N, P, and K fertilizer.
	ZnEDTA	1.1–2.2	Broadcast	
Corn, snapbean, vegetables	ZnSO$_4$	2.2–4.5	Band	Band with N, P, and K fertilizer.
	ZnEDTA	0.6–1.1	Band	
Flax	ZnSO$_4$	11.0	Broadcast	Incorporate by plowing.
Corn, bean, grain, sorghum, sweet corn	ZnSO$_4$	11.0–17.0	Broadcast	DTPA extractable Zn, <0.5 mg kg^{-1}.
		5.6–11.0	Broadcast	DTPA extractable Zn, 0.5 to 1.0 mg kg^{-1}.
Corn, bean, grain, sorghum, sweet corn	ZnSO$_4$	1.1–2.2	Band	DTPA extractable Zn, <0.5 mg kg^{-1}; band with starter fertilizer.
	Na$_2$ZnEDTA	<1.1	Band	
Corn, sorghum	ZnSO$_4$	11.0	Broadcast	DTPA extractable Zn, <0.40 mg kg^{-1}.
		5.5	Broadcast	DTPA extractable Zn, 0.41 to 0.81 mg kg^{-1}.
Corn, bean, flax, potato, sorghum, soybean	ZnSO$_4$	5.6–11.0	Broadcast	Reapply every 2 to 4 yr.
		1.0–5.5	Band	

Source: As summarized by Martens and Westermann (1991).

7.4.3 Placement of Phosphorus

Method of application of P fertilizer can have significant effects on the magnitude of P–Zn interaction. When supplemental P is required, placement of P close to seeds at planting has been recognized as a more highly effective method for inducing P absorption during early plant development than are preplant broadcast applications that are incorporated into the soil by tillage. Higher P concentrations are frequently noted in plants from banded P. This difference in early season P uptake is particularly pronounced when soil conditions are relatively cool and wet after seeding.

Since Zn uptake is depressed under cool, wet soil conditions, banded placement of P tends to intensify P–Zn interaction, especially when Zn is deficient or near-deficient in soil. Banded application of P near the seeds has been observed to depress Zn uptake by corn (Adriano and Murphy, 1970; Grant and Bailey, 1993; Stukenholtz et al., 1966). Strong evidence for the improved uptake of P from banded applications in cool, wet spring with concurrent depression of Zn concentrations in corn leaves were demonstrated in Kansas when P was applied without Zn and when P was banded (Adriano and Murphy, 1970). When the Zn deficiency was eliminated, the P–Zn interaction became positive, resulting in higher grain yields.

7.5 Plant Factors

Crop cultivars differ in their ability to utilize nutrients. Intercultivar variation in micronutrient uptake has been observed with corn, rice, barley, wheat, and cocoa (Adriano, 1986). Similar differences have been reported in the literature in relation to the abilities of sorghum and soybeans to absorb Fe. The substantial differences in Zn uptake among cultivars of lettuce and corn in Zn nutrition have been demonstrated (Adriano, 1986). Similarly, cultivars of soybeans and wheat reacted differently to P–Zn interaction (Paulsen and Rotimi, 1968; Sharma et al., 1968). Obviously, lines that have a poor ability to utilize micronutrients present in their root environment will be more subject to P-induced deficiencies. Brown and Tiffin (1962) reported that Zn deficiency is also dependent on plant species. Crops sensitive to Zn deficiency in the United States are beans, potatoes, and peaches in the northwest; cotton and rice in the southwest; corn, flax, and soybeans in the Great Plains; corn and beans in the north-central region; and corn and citrus in the southeastern region. In general, cereal crops have not been found to be sensitive to Zn deficiency (Viets et al., 1954).

Zinc-efficient genotypes have been generalized to typically contain lower Zn concentrations in roots and higher concentrations in young leaf blades (e.g., oilseed rape), indicating a better transport and utilization of Zn (Erenoglu et al., 1999; Graham and Rengel, 1993; Grewal et al., 1997; Khan et al., 1998).

8 Sources of Zinc in the Environment

Other than Zn arising from geogenic processes, all other sources of Zn in soils are anthropogenic. The most common sources of Cd and Pb pollution also produce high inputs of Zn into the environment. For example, Cd and Zn, and to some extent Pb coexist in sewage sludge and smelting emissions.

8.1 Zinc Fertilizers

In agriculture, Zn-carrying fertilizers are by far the largest source of Zn. There are four main classes of Zn sources: inorganic, synthetic chelates, natural organic complexes, and inorganic complexes (Mortvedt and Gilkes, 1993). Primary inorganic sources include $ZnSO_4$, the most commonly used Zn fertilizer, ZnO, $ZnCO_3$, $Zn(NO_3)_2$, $ZnCl_2$, and zinc–ammonia ($Zn–NH_3$) complex (see Table 16.10). Primary organic sources include ZnEDTA, zinc lignosulfonates, and zinc polyflavonoids. The last two are wood by-products in the paper industry. Zinc applied for crop production can have residual effects for up to 2 or more yr (Boawn, 1974; Brown et al., 1964; Follet and Lindsay, 1971).

Recommended rates of Zn for various crops range from <1 kg ha^{-1} Zn as chelates to as high as 22 kg ha^{-1} as $ZnSO_4$ (Murphy and Walsh, 1972). The rate of application depends upon the crop species, soil or leaf test recommendation, method of placement (foliar, banded, or broadcast), and carrier of Zn (chelate vs. inorganic). While foliar application and seed treatments can be satisfactory in supplying Zn to crops, it is generally accepted that there is no better substitute than soil application. Foliar application should be followed as a supplement only during critical periods. Very high rates of Zn applied to soil can be tolerated by crops, however. In a field experiment on a loamy sandy soil, rates up to 363 kg Zn ha^{-1} as $ZnSO_4$ failed to affect the yields of snap beans, cucumber, or corn deleteriously even if Zn concentrations in crop tissue exceeded 350 ppm (Walsh et al., 1972).

8.2 Municipal Sewage Sludge

Other than Zn fertilizers that are applied to soils according to soil test recommendations, sewage sludge has gained more acceptance as a soil conditioner and fertilizer source. Because most sludge studies conducted in the United States, Canada, and western Europe are experimental in nature, application rates are usually in tens or sometimes in hundreds of tonnes, giving much larger amounts of Zn than commercial fertilizers. In resource-limited Far Eastern countries (e.g., China), raw sewage has been applied directly to soils as a source of fertilizer for many centuries.

The review by Page (1974) indicates that metal concentrations of plants grown on sludge-amended soils are dependent on loading rates, soil properties (particularly pH), and plant species. There is a wide range of metal concentrations in sewage sludge and the chemical forms of these elements differ among sludges from different treatment plants. Sludge that contains large amount of metals may adversely affect soil productivity and lead to food chain accumulation by some metals. Based on these expectations, the U.S. Environmental Protection Agency (EPA) has set numerical limits for permissible concentrations of toxic metals in sewage sludge and in sludged soils (U.S. EPA, 1993). These limits are based on intensive research on the potential adverse effects of metals from sludge and potential transfer into the food chain (see Chapter 4).

To date, results on the effect of sewage sludge on crop growth, soil productivity, and food chain quality are quite conflicting, signaling the complexity of the issue. Increases in yield can be expected due to the N and P in the material. However, utilization of sludge on land often results in significant increases of toxic metals in soils, with Zn as one of the metals likely to cause phytotoxicity, particularly in acid soils.

Numerous findings on the accumulation of trace metals in tissue of edible vegetables grown on sludge-amended soils have been reported: substantial increases in levels of Zn in beets and Swiss chard (Chaney et al., 1977; Ryan, 1977); edible fruit, tuber, and root tissues of vegetable crops (Dowdy and Larson, 1975); and edible portion of several vegetables (Giordano and Mays, 1977). Zinc concentrations generally were higher in the leaf or vegetative tissue than in the storage tissue. The residual effects of large amounts of sewage sludge applied to soils in elevating plant tissue Zn content may linger. This is expected, since Zn in soil accumulates primarily in the root zone of crops and, unlike the major nutrients, is not easily depleted by plant uptake.

In general, sludge-borne metals do not represent a serious limitation for plant growth. Phytotoxicity is rarely observed from sludge applications even though large amounts of metals are added into the soil (Juste and Mench, 1992). Several hypotheses can be offered to account for the relatively low phytotoxic effect of sludge-borne metals. These include increased soil pH, formation of insoluble salts (e.g., phosphate, sulfate, and silicate salts), metal sorption by Fe and Mn oxides, or OM. The antagonistic interactions among the sludge metals may be another reason for their low phytotoxic effect.

Long-term field experiments in several parts of the world demonstrate that Zn is the most bioavailable metal in sludge-treated soils. In an experiment in Bordeaux, France the Zn concentrations in the corn's 6th leaf obtained from the high-rate sludge plots were sharply higher than that in the control since the beginning of the study (Juste and Mench, 1992). Except for 1976 and 1977, the mean Zn concentrations of leaves progressively increased year after year and reached 600 mg kg^{-1} in 1986. But corn containing this level of Zn did not exhibit any phytotoxicity symptoms. The phytotoxicity threshold of 300 mg kg^{-1} quoted by Hinesly et al. (1977) perhaps underestimated plant tolerance of Zn. The Zn concentration of grains was significantly higher than the control and fluctuated yearly up to 1982. It leveled off thereafter at approximately 60 mg kg^{-1}, indicating a higher mobility in plants than other metals. Barley grown in sludge-treated soils also accumulated Zn, and metal concentration in barley leaves and grains peaked as high as 820 and 250 mg kg^{-1}, respectively.

Results from experiments conducted at Woburn, United Kingdom and in Germany showed that Zn concentration of sugar beet leaves could be as high or higher than Zn concentrations of the soils in which they were grown. Sauerbeck and Styperek (1985) reported that signs of Zn toxicity were apparent in the sugar beets grown on soils with high levels of Zn. The Zn accumulation in potato tubers and carrot roots however, was significantly lower than in beets. The Zn concentration in tobacco levels was higher than that in soybeans, although the crops were grown on different sites with different sludges. Soils receiving lime-treated sludges had higher soil pH and produced plants that accumulated lower Zn.

Synthesis of data from long-term investigations on bioavailability of sludge-borne metals in soils reveals the following (Juste and Mench, 1992):

1. Sludge-borne metals in soils are accumulated in the surface layer. The lack of downward movement and the lateral displacement of metals caused by cultivation contributed to the varying recovery percentages with respect to time and metal species. The potential for metals to be transported outside of the experimental plots requires the attention of researchers and casts uncertainty on evaluating data of long-term experiments.
2. Sludge-borne metals do not cause serious toxicity to the majority of crops even if large amounts of metals are added to the soil.
3. Zinc is the most bioavailable sludge-borne metal. Cadmium and nickel are also bioavailable, but the ability for the absorbed Cd and Ni to move from vegetative parts into grains is substantially less than that for Zn.
4. Despite results provided by long-term field experiments, information is needed to evaluate long-term effects on factors controlling metal bioavailability in soils, and to determine long-term effects on soil properties as a result of metal accumulation.

8.3 Mining and Smelting

Mining and smelting activities have released substantial amounts of Zn into the environment. The hazards from mining arise from dispersal of spoils and mine drainage effluent, while atmospheric discharge of metals and other pollutants is the main concern in smelting. With smelting, the impacts from aerial discharges appear to diminish substantially beyond the 10-km radius (John, 1975; Whitby and Hutchinson, 1974). Hogan and Wotton (1984) found that in a boreal forest in Manitoba, Canada that had been exposed to smelter deposition for over 50 yr, significant deposition of metals occurred at sites up to 35 km from the stacks. The deposition of Zn from the source declines progressively with distance. The most heavily impacted sites were within the 6-km distance, where significant accumulation of metals at lower soil depths has occurred. Soils in the 10- to 15-cm depth at these sites averaged 260 ppm total Zn compared with 80 ppm total Zn in soils from the same depth, 35 km farther away.

TABLE 16.11. Zinc concentrations (ppm DW) in surface soils in the vicinity of mines and smelters.

Source of contamination	Range	Mean	Location
Zn–Pb smelting	11–10,500	205	Poland
Cu smelting	49–554	–	Poland
Zn–Pb mining and smelting	238–472	349	United Kingdom
Cu smelting	7–910	–	Poland
Zn–Pb mining	2040–50,000	14,970	United Kingdom
Zn–Pb mining	10–49,390	728	United Kingdom
Pb mining	11–641	96	United Kingdom
Zn–Pb mining	400–4245	–	Russia
Zn smelting	1340–180,000	–	Belgium
Zn–Pb smelting	180–3500	1,050	Zambia
Cu smelting	430–1370	–	United States
Zn smelting	110–60,700	–	Canada
Ag–Cu–Pb–Zn mining and smelting		554	Japan
Zn smelting		2,900	United States (Palmerton, PA)[a]

Normal Zn soil levels: 15–100

[a]Beyer and Storm (1995).
Source: Modified from Dudka and Adriano (1997).

Surface soils in the vicinity of smelters contain elevated levels of metals (Dudka and Adriano, 1997). Zinc in soils collected near smelters ranged from <10 to >100,000 ppm (Table 16.11). As mentioned earlier, the abundance and diversity of soil biota are generally reduced by smelter emissions, with a consequence of reduced soil fertility due to disrupted C and N biogeochemical cycles.

8.4 Diffuse Sources: A Case Study

A holistic view of metal input into an entire landscape is exemplified by the diffuse aqueous emissions to the Rhine River (Table 16.12). The data were estimated for the peak period of late 1960s to early 1970s with emissions in 1988 (Stigliani et al., 1993). Similar to the trends in air emissions, there have been significant declines over the previous 20 yr. For Zn, emissions in 1988 were 30% of emissions during the peak period. Emissions came from both point and diffuse sources. Although there have been significant declines in both types of source in relative terms, emissions from point-source emissions have declined more.

Several factors resulted in reductions in point-source emissions: (1) the installation of water pollution control equipment in compliance with increasingly stringent regulations; regulations however, generally lagged behind those for controlling air pollution by several years; thus, the Rhine River

TABLE 16.12. Estimated amounts of diffuse aqueous emissions to the Rhine River of zinc from urban and agricultural areas, by source (tonnes yr^{-1}).

Source	Late 1960s/early 1970s	1988
Urban areas		
Deposition (atmospheric)	868	127
Corrosion	2047	575
Traffic	191	308
Leaching	809	391
Agricultural areas		
Runoff	388	71
Erosion	390	426
Natural deposits	530	530
Total diffuse	5223	2428

Source: Extracted from Stigliani et al. (1993).

continued to be highly polluted by these chemicals throughout the 1970s; (2) the implementation of "good housekeeping" practices—for example, two thermal Zn smelters located by the river were the source of nearly half of all aqueous Cd emission during the 1970s; (3) the substantial increase in the recycling of industrial sludges and solid wastes by the metal-producing industries during the 1970s and 1980s—for example, wastes containing Cd and Zn from the iron and steel industry were shipped to zinc–cadmium refineries, which reduced emissions of these metals; and (4) the structural changes in the major polluting industries, where cleaner industrial processes replaced older technologies.

Diffuse sources of emissions in the Rhine Basin are dominated by storm runoff from urban and arable areas. Inputs of heavy metals on urban surfaces originate from vehicle exhausts, tire wear, corrosion of building materials, and atmospheric deposition. In arable areas, stormwater runoff, erosion, and seepage from natural geochemical deposits are the major sources of diffuse emissions. Anthropogenic inputs to arable soils occur from both atmospheric deposition and the application of agrochemicals. The reductions in atmospheric emissions as noted above have been a major factor in the observed overall decreases in diffuse-source emissions to the river.

Corrosion of galvanized steel has been a major source of emission of Zn, accounting for nearly 40% of the total aqueous diffuse-source emissions to the river during the peak period. By 1988, corrosion-derived emissions of Zn declined both in absolute amounts (28% of the peak year emission) and as a share of the total emissions (24%). The decrease coincided with the decline in urban concentrations of corrosive air pollutants such as SO_2. Until the late 1960s, Zn was produced mainly from thermal smelters in which trace Cd was also present in concentrations of 0.15 to 0.50%. Since that time, Zn has been produced mainly by the electrolytic method, in which Cd concentrations are ≤0.02%.

References

Adriano, D.C. 1986. *Trace Elements in the Terrestrial Environment*. Springer-Verlag, New York.

Adriano, D.C., and L.S. Murphy. 1970. *Agron J* 62:561–567.

Adriano, D.C., G.M. Paulsen, and L.S. Murphy. 1971. *Agron J* 63:36–39.

Ainsworth, C.C., and D. Rai. 1987. In EPRI Rep EA-5321, Electric Power Res Inst, Palo Alto, CA.

Albasel, N., and A. Cottenie. 1985. *Soil Sci Soc Am J* 49:386–390.

Alley, M.M., D.C. Martens, M.G. Schnappinger, and G.W. Hawkins. 1972. *Soil Sci Soc Am Proc* 36:621–624.

Archer, F.C. 1980. Minist Agric Fish Food (Great Britain) 326:184–190.

Bar-Tal, A., B. Bar-Yosef, and Y. Chen. 1988. *Soil Sci* 146:367–373.

Bauer, A. 1971. *Commun Soil Sci Plant Anal* 2:161–193.

Baveye, P., M.B. McBride, et al. 1999. *Sci Total Environ* 227:13–28

Bell, P.F., B.R. James, and R.L. Chaney. 1991. *J Environ Qual* 20:481–486.

Benson, N.R. 1966. *Soil Sci* 101:171–179.

Berrow, M.L., and G.A. Reaves. 1984. In *Proc Intl Conf Environmental Contamination*. CEP Consultants, Edinburgh.

Berthelsen, B.O, E. Steinnes, W. Solberg, and L. Jingsen. 1995. *J Environ Qual* 24:1018–1026.

Beyer, W.N., and G. Storm. 1995. In *Handbook of Ecotoxicology*. CRC Pr, Boca Raton, FL.

Bidwell, A.M., and R.H. Dowdy. 1987. *J Environ Qual* 16:438–442.

Bingham, F.T., A.L. Page, and J.R. Sims. 1964. *Soil Sci Soc Am Proc* 28:35 1354.

Boardman, R., and D.O. McGuire. 1990a. *Forest Ecol Manage* 37:167–205.

Boardman, R., and D.O. McGuire. 1990b. *Forest Ecol Manage* 37:207–218.

Boawn, L.C. 1971. *Commun Soil Sci Plant Anal* 2:31–36.

Boawn, L.C. 1974. *Soil Sci Soc Am Proc* 38:800–803.

Boawn, L.C., and J.C. Brown. 1968. *Soil Sci Soc Am Proc* 32:94–97.

Boawn, L.C., F.G. Viets, Jr., and C.L. Crawford. 1954. *Soil Sci* 78:1–7.

Boawn, L.C., F.G. Viets, Jr., and C.L. Crawford. 1957. *Soil Sci* 83:219–227.

Bosmans, H., and J. Paenhuys. 1980. *Pedologie* 30:191–223.

Bowen, H.J.M. 1979. *Environmental Chemistry of the Elements*. Academic Pr, New York.

Bradford, G.R., R.J. Arkley, P.F. Pratt, and F.L. Bair. 1967. *Hilgardia* 38:541–556.

Brennan, R.F., J.D. Armour, and D.J. Reuter. 1993. In A.D. Robson, ed. *Zinc in Soils and Plants*. Kluwer, Dordrecht, Netherlands.

Brinkerhoff, F., B.G. Ellis, J. Davis, and J. Melton. 1966. *Quart Bull Mich Agric Exp Sta* 49:262–275. East Lansing, MI.

Brown, A.L., and B.A. Krantz. 1960. *Calif Agric* 14:8–9.

Brown, A.L., and B.A. Krantz. 1966. *Soil Sci Soc Am Proc* 30:86–89.

Brown, A.L., B.A. Krantz, and P.E. Martin. 1964. *Soil Sci Soc Am Proc* 28:236–238.

Brown, A.L., J. Quick, and J.L. Eddings. 1971. *Soil Sci Soc Am Proc* 35:105–107.

Brown, J.C., and L.O. Tiffin. 1962. *Agron J* 54:356–358.

Bruemmer, G.W., J. Gerth, and K.G. Tiller. 1988. *J Soil Sci* 39:37–52.

Buchauer, M.J. 1973. *Environ Sci Technol* 7:131–135.

Cakmak, I., and H. Marschner. 1987. *Physiol Plant* 70:20–30.

Cavallaro, N., and M.B. McBride. 1984. *Soil Sci Soc Am J* 48:1050–1054.

Cernik, M., P. Federer, M. Borkovec, and H. Sticher. 1994. *J Environ Qual* 23:1239–1248.

Chaney, R.L., S.B. Hornick, and P.W. Simon. 1977. In R.C. Loehr, ed. *Land as a Waste Management Alternative*. Ann Arbor Sci, Ann Arbor, MI.

Chaney, R.L., W.N. Beyer, C.H. Gifford, and L. Sileo. 1988. *Trace Subs Environ Health* 22:263–280.

Chang, A.C., A.L. Page, K.W. Foster, and T.E. Jones. 1982. *J Environ Qual* 11:409–412.

Chang, A.C., J.E. Warneke, A.L. Page, and L.J. Lund. 1984. *J Environ Qual* 13:87–91.

Chapman, H.D., ed. 1966. *Diagnostic Criteria for Plants and Soils*. Quality Printing, Abilene, TX.

Chaudhry, F.M., F. Hussain, and A. Rashid. 1977. *Plant Soil* 47:297–302.

Chen, J., F. Wei, C. Zheng, Y. Wu, and D.C. Adriano. 1991. *Water Air Soil Pollut* 57–58:699–712.

Chesnin, L. 1963. *J Agric Food Chem* 11:118–122.

Chino, M. 1981a. In K. Kitagishi and I. Yamane, eds. *Heavy Metal Pollution in Soils of Japan*. Japan Sci Soc Pr, Tokyo.

Chino, M. 1981b. In K. Kitagishi and I. Yamane, eds. *Heavy Metal Pollution in Soils of Japan*. Japan Sci Soc Pr, Tokyo.

Chlopecka, A., J.R. Bacon, M.J. Wilson, and J. Kay. 1996. *J Environ Qual* 25:69–79.

Connell, W.E., and W.H. Patrick, Jr. 1968. *Science* 159:86–87.

Crowley, D.E., W. Smith, B. Faher, and J.A. Manthey. 1996. *Hort Sci* 31:224–229.

Dang, Y.P., R.C. Dalal, D.G. Edwards, and K.G. Tiller. 1993. *Plant Soil* 155/156:247–250.

Domingo, L.E., and K. Kyuma. 1983. *Soil Sci Plant Nutr* 29:439–452.

Dowdy, R.H., and W.E. Larson, 1975. *J Environ Qual* 4:278–282.

Dowdy, R.H., J.J. Latterell, T.D. Hinesly, R.B. Grossman, and D.L. Sullivan. 1991. *J Environ Qual* 20:119–123.

Dudka, S., and D.C. Adriano. 1997. *J Environ Qual* 26:590–602.

Duker, A., A. Ledin, S. Karlsson, and B. Allard. 1995. *Appl Geochem* 10:197–205.

Edwards, J.H., and E.J. Kamprath. 1974. *Agron J* 66:479–482.

Elgabaly, M.M. 1950. *Soil Sci* 69:167–174.

Ellis, R., Jr., J.F. Davis, and D.L. Thurlow. 1964. *Soil Sci Soc Am Proc* 35:935–938.

Elrashidi, M.A., and G.A. O'Connor. 1982. *Soil Sci Soc Am J* 46:1153–1158.

Erenglu, B., I. Cakmak, V. Romheld, R. Derici, and Z. Rengel. 1999. *Plant Soil* 209:245–252.

Fenn, L.B., H.L. Malstrom, T. Riley, and G.L. Horst. 1990. *J Am Soc Hort Sci* 115:741–744.

Follett, R.H., and W.L. Lindsay. 1970. In *Profile Distribution of Zinc, Iron, Manganese, and Copper in Colorado Soils*. Colorado Agric Exp Sta Tech Bull 110. Fort Collins, CO.

Follett, R.H., and W.L. Lindsay. 1971. *Soil Sci Soc Am Proc* 35:600–603.

Food and Nutrition Board. 1989. In *Recommended Dietary Allowances* ed 10. National Research Council, Washington, DC.

Forno, D.A., S. Yoshida, and C.J. Asher. 1975. *Plant Soil* 42:537–550.

Frank, R., K. Ishida, and P. Suda. 1976. *Can J Soil Sci* 56:181–196.

Friedland, A.J., A.H. Johnson, and T.G. Siccama. 1984. *Water Air Soil Pollut* 21:161–170.

Friesen, D.K., A.S.R. Juo, and M.H. Miller, 1980. *Soil Sci Soc Am J* 44:1221–1226.

Gangwar, M.S., and S.K. Chandra. 1976. *Commun Soil Sci Plant Anal* 7:295–310.

Ganiron, R.B., D.C. Adriano, G.M. Paulsen, and L.S. Murphy. 1969. *Soil Sci Soc Am Proc* 33:306–309.

Giller, K.E., S.P. McGrath, and P.R. Hirsch. 1989. *Soil Biol Biochem* 21:841–848.

Gillman, G.P., and R.L. Fox. 1980. *Soil Sci Soc Am J* 44:934–938.

Giordano, P.M., and D.A. Mays. 1977. In H. Drucker and R. Wildung, eds. *Biological Implications of Metals in the Environment.* CONF-750929. NTIS, Springfield, VA.

Giordano, P.M., and J.J. Mortvedt. 1978. *Agron J* 70:531–534.

Golley, F.B., J.T. McGinnis, R.G. Clements, G.I. Child, and M.J. Duever. 1975. *Mineral Cycling in a Tropical Moist Forest Ecosystem.* Univ Georgia Pr, Athens, GA.

Golovina, L.P., M.N. Lysenko, and T.I. Kisel. 1980. *Sov Soil Sci* 12:73–80.

Graham, R.D., and Z. Rengel. 1993. In A.D. Robson, ed. *Zinc in Soils and Plants.* Kluwer, Dordrecht, Netherlands.

Grant, C.A., and L.D. Bailey. 1993. *Can J Plant Sci* 73:17–29.

Grenwal, H., J. Stangoulis, T.D. Potter, and R.D. Graham. 1997. *Plant Soil* 191:123–132.

Gupta, U.C., F.W. Calder, and L.B. MacLeod. 1971. *Plant Soil* 35:249–256.

Hambidge, K.M., C.E. Casey, and N.F. Krebs. 1986. *Trace Elements in Human and Animal Nutrition,* 5th ed. Academic Pr, Orlando, FL.

Hamon, R.E., S.E. Lorenz, P.E. Holm, T.H. Christensen, and S.P. McGrath. 1995. *Plant Cell Environ* 18:749–756.

Harter, R.D. 1983. *Soil Sci Soc Am J* 47:47–51.

Harter, R.D. 1991. In J.J. Mortvedt et al., eds. *Micronutrients in Agriculture,* 2nd ed. SSSA 4. Soil Sci Soc Am, Madison, WI.

Havlin, J.L., and P.N. Soltanpour. 1981. *Soil Sci Soc Am J* 45:70–75.

Heckman, J.R., J.S. Angle, and R.L. Chaney. 1987. *J Environ Qual* 16:117–124.

Heinrichs, H., and R. Mayer. 1980. *J Environ Qual* 9:111–118.

Hickey, M.G., and J.A. Kittrick. 1984. *J Environ Qual* 13:372–376.

Hinesly, T.D., R.L. Jones, E.L. Ziegler, and J.J. Tyler. 1977. *Environ Sci Technol* 11:182–188.

Hogan, G.D., and D.L. Wotton. 1984. *J Environ Qual* 13:377–382.

Holm, P.E., T.H. Christensen, J.C. Tjell, and S.P. McGrath. 1995. *J Environ Qual* 24:183–190.

Hu, H., and D. Sparks. 1991. *Hort Sci* 26:267–268.

Ibekwe, A.M., J.S. Angle, R.L. Chaney, and P. Van Berkum. 1995. *J Environ Qual* 24:1199–1204.

Ichikura, T., Y. Doyama, and M. Maeda. 1970. *Bull Osaka Agric Res Center* 7:33–41. Osaka, Japan.

Iyengar, S.S., D.C. Martens, and W.P. Miller. 1981. *Soil Sci Soc Am J* 45:735–739.

Jeng, A.S., and B.R. Singh. 1993. *Soil Sci* 156:240–250.

John, M.K. 1972. *Soil Sci* 113:222–227.

John, M.K. 1974. *Can J Soil Sci* 54:125–132.

John, M.K. 1975. In *Proc Intl Conf Heavy Metals in the Environment.* Univ of Toronto, Ontario, Canada.

Jones, J.B., Jr. 1972. In J.J. Mortvedt, P.M. Giordano, and W.L. Lindsay, eds. *Micronutrients in Agriculture.* Soil Sci Soc Am, Madison, WI.

Jones, J.B. 1991. In J.J. Mortvedt et al., eds. *Micronutrients in Agriculture,* 2nd ed. SSSA 4, Soil Sci Soc Am, Madison, WI.

Jordan, M.J. 1975. *Ecology* 56:78–91.

Judy, W., J. Melton, G. Lessman, B.G. Ellis, and J. Davis. 1965. *Mich State Univ Agric Exp Sta Rep.* 33:8. East Lansing, MI.

Jugsujinda, A., and W.H. Patrick, Jr. 1977. *Agron J* 69:705–710.

Junus, M.A., and F.R. Cox. 1987. S*oil Sci Soc Am J* 51:678–683.

Juste, C., and M. Mench. 1992. In D.C. Adriano, ed. *Biogeochemistry of Trace Metals.* Lewis, Chelsea, MI.

Kanehiro, Y., and G.D. Sherman. 1967. *Soil Sci Soc Am Proc* 31:394–399.

Kang, B.T., and O.A. Osiname. 1985. *Fert Res* 7:131–150.

Kelly, J.J., and R.L. Tate. 1998. *J Environ Qual* 27:609–617.

Khan, A., and P.N. Soltanpour. 1978. *Agron J* 70:1022–1026.

Khan, A.A., and G.K. Zende. 1977. *Plant Soil* 46:259–262.

Khan, H.R., G.K. McDonald, and Z. Rengel. 1998. *Plant Soil* 198:1–9.

Kochian, L.V. 1993. In A.D. Robson, ed. *Zinc in Soils and Plants.* Kluwer, Dordrecht, Netherlands.

Krauskopf, K.B. 1979. *Introduction to Geochemistry,* 2nd ed. McGraw-Hill, New York.

Kuo, S., and A.S. Baker. 1980. *Soil Sci Soc Am J* 44:969–974.

Kuo, S., and B.L. McNeal. 1984. *Soil Sci Soc Am J* 48:1040–1044.

Laughlin, W.M., P.F. Martin, and G.R. Smith. 1974. *Am Potato J* 51:394–402.

LeClaire, J.P., A.C. Chang, C.S. Levesque, and G. Sposito. 1984. *Soil Sci Soc Am J* 48:509–513.

Lee, C.R., and G.R. Craddock. 1969. *Agron J* 61:565–567.

Lee, C.R., and G.R. Craddock. 1970. *Six Trace Elements in Soils.* Melbourne Univ Pr, Melbourne, Australia.

Lee, C.R., and N.R. Page. 1967. *Agron J* 59:237–240.

Legget, G.E., and D.T. Westermann. 1986. *Soil Sci Soc Am J* 50:963–968.

Lessman, G.M., and B.G. Ellis. 1971. *Soil Sci Soc Am Proc* 35:935–938.

Liang, Y., and R.B. Corey. 1993. *J Environ Qual* 22:1–8.

Liang, J., J.W.B. Stewart, and R.E. Karamanos. 1990. *Can J Soil Sci* 70:335–342.

Likens, G.E., and F.H. Bormann. 1970. In *Chemical Analyses of Plant Tissue from the Hubbard Brook Ecosystem in New Hampshire.* Yale School Forestry Bull 79. New Haven, CT.

Lindsay, W.L. 1979. *Chemical Equilibria in Soils.* Wiley, New York.

Lindsay, W.L., and W.A. Norvell. 1969. *Soil Sci Am Proc* 33:62–68.

Lindsay, W.L., and W.A. Norvell. 1978. *Soil Sci Soc Am J* 42:421–428.

Lindsay, W.L., J.F. Hodgson, and W.A. Norvell. 1967. In *Intl Soc Soil Sci Trans Comm H,* IV (Aberdeen, Scotland).

Lins, I.D.G., and F.R. Cox. 1988. *Soil Sci Soc Am J* 52:1681–1685.

Liu, Z., Q.Q. Zhu, and L.H. Tang. 1983. *Soil Sci* 135:40–46.

Loneragan, J.F., and M.J. Webb. 1993. In A.D. Robson, ed. *Zinc in Soils and Plants.* Kluwer, Dordrecht, Netherlands.

Loneragan, J.F., D.L. Grunes, R.M. Welch, and E.E. Cary. 1982. *Soil Sci Soc Am J* 46:345–352.

Lopez, A.S., and F.R. Cox. 1977. *Soil Sci Soc Am J* 41:742–747.

Lucas, R.E. 1967. In Ext Bull 486 Mich State Univ Coop Ext Service, East Lansing, MI.

MacLean, A.J. 1974. *Can J Soil Sci* 54:369–378.

Marsh, D.B., and L. Waters. 1985. *J Am Soc Hort Sci* 110:365–370.

Martens, D.C., and D.T. Westermann. 1991. In J.J. Mortvedt et al., eds. *Micronutrients in Agriculture*, 2nd ed. SSSA 4, Soil Sci Soc Am, Madison, WI.

Matsuda, K., and M. Ikuta. 1969. *Soil Sci Plant Nutr* 15:169–174.

McBride, M.B., and J.J. Blasiak. 1979. *Soil Sci Soc Am J* 43:866–870.

McGrath, S.P. 1993. In S.M. Ross, ed. *Toxic Metals in Soil-Plant Systems*. Wiley, Chichester, United Kingdom.

McGrath, S.P., P.C. Brookes, and K.E. Giller. 1988. *Soil Biol Biochem* 20:415–424.

McKeague, J.A., and M.S. Wolynetz. 1980. *Geoderma* 24:299–307.

Melton, J.R., B.G. Ellis, and E.C. Doll. 1970. *Soil Sci Soc Am Proc* 34:91–93.

Mortvedt, J.J., and R.J. Gilkes. 1993. In A.D. Robson, ed. *Zinc in Soils and Plants*. Kluwer, Dordrecht, Netherlands.

Murphy, L.S., and L.M. Walsh. 1972. In J.J. Mortvedt, P.M. Giordano, and W.L. Lindsay, eds. *Micronutrients in Agriculture*. Soil Sci Soc Am, Madison, WI.

Murthy, A.S.P. 1982. *Soil Sci* 133:150–154.

Nakos, G. 1983. *Plant Soil* 74:137–140.

Nelson, J.L., L.C. Boawn, and F.G. Viets, Jr. 1959. *Soil Sci* 88:275–283.

Norvell, W.A., and W.L. Lindsay. 1972. *Soil Sci Soc Am Proc* 36:778–783.

[NRC] National Research Council 1979. *Zinc*. Univ Park Pr, Baltimore, MD.

Ohki, K. 1975. *Physiol Plant* 35:96–100.

Ohki, K. 1977. *Agron J* 69:969–974.

Ohnesorge, F.K., and M. Wilhelm. 1991. In *Metals and Their Compounds in the Environment*. VCH Publ, Weinheim, Germany.

Oliver, D.P., R. Hannam, K.G. Tiller, and G.D. Cozens. 1994. *J Environ Qual* 23:705–711.

Olsen, S.R. 1972. In J.J. Mortvedt, P.M. Giordano, and W.L. Lindsay, eds. *Micronutrients in Agriculture*. Soil Sci Soc Am, Madison, WI.

Page, A.L. 1974. *Fate and Effects of Trace Elements in Sewage Sludge When Applied to Agricultural Lands*. EPA-670/2-74-005. United States Environmental Protection Agency, Cincinnati, OH.

Parker, D.R., J.J. Aguilera, and D.N. Thomason. 1992. *Plant Soil* 143:163–177.

Parker, G.R., W.W. McFee, and J.M. Kelly. 1978. *J Environ Qual* 7:337–342.

Pauli, A.W., R. Ellis, Jr., and H.C. Moser. 1968. *Agron J* 60:394–396.

Paulsen, G.M., and O.A. Rotimi. 1968. *Soil Sci Soc Am Proc* 32:73–76.

Payne, G.G., D.C. Martens, C. Wivarko, and N.F. Perea. 1988. *J Environ Qual* 17:707–711.

Peck, N.H., D.L. Grunes, R.M. Welch, and G.E. MacDonald. 1980. *Agron J* 72:528–534.

Pepper, I.L., D.F. Bezdicek, A.S. Baker, and J.M. Sims. 1983. *J Environ Qual* 12:270–275.

Ponnamperuma, F.N. 1972. *Adv Agron* 24:29–95.

Prasad, A.S., J.A. Halsted, and M. Nadimi. 1961. *Am J Med* 31:532–546.

Prasad, A.S., A. Miale, Z. Farid, and W.J. Darby. 1963. *Arch Int Med* 111:407–428.

Radjagukguk, B., D.G. Edwards, and L.C. Bell. 1980. *Aust J Agric Res* 31:1083–1096.

Randhawa, N.S., and F.E. Broadbent. 1965. *Soil Sci* 99:362–365.

Reddy, M.R., and H.F. Perkins. 1974. *Soil Sci Soc Am Proc* 38:229–231.

Reiners, W.A., R.H. Marks, and P.M. Vitousek. 1975. *Oikos* 26:264–275.

Reuter, D.J., J.F. Loneragan, A.D. Robson, and D. Plaskett. 1982. *Aust J Agric Res* 33:989–999.

Roberts, B.A. 1980. *Can J Soil Sci* 60:231–240.

Romheld, V., and H. Marschner. 1991. In J.J. Mortvedt et al., eds. *Micronutrients in Agriculture,* 2nd ed. SSSA 4. Soil Sci Soc Am, Madison, WI.

Rother, J.A., J.W. Millbant, and T. Thornton. 1983. *J Soil Sci* 34:126–127.

Russell, L.H., Jr. 1979. In F.W. Oehme, ed. *Toxicity of Heavy Metals in the Environment,* part 11. Marcel Dekker, New York.

Ryan, J.A. 1977. In *Proc Natl Conf. Disposal of Residues on Land.* Information Transfer, Rockville, MD.

Sakai, R., B.P. Singh, and A.P. Singh. 1982. *Plant Soil* 66:129–132.

Sanders, J.R., and M.I. El Kherbawy. 1987. *Environ Pollut* 44:165–176.

Sanders, J.R., S.P. McGrath, and T.M. Adams. 1987. *Environ Pollut* 44:193–210.

Sandstead, H.H., and G.W. Evans. 1984. In *Present Knowledge in Nutrition,* 5th ed. Nutrition Foundation, Washington, DC.

Sauerbeck, D.R., and P. Styperek. 1985. In *Processing and Use of Organic Sludge and Lignid Agricultural Wastes.* Reidel, Dordrecht.

Schulthess, C.P., and C.P. Huang. 1990. *Soil Sci Soc Am J* 54:679–688.

Sharma, K.C., B.A. Krantz, A.L. Brown, and J. Quick. 1968. *Agron J* 60:453–456.

Shuman, L.M. 1979. *Soil Sci* 127:10–17.

Shuman, L.M. 1991. In J.J. Mortvedt et al., eds. *Micronutrients in Agriculture,* 2nd ed. SSSA 4. Soil Sci Soc Am, Madison, WI.

Shuman, L.S. 1999. *J Environ Qual* 28:1442–1447.

Siccama, T.G., W.H. Smith, and D.L. Mader. 1980. *Environ Sci Technol* 14:54–56.

Silanpaa, M., and P.L.G. Vlek. 1985. *Fert Res* 7:151–167.

Sims, J.L., and W.H. Patrick, Jr. 1978. *Soil Sci Soc Am J* 42:258–262.

Singh, B.R., and J. Lag. 1976. *Soil Sci* 121:32–37.

Singh, J.P., R.E. Karamanos, and J.W.B. Stewart.1987. *Can J Soil Sci* 67:103–116.

Singh, M., and S.P. Singh. 1980. *Plant Soil* 56:81–92.

Smilde, K.W., B. Van Luit, and W. Van Driel. 1992. *Plant Soil* 143:233–238.

Smith, W.H. 1973. *Environ Sci Technol* 7:631–636.

Soltanpour, P.N. 1969. *Agron J* 61:288–289.

Soltanpour, P.N., and A.P. Schwab. 1977. *Commun Soil Sci Plant Anal* 8:195–207.

Soltanpour, P.N., J.O. Reuss, J.G. Walker, et al. 1970. *Am Potato J* 47:435–443.

Soon, Y.K. 1994. *Can J Soil Sci* 74:179–184.

Sparks, D. 1993. *Hort Sci* 28:1100–1102.

Sparr, M.C. 1970. *Commun Soil Sci Plant Anal* 1:241–262.

Stahl, R.S., and B.R. James. 1991a. *Soil Sci Soc Am J* 55:1291–1294.

Stahl, R.S., and B.R. James. 1991b. *Soil Sci Soc Am J* 55:1287–1290.

Staker, E.V., and R.W. Cummings. 1941. *Soil Sci Soc Am Proc* 6:207–214.

Stanton, D.A., and R.T. Burger. 1967. *Geoderma* 1:13–17.

Stewart, I., C.O. Leonard, and G. Edwards. 1955. *Florida State Hort Soc* 1955:8288.

Stewart, J.A., and K.C. Berger. 1965. *Soil Sci* 100:244–250.

Stigliani, W.M., P.R. Joffe, and S. Anderberg. 1993. *Environ Sci Technol* 27:786–793.

Strojan, C.L. 1978. *Oikos* 31:41–52.

Stukenholtz, D.D., R.J. Olsen, G. Gogan, and R.A. Olson. 1966. *Soil Sci Soc Am Proc* 30:759–763.

Suzuki, S., N. Djuangshi, K. Hyodo, and O. Soemarwoto. 1980. *Arch Environ Contam Toxicol* 9:437–449.

Swaine, D.J. 1955. *The Trace Element Content of Soils.* Commonwealth Bureau of Soil Sci (Great Britain), Tech Commun 48. Herald Printing Works, York, United Kingdom.

Swaine, D.J., and R.L. Mitchell. 1960. *J Soil Sci* 11:347–368.

Tan, K.H., L.D. King, and H.D. Morris. 1971. *Soil Sci Soc Am Proc* 35:748–752.

Tanner, J.T., and M.H. Friedman. 1977. *J Radioanal Chem* 37:529–538.

Tiller, K.G., and J.F. Hodgson. 1962. *Clays Clay Miner* 9:393–403.

Travis, D.O., and R. Ellis, Jr. 1965. *Trans Kansas Acad Sci* 68:457–460.

Trierweiler, J.F., and W.L. Lindsay. 1969. *Soil Sci Soc Am Proc* 33:49–54.

Underwood, E.J. 1973. In *Toxicants Occurring Naturally in Foods,* 2nd ed. National Academy of Sciences, Washington, DC.

[United States EPA] United States Environmental Protection Agency. 1993. *Fed Register* 58(32):9248–9415.

Van Campen, D.R. 1991. In J.J. Mortvedt et al., eds. *Micronutrients in Agriculture,* 2nd ed. SSSA 4. Soil Sci Soc Am, Madison, WI.

Van Hook, R.I., W.F. Harris, and G.S. Henderson. 1977. *Ambio* 6:281–286.

Viets, F.G., Jr., L.C. Boawn, and C.L. Crawford. 1954. *Soil Sci* 78:305–316.

Vinogradov, A.P. 1959. *The Geochemistry of Rare and Dispersed Chemical Elements in Soils.* Consultants Bureau, New York.

Wallace, A., and E.M. Romney. 1970. *Soil Sci* 109:66–67.

Walsh, L.M., D.R. Steevens, H.D. Seibel, and G.G. Weis. 1972. *Commun Soil Sci Plant Anal* 3:187–195.

Watanabe, F.S., W.L. Lindsay, and S.R. Olsen. 1965. *Soil Sci Soc Am Proc* 29:562–565.

Wear, J.I., and C.E. Evans. 1968. *Soil Sci Soc Am Proc* 32:543–546.

Webb, M.J., and J.F. Loneragan. 1990. *J Plant Nutr* 13:1499–1512.

Wedepohl, K.H., ed. 1978. *Handbook of Geochemistry.* Springer-Verlag, New York.

Welch, R.M. 1993. In A.D. Robson, ed. *Zinc in Soils and Plants.* Kluwer, Dordrecht, Netherlands.

Welch, R.M., M.I. Webb, and J.F. Loneragan. 1982. In *Proc 9th Intl. Plant Nutr. Coll,* Commonwealth Agricultural Bureau, Farnham Royal, United Kingdom.

Welch, R.M., W.H. Allaway, W.A. House, and J. Kubota. 1991. In J.J. Mordvedt et al., eds. *Micronutrients in Agriculture,* 2nd ed. SSSA 4, Soil Sci Soc Am, Madison, WI.

Whitby, L.M., and T.C. Hutchinson. 1974. *Environ Conserv* 1:191–200.

White, J.R., and C.T. Driscoll. 1987. *Environ Sci Technol* 21:211–216.

White, M.C., and R.L. Chaney. 1980. *Soil Sci Soc Am J* 44:308–313.

Whitney, D. 2000. Dept of Agronomy, Kansas State Univ, Manhattan, KS (personal communication).

Whitney, D.A., R. Ellis, Jr., L.S. Murphy, and G. Herron. 1973. Publ. L-360, Coop Ext Serv, Kansas State Univ, Manhattan, KS.

Whitton, J.S., and N. Wells. 1974. *NZ J Sci* 17:351–367.

[WHO] World Health Organization. 1983. In *Guidelines for Drinking Water Quality,* WHO, Geneva.

Williams, D.E., J. Vlamis, A.H. Pukite, and J.E. Corey. 1987. *Soil Sci* 143:124–131.

White, M.L. 1957. *Econ Geol* 52:645–651.

Xie, R.J., and A.F. MacKenzie. 1989. *J Soil Sci* 40:49–58.

Yoon, S.K., J.T. Gilmour, and B.R. Wells. 1975. *Soil Sci Soc Am Proc* 39:685–688.

Yoshida, S., J.S. Ahn, and D.A. Forno. 1973. *Soil Sci Plant Nutr* 19:83–93.

Young, H.E., and V.P. Guinn. 1966. *Tappi* 49:190–197.

Youngdahl, L.J., L.V. Svec, W.C. Liebhardt, and M.R. Teel. 1977. *Crop Sci* 17:66–69.

17
Nickel

1 General Properties of Nickel

Nickel (atom. no. 28) belongs in Group VIII of the periodic table, the so–
called iron–cobalt group of metals, and has an atom. wt. of 58.71, specific
gravity of 8.9, and melting pt. of 1453 °C. It is a silvery white, hard,
malleable, ductile, ferromagnetic metal that maintains a high luster and is
relatively resistant to corrosion. It is insoluble in water, but soluble in dilute
HNO_3, slightly soluble in HCl and H_2SO_4, and insoluble in NH_4OH.

Nickel has five stable isotopes in nature: ^{58}Ni (68.27%), ^{60}Ni (26.10%),
^{61}Ni (1.13%), ^{62}Ni (3.59%), and ^{64}Ni (0.91%). Nickel forms stable complexes
with many organic ligands; however, complexes with naturally occurring
inorganic ligands are formed only to a small degree, in the order OH^- >
SO_4^{2-} > Cl^- > NH_3. Under anaerobic conditions, sulfide may control the solubility of Ni.

Normally, Ni occurs in the 0 and II oxidation states, although the I and III
states can exist under certain conditions. The latter ions are not stable in
aqueous solution. Nickel is present as Ni^{2+} in common water-soluble Ni compounds such as the acetate, chloride, sulfate, bromide, fluoride, and nitrate
salts.

Nickel is closely related to Co in both its chemical and biochemical properties. Nickel can replace essential metals in metallo-enzymes and can cause
disruption of metabolic pathways.

2 Production and Uses of Nickel

Nickel is extracted from sulfide and oxide ores; the two economically
exploitable Ni ores are: pyrrhotite and pentlandite, a sulfide. The primary
mine producers of Ni in the world in decreasing order are: Russia, Canada,
New Caledonia, Indonesia, and Australia. World production of Ni in 1994
was approximately 0.9×10^6 tonnes. Canada is the world's largest producer.
All Canadian and some Russian and Australian Ni is produced from sulfide
ores. Others are produced from oxide ores.

The major uses of refined Ni in industry include electroplating, alloy
production and fabrication, the manufacture of Ni–Cd batteries and
electronic components, and the preparation of catalysts for hydrogenation
of fats and methanation.

Most Ni is used in alloys that are strong and corrosion resistant such as
stainless steel. Stainless steel production is the single largest application of
Ni and has been the most rapidly growing use in recent years. Thus, Ni is
found in a wide variety of commodities such as automobiles, batteries,
coins, jewelry, surgical implants, kitchen appliances, and sinks and utensils.
Nickel is also used in Ni–Fe (used for magnetic components in electrical

equipment), Ni–Cu (noted for resistance to corrosion especially in marine applications), Ni–Cr (used in heating elements for stoves and furnaces), and in Ni-Ag (used as a base for electroplated articles) alloys. Nickel–steel alloys are crucial in armor plating and in armaments. Other high-Ni alloys are used for such high technology applications as turbine blades, jet engine components, and in nuclear reactors. While markets for key Cd products have shrunk, demand for Ni–Cd batteries has grown and is expected to grow significantly during the next 5 yr, especially with the expansion of electric vehicles.

3 Nickel in Nature

Nickel constitutes about 80 ppm of the earth's crust making it the 23rd most common element. Most of this is in igneous rocks, of which Ni comprises approximately 100 ppm. Hence it is ubiquitous in the environment. Deposits that are commercially exploitable contain at least 10,000 ppm Ni. Nickel content varies greatly among rock types (Table 17.1). Sandstone has very low Ni content, whereas peridotite or serpentine (ultrabasic igneous rocks) averages as much as 2000 ppm. Serpentine rocks also contain high Cr, Mg, and Fe, but low Ca; serpentines have anomalously high Mg/Ca ratios (200,000:15) compared with the crustal average, and relatively low levels of silica. Thus, serpentine-derived soils are usually sparse in vegetation because of nutrient imbalance and phytotoxicity. The real cause of poor plant growth in these soils is still unclear, although the consensus is that Ni is more likely to be toxic than the relatively large concentrations of Co and Cr present. Thus, phytotoxicity due to Ni is confounded by the high Mg/Ca ratios causing Ca deficiency. In the tropics, Ni is concentrated with Fe in laterite, a hard red soil characterized by low silica but high Fe and Al contents. The Ni content of laterite may be as much as 60 times that of the parent rocks.

Laterite, a primary Ni oxide ore is mined mostly by open-pit technique in Australia, Indonesia, New Caledonia, the United States, and Russia. The Ni sulfide ore pentlandite is being mined underground in Canada and Russia. There are almost 100 minerals of which Ni is an essential constituent. The majority of Ni minerals are sulfides and arsenides. Pentlandite is a principal ore mineral of Ni and is a common accessory mineral in igneous rocks. The most important Ni ore deposits of commercial importance are the Ni sulfide minerals associated with mafic and ultramafic igneous rocks. The deposits in Sudbury (Ontario), Thompson (Manitoba), Norilsk and Pechenga (former Soviet Union), and Kambalda (Australia) are of this type.

The soil content of Ni is also variable (see Table 17.1), with the world's average reported around 20 ppm (Appendix Table A.24); normal soils contain from 5 to 500 ppm (Swaine, 1955). Relative to Ni, soils can be grouped

TABLE 17.1. Commonly observed nickel concentrations (ppm) in various environmental media.

Material	Average concentration	Range
Igneous rocks[a]	75	2–3600
Limestone[a]	20	—
Sandstone[a]	2	—
Shales and clays[a]	68	20–250
Black shales[a]	50	10–500
Sedimentary rocks[b]	20	—
Coal[c,d]	15	3–50
Coal ash[d]		
Fly ash	141	23–353
Bottom ash	216	<10–1067
FGD sludge	40	<5–145
Oil ash[d]	10,186	6180–13,750
Petroleum[e]	10	—
Soils (world)	20 (17–50)[f]	5–500[c]
Plants[h]	—	0.01–5
Herbaceous vegetables[e]	—	0.02–4
Common crops[g]	—	<0.1–4.0
Ferns[e]	1.5	—
Fungi[e]	1.8	—
Lichens[e]	—	1.5–3.8
Freshwater (μg L^{-1})[e]	0.5	0.02–27
Seawater (μg L^{-1})[e]	0.56	0.13–43

Sources: Extracted from
[a]Cannon (1978).
[b]Polemio et al. (1982).
[c]Trudinger et al. (1979).
[d]Ainsworth and Rai (1987).
[e]Bowen (1979).
[f]McGrath (1995). Values in parenthesis are range in median.
[g]Vanselow (1965).
[h]Uren (1992).

into two categories: (1) those derived from sandstones, limestones, or acid igneous rocks, containing <50 ppm and (2) those derived from argillaceous sediments or basic igneous rocks, containing from 5 to over 500 ppm of Ni. It is not uncommon for soils derived from ultrabasic igneous rocks to contain 5000 ppm or more of total Ni.

4 Nickel in Soils

4.1 Total Nickel in Soils

Igneous rocks, upon weathering, are the primary source of Ni in soils. The values for total Ni content of soils are quite variable. These variations are caused by several factors, including soil parent material. For world soils, an average concentration is likely around 20 ppm. Indeed, the average Ni concentration for U.S. surface soils is 20 ppm (range of <5 to 700 ppm,

TABLE 17.2. Concentrations of total nickel in surface soils of England and Wales according to soil texture.

| Soil texture | n | Minimum | Percentile | | | Maximum |
			25th	50th	75th	
Clayey	479	10.5	31.4	38.2	44.5	194.6
Fine loamy	2002	1.6	19.0	25.3	32.8	439.5
Fine silty	1063	2.3	20.7	28.2	36.2	298.8
Coarse silty	184	4.5	17.7	22.4	31.0	89.7
Coarse loamy	1143	0.8	10.8	15.8	21.9	436.4
Sandy	229	0.8	4.5	7.5	12.1	74.3
Peaty	557	0.8	4.5	6.6	12.7	123.9
All soils	5692	0.8	14.0	22.6	32.4	439.5

Source: Extracted from McGrath and Loveland (1992).

$n = 863$) (Shacklette et al., 1971). A value of 25 ppm Ni was reported for soils from various parts of the world (Berrow and Reaves, 1984), while an average of 22 ppm Ni was reported for tropical Asian paddy soils (Domingo and Kyuma, 1983). Values (in ppm) for several countries were reported as follow: the United States, 20; Canada, 20 (range, 5 to 50); Italy, 28 (range, 4.0 to 97.5); England and Wales, 26 (range, 4.4 to 228); and Sweden, 8.7. The most recent values for total Ni concentrations of surface soils from England and Wales are presented in Table 17.2.

Considerable variations in Ni content occur, especially in serpentine soils, with values of 100 to 3000 ppm for these soils reported from the United Kingdom and Scandinavia (Proctor, 1971). Reported average values (in ppm) for serpentine soils for other countries are as follow: Italy, 2600; former Soviet Union, 5000; Portugal, 4000; Zimbabwe, 7000; New Caledonia, 5400; western Australia, 700; and New Zealand, 3300. Nickel contents (in ppm) for typical Scottish surface soils developed from different parent materials are as follow: serpentine, 800; olivine gabbro, 50; andesite, 10; granite, 10; shale, 40; and sandstone, 15 (Mitchell, 1971). Noncultivated Swedish soils contained 7.8 ppm total Ni and cultivated soils contained 9.5 ppm Ni (Andersson, 1977).

4.2 Bioavailable Nickel in Soils

Since total Ni in soils is not a reliable index of plant bioavailability and mobility of Ni, the extractable fraction is commonly used for this purpose. As expected, there are considerable variations in the amount (as percentage of total concentration) of Ni removed by a particular extractant: 0.5 to 2.0% by DTPA extractant for some Canadian soils (Whitby et al., 1978), as much as 7.0% by DTPA extractant for some Nile Delta soils (Elrashidi et al., 1979), about 3% by NH_4OAc-EDTA extractant for some polluted Greek soils (Nakos, 1982), from 1.3 to 5.7% by DTPA-TEA extractant for some southern

California soils treated with sewage sludge (Valdares et al., 1983), and about 3.8% by HOAc extractant for soils in England and Wales (Archer, 1980). Danish soils treated with sewage sludge yielded higher Ni extractabilities with 0.01 M CaCl$_2$ than untreated soils and were also pH-dependent (Sanders et al., 1986). For example, at pH 6.5 extractability was 0.2 to 0.9% of total Ni from untreated soils vs. 0.2 to 2.3% for treated soils. At pH 4.5 however, the untreated soils produced 1.4 to 6.8% extractable Ni vs. 2.1 to 20.8% for treated soils. This extractant also proved to be more reliable in predicting plant uptake of Ni. Amounts of Ni extracted with water are usually very low, commonly <0.5% even for contaminated soils and serpentine soils.

Just like the other trace elements, several extractants to relate plant-available Ni to extractable soil Ni have been tested: mild acids, neutral salts, and chelating agents. Although some investigators found that a suite of extractants was unsatisfactory in predicting the availability of Ni to plants (Shewry and Peterson, 1976), some extractants, such as the DTPA solution, could predict plant uptake quite well (Valdares et al., 1983). In addition, the latter investigators found an excellent correlation between DTPA-extractable Ni and the total quantity of this element added to the soil (Fig. 17.1). In one case, the extractability of Ni is similar in the acid (pH 5.5) and neutral soils (pH 7.0), but is considerably lower in a calcareous soil (pH 7.7).

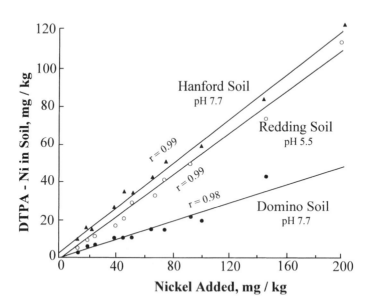

FIGURE 17.1. Relationship between DTPA-extractable nickel and total added nickel in soils. Nickel enrichment was via sewage sludge addition at various rates up to 4% (by wt.). Soils were planted to successive crops of lettuce and Swiss chard, after which extractable nickel was determined. (From Valdares et al., 1983.)

4.3 Nickel in Soil Profile

Reports on the distribution of Ni in the soil profile have been conflicting. The surface or the subsoil may be relatively higher or have similar Ni concentrations depending on the parent material, degree of contamination, or pedogenic processes. For nonpolluted soils, there is no discernible pattern in the distribution of Ni in the soil profile. Data for 20 arable soils in New Jersey indicate about equal concentrations of total Ni between the A (14.4 ppm) and B (16.5 ppm) horizons (Painter et al., 1953). However, additional data from New Jersey (Connor et al., 1957) indicate that for some Podzols, total Ni tends to increase with depth, from the A to B horizons, then to C horizon. Similar total Ni content of the surface soils (23 ± 4 ppm) and subsoils (27 ± 11) in 16 major soil series in Minnesota were found (Pierce et al., 1982). In a survey in California encompassing 49 soil profiles (Bradford et al., 1967), basically four patterns for Ni soil concentrations can be discerned: (1) uniform distribution throughout the profile (16 profiles), (2) increasing concentration with depth (20 profiles), (3) decreasing concentration with depth (6 profiles), and (4) increasing concentration, followed by a decreasing pattern (7 profiles). In Wales, similar total Ni contents were observed in surface soils (29 ppm) and subsoils (33 ppm) (Bradley et al., 1978). Elsewhere in the United Kingdom, observations for several profiles revealed similar patterns observed for California soils (Swaine and Mitchell, 1960).

In Swiss subalpine soils derived from serpentine material, total Ni and Mg generally increased with depth (Table 17.3). Exchangeable Ni was ≤ 10 mg kg^{-1}, considered still phytotoxic to most plants (Gasser et al., 1995). In arable soils in Canada, the distribution of total Ni in the profile was fairly similar (Mills and Zwarich, 1975; Whitby et al., 1978). Even for Canadian serpentine soils, the distribution of total Ni in the profile was remarkably uniform (Roberts, 1980). However, in arable muck soils in Canada, there was a declining trend in Ni concentration with depth (Hutchinson et al., 1974). The decline in concentration with depth was even more dramatic in virgin muck soils. In other parts of the world, the profile pattern may be uniform as in New Zealand (Quin and Syers, 1978), or variable as in Papua New Guinea soils (Bleeker and Austin, 1970). Considerable leaching of Ni in the soil profile could occur in instances where the soil has been severely acidified, such as in soils close to smelters (Hazlett et al., 1984).

In forest ecosystems, the Ni content of soils may tend to be higher in the mineral A horizon (0.52 ± 0.13 ppm) than in the underlying horizons (≤ 0.10 ppm) as indicated by data for extractable Ni in Nagoya University Forest (Memon et al., 1980). Conversely, Ni content tended to increase with mineral soil depth in the Solling Forest ecosystem in the Hartz Mountains, Germany (Heinrichs and Mayer, 1980).

TABLE 17.3. Total nickel, iron, calcium, and magnesium concentrations (mmol kg^{-1}) in serpentine soils.[a]

	Fe	Mg	Ca	Ni
Profile 1				
O	287	370	25	2
A	1218	1646	125	7
Bw1	1630	2139	125	10
Bw2	1146	6830	50	19
BC	1110	8517	25	29
C	1057	8846	25	34
Profile 2				
O	358	370	225	2
AE	824	905	399	3
Bw	1180	4320	374	15
BC	1200	6542	274	24
C	1003	7982	250	24
Profile 3				
O	519	3168	250	12
AC	1003	7653	274	32
Bw	1862	5596	599	20
BC	1379	7365	474	32
C	1236	7900	449	36

[a]Ranges: 16–105 g kg^{-1} Fe, 9–215 g kg^{-1} Mg, 1.24 g kg^{-1} Ca, and 0.1–2.1 g kg^{-1} Ni. Atomic mass (g mol^{-1}): Fe = 55.85, Mg = 24.31, Ca = 40.08, Ni = 58.70.
Source: Extracted from Gasser et al. (1995).

4.4 Forms and Chemical Speciation of Nickel in Soils

Nickel in the solid phase of soils occurs in several chemical forms, including occurrence on exchange sites, specific adsorption sites, adsorbed or occluded into the sesquioxides, fixed within the clay lattice, or bound in organic residues and microorganisms. In the aqueous phase (i.e., soil solution), Ni occurs in free ionic form (Ni^{2+}) and complexed form (i.e., with organic and inorganic ligands).

The most common technique of identifying operationally defined fractions (or forms) of metals in soils is by using sequential extraction with selective chemical reagents. As much as 50% of Ni in soils may be associated with the residual fraction (HF and $HClO_4$ soluble) and about 20% in the Fe–Mn oxide fraction, with much of the remainder in the carbonate fraction, and only a relatively small fraction in the exchangeable and organic fractions (Hickey and Kittrick, 1984).

Application of sewage sludge to soils may result in a shift in the proportion of metal forms in both phases. For example, Ni in aqueous extracts of sludge-amended soils has been found to be less than in the sludge itself (Lake et al., 1984). In three southern California soils treated with sewage sludge, most of the soil Ni exists in the residual form (87%), followed by the carbonate form (5.1%) (Table 17.4). The residual form is considered the most stable form of Ni because it requires the most drastic extracting

TABLE 17.4. Distribution (as %) of chemical forms of nickel in sewage sludge and sludge-treated soils at three sites in southern California.

Extractant	Chemical form	Sewage sludge	Treated soil		
			Holland (pH 5.8–6.4)	Ramona (pH 5.2–6.7)	Helendale (pH 7.4–8.0)
0.5 M KNO$_3$	Exchangeable	10.9	<0.1	<0.1	<0.1
H$_2$O	Adsorbed	0.5	<0.1	<0.1	<0.1
0.5 M NaOH	Organically bound	24.3	<0.1	<0.1	<0.1
0.05 M EDTA	Carbonate	31.9	3.4	3.9	8.1
4.0 M HNO$_3$	Sulfide/residual	26.4	79.8	93.5	87.1

Source: Emmerich et al. (1982a).

reagent for its removal. In another instance, Ni in a sandy loam soil treated for 7 yr with composted sludge was mostly in the sulfide/residual (41%), carbonate (34%), and organic (23%) forms (Chang et al., 1984). For soils heavily enriched with heavy metals, Ni was distributed as follows: residual ≫ Fe–Mn oxide > organic ≈ carbonate > exchangeable form (Hickey and Kittrick, 1984). However, in a polluted sediment, Ni was distributed as Fe–Mn oxide ≫ residual = carbonate > organic = exchangeable form. For contaminated soils and river sediments, Ni occurred primarily in the precipitated and 0.5 N HNO$_3$-extractable forms (Cottenie et al., 1979). Only very small amounts were in the water-soluble form.

Based on thermodynamic considerations, Ni ferrite (NiFe$_2$O$_4$) is the most probable solid phase that can precipitate in soils (Sadiq and Enfield, 1984a,b). Under acidic and reducing conditions, the sulfides of Ni are likely to control the solubility of Ni in the soil solution. Above pH 8, Ni(OH)$^+$ and Ni^{2+} are likely to predominate in the soil solution, while in acid soils Ni^{2+}, NiSO$_4^0$ and NiHPO$_4$ are important species, with the relative proportion depending on the levels of SO$_4^{2-}$ and PO$_4^{3-}$. However, Sposito and Page (1984) emphasized the importance of the bicarbonates (as NiHCO$_3^+$) and carbonates (as NiCO$_3^0$) in Ni speciation, especially in alkaline, calcareous soils.

Using the GEOCHEM computer program, the forms of Ni in the soil solutions for a soil treated with sewage sludge were predicted as follows: in the sludge layer—free ionic form, 52.5%; inorganic complex form, 22%; and organic complex form, 26.5%; in the mineral soil below the sludge layer—free ionic form, 68%; inorganic complex form, 26.5%; and organic complex form, 6.0% (Emmerich et al., 1982b).

Using another fractionation scheme, the following percentages for sediments in natural marsh ecosystems near Galveston, Texas were found: residual, 52.1%; moderately reducible, <7.8%; OM plus sulfide, 28.5%; easily reducible, 7.6%; and exchangeable plus water-soluble, <1.2% (Lindau and Hossner, 1982). However, the exchangeable form of Ni may constitute a significant portion of the total Ni for certain soils, such as in serpentine soils and some soils of Japan.

4.5 Sorption and Complexation of Nickel in Soils

Since Ni(II) is stable over a wide range of pH and redox conditions, this is the species of Ni expected to occur in most soils. Nickel is known to complex readily with a variety of inorganic and organic ligands. Nickel-halides and salts of oxo-acids are generally soluble in water, while Ni carbonate is fairly insoluble. In examining soils from arable and potential waste disposal sites, Bowman et al. (1981) found that sorption of Ni from 0.01 N $CaCl_2$ was extensive for all samples; >99% of added Ni was sorbed in some cases. In spite of this almost complete sorption, Ni sorption was not completely irreversible, as some Ni was desorbed by the chelating agent DTPA. However, in the presence of high levels of Ca^{2+} ions and small amount of the chelating agent EDTA in the equilibrating solution, Ni sorption by the soils was minimized. In this case, competition between Ca^{2+} ions and Ni^{2+} ions for adsorption sites prevailed over Ni–Cl complex formation, while the EDTA complexed with Ni, maintaining it in soluble form in the aqueous phase. Nickel may bind with a variety of synthetic chelates to the virtual exclusion of competing cationic species at typical soil pH (Sommers and Lindsay, 1979). In some instances, inorganic complexation of Ni may be significant in affecting its mobility. Increased Ni mobility or decreased sorption of Ni in the presence of Cl^- ions, apparently due to Cl^- complex formation, has been observed (Bowman and O'Connor, 1982; Doner, 1978). Similarly, SO_4^{2-} or NO_3^- may also have a complexing effect on Ni.

Nickel is retained in certain soils primarily by specific adsorption mechanisms at low (≤10 ppm) concentrations. The isotherm conforms to the Freundlich equation up to about the initial solution concentrations of 10 ppm Ni. Depending on sorption capacities, the Freundlich equation may describe the adsorption of Ni at concentrations well above 10 ppm Ni. However, Ni sorption by soil can also display a complex (i.e., multiphasic) isotherm (Harter, 1983). Adsorption of metals in soils (pH 4.3 to 5.6) was in the order Pb > Cu > Zn > Ni (Harter, 1983). Using soil columns, the mobility of metals in soils (pH 3.8 to 7.1) tended to be: Cd ≤ Ni ≤ Zn ≫ Cu (Tyler and McBride, 1982). In general, differences in sorption or mobility of the metals are affected by differences in physicochemical properties of the soil. Sorption of Ni in soils is largely pH-dependent.

5 Nickel in Plants

5.1 Essentiality of Nickel in Plants

Although Ni is toxic to organisms at elevated levels, trace amounts of this element are required for several biological processes. For example, Ni is a constituent in four enzymes: urease, methyl coenzyme M reductase, hydrogenase, and carbon monoxide dehydrogenase (Hausinger, 1992). Other essential biological roles for Ni in plants, animals, and microorganisms are known. The essentiality of Ni for higher plants has been established (Brown et al., 1987a,b; Eskew et al., 1983). Although little is known about the essential functions of Ni in plants, its role as an essential component of the enzyme urease is firmly established. Therefore, Ni deficiency could result in the disruption of N as well as amino acid metabolism in plants.

Nickel deficiency has been shown to produce a variety of physiological effects in plants. In soybeans, urea accumulates to toxic levels in leaflet tips of Ni-deficient plants as a result of depressed levels of urease activity (Eskew et al., 1983). Similarly, when grown under conditions of low Ni supply, tomato plants developed chlorosis in the youngest leaves leading to necrosis of their meristematic tissues. Some of the most striking effects of Ni deficiency have been observed in cereals (wheat, oats, and barley) where growth depression, premature senescence, decreased tissue Fe levels, and grain development inhibition were some of the consequences (Brown et al., 1987). In addition, Ni deficiency resulted in the disruption of malate and various inorganic anions in roots, shoot, and grain of barley (Brown et al., 1990).

5.2 Uptake and Translocation of Nickel in Plants

Since Ni is ubiquitous in the environment, it is a normal constituent of plant tissues. Nickel content of field-grown crops and natural vegetation ranges from 0.05 to 5.0 ppm in the dry matter (Vanselow, 1965). Subsequently, mean values of about 0.20 to 4.5 ppm for nearly 2000 specimens of field crops and natural vegetation from the United States have been reported (Connor et al., 1975). In normal plants, the following ranges of values have been reported (Kabata-Pendias and Pendias, 1992): grasses, 0.10 to 1.7 ppm DW; clovers, 1.2 to 2.7 ppm; vegetables, 0.20 to 3.7 ppm; cereals, average of 0.50 ppm. Generally, the level of Ni in most plant species that may produce phytotoxicity ranges from 10 to 100 ppm.

Nickel is usually absorbed in the ionic form Ni^{2+} from the soil or culture medium. Nickel is apparently more easily absorbed by plants when supplied in the ionic form than when chelated (Mishra and Kar, 1974). A possible explanation is that the charge on the chelated molecule inhibits its absorption by the roots, i.e., a molecule with no charge or a slight negative charge can be taken up, while a complexed molecule with strong negative charge cannot. High Ni content of the soil, particularly when the exchangeable form is also high, facilitates the absorption of this element by plants when other conditions are also favorable.

Metabolic control of Ni uptake by plants is most likely (Aschmann and Zasoski, 1987; Cataldo et al., 1978a). Following its absorption by the root, Ni is transported within the xylem as organic complexes and is highly mobile in plants, with leaves being the major sink in the shoot during vegetative growth (Cataldo et al., 1978b). At senescence, >70% of the Ni present in the shoot is remobilized to seeds. This behavior of Ni is shared by a number of micronutrients, i.e., Fe, Cu, Mn, and Zn. Nickel accumulating in seeds is primarily associated with the cotyledons. Because of the similar abundance of Ni and Cr in some soils, especially in serpentine-derived soils, their uptake deserves some attention. The fact that Ni is more mobile than Cr in soils indicates the likelihood for greater uptake and translocation of Ni in plants. Indeed, Ni was translocated much more rapidly than Cr from roots to other parts where it was accumulated (Ishihara et al., 1968). Chromium accumulated in fine roots and its translocation was much slower than that of Ni. Thus, it can be generalized that since Ni occurs in plants as chemically stable, soluble species it can accumulate readily. The behavior of Ni^{2+} in plants may be regulated by the same mechanisms operating for nutrient ions such as Cu^{2+} and Zn^{2+}, since Ni^{2+} has been shown to be a competitor for root transport sites for the other nutrients.

5.3 Hyperaccumulation of Nickel in Plants

Concentrations of Ni in plants growing in normal soils seldom exceed 5 ppm. However, this is not the case in environments with ultrabasic substrate, such as periodotite or serpentinite. The Ni content in dry matter of vegetation from serpentine areas is usually <50 ppm and seldom exceeds 100 ppm on dry weight basis (Brooks and Radford, 1978).

There are plant indicators or accumulators that bioaccumulate Ni at unusually high levels. Accumulators are plants whose mean content of a particular element is greater than the content of the same element in the substrate (Brooks et al., 1974). Severne (1974) more broadly defines accumulators as plants that have dry weight elemental concentrations greater than either the associated substrate or normal plants. Since there are unusual plants whose Ni content can exceed by a factor of 10 even the highest values observed for plants growing on ultramafic nickeliferous substrate, Brooks et al. (1977) ultimately used the term *hyperaccumulator* to denote such species. Leaves of such species should have at least 1000 ppm Ni on dry weight basis. Some of these species are indicated in Figure 17.2. Hyperaccumulators are not always restricted to Ni-rich ultrabasic substrates but are always associated with them when the tissue levels exceed 1000 ppm.

There are three reasons for interest in high Ni contents in plants (Brooks, 1980): (1) the challenge of reclaiming serpentine soils, i.e., making fertile an appreciable area of the earth covered by ultrabasic rocks, (2) the potential usefulness of using Ni levels in vegetation as an indication of the Ni contents of the corresponding soils, e.g., in biogeochemical prospecting, and (3) the phytochemical and geobotanical considerations. Serpentine soils are unsuitable to support normal vegetation because of their low nutrient contents, high Mg content (~20%) which can antagonize Ca uptake by plants, and direct toxicity of indigenous elements such as Ni and Cr. Application of biogeochemical prospecting has been relatively successful for Ni in New Zealand, Australia, New Caledonia, and the former Soviet Union. It may

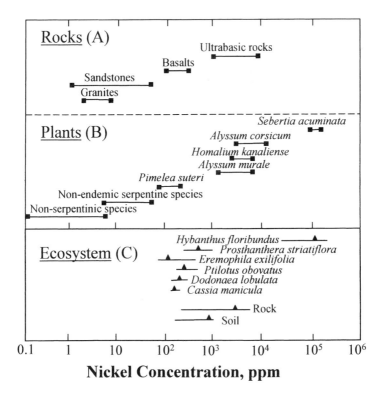

FIGURE 17.2. Ranges of nickel levels (ppm) in (A) rocks, (B) plants and (C) shrubs, rocks and soils from the Goldfields of western Australia. (Modified from Brooks, 1980; Severne and Brooks, 1972.)

also be applicable to U, Co, Cr, Pb, and Pt prospecting but tends to be unreliable for Cu and Zn. Hyperaccumulators are of special interest to geobotanists and phytochemists because of their adaptability to ultrabasic areas and the excessive levels of Ni in plant tissues. This phenomenon is considered a tolerance mechanism involving the complexation of Ni with malic and malonic acids in stems and leaves (Brooks et al., 1981). More recently, an amino acid, histidine has been demonstrated to be involved both in the mechanism of Ni tolerance and in the high rates of Ni transport into the xylem required for hyperaccumulation in shoot (Kramer et al., 1996). However, the Ni detoxification is arguably of primary importance in the root cell cytoplasm rather than in the xylem as the latter represents an apoplastic, extracellular phase, but effective flux of Ni from roots to shoot via the xylem is essential for achieving hyperaccumulation.

In Ni-olerant *Silene italica* from serpentine soils, addition of either Ni^{2+} or Cd^{2+} induced substantial increase of glucose-6-phosphate dehydrogenase, glutamate dehydrogenase, and malic enzyme than in nontolerant population, indicating the adaptation of the former population to toxic levels of metals

(Mattioni et al., 1997). To date, over 240 Ni-hyperaccumulating taxa have been reported of which 76 are in the family Brassicaceae. This family is largely confined to the genera *Alyssum* (48 taxa) and *Thlaspi* (23 taxa). Some of these species can accumulate up to 3% Ni in leaf dry matter (Reeves, 1992).

5.4 Nickel Interactions with Other Elements in Plants

In plant nutrition, elements can interact synergistically or antagonistically. Serpentine soils usually have elevated levels of Fe, Co, Cr, Ni, and Mg. Their high Mg contents can inhibit Ca uptake by plants. Therefore, plants growing on serpentine soils can be considered as highly adaptable to low available Ca. Since the uptake behavior of Ni by plants is characteristic of nutrient ions, Ni may interact with other nutrient ions. To this end, the rate of Ni^{2+} uptake by soybean plants in the presence or absence of Co^{2+}, Cu^{2+}, Fe^{2+}, Mg^{2+}, Mn^{2+}, and Zn^{2+} was measured (Cataldo et al., 1978a). Of the ions investigated, Mg^{2+} and Mn^{2+} did not inhibit Ni^{2+} absorption, while Co^{2+}, Cu^{2+}, Fe^{2+}, and Zn^{2+} did. In the presence of the last four elements, Ni^{2+} uptake was reduced by 25 to 42% of the rates found for control treatments. In addition, Cu^{2+} was found to be a competitive inhibitor of Ni^{2+} uptake and Cu^{2+} appears to be a better competitor of Ni^{2+} than is Zn^{2+}.

Several crops grown on serpentine soil did not exhibit any Ni toxicity symptoms when the Cu/Ni ratio was equal to or greater than 1 or when the Fe/Ni ratio was equal to or greater than 5 (Mizuno, 1968). Crops such as kidney beans, soybeans, cabbage, and alfalfa that contained low ratios of Fe/Ni and Cu/Ni were extremely susceptible to Ni toxicity, whereas crops such as potatoes and corn, which contained high ratios, were not affected. The severity of Ni toxicity symptoms in mulberry plants grown on serpentine soils can be alleviated by either topdressing the soil or foliar spraying with Fe and Zn compounds (Iizuka, 1975). Interactions among Cu, Ni, and Zn in young barley plants were insignificant and thus may be ignored when dealing with subtoxic accumulations (Beckett and Davis, 1978). However, at higher concentrations, the interactions became larger and more complex.

Because of the role of Fe in overcoming part of the phytotoxicity caused by some metals, trace metal–induced chlorosis is considered as a form of Fe deficiency. Indeed it is common to observe that excessive concentrations of metals applied to soils produce Fe–deficiency chlorosis concomitant with the secondary symptoms specific to the metal in question, with Ni being more pronounced than other metals. Nickel as well as Cu can cause an actual Fe deficiency by inhibiting its translocation from roots to shoot (Foy et al., 1978).

Increasing Fe level in the culture solution produced a reduction in Ni toxicity symptoms (Crooke et al., 1954). Furthermore, a correlation of necrotic symptoms with the Ni content and with the Ni/Fe ratio in the plant was noted (Fig. 17.3). Nickel consistently reduced Fe contents of the roots and

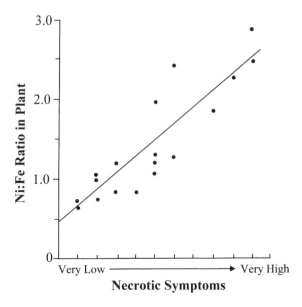

FIGURE 17.3. Relationships between Ni/Fe ratio in the plant and the degree of necrotic symptoms. (From Crooke, 1955.)

shoots of oats and tomato plants (Crooke, 1955). In contrast, excess Co^{2+}, Cu^{2+}, Mn^{2+}, and Zn^{2+} reduced Fe translocation to the shoot of barley, while excess Ni^{2+}, which was the most toxic, did not (Agarwala et al., 1977).

6 Ecological and Health Effects of Nickel

6.1 Phytotoxicity of Nickel

In general, when Ni concentrations in vegetative tissue of plants exceed 50 ppm DW, plants may suffer from excess Ni and manifest toxicity symptoms. Exceptions are the endemic species of serpentine soils or the so-called Ni accumulators and hyperaccumulators, which may contain several thousands ppm of Ni (see Fig. 17.2).

A wide range of levels of Ni in the growth media and in plant tissues has been reported to be toxic to plants. In nutrient solution, Ni may be toxic to plants at levels as low as <1 ppm to as high as 300 ppm, depending on plant species (Mishra and Kar, 1974). For a wide range of agronomic and horticultural crops, the following concentrations (ppm DW) in soils (by 0.5 M HOAc extraction) and plant tissues were reported for normal and toxic levels: in soils—normal, 1 to 5 ppm; toxic, 15 to 25 ppm; and in plants—normal, 15 to 25 ppm; phytotoxic, >50 ppm (Macnicol and Beckett, 1985).

With trees, levels ≥10 ppm Ni in the nutrient solution substantially reduced the growth of white pine and white spruce seedlings (Lozano and Morrison, 1982). In soil, Ni levels can be much higher before phytotoxicity occurs. For example, no yield depressions and the absence of chlorotic symptoms in ryegrass were noted until at about the 90 ppm level in a noncalcareous soil (Khalid and Tinsley, 1980).

In plants, the following toxic levels (ppm DW) in the foliage have been found: rice, 20 to 50; ryegrass, 154; barley, 26; hardwood species, 100 to 150; and citrus, 55 to 140. These results indicate that plants are divergent in their sensitivity to Ni. The phytotoxicity symptoms produced by Ni are similar to those produced by other heavy metals and consist of chlorosis caused by Ni-induced Fe deficiency, and specific effects of the metal itself. A typical toxicity symptom produced by Ni is the chlorosis or yellowing of leaves followed by necrosis. Other symptoms include stunted growth of roots and shoot, deformed plant parts, unusual spotting, and in severe cases, death of the whole plant. With sorghum, the symptoms of excessive Ni are similar to those for excessive Co, which in turn resemble Fe deficiency, except that the yellow streaks of the Co-induced symptoms did not extend as readily to the leaf tips as in Fe deficiency (Clark et al., 1981). Roots were stubby, short, thick, dark-colored, and brittle. In rice, interveinal chlorosis appears in new leaves, and the numbers of leaves and tillers are decreased (Chino, 1981). Root growth is severely depressed. In potato and tomato, Ni toxicity symptoms resemble those of Mn deficiency. Excessive Ni can cause mitotic disturbances in root tips of some plants.

Several reports have compared the phytotoxicity of other metals with Ni. In barley, the effectiveness of metals in inducing visible toxicity symptoms was in the order Ni > Co > Cu > Mn > Zn (Agarwala et al., 1977). In rice, the order was Cu > Ni > Co > Zn > Mn, and Hg > Cd > Zn (Chino, 1981); in oats, the order was Ni > Cu > Co > Cr > Zn > Mo > Mn (Hunter and Vergnano, 1953). In corn however, Cd proved more toxic than Ni (Traynor and Knezek, 1974). In corn and sunflower, the toxic effect of metals on photosynthesis and growth was in the order Tl > Cd > Ni > Pb (Carlson et al., 1975).

6.2 Toxicity of Nickel in Freshwater Fish and Invertebrates

Extensive reviews of the toxicity of Ni to freshwater fish indicate the similarity of response for all species and the mitigating influence of water hardness (Mance, 1987). In short-time exposures, both salmonid and nonsalmonid species are equally sensitive to Ni especially in soft water. Total Ni content in the muscle tissue of fish, both freshwater and marine, are normally low, <0.5 mg kg^{-1} WW (Moore, 1991). However, in highly contaminated ecosystems, Ni levels in fish can be substantially higher. For example, six freshwater species from the Sudbury, Ontario region had 9.5 to 13.8 mg Ni kg^{-1} WW (Hutchinson et al., 1975). The specimens were collected from an area where Ni-bearing ore deposits were being mined and smelted.

Exposure of fish to low levels of Ni over extended periods may result in a number of toxicological effects, including reduced skeletal calcification and diffusion capacity of gills, which leads to asphyxiation (Moore, 1991). In general, Ni^{2+} is relatively nontoxic to fish in both freshwater and marine systems.

It is also apparent that water hardness has a mitigating effect on Ni toxicity to freshwater invertebrates as indicated by results for *Daphnia magna* and *T. tubifex*. In general, Ni^{2+} is considered as one of the least toxic inorganic contaminants to invertebrates.

6.3 Effects of Nickel on Human Health

No toxicological incidents from Ni and its salts been recorded in the early 1900s (Sunderman, 1992). In fact, Ni salts had some medicinal value for the treatment of epilepsy, chorea, migraine, and neuralgia. Nickel was also used as a sedative and tonic of peculiar and elective power in controlling the damaging effects of sexual vice on the nervous system. However, since World War II, the use of Ni for medicinal purposes has been completely abandoned and it has been widely recognized that in addition to nickel carbonyl [$Ni(CO)_4$], exposure to Ni and other Ni compounds may adversely affect human health (Table 17.5). The physiological response in humans to Ni depends on the following factors: physicochemical characteristics of the Ni compound, the concentration, the length and type of exposure, and the sensitivity and well-being of the exposed individual. In addition, the route of exposure (see Table 17.5) and the target tissue are important.

Individuals may be exposed to Ni in the workplace or through contact with everyday items such as Ni-containing jewelry, cooking utensils, stainless steel kitchens, and clothing fasteners. Contact dermatitis, consisting of itching of the fingers, wrists, and forearms, is the most common effect in humans from chronic (long-term) Ni exposure. Respiratory effects, such as asthma

---

TABLE 17.5. Types of nickel poisoning in humans.

Inhalation [$Ni(CO)_4$, Ni, Ni_3S_2, NiO, Ni_2O_3]
Acute: Pneumonitis with adrenal cortical insufficiency; hyaline membrane formation; pulmonary edema and hemorrhage; hepatic degeneration; brain and renal congestion
Chronic: Cancer of respiratory tract; pulmonary eosinophilia (Loeffler's syndrome); asthma
Oral
Food and beverages; drugs
Skin contact
Primary irritant dermatitis; allergic dermatitis (nickel itch); eczema
Parenteral (Prosthetic Implantations)
Allergenic reactions; osteomyelitis; osteonecrosis; malignant tumors

Source: Sunderman (1992).

and an increased risk of chronic respiratory infections, have also been reported in humans from inhalation exposure to Ni.

Occupational exposure to Ni and its compounds occurs in both the Ni-producing and Ni-using industries. In recent years, the carcinogenicity of Ni and its compounds has become of increasing concern. The high incidence of pulmonary (respiratory) cancer occurring in Ni workers was first recognized in 1937 (Sunderman, 1992). The carcinogenicity of Ni has probably been more closely scrutinized than that of most other elements. However, no evidence of carcinogenicity exists for ingestion of Ni via food or water (Moore, 1991). Because occupational exposure is the most relatively consequential exposure to Ni, several countries have occupational exposure limits for Ni (Table 17.6). The ACGIH guideline values can be surpassed in a number of industries, including electrowinning and electrorefining processes, Ni–Cd battery manufacture, Ni power production, high-Ni alloy production, and Ni salt production (Nieboer, 1992).

Inhalation and ingestion are the major routes of Ni intake in humans (Nieboer et al., 1992). Although absorption through the skin (percutaneous) does not appear to be a major source of Ni in humans, it is of great importance in Ni dermatitis. In occupational settings, the most important route of Ni uptake is respiratory absorption. Current industrial exposures

TABLE 17.6. Occupational exposure limits for nickel (mg Ni m^{-3}).[a]

Country	Ni metal, sparingly soluble compounds	Soluble Ni compounds	Nickel carbonyl
Czech Republic	0.05	0.05	0.0034 (1.4 ppb)[f]
Denmark	0.5	1	—
Former Soviet Union	0.05[c]	0.005[c]	0.0005[c]
Germany	0.5[b]	0.05	0.24 (100 ppb)[f]
Japan	1	1	0.0024 (1 ppb)[f]
Netherlands	1	0.1	—
Norway	0.1	0.1	—
Sweden	0.5[d]	0.1[e]	0.0024 (1 ppb)[f]
United Kingdom	1	0.1	0.12 (50 ppb)[f]
United States (ACGIH)	1	0.1	0.12 (50 ppb)[f]
United States (OSHA)	1	1	0.0024 (1 ppb)[f]
United States (NIOSH)	0.015[g]	0.015[g]	0.0024 (1 ppb)[f]

[a]Abbreviations: ACGIH, American Conference of Government and Industrial Hygienists; NIOSH, National Institute of Occupational Safety and Health; OSHA, Occupational Safety and Health Administration.
[b]Respirable, technical guideline only.
[c]Aerosols, ceiling value.
[d]Nickel metal only; for nickel subsulfide, 0.01.
[e]Nickel compounds.
[f]Refers to Ni(CO)$_4$ in units of nanoliter per liter: 1 ppb (v/v) = 0.007 mg Ni(CO)$_4$/m^3 = 0.0024 mg Ni/m^3.
[g]Recommended TLV.
Source: Modified from Nieboer (1992).

are generally well below the ACGIH guidelines of 1 mg m^{-3} for sparingly water-soluble compounds (e.g., Ni metal, oxides and sulfides) and 0.10 mg m^{-3} for water-soluble salts (see Table 17.6). By comparison, ambient Ni levels are very low even in urban areas.

Human and animal studies have reported an increased risk of lung and nasal cancers from exposure to Ni refinery dusts and to Ni subsulfide. The U.S. Environmental Protection Agency (EPA) has classified Ni refinery dusts and Ni subsulfides as human carcinogens; the U.S. EPA has calculated inhalation unit risk estimates of 2.4×10^{-4} (μg/m^3)$^{-1}$ for Ni refinery dusts and 4.8×10^{-4} (μg/m^3)$^{-1}$ for Ni subsulfide (U.S. EPA, 1993; ATSDR, 1993). Nickel carbonyl has been reported to produce lung tumors in animal studies; thus the U.S. EPA has classified Ni carbonyl as a probable human carcinogen. Nickel carbonyl is the most acutely toxic form of Ni in humans, with the lungs and the kidneys as the target organs. Symptoms such as headache, vomiting, chest pains, dry coughing, and visual disturbances have been reported from acute inhalation exposure in humans.

The U.S. EPA has not established a reference concentratrion (RfC) for any form of Ni, but it has established a reference dose (RfD) for Ni (soluble salts) at 20 μg/kg/day. The U.S. EPA estimates that consumption of this dose or less, over a lifetime, would not likely result in the occurrence of chronic, noncancer effects (U.S. EPA, 1993).

6.3.1 Dietary Exposure to Nickel

Nickel has been demonstrated to be an essential element in some animal species; accordingly it has been suggested that Ni may be essential also for human nutrition. A daily requirement of 50 μg of Ni has been estimated (U.S. EPA, 1985, 1993). Food is the major source of Ni exposure with an average intake for adults estimated to be approximately 100 to 300 μg day^{-1} (U.S. EPA, 1985).

Total Ni concentrations in drinking water are generally low (<0.05 mg L^{-1}). The highest level reported for a drinking water supply was 0.49 mg L^{-1}. Clemente et al. (1980) reported that the Ni content of most mineral and deep spring waters in Italy is below 10 ppb. A few water samples had about 100 ppb.

Nickel also occurs at low concentrations in foods. Based on limited information available, diet is the major source of Ni exposure in the general population, with drinking water making only a minor contribution. The typical western diet yields 100 to 900 μg Ni day^{-1} (average of 160 to 500 μg day^{-1}) for a 70-kg reference human. The Ni content of representative Canadian diets are presented in Table 17.7, which indicates that the primary sources of Ni in diets are meat, fish, and poultry, followed by grain and cereal. Because of the divergent nature of Ni concentrations in various food items and differences in eating habits, intakes of Ni vary considerably. For example, the levels of Ni in diets of institutionalized children, ages

TABLE 17.7. Estimated Canadian dietary intake and concentration of nickel by food class.

Food class	Ni intake		Ni concentration (ppm WW)
	µg Ni day^{-1}	% of total diet	
Dairy products	44	9.5	0.09
Meat, fish, and poultry	98	21.1	0.36
Grain and cereal	77	16.7	0.41
Potatoes	50	10.8	0.26
Leafy vegetables	14	3.0	0.29
Legume vegetables	24	5.2	0.74
Root vegetables	8	1.7	0.18
Garden fruits	42	9.1	0.50
Fruits	30	6.5	0.16
Oils and fruits	16	3.5	0.61
Sugars and adjuncts	46	10.0	0.32
Beverages	13	2.8	0.22
Total	462		

Source: Based on a survey by Kirkpatrick and Coffin (1974) in Halifax and Vancouver, Canada.

9 to 12, from 28 U.S. cities ranged from 0.14 to 0.32 ppm (Murthy et al., 1973). The intakes of Ni in Italy by adults range from 100 to 700 µg day^{-1}, comparable with range values of 150 to 800 µg day^{-1} in the United States (Clemente et al., 1980). The Canadian dietary intake of Ni (see Table 17.7) ranged from 347 to 576 µg day^{-1} for Halifax and Vancouver, respectively (Kirkpatrick and Coffin, 1974). An intake of <300 µg day^{-1} for the United Kingdom has been reported (Hamilton and Minski, 1972); that of the International Commission on Radiological Protection is about 400 µg day^{-1}. Because of the usual low levels of Ni in drinking water, the contribution of drinking water in intake is negligible. Consequently, the WHO has not recommended any drinking water guideline for Ni.

Absorbed Ni averaged $27 \pm 17\%$ of the dose ingested from water, compared with only $0.7 \pm 0.4\%$ of the same dose in food (Nieboer et al., 1992). In general, about 1% of dietary Ni is absorbed, whereas absorption from inhalation exposure to Ni carbonyl is both rapid and extensive (Sunderman et al., 1989). Nickel is transported in the body via plasma where it is bound to high-molecular-weight carriers (e.g., proteins and amino acids).

7 Factors Affecting Mobility and Bioavailability of Nickel

Bioavailable and soil solution concentrations of metals depend on a number of factors including total metal levels in soil, soil pH, and the soil's CEC, which reflects the OM, hydrous oxides content, and texture and mineralogy of the soil.

7.1 pH

The solubility in soils and bioavailability of most heavy metals to plants are known to be inversely related to pH. The effects of pH on Ni chemistry in soils have been demonstrated in soil sorption, sewage sludge application, reclamation of serpentine soils, and plant uptake studies. In sorption studies, the amount of Ni retained was dependent upon the pH of the soil, with retention increasing with increasing pH. For example, Ni extracted with 0.01 N HCl averaged about 70 to 80% of the amount retained by soils from pH <5.0 to ~8.0 (Harter, 1983). However, the extractability decreased to <75% of the Ni retained at pH 7 or above. Also pH influences the precipitation of Ni with other compounds such as phosphates. Pratt et al. (1964b) reported that formation of Ni-P complexes could occur at pH values ≥7.0, thereby reducing Ni toxicity in soils.

In lands receiving sewage sludge high in metal contents, the liming of soils to elevate or maintain pH at 6.5 or above should minimize metal phytotoxicity. Liming sludged soils is important since sludge high in ammoniacal N tends to acidify the soil through the production of H^+ ions from the mineralization and nitrification of sludge N. Indeed, depressed plant tissue Ni concentrations were associated with sludge–lime interaction (John and Van Laerhoven, 1976). Yield reductions in crops growing on sludged soils may occur if the soil pH is 5.5 or lower (Sanders et al., 1986; Smith, 1994). Nickel phytotoxicity effects can also be alleviated by increasing the levels of Ca^{2+} in the growth medium, which can be induced under field conditions by liming.

Liming serpentine soils also has dramatic effects on Ni uptake by plants. Elevating the pH of these soils can reduce the amount of exchangeable Ni and therefore Ni toxicity to plants. The drastic effect of elevated pH on such soils on Ni uptake by potato, alfalfa, and oats is demonstrated in Figure 17.4. Others have also demonstrated the effect of pH on Ni uptake by plants. Bingham et al. (1976) indicated that the critical soil concentration of Ni was much lower for acid than for calcareous soils. For example, in wheat, the Ni level in soil that produced 50% yield decrement was 195 ppm in acid soil, compared with 510 ppm in calcareous soil; with romaine lettuce the values were 110 and 440 ppm, respectively, for acid and calcareous soils. Approximately 20% more Ni was removed by alfalfa from plots having pH values lower than 6.8 than from plots having higher pH values (Painter et al., 1953). In general, liming the soil is the most effective means of mitigating plant uptake of Ni.

7.2 Organic Matter

Depending on its nature, OM can immobilize or mobilize metals. Solid OM (e.g., OM particulate, chips, etc.) binds Ni rendering it less available to plants. It is a common practice to reclaim serpentine soils by liming or adding OM, or both. This is because OM can add to the sorption capacity of

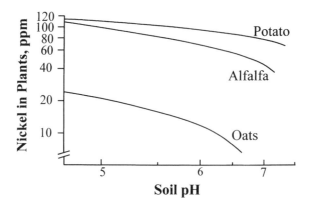

FIGURE 17.4. Relationships between serpentinic soil pH and nickel content (ppm) in selected crops. (Modified from Mizuno, 1968.)

soils for metals. For example, in comparing the addition of heavy metals to soil as inorganic salts with the addition of sludge spiked with metals, Cunningham et al. (1975) found that Cu, Cr, and Zn were less bioavailable to plants when applied in the latter form. Similar results were obtained when approximately equivalent amounts of metals (Zn, Cu, Cd and Ni) were added in the form of sewage sludge or salts (Tadesse et al., 1991). The binding mechanism in sludge OM involves acidic functional groups. A large portion (50 to 80%) of OM of stabilized sewage sludge is fairly resistant to decomposition in the soil and is similar to humified OM. The average rate of decrease in soil OM from sludge application was estimated at 10 to 18% per year (Latterell et al., 1978). The OM decomposition rate is expected to decelerate with time. Whether all the OM from applied biosolids will completely decompose and become soluble is still unknown.

In contrast, metals can be complexed by dissolved OM and therefore rendered more mobile. The capacity for chelation of Ni by OM is approximately 2.2% by weight. For example, a soil with 1% OM could complex at least 220 μg Ni g^{-1} (Pratt et al., 1964a). Metals can be mobilized in association with colloidal humified OM and as soluble complexes in soil solution (Bloomfield et al., 1976). Also during the decomposition of OM, a number of organic acids are produced by microbial activity. These acids can act as chelating agents and form complexes with Ni. The organic acids most effective in complexing Ni are dicarboxylic and tricarboxylic acids. Thus, the movement of metals to lower horizons or deeper strata is facilitated by organic compounds having complexing ability.

7.3 Other Factors

Although data are meager, Ni content in various soil fractions tends to increase with decreasing particle size (i.e., consistent with other metals) and can be generalized as sand < silt < clay. This indicates the role of CEC in retaining Ni in soils, which as a rule of thumb, increases with decreasing particle size. The effects of parent material on Ni content of soils was discussed previously in Section 3.

7.4 Plant Factors

Like the other metals, uptake of Ni by plants varies according to species and cultivars. In addition, Ni accumulates differentially between plant parts and tends to accumulate higher in younger leaves. Agronomic crops can be adapted to relatively high-Ni soils by selecting species that possess low influx of ions into roots and low transport from roots to shoot, the so-called inefficient species or cultivars (Yang et al., 1996). For example, white clover is more tolerant of high Ni in such substrate than is ryegrass because of the high influx and transport of Ni in the latter.

8 Sources of Nickel in the Environment

8.1 Mining and Smelting

Mining and smelting of Ni-bearing ores substantially contribute to the cycling of Ni in the environment. Release of air pollutants from metal-ore smelting—chiefly SO_2 and heavy metals primarily Ni, Cu, Co, and Fe—has been implicated as the cause of extensive damage to terrestrial vegetation and soils in the Sudbury, Ontario area (Freedman and Hutchinson, 1980; Hutchinson and Whitby, 1974). Three major air pollution processes have been identified in the Sudbury region. In the outermost zone, pollution takes the form of acid deposition, merged with the larger background of atmospheric acidity over eastern North America. Closer to the smelters, acid deposition intensifies, and direct injury to vegetation may occur due to SO_2 fumes. Nearest to the smelters, still more intensified acid aerosol washout and SO_2 fumes occur, augmented by substantial deposition of particulates laden with Ni, Cu, and to a lesser degree, Fe and Co (Dudka and Adriano, 1997).

In the Copper Cliff Coniston Ni smelters near Sudbury, large amounts of Ni were released in the 1960s and 1970s, depositing substantial amounts in soils. The highest levels of metals in soils in the Sudbury region occurred in surface soils close to the smelters (levels up to 3000 to 5000 ppm Ni), but elevated levels of Ni and Cu in soils have been reported from as far away as 50 km from the smelters (Hutchinson and Whitby, 1974). Concentrations of Ni and Cu in roots, stems, and leaves of lowbush blueberry (*Vaccinium angustifolium*) declined logarithmically with increasing distance from the smelter in Sudbury (Fig. 17.5). The impacts of the smelting activity are also detectable in lakes within a 30-km radius of the smelting complex. In fact,

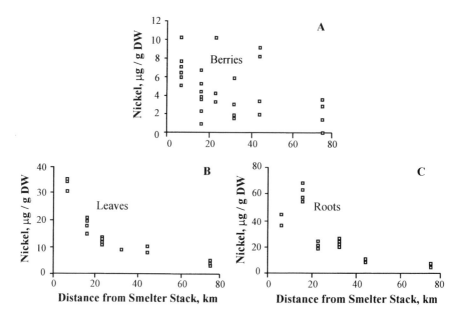

FIGURE 17.5. Concentrations of nickel in tissues of *V. angustifolium* (lowbush blue berry) as a function of distance from a smelter stack, Sudbury, Ontario, Canada. (Extracted from Bagatto and Shorthouse, 1991.)

the enrichment factors in surficial sediments and deposition rates for Ni and Cu are among the highest ever recorded anywhere in the world (Nriagu et al., 1982). In response to this exposure to high Ni over the past 70 years, some Ni-tolerant populations of plants and animals have evolved. Nickel pollution from metal smelting has also been reported in Australia.

The effects of Ni emission on the concentration of Ni in crops and feed-stuff are indicated in Table 17.8. In general, aside from smelters, Ni in air comes primarily from fuel oil and coal combustion since petroleum contains Ni and there is a high correlation between V and Ni in air (Schroeder, 1971). However, Swaine (1980) reported that coal combustion does not contribute as much Ni to the atmosphere as oil combustion. The former emits only 0.66×10^6 kg Ni yr^{-1} vs. 26.7×10^6 kg Ni yr^{-1} for the latter.

8.2 Biosolids

As sewage sludge may contain up to 0.5% total Ni, this biosolid applied on land can cause soil enrichment with Ni. Of the metals in sludge, Cu, Ni, and Zn are considered the ones most likely to cause phytotoxicity while Cd, Cr, Mn, Hg, and Pb are regarded as less of a potential problem in this regard. Although considerable information has indicated increased uptake of metals due to sludge application, long-term studies show that while Ni has

TABLE 17.8. Nickel concentrations (ppm DW) of crops and feedstuff with and without the influence of nickel emission[a].

Plant	Average	
	Without emission	With emission
Onions	0.83	2.06
Potatoes	0.56	1.03
Carrots	0.50	1.41
Turnips	0.49	2.20
Dwarf beans	3.07	8.22
Parsley	1.72	11.14
Lettuce	1.40	10.69
Cabbage	0.84	4.65
Kohlrabi	0.62	1.72
Tomatoes	0.57	1.91
Apples	0.43	1.27
Oats	0.71	1.46
Winter wheat	0.30	1.26
Winter rye	0.26	1.04
Summer barley	0.25	0.39
Turnip leaves	1.46	5.76
Pasture grass	1.00	9.00
Green maize	1.05	2.81

[a]Samples were taken from four different areas in eastern Germany without Ni emission and within the radius (20 km) of a Ni-processing plant that began operation 15 yr prior to sampling. *Source:* Anke et al. (1983).

accumulated substantially in soils, uptake by plants of this element has not been sufficient to be of concern in the food chain.

8.3 Other Sources

Since metals are found in phosphate rocks in varying amounts, application of phosphatic fertilizers in agricultural lands may elevate the levels of some metals in soils. In general, phosphate rocks from the western U.S. contain higher concentrations of most heavy metals than do rocks in the eastern states. Selected U.S. phosphate rocks can have as much as 64 ppm Ni; a range of 1 to 10 ppm Ni for Australian superphosphates has been reported (Williams, 1977). Although long-term use of phosphate fertilizers may raise the Cd levels in soils, plant uptake of Ni from fertilizers should not cause concern in the food chain.

Coal contains varying amounts of Ni (<2 to 150 ppm in European and Canadian coals; <1 to 90 ppm in Australian bituminous coals; <2 to >300 ppm in bituminous coals from the Appalachian region). Upon combustion, Ni is concentrated in the fly ash. Nickel is distributed between the fly ash and the bottom slag, with about 80% of the Ni being in the fly ash. Swaine (1980) reported that fly ash from several countries can have Ni from <2 to 960 ppm. Despite possible high levels of Ni in some fly ashes, it should not pose any threat to the growth and quality of plants unlike Se, Mo, and B (Adriano et al., 1980; Carlson and Adriano, 1993).

References

Adriano, D.C., A.L. Page, A.A. Elseewi, A.C. Chang, and I. Straughan. 1980. *J Environ Qual* 9:333–344.

Agarwala, S.C., S.S. Bisht, and C.P. Sharma. 1977. *Can J Bot* 55:1299–1307.

Ainsworth, C.C., and D. Rai. 1987. In EPRI Rep EA–5321. Electric Power Res Inst, Palo Alto, CA.

Andersson, A. 1977. *Swed J Agric Res* 7:7–20.

Anke, M., M. Gron, B. Groppel, and H. Kroneman. 1983. In B. Sarkar, ed. *Biological Aspects of Metals and Metal Related Diseases*. Raven Pr, New York.

Archer, F.C. 1980. *Minist Agric Fish Food* (Great Britain) 326:184–190.

Aschmann, S.G., and R.J. Zasoski. 1987. *Physiol Plant* 71:191–196.

[ATSDR] Agency for Toxic Substances and Disease Registry. 1993. *Toxicological Profile for Nickel*. U.S. Dept Health and Human services, Atlanta, GA.

Bagatto, G., and J.D. Shorthouse. 1991. *Can J Bot* 69:1483–1490.

Beckett, P.H.T., and R.D. Davis. 1978. *New Phytol* 81:155–173.

Berrow, M.L., and G.A. Reaves. 1984. In *Proc Intl Conf Environmental Contamination*. CEP Consultants, Edinburgh.

Bingham, F.T., G.A. Mitchell, R.J. Mahler, and A.L. Page. 1976. In *Proc Intl Conf Environ Sensing and Assessment,* vol 2. Inst Elect and Electron Engrs, New York.

Bleeker, P., and M.P. Austin. 1970. *Aust J Soil Res* 8:133–143.

Bloomfield, C., W.I. Kelso, and G. Pruden. 1976. *J Soil Sci* 27:16–31.

Bowen, H.J.M. 1979. *Environmental Chemistry of the Elements*. Academic Pr, New York.

Bowman, R.S., and G.A. O'Connor. 1982. *Soil Sci Soc Am J* 46:933–936.

Bowman, R.S., M.E. Essington, and G.A. O'Connor. 1981. *Soil Sci Soc Am J* 45:860–865.

Bradford, G.R., R.J. Arkley, P.F. Pratt, and F.L. Bair. 1967. *Hilgardia* 38:541–556.

Bradley, R.I., C.C. Rudeforth, and C. Wilkins. 1978. *J Soil Sci* 29:258–270.

Brooks, R.R. 1980. In J.O. Nriagu, ed. *Nickel in the Environment*. Wiley, New York.

Brooks, R.R., and C.C. Radford. 1978. *Proc Royal Soc Lond* [*Biol*] 200:217–224.

Brooks, R.R., J. Lee, and T. Jaffre. 1974. *J Ecol* 62:493–499.

Brooks, R.R., J. Lee, R.D. Reeves, and T. Jaffre. 1977. *J Geochem Explor* 7:49–57.

Brooks, R.S., S. Shaw, and M.A. Asensi. 1981. *Physiol Plant* 51:167–170.

Brown, P.H., R.M. Welch, and E.E. Cary. 1987a. *Plant Physiol* 85:801–803.

Brown, P.H., R.M. Welch, E.E. Cary, and R.T. Checkai. 1987b. *J Plant Nutr* 10:1442–1455.

Brown, P.H., R.M. Welch, and J.T. Madison. 1990. *Plant Soil* 125:19–27.

Cannon, H.L. 1978. *Geochem Environ* 3:17–31.

Carlson, C., and D.C. Adriano. 1993. *J Environ Qual* 22:227–247.

Carlson, R.W., F.A. Bazzaz, and G.L. Rolfe. 1975. *Environ Res* 10:113–120.

Cataldo, D.A., T.R. Garland, and R.E. Wildung. 1978a. *Plant Physiol* 62:563–565.

Cataldo, D.A., T.R. Garland, R.E. Wildung, and H. Drucker. 1978b. *Plant Physiol* 62:566–570.

Chang, A.C., A.L. Page, L.E. Warneke, and E. Grgurevic. 1984. *J Environ Qual* 13:33–38.

Chino, M. 1981. In K. Kitagishi and L. Yamane, eds. *Heavy Metal Pollution in Soils of Japan*. Japan Sci Soc Pr, Tokyo.

Clark, R.B., P.A. Pier, D. Knudsen, and J.W. Maranville. 1981. *J Plant Nutr* 3:357–374.

Clemente, G.F., L.C. Rossi, and G.P. Santaroni. 1980. In J.O. Nriagu, ed. *Nickel in the Environment*. Wiley, New York.

Connor, J., N.F. Shimp, and J.C.F. Tedrow. 1957. *Soil Sci* 83:65–73.

Connor, J.J., H.T. Shacklette, R.J. Ebens et al. 1975. In USGS Prof Paper 574-F. U.S. Geological Survey, Washington, DC.

Cottenie, A., R. Camerlynck, M. Verloo, and A. Dhaese. 1979. *Pure APAI Chem* 52:45–53.

Crooke, W.M. 1955. *Ann Appl Biol* 43:465–476.

Crooke, W.M., J.G. Hunter, and O.Vergnano. 1954. *Ann Appl Biol* 41:311–324.

Cunningham, J.D., D.R. Keeney, and J.A. Ryan. 1975. *J Environ Qual* 4:455–459.

Domingo, L.E., and K. Kyuma. 1983. *Soil Sci Plant Nutr* 29:439–452.

Doner, H.E. 1978. *Soil Sci Soc Am J* 42:882–885.

Dudka, S., and D.C. Adriano. 1997. *J Environ Qual* 26:590–602.

Elrashidi, M.A., A. Shehata, and M. Wahab. 1979. *Agrochimica* 23:245–253.

Emmerich, W.E., L.J. Lund, A.L. Page, and A.C. Chang. 1982a. *J Environ Qual* 11:178–181.

Emmerich, W.E., L.J. Lund, A.L. Page, and A.C. Chang. 1982b. *J Environ Qual* 11:182–186.

Eskew, D.L., R.M. Welch, and W.A. Norvell. 1983. *Science* 222:621–623.

Foy, C.D., R.L. Chaney, and M.C. White. 1978. *Ann Rev Plant Physiol* 29:511–566.

Freedman, B., and T.C. Hutchinson. 1980. *Can J Bot* 58:1722–1736.

Gasser, U.G., S.J. Juchler, W.A. Hobson, and H. Sticher. 1995. *Can J Soil Sci* 75:187–195.

Hamilton, E.I., and M.J. Minski. 1972. *Sci Total Environ* 1:375–394.

Harter, R.D. 1983. *Soil Sci Soc Am J* 47:47–51.

Hausinger, R.P. 1992. In E. Nieboer and J.O. Nriagu, eds. *Nickel and Human Health*. Wiley, New York.

Hazlett, P.W., G.K. Rutherford, and G.W. Van Loon. 1984. *Geoderma* 32:273–285.

Heikal, M.M.D., W.L. Berry, A. Wallace, and D. Herman. 1989. 147:413–415.

Heinrichs, H., and R. Mayer. 1980. In J.O. Nriagu, ed. *Nickel in the Environment*. Wiley, New York.

Hickey, M.G., and J.A. Kittrick. 1984. *J Environ Qual* 13:372–376.

Hunter, J.G., and O. Vergnano. 1953. *Ann Appl Biol* 40:761–777.

Hutchinson, T.C., and L.M. Whitby. 1974. *Environ Conserv* 1:123–132.

Hutchinson, T.C., M. Czuba, and L.M. Cunningham. 1974. *Trace Subs Environ Health* 8:81–93.

Iizuka, T. 1975. *Soil Sci Plant Nutr* 21:47–55.

Ishihara, M., Y. Hase, H. Yolornizo, S. Konno, and K. Sato. 1968. Engei Shikenio Hokoku (ser. A, 1968) 7:39–54.

John, M.K., and C.J. Van Laerhoven. 1976. *J Environ Qual* 5:246–251.

Kabata–Pendias, A., and H. Pendias. 1992. *Trace Elements in Soils and Plants*. CRC Pr, Boca Raton, FL.

Khalid, B.Y., and J. Tinsley. 1980. *Plant Soil* 55:139–144.

Kirkpatrick, D.C., and D.E. Coffin. 1974. *J Inst Can Sci Technol Aliment* 7:56–58.

Kramer, U., J.D. Cotter-Howells, J.M. Charnock, A.J.M. Baker, and J.A.C. Smith. 1996. *Nature* 379:635–638.

Lake, D.L., P.W.W. Kirk, and J.N. Lester. 1984. *J Environ Qual* 13:175–183.

Latterell, J.J., R.H. Dowdy, and W.E. Larson. 1978. *J Environ Qual* 7:435–440.

Lindau, C.W., and L.R. Hossner. 1982. *J Environ Qual* 11:540–545.

Lozano, F.C., and L.K. Morrison. 1982. *J Environ Qual* 11:437–441.

Macnicol, R.D., and P.H.T. Beckett. 1985. *Plant Soil* 85:107–129.

Mance, G. 1987. *Pollution Threat of Heavy Metals in Aquatic Environments*. Elsevier, London.

Mattioni, C., R. Gabbrielli, J. Vangronsveld, and H. Clijsters. 1997. *J Plant Physiol* 150:173–177.

McGrath, S.P. 1995. Chromium and Nickel. In B.J. Alloway, ed. *Heavy Metals in Soils*. Blackie, London.

McGrath, S.P., and P.J. Loveland. 1992. In *The Geochemical Atlas of England and Wales*. Blackie, London.

Mills, J.G., and M.A. Zwarich. 1975. *Can J Soil Sci* 55:295–300.

Memon, A.R., S. Ito, and M. Yatazawa. 1980. *Soil Sci Plant Nutr* 26:271–280.

Mishra, D., and M. Kar. 1974. *Bot Rev* 40:395–452.

Mitchell, R.L. 1971. In Minist Agric Fish Fed Tech Bull 21. Her Majesty's Sta Office, London.

Mizuno, N. 1968. *Nature* 219:1271–1272.

Moore, J.W. 1991. *Inorganic Contaminants in Surface Water*. Springer-Verlag, New York.

Murthy, G.K., U.S. Rhea, and J.T. Peeler. 1973. *Environ Sci Technol* 7:1042–1045.

Nakos, G. 1982. *Plant Soil* 66:271–277.

Nieboer, E., and J.O. Nriagu. 1992. *Nickel and Human Health*. Wiley, New York.

Nieboer, E., W.E. Sanford, and B.C. Stace. 1992. In E. Nieboer and J.O. Nriagu, eds. *Nickel and Human Health*. Wiley, New York.

Nriagu, J.O., H.K.T. Wong, and R.D. Coker. 1982. *Environ Sci Technol* 16:551–560.

Painter, L.I., S.J. Toth, and F.E. Bear. 1953. *Soil Sci* 76:421–429.

Pierce, F.J., R.H. Dowdy, and D.F. Grigal. 1982. *J Environ Qual* 11:416–422.

Polemio, M., S.A. Bufo, and N. Senesi. 1982. *Plant Soil* 69:57–66.

Pratt, P.F., F.L. Bair, and G.W. McLean. 1964a. 8th Intl Cong Soil Sci 3:243–248, Bucharest, Romania.

Pratt, P.F., F.L. Bair, and G.W. McLean. 1964b. *Soil Sci Soc Am Proc* 28:363–365.

Proctor, J. 1971. *J Ecol* 59:827–842.

Quin, B.F., and J.K. Syers. 1978. *NZ J Agric Res* 21:435–442.

Reeves, R.D. 1992. In *The Vegetation of Ultramafic (Serpentine) Soils*. Intercept, Andover, United Kingdom.

Roberts, B.A. 1980. *Can J Soil Sci* 60:231–240.

Sadiq, M., and C.G. Enfield. 1984a. *Soil Sci* 138:262–270.

Sadiq, M., and C.G. Enfield. 1984b. *Soil Sci* 138:335–340.

Sanders, J.R., S.P. McGrath, and T.M. Adams. 1986. *J Sci Food Agric* 37:961–968.

Schmidt, J.A., and A.W. Andren. 1980. In J.O. Nriagu, ed. *Nickel in the Environment*. Wiley, New York.

Schroeder, H.A. 1971. *Environment* 13:18–32.

Severne, B.C. 1974. *Nature* 248:807–808.

Shacklette, H.T., J.C. Hamilton, J.G. Boerngen, and J.M. Bowles. 1971. In USGS paper 574–D, U.S. Geological Survey, Washington, DC.

Shewry, P.R., and P.J. Peterson. 1976. *J Ecol* 64:195–212.

Smith, S.R. 1994. *Environ Pollut* 85:321–327.

Sommers, L.E., and W.L. Lindsay. 1979. *Soil Sci Soc Am J* 43:39–46.

Sposito, G., and A.L. Page. 1984. In *Circulation of Metal Ions in the Environment: Metal Ions in Biological Systems*. Marcel Dekker, New York.

Sunderman, F.W. 1992. In E. Nieboer and J.O. Nriagu, eds. *Nickel and Human Health*. Wiley, New York.

Sunderman, F.W., S.M. Hopfer, K.R. Sweeney, A.H. Marcus, B.M. Most, and J. Cuanson. 1989. *Proc Soc Biol Med* 191:5–11.

Swaine, D.J. 1955. *The Trace Element Content of Soils*. Commonwealth Bureau Soil Sci(EB), Tech Commun 48. Herald Printing Works, York, United Kingdom.

Swaine, D.J. 1980. In J.O. Nriagu, ed. *Nickel in the Environment*. Wiley, New York.

Swaine, D.J., and R.L. Mitchell. 1960. *J Soil Sci* 11:347–368.

Tadesse, W., J.W. Shuford, R.W. Taylor, D.C. Adriano, and K.S. Sajwan. 1991. *Water Air Soil Pollut* 55:397–408.

Traynor, M.F., and B.D. Knezek. 1974. *Trace Subs Environ Health* 7:75–81.

Trudinger, P.A., D.J. Swaine, and G.W. Skyring. 1979. In P.A. Trudinger and D.J. Swaine, eds. *Biogeochemical Cycling of Mineral-Forming Elements*. Elsevier, Amsterdam.

Tyler, L.D., and M.B. McBride. 1982. *Soil Sci* 134:198–205.

[U.S. EPA] United States Environmental Protection Agency 1985. *Health Assessment Document for Nickel*. EPA/600/8–83/012F. Natl Environ Res Centre, Cincinnati, OH.

U.S. EPA. 1993. *Integrated Risk Information System (IRIS) on Nickel*. Natl Environ Res Center, Cincinnati, OH.

Valdares, J.M.A.S., M. Gal, U. Mingelgrin, and A.L. Page. 1983. *J Environ Qual* 12:49–57.

Vanselow, A.P. 1965. In H.D. Chapman, ed. *Diagnostic Criteria for Plants and Soils*. Quality Printing, Abilene, TX.

Whitby, L.M., J. Gaynor, and A.J. MacLean. 1978. *Can J Soil Sci* 58:325–330.

Williams, C.H. 1977. *J Aust Inst Agric Sci* (Sept–Dec):99–109.

Yang, X., V.C. Baligar, D.C. Martens, and R.B. Clark. 1996. *J Plant Nutr* 19:73–85.

18
Selenium

1 General Properties of Selenium

Selenium (atom. no. 34) is a member of Group VI-A, also known as the S family in the periodic table. It has a specific gravity of 4.79 for the metallic (gray) form or 4.28 for the vitreous (black) form. It has an atom. wt. of 78.96 and melting pt. of 217 °C (gray form). Selenium is classified as a metalloid having properties of both a metal and a nonmetal. Because of its chemical similarity to S, it resembles S both in its form and compounds, and this accounts for their many interrelations in biology. It has six stable isotopes in nature with the following composition (in %): ^{74}Se, 0.87; ^{76}Se, 9.02; ^{77}Se, 7.58; ^{78}Se, 23.52; ^{80}Se, 49.82; and ^{82}Se, 9.19. The most important oxidation states of Se are $-$II, 0, IV, and VI. Elemental Se (Se^0) is often associated with S in compounds such as selenium sulfide (Se$_2$S$_2$) and polysulfides. Selenite (SeO$_3^{2-}$) and selenate (SeO$_4^{2-}$) are common ions in soils and natural waters and are highly water soluble; Se^0 is much less soluble in water. Selenium can be easily oxidized from elemental Se [Se(0)], to SeO$_3^{2-}$ [Se(IV)], and to SeO$_4^{2-}$ [Se(VI)]. Some of the better-known commercial Se compounds include H$_2$Se, metallic selenides [Se($-$II)], SeO$_2$, H$_2$SeO$_3$, SeF$_4$, Se$_2$Cl$_2$, and H$_2$SeO$_4$ (selenic acid). Selenium also forms a large number of organic compounds that are similar to those of S.

2 Production and Uses of Selenium

Selenium is a rare element, there being no large deposits anywhere. It is obtained commercially as a by-product of the electrolytic refining of Cu. The leading producers of Se include Canada, Japan, the United States, the former Soviet Union, Mexico, Sweden, and Belgium—yielding a combined output of over 1600 tonnes of Se annually.

Compounds of Se are used in photoelectric cell devices (e.g., in photographic exposure meters, photometers, counting devices, and light-controlled switches) as well as xerography; as maroon and orange pigments, in combination with Cd sulfide, for plastics and ceramics; in increasing the resistance of rubber to heat, oxidation, and abrasion; in glass making; and as lubricants to increase the machinability of stainless steel. Glass making (to counteract coloration due to iron oxides and to color the glass red) constitutes about 20% of its overall industrial use. Selenium is an antioxidant, which makes it useful to include in mineral and vegetable oils, lubricants, and inks. Selenium sulfide is used as a therapeutic agent in shampoos for certain scalp conditions. Selenium is also used in coloring copper and copper alloys, and as insecticides and fungicides.

3 Selenium in Nature

The abundance of Se in the earth's crust ranges from 0.05 to 0.09 ppm, which is about 1/6000 of the abundance of S and 1/50 of the abundance of As. It is ranked 68th in crustal abundance among elements, ahead of Hg. Although Se is inconsistently dispersed in geologic materials, it is detectable in most earth materials (Table 18.1) and is associated with various types of sedimentary rocks at concentrations ranging from 0.10 to 675 ppm; it is

TABLE 18.1. Commonly observed selenium concentrations (ppm) in various environmental media.

Material	Average concentration	Range
Igneous rocks[a]	0.05	—
Limestone[b]	0.08	0.1–6
Sandstone[a,b]	0.05	—
Shale[a]	0.60	<1–675
Black shales[f]	24.1	—
Crude oil[f]	0.17	0.06–0.35
Coal[c]	3.4	0.05–11
Coal ash[d]:		
Fly ash	14	5.5–47
Bottom ash	4.1	<1.5–10
FGD sludge	36	<2–162
Oil ash[d]	29	23–34
Phosphate rocks[a,b]	mostly <20	1–300
Superphosphate[a]	mostly <4	up to 25
Soils (world)[a]	<0.1–2	up to 5000
Seleniferous soils	>2.0	—
Sediments (stream)[f]	<5	—
Se-deficient soils[f]	<0.3	—
Grasses[e]	0.26	<0.01–9.0
Plants[a]	0.1–15	some up to 1200
Se accumulator plants[b]	—	100–10,000
Vegetables and fruits[e]	0.05	0.01–0.20
Freshwater[a]	0.0002	0.0001–0.4
Seawater[c]	0.0002	0.0001–0.0002

Sources: Extracted from
[a]Swaine (1987).
[b]Gamboa-Lewis (1976).
[c]Bowen (1979).
[d]Adriano et al. (1980); Ainsworth and Rai (1987).
[e]Cannon (1974).
[f]Ihrat (1989).

frequently enriched in black shales. It tends to concentrate in carbonaceous debris in sandstones; it is also enriched in phosphate rocks (1 to 300 ppm) and thus may contaminate soils from applications of phosphatic fertilizers. In contrast, igneous rocks contain small amounts of Se, usually <0.20 ppm, while volcanic ashes and tuffs can contain much higher content. Calcareous rocks likewise contain small amounts. Coal and crude oil generally contain much more Se than igneous rocks, presumably because of their higher content of OM. Combustion of both forms of fossil fuel can release substantial amounts of Se into the environment.

The primary sources of Se in nature are volcanic eruptions and metallic sulfides associated with igneous activity; secondary sources are biological pools in which Se has accumulated (NAS, 1974). Examples of primary sources are western shales of the Cretaceous Age and the sandstone ores of the Colorado Plateau. In general, shales have the highest concentrations of Se and are the primary sources of high-Se soils in the Great Plains and Rocky Mountain foothills in the United States (Lakin, 1961). The higher contents in shales can be attributed partly to biological activity. Although about 50 Se minerals are known, Se is commonly associated with heavy metal (Ag, Cu, Pb, Hg, Ni, etc.) sulfides, where it occurs as a selenide, or as a substitute ion for S in the crystal lattice. Neither Se nor S is an essential constituent of rock-forming silicate minerals. The close similarity of the ionic radii of S (1.84 Å) and Se (1.91 Å) indicates that Se can readily enter sulfide minerals by isomorphous substitution. The highest Se content of sulfides is associated with uranium ores in sandstone-type deposits in the western United States. The maximum levels of Se for minerals are: pyrite, 5%; marcasite, 0.65%; and chalcocite, 5% (Coleman and Delevaux, 1957). The highest resources of Se are reported for chalcopyrite-pentlandite-pyrrhotite deposits geologically related to basic and ultrabasic rocks of the Precambrian Age.

Certain regions in some countries have soils that contain Se well above the normal levels. These soils, generally known as seleniferous soils are located in arid and semiarid areas of China, Mexico, and Colombia, and in Hawaii and the western interior of the United States and Canada. These soils are known for having high bioavailable Se that frequently supports vegetation characteristically high in Se. In arid and semiarid areas of the United States having mean annual rainfall of <50 cm, soils have been reported to produce seleniferous vegetation (Trelease and Beath, 1949). In general, high-Se soils do not occur in humid regions.

4 Selenium in Soils/Sediments

4.1 Total Selenium in Soils

Soil Se may provide some indication as to Se sufficiency to plants and animals. Most noncontaminated world soils contain 0.10 to 2 ppm of total

Se (Appendix Table A.24). Another report indicates a range of 0.03 to 2.0 ppm (mean of 0.40 ppm) for world soils ($n = 1623$) (Ure and Berrow, 1982). From extensive U.S. soil surveys, Se varied from trace amounts to a maximum of 82 ppm; Se levels of 1 to 6 ppm are about normal for seleniferous soils (Byers et al., 1938). Seleniferous soils of North America extend into Alberta, Manitoba, and Saskatchewan, west to the Pacific coast, and south into Mexico. In the United States, the region extends primarily from North Dakota to Texas (i.e., the Great Plains) and west to the Pacific coast (excluding the Pacific northwest). The soils in the Great Plains originate from the Cretaceous-age shale, tend to be alkaline, and have Se levels in the 6 to 28 ppm range that have caused Se toxicity in livestock.

Seleniferous soils in the three Canadian provinces showed Se values of 0.10 to 6 ppm (Byers and Lakin, 1939). Some seleniferous soils in the United Kingdom have Se contents of 0.90 to 91.4 ppm (Nye and Peterson, 1975). In Wales and Ireland, Se has leached from Avonian shales enriching the low-lying, high-OM soils with Se in the 30 to 300 ppm range. However, normal soils for the United Kingdom only average 0.60 ppm, with values ranging from 0.20 to 1.8 ppm ($n = 114$) (Archer, 1980).

Selenium deficiency in livestock has been observed in the United States, Canada, New Zealand, Australia, Scotland, Finland, Denmark, Sweden, the United Kingdom, France, Germany, Turkey, and the former Soviet Union. In contrast, Se-toxic soils have been discovered in the United States, Canada, Colombia, Ireland, South Africa, Australia, the United Kingdom, Mexico, and the former Soviet Union. A ranking for Se in New Zealand soils has been proposed as follows: <0.30, very low; 0.30 to 0.50, low; 0.50 to 0.90, average; 0.90 to 1.5, high; and >1.5 ppm, very high (Wells, 1966). Soils having more than 2 ppm total Se are likely to be seleniferous.

Selenium can be expected to be present in excessive amounts only in semiarid and arid regions in soils derived from Cretaceous shales (Rosenfeld and Beath, 1964), but in humid climates or in irrigated areas, most of the Se is leached from soils of this origin. In some cases, however, Se may be insufficient in soils relative to animal nutrition, particularly in soils originating from igneous rocks; Se in the surface layer may be further depleted by intensive irrigation (Oldfield, 1972).

In general, Se concentration in soils mostly depends on the parent rock, the topography and age of the soil, and climate.

4.2 Bioavailable Selenium in Soils

Total Se in soils is not a reliable index of plant-available Se. For this reason, several extracting solutions have been evaluated to predict bioavailability of Se in plants. Normal soils contain <50 ppb of water-soluble Se (Workman and Soltanpour, 1980). Soils supporting vegetation containing toxic quantities of Se can be expected to have water-soluble Se in the range of a tenth of a ppm. Water-soluble levels of Se in U.S. soils ranged from 0.10 ppm to 38 ppm (Byers et al., 1938). However, in about 80% of the samples, Se did not exceed 0.10 ppm. The most common water-soluble Se has been identified as the selenate form in soils from Wyoming and the midwestern United States (Rosenfeld and Beath, 1964), Ireland (Fleming and Walsh, 1957), and Canada (Levesque, 1974b). Water-soluble Se varied from 2.0 to 7.0% of the total Se (0.19 to 0.74 ppm) for some Canadian soils (Levesque, 1974b) and from 0.33 to 2.90% of total Se (20 to 850 ppm) for some Irish soils (Fleming and Walsh, 1957). Water-extractable Se is of doubtful utility as a plant-available indicator in Se-deficient soils, but can be a useful criterion for assessing the behavior of Se applied to soils.

Other extractants that have been tested in predicting plant-available Se include hot water, NH_4HCO_3–DTPA, $CaCl_2$, $Ca(NO_3)_2$, and K_2SO_4 solutions. In low-Se soils of Scandinavia, total Se content in plants was not correlated with either the water-soluble or total Se in surface soils (Hamdy and Gissel-Nielsen, 1976; Lindberg and Bingefors, 1970). In contrast, Sippola (1979) found that Se in timothy grass growing in different field sites throughout Finland reported a better correlation with total than with extractable Se. Similarly, significant correlations between total Se in some British soils and plant content were observed (Nye and Peterson, 1975) and total Se in Belgian soils and ryegrass content (Robberecht et al., 1982). Hot water and NH_4HCO_3–DTPA extracted approximately equal amounts of Se from soil (Fig. 18.1) (Soltanpour and Workman, 1980). Significantly high correlation ($r^2 = 0.96$) was found between the amounts of Se extracted by NH_4HCO_3–DTPA and Se uptake by alfalfa plants. The DTPA-extractable soil Se levels of over 100 ppb produced alfalfa plants containing levels of Se above the 5 ppm level. The inconsistencies in the results of various

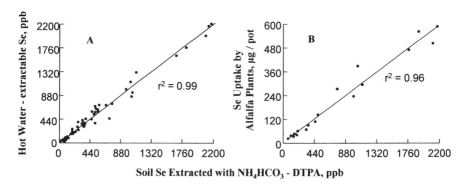

FIGURE 18.1. (A) Relationships of hot water-extractable vs. NH_4HCO_3–DTPA–extractable selenium in soils before planting and after harvest of alfalfa. (B) Selenium uptake by two cuttings of alfalfa as a function of NH_4HCO_3–DTPA–extractable selenium in soils. (From Soltanpour and Workman, 1980.)

investigators indicate the complex chemistry of Se in soils and its differing bioavailability to plants, which in turn is affected by a number of factors discussed later.

4.3 Selenium in Soil Profile

The distribution of Se in the soil profile, which gives clues as to the behavior of this element during the processes of soil formation, is influenced by several factors including the parent material, OM, pH, Fe oxide, and to some extent, rainfall. In a survey involving 54 Canadian profiles with possible animal nutrition problems due to low Se contents in crops, the highest Se values (0.94 ppm) were obtained typically from the podzolic B horizon and surface organic layers (Levesque, 1974a). This indicates the importance of OM in retaining Se in surface soil. Selenium distribution in the profile varied with the soil order. For Podzols, the surface layer in the A horizon gave a total Se content of 0.084 ppm compared with a content of 0.073 ppm in the C horizon. Among the mineral horizons, the B horizon produced the highest Se content (0.52 ppm). In Gleysolic soils, marked accumulations occurred in the A horizon (0.48 ppm) and in the B horizon (0.25 ppm) as compared with only 0.10 ppm in the C horizon. In Brunisolic soils, there was a decreasing pattern of Se concentration with depth, 0.64 ppm Se in the surface layer vs. 0.11 ppm in the C horizon. It is clear from this survey that Se is closely associated with organic C in the upper horizon and with Fe oxide in the lower horizon.

In New Zealand soils with little profile development (i.e., A-C soils), the Se content of surface soil depends upon the nature of the parent material (Wells, 1967); it is very low in granites and rhyolitic pumices, low in mica schist and graywacke (excluding volcanic graywacke), but is high in andesitic and basaltic ashes and in argillite. However, high amounts of Se occur in strongly weathered surface soils from which the clay and Fe compounds have not been eluviated. Low-Se surface soils were found in the following soil groups: brown-gray earths, yellow-gray earths away from the andesitic ash zone, high country yellow-brown earths, Podzols, pumice soils, and southern steepland soils (Wells, 1967). In contrast, high-Se surface soils were found in northern yellow-brown earths, rendzinas, and some andesitic and basaltic soils. These findings are similar to those from Canada in that the levels of Se are very high in well-developed B2 horizons (1.4 ppm) where clay accumulation occurs and at a maximum in the concretion layer of ironstone soils and in the iron-pan layer of Podzols. For comparison, the A horizon has an average Se content of 0.60 ppm.

In Finnish soils, Se is enriched in the O-A1 horizon enriched in OM and in Fe-rich B horizon, but depleted in the eluviated A2 horizon (Koljonen, 1975). However, in areas of low rainfall and high temperature, Se moved up

TABLE 18.2. Forms of selenium in soil profiles (μg Se kg^{-1}) of a Swedish forest.

Site	Horizon	NaH$_2$PO$_4$-soluble SeO$_3^{2-}$	Na$_4$P$_2$O$_7$/ NaOH-soluble Se	Total SeO$_3^{2-}$
Skogaby	E	1	84	5
	Bs$_1$	5	280	24
	Bs$_2$	17	347	43
	Bs$_3$	23	280	51
	B/C	16	119	29
	C1	8	54	18
Ulriksdal	A/E/B	1	107	4
	Bs$_1$	5	173	14
	Bs$_2$	7	161	14
	Bs$_3$	6	118	12
	B/C1	3	44	8
	B/C2	1	17	5

Source: Extracted from Gustafsson and Johnsson (1992).

the soil profile, i.e., due to high evaporation, while in high rainfall areas it was leached down the profile. Indeed this is what happens to Se in arid high-temperature areas as in India (Singh and Kumar, 1976) and in parts of California (see Section 6.2.1).

In Swedish forest soils whose anthropogenic source of Se is atmospheric fallout, all soil OM fractions, particularly those in the B horizon, are considerably enriched in Se compared with Se in plant tissues (Table 18.2) (Gustafsson and Johnsson, 1992). Selenium present in organic fractions accounted for most of the Se extracted by Na$_4$P$_2$O$_7$/NaOH (to estimate organic Se compounds). Data on loss-on-ignition and oxalate treatment of soils indicate the considerable influence of OM, and Fe and Al oxide contents of soils on Se distribution in the soil profile.

In China, soils in the southern provinces (laterite and red earth) and arid regions (chestnut and gray desert soil), the tendency to retain Se increases with the profile depth, whereas for the temperate forest and forest-steppe soil (dark brown earth and black soil) the tendency decreases with depth (Tan et al., 1994).

4.4 Forms and Chemical Speciation of Selenium in Soils

The chemical form of Se in soils largely determines its mobility and bioavailability to plants and animals. Selenium in nature exists in four oxidation states: VI(selenate), IV(selenite), 0(elemental Se), and −II(selenide) (Appendix Table A.33). Although solubility data for Se minerals are still incomplete, precipitation–dissolution reactions can also play a role in determining Se concentrations in aqueous phase (Elrashidi et al., 1987). Selenium species in soil solution can be present primarily in three oxidation states:

1. Selenates(VI): SeO_4^{2-}, $HSeO_4^-$, and H_2SeO_4 (selenic acid)
2. Selenites(IV): SeO_3^{2-}, $HSeO_3^-$, and H_2SeO_3 (selenous acid)
3. Selenides(−II): Se^{2-}, HSe^-, H_2Se (hydrogen selenide), and R_2Se (organic selenides)

Available equilibrium reactions of Se from thermodynamic data indicate that in highly oxidizing conditions, SeO_4^{2-} is the predominant species in solution in both acid and alkaline soils (Elrashidi et al., 1987; 1989). In the medium redox range, both $HSeO_3^-$ and SeO_3^{2-} are the most important species in solution; the $HSeO_3^-$ species is more predominant in acid soils and SeO_3^{2-} is the main species under alkaline conditions (Fig. 18.2). More specifically, at a solution pH of 7 and a total Se concentration of 300 μg L^{-1}, SeO_4^{2-} and $CaSeO_4^0$ are the dominant species under highly oxidizing conditions (>400 mV); in moderately oxidizing conditions (0 to 400 mV), selenite stability is dominated by $CaSeO_3^0$ and $HSeO_3^-$; and under strongly reducing conditions (<0 mV) selenide stability is dominated by HSe^- (White

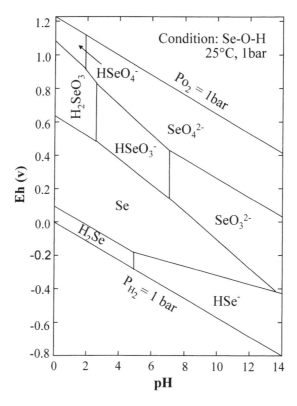

FIGURE 18.2. Predicted Eh–pH stability field for selenium. The activity of dissolved Se $= 10^{-6}$. For more details see Brookins (1988).

and Dubrovsky, 1994). Other selenate, selenite, or selenide species are not important in this redox range in soils. Under highly reducing conditions, HSe^- is the most important species in soil solution, especially in the alkaline pH range. Only under strongly acidic conditions does H_2Se^0 species contribute considerably to Se in solution. The species Se^{2-} is extremely low in solution and is not important for most soils.

In the oxidized alkaline soils of the arid western San Joaquin Valley, California Se(VI) is the dominant soluble form of Se, while Se(IV) represents the dominant adsorbed Se species (Fujii et al., 1988; Fio and Fujii, 1990). In addition, speciation of saturation extracts from selected San Joaquin Valley soils showed that Se was present as Se(VI), Se(IV), and hydrophobic organic Se (Fio and Fujii, 1990). The hydrophobic organic Se (<1 to 4 μg L^{-1}) was probably associated with the DOC of the saturation extract, which represents only a small fraction of the soil OM. Selenite concentrations in the extracts were also low (<1 to 4 μg L^{-1}) but 98% of the soluble Se was Se(VI) (4 to 640 μg L^{-1}) in most of the samples. The high concentrations of soluble Se and salinity, where they parallel in distribution in the soil profile, in shallow groundwaters, and soils in western San Joaquin Valley, are primarily due to evaporation from a shallow water table. Another Kesterson Reservoir study confirmed the predominance of Se(VI) in soil water that accounts for most of the soluble inventory, with Se(IV) accounting for the rest (Tokunaga et al., 1994). With the low pKa of H_2SeO_4 (pKa = 1.91), selenate occurs solely as SeO_4^{2-} in these neutral to alkaline soil solutions (Elrashidi et al., 1987). Selenite, with a pKa of 7.3, occurs as both $HSeO_3^-$ and SeO_3^{2-} in these solutions.

In evaluating the forms of Se in arable soils in the Panoche Creek alluvial fan area in western San Joaquin Valley, Fujii et al. (1988) estimated that 1 to 11% (of total soil Se) and <1 to 63% represent the adsorbed and soluble Se respectively. Most of the Se in the KCl extracts (i.e., soluble form) was selenate, whereas most of the Se in the phosphate extracts (i.e., adsorbed form) was selenite (Table 18.3). Detailed fractionation of soils and sediments at the reservoir (in the San Joaquin Valley) reveals the following primary Se associations: soluble Se(VI), soluble Se(IV), adsorbed Se, organic Se, carbonate Se, and refractory (recalcitrant) Se (Lipton, 1991; Tokunaga et al., 1991). Refractory Se dominates (50 to 72% of total Se) in surface layers (0 to 10 cm) while soluble Se was predominant (45 to 62% total Se) in deeper layers (45 to 55 cm deep) (Zawislanski and Zavarin, 1996). This unusual trend in the Se form distribution can be explained by the history of Se deposition at the Kesterson. The high percentage of insoluble Se (95 to 98%) in the surface soil is due to strongly reducing conditions during ponding periods (Fig. 18.3). Most of the total Se was immobilized in the top 15 cm of the soil, but as the ponds dried out and the surface became oxidized, Se was gradually solubilized and transported as selenates and selenites deeper into the soil profile. The second most common pool of Se is the organic

TABLE 18.3. Selenium forms extracted from soil samples from the 1-yr and 15-yr fields in western San Joaquin Valley, California.

Soil sample			Ground water Se (μg L^{-1})	Extract								
				Saturated soil paste			250 mol m^{-3} KCl			32 mol m^{-3} phosphate		
Field	Depth (m)	Total Se (mg kg^{-1})		Total Se, mg kg^{-1}	% SeO$_4^{2-}$	% SeO$_3^{2-}$	Total Se, mg kg^{-1}	% SeO$_4^{2-}$	% SeO$_3^{2-}$	Total Se, mg kg^{-1}	% SeO$_4^{2-}$	% SeO$_3^{2-}$
1-yr	0.9	1.4	6200	0.121 (150)[a]	98	2	0.273	98	2	0.0032	12	88
15-yr	2.4	0.8	610	0.061 (110)[a]	100	0	0.074	99	1	0.015	33	67

[a]Total Se in solution, μg L^{-1}

Source: Extracted from Fujii et al. (1988).

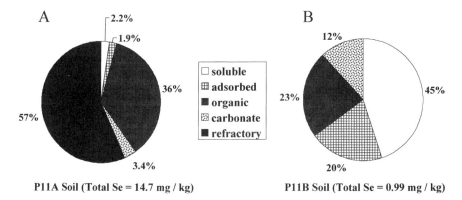

A   2.2%
    1.9%
    36%
57%
    3.4%

□ soluble
⊞ adsorbed
■ organic
☒ carbonate
■ refractory

B   12%
    23%   45%
    20%

P11A Soil (Total Se = 14.7 mg / kg)   **P11B Soil (Total Se = 0.99 mg / kg)**

FIGURE 18.3. Selenium fractions in (A) surface topsoil (P11A, 0–10 cm) and (B) bottom soil (P11B, 45–55 cm deep) from the Kesterson Reservoir, California. (Extracted from Zawislanski and Zavarin, 1996.)

fraction. Total Se concentrations are one to two orders of magnitude higher in surface soils than in subsoils, again conforming to the depositional history.

Another fractionation study evaluated the trend in the distribution of Se forms in the Kesterson area, including the shales (i.e., Se source), alluvial soils, drain sediment, and evaporation pond soils (i.e., Se sinks) (Martens and Suarez, 1998). The Se(VI) fraction is found only in the water extraction along with low levels of water-soluble Se(IV) (Table 18.4). Selenium was found in all samples as water-soluble, exchangeable, and very tightly held Se(IV). The sum of water-soluble and exchangeable Se was low ($<15\%$) compared with total Se. The extraction scheme indicates that a large proportion of the Se inventory ($>80\%$) is present in these materials as organic Se and as Se(0). Direct evidence for the existence of Se(0) was provided by X-ray absorption spectroscopy.

TABLE 18.4. Selenium speciation (mg kg^{-1}) in shales and alluvial soils in San Joaquin Valley, California.[a]

Material	Water Se[IV]	Water Se[VI]	P-buffer Se[IV]	P-buffer Se[–II]	NaoH Se[IV]	NaoH Se[–II]	HNO$_3$ Se[0]	Total Se
Kreyenhagen shale	0.03	0.47	1.35	0.36	3.98	2.98	1.77	10.94
Moreno shale	0.03	0.10	0.49	0.20	0.98	1.57	1.84	5.21
Tumey Gulch soil	ND[b]	0.09	0.02	0.02	0.04	0.33	0.56	1.06
Moreno Gulch soil	ND	0.05	0.01	0.02	0.01	0.18	0.31	0.58
Salt Creek soil	ND	0.01	0.01	0.01	0.01	0.10	0.27	0.41

[a]The shales and soils (5 g) were placed in 40 mL Teflon centrifuge tubes and sequentially extracted with water (25 mL), P buffer (25 mL), NaOH (25 mL; 90 °C), and HNO$_3$ (2.5 mL + 20 mL water; 90 °C); Se speciation in the extractants was determined by HGAAS; $n = 4$
[b]ND, not detectable.
Source: Extracted from Martens and Suarez (1998).

4.5 Sorption and Solubility of Selenium in Soils/Sediments

The solubility and mobility of Se in soils depend on such processes as adsorption, precipitation, and transformation, which in turn depend on Se chemical species, pH, redox, etc. Several species of Se may exist in soils and sediments depending on the pH and redox potential, but SeO_3^{2-} and SeO_4^{2-} are the predominant ones in oxidized condition. However, SeO_3^{2-} is the predominant mobile inorganic form of Se in soils of humid regions and is most likely adsorbed onto hydrous sesquioxide (Cary et al., 1967). Selenite is specifically adsorbed on gibbsite and goethite (Hingston et al., 1971); both amorphous and crystalline forms of Fe_2O_3 influence Se adsorption by New Zealand soils (John et al., 1976).

Extended X-ray absorption fine structure (EXAFS) studies provide direct evidence for inner-sphere, bidentate adsorption of Se(IV) on goethite (Hayes et al., 1987). Similarly, evidence of inner-sphere complexes of Se(VI) on goethite and hydrous ferric oxide by EXAFS has been obtained (Manceau and Charlet, 1994). The role of ferric hydroxide in precipitating selenite in acidic to neutral soils has been demonstrated (Lakin, 1961). However, Geering et al. (1968) presented evidence that selenite may form adsorption complexes with ferric oxides in soils rather than as crystalline ferric selenite. In some instances, the coprecipitation of selenite with $Fe(OH)_3$ is via adsorption and occlusion (Plotnikov, 1960, 1964). Using a fractionation scheme of Se in soil, Cary et al. (1967) indicated possible occlusion of Se in sesquioxides or as solid phase of selenites on sesquioxides.

Adsorption of selenite or selenate Se by soil and geologic materials has been described by either the Langmuir or the Freundlich isotherm (Adriano, 1986). Several studies have demonstrated that selenite is sorbed onto soil to a much greater extent than selenate (Gissel-Nielsen et al., 1984). Therefore, selenite is not as bioavailable as selenate for uptake by organisms. The soil's surface area and organic C, sesquioxide, and allophane contents are closely related to Se adsorption (John et al., 1976).

The practical importance of adsorption–desorption relates to Se removal from soil solution and groundwater that is highly dependent on speciation. Selenate generally adsorbs weakly onto clays, iron oxides, and other soil phases at neutral and alkaline pH. Selenite generally adsorbs more strongly onto soil substrates than selenate and is essentially independent of SO_4^{2-} concentration (White and Dubrovsky, 1994). This difference in sorption affinities explains why the selenate/selenite ratios are often higher in groundwater than in surface water. For example, in comparing SeO_4^{2-}/ SeO_3^{2-} ratio between surface drainage water and drainage water in the shallow alluvial aquifer beneath the Kesterson Reservoir, the high ratios (i.e., low selenite concentrations) in the groundwater resulted from a strong sorption affinity of SeO_3^{2-} for the aquifer substrate.

The low solubility of Se in acid or neutral soils can be attributed to its presence as selenite in combination with ferric iron or as basic ferric selenite. Ferric selenite has a solubility product of 10^{-33}, whereas $Fe_2(OH)_4SeO_3$ has a solubility product of about 10^{-63} (Geering et al., 1968); the solubility of selenite in soils that contain reactive Fe oxide may be explained by the formation of ferric selenite–ferric hydroxide adsorption complexes. In some soils, Se reacts with active hydroxides in a manner similar to interactions with active ferric hydroxide. Oxidation of selenite to selenate in alkaline soils is favored by a decrease in the stability of ferric hydroxide–selenite complexes.

Sorption studies using clays simulating prevailing Se concentrations (0 to 12 μmol L^{-1}) and pH (4 to 8) in Kesterson confirm the importance of adsorption as the governing process controlling Se concentration in that system (Bar-Yosef and Meek, 1987). The SeO_3^{2-} sorption kinetics on kaolinite indicate a rapid initial reaction for about 35 hr, which is usually attributed to adsorption, followed by a much slower process, probably occlusion, which accounted for <5% of the total sorption in 720 hr. The adsorption of SeO_3^{2-} decreased with elevated pH values, becoming negligible above pH 8. At pH above 7, SeO_3^{2-} adsorption by montmorillonite exceeded that of kaolinite; the reverse occurred at pH 5.5.

In batch adsorption studies using Panoche soil from the Kesterson, sorption was greater for selenite (highest) and selenomethionine than for selenate (lowest) (Alemi et al., 1991). Indeed, selenate was rapidly leached from the soil columns compared with other Se species, confirming its low affinity for soil constituents.

4.6 Transformation of Selenium in the Environment

A severe ecotoxicological event in the 1980s at the Kesterson Reservoir and Kesterson National Wildlife Refuge, located in the western part of California's San Joaquin Valley, has spurred great interest in recent years in the biogeochemistry of Se. The fate and effects of Se have virtually become the central focus of research, primarily orchestrated by scientists from the University of California system, U.S. Geological Survey, U.S. Fish and Wildlife, National Labs, and other state and federal agencies (Frankenberger and Benson, 1994; Frankenberger and Engberg, 1998). Intensive efforts have focused on transformation mechanisms and pathways in both soil and aquatic systems.

There are three major transformation mechanisms for Se: oxidation–reduction, methylation–demethylation, and volatilization. All can be microbially mediated. The net effects of these mechanisms include immobilization of Se (e.g., by reduction), mobilization (e.g., by oxidation), and eventual loss (e.g., by demethylation and volatilization). In other words, these mechanisms determine the partitioning of Se between the various phases in the environment (i.e., solid, aqueous, and gaseous).

Selenium tends to be reduced rather than oxidized in living systems (Milne, 1998; Oremland, 1994; Thompson-Eagle and Frankenberger, 1992). Microbial extracts of *Micrococcus lactilyticus, Clostridium pasteurianum,* and *Desulfovibrio desulfuricans* have reduced SeO_3^{2-} but not SeO_4^{2-} to Se^{2-}. However, a strain of *D. desulfuricans* is capable of reducing SeO_4^{2-} to Se^{2-} (Zehr and Oremland, 1987). Selenite can be reduced to Se^0 by aerobically grown *Salmonella heidelberg* as well as by cell extracts of *Streptococcus faecalis* and *S. faecium.* It is also possible to convert seleno-oxyanions to Se^0 by anaerobic bacterial transformation (Oremland et al., 1989). Under reducing conditions, microbial reduction of selenates and selenites has led to accumulation of Se^0 in the Kesterson sediments (Tokunaga et al., 1996). The various reduction reactions of Se in reducing environments are exhibited in Figure 18.4.

Documented cases of Se oxidation indicate that Se^0, SeO_3^{2-}, and copper selenide can be oxidized by bacterial cultures. *Aspergillus niger* has also been reported to oxidize SeO_3^{2-} to SeO_4^{2-} (Thompson-Eagle and Frankenberger, 1992). In examining Se oxidation potential of San Joaquin Valley soils, Losi and Frankenberger (1998) found that the reaction appears to be largely biotic in nature, occurring at relatively slow rates, and yielding either SeO_3^{2-}, or a combination of SeO_3^{2-} and SeO_4^{2-}. The capacity of soils to oxidize Se^0 was enhanced by prior exposure to Se. A number of microorganisms,

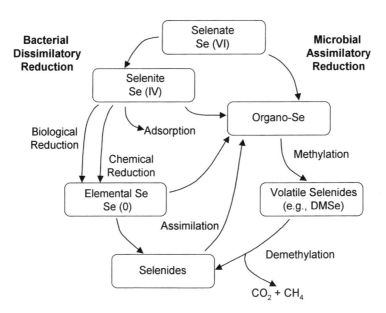

FIGURE 18.4. Proposed reductive pathways of selenium in the environment. (Reprinted from Oremland, 1994, p. 415, by courtesy of Marcel Dekker, Inc.)

including both fungi and bacteria, are capable of forming volatile, alkylated Se compounds from inorganic forms of Se.

Biomethylation of Se species including SeO_3^{2-}, SeO_4^{2-}, Se^0, and various organoselenium compounds into a less toxic, volatile form (e.g., DMSe) is believed to be a widespread conversion in seleniferous environments (Chau et al., 1976; Doran, 1982). Dimethylselenide (CH_3SeCH_3) is the major metabolite of Se volatilization although other products such as DMDSe ($CH_3Se_2CH_3$), methaneselenone [$(CH_3)_2SeO_2$], methaneselenol (CH_3SeH) and dimethylselenide sulfide (CH_3SeSCH_3), may also be produced (Doran, 1982; Oremland, 1994; Thompson-Eagle and Frankenberger, 1992). Microbially-mediated volatilization has been suggested as an effective means to dissipate Se from severely contaminated soils and sediments, i.e., bioremediation (Gao and Tanji, 1995; Karlson and Frankenberger, 1989; Thompson-Eagle and Frankenberger, 1992). Several strains of bacteria and fungi that are capable of volatilizing Se have already been isolated (Frankenberger and Karlson, 1994). Volatilization via methylation is thought to be a protective mechanism used by microorganisms to avoid Se toxicity in seleniferous environments.

Volatilization of Se from selenite by filamentous cyanophyte-dominated mat was via DMSe, DMDSe, and dimethylselenyl sulfide, with evidence that the precursors were methyl selenomethionine and methyl selenocysteine (Fan et al., 1998). Up to 75% of total Se in water (at the 20 to 10,000 µg L^{-1} Se level) was lost over a 50-day period.

The Kesterson Reservoir and Wildlife Refuge is an excellent setting to follow Se transformation due to depositional history of the element and the cyclic nature of environmental factors (e.g., ponding-drying, salinity, nutrients, etc.). In Kesterson, the Se in drain and pond waters is in the dissolved form primarily as SeO_4^{2-} (Tokunaga et al., 1996). Upon entering the anoxic pond bottoms, microbially mediated reduction of Se(VI) occurs, yielding selenite (as $HSeO_3^-$ and SeO_3^{2-}), elemental Se, and various forms of organic Se (Oremland et al., 1989; Tokunaga et al., 1996). Selenite becomes strongly adsorbed on soils, while elemental Se is essentially insoluble. Organic Se generally comprises <1% of the dissolved Se inventory.

Manipulation of redox potential and pH under controlled environmental conditions can shed more light onto the transformation sequence of Se. To this end, transformation of Se in Kesterson Reservoir sediments was evaluated according to changes in redox potential and pH (Masscheleyn et al., 1990). The sediments used were considerably reducing and characterized by a high pH (8.1), an OM content of 5.2 %, and a dark grayish color due the presence of iron monosulfides. Other minerals [e.g., thenardite (Na_2SO_4), calcite, and pyrite] were also present. The sediment had a total Se content of 9.06 ± 2.40 ppm DW ($n = 8$). Upon a 3-wk incubation at -200 mV, the solubility of Se was very low and Se($-II$, 0) were the only detectable species. At this redox condition, N, Mn, and Fe were present in a reduced

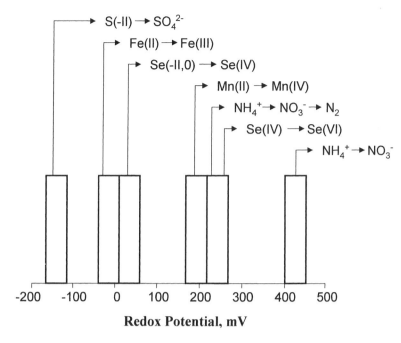

FIGURE 18.5. Sequential oxidation of several redox systems in the Kesterson Reservoir, California sediments. (Modified from Masscheleyn et al., 1990.)

form. Sulfide concentrations up to 74 ppm in sediment were measured. Oxidation of $Se(-II, 0)$ to $Se(IV)$ occurred at an Eh of about 0 mV (Fig. 18.5), where levels of dissolved Fe and sulfides decreased, and sulfate concentrations increased sharply, indicating oxidation of the iron sulfides.

The transformation of selenite to selenate began at an Eh of about 200 mV (see Fig. 18.5) and occurred at values corresponding with nitrification and denitrification. Concomitant decreases in concentrations of dissolved Mn, Fe, and $S(-II)$ also occurred. Over time, almost all selenite was oxidized, resulting in selenate as the dominant species in solution.

Subjecting the same Kesterson sediments to changing pH (6.5, natural, 8.5, and 9) and redox (-200, 0, 200, and 450 mV) conditions, Se solubility increased with increasing Eh at all pH levels (see Fig. 18.5). The pH had a major influence on the level and speciation of dissolved Se. In general, Se solubility was lowest at natural (unadjusted) pH. A decrease or increase in natural pH led to higher amounts of total dissolved Se. The highest total soluble Se levels occurred at pH 6.5; up to 67% of the total Se in the sediments was soluble at high Eh value (450 mV). This is caused by the higher solubility of iron sulfides, releasing Se. The oxidation of iron sulfides resulted in the formation of hydroxylated ferric oxides that can adsorb selenite and selenate to a much lesser extent.

It has been shown that oxyanions such as NO_3^- can facilitate the oxidation of Se (Wright, 1999) or inhibit the reduction of SeO_3^{2-} (Dungan and Frankenberger, 1998). Thus, the presence of high levels of NO_3^- in soils and groundwaters may contribute to the persistence of selenates in these media. This might well be a contributing factor in the high levels of SeO_4^{2-} in drainage effluents from agricultural drains at the San Joaquin Valley in California.

5 Selenium in Plants

5.1 Essentiality of Selenium in Plants

Selenium first came to attention in the 1930s when studies in the western and Great Plain regions of the United States showed many areas where levels of Se in plants were sufficiently high to be toxic to livestock. Selenium was the first element found to occur in native vegetation at levels toxic to animals (Allaway, 1968). Livestock consuming plants containing excessive amounts of Se can be afflicted with two maladies, commonly known as *alkali* disease and *blind staggers.* On the other hand, Se-deficient feeds has caused a disorder known as *white muscle* disease. The safe range between normal and toxic levels in feedstuff is very narrow, one reason why Se levels in plants have received critical attention from soil, plant, and animal scientists. The minimal dietary level required to prevent *white muscle* disease is from 0.03 to 0.10 ppm Se, depending upon the level of vitamin E and possibly other substances in the diet (Allaway et al., 1967). However, the generally accepted critical Se level for preventing Se disorder in livestock is 0.10 ppm DW. Concentrations above 3 to 4 ppm are considered toxic (Coles, 1974; Underwood, 1977).

Worldwide surveys on Se concentrations in crops and feedstuff indicate that areas producing crops with Se contents too low (<0.10 ppm) to meet animal requirements are more common than areas producing toxic levels (>2 ppm) of Se in crops (Gissel-Nielsen et al., 1984).

While Se is an important element in human and animal nutrition, it is known to be nonessential for plant growth. However, there are plant genera called indicator or accumulator plants where Se may be required for normal growth (Shrift, 1969). But since no beneficial effects were found in agronomic species when Se was added to purified medium in which plants were grown (Broyer et al., 1966), Se does not meet the established criteria for elemental essentiality for plants. In general, evidence that Se is essential to plants has not been conclusive.

5.2 Uptake and Bioaccumulation of Selenium in Plants

Research on plant uptake of Se has been reinvigorated due to Se toxicosis in organisms in Kesterson, California in the 1980s as well as a potential cleanup technique (i.e., phytoremediation) of highly Se contaminated environments. The Kesterson incident triggered extensive Se analysis of food crops raised in the San Joaquin Valley. Among the different crop species examined, broccoli, cotton, and cauliflower had the highest tissue Se concentrations (Burau et al., 1988). Broccoli and cauliflower are in the genus *Brassica* and are high S uptake species. Plants that contain high concentrations of S accumulate higher Se than those that do not. The lowest Se concentrations were found in corn grain, lettuce, sugar beet, and tomato. Within the valley, large variations in tissue Se concentration occurred. For example, Se level in broccoli ranged from 60 to 1530 ppb DW and in alfalfa from trace to 1460 ppb. The wide variations were caused by the sample locations having variable Se contents.

Selenium is absorbed by plants as selenate, selenite, or organic Se. Probably the volatile forms from soil are also absorbed by the foliage of plants. Uptake of added Se by plants is in the neighborhood of 1% of the total amount added (Gissel-Nielsen and Gissel-Nielsen, 1973; Levesque, 1974b). Selenate is taken up metabolically by plants, whereas selenite is largely taken up by passive processes and at lower levels (Peterson et al., 1981). Selenate is taken up through the same binding sites in the plant roots as sulfate (Leggett and Epstein, 1956). The two ions, analogues in biogeochemical reactions, are taken up by the same active uptake processes in competition with each other, while selenite is taken up through other sites.

Once selenate has entered the plants, it is metabolized by the same enzymes as the S assimilatory pathway (Terry and Zayed, 1994). This is due to the fact that Se has the ability to mimic S, thereby forming Se analogues of S compounds that are substrates for the S assimilation enzymes. Thus, the toxicity of Se to plants is essentially related to the competitive interactions between S compounds and their Se analogues. This view is supported by well-documented inhibition of plant growth caused by the addition of selenate that can be overcome by supplying sulfate. The primary cause of Se toxicity is likely to be the replacement of S by Se in the amino acids of proteins, thereby disrupting catalytic reactions.

Tall fescue grown in nutrient culture spiked with either selenate or selenite had shoot tissue Se concentrations much higher from the selenate than the selenite (83 vs. 142 mg kg^{-1}, respectively) (Wu, 1998). However, the roots from the selenite treated plants had slightly higher Se (433 vs. 384 mg kg^{-1}) than the selenate source. Apparently the greater growth inhibition by selenate was due to its greater translocation than the selenite to the shoot tissue. Overall, the data indicate that soil selenite was not significantly correlated with plant Se level. In contrast, soil selenate was highly correlated with tissue Se content.

Food crop plants apparently incorporate Se principally in their protein in a bound form when grown in soils high in Se. Those that normally contain high levels of S, such as cabbage tend to accumulate more Se than those that are normally low in S. Marked substitution of Se for S in normal cell constituents of nonspecialized plants does occur, which may account for their sensitivity to large amounts of Se in soil (Stadtman, 1974). The most likely biogeochemical changes in Se involved in its movement from soils through vegetation and eventually to domestic animals are depicted diagramatically in Figure 18.6.

Selenium content of crops receives widespread attention because of its importance in the food chain. Selenium-responsive diseases, such as *white muscle* disease in livestock, are most likely to occur in areas producing *feedstuff* low in Se. Accordingly, attempts were made to delineate soils in regions in many parts of the world as to deficient, sufficient, and excessive Se levels. For example, the United States was delineated into regions based on Se concentrations in forages, and the regions were further divided into smaller areas. The Pacific Northwest, northeastern, and southeastern regions are classified as low-Se areas (Kubota et al., 1975). The western half of the states of Washington and Oregon and part of northern California were identified as extremely low Se areas. In most of the west-central region, the Se contents of crops are mainly in the protective, but nontoxic range of Se concentrations. Other parts of the country are characterized by variable Se contents in plants.

Similarly, soils in the Atlantic coast region of Canada have been found to be very low in Se, producing forage crops generally Se-deficient ($\ll 0.10$ ppm DW) for poultry and livestock (Gupta and Winter, 1975; Winter and Gupta, 1979). Other countries that have reported low Se contents in plants include Finland, New Zealand, and Scandinavia. Serious Se deficiency in livestock in Australia and New Zealand was recognized in the early 1960s (Gissel-Nielsen et al., 1984). In Australia, Se deficiency is most pronounced in the western and southern parts of the country, while all of New Zealand is considered as a low Se area. Selenium deficiency in livestock in England and Scotland has also been reported.

Data on Se in crops from Europe, especially the northern part, are available. In general, Se levels in plants in northern Europe are considered inadequate. For example, cereal crops in Sweden, Norway, Finland, and Denmark contain low Se levels, usually <40 ppb (Gissen-Nielsen et al., 1984). However, pasture crops in general contain more Se, about double that for cereal crops. The low levels of Se in plants in this region can be attributed to strongly leached soils that were developed on glacier-deposited material.

The most severely Se deficient area of the world is the Keshan region in southeastern China, where a great number of children have died from a Se deficiency disease known as the *Keshan* disease (Tan et al., 1994).

Toxic levels of Se (up to 500 ppm Se in crops) in Europe are limited to a few spots in Wales and some areas in Ireland. Selenium in these soils, ranging from 30 to >300 ppm, originates from Se leached from Avonian shales and deposited in low-lying valleys. Thus, the Se level in soils and

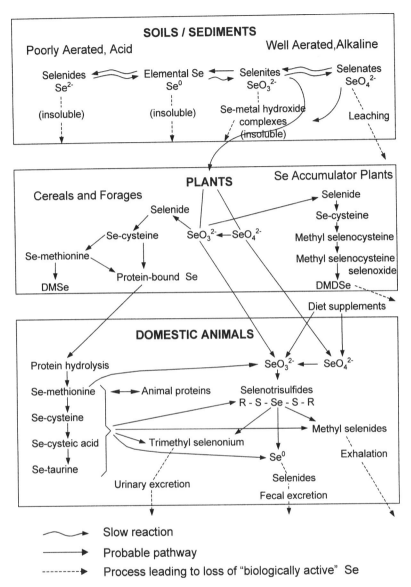

FIGURE 18.6. Proposed biogeochemical reactions and pathways of selenium in the soil–plant–animal system. (Modified and integrated from Allaway, 1973; Terry and Zayed, 1994. With input from Prof. Norman Terry, Univ of California, Berkeley, CA.)

plants is greatly influenced by the soil parent material and climatic conditions.

Certain plant species absorb soil Se very efficiently. These indicator or accumulator plants take up Se from geological formation, shale, or soil, and play a very important role in the cycling of Se. The accumulator plants, which are always associated with semiarid, seleniferous soils, biosynthesize organic Se compounds such as Se-methylselenocysteine and release these compounds to the soil upon decay. Accumulator plants have been defined as vegetation that can accumulate Se to levels 10^2 to 10^4 times greater than levels in normal vegetation (Lewis, 1976). These plants occur in at least 15 countries including Canada, Mexico, the United Kingdom, Ireland, Israel, Venezuela, Australia, and the United States.

A range of seleno-amino acids has been isolated from various plants (Peterson et al., 1981). Plant accumulators synthesize selenocystathionine and methyl selenocysteine, whereas nonaccumulators synthesize selenomethionine and methyl selenomethionine (see Fig. 18.6). It has been hypothesized that accumulator species may have evolved mechanisms to exclude, for example selenocysteine from their enzymes and other proteins as a detoxifying mechanism, whereas nonaccumulator species incorporate this amino acid into their proteins (Peterson and Butler, 1967).

Because of the extreme variations in ability to accumulate Se, plants have been divided into three groups:

1. Primary indicators—plants that contain large amount of Se (10^3 to 10^4 ppm DW) and seem to require Se for their growth. These include several species of *Astragalus* (Leguminosae), *Stanleya* (Cruciferae), *Haplopappus* (Compositae) and *Machaeranthera* (Compositae). Selenium is present in tissues primarily in inorganic form. Plants consumed by livestock can cause blind staggers, an acute form of Se poisoning.
2. Secondary indicators—plants that accumulate Se from about 100 to 1000 ppm when grown on soils of high bioavailable Se contents and do not appear to require Se for their growth. Included in this group are species of *Aster, Atriplex, Castilleja, Comandra, Machaeranthera, Mentzelia, Grayia, Grondelia,* and *Gutierrezia.* Most of the Se in these plants occurs as selenate, with lesser amounts in the organic form. These plants can cause acute or chronic *selenocosis* in livestock.
3. Nonaccumulators—most cultivated crop plants, cereals, and grasses generally accumulate low levels of Se (at <30 ppm under field conditions).

Selenium is associated primarily with plant proteins. Alkali disease, a milder form of Se poisoning in livestock, may result from consumption of these plants.

Indicator plants are also known as converter plants because of their unique ability to absorb Se from the soil in a form that is not available to other plants and recycle it in a form that is available. Since most of these plants are deep rooted, they can absorb bioavailable Se from the deep strata of soil profile and deposit it on the surface where it is more readily accessible and likely more available to other plants.

5.3 Phytotoxicity of Selenium

Selenium is toxic to plants at low concentrations in nutrient culture, and crops—including the forage—are sensitive to the addition of a small amount of Se in soil. As low as 0.025 ppm Se in nutrient solutions has decreased the yield of alfalfa (Broyer et al., 1966). Mild toxicity was displayed in bush beans at 8 ppm Se (as Na_2SeO_3) in solution culture (Wallace et al., 1980); severe toxicity was obtained at 80 ppm Se concentration, with plants characterized by stunting and brown spots. Leaves, stems, and roots of these plants had 901, 1058, and 1842 ppm of Se, respectively. In sand culture, the critical level of Se in solution was 5 ppm Se (as SeO_3^{2-}); the critical level in barley tissue was about 30 ppm (Davis et al., 1978). In culture solutions void of sulfate, concentrations of selenate-Se as low as 0.10 ppm produced minor injury to wheat plants (Hurd-Karrer, 1935), but this was probably due to S deficiency since injury symptoms did not occur in the presence of S. In soils, selenite-Se added at 20 ppm did not affect wheat germination but retarded plant growth (Beath et al., 1937). At the rate of 40 ppm, it reduced germination substantially and produced stunted and chlorotic plants. The characteristic toxic symptoms in wheat plants are chlorosis, stunting, and yellowing of the leaves.

For most cereal crops, snow-white chlorosis is a typical symptom of Se toxicity. Selenate is a more toxic source of Se than selenite, but this depends on S concentrations in the medium; at S concentrations above 30 ppm, selenite-Se could become more toxic (Hurd-Karrer, 1937). In sand culture, cotton growth was slightly affected at 20 ppm selenate-Se; at 50 ppm in the medium, growth was markedly depressed (Mason and Phillis, 1937). The Se contents of plants grown on soils containing 40 to 88 ppm of Se and showing toxicity symptoms were 117 ppm in meadow fescue, 30 ppm in meadow sweet, and 7.2 ppm in barley grain (Walsh and Fleming, 1952). Potted soils treated with Na_2SeO_3 and yielding about 2 ppm of soil Se extracted with NH_4HCO_3–DTPA solution, were highly toxic to alfalfa plants and resulted in plant concentrations of over 1000 ppm of Se (Soltanpour and Workman, 1980).

Visible symptoms of Se toxicity in cultivated crops under field conditions have never been observed (NAS, 1976). It appears then that under field conditions, bioavailable Se is not provided to plants at a level high enough to cause injury. Crops may not show any injury until they contain at least 300 ppm of the element, which is much higher than what plants normally contain even in seleniferous areas (Rosenfeld and Beath, 1964).

5.4 Phytovolatilization of Selenium

Nonaccumulator plants, such as cabbage and alfalfa, give off DMSe and other volatile Se compounds, and accumulator species, such as *Astragalus racemosus,* produce DMDSe when exposed to selenite (Lewis, 1976; Lewis et al., 1966, 1974; Evans et al., 1968).

Because of the potential of phytovolatilization in removing Se in contaminated environments, much research in this area has focused on the kinetics of volatilization and identifying plant species, both aquatic and terrestrial, that can maximize Se loss via this pathway. In evaluating the volatilization potential of various agronomic and horticultural species, Terry et al. (1992) classified the plants as: high—rice, broccoli, and cabbage; intermediate—carrot, barley, alfalfa, tomato, cucumber, cotton, eggplant, and maize; and low—sugar beet, beans, lettuce, and onion. The volatilization rates ranged from <250 µg Se kg^{-1} DW per day for low volatilizers to as high as 1500 to 2500 µg Se kg^{-1} per day for the high volatilizers. Plants that had high Se in their tissues tended to also have high volatilization rates. Similar intraspecies variation in Se volatilization, which accounted for 0.5% (for alfalfa) to 6.1% (for *Astragalus bisulcatus*) of the Se lost from the soil, has also been observed (Duckart et al., 1992).

Mechanisms of phytovolatilization in plants have been proposed (Terry and Zayed, 1994; 1998). Briefly, once SeO_4^{2-} has entered the plant it is activated by the enzyme ATP sulfurylase (for which sulfate is the normal substrate) to form adenosine 5'-phosphoselenate, which apparently undergoes reduction to selenite, which is then reduced to selenide, followed by the incorporation of selenide into Se-cysteine (see Fig. 18.6); the latter is converted to Se-methionine via Se-cystathionine and Se-homocysteine. Se-methionine is the most likely precursor, eventually leading to the formation of volatile Se, DMSe, which is the species of volatile Se produced by the majority of plants and microorganisms.

Similar attempts to identify aquatic microphytes that can dissipate water-borne Se have been conducted (Fan and Higashi, 1998). The role of microorganisms within the plant and those inhabiting the rhizosphere in phytovolatilization is also under scrutiny. It is now known that most of the Se produced by plants is volatilized from their roots (Terry and Zayed, 1994). Indeed, certain plants can volatilize Se in substantial amounts, up to 35.5 g Se ha^{-1} per day.

6 Ecological and Health Effects of Selenium

Although Se is not a confirmed essential element in plant nutrition (but it is a beneficial element for certain species), it is essential in animal and human nutrition. Therefore, Se levels in plants are still important to animal and human nutritionists because of the dietary role of plant-based Se. On a global scale, the concerns about low Se concentrations in soils and plants and Se deficiency in animals and humans far outweigh toxicity from the element. These extreme cases are highlighted in this section.

6.1 Essentiality of Selenium in Animal and Human Nutrition

The nutritional role of Se was initially recognized in 1957 when it was found to have a complementary role to vitamin E in preventing dietary hepatic necrosis and exudative diathesis in chicks and rats (Mayland, 1994). By the late 1960s it was firmly established as a specific nutritional requirement for chicks. Eventually Se was shown to be an essential constituent of biologically important glutathione peroxidase, a major cellular antioxidant enzyme, which can convert free radicals to peroxides, then to water and oxygen, whereas vitamin E scavenges the free radicals and neutralizes their potential damaging effects. Thus, low Se intake with vitamin E deficiency increases oxidative stress, which may enhance oxidation damages.

Clinical signs of Se deficiencies in fish, small animals, livestock, and humans include reduced growth and appetite, anemia, reduced reproductive fertility, and muscular weakness (Combs and Combs, 1984). Specific disorders include exudative diathesis and increased embryonic mortality in birds. Muscular dystrophy affects fish, birds, and other animals. Retained placenta can occur in Se-deficient cows, while mulberry heart disease is noted in pigs. Although Se toxicosis in ruminant and monogastric animals (i.e., alkali disease)—characterized by hair loss, deformation and sloughing of the hooves, erosion of the joints of the bones, anemia, and effects on the heart, kidney, and liver (ATSDR, 1989; U.S. EPA, 1986)—roaming seleniferous areas still occurs as well as in wildlife inhabiting areas where sediments and aquatic vegetation contain excess Se, *selenocosis* is generally not a major concern.

Severe nutritional deficiency is associated with endemic cardiomyopathy (i.e., *Keshan* disease)—characterized by heart failure, cardiac enlargement, electrocardiogram abnormalities, and cardiac shock—in youngsters from a large discrete region in China, as well as with endemic osteoarthrosis (i.e., *Kaschin-Beck* disease)—characterized by atrophy, degeneration, and necrosis of cartilage tissue—in juvenile (ages 5 to 13) Chinese.

6.2 Ecological Effects of Selenium

Selenium has served as a nutrient and poison in animals. As a nutrient, it is required in the diet of fish at concentrations of about 0.10 to 0.50 $\mu g\ g^{-1}$ DW (Lemly, 1998). At a dietary concentration of 7 to 30 times that required (i.e., >3 $\mu g\ g^{-1}$) Se may become toxic. Several cases of chronic and acute Se poisoning in fish, waterfowl, and other animals in habitats where Se has accumulated in sediments, vegetation, and fish-food organisms have been reported during the last 15 yr (Hamilton, 1998; Skorupa, 1998; Stephens and Waddell, 1998; Zhang and Moore, 1998). The best known example is the Kesterson Reservoir and Kesterson National Wildlife Refuge in the western San Joaquin Valley, California where massive poisoning of fish and wildlife has occurred (Frankenberger and Engberg, 1998; Lemly, 1998; Presser and Piper, 1998).

High levels of Se in the blood of waterfowl have been detected in other habitats. For example, up to 10 ppm WW of Se in blood samples from emperor geese (*Chen canagica*) at their breeding grounds in the Yukon-Kuskokwim Delta in western Alaska were found (Franson et al., 1999). There is also evidence that emperor geese are exposed to higher Se in the marine environment of their wintering and staging areas in the Alaskan Peninsula than at the breeding grounds. Similarly, adult male greater scaup (*Aythya marila*) and surf scoters (*Melannita perspicillata*) from the San Fransisco Bay area were found to have elevated hepatic Se levels (greater scaup = 67 ppm DW; surf scoters = 119 ppm DW) (Hoffman et al., 1998). Elevated enzymes related to glutathione metabolism and antioxidant activity have been associated with elevated Se concentrations that can adversely affect reproduction and neurological functions in wildlife (Hoffman and Heinz, 1998).

Teratogenic deformities in fish (i.e., congenital malformations that are a permanent pathological marker of Se poisoning) can occur rapidly when dietary Se exceeds 3 μg g^{-1} and increase rapidly when Se concentration in the fish eggs exceeds 10 μg g^{-1} (Lemly, 1998). Teratogenic thresholds in the diet range from 5 to 20 μg g^{-1}. Teratogenesis is induced when larval fish are relying on their attached yolk sac for nourishment and development. Once external feeding has begun, the potential for teratogenic effects declines and soon disappears. Thus, the teratogenic process is strictly an egg–larval phenomenon.

The severe Se-poisoning in organisms at the Kesterson prompted policymakers and environmental regulators in the state of California to establish guidelines on Se concentrations in key environmental media and biota as a tool in the environmental protection and management of similar Se-affected areas (Table 18.5). Measurement of Se has expanded from the traditional

TABLE 18.5. Selenium levels of concern for selected environmental indicators.[a]

Indicator	Normal background	Threshold (range)	Level of concern (range)	Toxicological and reproductive effects are certain
Water	<0.5–1.5	Avian protection, 2–3; Fish protection, 2–5	2–5	>5
Sediment	<2	2–4	2–4	>4
Food chain	Usually <2, rarely >5	2–4 in the diet	3–7	>7
Fish	Usually <2, rarely >5		4–12	>12
Avian eggs	Usually <3	4.2–9.7	3–8 (pop. hatchability)	>8

[a]Units for water μg L^{-1}; all other measurements are μg g^{-1} DW.
Source: Based on Engberg et al. (1998).

way of just monitoring the water column and now includes sediments, food-organisms (diet), fish, and avian eggs (Engberg et al., 1998).

Most documented areas of Se contamination have been due to leaching from Se-rich soils and drainage from coal fly ash ponds (Frankenberger and Engberg, 1998). Selenium exists in surface waters in inorganic (selenate and selenite) and organic forms. Organic forms (e.g., Se-methionine) are generally more bioaccumulative and toxic to aquatic organisms than either selenite or selenate (Besserb et al., 1993). The two principal exposure routes to Se for aquatic organisms are from the water and food. Dietary concentrations of <0.10 µg Se g^{-1} DW for rainbow trout can lead to Se deficiency symptoms; whereas at levels above 10 µg g^{-1}, toxicity may begin to manifest (Hodson and Hilton, 1983). Most bodies of water are not Se deficient, with food organisms generally containing between 0.10 and 4.7 µg Se g^{-1} DW in noncontaminated waters. The lowest adverse effects have been observed mainly with species at the top of the food chain (i.e., fish and aquatic birds) under long-term chronic exposure from food and water (Lemly, 1998). The current recommended chronic water quality criterion for Se is 5 µg L^{-1}, based primarily on results from contaminated sites such as Belews Lake, North Carolina (see Section 8.3).

6.2.1 The Kesterson Case Study

The public was alarmed when aquatic birds nesting at the Kesterson Reservoir in 1983 were found to have high rates of embryo deformities and mortalities (Ohlendorf and Santolo, 1994). The following year, adult birds were also found dead in unusually high numbers. Postmortem assays revealed that these effects were due to exceptionally high concentrations of Se in the birds' diet and their tissues. This bioaccumulation of Se occurred because of the high levels of Se in the subsurface agricultural drainage carried to Kesterson by the San Luis Drain. In 1986, reclamation measures stopped discharge of drainage effluent to Kesterson and in 1988 dewatered the reservoir, and filled all areas to at least 15 cm above the expected average seasonal rise of groundwater. Although not considered a Superfund site, ecological risk assessment was performed (Ohlendorf and Santolo, 1994).

Kesterson Reservoir is a 500-ha shallow impoundment (1 to 1.5 m deep) subdivided into 12 interconnected cells. Located at the terminus of the San Luis Drain, the reservoir served dual roles as an evaporation basin for agricultural drainage effluents and as a managed wetland intended to benefit fish and wildlife. Initially (1972 to 1978) the basin was receiving high-quality agricultural spill water, but by 1981 the basin started receiving highly saline subsurface drainage water (300 µg Se L^{-1}) (Skorupa, 1998). By 1982, federal biologists and resource managers noticed deterioration of the aquatic ecosystem. In 1983 to1985, detailed ecotoxicological research was conducted. Birds are the most sensitive species to chronic *selenocosis* because of reproductive effects and are the main focus of the risk assessment, along with the mammals. Waterborne Se concentrations of 10 μg L^{-1} have been associated with impaired hatchability of shorebirds and concentrations of 1 to 20 μg L^{-1} have been associated with teratogenic effects. There are reasonably good relationships between Se levels in water, invertebrates, and waterbird eggs, and also between these Se concentrations and reproductive effects in exposed bird populations. Se levels in eggs of birds inhabiting nonaffected reference sites in the United States average <3 μg g^{-1} DW (i.e., associated with normal rates of egg viability). The threshold for mean egg Se associated with impaired hatchability is 8 μg g^{-1}. Small mammals (e.g., rodents and shrews) have been monitored most consistently because they are less mobile than birds, nonmigratory, potentially affected by dietary Se, and widespread throughout the reservoir. These mammals are good bioindicators of local conditions because of their small home ranges.

Early studies conducted in 1983 to1985 showed high bioaccumulation in aquatic plants, invertebrates, fish, and birds. The most pronounced effects on wildlife were found in birds feeding regularly in the ponds (Fig. 18.7).

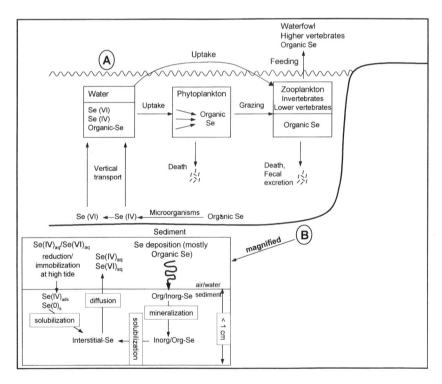

FIGURE 18.7. Proposed biogeochemical cycle of selenium in (A) water column and (B) sediment in freshwater systems. (Extracted and integrated from several sources.)

The abnormalities in aquatic birds were similar to those in chickens and mallards fed diets containing 7 to 10 µg Se g^{-1}. The greatest risk of Se exposure to terrestrial wildlife appears to be from the consumption of mushrooms, which contain the highest levels of Se currently found at the site.

Selenium concentrations in several key environmental media in the San Joaquin Valley as well as Se levels of concern are indicated in Table 18.6. These values are compared with values from noncontaminated areas.

6.3 Effects of Selenium on Human Health

The concern for Se in the general population is its deficiency in human nutrition. There are large areas in North America, China, and Europe, especially Scandinavia, where soils are too deficient in Se to produce Se-nutritious food crops. The cases in Finland and China will be the focus here.

In Finland, the Se content of arable soils is relatively low (0.20 to 0.30 mg kg^{-1}) (Aro et al., 1998). The climatic conditions, low pH, and high Fe content of the soil favor the immobilization of Se in soils, and hence low Se uptake by crops. A large study of the mineral composition of Finnish foods conducted in the late 1970s indicated that all agricultural products grown and produced in Finland contained very low amounts of Se.

Since 1969, commercial animal feeds in Finland have been supplemented with 0.10 mg of sodium selenite per kg DW of product. This reduced the incidence of Se deficiency problems in animals but did not entirely eliminate the need for Se medication of domestic animals and did not appreciably affect human nutrition. During the 1970s, the dietary Se intake was relatively stable, estimated at as low as 25 µg Se/2400 Kcal. During the 1980s when supplementary high-Se wheat was imported from North America, the Se content of cereal products increased. Subsequently, dietary Se intakes and serum Se levels increased in the early 1980s.

Although no apparent disorders in the Finnish population caused by Se deficiency have been observed, low Se intakes or low Se levels in the serum are believed a causal factor in the increased risk of cardiovascular death and certain types of cancer (Salonen et al., 1982, 1984). To abate the concern on insufficient supply of Se via the diets of domestic animals and humans, the Finnish government in 1984 started supplementing mineral fertilizers with Se as sodium selenate. Direct enrichment of foods was considered hazardous because the window of safe intake for Se is very narrow. This monumental decision was based on research conducted in Denmark and Finland. The fertilizers for grain production were supplemented with 16 mg Se kg^{-1} of fertilizer (Se as selenate) and those for the production of hay and fodder with 6 mg kg^{-1}. The goal was to elevate the Se level of cereal grains tenfold, and yield food intake in the sufficient range of 0.05 to 0.20 mg day^{-1}. This supplementation practice still continues today, except that all fertilizers irrespective of land use receive 6 mg Se kg^{-1} of fertilizer (Varo et al., 1994).

TABLE 18.6. Selenium concentrations, levels of concern, and loads for environmental components (source rock, soil sediment, water, and biota) in San Joaquin Valley, California, and selected areas.[a]

Environmental components		Se content (mean/range)
SJV source rock ($\mu g\ g^{-1}$)	Seleniferous rocks (e.g., Kreyenhagen, Monterey, and Moreno formations)	8.9
	Nonseleniferous rocks (e.g., Panoche and Lodo formations)	1.1
Soil ($\mu g\ g^{-1}$)	San Joaquin Valley (western slope)	0.14
	San Joaquin Valley (Panoche fan)	0.68
	Western United States	0.34
	Conterminous United States	0.26
Sediment ($\mu g\ g^{-1}$)	Kesterson Reservoir	
	Top 15 cm	5.0
	Top 5 cm "muck"	55
	Organic detritus	165
	Cleanup criterion	4.0
	San Luis drain	84
	Tulare evaporation ponds	0.05–15
	Grassland wetlands	<2.0
	San Francisco Bay	0.1–0.6
	Deep sea	0.17
Water ($\mu g\ L^{-1}$)	Westlands drainage sumps	140–1400
	Grassland drainage sumps	8–4200
	San Luis drain (inflow to pond 2)	330
	Tulare evaporation ponds	<1–6300
	Mud and salt sloughs	16/2–100
	Panoche Creek	2–60
	Coast ranges (seeps/ephemeral streams)	<2–3500
	U.S. rivers	0.2
	San Jaoquin River water quality objective	5.0
	San Francisco Bay	0.07–0.4
	Pacific Ocean	0.13
	Aquatic maximum (drinking water standard)	50
Concern level–solids ($\mu g\ g^{-1}$)[b]	Sediment	2.0–4.0
	Aquatic vegetation and invertebrates	3.0–7.0
	Fish (whole body)	4.0–12.0
	Avian eggs	3.0–8.0
Concern level–liquids ($\mu g\ L^{-1}$)[b]	Aquatic level (for protection of wildlife)	<2.3–5.0
Hazardous waste	Solid ($\mu g\ g^{-1}$)	100
	Liquid ($\mu g\ L^{-1}$)	1000
Load level (kst)[a]	San Luis drain (bed sediment)	0.26–0.85
	San Luis drain (reuse annual target)	0.38[c]
	San Joaquin River (total, 1986–1995)	5.85
	Kesterson Reservoir (total, 1981–1985)	1.0
	Grassland wetlands (cumulative loss, 1986–1995)	0.95
	Annual human dietary intake	10^{-8}–$10^{-8.6}$

[a]Concentrations for solids are reported on a dry weight basis. Load levels are in Kestersons (kst): 1 kst = 7900 kg Se.

[b]Toxicity levels, although not listed, are represented by the upper limit of the concern levels.

[c]This amount (i.e., 0.38 kst = 3000 kg) represents the agreed-on annual load limit at compliance point B (USBR, 1995).

Source: Modified from Presser and Piper (1998).

Before the Se supplementation of fertilizers in Finland in the mid-1970s, the daily intake was about 25 µg Se/2400 Kcal per person (corresponding serum Se concentration of 0.63 to 0.76 µmol L^{-1}, which is among the lowest in the world), well below the recommended minimum intake by the U.S. National Academy of Sciences (Fig. 18.8). Fertilizer supplementation with Se caused a plateau of 110 to 120 µg/2400 Kcal in 1987 (Fig. 18.8), which remained constant until 1990. In 1991, the mean Se intake was 95 µg/day^{-1} (120 µg/day^{-1} in males vs. 85 µg/day^{-1} in females). This decrease was concomitant with the reduction in Se supplementation of fertilizers for grain production by 60% in 1990. No further substantial decreases were noted in 1995 to 1996, indicating that the Se intake has reached a new plateau. This practice has proven to be an effective, safe, and precise way of delivering the Se intake to the general Finnish population at the recommended level.

In China, two endemic human diseases, the *Keshan* disease and *Kaschin-Beck* disease are distributed mainly in a distinct wide belt, running from the northeast to the southwest of the country. The disease belt is characterized by temperate forest and forest-steppe soils that belong to the brown-drab

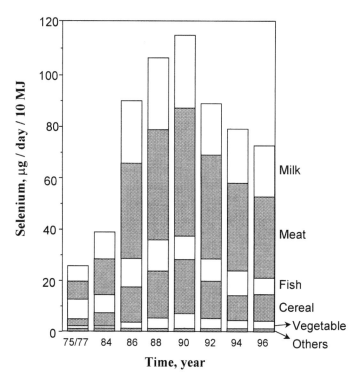

FIGURE 18.8. Estimated dietary selenium intakes (energy of 10 MJ or 2400 Kcal) in Finland before (1984) and during selenium fertilization. (Reprinted from Frankenberger and Engberg, 1998, p. 87, by courtesy of Marcel Dekker, Inc.)

soil series. Flanking it on both sides, to the southeast and the northwest, are two relatively high Se belts. From an extensive survey of the soils of China, Se categories to delineate the low from high Se levels in soils, grain, and human hair were established (Table 18.7). The threshold values are: 0.15 μg total Se g^{-1} soil, or 0.003 μg water-soluble Se g^{-1} soil, 0.025 μg Se g^{-1} for cereal grain, and 0.20 μg Se g^{-1} for human hair. In the disease belt area in China, Se intake in the order of 11 μg Se/person per day is inadequate; adequate Se intake is in the range of 110 μg/person per day; and excessive Se intake is up to 750 μg/person per day. Chronic Se poisoning may be associated with intake of 5 mg Se day^{-1}.

Food is the primary source of exposure to Se in humans. Because of the differences in dietary habits, human Se intake varies considerably. The greatest variation in Se intake occurs where the intake of plant food predominates. For example, the daily intake of Se in China ranges from <10 μg where soil Se is low to >500 μg where food plants are grown on seleniferous soils (Gissel-Nielsen et al., 1984). The estimated Se intake for the U.S. population ranges from 71 to 152 μg day^{-1} (ATSDR, 1989), much higher than the recommended dietary allowance for U.S. adults (Table 18.8). However, a much lower estimate of 62 μg day^{-1} had been reported for the northeastern United States (Schroeder et al., 1970). The reported values (μg Se day^{-1}) for other countries are: Canada, 113 to 220; the Netherlands, 110; Italy, 141; France, 166; New Zealand, 25; Japan, 208; the United Kingdom, 200; and Germany, 60. Values for the average British diet (∼60 μg day^{-1}) and for the Italian diet (14 μg day^{-1}) have also been reported. The U.S. Department of Agriculture has recommended that the daily intake of Se should range from 100 to 200 μg day^{-1} (Shamberger, 1981).

In U.S. adults, approximately 99% of ingested Se comes from cereals, grains, fish, meat, and poultry. In average British diets, about 50% of Se ingested comes from cereals and cereal products and another 40% from meat and fish (Thorn et al., 1978). Milk, table fats, fruits, and vegetables provided little or no Se. In Canadian diets, cereals provided the most Se (62 to 112 μg day^{-1}), followed by meat, poultry, and fish (25 to 90 μg day^{-1}), and dairy products (5 to 25 μg day^{-1}) (Thompson et al., 1975). In general,

TABLE 18.7. Selenium concentration categories in soils, cereal grains, and human hair in China.

Se category	Total in surface soil	Water-soluble in surface soil	Food grains	Hair	Effect
Deficient	<0.125	<0.003	<0.025	<0.200	Se–responsive diseases
Marginal	0.125–0.175	0.003–0.006	0.025–0.04	0.200–0.25	Potential Se deficiency
Moderate to high	0.175–0.40	0.006–0.008	0.040–0.070	0.25–0.500	Sufficiency range
Excessive	≥3.0	≥0.02	≥1.0	>3.0	Se poisoning

Source: Modified from Tan et al. (1994).

TABLE 18.8. Recommended dietary allowances for selenium.

Category	Age (yr)	Se (μg day^{-1})
Infants	0.0–0.5	10
	0.5–1.0	15
Children	1–6	20
	7–10	30
Males	11–14	40
	15–18	50
	19+	70
Females	11–14	45
	15–18	50
	19+	55
Pregnant		65
Lactating		75

Source: Food and Nutrition Board (1989).

seafoods, meat, high-S vegetables, and cereals contain relatively high Se; fruits tend to have the lowest. Drinking water usually contains Se at very low levels (usually <10 μg L^{-1}). However, occasionally, higher levels of Se may be found in drinking water, usually in areas where high levels of Se in soils contribute to the Se content of the water (ATSDR, 1989).

In epidemiological studies of populations exposed to high levels of Se in food and water, discoloration of the skin, pathological deformation and loss of nails, loss of hair, excessive tooth decay and discoloration, lack of mental alertness, and listlessness were reported (ATSDR, 1989; U.S. EPA, 1986). The U.S. Environmental Protection Agency (EPA) has not yet established a reference concentration (RfC) for Se, but it has established a reference dose (RfD) which is 5 μg/kg/day. The U.S. EPA estimates that consumption of this dose or less, over a lifetime, would not likely result in the occurrence of chronic noncancer effects.

7 Factors Affecting Mobility and Bioavailability of Selenium

The chemical form, solubility, and bioavailability of Se in soils and sediments are influenced by factors such as pH, redox potential, OM, oxide content, as well as plant-related factors.

7.1 Parent Material

Selenium content of plants varies with the geologic substrate on which they grow. Furthermore, close correlations have been found between the parent rocks from which the soils have developed and the amount of Se found in soils (Byers and Lakin, 1939). Igneous rocks, being inherently low in Se, give rise to a great many Se-deficient soils. In contrast, parent materials of sedimentary origin produce soils high in Se, with the shales of particular importance. Shales, particularly the Cretaceous type, generally have the highest concentrations of Se and are the major sources of high-Se soils in the Great Plains and Rocky Mountain regions of the United States and vast areas in Alberta and Saskatchewan (Canada), South Africa, Australia, Ireland, and the United Kingdom.

High-Se soils in Israel are derived from limestone alluvium, while those in South America are from black slate alluvium (Peterson et al., 1981). In the west-central area of Saskatchewan, wheat plants have the highest Se concentrations when grown over lacustrine clay and glacial till; the Se concentrations were intermediate in plants grown on lacustrine silt; and the Se concentrations were lowest in plants grown on aeolian sand.

The following summary describes Se interrelationships among parent rocks, soils, and plants:

1. Where parent material with a high Se content weathers to form well-drained soils in dry areas (<500 mm annual rainfall), selenides and other insoluble forms will convert to selenates and organic Se compounds. The vegetation will be potentially toxic to livestock.
2. Where parent material with a high Se content weathers to form soils in humid areas, relatively insoluble complexes of ferric oxide or hydroxide and selenite will be formed. These soils will be slightly to strongly acidic, and the vegetation will probably contain sufficient Se to protect livestock from Se deficiency.
3. Where parent material with a high Se content weathers to form poorly drained soils or where such soils are enriched by alluvial action and the conditions are alkaline, toxic vegetation will probably be produced.
4. Where parent material with a low Se content weathers to form soils under either humid or dry conditions, the vegetation is likely to contain insufficient Se to protect animals from Se deficiency. The more humid the area and the more acid the soil, the greater the likelihood of extremely low Se concentrations in the vegetation.

7.2 pH

It is known that Se is quite bioavailable to plants growing on well-aerated, alkaline soils where Se primarily occurs as selenates. In acid to neutral soils however, the ferric oxide–selenite complex is formed, which is only sparingly soluble and is unavailable to plants. The solubility of Se in soils is low when the soil is slightly acidic to neutral. In general, selenates predominate in high-Se alkaline soils as in North Dakota, while selenites predominate in high-Se acidic soils as in Ireland (Peterson et al., 1981). Thus, liming the soil could enhance Se uptake by plants. For example, liming of soil to raise the pH from 6.0 to 6.9 has been shown to increase the uptake of added Se by alfalfa (Cary et al., 1967). Cary and Allaway (1969) further observed increased uptake of added Se from several sources, including elemental Se by alfalfa, when grown on higher pH treatment. Likewise, Gissel-Nielsen (1971b) found higher uptake of added selenite by ryegrass in limed soils. Based on this trend, applying Se to alkaline soils is likely to be more hazardous in producing feedstuff for livestock than applying the element to acidic soils. The pH affects not only the solubility and bioavailability of Se but also the biomethylation processes. The optimal pH for Se biomethylation and subsequent volatilization from a seleniferous Kesterson sediment was 8.0 (Frankenberger and Karlson, 1994). Elevating the pH of a sandy soil from 6.0 to 7.0 increased Se volatilization by 20%.

7.3 Chemical Species of Selenium

In general, increasing the level of soil Se increases Se uptake by plants. Selenate is almost certainly absorbed most efficiently, selenite much less efficiently, and organic Se absorbed the least. Therefore, selenate and selenite are the primary Se forms that can be toxic to plants. Selenate added to soils is readily taken up by the plants and can remain bioavailable for some time in some soils. Uptake of Se by plants from Danish soils (pH 4.8 to 7.6) was on an average 8 times greater with selenate than with selenite (Bisbjerg and Gissel-Nielsen, 1969). When selenite is added to soils, it can be immobilized to insoluble forms causing limited uptake by plants. These two forms of Se are differentially translocated, with Se accumulating higher in the roots when supplied as selenite; the reverse trend happened with selenate as the source (Gile and Lakin, 1941).

Using limed and unlimed potted soils, Carlson et al. (1991) found that selenate had a much greater toxic effect on a forage sorgrass than selenite, reducing plant biomass by as much as 97% and increasing tissue Se concentrations over the 1000 ppm DW level (Fig. 18.9). In contrast, selenite additions (up to 4 mg kg^{-1} soil) generally did not affect plant biomass due to much lower Se accumulation. Between the two soil types used (Blanton sand vs. Orangeburg loamy sand), Se concentrations were higher in plants grown in the soil with lower clay and hydrous oxide (i.e., Blanton soil) contents. Liming resulted in lower tissue Se concentrations in plants grown on either soil treated with selenate. Liming also reduced Se concentrations in plants grown on the Blanton soil treated with selenite. In contrast, others have reported increased plant Se concentrations with liming due to increased

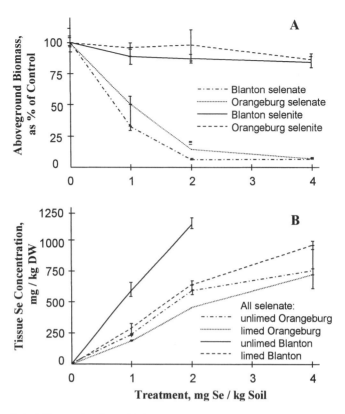

FIGURE 18.9. Effects of soil-applied selenium added as selenate or selenite on the (A) biomass and (B) tissue Se content of plants grown in limed sandy Blanton and loamy sand Orangeburg soils. The values given represent means ± 1 SE. (Modified from Carlson et al., 1991.)

availability of soil Se as soil pH increased (Gissel-Nielsen, 1971; Gupta et al., 1982; Mikkelsen et al., 1989). Only the results from the Orangeburg soil amended with selenite in this study were consistent with this earlier work. Careful examination of previous studies revealed that all of the reported increases in plant Se associated with liming involved the selenite form. In one instance where the effect of liming on the availability of the selenate form was reported, Ylaranta (1983) found that liming slightly increased or decreased the extractability of soil selenate, depending on the extract used and the soil type. It is apparent that literature reports of increased Se accumulation with liming have been inappropriately generalized from studies using only the selenite form. The effect of liming on the selenate form has not received much study.

The effect of liming on tissue Se concentrations in the study by Carlson et al. (1991) may be partially explained by differences in extractable SO_4^{2-}

between the limed and unlimed soils. Extraction of SO_4^{2-} from the Blanton and Orangeburg soils gave the following results: Blanton unlimed, 12.7 mg SO_4^{2-} kg^{-1}; Blanton limed, 22.6 mg SO_4^{2-} kg^{-1}; Orangeburg unlimed, 60.7 mg SO_4^{2-} kg^{-1}. In the Blanton soil, the sulfate concentration was higher after liming, while in the Orangeburg soil, it was slightly lower. Sulfate addition has been shown to reduce plant accumulation of Se in a variety of studies (Girling, 1984; Gissel-Nielsen, 1973; Mikkelsen et al., 1989a;b). Thus it may be that the decreased Se accumulation observed in the limed Blanton soil was a result of a higher available sulfate interacting antagonistically with Se. In the case of the Orangeburg soil, available sulfate was slightly reduced in the limed soil and was therefore not a factor in determining the availability of Se to the plants. In this soil, liming increased plant accumulation of Se from the selenite treatments due to increased availability of selenite at the higher pH, presumably resulting from desorption of selenite from soil hydrous oxides. Some indicator species can absorb comparatively larger amounts of organic Se from the soil. In some vegetable species, uptake of Se was greater when plants were grown on soils containing organic Se than when grown on selenate-enriched soils (Hamilton and Beath, 1964).

The form of Se added to increase plant Se concentration to a desired level may well be tailored to the soil pH. Selenites may be more useful than selenates as slowly available Se additives on alkaline soils, since selenates added to these soils can produce plants with Se concentrations toxic to livestock. To this end, it was suggested that soluble selenites may be the most suitable form of Se to add to neutral or acidic soils to produce crops low in Se (Cary and Allaway, 1969).

Total uptake of Se (as percentage of added Se) from a 2-yr field experiment using mustard yielded values of 0.01% from Se^0, 4% from SeO_3^{2-}, and 30% from SeO_4^{2-} (Gissel-Nielsen and Bisbjerg, 1970). With lucerne, barley, and sugar beets, the uptakes were one-third or less of these values. Thus, the use of Se^0 is impractical as a Se source because of its insolubility. With potted cowpeas, a plant used as a fodder crop in India as an indicator, inhibition of plant growth from soil-added Se was in the order $SeO_4^{2-} > H_2SeO_3 > SeO_3^{2-} > Se^0$ (Singh and Singh, 1979). Addition of $BaCl_2$ to some alkaline seleniferous soils reduced Se uptake by lucerne plants. Presumably the sparingly soluble $BaSeO_4$ was formed, resulting in the low levels of soluble Se (Ravikovitch and Margolin, 1959). Because it is fairly insoluble, $BaSeO_4$ can serve as a slowly available Se source in low Se alkaline soils.

In summary, Se in its highest oxidation state [i.e., Se(VI)] is readily available for plant uptake, while Se(IV) is much less available due to its higher sorption affinity for soil constituents. Elemental Se is unavailable for plant uptake while Se($-$II) are generally metal selenides of extremely low solubility.

7.4 Soil Texture and Clay Mineral

In general, uptake of Se by plants from mineral soils can be expected to be inversely related to the clay content of the soil. Bisbjerg and Gissel-Nielsen (1969) observed that Se concentrations in plants (red clover, barley, and white mustard) were sometimes twice as high on sandy than on loamy (more clay) soils. In addition, Singh et al. (1981) noted that in soil sorption studies of selenite and selenate, the degree of sorption was influenced positively by the clay content. Thus if selenites were to be applied to farmlands low in Se, smaller amounts should be used for sandy than for fine-textured soils.

Loamy and clayey soils were found to contain greater Se than Podzols and sandy soils in Ontario, Canada (Frank et al., 1979), while in Finnish mineral soils, the mean total Se content decreased with increasing particle size (Sippola, 1979). Sandy soils that characteristically contain less Al and Fe oxides can be expected to release Se more easily for plant uptake (Carlson et al., 1991).

Frost and Griffin (1977) found that selenite adsorption was greater in montmorillonite than in kaolinite; SeO_4^{2-} has lower affinity for kaolinite than SeO_3^{2-} (Bar-Yosef and Meek, 1987).

7.5 Organic Matter

In nature, Se is associated with the OM in soils via the cycling of the element through decay of plant material. Thus, Se levels in surface soils can be expected to be higher than in subsoils due to higher OM content in the former. In soil sorption studies, sorption of Se was influenced positively by OM content, indicating its complexing role on Se (Singh et al., 1981). Levesque (1972, 1974b) also indicated that Se was presumably bound to organometallic complexes. These findings however, are not consistent with plant uptake studies since organic soils, such as peats, failed to inhibit Se uptake by plants (Davies and Watkinson, 1966b; Gile and Lakin, 1941).

In Se transformation in soils, OM apparently plays an important role. Conversion of inorganic forms of Se, both added and indigenous, to volatile products by soil microorganisms may be enhanced if soils are amended with organic material (Doran and Alexander, 1977). Additions of OM to soils are known to stimulate microbial activity, thus increased uptake and metabolism of Se can be expected. This can then affect the various transformation reactions, including reduction and methylation. For example, Karlson and Frankenberger (1988, 1989) found that C amendment of Se(IV)-affected soils accelerated the production of alkylselenides. Apparently, the addition of C stimulated the methylating organisms in soil. For example, pectin addition accelerated Se evolution 2- to 130-fold, which is as pronounced with Se(VI) as with Se(IV). The total Se volatilized ranged from 11.3 to 51.4% of the added Se. Fungal growth is generally favored by the addition of C source, as they are known to be the primary producers of methyl-Se.

However, the stimulation by OM depends on the type of C source. For example, in the above study, pectin and galacturonic acid caused the greatest stimulation; cellulose has minimal effect because its long-chain structure is disliked by microorganisms. However, addition of compost manure of gluten (a pure protein) to soils may aid in the persistence of DMSe since the microorganisms involved prefer the C source contained in this organic material (Guo et al., 1999).

It is well known that addition of OM lowers the soil redox potential under flooded conditions. Since Se species are redox sensitive, ponded soils amended with OM can affect the speciation of Se considerably, and hence its uptake by organisms. In a pot experiment, flooding the Se-spiked soil (-390 mV) amended with OM (as cellulose) decreased Se(VI) concentration in the soil solution at a much faster rate than in unamended soil (Mikkelsen et al., 1989). It was hypothesized that the disappearance of Se(VI) from the soil solution was due to the reduction of selenate to selenite, and eventually to Se^0. In addition, Se was accumulated to phytotoxic levels in all parts of the rice plants when grown under unflooded conditions or in flooded soil in the absence of OM. Similarly, Se in incubated Kesterson sediments remained in the Se(0) state when amended with OM (Tokunaga et al., 1996), confirming the important role of OM in Se transformation.

7.6 Redox Potential

The effect of redox potential on Se speciation and transformation has been discussed in Sections 4.4 and 4.6. Briefly, at high redox potential in well-aerated alkaline soils, the highly soluble selenates are the predominant species, and at circumneutral pH the selenites occur at approximately equal concentrations. In reducing conditions, the oxidized species may undergo reduction reactions yielding elemental Se, further converting to the volatile selenides at much reducing conditions.

Selenate is stable in oxidized environments and is the species in which Se is most readily taken up by plants. In most soils, the selenates and selenites are the predominant Se species.

7.7 Competing Ions

The biogeochemical similarity between S and Se and the fact that almost all S compounds are known to have Se analogues led to the concept that Se functions in animal and plant seleno-amino acids (Frost, 1972). In biological systems, enzymes responsible for assimilation of the S ions can also use the Se ions. Because of their similarity, Se and S interactions in plant metabolism can be expected. However, results are rather conflicting, although evidence is more overwhelming toward the antagonistic effect of S on Se nutrition. Plant uptake of applied Se was highest where a fertilizer low in S was used (Gissel-Nielsen, 1971a). In some instances however, addition of S to soil may fail to inhibit the uptake of Se by plants (Gupta and Winter, 1975) particularly in seleniferous soils. Carter et al. (1969) found that uptake of SeO_4^{2-} by alfalfa from $BaSeO_4$ applied to an alkaline silt loam in the field was enhanced by the application of S, particularly when the S/Se ratio approached 10.

The interactive nature of SO_4^{2-} and SeO_4^{2-} ions depends to a large extent on plant species. For example, the primary accumulator *A. bisulcatus* exhibits preferential uptake of SeO_4^{2-} over SO_4^{2-} such that high shoot Se concentrations can be expected even in plants growing in gypseferous (i.e., high-SO_4^{2-}) soils (Bell et al., 1992). In another instance, the application of gypsum ($CaSO_4 \cdot 2H_2O$) to soil reduced the uptake of Se by forbs and grasses grown on a soil-capped, fly ash landfill (Woodburg et al., 1999), indicating again the interactive nature between the SO_4^{2-} and SeO_4^{2-} ions.

By comparing plant molar Se/S ratios (whole plant basis) with nutrient solution SeO_4^{2-}/SO_4^{2-} ratios, Bell et al. (1992) demonstrated that while alfalfa discriminates against Se, *A. bisulcatus* exhibits preferential uptake for Se relative to S (Fig. 18.10). The solid line indicates the equality of the two

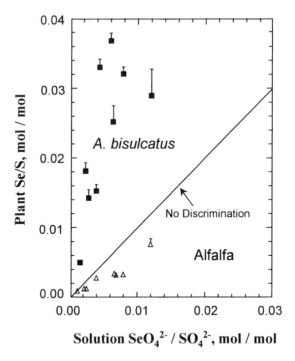

FIGURE 18.10. Plant molar Se/S ratio vs. the SeO_4^{2-}/SO_4^{2-} ratio in nutrient solution used to grow *Astragalus bisulcatus* (a primary Se accumulator) or alfalfa (a nonaccumulator). Plant ratios were computed on the basis of whole plant (roots plus shoot) uptake of Se and S. Solid line (slope = 1) denotes no plant discrimination between the analogues SeO_4^{2-} and SO_4^{2-}. (From Bell et al., 1992.)

molar ratios which represent no discrimination in the transport and assimilation of SeO_4^{2-} and SO_4^{2-}. Indeed a summary of published results indicates that primary Se accumulators are uniquely able to accumulate SeO_4^{2-} in the face of competition with SO_4^{2-}, probably as a consequence of their specialized Se metabolism, which is lacking in other plant species. Most nonaccumulators such as alfalfa, seem to discriminate against SeO_4^{2-} relative to SO_4^{2-}. Hence, arable soils in the western San Joaquin Valley, which contain elevated levels of both SO_4^{2-} and SeO_4^{2-} rarely produce agronomic crops with elevated Se concentrations. However, certain plant species, such as upland rice and members of the genus *Cruciferae,* can be expected to more readily accumulate SeO_4^{2-} from such soils (Banuelos et al., 1993).

Selenium-S interaction has also been observed in soil systems. Sulfate ions (as $BaSO_4$) added to the soil aided in leaching Se from soil (Brown and Carter, 1969). Furthermore, when a gypsum ($CaSO_4 \cdot H_2O$) solution was used for leaching, more Se was removed from the soil columns than when water alone was used. Apparently the sulfate ions induced desorption of Se ions from these alkaline soils (pH=7.8). Thus, application of SO_4^{2-}, especially to an alkaline field soil, can increase Se uptake by plants. In addition, sulfate and phosphate ions were shown to desorb Se from soils (Singh et al., 1981). Such desorption can be explained by the replacement of Se ions by these competing ions from the exchange complex since it has been demonstrated that SeO_3^{2-} and SeO_4^{2-} are retained in soils through one of the mechanisms by which phosphate and sulfate are retained (Rajan and Watkinson, 1976).

Selenate has been shown time and again to behave much like sulfate (Goldberg and Glaubig, 1988; Neal and Sposito, 1989) in that its sorption by soils is generally low and its mobility is high, whereas selenite appears to behave in a manner similar to that of phosphate (Goldberg and Sposito, 1985; Neal et al., 1987), being sorbed to a much greater extent than selenate by most soils. Thus SO_4^- and PO_4^{2-} ions can compete with selenate and selenite ions, respectively, for the same exchange complex in soils.

Although the competition between SeO_4^{2-} and SO_4^{2-} in plant nutrition is well documented under controlled conditions, it has been a scientific curiosity whether such relationships are valid under field conditions. To this end, a landscape study in the San Joaquin Valley was undertaken where alfalfa tissues and 721 soil samples were collected (Severson and Gough, 1992). Results of this survey indicate the interrelationships between total Se and S in soils and in alfalfa tissues. At relatively low total Se in soil (<0.50 mg kg^{-1}), increasing soil S had only minimal effect on Se accumulation in alfalfa. The Se level in alfalfa remained constant at about 0.20 mg kg^{-1}. As total soil Se increased from 0.50 to 1.7 mg kg^{-1}, alfalfa accumulation of Se increased drastically from 0.20 to about 1.0 mg kg^{-1}. At the highest level of total Se in soil (>1.7 mg kg^{-1}), the accumulation of Se in alfalfa remained constant at about 1.0 mg kg^{-1} as total S in soil increased. This plateau in Se accumulation at the higher total Se levels in soil may be partly due to SeO$_4^{2-}$-SO$_4^{2-}$ competition, but since this is under field conditions other biogeochemical processes in the soil–plant system may also have contributed to this observation.

Like the Se-S interaction, interactions between Se and P in plants have also been demonstrated. The pattern has also been inconsistent, although more results indicate that the effect of P tended to be more positive (Carter et al., 1972; Levesque, 1974b). Two explanations for the Se-P interaction are possible: (1) Se and P ions compete for the same reaction sites in soils; and (2) P applications may stimulate plant growth, including greater root proliferation, enabling the plants to absorb more Se (Carter et al., 1972). The more extensive the root system, the greater the root–soil contact and the capacity to absorb ions.

7.8 Salinity

In saline soils, total salinity and other ions may interfere with Se metabolism in plants. For example, in the San Joaquin Valley, SO$_4^{2-}$ and Cl$^-$ are major inorganic anions in the shallow groundwater. Tall fescue from the field irrigated with water low in salinity had higher Se concentrations than plants from the field irrigated with water high in salinity (Wu and Huang, 1991). Similarly, alfalfa grown in potted soils accumulated Se (from SeO$_4^{2-}$) in low-SO$_4^{2-}$ soils (i.e., 10 to 20 times greater) than plants grown in high-SO$_4^{2-}$ soils (Wan et al., 1988).

A fundamental premise in plant nutrition research in the western part of the San Joaquin Valley is that high Se levels in soils and drainage waters are almost always accompanied by high levels of soluble salts and B (Parker et al., 1991; Parker and Page, 1994). Salinity levels in the valley can be exceptionally high in evaporation pond sediments, in soils adjacent to evaporation ponds, and in salt-affected and unreclaimed soils. The soluble salts in the area are dominated by Na$^+$ and SO$_4^{2-}$ ions, although significant quantities of Ca^{2+}, Mg^{2+}, and Cl$^-$ are also typically present. In some soils, soluble B levels are also quite high, with concentrations in excess of 10 mg L^{-1} being common. Plant tolerance to these saline constituents is crucial in selecting plant species for bioremediating highly affected soils and sediments. In other words, since high Se coexists with those saline constituents, especially Cl$^-$ and B, plants should possess co-tolerance properties in such environment.

7.9 Plant Factors (Species, Cultivars, Parts)

There are considerable differences in Se accumulation among plant species, particularly in soils high in bioavailable Se (Fig. 18.11). Under

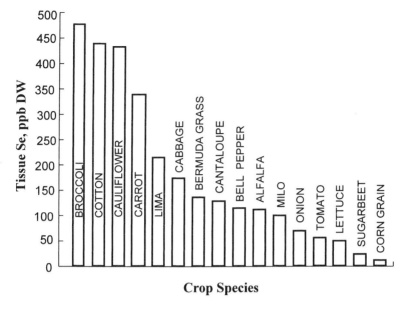

FIGURE 18.11. Ranked geometric mean values of selenium concentrations in edible tissues of crops sampled from the west side of the San Joaquin Valley, California. (From Burau et al., 1988.)

these conditions, tenfold differences or more are common. However, there is only meager information on plant species differences in soils indigenously low in bioavailable Se or in typical agricultural soils. In general, differences among plant species in Se uptake from soils low in Se are small.

Surveys on Se concentrations in Danish forages and San Joaquin Valley crops indicate that Se levels in the same species vary considerably within a small area due to variations in soil Se content (Gissel-Nielsen et al., 1984; Wu, 1998). Similarly, differences in tissue Se content among species can be very substantial (see Fig. 18.11). Even among S-accumulating vegetables, Se uptake can vary considerably among them as well as between parts in a particular species. The composite leaf tissue Se contents of Swiss chard and collards (0.45 ppm DW) are higher than in similar tissues of cabbage and broccoli (0.35 ppm) (Banuelos et al., 1989). However, the broccoli florets contained the highest Se levels (1.39 ppm). Generally, Se is highest in the leaves and lowest in the stems.

Stage of growth and harvesting can also determine the Se concentration in plants. Selenium concentrations in plant tissues usually decline with maturity or with subsequent cuttings of pasture or forage crops. Exceptions for this trend can occur in low-Se soils.

8 Sources of Selenium in the Environment

8.1 Geologic Material

Highly Se-contaminated areas that have produced *selenocosis* in animals derived the Se from geologic formation. The common source is shales (e.g., Cretacean shales) that inherently contain high levels of Se. High accumulation of Se occurs over time in low-lying areas or where there are clay pans that receive Se from leachates or drainage waters from overlying areas. A good example of such phenomenon is the Kesterson Reservoir in California that has been receiving drainage waters from higher-elevated arable fields in the western San Joaquin Valley. High evaporation rates also aided in the selenification of affected soils (see Section 6.2.1).

8.2 Agricultural Practices

Although low-Se soils may produce forage having insufficient Se for normal nutrition required by livestock (i.e., 0.03 to 0.10 ppm Se in the diet, depending on vitamin E and other factors), Se compounds can be added to these soils or plants. There are three conventional means of increasing the content of Se in plants: soil application, foliar spray, and seed treatment. For soil application, rates of application may vary from tens of grams to a few kilograms per hectare, depending on soil type, type of crops, and other factors. In New Zealand topdressing pasture with about 70 g Se ha^{-1} as Na_2SeO_3 or Na_2SeO_4 produced Se concentrations toxic to livestock for periods of more than a year after application (Grant, 1965). The high Se levels in plant tissues were attributed to foliar absorption and resulted in uptake that was greater than if the Se had been applied directly to the soil. Since 1982, New Zealand has permitted fertilizer enrichment with Se as sodium selenate in granules containing 1% Se at a rate of 1 kg of the granules per hectare.

Slow-release fertilizers containing barium selenate have also been used. Alfalfa grown in the field on acidic soils treated with 1.12 kg Se ha^{-1} as Na_2SeO_3 contained up to 2.7 ppm, within the considered safe limit for livestock (Allaway et al., 1966). In another field trial, barley and timothy grown on an acid soil treated with up to 2.24 kg Se ha^{-1} as Na_2SeO_3 gained adequate levels of Se for up to 5 yr (Gupta and Winter, 1981). Elemental Se applied at rates as low as 0.25 to 0.5 kg ha^{-1} can provide adequate Se for up to two crops of alfalfa (Carter et al., 1969). However, it may be an impractical carrier of Se because of its short and low recovery by plants.

Field application of Se should be implemented with caution when Se is topdressed or when the soil is limed because of the greater availability of Se with increasing pH. Foliar application of Se is potentially efficient and safe for increasing Se in forages and feeds (Cary and Rutzke, 1981; Gupta et al., 1983a). Seed treatment of Se has been shown to be effective also in increasing Se concentrations in plants (Gupta et al., 1983b). Application of Se as Na_2SeO_3 solution to seeds at 50 to 200 g ha^{-1} produced forages containing greater than 0.10 ppm Se for at least three harvests. A big advantage of foliar application over soil application is that it avoids the influence that soil conditions can exert on plant uptake.

Macronutrient fertilizers are another source of Se in agriculture, especially phosphatic fertilizers. Selenium occurs in phosphate rocks, ranging from trace amounts to several hundreds ppm. Selenium concentrations in seven Florida phosphate rocks ranged from 0.70 to 7.0 ppm (Robbins and Carter, 1970); the range was from 1.4 to 178 ppm in seven samples of phosphorus fertilizer produced from western United States phosphate mines. Therefore, it seems prudent to expect that normal phosphate fertilization could provide the required Se amounts for livestock, provided the fertilizer contains sufficient Se. In general however, the contribution of fertilizers, unless process-enriched with Se, to the total Se content of plants is negligible.

8.3 Coal Combustion and Its Residues

Burning of fossil fuels, including coal and fuel oils, can also be an important source of Se in the atmosphere (Hashimoto et al., 1970). Because large amounts of coal residues are being produced yearly ($>100 \times 10^6$ tonnes yr^{-1}) by coal-fired power plants in the United States, this solid waste looms as a potential source of Se in the environment. Selenium from fly ash is quite available to plants and could place some constraints on land application of this by-product (Adriano et al., 1980; Adriano and Weber, 1998; Carlson and Adriano, 1993). It can accumulate in plant tissues in the range of several hundred ppm and up to two orders of magnitude of that in the growth medium. A severe case of fish poisoning at Belews Lake, North Carolina in the late 1970s occurred due to Se accumulation in the reservoir from coal fly ash related contamination (Skorupa, 1998). Ash effluents entering the lake had 150 to 200 μg Se L^{-1}. Although an average value of 10 μg Se L^{-1} of the lake water (8 to 22 μg L^{-1} range) has been reported, high rates of teratogenic fish primarily attributed to Se were observed. Of the fish species initially present, only four remained by 1978.

The solubility and subsequent leaching of Se from coal fly ash can result in toxic elevations of Se in aqueous media. For example, bioaccumulation of Se in fish at the Pigeon River/Pigeon Lake in Michigan affected by coal fly ash effluents was significantly higher than in fish populations upstream of the ash disposal facility (Besser et al., 1996). Se concentrations in fish from the lower reach of the river/lake exceeded LOAECs (lowest observable adverse effect concentrations) for dietary Se exposure for sensitive species of birds and mammals. Similarly, Hopkins et al. (1999) obtained 32% higher mean standard metabolic rates in snakes from a site contaminated by coal combustion residue compared with snakes collected from reference sites.

This indicates that the elevated levels of trace elements, including Se and As, have been stressing the animals to the extent that less energy remains available for growth, reproduction, and storage.

8.4 Atmospheric Deposition

Both natural and anthropogenic sources of Se contribute to the release of Se into the atmosphere, (i.e., both particulate and gaseous forms) which eventually falls on the earth's surface. The importance of deposition on the enrichment of soils and vegetation has been evaluated for trace elements including Se, in the United Kingdom (Haygarth, 1994). The results are summarized below:

1. Deposition has contributed to a 15% increase in soil Se concentration in the last century.
2. Deposition has contributed between 33 and 82% of plant leaf Se uptake.
3. Deposition is heavily influenced by geographic proximity to an emission source, with highest levels associated with industrial and coastal zones.
4. The deposition process is considerably higher during the winter months, probably reflecting the prevalence of wet removal processes combined with climatological characteristics, temperature inversions, and occult deposition with higher combustion and emission rates.

References

Adriano, D.C., 1986. *Trace Elements in the Terrestrial Environment*. Springer-Verlag, New York

Adriano, D.C., and J. Weber. 1998. *Coal Ash Utilization for Soil Amendments to Enhance Water Relations and Turf Growth*. EPRI Rep TR-111318. Electric Power Res Inst, Palo Alto, CA.

Adriano, D.C., A.L. Page, A.A. Elseewi, A.C. Chang, and I. Straughan. 1980. *J Environ Qual* 9:333–344.

Ainsworth, C.C., and D. Rai. 1987. *Chemical Characterization of Fossil Fuel Combustion Wastes*. EPRI Rep EA-5321. Electric Power Res Inst, Palo Alto, CA.

Alemi, M.H., D.A. Goldhamer, and D.R. Nielsen. 1991. *J Environ Qual* 20:89–95.

Allaway, W.H. 1968. *Trace Subs Environ Health* 2:181–206.

Allaway, W.H. 1973. *Cornell Vet* 63:151–170.

Allaway, W.H., D.P. Moore, J.E. Oldfield, and O.H. Muth. 1966. *J Nutr* 88:411–418.

Allaway, W.H., E.E. Cary, and C.F. Ehlig. 1967. In O.J. Muth, ed. *Selenium in Biomedicine*. AVI Publ, Westport, CT.

Archer, F.C. 1980. *Minist Agric Fish Food* (Great Britain) 326:184–190.

Aro, A., G. Alfthan, P. Ekholm, and P. Varo. 1998. In W.T. Frankenberger and R.A. Engberg, eds. *Environmental Chemistry of Selenium*. Marcel Dekker, New York.

[ATSDR] Agency for Toxic Substances and Disease Registry. 1989. *Toxicological Profile for Selenium*, U.S. Dept Health and Human Services, Atlanta, GA.

Banuelos, G.S., and D.W. Meek. 1989. *J Plant Nutr* 12:1255–1272.

Banuelos, G.S., G. Gordon, B. Mackey, et al. 1993. *J Environ Qual* 22:786–792.

Bar-Yosef, B., and D. Meek. 1987. *Soil Sci* 144:11–18.

Beath, O.A., H.F. Eppson, and C.S. Gilbert. 1937. *J Am Phar Assoc* 26:394–405.

Bell, P.F., D.R. Parker, and A.L. Page. 1992. *Soil Sci Soc Am J* 56:1818–1824.

Besser, J.M., T.J. Canfield, and T.W. LaPoint. 1993. *Environ Toxicol Chem* 12:57–72.

Besser, J.M., J.P. Giesy, R.W. Brown, J.M. Buell, and G.A. Dawson. 1996. *Ecol Environ Safety* 35:7–15.

Bisbjerg, B., and G. Gissel-Nielsen. 1969. *Plant Soil* 31:287–298.

Bowen, H.J.M. 1979. *Environmental Chemistry of the Elements.* Academic Pr, New York.

Brookins, D.G. 1988. *Eh–pH Diagrams for Geochemisty.* Springer-Verlag, New York.

Brown, M.J., and D.L. Carter. 1969. *Soil Sci Soc Am Proc* 33:563–565.

Broyer, T.C., D.C. Lee, and C.J. Asher. 1966. *Plant Physiol* 41:1425–1428.

Burau, R.G., A. McDonald, A. Jacobson, D. May and V. Rendig. 1998. In *Selenium Contents in Animal and Human Food Crops Grown in California.* ANR Spec Publ 3330, Sacramento, CA.

Byers, H.G., and H.W. Lakin. 1939. *Can J Res* 17-B:364–69.

Byers, H.G., J.T. Miller, K.T. Williams, and H.W. Lakin. 1938. In USDA Tech Bull 601. U.S. Dept Agriculture, Washington, DC.

Cannon, H.L. 1974. In P.L. White and D. Robbins, eds. *Environmental Quality and Food Supply.* Futura Publ, Mt Kisco, NY.

Carlson, C., and D.C. Adriano. 1993. *J Environ Qual* 22:227–247.

Carlson, C.L., D.C. Adriano, and P.M. Dixon. 1991. *J Environ Qual* 20:363–368.

Carter, D.L., M.J. Brown, and C.W. Robbins. 1969. *Soil Sci Soc Am Proc* 33:715–718.

Carter, D.L., C.W. Robbins, and M.J. Brown. 1972. *Soil Sci Soc Am Proc* 36:624–628.

Cary, E.E., and W.H. Allaway. 1969. *Soil Sci Soc Am Proc* 33:571–574.

Cary, E.E., and M. Rutzke. 1981. *Agron J* 73:1083–1085.

Cary, E.E., G.A. Wieczorek, and W.H. Allaway. 1967. *Soil Sci Soc Am Proc* 31:21–26.

Chau, Y.K., P.T.S. Wong, B.A. Silverberg, P.L. Luxon, and G.A. Bengert. 1976. *Science* 192:1130–1131.

Coleman, R.G., and M.H. Delevaux. 1957. *Econ Geol* 52:499–527.

Coles, L.E. 1974. *J Assoc Public Anal* 12(3):68–72.

Combs, G.F., and S.B. Combs. 1984. *Ann Rev Nutr* 4:257–280.

Davies, E.B., and J.H. Watkinson. 1966b. *NZ J Agric Res* 9:641–652.

Davis, R.D., P.H.T. Beckett, and E. Wollan. 1978. *Plant Soil* 49:395–408.

Doran, J.W. 1982. *Adv Microb Ecol* 6:1–32.

Doran, J.W., and M. Alexander. 1977. *Soil Sci Soc Am J* 41:70–73.

Duckart, E.C., L.J. Waldron, and H.E. Doner. 1992. *Soil Sci* 53:94–99

Dungan, R.S., and W.T. Frankenberger. 1998. *J Environ Qual* 27:1301–1306.

Elrashidi, M.A., D.C. Adriano, S.M. Workman, and W.L. Lindsay. 1987. *Soil Sci* 144:141–152.

Elrashidi, M.A., D.C. Adriano, and W.L. Lindsay. 1989. In L.W. Jacobs, ed. *Selenium in Agriculture and the Environment.* SSSA 23, Madison, WI.

Engberg, R.A., D.W. Westcot, M. Delamore, and D.D. Holtz. 1998. In W.T. Frankenberger and R.A. Engberg, eds. *Environmental Chemistry of Selenium.* Marcel Dekker, New York.

Evans, C.S., C.J. Asher, and C.M. Johnson. 1968. *Aust J Biol Sci* 21:13–20.

Fan, T.W., and R.M. Higashi. 1998. In W.T. Frankenberger and R.A. Engberg, eds. *Environmental Chemistry of Selenium.* Marcel Dekker, New York.

Fan, T.W., R.M. Higashi, and A.N. Lane. 1998. *Environ Sci Technol* 32:3185–3193.

Fio, J.L., and R. Fujii. 1990. *Soil Sci Soc Am J* 54:363–369.

Fleming, G.A., and T. Walsh. 1957. *Royal Irish Acad Proc* 58B:151–165.

Food and Nutrition. Board. 1989. *Recommended Dietary Allowances.* National Research Council, Washington, DC.

Francis, A.J., J.M. Duxbury, and M. Alexander. 1974. *Appl Microbiol* 28:248–250.

Frank, R., K.I. Stonefield, and P. Suda. 1979. *Can J Soil Sci* 59:99–103.

Frankenberger, W.T., and S. Benson, eds. *Selenium in the Environment.* Marcel Dekker, New York.

Frankenberger, W.T., and R.A. Engberg, eds. 1998. *Environmental Chemistry of Selenium.* Marcel Dekker, New York.

Frankenberger, W.T., and U. Karlson. 1994. In W.T. Frankenberger and S. Benson, eds. *Selenium in the Environment.* Marcel Dekker, New York.

Frankenberger, W.T., and U. Karlson. 1994. *Geomicrobiol J* 12:265–278.

Franson, J.C., J.A. Schmutz, L.H. Creekmore, and A.C. Fowler. 1999. *Environ Toxicol Chem* 18:965–969.

Frost, D.V. 1972. *CRC Critical Rev Toxicol* 1:467–514.

Frost, R.R., and R.A. Griffin. 1977. *Soil Sci Soc Am J* 41:53–57.

Fujii, R., S.J. Deverel, and D.B. Hatfield. 1988. *Soil Sci Soc Am J* 52:1274–1283.

Gamboa-Lewis, B.A. 1976. In J.O. Nriagu, ed. *Environmental Biogeochemistry,* vol. 1. Ann Arbor Science, Ann Arbor, MI.

Gao, S., and K.K. Tanji. 1995. *J Environ Qual* 24:191–197.

Geering, H.R., E.E. Cary, L.H.P. Jones, and W.H. Allaway. 1968. *Soil Sci Soc Am Proc* 32:35–40.

Gile, P.L., and H.W. Lakin. 1941. *J Agric Res* 63:559–581.

Gissel-Nielsen, G. 1971a. *J Agric Food Chem* 19:564–566.

Gissel-Nielsen, G. 1971b. *J Agric Food Chem* 19:1165–1167.

Gissel-Nielsen, G., and B. Bisbjerg. 1970. *Plant Soil* 32:382–396.

Gissel-Nielsen, G., and M. Gissel-Nielsen. 1973. *Ambio* 2:114–117.

Gissel-Nielsen, G., U.C. Gupta, M. Lamand, and T. Westermarck. 1984. *Adv Agron* 37:397–460.

Girling, C.A. 1984. *Agric Ecosys Environ* 11:37–65.

Goldberg, S., and R.A. Glaubig. 1988. *Soil Sci Soc Am J* 52:954–958.

Goldberg, S., and G. Sposito. 1985. *Commun Soil Sci Plant Anal* 16:801–821.

Grant, A.B. 1965. *NZ J Agric Res* 8:681–690.

Guo, L., W.T. Frankenberger, and W.A. Jury. 1999. *Environ Sci Technol* 33:2934–2938.

Gupta, U.C., and K.A. Winter. 1975. *Can J Soil Sci* 55:161–166.

Gupta, U.C., and K.A. Winter. 1981. *J Plant Nutr* 3:493–502.

Gupta, U.C., K.B. McRae, and K.A. Winter. 1982. *Can J Soil Sci* 62:145–154.

Gupta, U.C., H.T. Kunelius, and K.A. Winter. 1983a. *Can J Soil Sci* 63:455–459.

Gupta,U.C., H.T. Kunelius, and K.A. Winter. 1983b. *Can J Soil Sci* 63:641–643.

Gustafsson, J.P., and L. Johnsson. 1992. *J Soil Sci* 43:401–472.

Hamdy A.A., and G. Gissel-Nielsen. 1976. In Rizo Rep 349, Roskilde, Denmark.

Hamilton, J.W., and O.A. Beath. 1963. *J Range Manage* 16:261–264.

Hamilton, J.W., and O.A. Beath. 1964. *J Agric Food Chem* 12:371–374.

Hamilton, S.J. 1998. In W.T. Frankenberger and R.A. Engberg, eds. *Environmental Chemistry of Selenium*. Marcel Dekker, New York.

Hashimoto, Y., J.Y. Hwang, and S. Yanagisawa. 1970. *Environ Sci Technol* 4:157–158.

Hayes, K.F., A.L. Roe, G.E. Brown, K.O. Hodgen, J.O. Leckie, and G.A. Parks. 1987. *Science* 238:783–786.

Haygarth, P.M. 1994. In W.T. Frankenberger and S. Benson, eds. *Selenium in the Environment*. Marcel Dekker, New York.

Hingston, F.J., A.M. Posner, and J.P. Quirk. 1971. *Disc Farraday Soc* 52:334–342.

Hodson, P.V., and J.W. Hilton. 1983. *Ecol Bull* 35:335–347.

Hoffman, D.J., and G.H. Heinz. 1998. *Environ Toxicol Chem.* 17:161–166.

Hoffman, D.J., H.M. Ohlendorf, C.M. Marn, and G.W. Pendelton. 1998. *Environ Toxicol Chem* 17:167–172.

Hopkins, W.A., C.L. Rowe, and J.D. Congdon. 1999. *Environ Toxicol Chem* 18:1258–1263.

Hurd-Karrer, A.M. 1935. *J Agric Res* 50:413–427.

Hurd-Karrer, A.M. 1937. *Am J Bot* 24:720–728.

Ihnat, M. 1989. *Occurrence and Distribution of Selenium*. CRC Pr, Boca Raton, FL.

John, M.K., W.M.H. Saunders, and J.W. Watkinson. 1976. *NZ J Agric Res* 19:143–151.

Karlson, U., and W.T. Frankenberger. 1988. *Soil Sci Soc Am J* 52:1640–1644

Karlson, U., and W.T. Frankenberger. 1989. *Soil Sci Soc Am J* 53:749–753.

Koljonen, T. 1975. *Ann Agric Fenn* 14:240–252.

Kubota, J., E.E. Cary, and G. Gissel-Nielsen. 1975. *Trace Subs Environ Health* 9:123–130.

Lakin, H.W. 1961. In *Selenium in Agriculture*. USDA Handb 200. U.S. Dept Agriculture, Washington, DC.

Leggett, J.E., and E. Epstein. 1956. *Plant Physiol* 31:222–226.

Lemly, A.D. 1998. In W.T. Frankenberger and R.A. Engberg, eds. *Environmental Chemistry of Selenium*. Marcel Dekker, New York.

Levesque, M. 1972. *Soil Sci* 113: 346–353.

Levesque, M. 1974a. *Can J Soil Sci* 54:63–68.

Levesque, M. 1974b. *Can J Soil Sci* 54:205–214.

Lewis, B.G., C.M. Johnson, and C.C. Delwiche. 1966. *J Agric Food Chem* 14:638–640.

Lewis, B.G., C.M. Johnson, and T.C. Broyer. 1974. *Plant Soil* 40:107–118.

Lindberg, P., and S. Bingefors. 1970. *Acta Agric Scand* 20:133–136.

Lipton, D.S. 1991. *Associations of Selenium in Inorganic and Organic Constituents of Soils from a Semi-arid Region*. PhD dissertation, Univ. California, Berkeley, CA.

Lo, M.T., and E. Sandi. 1980. *J Environ Pathol Toxicol* 4:193–218.

Losi, M.E., and W.T. Frankenberger. 1998. *J Environ Qual* 27:836–843.

Manceau, A., and L. Charlet. 1994. *J Colloid Interf Sci* 168:87–93.

Martens, D.A., and D.L. Suarez. 1998. In W.T. Frankenberger and R.A. Engberg, eds. *Environmental Chemistry of Selenium*. Marcel Dekker, New York.

Mason, T.G., and E. Phillis. 1937. *Emp Cotton Grow Rev* 14:308–309.

Masscheleyn, P.H., R.D. Delaune, and W.H. Patrick. 1990. *Environ Sci Technol* 24:91–96.

Mayland, H.F. 1994. In W.T. Frankenberger and S. Benson, eds. *Selenium in the Environment*. Marcel Dekker, New York.

Mikkelsen, R.L., D.S. Mikkelsen, and A. Abshahi. 1989a. *Soil Sci Soc Am J* 53:122–127.

Mikkelsen, R.L., G.H. Haghnia, and A.L. Page. 1989b. In L.W. Jacobs, ed. *Selenium in Agriculture and the Environment*. SSSA 23, Madison, WI.

Milne, J.B. 1998. In W.T. Frankenberger and R.A. Engberg, eds. *Environmental Chemistry of Selenium*. Marcel Dekker, New York.

[NAS] National Academy of Sciences. 1974. *Geochem Environ* 1:57–63.

[NAS] National Academy of Sciences. 1976. *Selenium*. NAS, Washington, DC.

Neal, R.H., and G. Sposito. 1989. *Soil Sci Soc Am J* 53:70–74.

Neal, R.H., G. Sposito, K.M. Holtzclaw, and S.J. Traina. 1987. *Soil Sci Soc Am J* 51:1165–1169.

Nye, S.M., and P.J. Peterson. 1975. *Trace Subs Environ Health* 9:113–121.

Ohlendorf, H.M., and G.M. Santolo. 1994. In W.T. Frankenberger and S. Benson, eds. *Selenium in the Environment*. Marcel Dekker, New York.

Oldfield, J.E. 1972. *Geol Soc Am Bull* 83:173–180.

Oremland, R.S. 1994. In W.T. Frankenberger and S. Benson, eds. *Selenium in the Environment*. Marcel Dekker, New York.

Oremland, R.S., J.T. Hollibaugh, A.S. Maest, T.S. Preser, L.G. Miller, and C.W. Culbertson. 1989. *Appl Environ Microbiol* 55:2333–2343.

Parker, D.R., and A.L. Page. 1994. In W.T. Frankenberger and S. Benson, eds. *Selenium in the Environment*. Marcel Dekker, New York.

Parker, D.R., A.L. Page, and D.N. Thomason. 1991. *J Environ Qual* 20:157–164.

Peterson, P.J., and G.W. Butler. 1967. *Nature* 213:599–600.

Peterson, P.J., L.M. Benson, and R. Zieve. 1981. In N.W. Lepp, ed. *Effect of Heavy Metal Pollution in Plants, vol. 1: Effect of Trace Metals on Plant Function*. Applied Science Publ, London.

Plotnikov, V.I. 1960. *Russ J Inorg Chem* 5:351–354.

Plotnikov, V.I. 1964. *Russ J Inorg Chem* 9:245–247.

Presser, T.R., and D.Z. Piper. 1998. In W.T. Frankenberger and R.A. Engberg, eds. *Environmental Chemistry of Selenium*. Marcel Dekker, New York.

Rajan, S.S.S., and J.H. Watkinson. 1976. *Soil Sci Soc Am J* 40:51–54.

Ravikovitch, S., and M. Margolin. 1959. *Emp J Exp Agric* 27:235–240.

Robberecht, H., D. Vanden Berghe, H. Deelstra, and R. Van Grieken. 1982. *Sci Total Environ* 25:61–69.

Robbins, C.W., and D.L. Carter. 1970. *Soil Sci Soc Am Proc* 34:506–509.

Rosenfeld, I., and O.A. Beath. 1964. *Selenium*. Academic Pr, New York.

Salonen, J.T., G. Alfthan, J.K. Huttunen, J. Pikkarainen, and P. Puska. 1982. *Lancet* 11:125–179.

Salonen, J.T., G. Alfthan, J.K. Huttunen, and P. Puska. 1984. *Am J Epidemiol* 120:342–349.

Schroeder, H.A., D.V. Frost, and J.J. Balassa. 1970. *J Chron Dis* 23:227–243.

Severson, R.C., and L.P. Gough. 1992. *J Environ Qual* 21:353–358.

Shamberger, R.J. 1981. *Sci Total Environ* 17:59–74.

Shrift, A. 1969. *Ann Rev Plant Physiol* 20:475–494.

Singh, M., and P. Kumar. 1976. *J Indian Soil Sci* 24:62–71.

Singh, M., N. Singh, and P.S. Relan. 1981. *Soil Sci* 132:134–141.

Sippola, J. 1979. *Ann Agric Fenn* 18:182–187.

Skorupa, J.P. 1998. In W.T. Frankenberger and R.A. Engberg, eds. *Environmental Chemistry of Selenium*. Marcel Dekker, New York.

Soltanpour, P.N., and S.M. Workman. 1980. *Commun Soil Sci Plant Anal* 11:1147–1156.

Stadtman, T. 1974. *Science* 183:915–922.

Stephens, D.W., and B. Waddell. 1998. In W.T. Frankenberger and R.A. Engberg, eds. *Environmental Chemistry of Selenium*. Marcel Dekker, New York.

Swaine, D.J. 1987. *Trace Subs Environ Health* 12:129–134.

Tan, J.A., W.Y. Wang, D.C. Wang, and S.F. Hou. 1994. In W.T. Frankenberger and S. Benson, eds. *Selenium in the Environment*. Marcel Dekker, New York.

Terry, N., and A.M. Zayed. 1994. In W.T. Frankenberger and S. Benson, eds. *Selenium in the Environment*. Marcel Dekker, New York.

Terry, N., and A.M. Zayed. 1998. In W.T. Frankenberger and R.A. Engberg, eds. *Environmental Chemistry of Selenium*. Marcel Dekker, New York.

Terry, N., C. Carlson, T.K. Raab, and A.M. Zayed. 1992. *J Environ Qual* 21:341–344.

Thompson, J.N., P. Erdody, and D.C. Smith. 1975. *J Nutr* 105:274–277.

Thompson-Eagle, E.T., and W.T. Frankenberger. 1992. *Adv Soil Sci* 17:261–310.

Thorn, J., J. Robertson, D.H. Buss, and N.G. Bunton. 1978. *Br J Nutr* 39:391–396.

Tokunaga, T.K., D.S. Lipton, S.M. Benson, et al. 1991. *Water Air Soil Pollut* 57-58:31–41.

Tokunaga, T.K., P.T. Zawislanski, P.W. Johannis, S. Benson, and D.S. Lipton. 1994. In W.T. Frankenberger and S. Benson, eds. *Selenium in the Environment*. Marcel Dekker, New York.

Tokunaga, T.K., I.J. Pickering, and G.E. Brown. 1996. *Soil Sci Soc Am J* 60:781–790.

Trelease, S.F., and O.A. Beath. 1949. *Selenium*. Champlain Printers, Burlington, VT.

Trelease, S.F., A.A. DiSomma, and A.L. Jacobs. 1960. *Science* 132:618.

Underwood, E.J. 1977. *Trace Elements in Human and Animal Nutrition*. Academic Pr, New York.

[U.S. EPA] United States Environmental Protection Agency 1986. *Final Drafts for Drinking Water Criteria for Selenium*. U.S. EPA, Office of Drinking Water, Washington, DC.

U.S. EPA. 1987. *Ambient Water Quality Criteria for Selenium, 1987*. EPA 440/5-87/006, U.S. EPA, Washington, DC.

Varo, P., G. Alfthan, J.K. Huttunen, and A. Aro. 1994. In R.F. Burk, ed. *Selenium in Biology and Human Health,* Springer-Verlag, New York.

Wallace, A., R.T. Mueller, and R.A. Wood. 1980. *J Plant Nutr* 2:107–109.

Walsh, T., and G.A. Fleming. 1952. *Int Soc Soil Sci Trans* (Comm II and IV) 2:178–183.

Wan, H.F., R.L. Mikkelsen, and A.L. Page. 1988. *J Environ Qual* 17:269–272.

Wells, N. 1967. *NZ J Sci* 10:142–179.

White, A.F., and N.M. Dubrovsky. 1994. In W.T. Frankenberger and S. Benson, eds. *Selenium in the Environment*. Marcel Dekker, New York.

Winter, K.A., and U.C. Gupta. 1979. *Can J Anim Sci* 59:107–111.

Woodbury, P.B., M.A. Arthur, G. Rubin, L.H. Weinstein, and D.C. McCune. 1999. *Water Air Soil Pollut* 110:421–432.

Workman, S.M., and P.N. Soltanpour. 1980. *Soil Sci Soc Am J* 44:1331–1333.

Wright, W.G. 1999. *J Environ Qual* 28:1182–1187.

Wu, L. 1998. In W.T. Frankenberger and R.A. Engberg, eds. *Environmental Chemistry of Selenium.* Marcel Dekker, New York.

Wu, L., and Z.Z. Huang. 1991. *Crop Sci* 31:114–118.

Ylaranta, T. 1983. *Ann Agric Fenn* 22:29–39.

Zawislanski, P.T., and M. Zavarin. 1996. *Soil Sci Soc Am J* 60:791–800.

Zehr, J.P., and R.S. Oremland. 1987. *Appl Environ Microbiol* 53:1365–1369.

Zhang, Y.Q., and J.N. Moore. 1998. In W.T. Frankenberger and R.A. Engberg, eds. *Environmental Chemistry of Selenium.* Marcel Dekker, New York.

19
Other Trace Elements

1 Antimony

The bluish white, lustrous, very brittle metal (atom. no. 51, atom. wt. 121.75, specific gravity 6.69, melting pt. 631 °C) is found in nature in more than 100 minerals. However, only about one dozen ores are commercially important, such as Sb oxides and sulfides, and complex Cu–, Pb–, and Hg–Sb sulfides, the most important of which is stibnite (Sb_2S_3). It also occurs with sphalerite, pyrite, and galena and is found in minor amounts in Hg deposits. In nature, it occurs primarily in the III oxidation state.

The major commercial sources of Sb are stibnite and antimonial Pb ores with minor contribution from antimonoxides. Antimony is used about equally between nonmetal and metal products. For nonmetal products, Sb is used in paints and lacquers, rubber compounds, ceramic enamels, glass and pottery, and abrasives. For metallic products, it is alloyed with Pb and is used in the manufacture of bearings, battery parts, sheet and pipe, tubes and foil, and ammunition. A Pb–Sb alloy, antimonial Pb, is used for storage batteries.

Antimony is 62nd in crustal abundance, slightly higher than Cd (Krauskopf, 1979). In soils, a median value of 1 ppm (range of 0.20 to 10 ppm) has been reported. Surficial soils of the United States have a geometric mean of 0.48 ppm, with a range of <1 to 10 ppm (Shacklette and Boerngen, 1984) (Appendix Table A.24). Reported values for world soils indicate a range of 0.05 to 2.3 ppm Sb (Kabata-Pendias and Pendias, 1984). Thus, most of world soils appear to have mean Sb values higher than the considered crustal mean of 0.20 ppm (Mitchell and Burridge, 1979). Surface soils from Ontario, Canada were reported to contain a mean of 0.24 ppm Sb with a range of 0.05 to 1.64 ppm (Frank et al., 1979). Norwegian surface soils have Sb content in the range of 0.17 to 2.20 ppm (Allen and Steinnes, 1979). The trend of metal concentrations where areas in southern Norway had several times greater metal contents than areas farther to the north indicates that the metals were deposited via long-range aerial transport.

Since the geochemical behavior of Sb is similar to that of As, it is commonly associated with nonferrous deposits and therefore, is emitted to the environment during the smelting of these ores. For example, the largest anthropogenic source of As and Sb to Puget Sound is a large Cu smelter located near Tacoma, Washington (Crecelius et al., 1975). This particular smelter releases Sb along with As into environment in three ways:

1. As stack dusts into the air—about 20×10^3 kg yr^{-1} of Sb oxides.
2. As dissolved Sb species in liquid effluent discharged directly into Puget Sound—about 2×10^3 kg Sb yr^{-1}.
3. As crystalline slag particles dumped directly into the Sound—about 1.5×10^6 kg Sb yr^{-1}.

On the other hand, only 250 kg yr^{-1} of Sb are discharged into the Sound in the form of sewage treatment plant effluents. Surface soils (0 to 3 cm) within 5 km of the Cu smelter can be enriched with up to 200 ppm Sb (Crecelius et al., 1974). Similarly, in a Sb-smelter setting, the levels of Sb in soils and plants are a function of distance (Ainsworth et al., 1990a; Asami and Kubota, 1993). Data indicate that contribution of root uptake to plant Sb content is insignificant, rather aerial deposition predominates over any other pathway. Evidence also indicates that the mobility of Sb in the food chain is low, with ground-living invertebrates containing lower Sb concentrations than their diets, and the tissues of small mammals being low in Sb compared with dietary intake (Ainsworth et al., 1990b).

Our present knowledge of Sb behavior in soils is insignificant. If present in soils in a soluble form, it probably occurs as antimonate and could be adsorbed by the same soil constituents that bind phosphate and arsenate. Indeed, Crecelius et al. (1975) found that chemical extraction of sediments from Puget Sound showed that about 50% of the As and Sb was associated with extractable Fe and Al compounds. Strong adsorption of Sb(V) by hematite occurred below pH 7 with a rapid reduction in adsorption at higher pH (Jones et al., 1990). Adsorbed Sb ions were not desorbed at the pH range of 2 to 5; the $Sb(OH)_6^-$ species predominated in the solution.

In general, potential phytotoxicity of Sb has been described as moderate (Brooks, 1972), although there are no reported environmental cases. Plants are more tolerant to Sb than As. The following normal levels have been quoted for terrestrial flora: 100 ppb for natural vegetation; 50 ppb for pasture grasses, herbage, and herbaceous vegetables, grain and cereal products; and 5 ppb for leguminous and root vegetables, and garden fruits (Jones et al., 1990).

2 Barium

The silvery white metal (atom. no. 56, atom. wt. 137.3, specific gravity 3.5, melting pt. 725 °C) belongs to the alkaline-earth group of elements, has a II oxidation state, and geochemically resembles Ca and Sr. This element is used in the form of Ba compounds primarily as barite ($BaSO_4$), which in turn is produced from high-grade ore (75 to 89%), often in association with granite and shale. The main exporters of Ba are China, Morocco, and Ireland, while the United States, the United Kingdom, and Germany are the main importers (Moore, 1991).

Most of commercial barite is used extensively as a weighting agent in oil- and gas-drilling fluids. Barite is circulated down in the well to lubricate and cool the bit and string, removing cuttings from the hole and controlling formation pressures. The other major uses for barite are in the production of Ba chemicals and in glass, paint, and rubber industries.

At 300 to 500 ppm in the earth's crust, it ranks 14th among the elements in order of abundance (Krauskopf, 1979). Bowen (1979) reported that in soils, Ba concentrations range from 100 to 3000 ppm, the median being about 500 ppm (Appendix Table A.24). Surficial soils in the United States have a geometric mean of 440 ppm and a range of 10 to 5000 ppm (Shacklette and Boerngen, 1984). Barium contents of 80 to 1700 ppm for soils of France and Italy were reported (Bertrand and Silberstein, 1928). For Russian soils, Vinogradov (1959) reported a range of 100 to 1500 ppm. Soils near barite deposits may contain up to 3.7% Ba.

Barium can accumulate in Mn oxides in soil; thus, Ba content of sedimentary manganese deposits is usually high (Peterson and Girling, 1981). Similarly, binding of Ba by ferromanganese nodules in the ocean's bottom is considered one of the major sinks for this element (Van der Weijden and Kruissink, 1977). In soil minerals as in micas and feldspars, Ba can substitute for K in the lattice structure.

Because it can be spilled and spread over sediments arising from drilling-related discharges (Phillips et al., 1998), substantial amounts of Ba can be dissolved and released to the water column under a combination of low pH and highly anaerobic conditions (Carbonell et al., 1999). For example, using controlled microcosm technique, an order of magnitude higher release of Ba from Mississippi River sediments was noted under acidic and anaerobic conditions than when under alkaline and anaerobic or aerobic conditions (i.e., 4.4 vs. 0.30% of total native Ba for acidic and alkaline conditions, respectively).

Early interest in Ba in plant nutrition was spurred by the known toxicity of soluble Ba salts to animals. Barium was believed to be taken up by some pasture plants known as locoweeds (*Astragalus* spp.) which can be lethal to cattle (Vanselow, 1965a). As it turned out, it was probably the Se and certain alkaloids in these plants that caused the toxicity. Norrish (1975) noted that barite is soluble enough to be taken up by plants growing on Ba-rich soils and accumulates at levels potentially toxic to animals. Other than this special scenario, Ba toxicity to animals from ingestion of plant-enriched Ba is unlikely.

Vanselow (1965a) indicates that vegetative tissues of most agronomic crops contain Ba in the range of 10 to 100 ppm, with some values approaching 2000 ppm. The locoweeds contained 100 to 290 ppm Ba. Because of the chemical similarities between Ba, Sr, and Ca, competition among these ions for plant uptake is likely. In solution culture, Ca decreased both Sr and Ba uptake by bush beans, with Ba being preferentially taken up over Sr (Wallace and Romney, 1971). Indeed, Menzel (1954) showed that uptake of Sr and Ba was inversely proportional to exchangeable soil Ca.

Under experimental conditions, Ba toxicity to bush beans and barley was induced when the soil was spiked with 2000 ppm Ba as $Ba(NO_3)_2$, resulting in 1 to 2% Ba levels in leaf tissue (Chaudry et al., 1977). Applications of lime or sulfur to soil tend to immobilize Ba due to formation of sparingly soluble $BaCO_3$ and $BaSO_4$, respectively.

Although Ba is passed through the food chain, evidence suggests that it is not concentrated in either aquatic or terrestrial organisms and should generally be considered to have low toxicity (Hope et al., 1996; Moore, 1991).

3 Beryllium

The silvery gray metal (atom. no. 4, atom. wt. 9.01, specific gravity 1.85, melting pt. 1278 ± 5 °C) is a very light metal but has a very high melting point. In nature, it occurs in the II oxidation state. The most important commercial sources of the element and its compounds are beryl ($3BeO \cdot Al_2O_3 \cdot 6SiO_2$) and bertrandite ($4BeO \cdot 2SiO_2 \cdot H_2O$). Other Be primary minerals are chrysoberyl ($BeAl_2O_4$) and phenacite (Be_2SiO_4). Beryllium and its salts are fairly toxic to humans, resulting in both acute and chronic effects.

Beryllium's major uses include: as a hardening agent in alloys, primarily as Be–Cu, Be–Zn, and Be–Ni; as a neutron moderator in nuclear reactors and cladding material for uranium (U); in aircraft brakes; in inertial guidance systems in missiles and aircraft; and in structural components of space vehicles. The United States and Russia are the major users of the metal—a reflection of the high demand from the reactor, weapons, and aerospace industries.

The average crustal abundance of Be (45th among elements) is estimated at around 3 ppm (Krauskopf, 1979), primarily in association with silicate minerals rather than sulfides; thus, it is ubiquitous in nature but in only trace amounts. However, values as high as 10 ppm may occur infrequently. Mitchell (1971) noted that for some typical Scottish surface soils, Be levels are usually below 5 ppm for soils derived from sandstone, shale, andesite, olivine-gabbro, serpentine, and granite. Soils derived from trachyte, mica schist, and quartzite may have slightly higher Be contents. For U.S. surface soils, Be ranged from <1 to 15 ppm (mean = 0.63 ppm) (Shacklette and Boerngen, 1984); for Russian surface soils, it ranged from 1.2 to 13 ppm (Bakulin and Mokiyenko, 1966); and for Canadian surface soils, it ranged from 0.10 to 0.89 ppm (Frank et al., 1979). Bowen (1979) reported a median value of 0.30 ppm Be for world soils (Appendix Table A.24).

The primary environmental source of Be is coal combustion. Estimates of total Be emissions in the United States are around 193 tonnes yr^{-1}, of which 180 come from coal (Moore, 1991). Coal-burning discharges of Be into aquatic systems are considered major sources of Be contamination of surface waters. Beryllium in coal, where it was shown to be primarily bound to the OM of the coal, varies from <0.40 to 8 ppm (mean = 1.5 ppm) in Australian bituminous coal (Swaine, 1979), and from 0.40 to 3 ppm (mean = 1.8 ppm) in British coal (Lim, 1979). Coal ashes have been known to contain as much as 100 ppm Be.

In soils, Be occurs as a divalent cation Be^{2+}; in alkaline soils it can be expected to occur as BeO (Fig. 19.1). Since Be readily complexes with organic compounds, it may accumulate in the surface layer of the soil profile and remain relatively unavailable for plant uptake. However, its inorganic complexes such as $BeCl_2$ and $BeSO_4$ can be fairly mobile and therefore readily available for plant uptake.

Beryllium data for plants are meager. Its typical concentrations in plants are at levels >0.01 and <2 ppm in ashes of plant species collected over nonmineralized areas (Peterson and Girling, 1981). Accumulator species such as *Vaccinium myrtillus, Vicia sylvatica, Aconitum excelsum,* and *Calamagrostis arundinacea* however, may accumulate Be exceeding 10 ppm levels. Newland (1982) reported that hickory is an accumulator of Be, containing as much as 1 ppm DW. Most results indicate that leaves contain more Be than either twigs or fruits, although some desert species contain more Be in the twigs.

Although very phytotoxic, availability of Be in low amounts may have beneficial effects on plants. These effects were manifested as a stimulation in growth of *Nicotiana tabacum* grown in nutrient culture containing 1 ppm Be as $BeCl_2$ (Tso et al., 1973), as improved growth of timothy plants grown in sandy loam soil spiked with 10 ppm Be (Ruhland, 1958), and in activation

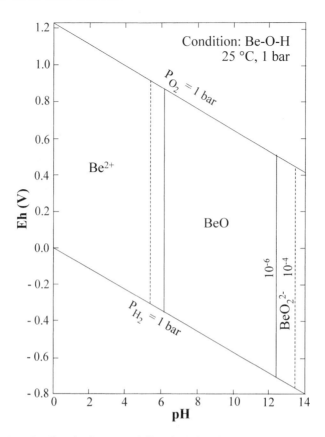

FIGURE 19.1. Predicted Eh–pH stability field for beryllium. Assumed activity of dissolved Be $= 10^{-6,-4}$. (From Brookins, 1988.)

of certain enzymes such as acid phosphatase and inorganic pyrophosphatase, but Be seems to inhibit alkaline phosphatase (Horovitz and Petrescu, 1964). The inhibition of the last enzyme was shown to be due to Be substitution for Mg (Tepper, 1980).

Soluble Be has been shown to reduce the growth of various crops at concentrations greater than 2 ppm in nutrient solution or at concentrations equivalent to 2 to 4% of CEC of the soil (Romney and Childress, 1965). Beryllium in plants accumulates primarily in roots, and upon translocation, has a tendency to accumulate more in leaves than in fruits and stems. This metal apparently interferes in plant nutrition, not only through inhibition of certain enzymes, but also through antagonism of Ca, Mg, and Mn nutrition. Since Be phytotoxicity can be prevalent only in acidic soils, liming the soil would render this element immobile (Williams and Riche, 1968).

The U.S. Environmental Protection Agency (1976) has issued several water quality criteria for Be, including limiting Be content to 100 ppb in water intended for continuous irrigation of soils, except that for neutral to alkaline fine-textured soils, the limit can be as high as 500 ppb.

Bioaccumulation of Be in aquatic organisms does occur, but does not biomagnify in trophic levels (Moore, 1991). Thus, it does not appear to be a concern in the consumption of fish and other seafood products. Because of the known toxicological nature of Be and the likelihood of its increased releases into the environment with intensification of coal combustion, this metal, along with T1 and V, should be viewed as potential problem metals of the future.

4 Cobalt

Cobalt (atom. no. 27, atom. wt. 58.93, specific gravity 8.9) is a silvery white metal with a pink or blue tinge when polished. It is resistant to corrosion and to alkalis but is soluble in acids. It is harder but more brittle than Fe or Ni. It is geochemically similar to Ni and exhibits an oxidation state of II or III.

Cobalt occurs in the minerals cobaltite, smaltite, and erythrite, and is often associated with Ag, Ni, Pb, Cu, and Fe ores from which it is commonly obtained as a by-product. The major ore deposits of Co occur as sulfides. Large deposits are found in Zaire, Morocco, and Canada.

Cobalt is used in the production of high-grade steels, alloys, superalloys, and magnetic alloys. More than 75% of the world's production of Co is used in the manufacture of alloys. It is also used in smaller quantities as a drying agent in paints, varnishes, enamels, and inks; as a pigment; and as a glass decolorizer. It is also an important source of catalysts in the petroleum industry.

Values for crustal abundance of Co range from 20 ppm (Bowen, 1979) to 27 ppm (Carr and Turekian, 1961); Krauskopf (1979) gave 22 ppm, ranking Co 30th among the elements. Soils developed from ultrabasic rocks are usually enriched in Co, Ni, and Cr (Anderson et al., 1973). Mitchell (1945) noted that soils derived from basic igneous rocks or argillaceous sediments may contain 20 to 100 ppm total Co, whereas soils derived from sandstones, limestones, and acid igneous rocks may have <20 ppm total Co.

Cobalt occurs in high amounts in ultrabasic rocks where it is associated with olivine minerals. Serpentine soils in New Zealand contain up to 1000 ppm Co with mean values around 400 ppm (Lyon et al., 1968, 1971). Kidson (1938) reported that Co content of a wide variety of soils is related to Mg content of their parent rocks, e.g., serpentines that are enriched in Mg give soils high in Co whereas soils derived from granite have low Co content. Total Co concentration of world soils has been summarized by Young (1979) and Proctor and Woodell (1975) showing that serpentine soils generally contain 100 ppm or more Co. Total Co in soils may range from about 0.30 ppm in severely Co-deficient areas to 1000 ppm in mineralized areas; but most soils have Co levels within a range of 2 to 40 ppm (Young, 1979).

In the conterminous United States, mean total Co content of soils and other surficial materials is 6.7 ppm (Shacklette and Boerngen, 1984); for world soils, the median total content appears to be around 8 ppm (Appendix Table A.24); the average for Asian paddy soils was 56 ppm (Domingo and Kyuma, 1983). A mean value of 4.7 ppm Co was reported as background value for soils of Poland that are derived from sedimentary rocks of glacial origin (Czarnowska and Gworek, 1990).

There is a narrow range for extractable Co determined by common soil extractants. In Co-deficient farmlands of Scotland, 2.5% HOAc generally extracted <0.25 ppm Co, whereas for fertile soils, values were above 0.30 ppm (Mitchell, 1945). Alban and Kubota (1960) found an increase in Co content of black gum (*Nyssa sylvatica*) leaves with HOAc-extractable Co. In general, extractable Co (as a percentage of total Co) shows a very wide range of 1 to 93% but is usually between 3 and 20% (Young, 1979). In terms of concentration, extractable Co varies from about 0.01 to 6.8 ppm but is usually in the range of 0.10 to 2 ppm.

Using ^{60}Co, Gille and Graham (1971) observed that 0.1 N HCl solution would give a good estimate of the quantity of Co in soil and that 0.1 M CaCl$_2$ would give the best indication of potential concentration of Co in soil solution.

In typical soils of the Hawaiian Islands, Co content is highest in A horizon and appears to be correlated with Fe oxide content of soil (Fujimoto and Sherman, 1950). In the eastern United States, higher Co was dithionite-extractable from A rather than from B horizon of red-yellow Podzols (Kubota, 1965).

Several factors are known to influence the sorption of Co by soils. Graham and Killion (1962) found that uptake of Co by several plants from soil constituent was in the order: illite > kaolinite > Putnam clay > sedimentary peat > montmorillonite. The peats were found to adsorb more Co than any of the clays. Using pot and field experiments, Adams et al. (1969) found that uptake of Co by clover was inversely related to total Mn content of soil. Tiller et al. (1969) found that the Co sorption capacity of soils was highly correlated with Co content and surface area and to a lesser extent with Mn and clay contents and pH. Taylor and McKenzie (1966) have shown that in some cases almost all of the Co in soils could be accounted for by that present in Mn minerals, indicating that these particular minerals can be an important sink for Co in soil. Sorption of Co by Fe and Mn oxides as a function of pH is shown in Figure 19.2. Cryptomelane (K$_2$Mn$_8$O$_{16}$), with a point zero charge below 3 and a high surface area (i.e., 200 m^2 g^{-1}) sorbed significant amounts of Co even at relatively low pH. At the other extreme, goethite, which had a relatively small surface area (i.e., 90 m^2 g^{-1}) and a point zero charge of 8.7, only showed significant sorption of Co at pH values above 6.0. Hodgson (1960) identified two forms of bound Co in montmorillonite. One form, characterized as being slowly dissociable, appears to be bound in a monolayer by chemisorption and would exchange

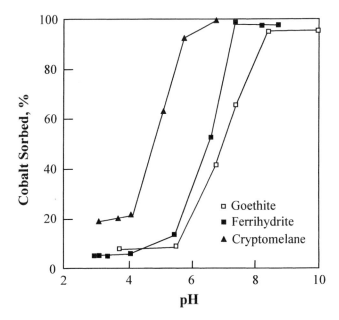

FIGURE 19.2. Sorption of Co(II) onto Fe and Mn oxides as a function of pH (initial Co concentration = 5×10^{-4} M). (Modified from Backes et al., 1995.)

with Zn^{2+}, Cu^{2+}, or other Co^{2+} ions but not with a Ca^{2+}, Mg^{2+}, or NH_4^+ ions. A second form of Co is not dissociable and is believed to either enter the crystal lattice or become occluded in the precipitates of another phase.

Cobalt has not been demonstrated to be essential for the growth of higher plants, although it has been shown to be required by certain blue-green algae. Cobalt is essential for N_2 fixation by free-living bacteria, blue-green algae, and symbiotic systems (Palit et al., 1994; Reisenauer, 1960; Welch et al., 1991). For example, Co is required by *Rhizobium,* the symbiotic bacterium that fixes N_2, in root nodules of legumes. In higher plants, Co supplements have been reported to increase growth of rubber plants and tomatoes and to have increased the elongation of pea stems.

Although Co is not essential to plants, its level in plant tissue is of general concern because of its essentiality in animal nutrition. It is a constituent of vitamin B_{12}, which is required by all animals (Miller et al., 1991). Low levels of Co in feedstuff can cause nutritional diseases in sheep and cattle known as bush sickness in New Zealand, coast disease or wasting disease in Australia, pining in United Kingdom, nakumitis in Kenya, and Lecksucht in Germany and the Netherlands. Young (1979) indicated that a level of about 0.07 ppm of Co in feedstuff is essential to maintain the normal health of animals, although the critical level varies slightly among countries and among animals. Ruminants consuming diets containing <0.07 to 0.11 ppm Co DW show loss of appetite, reduced growth, or loss in body weight, followed by extreme emaciation, listlessness, anemia, and eventually, death (Miller et al., 1991). Tissues of higher plants are generally reported to contain <1 ppm Co on dry weight basis. Leafy plants, such as lettuce, cabbage, and spinach have a relatively high Co content, whereas Co is low in grasses and cereals (Kipling, 1980). Compared with grasses, legumes take up more Co and are more definitive of the Co status of the soil (Kubota, 1980).

As a general rule, plants growing on soils having low Co concentrations would have low Co contents. For example, in the eastern United States, coarse-textured Podzols have been associated with Co deficiency of ruminants (Kubota, 1965). Cobalt deficiency in ruminants most commonly occurs on acidic, highly leached sandy soils; on soils derived from granites; and on some highly calcareous soils and some peaty soils (Peterson and Girling, 1981). Plants growing on such soils generally contain <0.07 to 0.08 ppm Co and serious deficiency conditions commonly occur below 0.04 ppm.

In United States, Kubota (1980) has delineated a geographic pattern of low and adequate Co areas. The lower Atlantic Coastal Plain has fairly low Co contents. Problem areas abound in sandy Podzols that have 1 ppm or less total Co, with grasses and legumes grown on these soils not having quite enough Co to meet animal needs. The glaciated region of the northeast is also characteristically low in total Co—legumes containing <0.07 ppm Co were found in some areas. In United Kingdom, low Co contents can be found in soils derived from sandstones, sands, limestones, certain shales, and acid igneous rocks in western and northern areas of England and Wales (Thornton and Webb, 1980).

Some plants appear to accumulate Co and several hyperaccumulator species have been reported to inhabit mineralized areas (Malaisse et al., 1979). The hyperaccumulators could contain Co in their tissues in excess of 1000 ppm. Cobalt concentrations over 20 ppm in wheat grasses (*Agropyron* spp.) in northern Nevada ranges have been reported, although they are not known accumulators of Co (Lambert and Blincoe, 1971).

As with most metals, Co in plants is mostly associated with roots but it can be translocated to foliage. Patel et al. (1976) noted that Co concentrations in leaves and stems of chrysanthemum significantly increased with increasing levels of Co in nutrient solution and caused reduction in levels of Fe, Mn, and Zn in plant tissue. Toxicity symptoms due to excess Co include the typically Fe deficiency–induced chlorosis and necrosis. In solution culture, small amounts of Co, sometimes as low as 0.10 ppm, can produce adverse effects on many crops (Vanselow, 1965b).

Several factors can influence the availability of Co to plants. The following soil types are prone to produce Co deficiency in plants (Vanselow, 1965b): highly leached, acid, sandy soils; soils derived from granites; some highly calcareous soils; Atlantic Coastal Plain soils; and some peaty soils. In addition, the following soil management and other practices can affect Co status of soils: (1) cobalt content of soils, in general, is related to Mg content of the parent rocks; common fertilizer treatments over many years should have no effect on Co status; (2) soil acidification increases the availability of Co as Co^{2+} (Fig. 19.3); (3) liming soils can reduce uptake of Co by plants, whereas application of gypsum to sandy loam could increase Co uptake; and (4) leaching of soils reduces available Co.

Several investigators (Filipovic et al., 1961; Price et al., 1955; Wright and Lawton, 1954) have reported decreased uptake of Co by plants with increasing soil pH. For example, lime appears to have a depressing effect on plant uptake of Co (Adriano et al., 1977; Askew and Dixon, 1937). Prokhorov et al. (1979) noted that of all properties they studied, soil pH had the greatest effect on Co sorption by soil. The dramatic inverse relationship of plant uptake of Co and soil pH is demonstrated in Figure 19.4.

Various investigators found that Co accumulates in hydrous oxides of Fe and Mn in soils. Kubota (1965) indicated that Co was associated with Fe oxides; similarly, Mn oxides play a key role in the chemistry of soil Co (McKenzie, 1975, 1978). However, Hodgson et al. (1969) found that

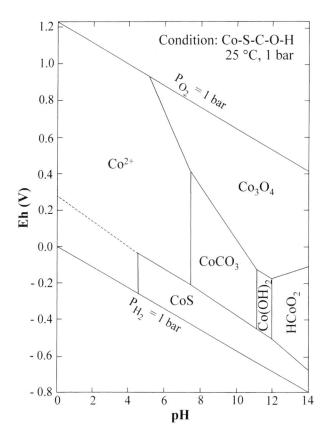

FIGURE 19.3. Predicted Eh–pH stability field for cobalt. Assumed activities for dissolved species are: $Co = 10^{-6}$, $C = 10^{-3}$, $S = 10^{-3}$. (From Brookins, 1988.)

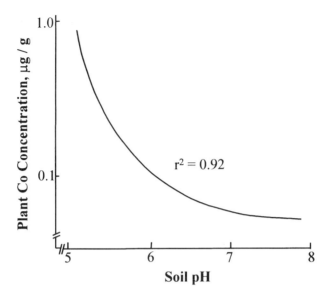

FIGURE 19.4. Relationship between cobalt concentration in ryegrass and soil pH. (Modified from McLaren et al., 1987.)

adsorption of Co by certain soils was increased by removal of Fe from the soil. In that case, Fe removal exposed clay mineral surfaces that were more reactive than previously exposed Fe oxide surfaces.

The following factors can affect the Co status of plants: soil properties, including levels of Mn and Fe; soil pH and moisture; and plant factors such as stage of growth, plant species, and type of organ (Peterson and Girling, 1981). Cobalt is relatively nontoxic to animals and man. A deficiency of Co is of far greater concern than potential toxic levels in plants.

Analyses of water samples from the Aare and Rhine rivers and the Biel and Lucerne lakes, Switzerland indicate total dissolved Co concentrations are in the range of 0.5 to 6.5 nM and Co^{2+} is in the range of 0.05 to 0.5 nM; organic complexes of Co are predominant in most samples with 80 to 96% of total dissolved Co (Qian et al., 1998).

5 Fluorine

Fluorine (atom. no. 9, atom. wt. 18.99) is the most electronegative and chemically reactive of all halides. It is a pale yellow, corrosive gas that reacts with practically all organic and inorganic substances. In nature, it does not exist in the free state but always forms compounds in an oxidized state, −I.

Fluorine occurs mainly in the silicate minerals of the earth's crust at a concentration of about 650 ppm and ranks 13th among the elements in crustal abundance. The most important F-bearing minerals are fluorspar (CaF_2), fluorapatite [$Ca_5(PO_4)_3F$], and cryolite (Na_3AlF_6). The largest known deposits of fluorite (fluorspar or calcium fluoride) are located in the United States, England, and Germany.

The largest use of calcium (or sodium) fluoride is in steel production to prolong the fluidity of the ingot, thus facilitating the escape of undesirable gaseous products. Fluorides are also used as a flux in smelting of Ni, Cu, Au, and Ag, as rodenticides, and as catalysts for certain organic reactions.

Fluorapatites are used industrially for preparation of phosphoric acid, fertilizers, and mineral feeds. Cryolite is used as an insecticide and in molten form, as an electrolyte in the production of Al from alumina. Fluorine and its compounds are used in producing U (from the hexafluoride form, UF_6) for the nuclear industry, high-temperature plastics, air conditioning, and refrigeration. It is also used in electroplating, wood preservation, paper production, and manufacturing of petroleum products.

Fluorine is ubiquitous in the environment and is always present in plants, soils, and phosphatic fertilizers, on the order of 3×10^0, 3×10^2, and 3×10^4 ppm F, respectively (Larsen and Widdowson, 1971). Fluorine is a common constituent of rocks and soils. It is an essential element in the following minerals: fluorite, apatite, cryolite, topaz, phlogophite, and lepidolite.

Very common soil minerals such as biotite, muscovite, and hornblende, may contain as much as several percent of F and therefore would seem to be the main source of F in soils. However, Steinkoenig (1919) indicated that micas, apatite, and tourmaline in the parent materials were the original sources of F in soils. It appears therefore, that the F content of soils is largely dependent on mineralogical composition of soil's inorganic fraction.

Total F concentrations in U.S. soils range from <100 ppm to over 1000 ppm (NAS, 1971). For world soils, the median value reported ranged from 200 to 300 ppm, whereas an average of 430 ppm was reported for U.S. surface soils (Appendix Table A.24). The total F concentration in normal arable soils ranges from 150 to 360 ppm, but can reach up to 620 ppm (McLaughlin et al., 1996). The F contents of 30 soil profiles ($n = 137$) of representative soils of varied texture, parent material, and geographic distribution varied from trace amounts to 7070 ppm for an unusual Tennessee soil containing phosphate rocks (Robinson and Edgington, 1946). The average for surface samples of these soils was 292 ppm. In general, F was concentrated in colloidal material and increased with depth of soil.

Nömmik (1953) observed that natural F content of soil increases with increasing depth, and that only 5 to 10% of total F content of soil is water-soluble. Under natural conditions, the concentrations of F in soil solutions are usually below 1 ppm, but under severely F-polluted soils, they may reach levels of about 10 ppm (Polomski et al., 1982). In soils in humid temperate climate, F could be readily lost from minerals in acid horizons (Omueti and Jones, 1980). A substantial amount of this F is retained in subsoil horizons, where it is complexed with Al that is most likely associated with phyllosilicates. The extent of F retention depends on the amount of clay and pH; therefore, the pattern of F distribution with depth follows the clay pattern closely. Fluorine occurs only in trivial amounts in the OM fraction of these soils (Omueti and Jones, 1980).

Several investigators observed that solubility of F in soils is highly variable and has the tendency to be higher at pH below 5 and above 6 (Fig. 19.5). The solubility of F tends to be lowest in the pH range of 5 to 6.5, which coincides with the greatest F sorption (Arnesen and Krogstad, 1998; Wenzel and Blum, 1992). Fluorine solubility in soil is complex and may be controlled by solid phases even more insoluble than CaF_2. In addition, F solubility may be related to solubility of Al or other ionic species with which it form complexes. At lower pH, sorption declines due to formation of soluble Al–F species, such as $(AlF)^{2+}$ and $(AlF_2)^+$ complexes (Barrow and Ellis, 1986; Wenzel and Blum, 1992). This may explain positive correlations in Al and F contents in white clover and ryegrass grown in F-contaminated soils (Arnesen, 1997). Soils having high pH and low levels of amorphous Al species, clay surface area, and OM generally sorb little F (Kau et al., 1997; Omueti and Jones, 1977a). In alkaline pH, the increased negative surface charge results in the repulsion of anionic F. Thus, it appears that the predominant retention mechanism is that of F exchange with OH group of amorphous materials, such as Al-hydroxides. In this case, the crystal lattice OH of clay minerals is replaced by F, resulting in a simultaneous release of Al and Fe. Another F retention mechanism involves F precipitation as CaF_2, as in calcareous soils (Slavek et al., 1984).

Fluorine occurs in soils as the singly charged fluoride ion, F^-, or occasionally as a component of such complex anions as $(BF_4)^-$, $(AlF_6)^{3-}$, and $(SiF_6)^{2-}$ (Hopkins, 1977). At neutral to alkaline pH, F exists predominantly as the free F ion (F^-), and at pH <5.5, F^- is complexed with Al (Wenzel and Blum, 1992). Indeed, F^- in soil solution may induce enhanced leaching of DOC, Al, and metals from soils (Totsche et al., 2000). The F^- increased the solubility of soil OM, which in turn facilitated the formation of organo-metal complexes. Likewise, fluoro complexation of Al occurred.

Fluorine is not an essential plant element but is essential for animals. However, continuous ingestion by animals of excessive amounts of F can lead to the disorder *fluorosis,* and suboptimal levels in the diet may have an equally damaging effect. Therefore, plant content of F is of interest to livestock producers. In endemic *fluorosis* areas, the main F source is either water or dust-contaminated feedstuff (including pastures), or both (Underwood, 1977). In other areas, the main source of F is naturally fluoridated drinking water derived from deep wells, which commonly contain 3 to 5 ppm F, and sometimes as high as 10 to 15 ppm F. The chronic *fluorosis* of sheep, horses, and cattle locally known as darmous that occurs in parts of North Africa is caused by contamination of water supplies and feedstuff with dusts originating from rock phosphate deposits and mines (Underwood, 1977).

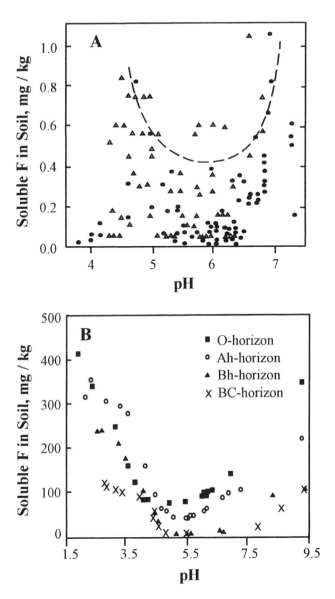

FIGURE 19.5. Relationships of soluble fluoride and pH. (A) F was extracted with 0.01 M CaCl$_2$. Dots represent the data of Larsen and Widdowson (1971); triangles are from the data of Gilpin and Johnson (1980); dashed line represents the proposed upper limit of F solubility. (B) F concentration in suspensions of contaminated soils after adjusting pH with either HCl or NaOH (Arnesen and Krogstad, 1998.)

The maximum recommended limit for F content in hay and pasture grass in Norway is 30 ppm (Arnesen, 1997).

Small amounts of F are often added to toothpaste or municipal water systems to prevent dental caries in humans. However, excessive exposure to F may cause weakening of bones and skeletal deformities in humans, livestock, and wildlife. Surface waters such as those used for drinking and cooking in most areas generally contain <1 ppm F, or even 0.10 ppm or less (Underwood, 1977). In regions where water comes from deep wells or artesian bores that are inherently high in F, *fluorosis* in humans has occurred as in southern India and South Africa, where concentrations of F as high as 20 to 40 ppm have been reported.

Background concentrations of F in plants are usually <10 ppm (Arnesen, 1997). Kumpulainen and Koivistoinen (1977) reported that cereals usually contain <1 ppm F, where F tends to accumulate in the outer layer of the grain and in embryo. The F contents of both leafy and root vegetables do not differ appreciably from those of cereals, with the exception of spinach, which is unusually enriched in F. Potato peelings can contain as much as 75% of total F in the whole tuber. Tea is one of the most F-enriched beverages with about two thirds of the F in tea leaves being soluble in the beverage itself.

Brewer (1965) reported that availability of soil F to plants is controlled by the following factors: pH, soil type and amount of clay, Ca, and phosphate. Liming acid soils to around pH 6 to 6.5 would decrease the levels of soluble F compounds present in or added to soil. The use of phosphate fertilizers having low F concentrations can also be beneficial as it can reduce F toxicity presumably through ion competition, at low pH values as long as the F level does not exceed 180 ppm.

Several anthropogenic sources of F can enrich soils with this element. Phosphatic fertilizers, especially the superphosphates, are the single most important source of F in arable lands. Rock phosphates generally contain around 3.5% F; phosphatic fertilizers commonly contain between 1.5 and 3.0% F (McLaughlin et al., 1996). Repeated applications of rock phosphates containing several percent of F was shown to significantly elevate the F content of soils (Omueti and Jones, 1977b). Typical additions of phosphate fertilizers (50 to 100 kg P_2O_5 ha^{-1} yr^{-1}) could elevate soil F content by 5 to 10 ppm yr^{-1} (Gilpin and Johnson, 1980). Other sources that may affect soil status of F are precipitation, fallout of combustion products of coal and other industrially polluted air, and F-bearing insecticides (Adriano and Doner, 1982).

The effect of F-bearing fertilizers on the F content of plants is usually insignificant. However, plants and soils growing in the vicinity of industries such as aluminum smelting can have substantially increased F contents. For example, Arnesen et al. (1995) found that F pollution of soils can be traced to more than 30 km from one of the Al smelters in Norway.

The atmospheric release of F by industries is of ecological importance. The steel industry is a major source of atmospheric F in the United States and the third largest source in Canada (NRCC, 1977). The primary aluminum production industry is the third largest source of atmospheric F in the United States and the largest source in Canada.

Fluoride can be toxic in plant and animal nutrition. Chronic emission of F from Al smelters can have drastic ecological effects, especially in adjacent areas. Plants take up F mostly through stomates in the form of HF gas, but it can also be taken up through the roots (Arnesen and Krogstad, 1998). Fluoride may decrease the microbial activity in soil (Van Wensem and Adena, 1991) and may also bioaccumulate in isopods and earthworms (Walton, 1987).

Petroleum refining produces a variety of wastes including HF, which is used as a catalyst in gasoline production. Landfarming is frequently used by petroleum refineries as an economical means of disposing of oily sludges that may contain F (Schroder et al., 1999). Because F can accumulate in landfarmed areas, it can be an exposure source of elevated F that may cause *fluorosis* in wildlife. Indeed, 80% of cotton rats (*Sigmodon hispidus*) collected from a landfarm developed dental *fluorosis* (Schroder et al., 1999). The mean bone F (1515 ppm) and mean total soil F (1954 ppm) from the landfarm were substantially greater than the bone F (121 ppm) and total soil F (121 ppm) from a control site. Similarly, elevated levels of F were detected in bones of game wildlife inhabiting the vicinity of Al smelters in Norway (Arnesen, 1997).

6 Silver

The lustrous, brilliant white metal (atom. no. 47, atom. wt. 107.87, specific gravity 10.5, melting pt. 962 °C) is very ductile and malleable. Metallic Ag occurs naturally and in ores such as argentite (Ag_2S) and hornsilver (AgCl); Pb, Pb–Zn, Cu, Ag, and Cu–Ni ores are also important commercial sources. Silver is also recovered during the electrolytic refining of Cu and smelting of Ni ores. The world's leading mine producers are Mexico, Peru, the former Soviet Union, the United States, and Canada.

According to the Silver Institute's World Silver Survey (Silver Institute, 1995) photographic manufacturing accounted for slightly over 50% of U.S. industrial demand. The photographic uses of Ag in United States, Japan, and western Europe together accounted for ~40% of the world's total Ag demand in 1990. The next largest user category was electrical contacts and conductors (15% of total demand), followed by brazing alloys and solders (7%), catalysts (3%), batteries (3%), sterlingware (3%), jewelry (3%), silverplate (3%), mirrors (1%), and miscellaneous uses (8%).

Silver is found in nature in four oxidation states: 0, I, II and III, with 0 and I states being the most common; II and III species are seldom detected naturally in the environment.

Silver occurs as a minor constituent of the earth's crust with an average of about 0.10 ppm (Purcell and Peters, 1998). In soils, Ag is found primarily as sulfides in association with Fe, Pb, Au, or tellurides. Smith and Carson (1977b) reported that the usual range of Ag in normal soils is <0.10 to 1 ppm, with unusual values exceeding 5 ppm. For the conterminous United States, Shacklette and Boerngen (1984) found that surficial mineral averaged 0.70 ppm Ag, while soils high in OM had Ag content in the range of 2 to 5 ppm. Mitchell (1944) reported up to 2 ppm Ag in some Scottish soils. The Ag content of the surface soils in arable areas of Ontario, Canada had a mean of 0.44 ppm Ag (range of 0.04 to 1.81 ppm) (Frank et al., 1979). In some southern California soils, Vanselow (1965c) found Ag in the range of 0.20 to 0.70 ppm.

Soil data collected from noncontaminated sites and for areas contaminated by fluvial and aerial processes from derelict 19th century mine operations in the United Kingdom indicate that soils derived from black shales (usually 0.50 ppm Ag), or high in OM (0.50 ppm) are indigenously richer in Ag than the sandstones (0.05 ppm) or limestone-derived (0.07 ppm) soils (Jones et al., 1984). Contaminated soils in these areas may contain up to 10 ppm Ag, while mine spoils may contain Ag in the range of 4 to 65 ppm.

The typical level of Ag in plant tissue is <1 ppm (Smith and Carson, 1977b; Vanselow, 1965c). Silver accumulates in plant roots and is poorly translocated to the shoot (Teller and Klein, 1974; Ward et al., 1979). When added to soil as $AgNO_3$ at 75 ppm level, only 0.50 ppm Ag accumulated in the leaves of citrus plants (Vanselow, 1965c). Soil pH does not influence much the plant uptake of Ag.

Peterson et al. (1976) found that Ag was not appreciably translocated to the shoot of cucumber, tomato, and wheat plants grown in nutrient solution containing 10 μM $AgNO_3$. Suzuki (1958) found only trace amounts of Ag taken up by the mint plant. Based on experiments with wheat, maize, and soybeans grown on sandy and loamy soils spiked with AgI and $AgNO_3$, even large amounts of Ag as AgI in soils had practically no effect on plant growth (Weaver and Klarich, 1973). Silver from $AgNO_3$ was apparently more available to plants than Ag from AgI (Teller and Klein, 1974).

Under field conditions, Horovitz et al. (1974) found a range of 0.01 to 16 ppm Ag in the dry matter of 35 plant species, with lower plant species having higher Ag contents. Klein (1979) concluded that Ag tends to accumulate mainly in roots, and that this element is excluded in plant foliage even when present in soil at high concentrations. Ward et al. (1979) found that as much as 90% of Ag added to a nutrient solution can be immobilized in roots of *Lolium perenne* and *Trifolium repens*.

Data on phytotoxicity of Ag are scarce. Wallace et al. (1977) observed that Ag was very lethal to bush beans grown in solution culture containing 11 ppm Ag as $AgNO_3$; even at 1.1 ppm level, yield was still significantly reduced, but plants exhibited no symptoms and produced Ag concentrations of 5.8, 5.1, and 1760 ppm in leaves, stems, and roots, respectively. Davis et al. (1978) noted that 0.50 ppm of Ag as $AgNO_3$ in nutrient culture produced a critical level of about 4 ppm in foliage tissue of barley. In short, Ag added to soils (e.g., from sewage sludge) should not adversely affect the growth of crops (Hirsch, 1998).

Because of the considerable unavailability of Ag to plants, environmental concern about Ag arises from anthropogenic sources that could enrich aquatic ecosystems. Although there is no evidence that terrestrial animals biomagnify Ag from anthropogenic sources, aquatic species such as fish, mollusks, and crustaceans could be adversely affected. Silver ions are considered one of the most toxic heavy metal ions to microorganisms, particularly the heterotrophic bacteria (Ratte, 1999). However, since Ag^+ easily forms insoluble compounds, it is considered relatively harmless in terrestrial environments.

An estimate of 2×10^6 kg of Ag were released into the environment in 1978 in the United States (Scow et al., 1981). This amount of emission was partitioned into 4, 39, and 68% released to air, discharged to water, and disposed as solid waste onto land, respectively. Of the total amounts of Ag released to water, more than 62% came from natural sources (438×10^3 kg), 17% came directly from photographic uses, 10% from sewage treatment plant discharges, and an additional 10% from urban runoff. Because of the magnitude of Ag releases to aquatic systems, the bioavailability and toxicity of Ag to aquatic life has recently received much attention (Bury et al., 1999; Hogstrand and Wood, 1998; Ratte, 1999; Shaw et al., 1998).

Reported concentrations of Ag^+ in natural waters are very low (<0.20 μg L^{-1}) (Ferguson and Hogstrand, 1998). However, when present in substantial amounts in the free ionic form (i.e., Ag^+), it is one of the most toxic metals to aquatic organisms, with a median lethal concentration (LC_{50}) value for freshwater fish in the range of 6.5 to 70 μg L^{-1} (Fig. 19.6). The value for seawater is much higher at 330 to 2700 μg L^{-1}.

The most important factor governing bioavailability and toxicity of Ag in aquatic organisms is chemical speciation (Bury et al., 1999; Hogstrand and Wood, 1998; Shaw et al., 1998). Ag^+ is very reactive and forms stable complexes with negative binding sites in suspended solids, sediments, as well as dissolved organic particles and inorganic anions (Fig. 19.7). The toxicity of Ag to freshwater fish can be influenced by water hardness, pH, alkalinity, Cl^-, and DOC (Campbell, 1995; Morel and Hering, 1993). Ligands that form Ag complexes, such as Cl^- and DOC have been shown to ameliorate $AgNO_3$ toxicity in rainbow trout (Bury et al., 1999); $AgNO_3$ is a salt that readily dissociates to Ag^+ and NO_3^- in freshwater.

Several investigators noted that dissolved $AgCl_n$ (e.g., $AgCl_{aq}$, $AgCl_2^-$, $AgCl_3^{2-}$, and $AgCl_4^{3-}$) was at least 10 times less toxic to rainbow trout than $AgNO_3$, leading to a conclusion that the concentration of Cl^- in water column is a primary modulator of Ag toxicity (Galvez and Wood, 1997; Hogstrand et al., 1996). Similarly, Ag thiosulfates have been shown to cause low toxicities. In seawater, $AgCl_n$ species are expected to predominate while Ag^+ concentrations should diminish.

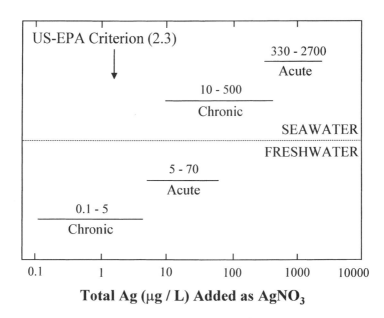

FIGURE 19.6. Toxicity ranges of silver (μg Ag L^{-1}) added as AgNO$_3$ to fish in freshwater and seawater. The EPA criterion is the current U.S. EPA acute water quality criterion for silver in seawater (2.3 μg Ag L^{-1}). For freshwater systems, the EPA limit depends on water hardness. There are no chronic criteria for silver in effect. (Reprinted with permission from Hogstrand and Wood, *Environmental Toxicology and Chemistry* 17:547–561. Copyright © 1998 SETAC, Pensacola, Florida, USA.)

The WHO has not recommended any guideline for Ag in drinking water (Table 19.1). Several nations use a provisional drinking water standard of 0.05 mg Ag L^{-1} to prevent argyria. However, Ag is relatively nontoxic to humans and is normally found at low concentrations in drinking water (Moore, 1991).

7 Thallium

Thallium (atom. no. 81, atom. wt. 204.4, specific gravity 11.85, melting pt. 303 °C) is a gray-white, soft, ductile metal. Although Tl does not occur in the free state in nature, several minerals contain Tl as a major constituent (16 to 60%), e.g., crooksite (Cu,Tl, Ag)$_2$-Se, lorandite (TlAsS$_2$), and hutchinsonite (Pb,Tl)$_2$(Cu, Ag)As$_5$S$_{10}$. However, there are no known commercial deposits (Howe and Smith, 1970). Most ores containing relatively high levels of Tl occur in Switzerland, Yugoslavia, and central Asia (Smith and Carson, 1977a).

In nature, the primary oxidation state of Tl is I (thallous species), and Tl occurs in soil via its isomorphous substitution with K$^+$ in feldspars and silicates. Although it can also occur in III oxidation state (thallic species),

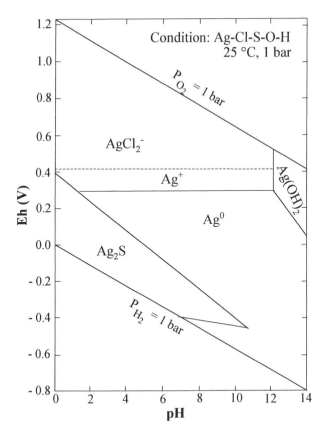

FIGURE 19.7. Predicted Eh–pH stability field for silver. The assumed activities for dissolved species are: $Ag = 10^{-8}$, $S = 10^{-3}$, $Cl = 10^{-3.5}$. (From Brookins, 1988.)

the lower oxidation state may predominate in aqueous systems. In freshwater systems however, Tl(III) could comprise $68 \pm 6\%$ of the total dissolved Tl (Lin and Nriagu, 1999). Its geochemical behavior is similar to (Rb) and other alkali metal cations. It is concentrated in certain sulfides, and deposits characteristically high in As are usually also high in Tl.

Thallium is produced mainly as a by-product of Zn smelting and refining. Many Zn ores contain >2 ppm Tl, which becomes concentrated in flue dusts and residues, and these provide the commercial source for Tl production. Due to the volatile nature of most Tl compounds, the element becomes enriched on the flue dusts during the processing of ores. Practically all of the Tl produced in the world today at an annual rate of 17 tonnes (in 1988) is through Tl extraction of such dusts (Nriagu, 1998).

When Tl use in rodenticides and insecticides was banned in the early 1970s in the United States, its commercial application became rather limited. Today, Tl is used in Pb-, Ag-, or Au-based bearing alloys, which have very

TABLE 19.1. Selected regulations controlling silver in the workplace and environment.[a]

Country	Location	Limit
United States	Workplace: metal dust, soluble	0.010 mg m^{-3a}
	Drinking water	0.10 mg L^{-1} (total)[b]
	Surface water: ambient water quality criteria	2.3 μg L^{-1} (total metal) chronic[c]
	Solid waste	5.0 mg kg^{-1d}
Germany	Workplace	0.010 to 0.10 mg m^{-3}
	Drinking water	0.010 mg L^{-1}
	Sewage treatment discharge	0.10 to 3.0 mg L^{-1}
Canada	Fresh surface water	0.10 μg L^{-1} (total)
	Drinking water	None
Austria	Surface water	1 mg L^{-1}
South Korea	Processing effluents	No high silver discharges: 3 mg L^{-1} in washwaters

[a]PEL, permissible exposure limit; TWA, time-weighted average.
[b]SMCL, secondary maximum contaminant level.
[c]For freshwater systems, the U.S. EPA criterion depends on water hardness according to the equation: max total recoverable Ag (μg L^{-1}) = e (1.72 [ln hardness]$^{-6.52}$).
[d]TCLP, toxicity characteristic leaching procedure.
Source: Modified from Purcell and Peters (1999).

low friction coefficients, high endurance limits, and higher resistance to acids. However, the growing concern about its toxicity is resulting in it being phased out from a number of these alloys. The metal is also employed in the electrical and electronic industry, Hg vapor lamps, and deep-temperature thermometers. In the United States, the semiconductor industry accounts for 60 to 70% of current Tl consumption, the remainder going to the manufacture of highly refractive optical glass and cardiac imaging. It is also used as a catalyst in the synthesis of olefins and hydrocarbons and other organic compound–related reactions.

The crustal abundance of Tl has been estimated at about 0.80 ppm (Nriagu, 1998); the element is widely distributed in the environment. Interest in its environmental significance has been spurred by its known toxicity to living organisms in small amounts, potential biomethylation of Tl in environment, and its release into environment from the combustion of coal and from the cement industry (Nriagu, 1998; Peterson and Girling, 1981).

Combustion of coal appears to be a larger environmental source of Tl than smelting (Kaplan and Mattigod, 1998; Sager, 1998). In coal, Tl occurs as sulfides rather than in complexed form with OM. This explains its low concentrations in anthracite coal. Bituminous coal was reported to contain 0.70 ppm Tl (Voskresenskaya, 1968) and as high as 76 ppm Tl could be present in coal fly ash (Davidson et al., 1976). Smith and Carson (1977a) estimated that in the United States alone, nearly 1000 tonnes of Tl are released into the environment annually in the form of vapor and dusts (300 tonnes), nonferrous metals (60 tonnes), and aqueous and solid wastes (>500 tonnes). In Germany, Tl has been identified as one of nine priority inorganic pollutants in soils (Sager, 1998).

Soil Tl data are minimal even on polluted soils. Bowen (1979) reported a median soil content of 0.20 ppm (0.10 to 0.80 ppm range). The highest value reported for U.S. soils was 5 ppm (Smith and Carson, 1977a). Soils in France have a median Tl value of 0.29 ppm (mean = 1.51 ppm) (Table 19.2). Soils from other countries had Tl contents of 0.46 (Germany), 0.30 (Austria), 0.05 (Switzerland), and 0.58 ppm (China). Soils sampled from the vicinity of a cement plant near Leimen in southwestern Germany had average Tl contents of 3.6, 0.70, and 0.10 ppm in the 0 to 10, 40 to 50, and 60 to 70 cm depths, respectively (Schoer, 1984). This indicates that although Tl is somewhat mobile in soils it does not migrate substantially in the soil profile.

Water samples can be expected to only have trace amounts of Tl, whereas agricultural products can be expected to have Tl concentrations in the ppb to ppm level (Table 19.2 and 19.4). Tl^+ is the predominant Tl species expected to occur in soils and natural water systems (Table 19.3). Although Tl^+ is the predominant species expected to occur in natural surface waters (Fig. 19.8), Tl can also form stable complexes with fulvic acid and chloride as $TlCl^0$. Contrary to thermodynamic prediction that Tl(I) is favored in natural water, samples taken from lakes Michigan, Huron, and Erie had Tl(III) dominating ($68 \pm 6\%$) the total dissolved Tl (Lin and Niagu, 1999).

TABLE 19.2. Thallium contents of selected environmental samples.

Sample	n	Median	Mean	Country
Surface soils (mg kg^{-1} soil)				
	51	0.46	—	Germany
	460	—	0.30	Austria
	369	0.05	—	Switzerland
	853	—	0.58	China
	244	0.29	1.51	France
Water (ng L^{-1})				
Tap water			5.1–24	Poland
River water			5.0–17	Poland
			2.0–75	Germany
Lake water			1.2–26	Canada
Seawater			61–82	Poland/Russia
Agricultural products (ng g^{-1})				
Animal feedstuff			0.02–0.08	
Vegetables (edible part)			0.03–0.20	
Red wine			0.09–0.48	
White wine			0.04–0.30	
Milk			0.01–0.033	
Bream (freshwater fish)			0.8–8.6	

Sources: Primarily from Tremel et al. (1997); Sager (1998).

TABLE 19.3. Predicted distribution of Tl(I) species (% of total) in natural water systems.[a]

Species	Groundwater	River water	Lake water (eutrophic)	Bog water	Seawater
Tl^+	90.4	82.7	76.8	32.4	51.9
$TlHCO_3^0$	4.4	1.2	2.0	—	0.5
$TlCO_3^-$	—	—	—	—	0.1
$TlSO_4^-$	3.6	0.4	0.8	—	11.2
$TlCl^0$	0.1	0.1	0.1	—	30.7
$Tl(Cl)_2^-$	—	—	—	—	5.4
Tl-Fulvate	1.4	15.6	20.3	67.6	0.2

[a]Concentrations of $TlOH^0$, TlF^0, $TlNO_3^0$, $TlH_2PO_4^0$, $TlHPO_4^{2-}$ and $TlPO_4^{2-}$ species were assumed negligible in all systems.
Source: Kaplan and Mattigod (1998).

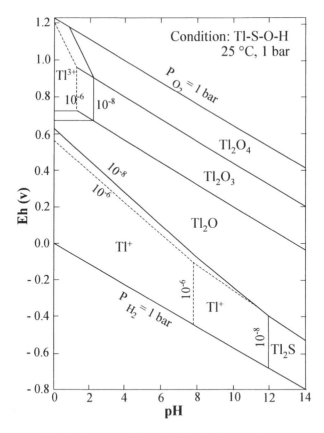

FIGURE 19.8. Predicted Eh–pH stability field for thallium. The assumed activities for dissolved species are: Tl = $10^{-8,-6}$, S = 10^{-3}. (From Brookins, 1988.)

Schoer (1984) concluded that the solubility and availability of Tl to organisms depend to a large extent on its source. Thallium present in flue dust emanating from the cement industry is 75 to 100% soluble in water but can become rapidly immobilized once incorporated in the soil. Soils contaminated by the flue dusts could yield as much as 40% of their total Tl content when extracted with 1 N NH$_4$OAc in 0.01 M EDTA but could yield only 0.5 to 5% of the total Tl once the metal was incorporated in tailing material.

Plant uptake data indicate that the amount of T1 that may be taken up by plants is proportional to Tl levels in soils (Sager, 1998; Schoer, 1984). These results further indicate that the concentration ratio (i.e., calculated as the ratio of element concentration in plant tissue and element concentration in soil, on DW basis) via root uptake may reach as high as 10, which is unusually high for metals.

Thallium values for plant material are relatively sparse. Smith and Carson (1977a) reported that terrestrial plants contain from 0.01 to 3800 ppm Tl, on ash wt. basis. However, they noted that an ash value of around 0.50 ppm is probably more typical for a wide range of plants. Assuming that ash constitutes approximately 10% of the plant's dry matter, then Tl concentration in the dry matter would be around 0.05 ppm. This extrapolated value appears realistic in view of the finding of Geilmann et al. (1960) that most crops grown in noncontaminated environments had Tl contents ranging from 0.02 to 0.04 ppm DW, with some crops such as leek, endive, and kale exhibiting higher values (0.075 to 125 ppm). King (1977) considers values of 2 to 7 ppm Tl (ash wt. basis) as anomalous and values >10 ppm as highly anomalous. These anomalous levels could be observed in plants growing in soils containing 3 to 5 ppm total Tl (Table 19.4).

Highest contents in plants have been obtained from areas of unusual Tl enrichment, such as in the Alsar region of Yugoslavia, where animals grazing on native vegetation had been intoxicated (Zyka, 1970). A case study on environmental pollution by Tl from a cement plant in Lengerich, Germany indicates that foodstuff grown within a 2-km radius of the plant (the center being ~500 m east of emission stack) could contain Tl levels exceeding the ingestion limit of 0.50 ppm WW in Germany (Schoer, 1984). The Tl contents of some vegetables (fresh wt.) were excessively high: kale, 45 ppm; kohlrabi, savoy, and white cabbage, 1 to 10 ppm.

Thallium uptake and accumulation in crops are highly dependent on soil pH and plant species. Using rape, bush beans, and ryegrass, Tl uptake was about 50% greater at pH 5.6 than at pH 6.2 (Sager, 1998). Rape bioaccumulated substantially more Tl than did bush beans and ryegrass. Similarly, when grown on T1-contaminated soils, white cabbage, kohlrabi, and green salad bioaccumulated Tl more substantially than carrot or celery. For this reason, rape and cabbage (and their member species) are not recommended to grow on Tl-contaminated areas. As a safety precaution, 1 ppm DW for rape and 0.10 ppm FW for fruits and vegetables have been recommended as the upper limit by the German Bundesgesundheitsamt (Sager, 1998).

TABLE 19.4. Thallium contents in various plant parts and their corresponding rhizosphere soils.

Plant	Rhizosphere soil (ppm)	Plant parts	ppm DW
Kohlrabi	3.7	Leaves	35.0
		Tuber	0.10
Turnip	9.4	Leaves	5.90
		Roots	0.40
Zucchini squash	5.2	Leaves	0.90
		Stalk	0.02
Cucumber	5.4	Leaves	0.70
		Fruits	0.10
Beets	5.2	Leaves	2.40
		Roots	0.60
Carrot	3.5	Leaves	0.30
		Roots	0.10
Onion	0.9	Stalk	0.10
		Tuber	0.01

Source: Hoffman et al. (1982).

Most plants showed phytotoxic symptoms when grown in media having relatively low levels of Tl. Carson and Smith (1977) indicated that at about 7 ppm total Tl in soils, many crops may exhibit injury symptoms. Tobacco plants are especially sensitive, showing toxic effects at 1 ppm level in soil or 0.04 ppm in nutrient culture. Similarly, wheat growing in a sandy loam soil was injured with a concentration of 1.4 ppm in soil, and at 28 ppm, death occurred. Significant reductions in photosynthesis of corn and sunflower were noticed when these plants were cultured in a nutrient solution containing 2 ppm Tl (Carlson et al., 1975). Excess Tl in growth medium has produced the following abnormalities in plants (Schoer, 1984): impairment of chlorophyll synthesis and seed germination, reduced transpiration due to interference of Tl in stomatal processes, growth reduction, stunting of roots, and leaf chlorosis. Thallium-affected soybeans grown in solution culture had fewer nodes than normal plants (Kaplan et al., 1990). In addition, the roots became severely stunted with only few lateral branches, and leaves became chlorotic. Nutritionally, the affected plants also suffered from nutrient imbalance, with Ca, Mg, and Mn being deficient. Indications are that Tl may be more toxic to plants than are Cd and Ni.

In general, our knowledge of the environmental biogeochemistry of Tl is limited, particularly Tl in the food chain pathway.

8 Tin

The white-silvery, soft, pliable metal (atom. no. 50, atom. wt. 118.7, specific gravity 5.75, melting pt. 232 °C) exhibits in its geochemical occurrence two oxidation states, II and IV (the prevalent forms in aquatic systems).

At 2.5 ppm crustal abundance, Sn is ranked 49th in its abundance among elements (Krauskopf, 1979). There are several known Sn-bearing minerals but only cassiterite (SnO_2) is commercially important. The main producers of refined Sn are Malaysia, Mexico, Brazil, the United Kingdom, and Indonesia, with Malaysia producing about 50% of the total amount produced. About 50% of that quantity is consumed in the United States.

Major uses of Sn are as a protective coating agent and as an alloying metal. About 80% of the total Sn produced is consumed in tinplate, bronze (1 part Sn to 9 parts Cu) and solder. Of the total amount used for tinplate, about 90% is used for cans, of which about 60% is used for food packaging, the rest for nonfood products. The Zn–Sn and Cd–Sn alloys are used as coatings in hydraulic brakes, aircraft landing gear equipment, and engine parts in the automotive and aerospace industries. Butyltin compounds, primarily monobutyltin (MBT), dibutyltin (DBT), and tributyltin (TBT), have been used as biocides or as stabilizers in a wide range of industrial and agricultural applications. Among the primary uses of butyltins, heat stabilization of polyvinyl chloride (PVC) polymers by MBT and DBT represents more than 50% of the global usage (Maguire et al., 1993). In addition, MBT and DBT are used as catalysts in the manufacture of silicone and polyurethane, and TBT is used mainly as an antifouling agent in paints used in ships, boats, aquaculture nets, and docks, and as a slimicide in cooling towers, and in wood preservation (Fent, 1996; Luijten, 1972). In agriculture, organotin compounds were used as fungicides for fungus control in potatoes and sugar beets (Barnes et al., 1971).

Tin is present in soils and plants in trace amounts. A common range of Sn in soil is 1 to 10 ppm (Peterson and Girling, 1981). Extensive analyses of surficial soil samples in the United States indicates a mean of 0.89 ppm, with a range of <0.10 to 10 ppm (Shacklette and Boerngen, 1984). Laycock (1954) reported 1.3 to 3 ppm Sn in tea soils of Nyasaland, while Prince (1957) found 1 to 11 ppm Sn in New Jersey soils.

Wallihan (1965) reported a narrow representative range of <0.10 to 3.0 ppm Sn in plants. A nonessential element in plant nutrition, it is fairly immobile in typical arable soils, especially those having neutral pH. Plants tend to accumulate Sn in roots and it is poorly translocated to the foliage. Bean plants grown in soils spiked with up to 500 ppm Sn as $SnCl_2$ showed only about 1 ppm Sn (DW) in leaves (Romney et al., 1975). In acidic soils however, it can be fairly bioavailable and phytotoxic.

A survey of several vegetables and cereal grains indicate a usual range of 0.10 to 2 ppm (Schroeder et al., 1964). A range of 0.80 to 6.8 ppm Sn in plant materials has been reported (Kick et al., 1971). Connor and Shacklette (1975) reported the Sn contents (ppm, ash wt.) of the following vegetables: carrot, 20; corn grain, 30; and beet tuber, 20. A fern, *Gleichenia linearis,* growing in Malaysia contained about 32 ppm Sn (DW), believed to be the highest recorded value for a terrestrial plant not contaminated by smelter operation (Peterson et al., 1976).

Because TBT has been widely used in antifouling paints for over 35 years, there is genuine concern on its environmental biogeochemistry and its ecotoxicological aspects in aquatic systems. Despite the partial restriction imposed in most countries, it is estimated that around 1200 tonnes TBT yr^{-1} are used for the protection of ship hulls (Davies et al., 1998). High contamination of port, marina, and harbor waters has often been reported. Consequently, TBT residues are often high in fish collected from those waters (Moore, 1991). In addition, notable concentrations of TBT have been measured in Tokyo Bay and in the Strait of Malacca where ship traffic is heavy, whereas concentrations elsewhere in the open sea have remained below the analytical threshold of 0.10 ng L^{-1} (Hashimoto et al., 1998).

Contamination of marine mammals represents an indication of TBT presence in Atlantic and Pacific waters, as did the bioaccumulation in squid livers from the contamination of waters in the Northern and Southern Hemispheres (Iwata et al., 1995; Yamada et al., 1997). In the Mediterranean Sea, total butyltin concentrations ranged from 1200 to 2000 ng g^{-1} in dolphin liver (Kannan et al., 1996). Coastal waters in the northwestern Mediterranean Sea are known to be highly contaminated by TBT (Michel and Averty, 1999). This is a function of the density of marinas along the French, Italian, and Spanish coasts, as well as considerable naval and commercial shipping traffic. More alarmingly, data on TBT in deep sea organisms collected from Suruga Bay, Japan infers that butyltin contamination has already reached deep waters (Takahashi et al., 1997).

Because of the known toxicological importance of organotin compounds to aquatic biota (Cardwell et al., 1999) and the possibility for Sn to be methylated (Moore, 1991; Summers and Silver, 1978), their biogeochemical behavior in the environment should be further assessed. More importantly, recent detection of butyltin compounds in human blood infers its potential effect on human health (Kannan et al., 1999).

In aquatic systems, TBT is known as the most toxic of all butyltin compounds (Bryan and Gibbs, 1991; Morcillo et al., 1991). It exerts chronic effects on aquatic organisms at concentrations as low as $\sim$10 to 20 ng L^{-1}. Thus, the U.S. EPA proposed 10 ng Sn L^{-1} as the chronic (marine) water quality criterion, which is protective of at least 95% of aquatic organisms (U.S. EPA, 1997). Following the restrictions on TBT usage (in the late 1980s in the United States), environmental monitoring indicates declining TBT concentrations in the United States, Japan, and Europe (Evans et al., 1995; Lange, 1996; Russell et al., 1996). For example, since passage of legislation controlling the use of TBT in the United States, median TBT levels in U.S. marine waters averaged less than 10 ng L^{-1}. Monitoring of TBT from pre-1989 to 1996 at saltwater and freshwater sites in the United States indicates declines in chronic risks from a high of 25% prior to the 1988 law restricting TBT usage, to <1 to 14% in 1996 (Cardwell et al., 1999). Most of the decline has been associated with marinas, reflecting limited TBT use on pleasure vessels.

Because of the inhibitory effect of TBT on fish growth and the behavior in fish larvae, there is also concern in other reproductive aspects (Fent and Meier, 1992; Pinkney et al., 1990). Indeed, transgenerational toxicity due to TBT was detected in the Japanese medaka (*Oryzias latipes*) (Nirmala et al., 1999).

Recent findings indicate the presence of butyltins in human blood, indicating the widespread exposure of humans to butyltin compounds from a variety of sources (Kannan et al., 1999). The toxicological implications of the concentrations of butyltins found in human blood are currently unknown, but their presence in the blood may have direct effects on hematological parameters and the potential to be distributed and accumulated in other body tissues.

9 Vanadium

Vanadium (atom. no. 23, atom. wt. 50.94, specific gravity 6.1) compounds in nature have oxidation states of II, III, IV, or V. The pentavalent form is the most soluble and is the predominant form in surface waters (Moore, 1991). Its tendency to form various oxyanions is a property it shares with Mo, As, W, and P. In nature, it is found in about 65 different minerals, the most important commercial sources of the metal being carnotite, roscoelite, vanadinite, and partonite. It is also found in phosphate rocks and in some Fe ores, as well as in crude oils as organic complexes. The metal is bright white, soft, and ductile. Deposits are being mined mainly in the United States and Africa.

Vanadium is used in manufacturing steel for high-speed tools that require rust resistance. About 80% of the V now produced is used as ferrovanadium or as a steel additive. Vanadium pentoxide is used in ceramics and as a catalyst in the production of certain chemicals. The metal is a major constituent in high-strength Ti alloys. Its compounds are used as mordants in the dyeing and printing of cotton and for fixing aniline black on silk.

Vanadium is ubiquitous in nature, ranking 20th in abundance (110 ppm) among the elements in the earth's crust (Krauskopf, 1979). Its compounds commonly occur in the trivalent oxidation state. Zenz (1980) reported that the average concentration of V in the earth's crust is about 150 ppm. Natural V occurs in igneous rocks, titaniferrous magnetites, certain deposits of phosphate rock, shales, some uranium ores, and in asphaltic deposits (Schroeder et al., 1963). The V contents of igneous rock, shale, sandstone, and limestone are 135, 130, 20, and 20 ppm, respectively (NAS, 1974). In general, the total V contents of soils derived from sands and sandstones are low compared with soils developed from shales and clays.

Pratt (1965a) indicated the V content of some soils ranging from 3 to 230 ppm, while Bowen (1979) reported a range of 3 to 500 ppm (mean = 90 ppm). Typical Scottish soils derived from shale and olivine-gabbro contained around 200 ppm, while soils derived from serpentine, sandstone, and granite had 100, 60, and 20 ppm, respectively (Thornton and Webb, 1980). Surface soils in the United States contain about 58 ppm V on the average (Shacklette and Boerngen, 1984). In China, the geometric mean value for V is 77 ppm ($n = 3874$) (Chen et al., 1991). Among the soil orders, V tends to be highest in Lithosols (114 ppm) and lowest in Oxisols (57 ppm). The mean concentration of V for Asian paddy soils is 166 ppm (Domingo and Kyuma, 1983).

In nature, V occurs as vanadates and in various organic complexes (Peterson and Girling, 1981). The vanadyl ion forms stable complexes with humic acids (Moore, 1991). In sediments and soils, V is known to occur primarily as oxovanadium (IV) in organic chelates (Yen, 1972).

In soils, V can substitute for Fe and can be sorbed onto Fe oxides, which explains the usually high correlations between the Fe and V contents in soils (Norrish, 1975). Although V tends to be more enriched in the A horizon, it is rather uniformly distributed throughout the soil profile (Vinogradov, 1959). However, in cases where the B horizon contains higher amounts of clay and Fe oxides, V can be expected to accumulate more in this horizon (LeRiche and Weir, 1963).

Vanadium is not considered to be an essential element for higher plants. Welch and Huffman (1973) did not obtain any growth response of lettuce and tomato plants to added V. The plants were grown in purified nutrient solution with and without 50 ng V ml^{-1}. All plants were healthy with similar growth and development patterns. However, some evidence exists to suggest that it is essential for the growth of certain bacteria and algae. The green algae *Scenedesmus obliquus* apparently requires it for their growth (Arnon and Wessel, 1953; Arnon, 1954). Bertrand (1950) found that trace amounts of V in the growth medium served as a growth stimulant for *Aspergillus niger*. Recently, it has been shown that during the fixation of N_2 by *Azotobacter*, V can replace Mo in the nitrogenase enzyme, but was not as effective as Mo-containing enzyme in fixing N_2 (Nicholas, 1975).

Vanadium is widely distributed in most biological materials. The most important forms in biological systems are the vanadate, VO_3^- (pentavalent) and the vanadyl ion, VO^{2+} (tetravalent) (Fox, 1987). Vanadate ion is a potential inhibitor of the ATPase system as well as an inhibitor or stimulator of many enzymes. Bengtsson and Tyler (1976) indicated that it is not unreasonable to expect V concentrations to fall in the range of 0.50 to 2 ppm DW in plants growing in noncontaminated areas. Bertrand (1950) reported a range of 0.27 to 4.2 ppm V in 62 plant materials (mean = 1 ppm) and found that plants growing on soils having higher V contents accumulated greater amounts of V than those growing on soils having less V. The roots of these plants contained higher V than the aerial tissues. Cannon (1963) gave V values (ppm DW) for the following plants: grasses, 1.4; legumes, 0.84; forbs, 1.2; deciduous trees, 1.65; deciduous shrubs, 2.7; ferns, 1.28; fungi, 0.22; and lichens, 8.6. Vegetable crops generally contain

<0.50 ppm V (Bengtsson and Tyler, 1976; Schroeder et al., 1963). Among the vegetables, radish and parsley tend to accumulate more V (Söremark, 1967).

Thirty-five plant species collected from 36 different locations in the Rocky Mountain region had V contents ranging from nondetectable to 6.6 ppm, whereas soil content ranged from 0.65 to 98 ppm (Parker et al., 1978). In this region, increased plant and soil residues were obtained at locations near phosphate rock, ore smelting, and coal burning facilities. In general, there was no relationship between the V content in plants and soils and the geologic deposit of V. It was concluded that soils possessing high V levels should not pose any risk from V bioaccumulation in the food chain.

Plant uptake of V can depend largely on soil type (Peterson and Girling, 1981). Plant accumulation of V may occur in seleniferous soils or river alluvium, while calcareous soils may restrict V movement from roots to shoot. In roots, V is thought to precipitate as calcium vanadate.

Under field conditions, V toxicity is virtually nonexistent. However, when added to nutrient solution at concentrations of 0.50 ppm or greater, V could be phytotoxic (Davis et al., 1978; Pratt, 1965a). Excess V interferes with chlorophyll synthesis and photosynthetic electron transport (Brauer and Stitt, 1990), inhibits the plasma membrane ATPase and acid phosphatases (Amodeo et al., 1992; Faraday and Spanswick, 1992; Poder and Penot, 1992), and inhibits the uptake of Ca, K, Mg, Cl^-, and PO_4^{2-} (Amodeo et al., 1992; Kaplan et al., 1990; Poder and Penot, 1992). Since V is poorly translocated, analysis of vegetative tissue may not be a useful diagnostic tool for V toxicity. Some plant species suffering from excess V contain only 1 to 2 ppm in their vegetative tissues while exhibiting several-fold concentrations in their roots (Davis et al., 1978; Pratt, 1965a; Warington, 1955, 1956). Berry (1978) concluded that V, along with Cd, is one of the most toxic trace elements in nutrition of higher plants grown in solution culture.

Recent interest in V content of plant materials has been spurred by the essentiality of V for animal nutrition and its emission in large amounts to the atmosphere from the combustion of fossil fuels. In the United States, practically all of the coal burned contains high V with the average content of about 30 ppm (Zenz, 1980). Australian bituminous coals have V contents ranging from 4 to 90 ppm (mean = 20 ppm) (Swaine, 1977). Fly ashes from bituminous and lignite coals contain 290 (99 to 652) and 209 (<25 to 268) ppm V, respectively (Ainsworth and Rai, 1987).

About 1750 tonnes of V were emitted to the atmosphere in the United States in 1969 as a result of the combustion of bituminous coal, with an additional 375 tonnes emitted from the combustion of anthracite coal (Zenz, 1980). In addition to releases of V from coal combustion, ash from petroleum-fired facilities may contain up to 38% V in the pentoxide form but concentration varies considerably, depending upon the source of the crude oil. Oil ash from the United States contain as much as 2.6% V, on the average (Ainsworth and Rai, 1987). Crude oils were found to contain as much as 1400 ppm V, with the Venezuelan oils characteristically enriched in this metal (Bengtsson and Tyler, 1976). Adoption of pollution abatement measures by electric power utilities has reduced by as much as 85% the release of V to the environment from fly ash. Since the critical level (25 ppm of total diet) of V for livestock would unlikely be exceeded under usual field conditions, the environmental concerns about V primarily arise from the air pollution aspect. Although little is known about the ecotoxicological effects of V in aquatic systems, V is known to have low toxicity to fish (Moore, 1991). Information available on the environmental biogeochemistry of V is generally insignificant.

References

Adams, S.N., J.L. Honeysett, K.G. Tiller, and K. Norrish. 1969. *Aust J Soil Res* 7:29–42.

Adriano, D.C., and H.E. Doner. 1982. In A.L. Page et al., eds. *Methods of Soil Analysis,* part 2. Am Soc Agron, Madison, WI.

Adriano, D.C., M. Delaney, and D. Paine. 1977. *Commun Soil Sci Plant Anal* 8:615–628.

Alban, L.A., and J. Kubota. 1960. *Soil Sci Soc Am Proc* 24:183–185.

Allen, R.O., and E. Steinnes. 1979. In *Proc Intl Conf Heavy Metals in the Environment.* CEP Consultants, Edinburgh.

Ainsworth, C., and D. Rai. 1987. In *Chemical Characterization of Fossil Fuel Combustion Wastes.* EPRI Rep EA-5321. Electric Power Res Inst, Palo Alto, CA.

Ainsworth, N., J.A. Cooke, and M.S. Johnson. 1990a. *Environ Pollut* 65:65–77.

Ainsworth, N., J.A. Cooke, and M.S. Johnson. 1990b. *Environ Pollut* 65:79–87.

Amodeo, G., A. Srivastava, and E. Zeiger. 1992. *Plant Physiol* 100:1567–1570.

Anderson, A.J., D.R. Meyer, and F.K. Mayer. 1973. *Aust J Agric Res* 24:557–571.

Arnesen, A.K.M., 1997. *Plant Soil* 191:13–25.

Arnesen, A.K.M., and T. Krogstad. 1998. *Water Air Soil Pollut* 100:357–373.

Arnesen, A.K.M., G. Abrahamsen, G. Sandvik, and T. Krogstad. 1995. *Sci Total Environ* 163:39–53.

Arnon, D.I. 1954. *Intl Cong Bot (8th) Proc (Paris)* 11:73–80.

Arnon, D.I., and G. Wessel. 1953. *Nature* 172:1039–1040.

Asami, T., and M. Kubota. 1993. *Proc. Intl. Conf. Heavy Metals in the Environment,* vol. 2. Toronto.

Askew, H.O., and J.K. Dixon. 1937. *NZ J Sci Technol* 18:688–693.

Backes, C.A., R.G. McLaren, A.W. Rate, and R.S. Swift. 1995. *Soil Sci Soc Am J* 59:778–785.

Bakulin, A.A., and V.F. Mokiyenko. 1966. *Pochrovedenie* 4:66.

Barnes, R.D., A.T. Bull, and R.C. Poller. 1971. *Chem Ind (Lond)* 7:204.

Barrow, N.J., and A.S. Ellis. 1986. *J Soil Sci* 37:287–293.

Bengtsson, S., and G. Tyler. 1976. In *Vanadium in the Environment.* MARC Tech Rep 2. Monit Assess Res Center, Univ London, London.

Berry, W.L. 1978. In D.C. Adriano and I.L. Brisbin, eds. *Environmental Chemistry and Cycling Processes.* Conf. 760429. NTIS, Springfield, VA.

Bertrand, D. 1950. *Bull Am Mus Nat Hist* 94:405–455.

Bertrand, D., and L. Silberstein. 1928. *Compt Rend Acad Sci (Paris)* 186:335–338.

Bowen, H.J.M. 1979. *Environmental Chemistry of the Elements.* Academic Pr, New York.

Brauer, M., and M. Stitt. 1990. *Physiol Plant* 78:568–573.

Brookins, D.G. 1988. *Eh–pH Diagrams for Geochemistry.* Springer-Verlag, New York.

Brooks, R.R. 1972. *Geobotany and Biogeochemistry in Mineral Exploration.* Harper & Row, New York.

Brewer, R.F. 1965. In H.D. Chapman, ed. *Diagnostic Criteria for Plants and Soils.* Quality Printing, Abilene, TX.

Bryan, G.W., and P.E. Gibbs. 1991. In M.C. Newman and A. McIntosh, eds. *Metal Ecotoxicology: Concepts and Applications.* Lewis Publ, Chelsea, MI.

Bury, N.R., J.C. McGeer, and C.M. Wood. 1999. *Environ Toxicol Chem* 18:49–55.

Campbell, P.G.C. 1995. In *Metal Speciation and Bioavailability in Aquatic Systems.* Wiley, New York.

Cannon, H.L. 1963. *Soil Sci* 96:196–204.

Carbonell, A.A., R. Pulido, R.D. DeLaune, and W.H. Patrick. 1999. *J Environ Qual* 28:316–321.

Cardwell, R.D., M.S. Brancato, and L. Tear. 1999. *Environ Toxicol Chem* 18:567–577.

Carlson, R.W., F.A. Bazzaz, and G.L. Rolfe. 1975. *Environ Res* 10:113–120.

Carr, M.H., and K.K. Turekian. 1961. *Geochim Cosmochim Acta* 23:9–60.

Carson, B.L., and I.C. Smith. 1977. In Midwest Res Inst Tech Rep 5. Kansas City, MO.

Chaudry, F.M., A. Wallace, and R.T. Mueller. 1977. *Commun Soil Sci Plan Anal* 8: 795–797.

Chen, J., F. Wei, C. Zheng, Y. Wu, and D.C. Adriano. 1991. *Water Air Soil Pollut* 57–58:699–712.

Connor, J.J., and H.T. Shacklette. 1975. In USGS Paper 574-F. U.S. Geological Survey, Denver, CO.

Crecelius, E.A., C.J. Johnson, and G.C. Hofer. 1974. *Water Air Soil Pollut* 3:337–342.

Crecelius, E.A., M.H. Bothner, and R. Carpenter. 1975. *Environ Sci Technol* 9:325–333.

Czarnowska, K., and B. Gworek. 1994. *Environ Geochem Health* 12:289–290.

Davidson, R.L., D.F.S. Natusch, C.A. Evans, and P. Williams. 1976. *Science* 191:852–854.

Davies, I.M., S.K. Bailey, and M.J. Harding. 1998. *J Mar Sci* 55:34–42.

Davis, R.D., P.H.T. Beckett, and E. Wollan. 1978. *Plant Soil* 49:395–408.

Domingo, L.E., and K. Kyuma. 1983. *Soil Sci Plant Nutr* 29:439–452.

Evans, S.M., T. Leksono, and P.D. McKinnell. 1995. *Mar Pollut Bull* 30:14–21.

Faraday, C.D., and R.M. Spanswick. 1992. *J Exp Bot* 43:1583–1590.

Fent, K. 1996. *Crit Rev Toxicol* 26:1–117.

Fent, K., and W. Meier. 1992. *Arch Environ Contam Toxicol* 22:428–438.

Ferguson, E.A., and C. Hogstrand. 1998. *Environ Toxicol Chem* 17:589–593.

Filipovic, Z., B. Stankovic, and Z. Dusic. 1961. *Soil Sci* 91:147–150.

Fox, M.R. 1987. *J Anim Sci* 65:1744–1752.

Frank, R., K.L. Stonefield, and P. Suda. 1979. *Can J Soil Sci* 59:99–103.

Fujimoto, G., and G.D. Sherman. 1950. *Agron J* 42:577–581.

Galvez, F., and C.M. Wood. 1997. *Environ Toxicol Chem* 16:2363–2368.

Geilmann, W., K. Beyermann, K.H. Neeb, and R. Neeb. 1960. *Biochem Z* 333: 6270.

Gille, G.L., and E.R. Graham. 1971. *Soil Sci Soc Am Proc* 35:414–416.

Gilpin, L., and A.H. Johnson. 1980. *Soil Sci Soc Am J* 44:255–258.

Graham, E.R., and D.D. Killion. 1962. *Soil Sci Soc Am Proc* 26:545–547.

Hashimoto, S., M. Watanabe, and A. Otsuki. 1998. *Mar Environ Res* 45:169–177.

Hirsch, M.P. 1998. *Environ Toxicol Chem* 17:610–616.

Hodgson, J.F. 1960. *Soil Sci Soc Am Proc* 24:165–168.

Hodgson, J.F., K.G. Tiller, and M. Fellows. 1969. *Soil Sci* 108:391–396.

Hoffman, G., P. Schweiger, and W. Scholl. 1982. *Landwirtsch Forschung* 35:4554.

Hogstrand, C., and C.M. Wood. 1998. *Environ Toxicol Chem* 17:547–561.

Hogstrand, C., F. Galvez, and C.M. Wood. 1996. *Environ Toxicol Chem* 15:1102–1108.

Hope, B., C. Loy, and P. Miller. 1996. *Bull Environ Contam Toxicol* 56:683–689.

Hopkins, D.M. 1977. *J Res USGS* 5:589–593.

Horovitz, C.T., and O. Petrescu. 1964. In *Trans 8th Intl Cong Soil Sci* 4:1205–1213.

Horovitz, C.T., H.H. Schock, and L.A. Horovitz-Kisimova. 1974. *Plant Soil* 40:397–403.

Howe, H.E., and A.A. Smith. 1970. *J Electrochem Soc* 97:167C–170C.

Iwata, H., S. Tanabe, T. Mitzuno, and R. Tatsukawa. 1995. *Environ Sci Technol* 29:2959–2973.

Jones, K.C., P.J. Peterson, and B.E. Davies. 1984. *Geoderma* 33:157–168.

Jones, K.C., N.W. Lepp, and J.P. Obbard. 1990. In *Heavy Metals in Soils*. Blackie, Glasgow.

Kabata-Pendias, A., and H. Pendias. 1984. *Trace Elements in Soils and Plants*. CRC Pr, Boca Raton, FL.

Kannan, K., K. Senthilkumar, and J.P. Giesy. 1999. *Environ Sci Technol* 33:1776–1779.

Kaplan, D.I., and S.V. Mattigod. 1998. In *Thallium in the Environment*. Wiley, New York.

Kaplan, D.I., D.C. Adriano, C.L. Carlson, and K.S. Sajwan. 1990. *J Environ Qual* 19:359–365.

Kaplan, D.I., D.C. Adriano, C.L. Carlson, and K.S. Sajwan. 1990. *Water Air Soil Pollut* 49:81–91.

Kau, P.M.H., D.W. Smith, and P. Binning. 1998. *Geoderma* 84:89–108.

Kick, H., R. Nosbers, and J. Warnusz. 1971. In *Proc Intl Symp Soil Fertility Evaluation (New Delhi)* 1:1039–1045.

Kidson, E.B. 1938. *J Soc Chem Ind (Lond)* 57:95–96.

King, H. 1977. USGS Bull 1466. U.S. Geological Survey, Washington, DC.

Kipling, M.D. 1980. In H.A. Waldron, ed. *Metals in the Environment*. Academic Pr, New York.

Klein, D.A. 1979. In D.A. Klein, ed. *Environmental Impacts of Artificial Ice Nucleating Agents*. Dowden, Hutchinson & Ross, Stroudsburg, PA.

Krauskopf, K.B. 1979. *Introduction to Geochemistry*, 2nd ed. McGraw-Hill, New York.

Kubota, J. 1965. *Soil Sci* 99:166–174.

Kubota, J. 1980. In B.E. Davies, ed. *Applied Soil Trace Elements*. Wiley, New York.

Kumpulainen, J., and P. Koivistoinen. 1977. *Residue Rev* 68:37–57.

Lambert, T.I., and C. Blincoe. 1971. *J Sci Fed Agric* 22:8–9.

Lange, R. 1996. In *Proceedings of Present Status of TBT-Co-Polymer Antifouling Paints*. The Hague, Netherlands.

Larsen, S., and A.E. Widdowson. 1971. *J Soil Sci* 22:210–221.

Laycock, D.H. 1954. *J Sci Food Agric* 5:266–269.

LeRiche, H.H., and A.H. Weir. 1963. *J Soil Sci* 14:225–235.

Lim, M.Y. 1979. In *Trace Elements from Coal Combustion Atmospheric Emissions*. Intl Energy Agency, Coal Res, London.

Lin, T.S., and J. Nriagu. 1999. *Environ Sci Technol* 33:3394–3397.

Luijten, J.G.A. 1972. In *Organotin Compounds*. Marcel Dekker, New York.

Lyon, G.L., R.R. Brooks, P.J. Peterson, and G.W. Butler. 1968. *Plant Soil* 219:225–240.

Lyon, G.L., R.R. Brooks, P.J. Peterson, and G.W. Butler. 1971. *J Ecol* 49:421–429.

Maguire, R.J., G. Long, M.E. Meek, and S. Savard. 1993. In *Canadian Environmental Protection Act Priority Substances List Report—Nonpesticidal Organotin Comp*. Dept Environment, Ottawa.

Malaisse, F., J. Gregoire, R.S. Morrison, R.R. Brooks, and R.D. Reeves. 1979. *Oikos* 33:472–478.

McKenzie, R.M. 1975. In D.J.D. Nicholas and A.R. Egan, eds. *Trace Elements in Soil-Plant-Animal Systems*. Academic Pr, New York.

McKenzie, R.M. 1978. *Aust J Soil Res* 16:209–214.

McLaren, R.G., D.M. Lawson, and R.S. Swift. 1987. *J Sci Food Agric* 39:101–112.

McLaughlin, M.J., K.G. Tiller, R. Naidu, and D.P. Stevens. 1996. *Aust J Soil Res* 34:1–54.

Menzel, R.G. 1954. *Soil Sci* 77:419–425.

Michel, P., and B. Averty. 1999. *Environ Sci Technol* 33:2524–2528.

Miller, E.R., X. Lie, and D.E. Ullrey. 1991. In J.J. Mortvedt et al., eds. *Micronutrients in Agriculture*, 2nd ed. SSSA 4, Soil Sci Soc Am, Madison, WI.

Mitchell, R.L. 1944. *Proc Nutr Soc* 1:183–189.

Mitchell, R.L. 1945. *Soil Sci* 60:63–70.

Mitchell, R.L. 1971. In Tech Bull 21. Her Majesty's Sta Office, London.

Mitchell, R.L., and J.C. Burridge. 1979. *Phil Trans R Soc (London)* B288:1524.

Moore, J.W. 1991. *Inorganic Contaminants in Surface Waters*. Springer-Verlag, New York.

Morcillo, Y., A. Albalat, and C. Porte. 1999. *Environ Toxicol Chem* 18:1203–1208.

Morel, F.M.M., and J.G. Hering. 1993. *Principles and Applications of Aquatic Chemistry*. Wiley, New York.

[NAS] National Academy of Sciences. 1971. *Fluorides*. In *Comm Biol Effects Air Pollut*. NAS, Washington, DC.

[NAS] National Academy of Sciences. 1974. *Vanadium*. NAS, Washington, DC.

Newland, L.W. 1982. In O. Hutzinger, ed. *The Handbook of Environmental Chemistry, vol. 3(B): Anthropogenic compounds*. Springer-Verlag, Berlin.

Nicholas, D.J.D. 1975. In D.J.D. Nicholas and A.R. Egan, eds. *Trace Elements in Soil-Plant-Animal Systems*. Academic Pr, New York.

Nirmala, K., Oshima, R. Lee, and K. Kobayashi. 1999. *Environ Toxicol Chem* 18:717–721.

Nömmik, H. 1953. *Acta Polytech* 127:1–121.

Norrish, K. 1975. In D.J.D. Nicholas and A.R. Egan, eds. *Trace Elements in Soil-Plant-Animal Systems*. Academic Pr, New York.

Nriagu, J.O. 1998. In *Thallium in the Environment*. Wiley, New York.

Omueti, J.A.I., and R.L. Jones. 1977a. *J Soil Sci* 28:564–572.

Omueti, J.A.I., and R.L. Jones. 1977b. *Soil Sci Soc Am J* 41:1023–1024.

Omueti, J.A.I., and R.L. Jones. 1980. *Soil Sci Soc Am J* 44:247–249.

Parker, R.D.R., R.P. Sharma, and G.W. Miller. 1978. *Trace Subs Environ Health* 12:340–350.

Patel, P.M., A. Wallace, and R.T. Mueller. 1976. *J Am Soc Hort Sci* 101(5):553–556.

Palit, S., A. Sharma, and G. Talukder. 1994. *Bot Review* 60:149–181.

Peterson, P.J., and C.A. Girling. 1981. In N.W. Lepp, ed. *Effect of Heavy Metal Pollution on Plants,* vol. 1. Applied Science Publ, London.

Peterson, P.J., M.A.S. Burton, M. Gregson, S.M. Nye, and E.K. Porter. 1976. *Trace Subs Environ Health* 10:123–132.

Phillips, C., J. Evans, W. Hom, and J. Clayton. 1998. *Environ Toxicol Chem* 17:1653–1661.

Pinkney, A.E., L.L. Matteson, and D.A. Wright. 1990. *Arch Environ Contam Toxicol* 19:235–240.

Poder, D., and M. Penot. 1992. *J Exp Bot* 43:189–193.

Polomski, J., H. Flijhler, and P. Blaser. 1982. *J Environ Qual* 11:457–461.

Pratt, P.F. 1965a. In H.D. Chapman, ed. *Diagnostic Criteria for Plants and Soils*. Quality Printing, Abilene, TX.

Pratt, P.F. 1965b. In H.D. Chapman, ed. *Diagnostic Criteria for Plants and Soils*. Quality Printing, Abilene, TX.

Price, N.O., W.N. Linkous, and R.W. Engel. 1955. *J Agr Food Chem* 3:226–229.

Prince, A.L. 1957. *Soil Sci* 84:413–418.

Proctor, J., and S.R.J. Woodell. 1975. *Adv Ecol Res* 9:255–366.

Prokhorov, V.M., L.P. Moskevich, and V.A. Kudyashov. 1979. *Sov Soil Sci* 11:161–167.

Purcell, T.W., and J.J. Peters. 1998. *Environ Toxicol Chem* 17:539–546.

Purcell, T.W., and J.J. Peters. 1999. *Environ Toxicol Chem* 18:3–8.

Qian, J., H.B. Xue, L. Sigg, and A. Albrecht. 1998. *Environ Sci Technol* 32:2043–2050.

Ratte, H.T. 1999. *Environ Toxicol Chem* 18:89–108.

Reisenauer, H.M. 1960. *Nature* 186:375–376.

Robinson, W.O., and G. Edgington. 1946. *Soil Sci* 61:341–353.

Romney, E.M., and J.D. Childress. 1965. *Soil Sci* 100:210–217.

Romney, E.M., A. Wallace, and G.V. Alexander. 1975. *Plant Soil* 42:585–589.

Ruhland, W. 1958. In *Mineral Nutrition of Plants*. Springer-Verlag, Berlin.

Russell, D., M.S. Brancato, and H.J. Bennett. 1996. *Mar Sci Technol* 1:230–238.

Sager, M. 1998. In *Thallium in the Environment*. Wiley, New York.

Schoer, J. 1984. In O. Hutzinger, ed. *The Handbook of Environmental Chemistry,* vol. 3(C). Springer-Verlag, Berlin.

Schroeder, H.A., J.J. Balassa, and I.H. Tipton. 1963. *J Chron Dis* 16:1047–1071.

Schroeder, H.A., J.J. Balassa, and I.H. Tipton. 1964. *J Chron Dis* 17:483–502.

Schroder, J.L., N.T. Basta, D.P. Rafferty, and K. McBee. 1999. *Environ Sci Technol* 18:2028–2033.

Scow, K. 1981. *Exposure Risk Assessment for Silver*. U.S. Environmental Protection Agency, Office of Water, Washington, DC.

Shacklette, H.T., and J.G. Boerngen. 1984. *Element Concentrations in Soils and Other Surficial Materials of the Conterminous United States*. USGS Paper 1270. U.S. Geological Survey, Washington, DC.

Slavek, J., H. Farrah, and W.F. Pickering. 1984. *Water Air Soil Pollut* 23:209–220.

Shaw, D.M. 1952. *Geochim Cosmochim Acta* 2:118–154.

Shaw, J.R., C.M. Wood, W.J. Birge, and C. Hogstrand. 1998. *Environ Toxicol Chem* 17:594–561.

Silver Institute. 1995. *World Silver Survey*. Washington, DC.

Smith, L.C., and B.L. Carson. 1977a. *Trace Metals in the Environment, vol 1: Thallium*. Ann Arbor Science, Ann Arbor, MI.

Smith, L.C., and B.L. Carson. 1977b. *Trace Metals in the Environment, vol 2: Silver*. Ann Arbor Science, Ann Arbor, MI.

Söremark, R. 1967. *J Nutr* 92:183–190.

Steinkoenig, L.A. 1919. *J Indus Eng Chem* 11:463–465.

Summers, A.O., and S. Silver. 1978. *Ann Rev Microbiol* 32:637–672.

Suzuki, N. 1958. *Sci Rep Res Inst (Tohoku Univ)* 42:149–157.

Swaine, D.J. 1977. *Trace Subs Environ Health* 11:107–116.

Swaine, D.J. 1979. In *Trace Elements in Australian Bituminous Coals and Fly Ashes*. Proc Conf Combustion of Pulverized Coal—Effect of Mineral Matter, Newcastle, United Kingdom.

Takahashi, S., S. Tanabe, and T. Kubodera. 1997. *Environ Sci Technol* 31:3103–3111.

Taylor, R.M., and R.M. McKenzie. 1966. *Aust J Soil Res* 4:29–39.

Teller, H.L., and D.A. Klein. 1974. In *Oper Rep 3*. Colorado State Univ, Fort Collins, CO.

Tepper, L.B. 1980. In H.A. Waldron, ed. *Metals in the Environment*. Academic Pr, New York.

Thornton, I., and J.S. Webb. 1980. In B.E. Davies, ed. *Applied Soil Trace Elements*. Wiley, New York.

Tiller, K.G., J.L. Honeysett, and E.G. Hallsworth. 1969. *Aust J Soil Res* 7:43–56.

Totsche, K.U., W. Wilcke, M. Körber, J. Kobza, and W. Zech. 2000. *J Environ Qual* 29:454–459.

Tremel, A., P. Masson, T. Sterckeman, D. Baize, and M. Mench. 1997. *Environ Pollut* 95:293–302.

Tso, T.C., T.P. Sorokin, and M.E. Engelhaupt. 1973. *Plant Physiol* 51:805–806.

Underwood, E.J. 1977. *Trace Elements in Human and Animal Nutrition*. Academic Pr, New York.

[U.S. EPA] United States Environmental Protection Agency. 1976. In *Quality Criteria for Water*. U.S. EPA, Washington, DC.

U.S. EPA. 1997. In *Draft Ambient Aquatic Life Water Quality Criteria for Tributyltin*. U.S. EPA, Duluth, MN.

Van der Weijden, C.H., and E.C. Kruissink. 1977. *Mar Chem* 5:93–112.

Vanselow, A.P. 1965a. Barium. In H.D. Chapman, ed. *Diagnostic Criteria for Plants and Soils*. Quality Printing, Abilene, TX.

Vanselow, A.P. 1965b. Cobalt. In H.D. Chapman, ed. *Diagnostic Criteria for Plants and Soils*. Quality Printing, Abilene, TX.

Vanselow, A.P. 1965c. Silver. In H.D. Chapman, ed. *Diagnostic Criteria for Plants and Soils*. Quality Printing, Abilene, TX.

Van Wensem, J., and T. Adena. 1991. *Environ Pollut* 71:239–241.

Vinogradov, A.P. 1959. *The Geochemistry of Rare and Dispersed Chemical Elements in Soils.* Consultants Bureau, New York.

Voskresenskaya, N.T. 1968. *Geochem Intern* 5:158–168.

Wallace, A., and E.M. Romney. 1971. *Agron J* 63:245–248.

Wallace, A., G.V. Alexander, and F.M. Chaudry. 1977. *Commun Soil Sci Plant Anal* 8:751–756.

Wallihan, E.F. 1965. In H.D. Chapman, ed. *Diagnostic Criteria for Plants and Soils.* Quality Printing, Abilene, TX.

Walton, K.C. 1987. *Environ Pollut* 46:1–9.

Ward, N.I., E. Roberts, and R.R. Brooks. 1979. *NZ J Sci* 22:129–132.

Warington, K. 1955. *Ann Appl Biol* 41:1–22.

Warington, K. 1956. *Ann Appl Biol* 44:535–546.

Weaver, T.W., and D. Klarich. 1973. In *Final Res Rep 42.* Montana Agric Exp Sta, Montana State Univ, Bozeman, MT.

Welch, R.M., and E.W.D. Huffman, Jr. 1973. *Plant Physiol* 52:183–185.

Welch, R.M., W.H. Allaway, W.A. House, and J. Kubota. 1991. In J.J. Mortvedt et al., eds. *Micronutrients in Agriculture,* 2nd ed. SSSA 4, Soil Sci Soc Am, Madison, WI.

Wenzel, W.W., and W. Blum. 1992. *Soil Sci* 153:357–364.

Williams, R.J.B., and H.H. Riche. 1968. *Plant Soil* 29:317–326.

Wright, J.R., and K. Lawton. 1954. *Soil Sci* 77:95–105.

Yen, T.F. 1972. *Trace Subs Environ Health* 6:347–353.

Yamada, H., K. Tagayanagi, and K. Ikeda. 1997. *Environ Pollut* 96:217–229.

Young, R.S. 1979. *Cobalt in Biology and Biochemistry.* Academic Pr, London.

Zenz, C. 1980. In H.A. Waldron, ed. *Metals in the Environment.* Academic Pr, New York.

Zyka, V. 1970. *Sbornick Geol Ved Technol Geochem* 10:91–96.

Appendix A
Tables

APPENDIX TABLE A.1. Global atmospheric emissions of trace metals (thousand tonnes per year) from anthropogenic sources.[a]

Element	Energy production	Mining	Smelting and refining	Manufacturing processes	Commercial uses[b]	Waste incineration	Transportation	Total, range	Median value
Antimony	1.30	0.10	1.42	—	—	0.67	—	1.5–5.5	3.5
Arsenic	2.22	0.06	12.3	1.95	2.02	0.31	—	12–25.6	19
Cadmium	0.79	—	5.43	0.60	—	0.75	—	3.1–12.0	7.6
Chromium	12.7	—	—	17.0	—	0.84	—	7.3–53.6	31
Copper	8.04	0.42	23.2	2.01	—	1.58	—	19.9–50.9	35
Lead	12.7	2.55	46.5	15.7	4.50	2.37	248	289–376	332
Manganese	12.1	0.52	2.55	14.7	—	8.26	—	19.6–66.0	38
Mercury	2.26	—	0.13	—	—	1.16	—	0.9–6.2	3.6
Nickel	42.0	0.80	3.99	4.47	—	0.35	—	24.1–87.1	52
Selenium	3.85	0.16	2.18	—	—	0.11	—	1.8–5.9	6.3
Thallium	1.13	—	—	4.01	—	—	—	3.3–6.9	5.1
Tin	3.27	—	1.06	—	—	0.81	—	1.5–10.8	5.1
Vanadium	84.0	—	0.06	0.74	—	1.15	—	30.1–142	86
Zinc	16.8	0.46	72.0	33.4	3.25	5.90	—	70–193	132

[a] As of 1983.
[b] Including agricultural uses.
Source: Nriagu, J.O., and J.M. Pacyna. 1988. Nature 333:134–139.

APPENDIX TABLE A.2. Global inputs of trace metals (thousand tonnes per year) into soils.

Source	Element												
	Sb	As	Cd	Cr	Cu	Pb	Mn	Hg	Mo	Ni	Se	V	Zn
Agric. + animal wastes	4.9	5.8	2.2	82	67	26	158	0.85	34	45	4.6	19	316
Logging + wood wastes	2.8	1.7	1.1	10	28	7.4	61	1.1	1.6	13	1.6	5.5	39
Urban refuse	0.76	0.40	4.2	20	26	40	24	0.13	2.3	6.1	0.33	0.2	60
Muni. sewage + organic waste	0.18	0.25	0.18	6.5	13	7.1	8.1	0.44	0.43	15	0.11	1.3	39
Solid wastes from metal fabric	0.08	0.11	0.04	1.5	4.3	7.6	2.6	0.04	0.08	1.7	0.10	0.12	11
Coal ashes	12	22	7.2	298	214	144	1076	2.6	44	168	32	39	298
Fertilizers + peat	0.25	0.25	0.28	0.20	0.32	1.4	2.9	12	0.01	0.46	2.2	0.27	0.97
Discarded manufactured products[a]	2.4	38	1.2	458	592	292	300	0.68	1.9	19	0.15	1.7	465
Atmospheric fallout	2.5	13	5.3	22	25	232	27	2.5	2.3	24	2.0	60	92
Total range	4.7–47	52–112	6–38	484–1309	541–1367	479–1113	706–2633	1.6–15	30–145	106–544	6.0–76	43–222	689–2054
Median value	26	82	22	898	971	759	1669	8.3	87	294	41	128	1322

[a]Metals used for industrial installations and "durable" goods are assumed to have a definite life span and to be released into the environment at a constant rate.

Note: these inputs exclude mine tailing and slags at the smelter sites.

Source: Nriagu, J.O., and J.M. Pacyna. 1988. Nature 333:134–139.

APPENDIX TABLE A.3. Global inputs of trace metals (thousand tonnes per year) into aquatic systems.

Source	Element												
	Sb	As	Cd	Cr	Cu	Pb	Mn	Hg	Mo	Ni	Se	V	Zn
Domestic wastewaters	2.2	9.2	1.7	46	28	6.8	110	0.30	2.2	62	3.8	2.3	48
Electric power plants	0.18	8.2	0.12	5.7	13	0.72	11	1.8	0.65	11	18	0.30	18
Base metal mining and smelting	3.8	7.4	2.0	12	14	7.0	40	0.10	0.51	13	12	0.60	29
Manufacturing processes	9.3	7.0	2.4	51	34	14	21	2.1	4.2	7.4	4.3	0.55	85
Atmospheric fallout	1.1	5.6	2.2	9.1	11	100	12	2.0	0.95	10	0.82	26	40
Sewage discharges	1.5	4.1	0.69	19	12	9.4	69	0.16	2.9	11	2.0	3.5	17
Total range	3.9–33	12–70	2.1–17	45–239	35–90	97–180	109–414	0.3–8.8	1.8–21	33–194	10–72	2.1–21	77–375
Median value	18	42	9.1	143	112	138	263	6.5	11	114	41	33	237

Source: Nriagu, J.O., and J.M. Pacyna. 1988. Nature 333:134–139.

APPENDIX TABLE A.4. Industrial vs. natural mobilization of metals in the biosphere $(10^3 \text{ tonnes yr}^{-1})$.

Element	A: Mine production	B: Industrial emissions	C: Weathering mobilization	Biospheric enrichment factor $(A + B)/C$
As	45	105	90	1.7
Cd	19	24	4.5	9.6
Cr	6800	1010	810	9.6
Cu	8114	1048	375	24
Hg	6.8	11	0.9	20
Mo	98	98	15	13
Ni	778	356	255	4.4
Pb	3100	565	180	20
Sb	55	41	15	6.4
Se	1.6	76	4.5	17
Zn	6040	1427	540	14

Source: Nriagu, J.O. 1990. *Environment* 32:7–23.

APPENDIX TABLE A.5. Concentrations of trace elements (mg kg⁻¹ DW) in various soil-forming rocks and other natural materials.[a]

	Igneous rocks			Sedimentary rocks				
Element	Ultramafic igneous	Basaltic igneous	Granitic igneous	Shales and clays	Black shales	Deep-sea clays	Limestones	Sandstones
Arsenic	0.3–16 / 3.0	0.2–10 / 2.0	0.2–13.8 / 2.0	— / 10	—	13	0.1–8.1 / 1.7	0.6–9.7 / 2
Barium	0.2–40 / 1	20–400 / 300	300–1800 / 700	460–1700 / 700	70–1000 / 300	2300	10	20
Beryllium	—	1.0	2–3	3		2.6	—	—
Cadmium	0–0.2 / 0.05	0.006–0.6 / 0.2	0.003–0.18 / 0.15	0–11 / 1.4	<0.3–8.4 / 1.8	0.1–1 / 0.5	0.05	0.05
Chromium	1000–3400 / 1800	40–600 / 220	2–90 / 20	30–590 / 120	26–1000 / 100	90	10	35
Cobalt	90–270 / 150	24–90 / 50	1–15 / 5	5–25 / 20	7–100 / 10	74	0.1	0.3
Copper	2–100 / 15	30–160 / 90	4–30 / 15	18–120 / 50	20–200 / 70	250	4	2
Fluorine	—	20–1060 / 360	20–2700 / 870	10–7600 / 800	—	1300	0–1200 / 220	10–880 / 180
Iron	94,000	86,500	14,000–30,000	47,200	20,000	65,000	3800	9800
Lead	1	2–18 / 6	6–30 / 18	16–50 / 20	7–150 / 30	80	9	<1–31 / 12
Mercury	0.004–0.5 / 0.1?	0.002–0.5 / 0.05	0.005–0.4 / 0.06	0.005–0.51 / 0.09	0.03–2.8 / 0.5	0.02–2.0 / 0.4	0.01–0.22 / 0.04	0.001–0.3 / 0.05
Molybdenum	0.3	0.9–7 / 1.5	1–6 / 1.4	— / 2.5	1–300 / 10	27	0.4	0.2
Nickel	270–3600 / 2000	45–410 / 140	2–20 / 8	20–250 / 68	10–500 / 50	225	20	2
Selenium	0.05	0.05	0.05	0.6		0.17	0.08	0.05
Vanadium	17–300 / 40	50–360 / 250	9–90 / 60	30–200 / 130	50–1000 / 150	120	20	20
Zinc	— / 40	48–240 / 110	5–140 / 40	80–180 / 90	34–1500 / 100?	165	20	2–41 / 16

[a]The upper values are the usually reported range, the lower values the averages.
Source: Cannon, H.L. 1978. *Geochem Environ* 3:17–31.

APPENDIX TABLE A.6. Existing or proposed regulations limiting Cd content of fertilizers in Australia and in Australia's trading partners.

Country	Cd content (mg Cd kg^{-1} P)
Austria	275
Denmark	150 from July 1992
	110 from July 1995
France	None
Germany	200 voluntary
Japan	340
Netherlands	40
Norway	100
Sweden	over 5, V tax of 30 Swedish krona
	mg^{-1} Cd increase; 100 maximum
Switzerland	50 from 1992
United Kingdom	None
United States	None
Australia	
Tasmania	
Phosphatic fertilizers	450
Trace element fertilizers	80 (per kg product)
Soil amendments	10 (per kg product)
Victoria (proposed)	
Phosphatic fertilizers	350
Other fertilizers	10 (per kg product)

Source: Modified from McLaughlin, M.J., et al. 1995. *Aust J Soil Res* 34:1–54.

APPENDIX TABLE A.7. Concentrations of trace elements (mg kg^{-1} DW) in fertilizers.

Fertilizer	Zn	Cu	Mn	B	Mo	Co	Cr	Ni	Pb	V	Cd
Diammonium phosphate (20-48-0)[a]											
Reagent grade	1.0	1.6	0.6	—	—	—	0.2	1.1	0.5	0.3	0.9
Idaho phosphate rock	715	2.7	195	—	—	—	485	64	4.4	1600	50
North Carolina phosphate rock[b]	285	1.0	93	—	—	—	195	38	4.7	90	30
Rock phosphate[b]	187	32	975	72	555	109	184	—	962	—	—
Single superphosphate (0-16-0)[b]	165	15	890	132	335	77	87	—	488	—	—
Triple superphosphate (0-45-0)[b]	418	49	75	212	270	47	392	—	238	—	—
Diammonium phosphate[b]	112	7.2	307	396	75	16	80	—	195	—	—
Fluid fertilizer (10-15-0)[a]											
Idaho phosphate rock	673	1.1	125	—	—	—	344	8.0	9.1	1150	44
North Carolina phosphate rock	500	1.4	25	—	—	—	175	35	5.2	52	17
Urea (45-0-0)[b]	4.0	0.6	0.5	1.0	5.3	—	6.3	—	—	—	—
Calcium ammonium nitrate (25-0-0)[b]	7.6	2.8	25	9.0	56	6.6	8.5	—	116	—	—
Ammonium sulfate (21-0-0)[b]	11	0.8	3.5	—	6.0	24	4.0	—	—	—	—
Muriate of potash (0-0-60)[b]	10	3.1	3.5	16	26	22	13	—	117	—	—
N-P-K mixture (12-12-12)[b]	88	18	132	61	200	51	116	—	444	—	—
Superphosphate from apatite[c]	—	—	—	—	10	—	20	5	—	5	—

Sources:
[a]Mortvedt, J.J., and P.M. Giordano. 1977. In H. Ducker and R.E. Wildung, eds. Biological Implications of Metals in the Environment. CONF-750929. NTIS, Springfield, VA.
[b]Arora, C.L., et al. 1975. Indian J Agric Sci 45:80-85.
[c]Ermolenko, N.F. 1972. In Trace Elements and Colloids in Soils. NTIS, Springfield, VA.

APPENDIX TABLE A.8. Pesticide-containing metals recommended in Ontario, Canada, from period 1892–1975.

Chemical	Metal composition of product	Period of recommendation	Crops
Insecticides			
Copper aceto-arsenite (Paris green)	2.3% As, 39% Cu	1895–1920	Apples and cherries
		1895–1957	Vegetable and small fruit (foliar and bait)
Calcium arsenate	0.8–26% As	1910–1953	Fruit and vegetables
Lead arsenate	4.2–9.1% As	1910–1975	Apples
	11–26% Pb	1910–1971	Cherries
		1910–1956	Peaches
		1910–1955	Vegetables
Mercuric chloride	6% Hg	1932–1954	Cruciferous crops
Zinc sulfate	20–30% Zn	1939–1955	Peaches
Fungicides			
Copper sulfate-calcium salts	4–6% Cu	1892–1975	Fruit and vegetables
Fixed copper salts	2–56% Cu	1940–1975	Fruit and vegetables
		1948–1975	Fruit
Maneb	1–17% Mn	1947–1975	Fruit and vegetables
Mancozeb	16% Mn, 2% Zn	1966–1975	Fruit and vegetables
Methyl and phenyl mercuric salts	0.6–6% Hg	1932–1972	Seed treatment
Phenyl mercuric acetate	6% Hg	1954–1973	Apples
Zineb and ziram	1–18% Zn	1974–1975	Vegetables
		1957–1975	Fruit
Topkiller			
Calcium arsenite	30% As	1930–1972	Vegetables
Sodium arsenite	26% As	1920–1972	Vegetables

Source: Frank, R., et al. 1976. *Can J Soil Sci* 56:181–196.

APPENDIX TABLE A.9. Concentrations of trace elements (mg kg^{-1} DW) in sewage sludge.

Element	United States[a]		United Kingdom[b]		Canada[c] (mean)
	Mean	Range	Mean	Range	
Ag	—	—	32	5–150	—
As	14.3	3–30	—	—	—
B	37.0	22–90	70	15–1000	1950
Ba	621	272–1066	1700	150–4000	—
Be	<8.5	—	5	1–30	—
Bi	16.8	<1–56	34	<12–100	—
Cd	104	7–444	<200	<60–1500	38
Co	9.6	4–18	24	2–260	19
Cr	1441	169–14,000	980	<40–8800	1960
Cu	1346	458–2890	970	200–8000	1600
F	167	370–739	—	—	—
Hg	8.6	3–18	—	—	—
Mn	194	32–527	500	150–2500	2660
Mo	14.3	1–40	7	2–30	13
Ni	235	36–562	510	20–5300	380
Pb	1832	136–7627	820	120–3000	1700
Sb	10.6	2–44	—	—	—
Se	3.1	1–5	—	—	—
Sn	216	111–492	160	40–700	—
Ti	2331	1080–4580	2000	<1000–4500	—
V	40.6	15–92	75	20–400	15
W	20.2	1–100	—	—	—
Zn	2132	560–6890	4100	700–49,000	6140

Sources:
[a]Furr, K.A., et al. 1976. *Environ Sci Technol* 7:683–687; includes Atlanta, Chicago, Denver, Houston, Los Angeles, Miami, Milwaukee, Philadelphia, San Francisco, Seattle, Washington, DC, and five cities in New York.
[b]Berrow, M.L., and J. Webber. 1972. *J Sci Fed Agric* 23:93–100; includes 42 samples from different locations in England and Wales.
[c]Oliver, B.G., and E.G. Cosgrove. 1975. *Environ Letters* 9:75–90; from 10 sites in southern Ontario, Canada.

APPENDIX TABLE A.10. Concentrations of trace elements (mg kg⁻¹ DW) in animal wastes.

Waste	As	Ba	Be	Cd	Co	Cr	Cu	Hg	Mn	Mo	Pb	Sb	Se	Sn	Ti	V	Zn
Feedlot diet[a-1]	0.10	18	<0.03	0.05	0.10	0.75	3.0	<0.01	17	<2.5	0.36	<0.08	0.19	<0.8	8.1	0.57	20
Cattle manure (low fiber diet)[a-2]	0.88	105	<0.03	0.28	1.7	20	24	0.05	117	30	2.1	<0.08	0.35	4.7	55	3.2	115
Cattle manure (high fiber diet)[a-3]	2.2	305	<0.03	0.24	2.2	31	21	<0.03	161	49	3.3	<0.08	0.32	7.4	129	8.0	86
Processed cattle waste pellets[a-4]	0.60	70	<0.03	0.14	1.1	5	19	<0.09	100	16	3.3	<0.08	0.36	3.7	50	3.0	77
Poultry waste with litter[a-5]	0.57	54	<0.03	0.42	2.0	6	31	0.06	166	5.0	2.1	<0.08	0.38	2.0	12	3.9	155
Poultry waste w/o litter	0.66	57	<0.03	0.58	1.2	5	20	<0.04	242	7.2	3.4	0.10	0.66	4.1	27	4.3	158
Poultry waste[b]	—	—	—	—	5.0	10	4.4	—	187	42	90?	—	—	—	—	—	36
Swine waste[b]	—	—	—	—	11.0	14	13	—	168	34	168?	—	—	—	—	—	198
Farmyard manure[b]	—	—	—	—	11.0	12	2.8	—	69	21	120?	—	—	—	—	—	15
Cow manure[c]	4.0	268	—	0.8	5.9	56	62	0.2	286	14	16	0.5	2.4	3.8	2800?	43	71

a-1Typical feedlot diet = 70% corn, 3% hay, 5% beet pulp, 20% corn silage, and 2% mineral supplement.
a-2From feedlot heifers fed 59% corn, 2% alfalfa hay, 3% molasses, 33% corn silage, and 3% mineral supplement.
a-3From feedlot heifers fed 24% corn, 29% alfalfa hay, 3% molasses, 41% corn silage, and 3% mineral supplement.
a-4A commercial high-protein feedlot animal waste product similar to item 3 pelletized.
a-5Includes wood shavings; used layer ration.
bArora, C.L., et al. 1975. *Indian J Agric Sci* 45:80–85; high Pb results may be due to contamination of samples.
cFurr, K., et al. 1976. *Environ Sci Technol* 7:683–687.
Source: Capar, S.G., et al. 1978. *Environ Sci Technol* 7:785–790.

APPENDIX TABLE A.11. Total concentration of trace elements (µg g^{-1} DW) in fly ash, bottom ash, FGD sludge, and oil ash.

Element	Fly ash[a]		Bottom ash		FGD sludge		Oil ash	
	Mean	Range	Mean	Range	Mean	Range	Mean	Range
Al	113,000	46,000–152,000	101,000	30,500–145,000	30,300	8400–73,000	36,700	3500–87,300
As	156.2	7.7–1385	7.6	<5–36.5	25.2	<5–53.1	112.3	74.8–174
Ba	1880	251–10,850	1565	150–9360	370	<25–2280	601.6	148–938
Cd	11.7	6.4–16.9	<5	<5	<5	<5	<5	<5
Cr	247.3	37–651	585	<40–4710	72.7	<40–168	2411	364–4390
Fe	76,000	25,000–177,000	105,000	20,200–201,000	64,000	1300–138,000	241,800	46,200–521,000
Mn	357	44–1332	426	56–1940	138	37.3–312	707	113–1170
Mo	43.6	7.1–236	14.4	2.8–443	12.2	<4.0–52.6	375	22–779
Na	9087	1300–62,500	6188	814–41,300	3817	200–16,500	4800	4300–5200
Ni	141	22.8–353	216	<10–1067	40.4	<5–145	10,186	6180–13,750
Pb	170.6	21.1–2120	46.7	4.6–843	29.2	<4–181	508	226–1029
Sb	42.5	11–131	<10	<10	15.5	—	444	12–1072
Se	14.0	5.5–46.9	4.1	<1.5–9.96	35.8	<2–162	28.7	23.1–34.2
Sn	43.6	7.9–56.4	28.2	<9–90.2	29.9	<10–50.5	28.9	13.0–37.8
Ti	6644	1310–10,100	5936	1540–11,300	1599	120–5050	3640	360–5110
V	271.5	<95–652	176	<50–275	156	<50–261	26,023	13,800–50,100
Zn	449.2	27–2880	127	3.8–515	99.7	7.7–612	1140	404–2310

[a]The total numbers of observation were 39, 40, 11, and 3 for fly ash, bottom ash, FGD sludge, and oil ash, respectively. Dashes imply no data.
Source: Extracted from Ainsworth, C., and, D. Rai 1987. *Chemical Characterization of Fossil Fuel Combustion Wastes*. EPRI Rep EA-5321. EPRI, Palo Alto, CA.

APPENDIX TABLE A.12. Concentrations (mg kg^{-1}) of trace elements in surface soils impacted by smelters.

Distance from smelter (km)	Pb smelter[a]							Distance from smelter (km)	Cu and Ni smelter[b]							
	Cd	Sb	Ag	Pb	Zn	Se	As		Cu	Ni	Co	Zn	Ag	Pb	Mn	V
0.4	83	155	30	7900	13,000	4.6	100	1.1	2892	5104	199	96	7.9	82	255	103
1.1	25	5	9.3	3200	870	0.76	49	1.6	2416	1851	80	65	3.5	53	202	63
2.4	—	32	6.0	1700	970	—	69	2.2	2418	2337	92	82	7.8	58	174	115
3.2	32	260	31	6700	1400	5.1	94	2.9	1657	1202	41	50	3.3	48	143	25
3.7	—	28	2.7	2000	200	—	24	7.4	1371	1771	46	87	2.9	46	299	165
5.3	18	18	2.8	1000	940	—	36	10.4	287	282	54	72	2.3	28	364	137
8.1	—	20	3.6	300	320	—	53	13.5	233	271	42	100	4.3	23	602	151
12.6	—	20	3.7	890	804	—	24	19.3	184	306	24	61	ND[c]	28	264	55
19.0	—	40	10	2200	3000	—	37	24.1	45	101	18	46	5.5	19	207	33
								32.1	46	35	16	55	1.9	26	195	96
								38.6	2	39	29	62	ND	28	192	169
								49.8	26	35	22	83	1.0	20	168	23

Sources:
[a] In Kellogg, Idaho. Samples (0–2 cm surface soil) were not taken on a transect (Ragaini, R.C., et al., 1977 Environ Sci Technol 8:773–781).
[b] In Sudbury Basin in Ontario, Canada. Samples (top 10 cm soil) were taken along a transect from smelter (Hutchinson, T.C., and L.M. Whitby. 1974. Environ Conserv 1:123–132).
[c] Not detectable.

APPENDIX TABLE. A.13. Certain physicochemical properties of selected elements that influence their biogeochemical behavior.

Element	Ionic radius (Å)	Electronegativity (kcal/g atom)	Ionic potential (charge/radius)	Diameter of hydrated ion in aqueous solution, (Å)
K^+	1.7–1.6	0.8	0.6	3.0
Na^+	1.2–1.1	0.9	0.9	4.5
Cs^+	2.0–1.9	0.7	0.5	2.5
Rb^+	1.8–1.7	0.8	0.6	2.5
Ca^{2+}	1.2–1.1	1.0	1.8	6.0
Mg^{2+}	0.8	1.2	2.5	8.0
Sr^{2+}	1.4–1.3	1.0	1.5	5.0
Ba^{2+}	1.7–1.5	0.9	1.3	5.0
Pb^{2+}	1.6–1.4	1.8	1.9	4.5
Se^{3+}	0.8	1.3	3.7	9.0
Fe^{2+}	0.9–0.7[a]	1.8	2.6	6.0
Cu^{2+}	0.8	2	2.5	6.0
Ge^{4+}	0.5	1.8	8.3	—
Mo^{4+}	0.7	—	5.5	—
Mn^{2+}	1–0.8	1.5	2.0	6.0
Zn^{2+}	0.9–0.7	1.8	2.6	6.0
Fe^{3+}	0.7–0.6[a]	1.9	4.4	9.0
Co^{2+}	0.8–0.7	1.7	2.6	6.0
Cd^{2+}	1.03	—	—	—
Ni^{2+}	0.8	1.7	2.6	6.0
Cr^{3+}	0.7	1.6	4.3	9.0
Mn^{4+}	0.6	—	6.5	—
Li^+	0.8	1.0	1.2	6.0
Mo^{6+}	0.5	1.8	12.0	—
V^{5+}	0.5	—	11.0	—
Al^{3+}	0.6–0.5	1.5	5.6	9.0
Be^{2+}	0.3	1.5	5.7	8.0
Cr^{6+}	0.4	—	16.0	—
La^{3+}	1.4–1.3	1.1	2.3	9.0
Sn^{2+}	1.3	1.8	1.5	—
Si^{4+}	0.4	1.8	10.0	—
Ti^{4+}	0.7	1.5	5.8	—

[a]Values given for high and low spin, respectively.

APPENDIX TABLE A.14. Surface and sorption properties of several soil minerals.

Mineral	Total surface area (m² g⁻¹)	Cation exchange capacity (meq 100 g⁻¹)	Total sorption of cations (mM g⁻¹)	Sorption of cations (meq 100 g⁻¹)			
				Cd^{2+}	Mn^{2+}	Zn^{2+}	Hg^{2+}
Kaolinite	7–30	3–22	30–70	3.1	3.5	3.4	0.46
Montmorillonite	700–800	80–150	390–460	60–86	72–116	88–108	0.4–2.2
Illite	65–100	20–50	65–95	—	—	—	—
Vermiculite	700–800	100–150	—	98	92	98	9.7
Gibbsite	25–58	—	—	—	—	—	—
Goethite	41–81	—	51–300	—	—	—	—
Manganese oxide	32–300	150	200–1000	—	—	—	—
Zeolites	720–880	—	—	—	—	—	—
Allophane	145–900	5–50	—	—	—	—	—
Plagioclase	—	7	—	0.47	0.26	1.2	0.14
Quartz	—	7	—	—	—	—	—

Source: Modified from Kabata-Pendias, A., and H. Pendias. 1992. *Trace Elements in Soils and Plants*. CRC Pr, Boca Raton, FL.

APPENDIX TABLE A.15. Contaminants likely to be present at former gasworks sites.

Inorganics	Metals	Volatile aromatics	Phenolics	Polynuclear aromatic hydrocarbons
Ammonia	Aluminum	Benzene	Phenol	Acenaphthene
Cyanide	Antimony	Ethylbenzene	2-Methylphenol	Acenaphthylene
Nitrate	Arsenic	Toluene	4-Methylphenol	Anthracene
Sulfate	Barium	Total xylenes	2,4-Dimethylphenol	Benzo(a)anthracene
Sulfide	Cadmium			Benzo(a)pyrene
Thiocyanates	Chromium(VI)			Benzo(b)fluoranthene
	Copper			Benzo(g,h,i)perylene
	Iron			Benzo(k)fluoranthene
	Lead			Chrysene
	Manganese			Dibenzo(a,h)-anthracene
	Mercury			Dibenzofuran
	Nickel			Fluoranthene
	Selenium			Fluorene
	Silver			Indenol(1,2,3-cd)-pyrene
	Vanadium			Naphthalene
	Zinc			Phenanthrene
				Pyrene
				2-Methylnaphthalene

Source: Nakles, D.V., D.G. Linz, and T.D. Hays. 1993. In *Management of Manufactured Gas Plant Sites Technology Seminar,* pp 32–36. Remed Technol. Inc., Arlington, VA.

APPENDIX TABLE A.16. Description of common components of municipal solid waste.

Category	Description
Paper and paperboard	
High grade paper	Office paper and computer paper
Mixed paper	Mixed colored papers, magazines, glossy paper, and other paper not fitting the categories of high grade paper, newsprint, and corrugated
Newsprint	Newspaper
Corrugated	Corrugated boxes, corrugated and brown (kraft) paper
Yard waste	Branches, twigs, leaves, grass, and other plant material
Food waste	All food waste excluding bones
Glass	Clear and colored glass
Plastics	All types of plastics
Ferrous metals	Iron, steel, tin cans, and bi-metal cans
Nonferrous metals	Primarily aluminum, aluminum cans, copper, brass, and lead
Wood	Lumber, wood products, pallets, and furniture
Rubber	Tires, footwear, wire cords, gaskets
Textiles	Clothing, furniture, footwear
Leather	Clothing, furniture, footwear
Miscellaneous	Other organic and inorganic materials, including rock, sand, dirt, ceramics, plaster, bones, ashes, etc.

Source: Rhyner, C.R., et al. 1995. *Waste Management and Resource Recovery.* Lewis Publ, Boca Raton, FL.

APPENDIX TABLE A.17. Estimates of the percentage of municipal solid waste (MSW) landfilled in selected countries.

Country	Year	% landfilled (by wt.)
Denmark	1985	44
France	1983	54
Greece	1983	100
Ireland	1985	100
Italy	1983	85
Japan	1987	33
Netherlands	1985	56–61
Sweden	1985, 1987	35–49
Switzerland	1985	22–25
United Kingdom	1983	90
United States	1986	90[a]
Germany (FRG)	1985, 1986	66–74

[a]This figure refers to landfilling after recycling (e.g., of source-separated glass, paper, metals) has occured. For example, in U.S. landfills about 80% of all MSW, but about 90% of post-recycled MSW.
Source: Modified from Rhyner, C.R., et al. 1995. *Waste Management and Resource Recovery.* Lewis Publ, Boca Raton, FL.

APPENDIX TABLE A.18. Modes of sewage sludge disposal in selected European countries (10^3 tonnes of dry sludge/year).[a]

Country	Agriculture and other fertilizers	Landfill	Incineration	Sea	Total
Austria	57(28)[b]	67(33)	74	—	200
Belgium	8(28)	15(52)	6	0	29
Germany (FRG)	690(25)	1510(55)	400	0	2750
Denmark	57(43)	39(30)	35	0	131
England and Wales	519(52)	126(13)	76	263	987
France	234(28)	446(52)	170	0	850
Greece	30(15)	170(85)	0	0	200
Ireland	7	4	0	12	23
Italy	270(34)	440(55)	90	0	800
Luxembourg	12	3	0	0	15
Netherlands	150(54)	81(29)	28	22	280
Portugal	160(80)	25(13)	0	15	200
Spain	173(62)	28(35)	0	79	280
Sweden	108(60)	72(40)	0	0	180
Switzerland	113(45)	80(32)	57	—	250

[a]Best available data taken from a variety of sources from databases during the period 1984–1991.
[b]Values in parentheses indicate percentage of total.
Source: Based on Matthews, P. 1999. Anglian Water Intl Ltd, Cambridgeshire, UK. Personal communication.

APPENDIX TABLE A.19. Legislated maximum concentration limits for metals in soils treated with sewage sludge and in sewage sludge suitable for release in agriculture in EU countries.

Metals in sewage sludge (mg kg⁻¹ DW)

Element	86/278/EEC	Belgium (Flanders)	Germany (pH <6)	Denmark (30/06/2000)	Spain (pH <7 / pH >7)	France	Italy	Netherlands	Switzerland	UK
Cd	20–40	6	5 10	0.8 0.4	20 40	20/15/10	20	1.25	2	—
Cr	—	250 1500	900	100	1000 1500	1000	—	75	100	—
Cu	1000–1750	375 750	800	1000	1000 1750	1000	1000	75	600	—
Hg	16–25	5	8	0.8	16 25	10	10	0.75	2.5	—
Ni	300–400	50	200	30	300 400	200	300	30	50	—
Pb	750–1200	300	900	120	750 1200	800	750	100	100	—
Zn	2500–4000	900 2500	2000 2500	4000	2500 4000	3000	2500	300	800	—

Metals in soil (mg kg⁻¹ DW)

Element	86/278/EEC	Belgium (Flanders)	Germany (pH <6)	Denmark	Spain (pH <7 / pH >7)	France	Italy	Netherlands	Switzerland	UK (5.0 <pH >7)
Cd	1–3	1.2	1.5 1	0.5	1 3	2	2	0.8	0.4	3
Cr	—	78	100	30	100 150	150	—	100	30	400
Cu	50–140	109	60	40	50 210	100	100	36	40	80 200
Hg	1–1.5	5.3	1	0.5	1 1.5	1	1	0.3	0.3	1
Ni	30–75	55	50	15	30 112	50	75	35	30	50 110
Pb	50–300	120	100	40	50 300	100	100	85	40	300
Zn	150–300	350	200 150	100	150 450	300	300	140	75	200 300

Source: Courtesy of Dr. L. Marmo—DGXI.E.3, Commission of European Community, Brussels, Belgium.

APPENDIX TABLE A.20. Percentage of lakes in potentially acidified regions of the United States and Canada above or below specified reference values.

Location	pH ≤5.0	pH ≤6.0	ANC (μeq L^{-1}) ≤0	ANC (μeq L^{-1}) ≤200	DOC (μM) >417	Al[a] >100 μg/L
United States[b]						
Adirondacks	14.4	35.8	18.4	85.9	33.2	18.1
Maine	0.5	5.1	1.5	69.4	57.1	0.5
Upper Midwest	2.9	14.5	5.4	57.8	73.9	1.0
Southern Blue Ridge	0.0	1.2	0.0	70.4	1.7	0.0
Florida	34.8	67.7	46.0	87.1	62.7	4.5
Western Mountains	0.0	1.2	0.0	83.0	4.5	0.0
Canada[c]						
Northwest Ontario	0.0	4.0	0.0	39.7	81.2	n/a
Northeast Ontario	4.5	18.1	6.9	53.2	48.0	n/a
South-Central Ontario	1.7	29.8	2.8	83.9	31.9	n/a
Quebec	4.1	42.0	3.1	87.0	64.2	n/a
Labrador	0.4	16.6	0.0	95.1	52.5	n/a
New Brunswick	7.1	35.3	11.1	96.3	47.4	
Nova Scotia	25.8	75.1	27.5	97.1	54.6	n/a
Newfoundland	3.0	44.3	6.8	94.3	52.6	n/a

[a]Methyl-isobutyl-ketone extractable Al.

[b]Lakes with significant landuse disturbances have been excluded. Values represent weighted population estimates for lakes with surface area >4 ha and ≤2000 ha.

[c]Not all Canadian lakes have complete chemical characterization; therefore, percentages of lakes represent values based on unequal numbers of lakes for each variable. Caution should be used in interpreting the significance of the percentage of Canadian lakes above or below reference values, because it is uncertain how the sample represents the total population of Canadian lakes.

Source: Modified from Sparling, D.W. 1995. In D.J. Hoffman et al., eds. Handbook of Ecotoxicology. Lewis Publ, Boca Raton, FL.

APPENDIX TABLE A.21. Guideline values (mg L^{-1}) for trace elements in drinking water.

Constituent	Guideline value, mg L^{-1}	Remarks
Arsenic	0.05	Under review in view of massive As poisoning in Bangladesh and India
Barium	No guideline value set	
Beryllium	No guideline value set	
Cadmium	0.005	
Chromium	0.05	
Fluoride	1.5	Natural or deliberately added; local or climatic conditions may necessitate adaption
Lead	0.05	
Mercury	0.001	Methylated species are of most concern
Nickel	No guideline value set	
Selenium	0.01	
Silver	No guideline value set	

Source: World Health Organization (WHO). 1984. *Guidelines for Drinking-Water Quality, vol. 1: Recommendations.* Geneva.

APPENDIX TABLE A.22. Means of total metals in soils (mg kg^{-1}) of selected countries.

Country	Cu	Zn	Ni	Cr	Pb	Cd
Austria	17	65	20	20	150	0.20
Belgium	17	57	33	90	38	0.33
Denmark	11	7	7	21	16	0.24
France	13	16	35	29	30	0.74
Germany	22	83	15	55	56	0.52
Italy	51	89	46	100	21	0.53
Netherlands	18.6	72.5	15.6	25.4	60.2	1.76
Norway	19	60	61	110	61	0.95
Portugal	24.5	58.4	—	—	—	—
Spain	14	59	28	38	35	1.70
Sweden	8.5	182	4.4	2.3	69	1.20
England and Wales	15.6	78.2	22.1	44	48.7	0.70
Scotland	23	58	37.7	150	19	0.47
Calculated average	19.5	68	27	52.7	39	0.79
Holmgren et al. (1993) (USA)						
5 percentile	3.8	8.0	4.1	—	4.0	0.036
50 percentile	18.5	53	18.2	—	11	0.20
95 percentile	94.9	126	56.8	—	23	0.78
China	23	74	27	—	27	0.097
United States	25	60	19	—	19	—

Sources: Data for western Europe extracted from Angelone, M., and C. Bini. 1992. In D.C. Adriano, ed. *Biogeochemistry of Trace Elements.* Lewis Publ, Boca Raton, FL. Chinese soils from Chen, J., et al. 1991. *Water Air Soil Pollut* 57–58:699–712; U.S. soils from Shacklette, H.T., and J.G. Boerngen. 1984. In U.S. Geological Survey Prof Paper 1270. USGS, Washington, DC.

APPENDIX TABLE A.23. Background values[a] (μg kg^{-1} DW) for mobile (extractable)[b] trace metals in soils in Germany (DIN 19730).

Element	Soil pH[c]				
	4.5–5.0	5.5–6.0	6.0–6.5	6.5–7.0	>7.5
Silver	1.5	1.5	1.5	1.5	1.5
Arsenic	40	40	40	40	50
Beryllium	20	1	0.6	0.4	0.4
Bismuth	1	1	1	1	3
Cadmium	20	10	5	3	3
Cobalt	200	30	25	20	20
Chromium	15	10	10	12	15
Copper	250	250	250	300	400
Mercury	1	1	1	1	1
Manganese	25,000	15,000	10,000	5000	3000
Molybdenum	10	30	50	60	110
Nickel	600	250	200	200	200
Lead	150	15	10	6	3
Antimony	5	7	10	20	40
Thallium	20	12	10	12	15
Uranium	3	3	3	3	5
Vanadium	20	15	15	15	30
Zinc	3000	300	200	170	100

[a]Determined as 90 percentile values.
[b]Soils extracted with 1 M NH$_4$NO$_3$ solution.
[c]Usual pH range in agricultural soils = 5.5 to 7.5.
Source: Modified from Pruess, A. 1997. In R. Prost, ed. *Contaminated Soils.* INRA 85. INRA, Paris.

APPENDIX TABLE A.24. Median (M) or average (A) contents (mg kg⁻¹ DW) and ranges in contents reported for elements in soils and other surficial materials.

Element	Bowen (1979)[a]		Shacklette and Boerngen (1984)[b]		Alloway (1995)	Rose et al. (1979)[c]		Kabata-Pendias and Pendias (1992)
	Median	Range	Average	Range		Average or median		
Ag	0.05	0.01–8	—	—	0.01–8	—		—
As	6	0.1–40	7.2	<0.1–97	0.1–50	7.5	(M)	0.07–197
B	20	2–270	33	<20–300	—	29	(M)	—
Ba	500	100–3000	580	10–5000	—	300	(M)	—
Be	0.3	0.01–40	0.92	<1–15	—	0.5–4		—
Bi	0.2	0.1–13	—	—	—	—		—
Cd	0.35	0.01–2	—	—	0.01–2.4	—		0.01–2.53
Co	8	0.05–65	9.1	<3–70	1–40	10	(M)	0.1–122
Cr	70	5–1500	54	1–2000	5–1500	6.3	(M)	1–1100
Cs	4	0.3–20	—	—	—	—		—
Cu	30	2–250	25	<1–700	2–250	15	(M)	1–323
F	200	20–700	430	<10–3700	—	300	(M)	—
Hg	0.06	0.01–0.5	0.09	<0.01–4.6	0.01–0.3	0.056	(M)	0.0014–5.8
Mn	1000	20–10,000	550	<2–7000	20–10,000	320	(M)	7–8423
Mo	1.2	0.1–40	0.97	<3–15	0.2–5	2.5	(A)	0.2–17
Ni	50	2–750	19	<5–700	2–1000	17	(M)	0.2–450
Pb	35	2–300	19	<10–700	2–300	17	(M)	1.5–286
Sb	1	0.2–10	0.66	<1–8.8	0.05–260	2	(A)	0.05–4
Se	0.4	0.01–12	0.39	<0.1–4.3	0.01–2	0.3	(M)	0.005–4
Sn	4	1–200	1.3	<0.1–10	1–200	10	(A)	—
Sr	250	4–2000	240	<5–3000	—	67	(M)	—
Ti	5000	150–25,000	2900	70–20,000	—	—		—
Tl	0.2	0.1–0.8	—	—	0.03–10	—		—
V	90	3–500	80	<7–500	3–500	57	(M)	0.7–500
W	1.5	0.5–83	—	—	0.5–83	—		—
Zn	90	1–900	60	<5–2900	10–300	36	(M)	3–770

[a]Compilation from numerous sources.
[b]For conterminous USA.
[c]Elements useful in geochemical prospecting.

Sources: Alloway, B.J. 1995. Heavy Metals in Soils. Chapman & Hall, London. Kabata-Pendias, A., and H. Pendias. 1992. Trace Elements in Soils and Plants. CRC Pr, Boca Raton, FL.; Shacklette, H.T., and J.G. Boerngen. 1984. In USGS Prof. Paper 1270. USGS Printing Office, Washington, DC; Bowen, H.J.M. 1979. Environmental Chemistry of the Elements. Academic Pr, New York; Rose, A.W. et al., 1979. Geochemistry in Mineral Exploration. Academic Pr, London.

APPENDIX TABLE A.25. Cleanup goals (actual and potential) for total and leachable metals.

Description	As	Cd	Cr (Total)	Hg	Pb
	Total metal concentration (mg kg^{-1})				
Background (mean)	5	0.06	100	0.03	10
Background (range)	1 to 50	0.01 to 0.70	1 to 1000	0.01 to 0.30	2 to 200
Superfund site goals from TRD	5 to 65	3 to 20	6.7 to 375	1 to 21	200 to 500
Theoretical minimum total metals to ensure TCLP Leachate <threshold (i.e., TCLP × 20)	100	20	100	4	100
California total threshold limit concentration	500	100	500	20	1000
	Leachable metal concentration (μg L^{-1})				
TCLP threshold for RCRA waste (SW 846, Method 1311)	5000	1000	5000	200	5000
Extraction procedure; Toxicity test (EP Tox) (Method 1310)	5000	1000	5000	200	5000
Synthetic precipitate leachate procedure (Method 1312)	—[b]	—	—	—	—
Multiple extraction procedure (Method 1320)	—	—	—	—	—
California soluble threshold leachate concentration	5000	1000	5000	200	5000
Maximum contaminant level[a]	50	5	100	2	15
Superfund site goals from TRD[c]	50	—	50	0.05 to 2	50

[a]Maximum contaminant level = maximum permissible level of contaminant in water delivered to any user of a public system.
[b]No specified level and no example cases identified.
[c]TRD = EPA Technical Resource Document. U.S. EPA (1997).

APPENDIX TABLE A.26. Pathways for risk assessment for potential transfer of trace contaminants in biosolids to humans, livestock, or the environment, and the highly exposed individuals to be protected by a regulation based on the pathway analysis. Each pathway presumes 1000 t dry biosolids ha^{-1} and/or annual application of biosolids as N fertilizer.

Exposure pathway	Definition of highly exposed individuals
1. Sludge → Soil → Plant → Human	Individuals whose diets contain 2.5% of all food produced on biosolids-amended soils.
2. Sludge → Soil → Plant → Home gardener	Home gardeners who apply 1000 t ha^{-1} to strongly acidic soils and consume 60% "home garden" foods for 70 years.
3. Sludge → Soil → Child	Children ages 1–6 who ingest biosolid products (ingesting soil) at 200 mg day^{-1} for 5 yr.
4. Sludge → Soil → Plant → Animal → Human	Farm family members whose diets include meat and dairy products, 45% of which are produced using crops grown on soil with 1000 t ha^{-1}.
5. Sludge → Soil → Animal → Human	Same as Pathway 4, except annual surface applications allow soil ingestion.
6. Sludge → Soil → Plant → Animal	Livestock which consume feedstuffs that are grown on soils with 1000 t biosolids ha^{-1}.
7. Sludge → Soil → Animal	Livestock which graze $\frac{1}{3}$ of year on pastures with surface applied or incorporated (1000 t ha^{-1}) biosolids.
8. Sludge → Soil → Plant	"Crops" grown on strongly acidic soils amended with 1000 t biosolids ha^{-1}.
9. Sludge → Soil → Soil-biota	Earthworms, microbes, etc., in soil with 1000 t ha^{-1}.
10. Sludge → Soil → Soil-biota → Predator of soil-biota	Shrews or other small mammals which live on site and consume 33% earthworms in diet for whole life.
11. Sludge → Soil → Airborne dust → Human	Tractor operator for normal farm tillage.
12. Sludge → Soil → Contaminated surface water → Fish → Human	Individuals whose diets consist mainly of fish.
13. Sludge → Soil → Air → Human	Farm households.
14. Sludge → Soil → Groundwater → Human	Well water on farms where individuals rely solely on this source for their water supply.

APPENDIX TABLE A.27. Soil-washing removal efficiency for selected metals.

Site description	Location[a]	Metal	Concentration (mg kg⁻¹ DW)		
			Before	After	% Removal
Chemical plant	CA[1]	Pb	12,000	3800	68%
Wood preserving	CA[1]	As	89	27	70%
		Cr	63	23	63%
		Cu	29	13	55%
		Zn	345	108	69%
Wood preserving	NC[2]	As	289	64	78%
		Cr	195	51	74%
Wire drawing	NJ[3]	Cu	330	100	70%
		Ni	110	60	45%
		Ag	25	4	84%
Pharmaceutical	PA[4]	As	850	300	65%
Pesticide mfg.	CO[5]	As	68	18	74%
		As	150	20	87%
Battery recycling	VA[6]	Pb	14,506	1200	92%
		TCLP-Pb	1000	40–50	95%
Pharmaceutical	NJ[3]	As	1440	235	84%
		Pb	135	25	81%
		Hg	7	0.3	96%
		Se	2.5	0.3	88%

[a]1, 2, 3, 4, 5, 6 represent the states of California, North Carolina, New Jersey, Pennsylvania, Colorado, and Virginia, respectively.
Source: Bio Trol. 1993. Bio Trol, Inco, Chaska, MN. Personal communication.

APPENDIX TABLE A.28. Representative concentrations of cadmium in food crops field-grown in different regions of the United States.

Crop	Number of samples	Average concentration ($\mu g\ g^{-1}$)
Leafy vegetables		
Cabbage	24	0.093
Lettuce	40	0.420
Spinach	104	0.800
Swiss chard	16	0.470
Seed vegetables		
Sweet corn	268	0.016
Dry beans	35	0.110
Snap beans	42	0.024
Vegetable fruits		
Cucumber	22	0.093
Eggplant	2	0.380
Pepper	2	0.021
Tomato	231	0.270
Fruits		
Apple	36	0.034
Grapefruit	23	0.006
Grape	21	0.011
Orange	20	0.005
Peach	24	0.011
Pear	38	0.006
Plum	16	0.006
Field crops		
Wheat		
USA	288	0.048
Canada	—	0.045
Argentina	—	0.013
Australia	—	0.015
Japan	—	0.085
Barley	29	0.027
Rice	166	0.013
Field corn	277	0.014
Sorghum	36	0.033
Soybeans	322	0.064
Root and bulb crops		
Asparagus	10	0.032
Carrot	207	0.250
Onion	10	0.050
Potato	297	0.170
Peanut	320	0.090
Radish	17	0.310
Turnip	5	0.170

Source: Nriagu, J.O., and M.S. Simmons, eds. 1990. *Food Contamination from Environmental Sources*. Wiley, New York.

APPENDIX TABLE A.29. Intake of cadmium associated with the consumption of fish and seafood by populations of selected countries.

Country	Population (in millions)	Edible portion consumed[a]	Freshwater (biota)[c]	Seawater			Estimated Cd ingested ($\mu g \ day^{-1}$)[c]
				Fish[b]	Crustaceans[b]	Mollusks[b]	
Maldives	0.14	186(37.1)	—	186(100)	—	—	9.3
Japan	120	113(22.5)	7.5(7)	87(77)	5.0(4)	12(10)	10
Iceland	0.1	96(19.2)	3.0(3)	80(83)	12(13)	—	8.1
Bermuda	0.06	79(15.8)	—	69(87)	10(13)	—	6.4
Portugal	8.8	51(10.2)	—	69(87)	1.0(2)	1.5(3)	4.2
Former Soviet Union	260	47(9.4)	8.0(81)	38(81)	—	0.5(1)	2.8
Australia	13.8	17(3.4)	1.5(9)	12(68)	4.0(24)	—	1.9
United States	255	16(3.2)	1.0(6)	10(63)	3.5(22)	1.0(6)	2.0

[a]In grams living weight per day; equivalent grams of protein per day given in parentheses.

[b]Percentage in relation to total consumption is given in parentheses.

[c]Estimated assuming that the Cd concentrations in edible parts of fish, crustaceans, mollusks, and freshwater biota are 0.05, 0.30 and 0.10 $\mu g \ g^{-1}$, respectively.

Source: Modified from Nriagu, J.O., and M.S. Simmons, eds. 1990. Food Contamination from Environmental Sources. Wiley, New York.

APPENDIX TABLE A.30. Lead contents of agronomic and horticultural crops reported in the literature. To convert to dry wt. basis, use values of ~85 to 92% water in wet tissues. Unless noted, values may be considered as background values.

Crop	n	Pb contents (mg kg^{-1})				Country
		Min.	Max.	Mean	Median	
Lettuce	75	0.03	2.25	0.14	0.09	Netherlands
	31	<0.027	0.280	0.045	0.033	Denmark
	31	0.003	0.161	0.043		Sweden
	75	—	—	0.30	0.20	United Kingdom[a]
	150	0.001	0.078	0.013	0.008	United States
Tomato	40	0.002	0.08	0.010	0.008	Netherlands
	30	0.006	0.04	0.01		Sweden
	53	<0.018	0.10	0.02	0.01	Denmark
	231	<0.0001	0.025	0.002	0.002	United States
Cucumber	45	0.001	0.014	0.005	0.004	Netherlands
	17	0.006	0.044	0.013		Sweden
Spinach	82	0.01	0.29	0.09	0.08	Netherlands
	104	0.016	0.17	0.045	0.039	United States
	20	—	—	1.3	0.34	United Kingdom[a]
Endive	82	0.03	0.43	0.08	0.07	Netherlands
Curly kale	19	0.15	0.91	0.56	0.54	Netherlands
	33	0.53	5.9	1.7	1.4	Netherlands
	28	0.05	5.2	0.44	0.26	Denmark
	17	—	—	1.0	0.36	United Kingdom[a]
Cauliflower	84	0.002	0.38	0.012	0.007	Netherlands
	12	0.009	0.020	0.015		Sweden
	15			0.04		Germany
Cabbage	96	0.002	0.23	0.015	0.006	Netherlands
	41	0	0.04	0.01		Sweden
	42	<0.02	0.09	0.01	<0.01	Denmark
	66	—	—	0.46	0.22	United Kingdom[a]
	31	<0.01	0.54	0.09		United Kingdom

continued

APPENDIX TABLE A.30. (*continued*)

Crop	n	Min.	Max.	Mean	Median	Country
				Pb contents (mg kg^{-1})		
Carrot	100	0.011	0.21	0.05	0.04	Netherlands
	47	0.004	0.141	0.021		Sweden
	15	<0.01	0.12	0.04		United Kingdom
	207	0.001	0.125	0.009	0.0065	United States
	71	—	—	0.49	0.24	United Kingdom[a]
Onion	83	0.009	0.05	0.02	0.02	Netherlands
	21	0.005	0.05	0.01		Sweden
	39	<0.03	0.21	0.03	<0.01	Denmark
	230	<0.0002	0.054	0.005	0.004	United States
Potato	154	0.01	0.08	0.03	0.03	Netherlands
	63	0.01	0.06	0.02		Sweden
	20	0.01	0.03	0.02		Finland
	66	0.02	1.90	0.36	0.15	Germany
	62	—	—	0.16	0.13	United Kingdom[a]
	297	0.0002	0.370	0.009	0.005	United States
Wheat	84	0.03	0.65	0.16	0.14	Netherlands
	85	0.010	0.14	0.050		Finland
	44	0.05	0.40	0.126		Germany
	115	0.001	0.670	0.057		Austria
	288	<0.0008	0.716	0.037	0.0017	United States
Barley	45	0.08	0.71	0.27	0.24	Netherlands
	47	0.01	0.14	0.06		Finland
Oats	39	0.09	0.52	0.30	0.28	Netherlands
	31	0.030	0.110	0.061		Finland
Apple	100	0.03	0.12	0.06	0.005	Netherlands
	17	0.009	0.288	0.033		Sweden

[a]From a contaminated site in Shipham, UK (Alloway et al. 1988. *Sci Total Environ* 75:41–69).
Source: Wiersma D., B.J. Van Goor, and N.G. Van der, Veen 1986. *J Agric Food Chem* 34:1067–1074.

APPENDIX TABLE A.31. Mercury concentrations in the top ten types of fish consumed by the general U.S. population.

Fish	μg Hg/g WW	Comments
Tuna	0.206	From the average of mean concentrations measured in 3 types of tuna: albacore tuna (0.264 μg/g), skipjack tuna (0.136 μg/g), and yellowfin tuna (0.218 μg/g).
Shrimp	0.047	From the average of mean concentrations measured in 7 types of shrimp: royal red shrimp (0.074 μg/g), white shrimp (0.054 μg/g), brown shrimp (0.048 μg/g), ocean shrimp (0.053 μg/g), pink shrimp (0.031 μg/g), pink northern shrimp (0.024 μg/g), and Alaska (sidestripe) shrimp (0.042 μg/g).
Pollack	0.15	The U.S. FDA (1991/1992) reports the methyl Hg concentration in pollack in commerce as (0.04 μg/g).
Salmon	0.035	From the average of mean concentrations measured in five types of salmon: pink (0.019 μg/g), chum (0.030 μg/g), coho (0.038 μg/g), sockeye (0.027 μg/g), and chinook (0.063 μg/g).
Cod	0.121	From the average of mean concentrations in Atlantic cod (0.114 μg/g) and the Pacific cod (0.127 μg/g).
Catfish	0.088 to 0.16	From U.S. freshwater sources, including channel, largemouth, rock, striped, flathead, and white catfish. Not including raised catfish, which is the type of catfish predominantly consumed in the United States. The Hg content of farm-raised catfish may be significantly different than freshwater sources.
Clam	0.023	From the average of the mean concentrations measured in 4 types of clam: hard clam (0.034 μg/g), Pacific littleneck clam (0 μg/g), soft clam (0.027 μg/g), and geoduck clam (0.032 μg/g).
Flatfish (Flounder)	0.092	From the average of mean concentrations measured in nine types of flounder: Gulf (0.147 μg/g), summer (0.127 μg/g), southern (0.078 μg/g), four-spot (0.090 μg/g), windowpane (0.151 μg/g), arrowtooth (0.020 μg/g), witch (0.03 μg/g), yellowtail (0.067 μg/g), and winter (0.066 μg/g).
Crab	0.117	From the average of the mean concentrations measured in five types of crab: blue crab (0.140 μg/g), dungeness crab (0.183 μg/g), king crab (0.070 μg/g), tanner crab (*C. opilio*) (0.088 μg/g), and tanner crab (*C. bairdl*) (0.102 μg/g).
Scallop	0.042	From the average of the mean concentrations measured in four types of scallop: sea (smooth) scallop (0.101 μg/g), Atlantic bay scallop (0.038 μg/g), calico scallop (0.026 μg/g), and pink scallop (0.004 μg/g).

Source: Modified from the U.S. EPA (1996). *Mercury.* USEPA/R-96-001a.

APPENDIX TABLE A.32. Typical concentrations of mercury in fish muscle (ng g^{-1} WW) from studies over the last three decades.

Year reported	Total Hg mean	Total Hg range	Methyl Hg, mean	% Methyl Hg	Species	Location	Remark
1989	270	210–420	—	—	Sea bream	Ghana,	Coastal sites
1989	330	240–470	—	—	Yellow fin tuna	Africa	
1990	125	31–233	—	—	Yellow perch	Ontario	Uncontaminated
1990b	130	10–400	—	—	Roach	E. Anglia,	Two rivers,
1990	280	60–1400			Eels	UK	uncontaminated
1990	366	<10–2360	—	99	Yellow perch	Michigan, USA	Lake survey
1990	448	70–1164			Esox lucius		
1990	163	<10–590			White sucker		
1992	341	46–1969	—	—	Flounder	SE Irish Sea	Near UK NW estuaries
1992	—	87–364	—	99	Largemouth bass	Midwest lakes,	Uncontaminated
		160–260		99	Yellow perch	USA	
1993	1370	—	—	—	Rock bass	Virginia, USA	Contaminated
1993	520	40–2040	—	—	Largemouth	Florida lakes, USA	Uncontaminated
1995	—	—	685	96	Rainbow trout	N. Zealand	Volcanic zone
1995	720	50–1310	—	—	Juvenile bass	California reservoir	Abandoned Hg mine
1995	77	—	41	22–80	Redfish	Barrents Sea,	Background
1995	73	—	27	22–>100	Capelin	Arctic Circle	Concentration of Hg in fish
1995	67	—	41	11–>100	Haddock		
1996	255	57–501	223	87.5	Eel	R. Yare, UK	Previously contaminated
1996	55	28–121	47.5	86.4	Roach		
1996	104	15–285	84	82	Eels	Ormesby, UK	Uncontaminated, broad
1996	54	19–93	44	81	Roach		
1996	2000	250–410	—	—	Barbel	G'llego River, Spain	Downstream from a chlor-alkali plant
1996	740	300–360	—	—	Common carp		
1996	1730	1200–2610	—	—	Northern pike		

Source: Modified from Downs et al. 1998. *Water Air Soil Pollut* 108:149–189.

APPENDIX TABLE A.33. Common compounds of selenium in nature.

Species	Formula	Description/occurrence
Selenate (VI) Selenic acid	SeO_4^{2-} H_2SeO_4	Se(VI) is very soluble and is stable in oxidized environments, and very mobile in soils, hence easily available to plants. Slowly converted to more reduced forms: not as strongly adsorbed as Se(IV).
	$HSeO_4^-$	Most common form in soils.
Selenite (IV)	SeO_3^{2-}	Soluble form, common in mildly oxidizing conditions.
Trimethylselenonium (TMSe$^+$)	$(CH_3)_3Se^+$	TMSe$^+$ is an important urinary metabolite of dietary Se and is made rapidly unavailable to plants by fixation and volatilization.
Selenous acid	H_2SeO_3	Selenous acid is protonated in acid/neutral conditions. Se(IV) is easily reduced to Se^0 by ascorbic acid (vitamin C) or sulfur dioxide in acidic environments by microorganisms. Readily immobilized by Fe oxides, amorphous Fe hydroxides, and aluminum sesquioxides in soils.
Selenium dioxide	SeO_2	Gas SeO_2 formed as a product of fossil fuel combustion (sublimation temperature, 300 °C): dissolves in water to form selenous acid.
	$HSeO_3^-$	Common in soils.
Elemental selenium (0)	Se^0	Stable in highly reducing environments; (a) red crystalline alpha and beta monoclinic; (b) red glossy or black amorphous forms, all insoluble in water and oxidation/reduction very slow.
Selenides (−II)	Se^{2-}	Stable in highly reducing environments; forms insoluble metal complexes; highly immobile.
Dimethylselenide (DMSe)	$(CH_3)_2Se$	Gas formed by volatilization from soil bacteria and fungi.
Dimethyldiselenide (DMDSe)	$(CH_3)_2Se_2$	Gas formed by volatilization of plants.
Methaneselenone	$(CH_3)_2SeO_2$	Volatile metabolite, possibly formed as a final intermediate prior to reduction to DMSe.
Dimethylselenylsulfide	$(CH_3)_2SeS$	
Hydrogen selenide	H_2Se	Gas, unstable in moist air; decomposes to Se^0 in water.

Appendix B
Abbreviations, Acronyms, Symbols, and Terms

ACGIH	American Conference of Governmental Industrial Hygienists
ADI	acceptable daily intake
ADP	adenosine diphosphate
ALAD	δ-aminolevulinic acid dehydrogenase
ANC	acid-neutralizing capacity
Aqua regia	$HNO_3 + HCl$ (1:3 by volume)
ARARs	applicable or relevant and appropriate requirements
Arsine	AsH_3
ATP	adenosine triphosphate
ATSDR	Agency for Toxic Substances and Disease Registry (U.S.)
BG	background level
CA	cacodylic acid
CDC	Centers for Disease Control and Prevention (U.S.)
CDTA	trans 1,2-diaminocyclohexane-N,N,N',N'-tetraacetic acid
CEC	cation exchange capacity; also Commission of the European Communities
CERCLA	Comprehensive Environmental Response, Compensation, and Liability Act
CNS	central nervous system
CO_2	carbon dioxide
DBT	dibutyltin
DDT	dichlorodipthenyltrichloroethane
Dithizone	diphenylthiocarbazone
DMA	dimethylarsinic acid or dimethyl arsinate, $[(CH_3)_2AsOO^-]$
DMDSe	dimethyldiselenide, $[(CH_3)_2Se_2]$
DMSe	dimethylselenide, $[(CH_3)_2Se]$

DNA	deoxyribonucleic acid
2,4-D	2,4-dichlorophenoxyacetic acid
2,4-DNP	2,4-dinitrophenol
DOC	dissolved organic carbon
Double (mixed) acid extractant	0.05 N HCl + 0.025 N H_2SO_4
DSMA	disodium methanearsonate
DTPA	diethylenetriaminepentaacetic acid
DTPA-TEA	diethylenetriaminepentaacetic acid-triethanol-amine
DW	dry weight
EAE	environmentally acceptable endpoint
EDDHA	ethylenediaminedi(o-hydroxyphenylacetic) acid
EDTA	ethylenediaminetetraacetic acid
EGTA	ethyleneglycol-bis(β-aminoethylether)-N,N,N',N'-tetraacetic acid
Eh	redox potential
EU	European Union
EXAFS	extended X-ray absorption fine structure spectroscopy
FAO	Food and Agriculture Organization of the United Nations
GI tract	gastrointestinal tract
HBEF	Hubbard Brook Experimental Forest (New Hampshire, U.S.)
HEDTA	N-hydroxyethylenediaminetriacetic acid
HEI	highly exposed individual
HEIDA	N-(2-hydroxyethyl)iminodiacetic acid
HOAc	acetic acid
Hydrogen selenide	H_2Se
IAA	indoleacetic acid, a plant hormone (auxin)
ICRP	International Commission on Radiological Protection
IQ	intelligence quotient
KCN	potassium cyanide
K_d	solid-liquid partition coefficient
LC_{50}	lethal concentration 50; the concentration of a chemical in the air or water that will result in the death of 50 percent of the exposed species
LD_{50}	lethal dose 50 (median lethal dose); a statistically derived single dose of a chemical that will result in the death of 50 percent of the test species under a specific set of conditions

LOAEC	lowest observed adverse effect concentration
MBT	monobutyltin
MCL	maximum contaminant level
MCLG	maximum contaminant level goal
µeq	microequivalent
meq	milliequivalent
Methaneselenol	CH_3SeH
Methaneselenone	$(CH_3)_2SeO_2$
MMA	monomethylarsonic acid or monomethylarsenate, $(CH_3AsO_2OH^-)$
MSMA	monosodium methanearsonate
MSW	municipal solid waste
MTR	maximum tolerable risk
NAS	National Academy of Science (U.S.)
NIOSH	National Institute of Occupational Safety and Health (U.S.)
NOAEC	no observed adverse effect concentration
NO_X	nitrogen oxides
NPL	National Priority List (U.S.)
NRC	National Research Council (U.S.)
NTA	nitrilotriacetic acid
OAc	acetate
OC	organic carbon
OECD	Organization for Economic Cooperation and Development
OM	organic matter
Organic selenides	R_2Se
OSHA	Occupational Safety and Health Administration (U.S.)
PAH	polycyclic aromatic hydrocarbon
parts per billion (ppb)	$\mu g\ kg^{-1}$ or $\mu g\ L^{-1}$
parts per million (ppm)	$mg\ kg^{-1}$ or $\mu g\ g^{-1}$
parts per trillion (ppt)	$ng\ kg^{-1}$
PCB	polychlorinated biphenyl
PEL	permissible exposure limit
PMA	phenylmercuric acetate
PVC	polyvinyl chloride
RAOs	remedial action objectives
RCRA	Resource Conservation and Recovery Act (U.S.)
RDA	recommended dietary allowances
RfC	reference concentration (risk)
RfD	reference dose (risk)
RI/FS	remedial investigation/feasibility study

RNA	ribonucleic acid
ROD	record of decision
SARA	Superfund Amendments and Reauthorization Act (U.S.)
Selenic acid	H_2SeO_4
Selenous acid	H_2SeO_3
SO_X	sulfur oxides
TBT	tributyltin
TCLP	toxicity characteristic leaching procedure
TDI	tolerable daily intake
TMAO	trimethylarsine oxide, $[(CH_3)_3AsO^-]$
$TMSe^+$	trimethylselenonium, $[(CH_3)_3Se^+]$
UN	United Nations
UNESCO	United Nations Education, Scientific and Cultural Organization
U.S. DHHS	United States Department of Health and Human Services
U.S. DOD	United States Department of Defense
U.S. DOE	United States Department of Energy
U.S. EPA	United States Environmental Protection Agency
U.S. FDA	United States Food and Drug Administration
U.S. HUD	United States Department of Housing and Urban Development
USDA	United States Department of Agriculture
USGS	United States Geological Survey
WHO	World Health Organization
WW	wet (fresh) weight
XANES	X-ray absorption near-edge structure spectroscopy

Appendix C
Scientific Names for Plants and Animals

Plants

Alfalfa	*Medicago sativa*
Algae (green)	*Scenedesmus obliquus*
Apple	*Malus sylvestris*
Apricot	*Prunus armeniaca*
Artichoke (Jerusalem)	*Helianthus tuberosa*
Ash	*Fraxinus* spp.
Asparagus	*Asparagus officinallis*
Aspen	*Populus* spp.
Aster (heath)	*Aster ericoides*
Avocado	*Persea* spp.
Azalea	*Rhododendron* spp.
Banana	*Musa* spp.
Barley	*Hordeum vulgare*
Bean (bush)	*Phaseolus vulgaris*
Bean (common)	*Phaseolus vulgaris*
Bean (dwarf kidney)	*Phaseolus vulgaris*
Bean (French)	*Phaseolus vulgaris*
Bean (kidney)	*Phaseolus vulgaris*
Bean (lima)	*Phaseolus vulgaris*
Bean (navy)	*Phaseolus vulgaris*
Bean (pinto)	*Phaseolus vulgaris*
Bean (red mexican)	*Phaseolus vulgaris*
Bean (snap)	*Phaseolus vulgaris*
Bean (white pea)	*Phaseolus vulgaris*
Beech	*Fagus* spp.
Beech (American)	*Fagus grandifolia*
Beet (common and table)	*Beta vulgaris*
Beet (red)	*Beta vulgaris*
Beet (sugar)	*Beta saccharifera*
Bell pepper	*Capsicum annum grossum*
Bentgrass (creeping)	*Agrostis alba*
Bentgrass (velvet)	*Agrostis canina*
Bermudagrass	*Cynodon dactylon*
Berseem	*Trifolium alexandrinum*
Bilberry	*Vaccinium* spp.

Birch	*Betula* spp.
Birch (white)	*Betula alba*
Birch (yellow)	*Betula alleghaniensis*
Blackberry	*Rubus* spp.
Blueberry	*Vaccinium* spp.
Bluegrass	*Poa* spp.
Bluegrass (Kentucky)	*Poa pratensis*
Breadfruit	*Artocarpus altilis*
Broadbean	*Vicia faba*
Broccoli	*Brassica oleracea* var. *italica*
Bromegrass	*Bromus* spp.
Bromegrass (Japanese)	*Bromus japonicus*
Bromegrass (smooth)	*Bromus secalinus*
Brussels sprouts	*Brassica oleracea* var. *gemmifera*
Cabbage (Chinese)	*Brassica pekinensis*
Cabbage (common, savoy)	*Brassica oleracea* var. *capitata*
Calendula	*Calendula officinalis*
Cantaloupe	*Cucumis melo*
Carnation	*Dianthus caryophyllus*
Carrot	*Daucus carota*
Cassava	*Manihot esculenta*
Cauliflower	*Brassica oleracea* var. *botrytis*
Celery	*Apium graveolens*
Chard (swiss)	*Beta vulgaris* var. *cicla*
Cherry	*Prunus* spp.
Cherry (choke)	*Prunus virginiana*
Chrysanthemum	*Chrysanthemum* vars.
Clover (alsike)	*Trifolium hybridum*
Clover (crimson)	*Trifolium incarnatum*
Clover (ladino)	*Trifolium repens forma lodigense*
Clover (red)	*Trifolium pratense*
Clover (sweet)	*Melilotus* spp.
Clover (white)	*Trifolium repens*
Clover (white sweet)	*Melilotus alba*
Clover (yellow sweet)	*Melilotus officinalis*
Clover (zigzag)	*Trifolium medium*
Cocklebur	*Xanthium* spp.
Cocoa	*Theobroma* spp.
Coconut	*Cocos nucifera*
Coffee	*Caffea* spp.
Collard	*Brassica oleracea* var. *acephala*

Corn (including sweet)	*Zea mays*
Cotton	*Gossypium hirsutum*
Cowberry	*Vaccinium* spp.
Cowpea	*Vigna sinensis*
Crabgrass	*Digitaria* spp.
Cranberry (mountain)	*Vaccinium vitis-idaea*
Cress (curly)	*Lepidium sativum*
Crowberry	*Empetrum nigrum*
Cucumber	*Cucumis sativus*
Dallisgrass	*Paspalum dilatatum*
Deertongue	*Panicum clandestinum*
Dewberry	*Rubus caesius*
Dill	*Anethum* spp.
Dogwood	*Cornus* spp.
Eggplant	*Solanum melongens* var. *esculentum*
Elm	*Ulmus* spp.
Elm (American)	*Ulmus americana*
Endive	*Chichorium endiva*
Fescue	*Festuca* spp.
Fescue (meadow)	*Festuca elatior*
Fescue (red)	*Festuca rubra*
Fescue (tall)	*Festuca elatior*
Fig	*Ficus carica*
Fir (balsam)	*Abies balsamea*
Fir (Douglas)	*Pseudotsuga menziesii*
Fir (silver)	*Abies alba*
Flax	*Linum usitatissimum*
Foxtail	*Alopecurus* spp.
Gladiolis	*Gladiolus* spp.
Grapefruit	*Citrus paradisi*
Grapes	*Vitis* spp.
Grass (barnyard)	*Echinochloa crusgalli*
Gum (black)	*Nyssa sylvatica*
Hairgrass	*Deschampsia* spp.
Hairgrass (Tufted)	*Deschampsia cespitosa*
Heather	*Calluna* spp.
Hemlock (eastern)	*Tsuga canadensis*
Hickory	*Carya* spp.
Honeydew	*Cucumbis melo inodorus*
Hops	*Humulus* spp.
Johnsongrass	*Sorghum halepense*
Juniper	*Juniperus* spp.

Kale	*Brassica oleracea* var. *acephala*
Kale (curly)	*Brassica oleracea* var. *acephala*
Kohlrabi	*Brassica oleracea* var. *gongylodes*
Kola	*Cola* spp.
Larch	*Larix* spp.
Larkspur	*Delphinium* spp.
Lemon	*Citrus limon*
Lespedeza	*Lespedeza* spp.
Lespedeza (Korean)	*Lespedeza stipulacea*
Lespedeza (sericea)	*Lespedeza cuneata*
Lettuce (Romaine)	*Lactuca sativa*
Locoweeds	*Astragalus* spp.
Lovegrass (weeping)	*Eragrostis curvula*
Lucerne	*Medicago sativa*
Lupine	*Lupinus* spp.
Maize	*Zea mays*
Mangel	*Beta vulgaris* var. *macrorhiza*
Maple (mountain)	*Acer spicatum*
Maple (red)	*Acer rubrum*
Maple (silver)	*Acer saccharinum*
Maple (sugar)	*Acer saccharum*
Marigold	*Tagetes* spp.
Meadow sweet	*Spirea alba*
Millet (proso)	*Panicum miliaceum*
Milo	*Sorghum vulgare*
Mint	*Mentha* spp.
Moss (feather)	*Hylocomium* spp.
Mulberry	*Morus* spp.
Muskmelon	*Cucumis melo* var. *reticulatus*
Mustard	*Brassica juncea*
Mustard (white)	*Brassica hirta*
Needle and thread	*Stipa comata*
Nutsedge	*Cyperus* spp.
Oak	*Quercus* spp.
Oak (black)	*Quercus velutina*
Oak (chestnut)	*Quercus prinus*
Oak (northern red)	*Quercus rubra*
Oak (white)	*Quercus alba*
Oats	*Avena sativa*
Oats (wild)	*Uniola latifolia*
Olive	*Olea* spp.
Onion	*Allium cepa*
Orange (including sweet)	*Citrus sinensis*

Orchardgrass	*Dactylis glomerata*
Oxalis	*Oxalis* spp.
Palm	*Elaeis* spp.
Pangolagrass	*Digitaria* spp.
Pansy	*Viola* spp.
Parsley (common)	*Petroselinum crispum*
Parsley (curly)	*Petroselinum crispum* var. *latifolium*
Parsnip	*Pastinaca sativa*
Pea	*Pisum sativum*
Pea (black-eyed)	*Vigna sinensis*
Pea (field)	*Pisum arvense*
Pea (sweet)	*Lathyrus odoratus*
Peach	*Prunus persica*
Peanut	*Arachis hypogaea*
Peanut (spanish)	*Arachis hypogaea* var. *fastigiata*
Pear	*Pyrus communis*
Pecan	*Carya illinoensis*
Pepper (green)	*Capsicum annuum*
Peppermint	*Mentha piperita*
Persimmon (Japanese)	*Diospyros kaki*
Pigweed	*Amaranthus* spp.
Pine	*Pinus* spp.
Pine (eastern white)	*Pinus strobus*
Pine (loblolly)	*Pinus taeda*
Pine (Monterey)	*Pinus radiata*
Pine (red)	*Pinus resinosa*
Pine (Scotch)	*Pinus sylvestris*
Pine (shortleaf)	*Pinus echinata*
Pine (slash)	*Pinus elliotii*
Pine (Swiss stone)	*Pinus cembra*
Pine (western white)	*Pinus monticola*
Pineapple	*Ananas comosus*
Pistachio	*Pistacia* spp.
Plantain	*Musa x paradisiaca*
Plum	*Prunus domestica*
Poplar (yellow)	*Liriodendron tulipifera*
Poppy (California)	*Eschscholzia californica*
Potato (Irish)	*Solanum tuberosum*
Potato (sweet)	*Ipomoea batatas*
Pumpkin	*Cucurbita pepo*
Quackgrass	*Agropyron repens*
Queen-of-the-meadow	*Filipendula ulmaria*

Radish	*Raphanus sativus*
Ragweed	*Ambrosia* spp.
Rape	*Brassica napus*
Rapeseed	*Brassica* spp.
Raspberry	*Rubus* spp.
Red top	*Agrostis alba*
Reed (common)	*Phragmites communis*
Rhododendron	*Rhododendron* spp.
Rice	*Oryza sativa*
Rose	*Rosa* spp.
Rutabaga	*Brassica napobrassica*
Rye	*Secale cereale*
Ryegrass (Italian)	*Lolium multiflorum*
Ryegrass (perennial)	*Lolium perenne*
Safflower	*Carthamus tinctorius*
Sandbur	*Cenchrus* spp.
Sesame	*Sesamum indicum*
Sorghum	*Sorghum vulgare*
Sorrel	*Rumex* spp.
Soybean	*Glycine max*
Spearmint	*Mentha spicata*
Spinach	*Spinacea oleracea*
Spruce	*Picea* spp.
Spruce (black)	*Picea mariana*
Spruce (Norway)	*Picea abies*
Spruce (red)	*Picea rubens*
Spruce (white)	*Picea glauca*
Squash	*Cucurbita pepo*
Squash (Zucchini)	*Cucurbitata pepo* var. *medullosa* Alef.
Strawberry	*Fragaria* spp.
Sudangrass	*Sorghum sudanense* stapf.
Sugarcane	*Saccharum offinarum*
Sunflower	*Helianthus annuus*
Switchgrass	*Panicum virgatum*
Sycamore (American)	*Platanus occidentalis*
Taro	*Colocasia* spp.
Tea	*Camellia* spp.
Tea (Labrador)	*Ledum* spp.
Thyme	*Thymus* spp.
Timothy	*Phleum pratense*
Tobacco	*Nicotiana tabacum*
Tomato	*Lycopersicon esculentum*
Trefoil (birdsfoot)	*Lotus corniculatus*
Tulip poplar	*Liriodendron tulipifera*

Tung	*Aleurites fordi*
Tupelo	*Nyssa* spp.
Turnip	*Brassica rapa*
Vetch	*Vicia* spp.
Vetch (crown)	*Coronilla varia*
Vetch (hairy)	*Vicia villosa*
Vetch (milk)	*Astragalus* spp.
Violet	*Viola* spp.
Walnut	*Juglans* spp.
Water-fern	*Azolla* spp.
Watergrass	*Echinochloa crusgalli*
Watermelon	*Citrullus* spp.
Watermilfoil	*Myriophyllum* spp.
Wheat (common)	*Triticum aestivum*
Wheat (durum)	*Triticum turgidum*
Wheatgrass (crested)	*Agropyron cristatum*
Wheatgrass (range)	*Agropyron* spp.
Whortleberry (bog)	*Vaccinium myrtillus*
Woodsorrel	*Oxalis* spp.
Yam	*Dioscorea* spp.
Zinnia	*Zinnia* spp.

Animals

Barbel	*Barbus* spp.
Bass (largemouth)	*Micropterus salmoides*
Bass (rock)	*Amboplites rupestris*
Bass (striped)	*Morone saxatilis*
Beetle (Colorado potato)	*Leptinotarsa decemlineata*
Capelin	*Mallotus villosus*
Carp (common)	*Cyprinus carpio*
Catfish (channel)	*Ictalurus punctatus*
Catfish (flathead)	*Pylodictus olivarus*
Catfish (white)	*Ameiurus catus*
Clam (geoduck)	*Panope generosa*
Clam (Pacific littleneck)	*Protothaca staminea*
Clam (soft)	*Mya arenaria*
Cod (Atlantic)	*Gadus morhua*
Cod (Pacific)	*Gadus macrocephalus*
Crab (blue)	*Callinectes sapidus*
Crab (Dungeness)	*Cancer magister*
Crab (king)	*Paralithodes camtschatica*

Crab (snow)	*Chionoecetes opilio*
Crab (Tanner)	*Chionoecetes bairdl*
Eagle (bald)	*Haliaeetus leucocephalus*
Flounder (arrowtooth)	*Atheresthes stomias*
Flounder (fourspot)	*Paralichthys oblongus*
Flounder (gulf)	*Paralichthys albigutta*
Flounder (southern)	*Paralichthys lethostigma*
Flounder (summer)	*Paralichthys dentatus*
Flounder (windowpane)	*Scophthalmus aquosus*
Flounder (winter)	*Pseudopleurnectes americanus*
Flounder (witch)	*Glyptocephalus cynoglossus*
Flounder (yellowtail)	*Limanda ferruginea*
Gull (herring)	*Larus argentatus*
Haddock	*Melanogrammus aeglefinus*
Hog (domestic)	*Sus scrofa*
Jau	*Paulicea luetkeni*
Killifish (Banded)	*Fundulus diaphanus*
Mink	*Mustela vison*
Minnow (fathead)	*Pimephales promelas*
Moth (codling)	*Carpocapsa pomonella*
Muskrat	*Ondatra zibethica*
Mussel	*Lamellidans marginalis*
Northern pike	*Esox lucius*
Osprey	*Pandion haliaetus*
Otter (river)	*Lutra canadensis*
Oyster (eastern)	*Crassostrea virginica*
Oyster (Pacific)	*Crassostrea gigas*
Oyster (western, Olympia)	*Ostrea lurida*
Perch (yellow)	*Perca flavescens*
Piranha	*Serrasalmus nattereri*
Pollack	*Pollachius pollachius*
Redfish	*Sciaenops ocellatus*
Roach	*Rutilus rutilus*
Salmon (chinook)	*Oncorhynchus tshawytscha*
Salmon (chum)	*Oncorhynchus keta*
Salmon (coho)	*Oncorhyynchus kisutch*
Salmon (pink)	*Oncorhynchus gorbusca*
Salmon (sockeye)	*Oncorhynchus nerka*
Scallop (Atlantic bay scallop)	*Aequipecten irradians concentricus*

Scallop (calico scallop)	*Argopecten gibbus*
Scallop (sea scallop)	*Placopecten magellanicus*
Scallop (pink scallop)	*Chlamys* spp.
Shrew	*Sorex* spp.
Shrimp (Alaska; sidestripe)	*Pandalopis dispar*
Shrimp (brown)	*Penaeus aztecus*
Shrimp (ocean)	*Pandalus jordani*
Shrimp (pink)	*Penaeus duorarum*
Shrimp (pink northern)	*Pandalus borealis*
Shrimp (royal red)	*Pleoticus robustus*
Shrimp (white)	*Penaeus setiferus*
Snail	*Amnicola* spp.
Sucker (longnose)	*Catostomus catostomus*
Sucker (white)	*Catostomus commersoni*
Sunfish (bluegill)	*Lepomis macrochirus*
Sunfish (pumpkinseed)	*Lepomis gibbosus*
Trout (brook)	*Salvelinus fontinalis*
Trout (rainbow; steelhead)	*Oncorhynchus mykiss*
Tuna (albacore)	*Thunnus alalunga*
Tuna (skipjack)	*Euthynnus pelamis*
Tuna (yellowfin)	*Thunnus albacares*
Walleye	*Stizostedion vitreum*
Waterflea	*Daphnia magna*
Whitefish (lake)	*Coregonus clupeaformis*
Weevil (boll)	*Anthonomus grandis*

Index

About the Author

Domy C. Adriano is a professor of environmental soil science in the Department of Crop and Soil Sciences at the University of Georgia, Athens, and also a senior biogeochemical ecologist at the Savannah River Ecology Laboratory (SREL), Aiken, South Carolina. He received his BS in agricultural engineering (1961) from the Department of Agricultural Engineering at Central Luzon State, University, Nueva Ecija, Philippines, and MS (1965) and Ph.D. (1970) in agronomy (soil science) from the Department of Agronomy at Kansas State University, Manhattan. He was a Postdoctoral Fellow (1970–1972) in the Department of Soil and Environmental Sciences at the University of California, Riverside and served as an assistant professor (1973–1974) in the Department of Crop and Soil Sciences at Michigan State University, East Lansing, before joining the SREL (1975) as a research ecologist. He is a fellow of the Soil Science Society of America and the American Society of Agronomy and is a recipient of the prestigious Environmental Quality Research Award from the American Society of Agronomy (1999) and the Distinguished Service in Agriculture Award (1999) from his alma mater Kansas State University.

Dr. Adriano's research interests include biogeochemistry of trace metals in the soil-plant system, source term – bioavailability – risk relationships, risk reduction and management in metal-contaminated sites, waste minimization and recycling, and natural attenuation of contaminants. He has published more than 130 journal and proceedings articles and 52 book chapters and edited or coedited 12 books, including *Bioremediation of Contaminated Soils* (1999, American Society of Agronomy Monog. no. 37). He has presented numerous lectures, seminars, and keynote addresses in the United States and abroad and recently spent his sabbatical leave (1998–1999) as a distinguished guest professor at the Universität für Bodenkultur, Vienna, Austria, and Limburgs Universitair Centrum, Diepenbeek, Belgium. His other professional activities include serving as associate editor of the *Journal of Environmental Quality* (1980–1985) and of *Environmental Geochemistry and Health* (1999-present), division chair for soil chemistry (S-2) of the Soil Science Society of America (1990), founder and first organizer of the International Conference on the Biogeochemistry of Trace Elements (1990), Chair of subcommission on soil remediation of the International Union of Soil Science (1995–1998), chair of Global Soil Remediation Network (1998–present), and president of the International Society of Trace Element Biogeochemistry (1999–2003).